Metals in Antiquity

Edited by

Suzanne M. M. Young, A. Mark Pollard,
Paul Budd and Robert A. Ixer

BAR International Series 792
1999

Published in 2016 by
BAR Publishing, Oxford

BAR International Series 792

Metals in Antiquity

ISBN 978 1 84171 008 2

BAR Publishing is the trading name of British Archaeological Reports (Oxford) Ltd. British Archaeological Reports was first incorporated in 1974 to publish the BAR Series, International and British. In 1992 Hadrian Books Ltd became part of the BAR group. This volume was originally published by Archaeopress in conjunction with British Archaeological Reports (Oxford) Ltd / Hadrian Books Ltd, the Series principal publisher, in 1999. This present volume is published by BAR Publishing, 2016.

Printed in England

BAR
PUBLISHING

BAR titles are available from:

BAR Publishing
122 Banbury Rd, Oxford, OX2 7BP, UK
EMAIL info@barpublishing.com
PHONE +44 (0)1865 310431
FAX +44 (0)1865 316916
www.barpublishing.com

METALS IN ANTIQUITY

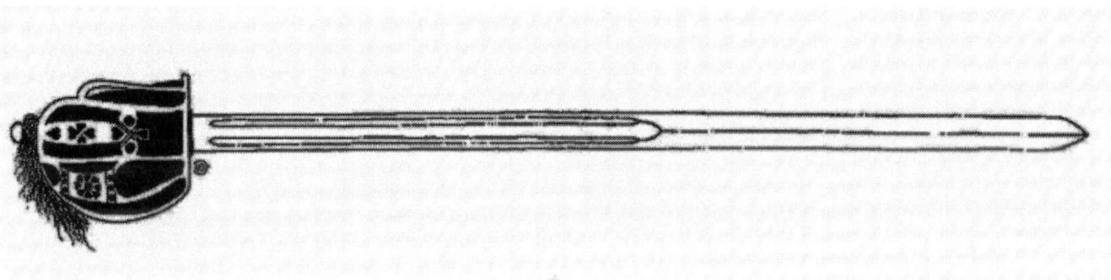

SUZANNE M. M. YOUNG[1], A. MARK POLLARD[2], PAUL BUDD[2], and ROBERT A. IXER[3]

[1]*Archaeometry Laboratories, Harvard University, Cambridge, USA*

[2]*Department of Archaeological Sciences, University of Bradford, Bradford, UK*

[3]*Department of Geology, Birmingham University, Birmingham, UK*

CONTENTS

Acknowledgements

This may read like those American movie or music awards where they thank everyone, everything, and the kitchen sink, but I am an extremely fortunate bluestocking and received much from many during the adventure of organizing this conference and this volume. Those people are great treasures to me, and so they shall be named.

Great praises and deep gratitude is extended to:

Nikolaas J. van der Merwe, for never clipping my wings, even when it meant I was running around doing a million or more things not *exactly* in my job description!

Carl C. Lamberg-Karlovsky (and ASPR) for much encouragement, both verbal and tangible (helping me arrange financial support towards the success of the conference).

A. Mark Pollard, for a billion details - small and practical - the sum total of which amounted to far greater assistance to me than any in the world gave towards the accomplishment of the conference and the organization of this volume.

Sheridan Bowman, for excellent and very helpful advice along the way.

VG Elemental, (innovator and manufacturer of many very fine elemental and isotopic mass spectrometers, especially all those that make my research both feasible and loads of fun), specifically thanks to Pete Brown, for financial support of the conference reception that brought us to the next level in decadent and delicious detail.

Marie Lorizio, for brilliance and good humour in handling all the accounts while I ran the conference - it was neither a small nor minor feat to handle my plunges in and out of debt and many a foreign transaction, not to mention dealing with me, who – unintentionally – specialised in constant *near* impossible requests!

Kelly Warner, for tons of time and help as my personal assistant during the months prior to the conference.

W.A. (Tinker) Green, for fantastic help during and after the conference as my personal assistant.

Joan Ryoo and Kit Marlow, for a terrific job helping during the conference and often going well beyond the call of duty for the sake of pleasing and helping conference attendees.

Barbara Wiberg, for great help in keeping the reception in the Peabody Museum splendid for all, and for very practical advice before, during, and after the reception.

Susan Williamson, a mathematician, friend and artist - my mentor from my university-years of many moons ago, whose presence at the archaeometallurgy conference meant more to me than words can say.

Olivia Lawrenson, Paul Budd's beloved wife, for putting up with all the visits, meetings, and work involved in our trans-Atlantic collaboration in organizing this conference and book and for helping us during the conference.

Rosie (Razzamatazz) Ixer, a jazz singer and star of Rob Ixer's life, who was his assistant while editing parts of this volume.

Marcia E. T. Young, for servicing as the grammer-police in much of this volume!

J. Chris Morton, my assistant while editing this volume, for all his computer expertise and help in truly uniting a volume edited by 4 very different people.

Geoffrey Dixon and his multimedia company Seven Stones, for the graphic work in the final production of this volume.

Paul Budd, for helping me ignite the flame that became this conference.

Rob Ixer, for his diplomatic talents, which made possible all that you see before you.

Conference participants, conference attendees, and volume contributors, for themselves! - because as my wise and wonderful pal, Ron Hancock, proclaimed of organising conferences: "We lay the foundation and frame, then its time to sit back and let the others build the house." This he told me during the conference - implying I could now chill out and enjoy a bit - that never happened (largely because those are close to impossible instructions for a complete control freak to follow!). But *now* that I have seen what a fine job all participants did in building the rest, perhaps next time there will be a drop of hope me!

Penultimately, very personally from me to G, for making the sunbeams warmer, the stars shine brighter, and the flowers smell sweeter in my world - for keeping a smile inside me no matter the struggles or trials along the way in the adventures (and misadventures) of producing this volume.

And lastly, the kitchen sink – for which you'll be all most grateful – because it means I have finished mushing and gushing!

I now pronounce you all safe to read and enjoy the rest of the volume in peace!

S.M.M.Y.

Preface and Dedication

From 10 to 13 September 1997, 170 archaeofolks - archaeological scientists, archaeometrists, archaeometallurgists, archaeologists, and even just archaeo-interested scientists - convened at Harvard University for the International Symposium ***Metals in Antiquity***. This conference was held in honour of Professor K. C. Chang of Harvard University and these proceedings are dedicated to him.

The conference aimed to promote an integrated and comprehensive understanding of ancient metallurgy. It sought to explore the distribution of metals in the natural environment, extractive metallurgy and fabrication processes, as well as the social context, use and deposition of artefacts, and to combine anthropology, archaeology and the earth sciences, involving archaeological, mineralogical, chemical and isotopic investigations of ancient metal production, use and provenance. This is a feat much easier to spell out than to carry out! It requires a traditional scholarship of the highest order, a focus on details, and an understanding of continents of information. And it requires the dissolution of walls separating disciplines that traditional education too often treats as distinct.

Increasing specialization is a natural response to the growth of a field of research, and those who resist the tendency longest while maintaining a high level of scholarship are rare. Professor K. C. Chang is such a scholar. While carrying out brilliant and seminal research in his chosen field he excluded nothing outside that field. The breadth of his knowledge and his keen interest in many disciplines are remarkable.

I have a special and very personal story to share with you all. I came to archaeometry nine years ago from organometallic synthesis and spectroscopy. In my first year I leapt into Bronze Age Nubia and studied swords using metallographic, chemical, and isotopic methods. Once my work was accepted for publication, I presented it at one of a series of colloquia in our department. K. C. Chang sat in the centre seat, second row. The next day in the hall he caught me and praised the talk and all the work and invited me to afternoon tea to discuss it some more. He was not meant to be my mentor. He was not my boss. I was never his responsibility in any way. We are in the same department, our offices along the same hallway; we could have smiled polite greetings to one another and no more for years. But that afternoon tea was to prove the first of many, and in it began some of the most important lessons in my life as an archaeometrist. Every time we met for tea I inquired what kind he wished, and every time it was the same – jasmine. I would make jasmine for two and rejoin him in his office. Each time he would wrap his fingers around the warm cup and hold it close so the steam could drift along his face for a moment. His lips would form a small smile, his bright eyes drift to the ceiling, and his mind look back across the horizon of time. He would say to me that jasmine tea is generally the lowest quality of leaves picked. They put the jasmine flowers in to hide that fact. But it is his favourite kind of tea, because whenever he smells the jasmine he remembers all the lovely ladies with beautiful jasmine flowers in their hair when he was young. He has a window into youth in a simple cup of tea, and he shared this same exact tale with me every time – part of the ritual. After his first sip all was serious again, down to business and research and archaeology. We talked long hours about people and times and cultures. He gently, but firmly, showed me the humanity needed in interpretations in archaeometry – in understanding cultures. He taught me to separate the mind-set associated with decades of training in the modern sciences from that of the creators of the objects under study – to put aside modern technical ways of thinking once the laboratory work is finished, and interpretation begins. He gave me assignments (suggestions! but I took it all deeply to heart) in social/cultural anthropological reading. Without making me feel we had done more than take a pleasant break from otherwise long hectic days, he gave me the gift of the knowledge of how to proceed on the path I had embarked upon, always with kindness and an interest in me and my work – no matter what region, what time, or what studies I was conducting.

K. C. Chang is the reason this conference took place at Harvard University. I hoped to create a meeting of minds with both shared and disparate interests, such that the whole would be far greater than the sum of its parts. In so doing I wanted to share with others the illumination K. C. shared with me. He is the reason such effort was put into the conference and the volume, and that all who cared to come were invited. The tea and Swiss chocolate truffles that were served were intended to create a lovely atmosphere fostering our exploration of knowledge, and to bring our field closer to its grandest goals.

The sessions of the meeting included the usual suspects for an archaeometallugical meeting:

- current research on ancient mining and archaeometallurgy
- the mineralogy and geochemistry of ore deposits and ancient extractive metallurgy,

• the characterisation of ore deposits for studies of provenance and technology,
• reconstructing ancient metal production processes,
• the social context of ancient metal production and use,
• theoretical aspects of ancient metallurgy,
• the ethnography of metallurgy.

But this conference was organized out of desire to promote an integrated and comprehensive understanding of ancient metallurgy. Even where sessions focused more on one method or discipline, the potential for future development at the interface between archaeology and the earth sciences was the intent. Greatest effort was given to bringing in those who have already succeeded in combining geological and archaeological approaches.

As the scientific investigation of ancient metal mining and production becomes more sophisticated and multi-disciplinary, the hypotheses adopted and conclusions drawn have to remain firmly rooted in archaeology and anthropology. The social organisation underlying early metallurgy and metalwork is of utmost importance. Thus this volume is organised differently than the conference sessions. It is arranged more or less chronologically and geographically. The role of metal making and use in antiquity was explored from both theoretical archaeological and ethnographic perspectives. This volume is arranged with social papers leading scientific studies done in the same region to provide greater cultural context wherever possible. Don Wagner's paper on China is a very well integrated study, and integration is the goal of this meeting. We break the chronological flow of the book in this case because it only makes sense to start the book with a fine chapter on Asia, the focus of much of K. C. Chang's own work.

Developing dialogue between social and analytical scientists is as difficult as the integration of the theoretical and the experimental in studies of ancient metallurgy. Archaeologists and archaeometrists do not often engage in academic discussion, and seldom read each other's literature. In order to make this archaeometallurgical volume most accessible to archaeologists, at the back of the volume can be found a map showing regions covered by each chapter, and two time lines – one showing the time periods covered in each chapter and the other showing the metal ages all over the world. This book contains papers on every continent save Antarctica.

Professor K. C. Chang rarely joins us at Harvard now, but I hope he is being served jasmine tea often. He is a mentor to all receptive to his wisdom and learning. He is a scholar greatly loved and revered, and it is my greatest hope that this congress and volume, intended to honour him, have done so to some small degree.

S. M. M. Y.

1

The earliest use of iron in China

DONALD B. WAGNER

Reverdilsgade 3, 1.th., DK–1701 Copenhagen V, Denmark

Until recently it was a reasonable hypothesis that the first use of iron in China was in the Southeast, perhaps in the 6th century BC. New finds, together with old finds just recently studied and published, have made this position untenable. It now seems likely that the technology of iron smelting diffused to China by the 8th century BC from the West via Scythian nomads in Central Asia.

I suggest as a working hypothesis that the craftsman of an indigenous Chinese tradition of making luxury weapons with meteoritic iron blades, which were probably better than bronze weapons, at some time learned bloomery smelting from steppe peoples and began substituting bloomery iron for meteoritic iron. These weapons were probably not a match for bronze weapons, but, by this time, they were probably intended for display rather than actual fighting.

In most of China the only uses for bronze had been for ritual objects and weapons; the 'barbarian' peoples of Southeast China were the first in the region of Chinese influence to use bronze agricultural implements to any great extent, and iron presumably provided a useful cheap substitute for bronze in this sort of application. It was probably here that iron casting was first developed, beginning with the carburization and melting of iron blooms in the kind of furnace that was used to melt bronze. The blast furnace probably developed here as well, as bloomeries were optimized for the purpose of providing iron for casting rather than forging.

Keywords: Meteoritic iron; Iron Age; China; Shang Period; Zhou Period; Central Asia; Scythians.

1. The state of the question

Over the centuries there have been numerous studies of when iron was first used in China, but until recently the only basis for a discussion of the question was in written texts. The terms of the debate were changed irrevocably by excavations in 1950–52 in Hui County, Henan.(Anon. 1956; An Zhimin 1990; Paul-David 1954; Wagner 1993: 199–206). Of 54 Warring States' graves excavated here, seven yielded a total of 161 iron artefacts, including decorative objects, weapons, and implements. In 1953 there was another major find, the site of an iron foundry of the third century BC in Xinglong County, Hebei (Zheng Shaozong 1956; Anon. 1980; Li Xueqin 1985: 327), where cast-iron moulds for casting iron implements were found. More surprises were to come as these and other iron artefacts were subjected to metallographic examination, but the finding of large amounts of iron securely dated as early as the fourth century BC was remarkable enough in itself.

With the publication of more, and more relevant, archaeological material, many scholars have been attracted to the question. Huang Zhanyue, in an important article published in 1976, brought order to a confusion of claims and counter-claims (Huang Zhanyue 1976). Using rigorous philological criteria he rejected, one by one, all of the arguments based on written sources. Considering then the archaeological evidence, he rejected several arguments on such grounds as methodological confusion, incorrect dating, or inadequate reporting of excavation details. He concluded that the first use of iron in China was in the south, for the earliest iron artefacts so far found in China were from Changsha, Hunan (in the ancient Chinese state of Chu) and Luhe, Jiangsu (in the state of Wu).

When in the mid-1980's I wrote my own contribution to the problem, in two chapters of a book entitled *Iron and Steel in Ancient China*, the available archaeological material appeared to provide a moderately clear picture (Wagner 1993: 51–146). Several artefacts of *meteoritic* iron from the late Shang and early Western Zhou found in north China seemed to be irrelevant to later developments; and if one ignored claims that were not properly documented with archaeological and metallurgical details, it seemed that in China the *smelting* of iron from ore began, as Huang Zhanyue had suggested, in the south, sometime before the beginning of the fifth century BC. I felt, however, that I could narrow this further, to the state of Wu.[1] From there, it spread within a century to the other major southern state, Chu, and within another century to the rest of the Zhou empire, almost entirely replacing bronze as the metal of choice for most practical implements and weapons. Earlier iron artefacts of non-Chinese nomadic peoples north of the Zhou empire (which I did not mention) appeared interesting but essentially irrelevant to the Chinese story: the rulers of the Zhou states had sufficient bronze for their purposes, and, if they were aware of the new metal used by their northern neighbours they found it to be a poor substitute.

Even as that book was being printed, new publications of archaeological material were undermining its story. What is available at the moment indicates that smelted iron definitely was in use in northwest China long before it was used in the south. A direct implication is that the technique of iron smelting came to northwest China from the West through Scythian intermediaries, for contacts with the iron-using nomadic peoples of Siberia were sufficiently intense that independent invention was hardly possible. Another surprising aspect of the new archaeological material is that it indicates that the early smiths may have used meteoritic and smelted iron interchangeably, so that the meteoritic iron artefacts have a significance which I had not earlier been willing to accept.

If iron-smelting technology came to China from the West, then the first iron smelting in China must have been done in

Figure 1: Map of China showing the places mentioned in text.

bloomeries rather than, as some other writers and I have assumed, in blast furnaces.

2. Iron on the northern borders of the Zhou empire

According to Chester S. Chard's useful survey of Northeast Asian prehistory, by about 700 BC 'there was marked cultural homogeneity . . . throughout the Eurasian steppe world, so that life in the Altai region has much in common with that of the Scythians of southern Russia' (Chard 1974: 153). In the upper Yenisei valley, 'during the fourth through the second centuries BC iron finally replaced bronze as the standard metal, a process that had taken place several centuries earlier among the Scythians. The time lag is probably explained by the very high quality of Altai bronze, which offered little inducement to change until iron-working had attained a comparable level locally.' I take this to mean that implicit or explicit evidence of a knowledge of iron is found here significantly earlier than the fourth century BC.

In Xinjiang some surprisingly early dates for iron have been published in recent years.[2] Iron is found in graves which for the most part show no sign at all of Chinese influence, and which have surprisingly early radiocarbon dates. Chen Ge gives a table of 35 radiocarbon dates (by three different laboratories) for graves in Xinjiang in which iron artefacts were found (Chen Ge 1989: 426–427). The dates can be summarized as follows:

earlier	1	
1000–901 BC	3	•••
900–801 BC	6	••••••
800–701 BC	7	•••••••
700–601 BC	3	•••
600–501 BC	5	•••••
500–401 BC	2	••
400–301 BC	2	••
300–201 BC	1	•
200–101 BC	2	••
later	3	

There seems to be little room for doubt that iron was known here by the eighth century BC, but I would tend to be wary of the dates which are earlier than this, for they are difficult to fit in with what we know of iron in Siberia and China proper. They are best explained as the tail of the Gaussian distribution of radiocarbon dates of the eighth century BC.

Xinjiang seems to have had very little contact with the Central Plain before the Han period. There are also some early dates for iron in parts of Gansu, perhaps again pointing to a date in the eighth century BC for iron (Pu Chaofu 1981; Pu Chaofu & Zhao Jianlong 1984; cf. Chen Ge 1989: 431), and here there are more signs of contact with the Central Plain.

. . .

The famous Frozen Tombs of Siberia, five barrows excavated at Pazyryk, in southern Siberia near the borders of Xinjiang and Mongolia, all contain iron artefacts (Rudenko 1970: 314, 315, 317, 318, 319, 321, 323, 326,

327). These include horse-bits, daggers, and a hairpin. All are of wrought iron, none of cast iron (Rudenko 1970: 207). A floating dendrochronology indicates that the tombs were constructed over a period of 48 years. Conventional dating methods suggest an absolute date in the late fifth century BC, and radiocarbon does not contradict this date. All of the tombs show clear Scythian influence, and most contain articles imported from China.

. . .

All of these dates for northern iron fit a pattern which requires no great strain of the reader's credulity: iron appears to have been introduced into Siberia by the Scythians, perhaps as early as the eighth century BC. Further east, in the Russian Maritime Province, it is something of a shock to find the following statement by Chard (1974: 94) without further discussion: 'Iron appears in the Vladivostok area in the eleventh to twelfth centuries BC, coinciding with the development of the so-called Shell Mound or Sidemi culture' (Chard 1974: 94). This is as early as the beginning of the Iron Age of Greece, and much earlier than iron anywhere else in Asia. Chard's date seems to be based on a single radiocarbon date, and should therefore not be considered reliable.[3]

We do not know whether the ancient northern iron artefacts were made locally or imported from further west. It seems *a priori* fairly likely that knowledge of the technology travelled with the people who used the artefacts, but there is not much in the way of real evidence one way or the other. There is some evidence that Korean iron technology may have come from the Scythians, and that this could have been as early as the eighth or seventh century BC (Yoon 1984: 35–46; Taylor 1989: 422, 427; 1990: 213–15; Barnes 1993: 214; Nelson 1993: 172–4; Kim 1978: 138–14).

3. Early Chinese meteoritic and smelted iron

The earliest evidence of the use of iron in the area of Shang and Zhou culture consists of bronze axeheads of several kinds with cast-in edges of meteoritic iron, all of which are likely to be from the Shang or early Zhou period:

> (1) Two bronze–iron weapons of the early Zhou period (Gettens *et al.* 1971), believed to come from a single tomb excavated in 1931 somewhere in the vicinity of Anyang, Henan. Analysis indicates that the iron parts of both contain significant amounts of nickel and cobalt. The distribution of nickel and cobalt in the microstructure of one indicates a Widmanstätten structure distorted by heating to a temperature above 600° C for a few minutes.
>
> (2) A bronze–iron axehead from a late Shang city-site at Taixicun in Gaocheng County, Hebei (Tang Yunming & Liu Shishu, 1973, pl. 1; Anon. 1977; Tang Yunming 1979; 1985). Analysis shows that the iron part contains significant amounts of nickel and cobalt, and their distribution in the microstructure suggests a Widmanstätten structure which has been severely distorted by hot forging (Li Chung 1979).
>
> (3) A bronze–iron axehead found in a Shang-period grave excavated at Liujiahe in Pinggu County, Beijing Municipality (Figure 2; Yuan Jinjing & Zhang Xiande, 1977, pl. 1.1; cf. Anon. 1990: 37–9). The iron part contains a significant amount of nickel, and the distribution of nickel in the microstructure is consistent with that in a Widmanstätten structure severely distorted by hot working (Zhang Xiande & Zhang Xianlu 1990).

The bronze parts of these artefacts are clearly Chinese. There is room for the supposition that the iron parts were imports from elsewhere, but very little reason for it. More likely there were Chinese smiths in the Shang and early Zhou who hot-forged meteoritic iron and produced small edges to be cast into bronze axes and other edged weapons. Meteoritic iron is usually fairly hard, and these bronze–iron weapons are likely to have been superior to weapons of bronze alone, but meteoritic iron is so rare that these must always have been uncommon luxury items. We do not know whether these smiths developed their techniques themselves or learned them from elsewhere. The probability of independent invention seems high, but I have no information at all on the use of meteoritic iron in early Central Asia.

3.1. Bronze–iron artefacts of the state of Guo

Evidence that the smithing of meteoritic iron is relevant to the later smithing of smelted iron in China is afforded by several iron artefacts from graves excavated in 1990–91 at Shangcunling in Sanmenxia, Henan, near the great bend of the Yellow River. An excavation report has not yet been published, but advance reports indicate that five bronze–iron artefacts were found in the excavations (Anon. 1992: 13–17; Zhong Shaoyi 1993). All five have been examined by metallurgists. A preliminary note on their findings indicates that three contained so much nickel that they clearly were made of meteoritic iron. In the other two, nickel was not detected, and their microstructures also appear to indicate that they were made of smelted iron: 'The microstructure and the existence of elongated but cracked inclusions in the rusted areas indicate that these two artefacts are likely to be fabricated of man-made iron; further study is in progress' (Han Rubin *et al.* 1994).

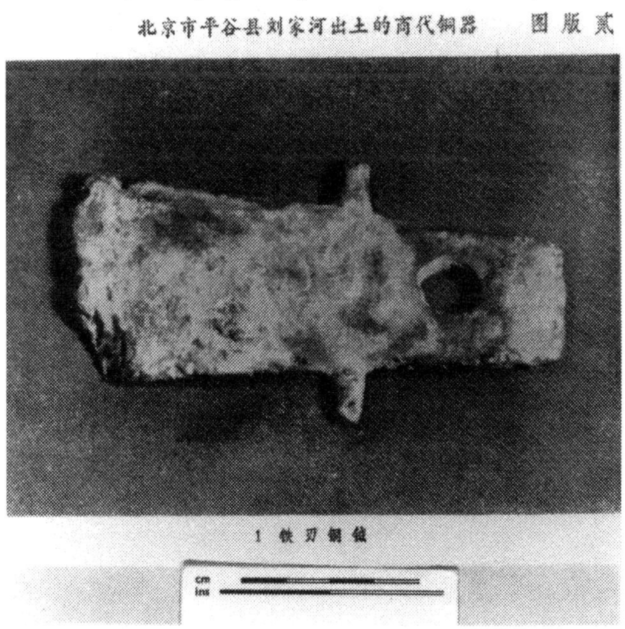

Figure 2: Photograph of a bronze-iron yue axehead found in a Shang-period grave excavated at Liujiahe in Pinggu County, Beijing Municipality, reproduced from Yuan Jinjing & Zhiang Xiande 1977, pl. 1.1; cf. Anon 1990: 37-9. Remaining dimensions 8.4 x 5 cm.

So little information is presently available, that it is not possible to make a definite statement on the dating of these

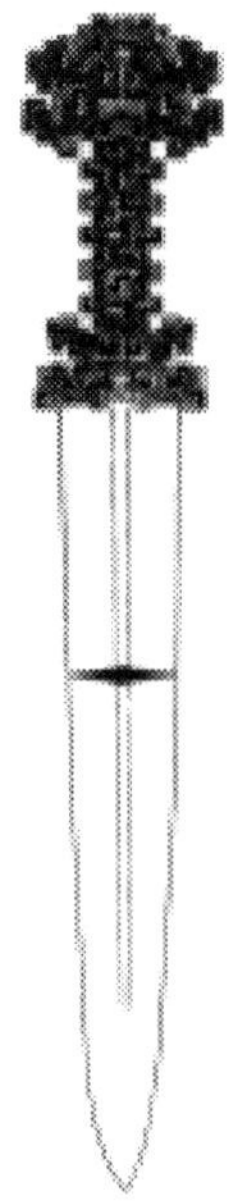

Figure 3.

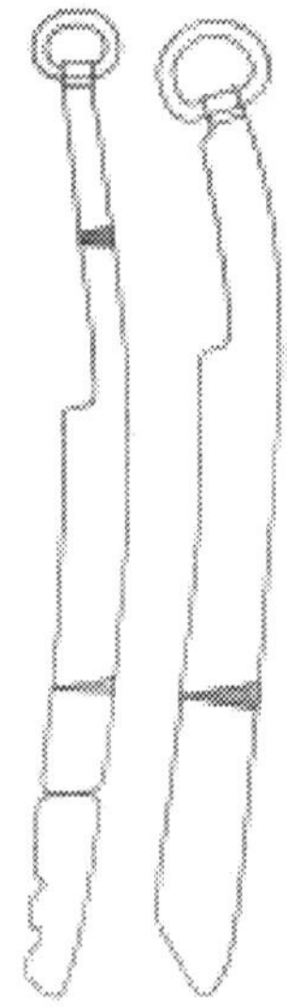

Figure 4.

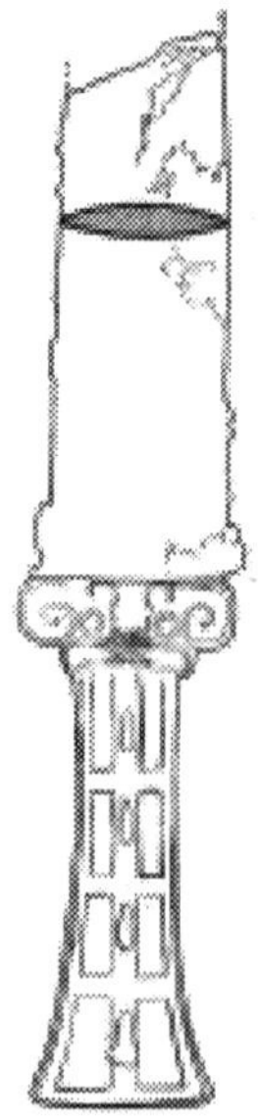

Figure 5.

Figure 3: Sketch of an iron short-sword with inlaid gold hilt from Grave number M2 at Yimencun in Baoji Municipality, Shaanxi (artefact number M2: 1), reproduced from Tian Renxiao 1993: 4, figure 7. Total length 35.2 cm, breadth of blade 4 cm.
Figure 4: Sketch of two knives with gold ring-heads from Grave number M2 at Yimencun in Baoji Municality, Shaanxi, reproduced from Tian Renxiao 1993: 6, figure 11. Right (artefact M2: 4): iron blade; total length 23.4 cm. Left (artefact M2: 18): bronze blade, fragment length 24 cm.
Figure 5: Sketch of a fragment of a bronze-iron short-sword from Tomb number M1 at Jingjiazhuang in Lingtai County, Gansu (artefact M1: 14), reproduced from Liu Dezhen and Zhu Jiantang 1981: 299. Length of bronze halt 8.5 cm, fragmentary iron blade 9 cm.

artefacts. Earlier excavations in the same place, in 1956–7, included 234 burials and four related horse-burials (Lin Shou-chin 1962; Lin Shoujin 1959; 1961; 1978; Li Xueqin 1985: 80–84; Anon. 1984: 283–5; Anon. 1991: 343). Bronze inscriptions indicate that this was a cemetery of the minor state of Guo, including what appears to be the grave of a Crown Prince. The newly excavated graves are said also to be of the state of Guo. Historical sources indicate that Guo was conquered in 655 BC, and this date is believed to be the *terminus ante quem* for the graves.

Until further information is forthcoming on the iron artefacts and their context we can probably assume that they date between the ninth and seventh centuries BC. An interesting aspect is a possible connection westward, to the vicinity of Baoji, where we shall see that other early iron swords have been found. Written sources indicate that there was a state named Guo near Baoji; whether the Guo of the Shangcunling graves was related to this Guo is less certain, but some art historians believe they can see elements of nomadic culture in the décor of the Shangcunling bronzes.[4]
. . .

Finding meteoritic and smelted iron in the same context, seemingly wrought by the same artisans, is surprising, nearly as surprising as finding smelted iron at all at such an early date in China. No doubt, more surprises are waiting for us, but it is clear that we must now look to the steppelands for the source of the earliest technology of iron smelting in China. Further evidence of this is afforded by a grave in Baoji, Shaanxi.

3.2. Gold–iron artefacts in a grave in Shaanxi

A grave in Baoji Municipality, Shaanxi, dated to the late Spring and Autumn period, contains some remarkable gold–iron artefacts, including three short-swords (e.g. Figure 3), thirteen ring-headed knives, two knives with gold ring-head, glass back, and iron edge (Figure 4), and two other knives (Tian Renxiao 1993; Li Xueqin 1993; Zhang Tian'en 1993). One of the short-swords has been examined by metallurgists, and was found to be of smelted, rather than meteoritic iron (Bai Chongbin 1994).

This is a typical Chinese small grave, but the grave-goods are highly unusual for a Chinese grave: there are no ceramic artefacts at all; there is a great deal of gold; and there are many artefacts which are more typical of the steppe cultures. It is difficult to date, but the date given by the excavators, the sixth century BC, seems secure enough.

With iron-using nomadic cultures close by, and clear signs of cultural influence from them, it is necessary to conclude that either the iron artefacts themselves, or the technology for their fabrication, came from the steppe peoples. The ring-handled knives might, in principle, have been made in China or anywhere in the steppelands. The origin of this design has long been a problem for archaeologists and art historians,[5] but the best guess at the moment is that its earliest development took place in Western Siberia.[6] It was a simple and useful design, and spread quickly; it became one of the typical Shang bronze artefact types, and is also found throughout the steppelands. The short-swords, on the other hand, are more specifically Chinese; the décor of their hilts is unusual, but similar sword-hilts have been found in China proper, and apparently nothing like them has been found outside China. While the sword-blades could have been made anywhere, the hilts were surely made by Chinese craftsmen, and it is only sensible to suppose that the blades were also made locally.

3.3. A Qin tomb in Gansu

A bronze–iron short-sword (Figure 5) comes from a tomb excavated in Lingtai County, Gansu (Liu Dezhen & Zhu Jiantang 1981). Lingtai is only about 60 km north of the ancient capital of Qin in Fengxiang County, Shaanxi, and comparison of burial styles, as well as artefact styles, shows clearly that this is a Qin tomb. The excavators date it to the eighth century BC, while Chen Ping argues for a slightly later date, in the seventh century BC (Chen Ping 1984a: 60).

A sample taken from the sword-blade by workers at the Beijing University of Science and Technology turned out to be totally corroded, and they felt that it was not possible to determine whether the iron was meteoritic or smelted.

3.4. The introduction of the bloomery

The evidence presented thus far suggests that iron-smelting techniques developed in the West had, by the eighth century BC, been brought by nomadic peoples of Central Asia all the way to the Pacific coast, and that these techniques diffused to the smiths of the Chinese states by way of various non-Chinese peoples of the northwest, in what is now Xinjiang. The Chinese smiths, now using smelted iron instead of meteoritic iron, continued to produce luxury articles such as swords and knives, with fittings of gold, jade, and bronze.

The earliest iron artefacts of this class are found in the northwest, but by the end of the sixth century BC, a few examples are found quite far away.[7] Perhaps the latest example, and the one found farthest afield, is a jade-hafted dagger found in what may have been a royal tomb in the northern part of Chu.[8] Almost nothing has been published about this interesting artefact, only a brief verbal description indicating that the blade has the shape of a willow leaf, with a rounded point. The total length is 22 cm, length of the cutting edge 12 cm, width 2.2 cm. It is entirely corroded, and no metallographic examination has been attempted. The excavators date the tomb to the beginning of the fifth century BC.

What part did these luxury edged weapons, of iron decorated with bronze, precious metals and jade, play in the later development of iron technology in China? We lack the technical information needed for a careful answer to this question, but there is some reason to suggest that they had no great importance at all. One metallographic examination of a relevant artefact has been published. This is one of the short-swords from Baoji.[9] The blade is of pure iron, with zero carbon, rather than steel, and this was not a better metal for the purpose than bronze. In Roman Europe, Gallic swords were often of pure iron, and were so soft that they bent in use, so that warriors had to fall back from battle and straighten them. The Gauls used soft iron because it was available, not because it was better than bronze; this cannot be the explanation for the use of iron in the luxury products we have seen here.

. . .

There is a possible exception to the pattern in which the early iron artefacts found in northwest China are all luxury wrought iron products. The tomb of a Duke of Qin, believed by many to be Duke Jing, who died in 537 BC, was excavated in Fengxiang County, Shaanxi, in the period 1978–86. In the fill of the tomb 'more than ten' iron digging implements were found. In a published photograph, one of these appears to be almost surely cast iron (*Guangming ribao* ("Daily", a newspaper) 1986.5.15: 4; reproduced, Wagner 1993: 93, fig. 2.41). At the moment, we are dependent on journalistic accounts for all information about this tomb.[10] It is therefore not certain that the tomb really is as early as this, and more importantly, until a proper excavation report is published, it will not be clear whether the digging implements were found in undisturbed earth or in robber tunnels of later times. This would not be the first time that iron digging implements have been found in ancient robber tunnels (e.g. Anon. 1982: 73, 113).

If, when the excavation report is published, it turns out that this tomb is indeed as old as some believe, and if the cast-iron digging implements are contemporary with the tomb, then the discussion that follows here will require radical revision, for I shall claim that the use of cast iron for practical implements began in south China, in a very different cultural context from that of Qin. It will be interesting to see how this question resolves itself.

4. Iron-casting and the blast furnace

The new evidence reviewed above definitely shows that wrought bloomery iron was the first smelted iron to be used in China. This has been a surprise, for not many years ago, the available evidence was consistent with the hypothesis that China started with *cast* iron, only later developing the techniques of the blacksmith (Barnard & Satö 1975: 67; Wagner 1993: 145–6). But while the use of wrought iron and the bloomery started in the northwest, and was a borrowed Western technique, iron casting was a Chinese invention. It seems to have started in the south.

If we restrict our view to properly published and reliably dated archaeological material, the earliest cast iron artefacts have been found in south China. Perhaps the most famous is a lump of cast iron from a grave dated to the early fifth century BC in Luhe, Jiangsu, in the territory of the ancient state of Wu. Other iron artefacts from Wu which are either uncertainly dated or not properly published may be older (Wagner 1993: 81–5).

In the territory of the ancient state of Chu, most of the evidence suggests a later date than this for the use of cast iron. The excavation of some 558 small and medium-sized graves at Yutaishan in Jiangling, Hubei, near the site of Ying, capital of Chu from 689 to 278 BC, has provided a chronological series for many aspects of Chu material culture.[11] Here iron artefacts do not appear until the early fourth century BC, and this suggests an approximate date for the widespread use of iron in Chu – though not, necessarily, for its *first* use there. The actual site of the capital has not yielded iron artefacts which are definitely earlier than this (Wagner 1993: 86). As yet, there have been no metallographic examinations of these artefacts, but all appear to be of cast iron.

We have already noted a luxury iron dagger, presumably wrought, found in a royal tomb in the northern part of Chu, dated to the end of the sixth century BC. Practical iron implements, apparently cast, which may be equally old, have been found at a Chu settlement site in Jingmen, Hubei, 75 km north of Jiangling (Li Zhaohua 1990: 17–18, 22–3).

. . .

While it is now difficult to maintain that iron-smelting itself was independently developed in south China, it still seems quite likely that iron casting, and perhaps also the blast furnace, were developed here. And it still seems quite reasonable to connect this development with the use of bronze for agricultural implements in the state of Wu.

The written evidence shows that the population of Wu was seen in the north as 'barbarian', and archaeological evidence shows that the material culture was indeed in significant ways different from that of the north.[12] They seem, in particular, to have had very different customs regarding the burial of the dead. These, together with some curious ritual animal-horns found in Wu contexts, suggest a completely different spiritual culture. On the other hand, they adopted some important elements of material culture from the north, among which was bronze technology.

Bronze production was, in the Chinese states of the north, concentrated in a few very large centres. Over the centuries, a narrow range of smelting and casting techniques was intensively developed, and the bronze-casting techniques developed in the Shang and Zhou foundries produced castings which are among the finest the world has known, even in modern times.

It was this highly sophisticated technology which the 'barbarians' of the south learned from their northern contacts – exactly when this occurred is a matter of debate, but not important here. In a new cultural context, this technology could be developed in new ways. New techniques were invented, and these techniques were applied to new uses, in particular to the production of agricultural implements (e.g. Figures 6, 7). Surprising numbers of these have been found, while comparatively few bronze implements, and none of the forms found here, are known from the north. Especially interesting is a number of bronze caps for digging implements, which are clearly the models for later iron implement-caps.

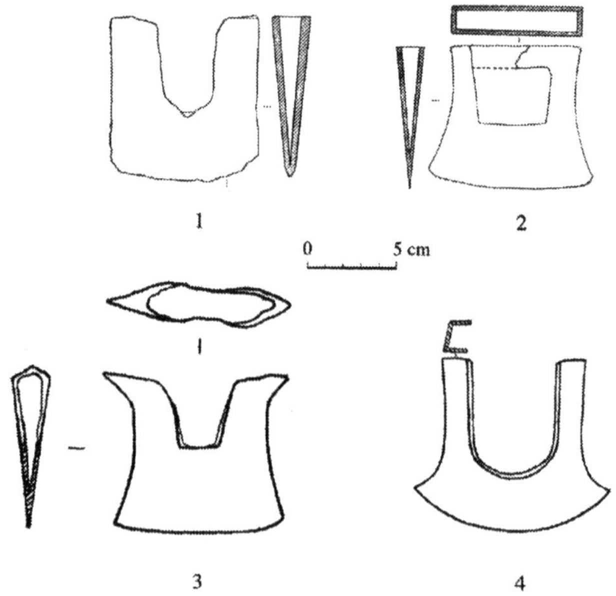

*Figure 7: Bronze caps for digging implements. Provenances: **1.** Tomb number 2 at Chengquiao in Luhe, Jiangsu, dated early 5th century BC, Anon. 1974: 119. On this tomb see Wagner 1993: 60-80. **2.** From a tomb at Jiulidun in Shucheng, Anhui, dated by the excavators, on the basis of bronze inscriptions, to the late Spring and Autumn period. Length 7.8 cm, breadth of edge 9.4 cm. Yang Jiuxia 1982: 237. **3.** Found in a scrap metal heap in Nanchang, Jiangxi, in 1975. Length 5.8 cm, breadth of edge 6.2 cm. Li Jiahe et at. 1977:60, 61; these authors date the artefact to the Shang period, but do not state the reason for this judgement. **4.** Found in the fill of tomb number M5 at Xi'eshan in Jiangling, Hubei, dated by the excavators to the fourth century BC. Length 9.8 cm, breadth of edge 11 cm. Yang Dingai 1984: 522.*

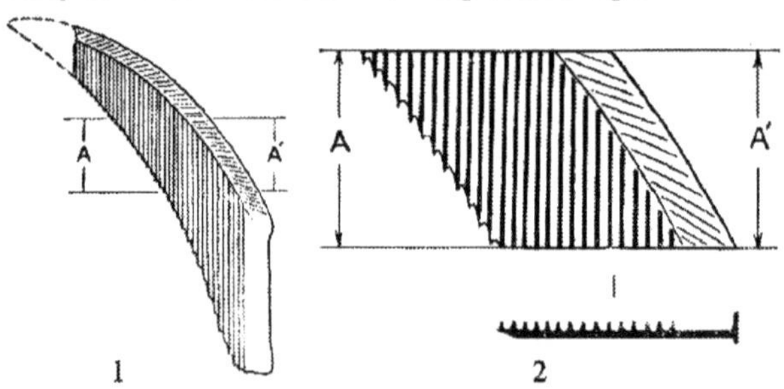

Figure 6: Saw-toothed sickle-blade, with magnified detail and cross-section, reproduced from Yun Hsiang 1985: 260, figure 4.

In the Wu region, copper deposits are small and scattered, and tin and lead deposits are rare; bronze agricultural implements must have been expensive. As agriculture in the region developed and became dependent on metal implements, it was necessary to find a cheaper metal. In one way or another, it was discovered that iron could be cast; the problems involved in casting white cast iron could be solved, at least adequately for castings which need not be beautiful, because the Wu ironfounders had at their command the highly sophisticated bronze-casting techniques developed in the north.

This hypothesis is strengthened somewhat by the fact that the earliest cast-iron implements known, the implement caps found in Chu graves of the fourth century BC and later, have clear bronze prototypes found in the Wu region. A weakness, on the other hand, is that no early iron implements of any kind excavated in the Wu region have yet been properly published.[13] An obvious answer to this objection is that Wu archaeology is still in its infancy; most finds of metal artefacts have been made by non-archaeologists, who might easily fail to recognize a badly corroded piece of iron as a valuable artefact, or even as an artefact at all (Wagner 1993: 83–5). Nevertheless, the possibility should be held open that the development described here occurred in Chu rather than Wu: it is in Chu that the earliest cast-iron implements were found, and a few bronze agricultural implements have also been found in Chu contexts.

. . .

My original suggestion was that the use in Wu of large copper-smelting furnaces led directly to the discovery of iron. In these furnaces, iron ores were used as a flux to reduce the melting point of the silica gangue of the copper ore. Much more iron ore than copper ore was used, and in normal operation the iron left the furnace harmlessly in the form of FeO in the slag. Improper furnace operation could lead to the production of metallic iron, and experimentation could have led in one way or another to the development of the iron blast furnace.[14]

That hypothesis might still explain the development of *cast* iron, but it has always suffered from the fact that it gives no explanation for the use of *wrought* iron in Wu just as early as cast iron (Wagner 1993: 146). It now seems more likely that bloomery iron smelting was introduced in Wu from the north, and that the first iron casting involved the carburization and melting of iron blooms in a cupola furnace of the sort used for melting bronze. Precisely this is described in a German manuscript of 1454, and it is possible that iron-casting began in Europe in this way (Johannsen 1910; 1913).

. . .

In whatever way cast iron was discovered, for most purposes it would not have been a good material. It would have been very low in silicon, and therefore be what is called *white* cast iron, which is difficult to cast and also very hard and brittle. The only advantage would have been that it was cheaper than bronze. In the north, the primary applications for bronze were weapons and symbolic objects of various kinds. Substitution of white cast iron for bronze in weaponry would have had disastrous effects, and casting difficulties would have made its use unlikely in fine ritual vessels.

The blast furnace and the finery were in use in China by the first century BC. If we accept that the earliest iron smelting in the south used the bloomery, and that iron casting began with the melting of iron blooms in a cupola furnace to cast agricultural implements, then the development of the blast furnace is quite easy to explain. Bloomery smelting is a delicately balanced operation, and can easily produce cast iron by mistake, as various modern experimenters have discovered. In the early West, until the advent of the iron cannon in Medieval times, no economically important use was known for cast iron, and the only intention in bloomery smelting was to produce wrought iron or steel. In some types of bloomery, great skill was required to limit the carbon content of the iron produced (e.g. Percy 1864: 326).

Lumps of cast iron produced in the bloomery would, however, melt just as easily in the cupola furnace as iron blooms (probably more easily), so that iron smelters who were engaged in producing iron for casting would have had no reason to care about the kind of iron they produced. Experience might easily have led, as many historians believe it did later in Medieval Europe, to the high bloomery or *Stückofen*, which often produced both wrought-iron blooms and flowing cast iron in the same operation. From there the development of the much more efficient blast furnace would have been straightforward. With the development of the finery to convert cast iron into wrought iron, the production of bloomery iron would have become uncompetitive, and in time, gone out of use.

We do not know where and when this development might have taken place, nor do we know anything about the actual furnaces used for iron production in China before late in the first century BC, so it would be unprofitable to speculate more here concerning details of the development of the blast furnace in China.

. . .

Recent studies of the origins of iron-smelting in the West stress that iron was not in the beginning a better metal than bronze. Iron seems to have been known here as early as 3000 BC, but first became an important metal about 1200 BC. It was originally put to use, for what amount to economic reasons, because of an increased demand or a decreased supply of either copper or tin. Once the iron had been put to use, the smiths, working intensively with it, gradually learned more about its properties and possibilities. In particular they discovered various ways of making steel by adding carbon into the iron. Once these discoveries had been made, developed, and widely diffused, iron became the material of choice for most weapons and implements.

There is, thus, a certain parallel between the Western story and the story told here for China. In the beginning, in China, in about the eighth century BC, smelted iron may have come into use as a cheap and inferior substitute for meteoritic iron in luxury weapons. As far as we know, there were no great improvements in the techniques of the smith until, in the south, perhaps in the sixth century BC, the founders of bronze agricultural implements discovered that they could cast iron and thus produce a useful product from a metal more readily available than bronze. How the cast-iron implements compared with bronze ones in practical use is difficult to know. The extreme hardness of white cast iron would have made it extremely wear-resistant, but also liable to break. The answer to this question is, however, not as important as the fact that in this application, iron was surely *economically* superior to bronze.

Experience with the new material, cast iron, led to great improvements in its mechanical properties. In particular, the development of what we now call *malleable cast iron* made cast iron a better material than bronze and at least competitive with wrought iron or steel.

References

AN ZHIMIN. 1990. Zhongguo kaoguxue de xinqidian – jinian Huixian fajue sishi nian (The starting point of Chinese archaeology: celebrating the fortieth anniversary of the Hui County excavations), *Wenwu tiandi* (All about cultural relics) 5: 30–34.

ANON. 1956. *Huixian fajue baogao*. (Excavations in Hui County, Henan). Beijing: Kexue Chubanshe.

1974. Jiangsu Luhe Chengqiao er-hao Dong Zhou mu (Eastern Chou Grave no. 2 at Chengqiao in Luhe County, Jiangsu) *Kaogu* (Archaeology) 2: 116–20 + plates 4–6.

1977. *Gaocheng Taixi Shang dai yizhi* (A Shang-period site at Taixi in Gaocheng County, Hebei). Beijing: Wenwu Chubanshe.

1980. *Hebei sheng chutu Wenwu xuanji* (Selected cultural relics unearthed in the province of Hebei). Beijing: Wenwu Chubanshe.

1982. Jiangling Tianxingguan yihao Chu mu (Chu Tomb no. 1 at Tianxingguan in Jiangling, Hubei). *Kaogu xuebao* (Acta archaeologia Sinica) 1: 71–116 + plates 11–24. English abstract p. 116.

1984. *Jiangling Yutaishan Chu mu* (Excavation report on 558 Chu graves at Yutaishan in Jiangling, Hubei). Beijing: Wenwu Chubanshe. English abstract pp. 192–4.

1986a. Surveys in the Qin capital. *China pictorial* (Beijing), 1: 14–17.

1986b. Ancient tomb adds new clue to history. *Beijing review*, 20: 6–7.

1990. *Beijing kaogu sishi nian* (Forty years of archaeology in Beijing). By Beijing Municipal Institute of Archaeology. Beijing: Yanshan Chubanshe. English abstract pp. 217–21.

1991. *Xichuan Xiasi Chunqiu Chu mu* (Chu tombs of the Spring – Autumn period at Xiasi, Xichuan). Beijing: Wenwu Chubanshe. English abstract, pp. 461–4.

1992. *Gems of China's cultural relics*. Beijing: Cultural Relics Publishing House. English and Chinese text. Chinese title: *Zhongguo wenwu jinghua*.

BAI CHONGBIN. 1994. Baoji shi Yimencun M2 chutu Chunqiu tie jian cankuai fenxi jianding baogao (Report on technical examination of a fragmentary iron sword from Tomb no. 2 at Yimencun in Baoji Municipality, Shaanxi). *Wenwu* (Cultural relics) 9: 82–5.

BARNARD, NOEL & SATŌ TAMOTSU. 1975. *Metallurgical remains of ancient China*. Tōkyō: Nichiōsha.

BARNES, GINA L. 1993. *China, Korea and Japan: The rise of civilization in East Asia*. London: Thames and Hudson.

CHARD, CHESTER S. 1961. First radiocarbon dates from the U.S.S.R. *Arctic anthropology* 1(1): 84-86.

1974. *Northeast Asia in prehistory*. Madison: University of Wisconsin Press.

CHEN FENGXIN. 1990. *Jiangling Yutaishan Chu mu fajue jianbao* (Excavation of Chu graves at Yutaishan in Jiangling, Hubei). *Jiang Han kaogu* 3: 1–7 + plate 1.

CHEN GE. 1981. Pamier gaoyuan gu mu (Ancient tombs on the Pamir Plateau). *Kaogu xuebao* (Acta archaeologia Sinica) 2: 199–216 + plates 1–4. English abstract p. 216.

1987. Guanyu Xinjiang xin shiqi shidai wenhua de xin renshi (New insights on the Neolithic cultures of Xinjiang). *Kaogu* (Archaeology) 4: 343–351 + 322.

1989. Xinjiang chutu de zaoqi tieqi – jiantan woguo kaishi shiyong tieqi de shijian wenti (Early iron artefacts unearthed in Xinjiang: With a note on the earliest use of iron in China). *Su Bingqi* 1989: 425–432.

1990. Guanyu Xinjiang diqu de qingtong shidai he zaoqi tieqi shidai wenhua (The Bronze Age and Early Iron Age culture in Xinjiang). *Kaogu* (Archaeology) 4: 366–374.
CHEN PING. 1984a–b. Shilun Guanzhong Qin mu qingtong rongqi de fenqi wenti (The periodisation of bronze vessels from Qin tombs of the central Shaanxi plain, parts 1–2). *Kaogu yu Wenwu* (Archaeology and cultural relics) [a:] 3: 58–73; [b:] 4: 63–73.
CHEN ZHENZHONG. 1985. Woguo gudai de qingtong xiaodao (Ancient Chinese bronze scraper-knives). *Kaogu yu Wenwu* (Archaeology and cultural relics) 4: 72–80 + 83.
CHEN ZHI. 1961. Changsha fajue baogao de jidian buzheng (Some corrections to Anon. 1956)). *Kaogu* (Archaeology) 5: 264.
CHERNYKH, E. N. 1992. *Ancient metallurgy in the USSR: The early metal age.* Translated by Sarah Wright. Cambridge: Cambridge University Press. (*New studies in archaeology*).
CONG DEXIN & CHEN GE. 1991. Xinjiang Luntai xian Qunbake muzang di er, san ci fajue jianbao (Second and third excavation seasons of graves at Qunbake in Luntai County, Xinjiang). *Kaogu* (Archaeology) 8: 684–703 + 736 + plates 1–3.
CRADDOCK, P. T. 1994. Review of Wagner 1993. *Antiquity* 1994, 68 (261): 886–890.
DEBAINE–FRANCFORT, C. 1989. Archéologie du Xinjiang des origines aux Han. IIème partie: L'âge du fer. *Paléorient* 15 (1), 183–213.
DIEN, ALBERT E., JEFFREY K. RIEGEL, & NANCY T. PRICE (ed.). 1985. *Chinese archaeological abstracts, vols. 2–4.* Institute of Archaeology, Los Angeles: University of California. (*Monumenta archaeologica*, vols. 9–11.) Cf. Rudolph 1978.
VON FALKENHAUSEN, LOTHAR. 1994. Review of Wagner 1993. *Chinese science*, 11: 103–7.
FONG, WEN (ed.). 1980. *The great Bronze Age of China: An exhibition from the People's Republic of China.* London: Thames and Hudson. Orig. New York: Metropolitan Museum of Art / Alfred A. Knopf.
GETTENS, RUTHERFORD J., ROY S. CLARKE, & W. T. CHASE. 1971. *Two early Chinese bronze weapons with meteoritic iron blades.* Washington, D.C.: Freer Gallery of Art. (Freer Gallery of Art, *Occasional papers*, vol. 4, no. 1).
HAN RUBIN, JIANG TAO, & HE SONGLIN. 1994. Meteoritic and man-made iron artifacts of 8–9c BCE, Sanmenxia, Henan, China. In *BUMA-3: The Third International Conference on the Beginning of the Use of Metals and Alloys, 18–23 April* 1994, Sanmenxia, pp. 69.
HAN WEI. 1987. In splendor laid. *China pictorial* 5: 14–17.
HUANG ZHANYUE. 1976. Guanyu Zhongguo kaishi yetie ho shiyong tieqi de wenti (On the problem of the first smelting of iron and use of iron implements in China). *Wenwu* (Cultural relics) 8: 62–70.
JOHANNSEN, OTTO. 1910. Eine Anleitung zum Eisenguss vom Jahre 1454. *Stahl und Eisen* 30 (32): 1373–1376.
1913. Die Bedeutung der Bronzekupolöfen für die Geschichte des Eisengusses. *Stahl und Eisen* 33 (26): 1061–1063.
KARLGREN, BERNHARD. 1945. Some weapons and tools of the Yin Dynasty. *Bulletin of the Museum of Far Eastern Antiquities* 17: 101–44 + plates 1–40.
KIM, JEONG-HAK. 1978. *The prehistory of Korea.* Translated by Richard J. Pearson & Kazue Pearson. Honolulu: University Press of Hawaii.
LI CHUNG. 1979. Studies on the iron blade of a Shang dynasty bronze *yüeh*-axe unearthed at Kao-ch'eng, Hopei, China. *Ars Orientalis*, 1979, 11: 259–89. Tr. of *Kaogu xuebao* (Acta archaeologia Sinica) 2: 17–34 + plates 1–8.
LI JIAHE (*et al.*). 1977. Jinnian Jiangxi chutu de Shang dai qingtong qi (Shang-period bronze artifacts unearthed in recent years in Jiangxi). *Wenwu* (Cultural relics) 9: 58–63.
LI WEIMING. 1988. Jianlun Shangdai qingtong dao (A brief study of the bronze knives of the Shang period). *Zhongyuan Wenwu* (Cultural relics from Central Plains) 2: 42–7.
LI XUEQIN. 1985. *Eastern Zhou and Qin civilizations.* Translated by K. C. Chang. New Haven & London: Yale University Press. (*Early Chinese civilizations series*). Orig. *Dong Zhou yu Qin dai wenming*, Beijing: Wenwu Chubanshe, 1984.
1993. Yimencun jin, yu qi wenshi yanjiu (The décor of the gold and jade artefacts in Tomb no. 2 at Yimencun in Baoji Municipality, Shaanxi). *Wenwu* (Cultural relics) 10: 15–19 + colour plate + plates 1–4.
LI ZHAOHUA. 1990. Jingmen shi Xianglinggang Dong Zhou yizhi yu mudi fajue jianbao (Excavation of an Eastern Chou site and cemetery at Xianglinggang in Jingmen Municipality, Hubei). *Jiang Han kaogu* (Jianghan archaeology) 4: 12–23 + 55 + plates 1–3.
LIN SHOU-CHIN [LIN SHOUJIN]. 1962. Culture of the state of Kuo. *China pictorial*, 3: 10–13.
LIN SHOUJIN. 1959. *Shangcunling Guo guo mudi: Huanghe shuikubaogao zhi san* (The cemetery of the state of Kuo at Shang Ts'un Ling: Archaeological excavations at the Yellow River Reservoirs, Report no. 3). Beijing: Kexue Chubanshe (*Zhongguo tianye kaogu baogao ji. Kaoguxue zhuankan*, D.10) English abstract, pp. 83–85.
1961. 'Shangcunling Guo guomudi' buji (Supplementary remarks to Lin Shoujin 1959). *Kaogu* (Archaeology) 9: 505–7.
1978. Shangcunling fajue de xueshu gongxian (Academic contributions of the excavation at Shang Ts'un Ling). *Xianggang Zhongwen Daxue Zhongguo Wenhua Yanjiusuo xuebao*, 9: (*shang*) 1–15 + plates 1–15. English summary, p. 15.
LIU DEZHEN & ZHU JIANTANG. 1981. Gansu lingtai xian Jingjiazhuang Chunqiu mu (Tombs of the Spring and Autumn period excavated at Jingjiazhuang in Lingtai County, Gansu). *Kaogu* (Archaeology), 4: 298–301 + plates 3–5.
LIU YIMAN. 1993. Yinxu qingtong dao (Bronze knives from the Yin Ruins in Anyang, Henan). *Kaogu* (Archaeology) 2: 150–66.
MURAKAMI KYŌTŌ. 1991. Dongbei Ya de zaoqi tieqi shidai (The Early Iron Age in Northeast Asia). *Beifang wenwu* (Northern cultural relics) 1: 94–101. Chinese tr. by Qiao Liang. Japanese orig. in *Kodai bunka* 1987, no. 9.
NELSON, SARAH MILLEDGE. 1993. *The archaeology of Korea.* Cambridge: Cambridge University Press. (*Cambridge world archaeology*).
OKLADNIKOV, A. P. 1965. *The Soviet Far East in antiquity: An archaeological and historical study of the maritime region of the U.S.S.R.* Translated by Dr. & Mrs. Stephen P. Dunn; ed. by Henry N. Michael. Toronto: University of Toronto Press. (Arctic Institute of North America, *Anthropology of the North: Translations from Russian sources* 6).
PAUL-DAVID, MADELEINE 1954. 'Les fouilles de Houei-hien.' *Arts asiatiques*, 1(2): 157–160.
PERCY, JOHN. 1861. *Metallurgy: The art of extracting metals from their ores, and adapting them to various purposes of manufacture. [Vol. 1:] Fuel; fire-clays; copper; zinc; brass; etc.* London: John Murray. Facs. repr. in 2 pts., Eindhoven: De Archaeologische Pers Nederland, n.d. [ca. 1985].
1864. *Metallurgy . . . [Vol. 2:] Iron; steel.* London: John Murray. Facs. repr. in 3 pts., Eindhoven: De Archaeologische Pers Nederland, n.d. [ca. 1983].
PU CHAOFU. 1981. Gansu Yongdeng Yushugou de Shajing muzang (Graves of the Shajing Culture at Yushugou in Yongdeng County, Gansu). *Kaogu yu Wenwu* (Archaeology and cultural relics) 4: 34–36 + plates 5–6.
PU CHAOFU & ZHAO JIANLONG. 1984. Gansu Yongchang Sanjiaocheng Shajing wenhua yizhi diaocha (Investigation of a site of the Shajing Culture at San-chiao-chhêng in Yongchang County, Gansu). *Kaogu* (Archaeology) 7: 598–601.
RUDENKO, SERGEI I. 1970. *Frozen tombs of Siberia: The Pazyryk burials of Iron Age horsemen.* Translated by M. W. Thompson. London: J. M. Dent. Orig. *Kultura Naseleniya Gornogo Altaya v Skifskoe Vremya*, Moscow / Leningrad: Academy of Sciences of the U.S.S.R., 1953.
RUDOLPH, RICHARD C. (ed.). 1978. *Chinese archaeological abstracts.* Los Angeles: Institute of Archaeology, University of California. (*Monumenta archaeologica*, vol. 6). Cf. Dien *et al.* 1985.
SU BINGQI. 1989. *Qingzhu Su Bingqi Kaogu wushiwu nian lunwenji*) (Festschrift for Su Bingqi's fifty-fifth anniversary in archaeology). Beijing: Wenwu Chubanshe.
SU MINSHENG. 1987. Qin excavations: Human sacrifices, tomb robbers & horse pits. *China reconstructs* 2: 31–33.
SUN BINGGEN & CHEN GE. 1987. Xinjiang Luntai Qunbake muzang di-yi-ci fajue jianbao (First excavation season of graves at Qunbake in Luntai County, Xinjiang). *Kaogu* (Archaeology) 11: 987–996.
1988. Xinjiang Hejing xian Chawuhu Goukou yihao mudi (Cemetery no. 1 at Chawuhu Pass, Hejing County, Xinjiang). *Kaogu xuebao* (Acta archaeologia Sinica) 1: 75–98 + plates 7–16. English abstract p. 99.
TANG JIGEN. 1993. Zhongguo yetieshu de qiyuan wenti (The question of the origin of iron smelting in China). *Kaogu* (Archaeology) 6: 556–65 + 553.
TANG YUNMING. 1979. Hebei Gaocheng Taixicun Shang dai yizhi fajue jianbao (Excavations at the Shang-period site at Taixicun in Gaocheng County, Hebei). *Wenwu* (Cultural relics) 6: 33–43 + plate 6. Summary tr. by Loretta Gibbs in Dien, Riegel, & Price 1985: 424–435.
1985. *Gaocheng Taixi Shang dai yizhi* (A Shang site at Taixi, Gaocheng). Beijing: Wenwu Chubanshe. English abstract, pp. 209–13.
TANG YUNMING & LIU SHISHU. 1973. Hebei Gaocheng Taixicun de Shang dai yizhi (A Shang-period site at Taixicun in Gaocheng County, Hebei). *Kaogu* (Archaeology) 5: 266–271 + plates 1–3.
TAYLOR, SARAH. 1989. The introduction and development of iron production in Korea: A survey. *World archaeology* 20 (3): 422–433.
1990. *Ploughshares into swords: The iron industry and social development in protohistoric Korea and Japan.* Ph.D. dissertation, Cambridge.
TIAN RENXIAO. 1993. Baoji shih Yimencun er-hao Chunqiu mu fajue jianbao (Excavation of Tomb no. 2 of the Spring and Autumn Period at Yimencun in Baoji Municipality, Shaanxi). *Wenwu* (Cultural relics) 10: 1–14 + colour plate + plates 1–4.
WAGNER, DONALD B. 1987. The dating of the Chu graves of Changsha: The earliest iron artifacts in China? *Acta Orientalia (Copenhagen)* 48: 111–156.
1993. *Iron and steel in ancient China.* Leiden: Brill. (*Handbuch der Orientalistik, vierte Abteilung: China*, no. 9).
WANG HUI. 1989. Lun Qin Jing gong (Duke Jing of Qin). *Shixue yuekan* (Journal of historical science) 3: 19–25 + 12.
WU EN. 1978. Guanyu woguo beifang de qingtong duanjian (Bronze short-swords of North China). *Kaogu* (Archaeology) 5: 324–33 + 360.

YANG JIUXIA. 1982. Anhui Shucheng Jiulidun Chunqiu mu (A Spring and Autumn period tomb at Jiulidun, Shucheng, Anhui). *Kaogu xuebao* (Acta archaeologia Sinica) 2: 229–42 + plates 17–22. English abstract p. 242.
YI XU. 1986. Tomb of the First Emperor's ancestor. *China reconstructs* 9: 58–9 + plates.
YOON, DONG-SUK. 1984. *Metallurgical study of the early Iron Age artifacts found in Korea.* Pohang Iron and Steel Co.
YUAN JINJING & ZHANG XIANDE. 1977. Beijing shi Pinggu xian faxian Shang dai muzang (Shang-period tombs excavated in Pinggu County, Beijing Municipality). *Wenwu* (Cultural relics) 11: 1–8 + plates 2–5.
YUN XIANG. 1985. Chiren tong lian chulun (A preliminary study of toothed bronze sickles), *Kaogu* (Archaeology) 3: 257–66.
ZHANG PING (a.o.). 1989. Xinjiang Hami Yanbulake mudi (The Yanbulake Cemetery in Hami, Sinkiang). *Kaogu xuebao* (Acta archaeologia Sinica) 3: 325–362 + plates 7–14. English abstract p. 362.
ZHANG TIAN'EN. 1993. Qin qi san lun – Yimencun Chunqiu mu jige wenti jianlun (Three notes on Qin artefacts in Tomb no. 2 at Yimencun in Baoji Municipality, Shaanxi). *Wenwu* (Cultural relics) 10: 20–27 + colour plate + plates 1–4.
ZHANG XIANDE & ZHANG XIANLU. 1990. Beijing Pinggu Liujiahe Shang dai tong yue tie ren de fenxi jianding (Laboratory studies of the iron edge of a Shang-period bronze *yue*-axehead from Liujiahe in Pinggu county, Beijing municipality). *Wenwu* (Cultural relics) 7: 66–71.
ZHANG YUZHONG. 1985. Wulumuqi shi Nanshan Kuangqu gu muzang (Ancient tombs in the Nanshan Mining Region, Urumchi Municipality, Xinjiang). Anon. 1985c: 256.
ZHENG SHAOZONG. 1956. Rehe Xinglong faxian de Zhanguo sheng-chan gongju zhufan (Moulds for casting production implements discovered at a Warring States period site in Xinglong County, Rehe). *Kaogu tongxun* (Archaeological bulletin) 1: 29–35 + plates 9–10.
ZHONG SHAOYI. 1993. Shishuo 'Yu Shu baojian' – Jianshuo zaoqi tieqi yingyong de yige xianxiang (On 'the precious sword of Yu Shu' – and on an aspect of the early use of iron in China). *Wenwu tiandi* (All about cultural relics) 3: 13–15.
ZHOU SHIRONG. 1990. Jiantan Hunan de Chu wenhua (On Chu culture in Hunan), *Zhongyuan wenwu* (Cultural relics from the Central Plains) 2: 85–8.
ZOU HOUBEN. 1984. Wu-xian Wufengshan shishi tudun mu (Stone-chamber tumulus sites at Wufengshan in Wu County, Jiangsu). Anon. 1984a, 105–6.

Notes

[1] In this I largely followed Huang Zhanyue. Since his article the new excavations of the Chu capital in Jiangling, Hubei, had discredited certain assumptions on which the dating of the Changsha material had been based; see Wagner 1987; Wagner 1993: 86–90; Zhou Shirong 1990. With the Changsha material essentially undatable, this put the use of iron in Chu at least a century later than in Wu. The most recent survey of the problem of the origin of iron smelting in China, by Tang Jigen (1993), accepts the greater part of Huang Zhanyue's conclusions, unfortunately including the now-obsolete dating of the Changsha material.
[2] See especially Chen Ge 1989; also 1981; 1987; 1990; Cong Dexin & Chen Ge 1991; Sun Binggen & Chen Ge 1987; 1988; Zhang Ping a.o. 1989; Zhang Yuzhong 1985; Debaine–Francfort 1989: 192, 194, 196–7, 198, 201, 206–7.
[3] Chard 1961: 85, item 21, no. RUL-165. I am grateful to Prof. Sarah M. Nelson for this reference. Okladnikov (1965: 184–142) discusses the Shell Mound culture in detail, but does not mention early iron artefacts. Murakami Kyötö (1991) mentions the iron artefacts of this culture, but does not take them seriously.
[4] E.g. Jenny F. So in Fong 1980: 252; note especially the mirror decorated with tigers and a horse, Lin Shoujin 1959: 27, fig. 21, pl. 40.2; Lin Shou-chin 1962: 11.
[5] Important studies include Karlgren 1945; Chen Zhenzhong 1985; Li Weiming 1988; Chernykh 1992: 268–70; Wu En 1978; Liu Yiman 1993.
[6] Chernykh (1992: 269), who however also notes that the Russian scholar who has carried out the most detailed investigation of the problem, N. L. Chlenova, suggests the opposite: that the design originated in China.
[7] See e.g. Tang Jigen 1993: 558; von Falkenhausen 1994: 104–5. Note however that many of the artefacts listed by these authors are insufficiently documented and rather uncertainly dated.
[8] Tomb M10 at Xiasi in Xichuan county, Henan, artefact no. M10:33. Excavation report Anon. 1991: 292, 347; cf. Wagner 1993: 90–91.
[9] Bai Chongbin 1994. The author does not, unfortunately, give the artefact number, so we cannot be sure which sword was examined. Probably it was the one which in its present state is least presentable in a museum, and that is probably M2:3.
[10] In English: Yi Xu 1986; Su Minsheng 1987; Han Wei 1987; Anon. 1986a; Anon. 1986b; *Xinhua News Agency news bulletin* (Hong Kong), 3, 5, 12, & 23 May 1986. In Chinese: *Guangming ribao* ('Guangming Daily'), 28 April 1986, 1; 2 May 1986, 1; 3 May 1986, 1; 4 May 1986, 1; 15 May 1986, 4; 24 May 1986, 1; 13 July 1986, 3; 6 August 1986, 1; *Renmin ribao* ('People's daily'), 3 May 1986, 3; 9 May 1986, 3; 25 May 1986, 3; 16 September 1986, 3. Discussion, Wang Hui 1989: 20–21; Wagner 1993: 91–4.
[11] Excavation report Anon. 1984; discussion Wagner 1993: 85–86, 152–7. Note also a later excavation of 73 more graves at Yutaishan, Chen Fengxin 1990.
[12] On the ancient state of Wu see especially Wagner 1993, ch. 3.
[13] Two 'iron artefacts' are reported very briefly from Wu graves in Wu County, Jiangsu, which are believed to be of Spring and Autumn period, but more details will be needed before this find can be taken seriously. Zou Houben 1984.
[14] This begs a number of technical questions raised by my friend Paul Craddock (1994: 887). The fact to be emphasised is that iron *could* be produced in a blast furnace of the same size and shape as the ancient copper-smelting furnaces, as traditional Chinese blast furnaces of the nineteenth and twentieth centuries demonstrate. The exact development route from copper to iron production would no doubt have been less straightforward than I suggested, but it would certainly have been possible.

2

Reconstructing the copper production process as practised among prehistoric mining/metallurgical communities in the Khao Wong Prachan Valley of central Thailand

VINCENT C. PIGOTT

Museum Applied Science Center for Archaeology (MASCA), University of Pennsylvania Museum, Philadelphia, PA 19104-6324, USA

Early in the second millennium BC the production of copper and bronze began in Southeast Asia. The Thailand Archaeometallurgy Project (TAP), conducted over the past decade, has revealed the presence of major copper mining complexes and associated production settlements, the latter among the largest such sites known in Asia. The combined evidence from these sites comprises much of the primary production data by which the so-called "Southeast Asian Metallurgical Tradition" can be characterized. Broadly speaking this "Tradition", defined in terms of the evidence from Thailand, arose under somewhat unusual conditions. In contrast to most of the major culture areas of the Old World where large scale production of metal, in particular bronze, is associated with complex societies and urban settings, in Southeast Asia current evidence suggests that such metallurgy developed in village contexts. These villages were active in extensive networks of inter-regional exchange, but exhibit only modest degrees of social ranking, at least during the early stages of metallurgy's appearance and development in the prehistoric period.

This paper presents an overview of the archaeometallurgical evidence for mining and copper production from recently excavated prehistoric sites in central Thailand. It is argued that production, which will be reviewed for the sites of Non Pa Wai, Nil Kham Haeng, and Non Mak La, was accomplished by a low investment, "low-tech.", very labour-intensive method, which was specialized only at the household-family level and which may have been only a dry season industrial activity. Despite this level of production organization, the evidence suggests that prior to the appearance of chiefdom levels of social complexity, and under conditions of economic decentralization, these village-based communities mined significant volumes of ores and produced large amounts of copper.

Keywords: Southeast Asia; Thailand; Archaeometallurgy; Production Organization; Copper; Economy.

Introduction

In the course of the last decade archaeology has made significant strides towards improving our understanding of the origin, development and social impact of copper-base metallurgy in mainland Southeast Asia. Models to explain these metallurgical developments are currently being tested, accepted and/or rejected. Among those being currently explored is the concept of Southeast Asia as what A.L. Kroeber (1948:423) might have termed a technological 'oikumene' - a region characterized by shared cultural traditions and defined, in part, by a set of technological traditions. Similarly, Joyce White (1988) suggests that Southeast Asia may be a distinct metallurgical province or perhaps more likely a mosaic of such provinces. She characterized the Southeast Asia metallurgical province as follows:

"Although there is site to site and sub-region to sub-region variation, in terms of typology and technology the configuration of the prehistoric metals in Southeast Asia shows both an internal coherence and, taken as a whole, considerable distinction from other major areas of early metallurgy. The internal variation within the region and over time suggests that metal technology was neither stagnant nor imitative. Instead it was probably adapted to local resources and responsive to temporal, geographic, and cultural conditions within the region."

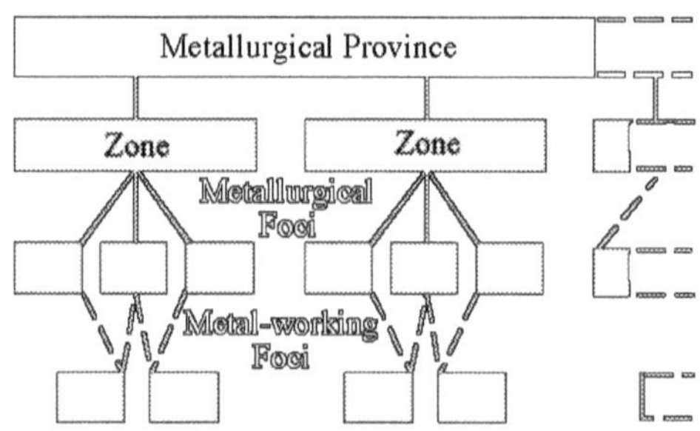

Figure 1. The structure of a metallurgical province (after Chernykh 1992, 9).

This follows on the concept of the metallurgical province (see Figure 1) defined by E.N. Chernykh (1980:320-1) as:

- "a system of kindred metallurgical and metal-working foci or centres, limited in space and time.
- The metallurgical foci were the principal production centres of a metallurgical province. Their production was characterized first by its metallurgy, by its types of alloys, and by the forms of metallic tools it produced.
- Metallurgical foci were, as a rule, located in copper ore zones where it was possible both to mine copper and to smelt the copper.
- Metal-working foci were mainly situated in zones without ore, and were dependent on the metallurgical foci for supplies of the metal. The craftsmen in metal-working foci very often imitated the forms of metallic

tools and decorations produced in the metallurgical foci.

- Metallurgical and metal-working foci were characterized by temporal and spatial limits.
- Foci of both types were associated with a definite archaeological culture or variant.
- At some later stages of development, however, metallurgical and metal-working foci often grew into super-cultural phenomena and embraced several cultures and their variants.
- Territorially and chronologically a metallurgical province was, as a rule, larger than even the largest prehistoric community.
- In the foci of such communities, metallurgical production was part of a metallurgical province only as a component of the more extensive system of production and its traditions."

Figure 2. An aerial view of the Khao Wong Prachan Valley during the rainy season. The heavily vegetated oval area to the left of the photograph is the prehistoric smelting site of Non Pa Wai.

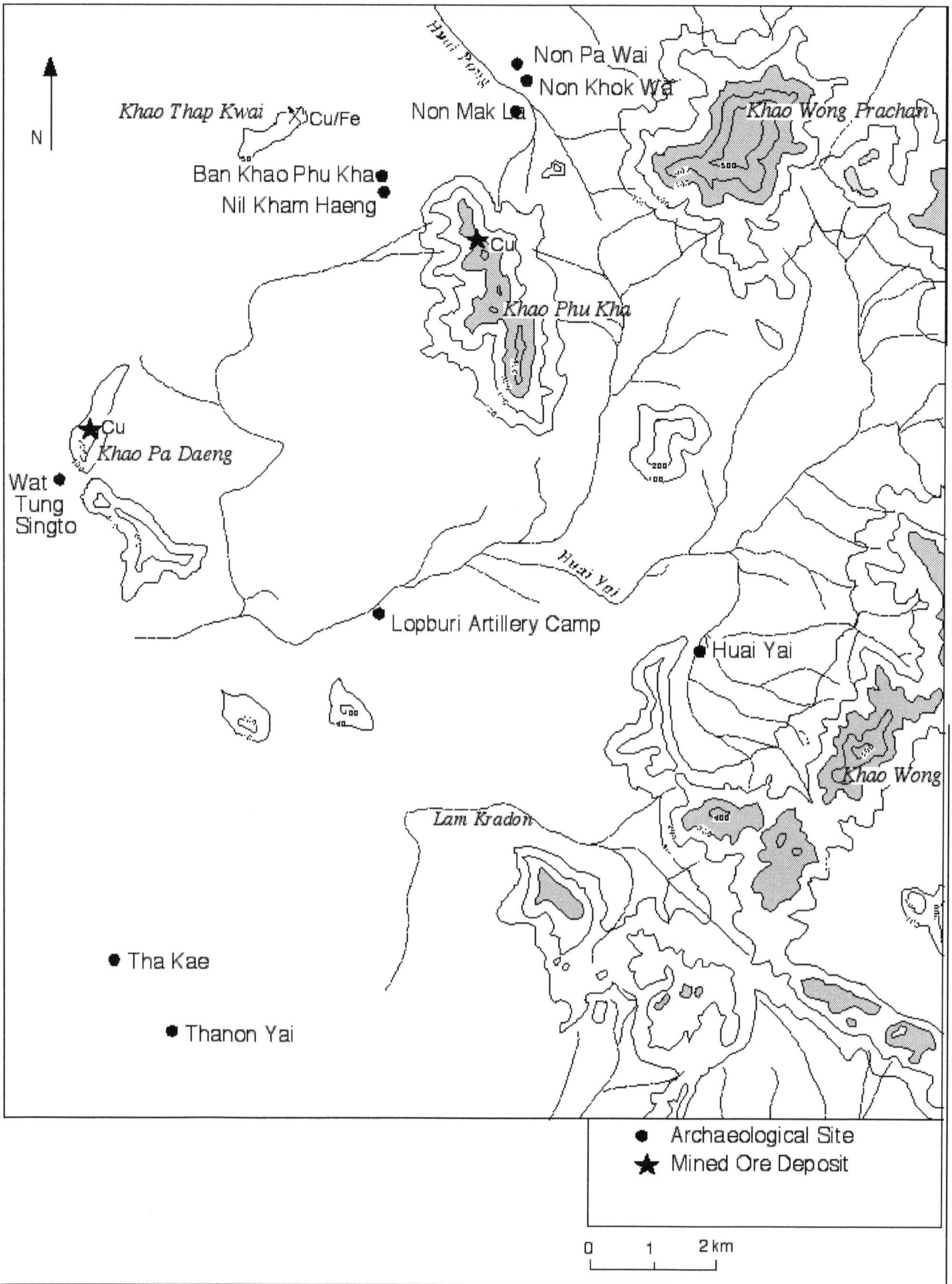

Figure 3. Map of Khao Wong Prachan Valley with archaeological sites and mined ore deposits indicated

Archaeology continues to reveal new material and gradually assembles the evidence that ultimately will support or dispel the validity of this and related concepts when applied to the context of prehistoric Southeast Asia. Between 1986 and 1994, through the focused efforts of the Thailand Archaeometallurgy Project, a small valley in central Thailand yielded evidence crucial in the continuing task of defining the concept of the metallurgical province. This valley, the Khao Wong Prachan Valley not far from the modern town of Lopburi was the prehistoric locus of a cluster of major copper producing settlements. In this cluster, we argue that craft specialization "largely developed within a community-based mode of production organization." This mode is defined by Costin (1991:8) as "autonomous individual or household-based production units, aggregated within a single community producing for unrestricted regional consumption." (see White and Pigott 1996:151). This discussion now turns to the archaeological and metallurgical evidence.

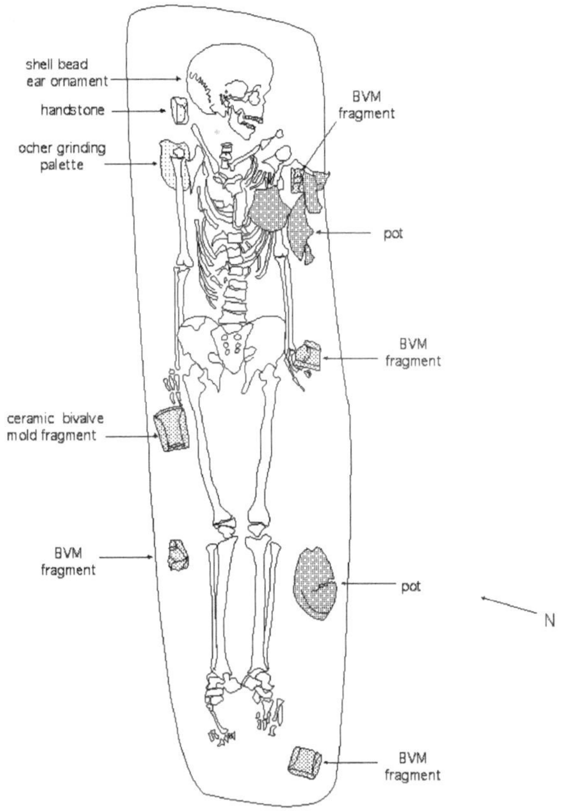

Figure 4. Drawing of the "Grave of the Metalworker" at Non Pa Wai (1986, Operation A).

Radiocarbon determinations date the earliest metallurgical activity to *ca.* 1500 BC in the sites clustered in the Khao Wong Prachan Valley (Figure 2). These sites are among the largest prehistoric metal production sites known in Asia and offer a valuable opportunity to study the organization and technology of intensive, centuries-long, production within its sociocultural context. This paper focuses primarily on the archaeometallurgical remains from our work at the site of Non Pa Wai relative to the neighbouring sites of Non Mak La and Nil Kham Haeng[1] (Figure 3). Non Pa Wai and Nil Kham Haeng are, in Chernykh's terms, metallurgical foci or principal production centres located within the same resource zone. A number of documented copper deposits are found in the Khao Wong Prachan Valley, most of which have indications of pre-modern, if not prehistoric, exploitation (Natapintu 1988: 114-15).

Prehistoric Non Pa Wai: a principal production Center

During the rainy season wild vegetation covers the five hectares of the Main Mound at Non Pa Wai because the density of concentrated industrial debris prevents cultivation. The site of Non Mak La lies across the stream. It is contemporaneous with Non Pa Wai and evidence suggests it is the settlement of the metalworkers whose centuries of effort produced the hectares of production remains at Non Pa Wai.

The industrial phase at Non Pa Wai dates between *ca.* 1500-700 BC The massive industrial layer accumulated several meters above the late 3rd millennium BC settlement encountered in the basal deposit. Numerous burials accompany this early settlement at Non Pa Wai, but none were found to contain metal.

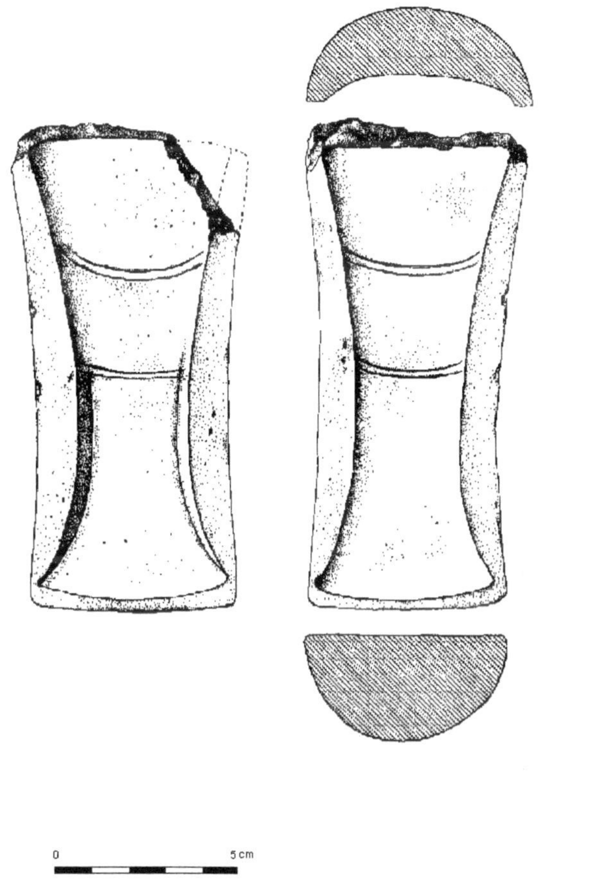

Figure 5. Drawing of a pair of ceramic bi-valve 'big axe' moulds from the second metalworker's burial at Non Pa Wai.

Four of the twenty burials found in the basal deposit and on its surface contained metallurgy-related artifacts. One burial contained an amorphous copper ornament and another a socketed axe while the other two were demonstrably those of metalworkers. This so-called "Grave of the Metalworker" (Figure 4) contains a tall adult male over 25 years of age buried holding fragments of ceramic casting moulds in his hands. Other mould fragments were distributed around the body, along with two pots, an ocher-stained hand stone and grinding palette, and a shell bead ear ornament. It appears that, at least in the earliest stages of metallurgical developments at Non Pa Wai, the people chose to live, smelt, die and be buried on the spot. The large bi-valve axe moulds from the "Grave of the Metalworker" (Figure 5) are a shape typical of the earliest period of metalworking at Non Pa Wai and may serve as an index fossil of sorts for this period.

During the initial period of intensive copper production at Non Pa Wai, it appears that industrial and some habitation activities occurred across the site contemporaneously. However, no burials from the industrial phase, other than the previously mentioned twenty in the basal deposit, were encountered. Despite the dense concentration of metalworking debris, very few identifiable metal artifacts have been recovered, nor any in situ metalworking installations. The metalworking debris did not occur on working or living surfaces. The production-related artifacts are almost all fragmentary and found in prehistoric secondary deposits, probably dumps. Considerable stratigraphic disturbance over time is apparent - some anthropogenic - and some from bioturbation due to animal burrows, tree roots and insect tunnels and nests.

Figure 6. Typical anvil stone for ore crushing from the surface of Non Pa Wai. Indentations form over time and are useful in holding larger ore fragments during crushing processes.

The metallurgical evidence from Non Pa Wai

In order to characterize the technology as a whole, we have outlined the various stages in the production process from mining to finished product (Figure 6). The major stages are discussed below.

Ore deposits

It is suggested that the deposit at Khao Tab Kwai was the local ore exploited by the early metalworkers at Non Pa Wai (Figure 3). It is mined today for iron, but it contains malachite, chrysocolla and chalcopyrite (the products of weathered copper sulfide ore bodies) and in prehistory was a source of copper as well. It may also have served as a greater regional source for red ocher, abundantly available there, and very common in burials in the prehistoric period across Thailand, if not Southeast Asia generally (Pigott *et al.*: in press).

Excavations at Non Pa Wai produced numerous copper ore samples from the industrial context. These, from expert hand specimen examination, physically resemble the copper ore typical of the deposit at Khao Tab Kwai. Geological study of the ores and the proximity of this substantial deposit suggests that Non Pa Wai metal workers were exploiting this deposit from the outset of copper production at the site, the earliest such production known in the Khao Wong Prachan Valley.

Non Pa Wai was an established settlement locus in the late 3rd millennium, prior to the advent of metalworking. The distance from Non Pa Wai to Khao Tab Kwai is not far, about a kilometre or so, and the amount of uphill climbing required to reach Khao Tab Kwai was minimal compared to other nearby ore sources. The rich, easy to smelt and readily available ores at Khao Tab Kwai were certainly a boon to an emerging metallurgical industry.

The process of producing copper

The analyses of Non Pa Wai slag by the late William Rostoker (unpublished, MASCA archive) consistently indicated a high matte, thus high sulfur content, which in turn suggests the use of sulfide ores. While it is recognized that oxide ores alone will smelt directly in a crucible, such a process requires reduction by carbon-rich gases and tends to generate little slag. In the Khao Wong Prachan Valley, at Non Pa Wai alone, slag volumes measure in the 100,000's of tonnes. The evidence suggest the co-smelting of oxide and sulfide ores, which can generate substantial amounts of slag (Rostoker *et al.* 1989).

Figure 7. Side view of Non Pa Wai chaff-tempered ceramic smelting crucible. The rim of the crucible has been systematically broken away to form a pouring spout. The crucible was probably recycled for remelting copper ingots and casting.

Ore crushing and beneficiation

Surface survey and excavation at Non Pa Wai have yielded both anvil stones and hand stones typically used in the ore crushing process. Clearly there was ample ore crushing going on at the site to reduce chunks of ore to smelting size (pea-sized gravel or smaller). Abundant samples of the ores used at Non Pa Wai were excavated and came in all sizes. It is presumed that rich ore fragments were brought from Khao Tab Kwai to Non Pa Wai for crushing. At Non Pa Wai, no discernible activity areas were excavated and the homogeneity of the Non Pa Wai industrial matrix was consistently striking. This homogeneity suggests that a fundamentally similar industrial activity, that of smelting, was practised uniformly across the site.

The smelting crucible

The evidence demonstrates that metalworkers at Non Pa Wai used large crucibles (Figure 7). to contain the smelting. Several 100,000s of rice-chaff tempered crucible sherds were excavated at Non Pa Wai, providing strong support for the suggestion that these were the primary smelting vessels used and that they were apparently used only once. This high volume of crucible sherds may be unprecedented in the Old World. It appears that these crucibles were locally manufactured, perhaps even on site, but there is no direct evidence of their production.

The typology of crucibles is presently incomplete. Shape can vary widely in crucibles and still serve the purpose of smelting. Why multiple shapes would exist can only be speculated. Two suggestions readily come to mind: first, a functional distinction i.e., between smelting and casting, and/or second, a typology relating to the production style of those who made the vessel. Moreover, if any number of family units were simultaneously practising the craft of smelting at Non Pa Wai, as suggested below, then a range of variability within this class of vessel can certainly be anticipated.

1)	$3Cu_2O + FeS \rightarrow FeO + SO_2 + 6Cu$	ΔF_R = -42.63 kcal
2)	$5CuO + CuFeS_2 \rightarrow FeO + 2SO_2 + 6Cu$	ΔF_R = -113.5 kcal
3)	$3CuO + FeS \rightarrow FeO + SO_2 + 3Cu$	ΔF_R = -65.23 kcal
4)	$2CuO + S \rightarrow SO_2 + 2Cu$	ΔF_R = -49.43 kcal

Figure 8. Equations for the direct reduction to copper metal by oxide/sulfide mineral interaction. The last reaction reflects the expectation that pyrite, commonly admixed with chalcopyrite, will dissociate and evolve sulfur vapour at about 700°C, leaving FeS (Hofmann 1914, 66; Rostoker et al. 1989, 73).

The smelting process

Rostoker's laboratory replication experiments (Rostoker *et al.* 1989) demonstrated that co-smelting of oxides and sulfides results in a direct one-step production of copper metal (Figure 8). Residual matte and substantial slag are by-products of this process (Figure 9). The matte would have been recycled along with some of the slag in each smelt as a flux. Rostoker found that slag piled atop the crucible charge served as a blanket to help contain the sulfur rich atmosphere.

The co-smelting process is technically simple. No fixed installations, nor any specific proportion to the oxide-sulfide mix, are needed. In fact, based on the stratigraphy of weathered copper sulfide ore bodies that typify the Khao Wong Prachan Valley, mixed oxide/sulfide ores may have been produced unintentionally rather than necessarily as a result of the deliberate choice and mixing of specific ores. Since smelting was conducted in crucibles or in shallow bowl furnaces, a large number of such operations could be undertaken simultaneously. Co-smelting in a crucible does not require charcoal as a fuel (Rostoker *et al.* 1989). Thus, there is no reason to suggest that charcoal preparation was a step in the production process. A high temperature (*ca.* 1250°C) and a sulfur-rich atmosphere is sufficient to drive the smelt. Such crucible smelting may have been driven by dry wood fuel.

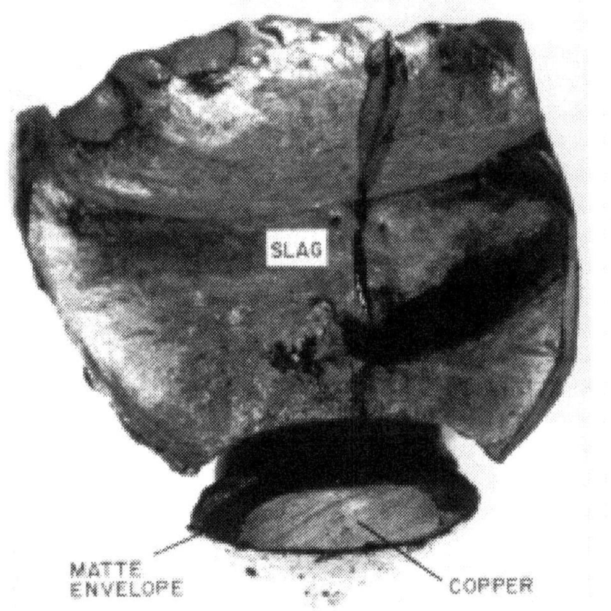

Figure 9. The products of Rostoker's smelting experiment #2 in cross section having been broken out of the crucible. The mass above is slag. The copper button (or ingot), surrounded by matte and positioned at the bottom, separated easily from the slag (Rostoker et al. 1989, 74). Note the large amount of slag for the copper produced.

Furnace chimneys

There was no indication of in situ furnace installations at Non Pa Wai. However, a second class of metallurgical ceramic artifacts has been identified that sheds further light on the smelting process. These have been termed 'furnace chimneys' (Figure 10). When fragments of these large, grit-tempered ceramic artifacts were encountered in the excavation at Non Pa Wai, it was although that they might constitute a portion of the smelting furnace. Subsequent discoveries supported this contention. However, interestingly it was from the neighbouring site of Nil Kham Haeng that reconstructible chimneys were excavated as part of the assemblage of grave goods among the burials. These portable chimneys must have required time and skill to make, and could have been used for a number of smelts before a new one was required. Although we have no petrographic data as yet, the macroscopic attributes of the temper (i.e., grit) and clay suggest the materials were selected for their refractory qualities.

When pieced together, the fragments form a circular chimney with a 20 cm internal diameter. At Non Pa Wai, we suggest the chimney was placed above a smelting crucible. Positioned in this way, the chimney would contain the upper portion of the furnace charge of ore and charcoal onto which the bellows blast was directed. The portion of the charge contained in the crucible or furnace bowl would smelt effectively under the high temperature, reducing conditions promoted by the low oxygen and high sulfur atmosphere. The holes in the wall of the chimney are presumed to have provided for the application of the air blast. Slagging frequently occurs in smelting furnaces at the juncture between the blast and kindled charge - i.e., the point of entry of the tuyére into the furnace. The absence of slagging on the interior of any of the chimney fragments examined to date suggests that the temperatures reached were insufficient to promote a slag forming reaction in this part of the installation. This may testify to the refractory nature of the chimney material, but it may be that slag formed elsewhere during smelting - most probably in the crucible below the chimney.

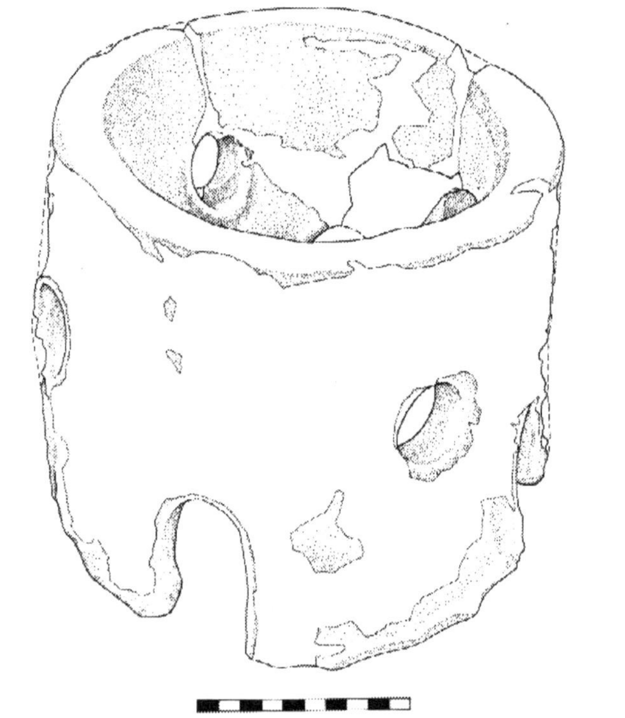

Figure 10. Drawing of a furnace chimney assembled from fragments excavated from 1990 operation 4, burial 1 at Nil Kham Haeng. Note the holes, presumably for a bellows or wind blast.

Another interesting point that merits discussion concerns the source of the draft for the smelt. It may be that the holes in the chimney accommodated a tuyére attached to a pair of piston bellows, although no tuyéres have yet been found. It is also unclear how, without a clay nozzle attached to a bamboo tube leading from the bellows, the blast could have been delivered to the hole in the chimney. When Non Pa Wai was excavated during the dry season, work was frequently stopped for hours due to the strong prevailing winds. These blew continuously enough to drive a smelting operation and the holes in a furnace chimney would have allowed sufficient draft to reach the charge. Furthermore, wind as draft could explain the absence at the smelting sites of both bellows and tuyéres for which there would have been no need.

Casting into cup and conical moulds

These ceramic ingot moulds, excavated by the tens of thousands at Non Pa Wai, range from about 1-10 cm in diameter. These ingot moulds appear to be unique to the Khao Wong Prachan Valley (Figure 11). Curiously, in the course of two full seasons of excavation at Non Pa Wai, nothing resembling an ingot was found. The rarity of metal at Non Pa Wai suggests that finished products were valuable and rarely lost and any miscast metal was probably recycled. Neighbouring Non Mak La, on the other hand, yielded finds which we believe are ingots cast in this type of moulds. Why such ingots would appear at that site will be discussed later.

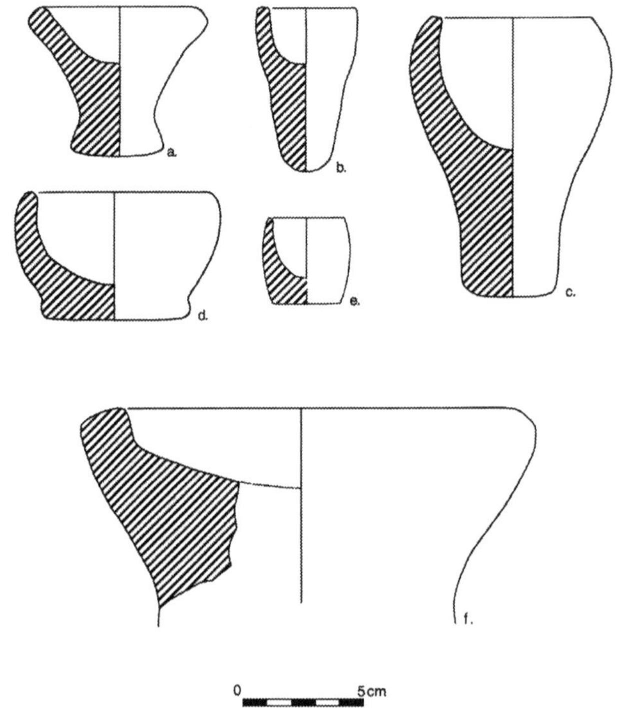

Figure 11. Illustration of the range of variation in ceramic ingot mould typology from Non Pa Wai (Armstrong 1994).

The small ingots produced in these moulds would have been suitable for counting, weighing, bagging, and transporting, all of which would facilitate the distribution of the metal. During the 1994 season we conducted an experimental casting at a local foundry in Lopburi. Conical moulds from Non Pa Wai were placed in soft surface soil, and a founder using a large crucible held in a metal "cradle" poured molten metal along the line of moulds. All these moulds had been baked in antiquity and no fracturing was seen as a result of thermal shock. After a short cooling period, the founder inverted each mould using a pair of pliers so that the ingot fell out. On cooling the ingots had contracted from the sides of the mould.

These raw copper ingots could be easily melted down in a crucible for casting, as they are small and have a substantial surface area exposed. This would have facilitated a "measuring" of proportions in the process of alloying with tin to make bronze. For example, among ingots of similar size, one ingot of tin to nine ingots of copper would yield a 10% tin-bronze.

Given the distinctive shape of the moulds, these ingots had characteristic shapes for which metal produced at Non Pa Wai might have been known intra-regionally, if not inter-regionally. Although Southeast Asian archaeology has not provided large numbers of metal artifacts and few have been subjected to analysis, finds of Southeast Asian metal known to be unalloyed copper are quite rare; most analyzed pieces are bronze. The significant amounts of copper produced in the Khao Wong Prachan Valley must have been leaving the valley or markets elsewhere. This observation is based on the fact that, in its characteristic ingot form, this metal was ideally suited as a portable and highly desirable trade good. Moreover, very little metal was excavated at Non Pa Wai and Non Mak La, while noticeable amounts of metal in a distinctly different form from that being produced at Non Pa Wai was found in burials at Nil Kham Haeng. Ingots from Non Pa Wai, produced in enormous numbers, are not to be found locally.

Casting into bi-valve moulds[2]

Bi-valve ceramic artifact moulds for casting large socketed axes (Figure 5) and arrow points, in addition to other shapes, have been excavated by the hundreds. But these quantities do not rival the huge quantities of cup and conical moulds, suggesting that ingot production was the primary focus of the Non Pa Wai industry. Nevertheless, the local metal producers must have needed copper-based tools, weapons, implements and ornaments for their own purposes. The bi-valve mould (BVM) industry may well represent local consumption of items for daily life, items which archaeology has yet to reveal outside of this valley.

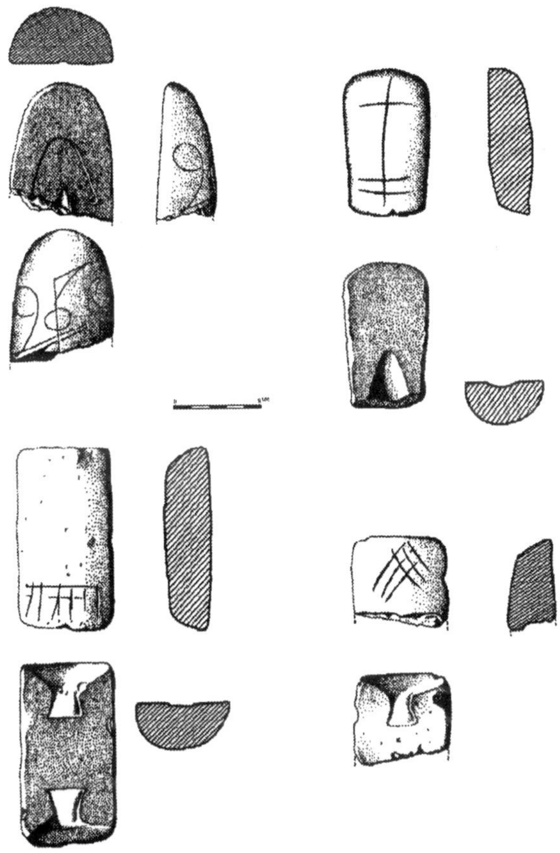

Figure 12. Examples of ceramic bi-valve casting moulds from Non Pa Wai with decorative incising on the exterior surfaces.

As mentioned earlier, the incidence of BVMs for the casting of large socketed axes may prove to be a diagnostic indicator of the earliest period of metalworking in the valley. Fragments of these moulds were found throughout the industrial deposit. It is possible that they are commonly distributed due to stratigraphic disturbances of the basal portion of the industrial deposit and were not in fact used in the latter phases of the industrial era at Non Pa Wai[3].

Exterior pattern decorating on BVMs

Many of the bi-valve moulds bore incised markings on their exterior surfaces (Figure 12). It is possible that these designs were a kind of maker's mark, a symbol of ownership, perhaps of the kin-based group responsible for the casting. Alternatively, the markings may indicate the type of artifact being cast, but thus far in our analysis no such relationships have been discerned.

Prehistoric Non Mak La: is this the village of the Non Pa Wai metalworkers?

At Non Pa Wai there appear to be no obvious spatial differentiation of activities of any type, whether they be domestic, industrial, or mortuary. Whereas excavations at the site of Non Mak La, just across the stream, revealed a site more consistent with a village-type occupation We believe that Non Mak La may be the settlement for the metalworkers from Non Pa Wai who, for reasons upon which we can only speculate (e.g., fire hazard, pollution) chose to settle away from their industrial work place. This suggestion, that Non Mak La is the metalworkers' settlement, is based on the following evidence.

The Mortuary evidence

Excavations at Non Pa Wai revealed no burials except at the start of the industrial phase. Non Mak La, on the other hand, occupied during the Non Pa Wai industrial phase, revealed the presence of burials throughout its occupational sequence. This sequence spans from *ca.* 2000 BC into the early centuries of the 1st millennium AD There was a high incidence of infant pot burials. Additional interesting patterns emerged during excavation, one group of burials was characterized by the presence of pottery anvils for the fashioning of ceramics while another, apparently contemporaneous, group of burials held only mussel shells (fresh water). This differentiation may represent a gender distinction. This suggestion will be explored in the analysis of physical anthropological data.

The Metallurgical evidence

As Non Pa Wai's habitation site, one would expect that Non Mak La would have metallurgical artifacts among its assemblage. Although little evidence for actual copper production exists at Non Mak La, almost one-fifth of the excavated materials were metallurgical artifacts. These included mould fragments for big axes. One particular fragment (Figure 13) is quite unusual because carved into its exterior surface is an impression of a cordiform implement. These artifacts (Figure 14) are known only from burials at the neighbouring site of Nil Kham Haeng. The presence of a single BVM bearing the impression of an artifact resembling a cordiform implement further connects Non Mak La to Non Pa Wai. Big axe moulds date to around 1500 BC (and possibly earlier) and cordiform implements from Nil Kham Haeng date between *ca.* 1100-300 BC Possibly the cordiform implement impression was carved into a discarded fragment of a big axe mould used at Non Pa Wai and subsequently carried to Non Mak La by a metalworker. The cordiform implement mould impression reveals no evidence that a mould plug to form a socket could have been accommodated. However, a sprue (an opening through which molten copper was poured into the mould) is carved into one end of the mould impression, a strong indication that castings were actually made in this small impression. Why make such an impression? Was this small cordiform implement impression used for casting small replicas of these points? One argument for this occurrence will be proposed in the closing discussion on cordiform implements at Nil Kham Haeng.

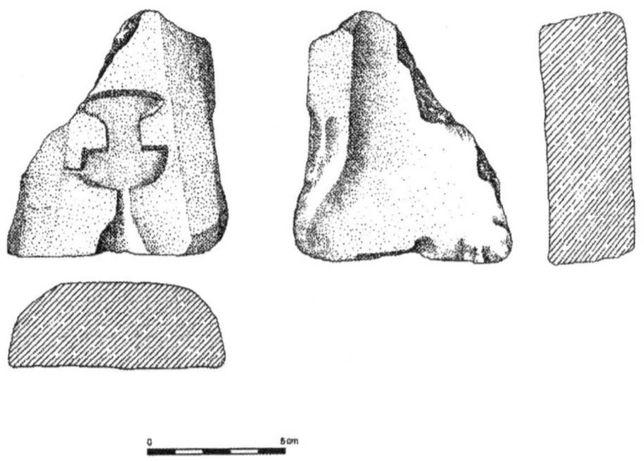

Figure 13. 'Big axe' ceramic mould fragment from Non Mak La. Note the impression of a cordiform implement carved on its exterior surface. The impression appears to have a sprue and may have been used for casting very small implements.

The Non Mak La excavation revealed what is believed to be cup mould ingots (Figure 15). It is possible that the daily ingot products at Non Pa Wai were brought home to Non Mak La by the 'commuting' resident metalworkers, where the ingots were stockpiled, perhaps in anticipation of future trading opportunities. The working assumption is that the prodigious production of ingots was undertaken specifically to produce items valuable for exchange in a network that was already established when metalworking began, or one that developed as a direct result of the availability of significant amounts of copper coming from this central Thailand source.

Prehistoric Nil Kham Haeng: a second principal production centre The discussion now turns to the other major production site in the Khao Wong Prachan Valley the site of Nil Kham Haeng. This site developed a few centuries after Non Pa Wai and in proximity to another major copper source, high on a nearby mountain, where ancient mine galleries have been found.

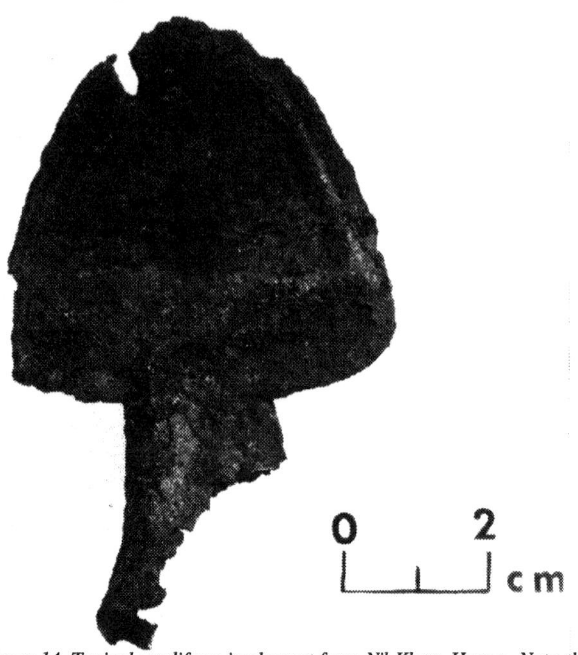

Figure 14. Typical cordiform implement from Nil Kham Haeng. Note the socket and what appears to be a casting flash along the outer edge.

Its stratigraphic character dramatically distinguishes Nil Kham Haeng from Non Pa Wai (Figure 15). The matrix of the entire site, over more than 4 hectares, consists of finely crushed ore host rock and crushed copper smelting slag that make up the thin strata.

Non Pa Wai and Nil Kham Haeng were both devoted to producing substantial amounts of copper and became enormous as a result of the process. However, by the time Nil Kham Haeng emerged as a production site, as much as four centuries after the beginning of Non Pa Wai's industrial phase, the means of copper production was undergoing significant change. This change was most likely a gradual transition developing over the centuries as metalworkers at Non Pa Wai slowly exhausted the readily available ores at local Khao Tab Kwai and turned to sources high up in the local mountains. The process of immeasurable man-hours of ore dressing and beneficiation resulted in the formation of the large mound of Nil Kham Haeng. Centuries of human effort were involved. The vast bulk of crushed and sorted ore was augmented by the process of crushing and recycling matte and matte-rich slag back into the smelting process. This served to extract additional copper and to flux the smelt The dramatic difference in stratigraphic composition so apparent between the two major production centres of Non Pa Wai and Nil Kham Haeng in the Khao Wong Prachan Valley - two sites in close proximity that share perhaps several centuries of contemporaneous production - may correlate with a shift in ore source exploitation. The early centuries of exploitation of the deposit at Khao Tab Kwai by Non Pa Wai metalworkers may have resulted in its gradual exhaustion. The shift in production may then relate to a major change in processing technology associated with the exploitation of ores increasingly rich in sulfur.

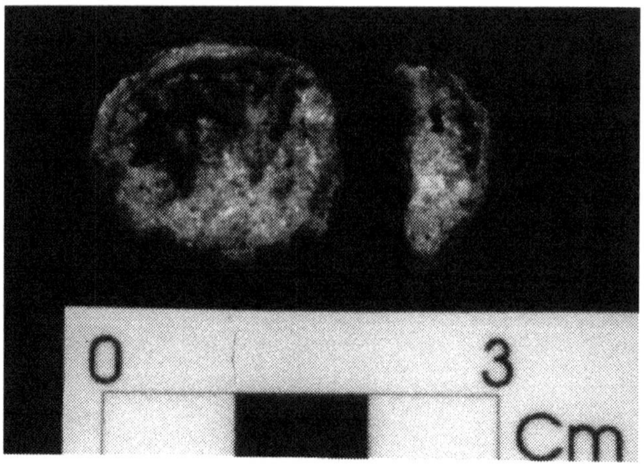

Figure 15. Copper-based ingot from Non Mak La. Ingots were probably cast at Non Pa Wai and taken to Non Mak La by the resident metalworkers.

Metalworkers may have already located additional deposits in neighbouring mountains such as that on Khao Phu Ka. Over time groups of families from Non Pa Wai, choosing not to hike the extra distance back to Non Mak La, may have decided to establish themselves near the base of Phu Ka at Nil Kham Haeng and gradually a new settlement arose. But why the intensive effort devoted to ore crushing? If the ores on Phu Ka were lower grade, they would require more time and effort devoted to beneficiation (i.e., crushing and sorting) to extract sufficient amounts of rich copper ore. Moreover, if the Phu Ka ores were more sulfidic than those at Khao Tab Kwai, then it would be necessary to crush and beneficiate oxidic ores to mix with the sulfidic ores for co-smelting. Finely crushed sulfidic ore would also roast faster and more easily, because it would be easier to drive off the sulfur from small ore fragments. When roasted, such ores would assist in the co-smelting process and/or smelt directly to copper.

Figure 16. The east balk at Nil Kham Haeng, 1990 operation 3. Note the finely layered stratigraphy of crushed materials. Lensing was probably due to torrential rains during the monsoon season distributing conical piles of crushed copper ore, host rock and slag.

The rather uniform size and lensed distribution of the crushed debris strongly suggest that it has been re deposited by water action. These lensed strata may result during the monsoon season from the redistribution by rainwater of numerous piles of crushed ore and slag that formed at crushing stations where an individual sat and worked. The lensing, therefore, may have been a site formation process driven primarily by nature rather than anthropogenic forces.

The remains at Nil Kham Haeng are much more of a mix of the industrial and the domestic than was apparent at Non Pa Wai. The lensing and the gravel matrix at Nil Kham Haeng allowed for the formation of living surfaces. Indications on living surfaces of domestic activity include a substantial ceramic assemblage, significant deposits of animal bone, post-hole patterns, evidence of shell and stone bracelets manufacture, and human burials .

Mortuary evidence for metallurgy

A tradition apparent from early Non Pa Wai appears to continue at Nil Kham Haeng. Both communities share the tradition of including tools of the metalworker's trade in burials. A particularly informative excavated burial was one quite rich in grave goods dating to *ca.* 500 BC. These grave goods included: sherds of complete vessels around and under the skeleton, five carnelian beads, iron bracelets covering both forearms, a cluster of ceramic furnace chimney fragments, and a cluster of eighteen cordiform implements in a pot. From this, and another burial, it was possible to reconstruct the furnace chimney discussed above. At Nil Kham Haeng, for the first time, metal occurred in the burials in large quantities. Clearly some amount of metal product was available and valuable as grave goods. Copper socketed points made up the bulk of the interred metal, with as many as sixty implements in one grave.

Other sites in the Khao Wong Prachan Valley have yielded very little metal in any context. An artifact similar to a Nil Kham Haeng cordiform implement was found at a site 60 km north of the Khao Wong Prachan Valley (S. Natapintu: pers. com.) and small socketed implements which appear to represent a cordiform implement variation were excavated in a mortuary context at Nong Nor, well to the south (C. Higham: pers. com.).

Furnace chimneys and the smelting process

At Nil Kham Haeng the smelting technology shifted from the crucible to the bowl furnace. The evidence supports this shift: crucible sherds were not common at Nil Kham Haeng. We believe that smelting was conducted instead in shallow bowl furnaces cut into the gravel lenses and lined with a refractory chaff-tempered clay. A furnace chimney was placed above the bowl. Such an approach would require that the furnace cool down before removing the copper smelted in the bowl. This copper would have had to have been remelted in the available crucibles (which could be used multiple times) and cast into the ingots and/or cordiform implements (see Figure 15) known to typify production at Nil Kham Haeng

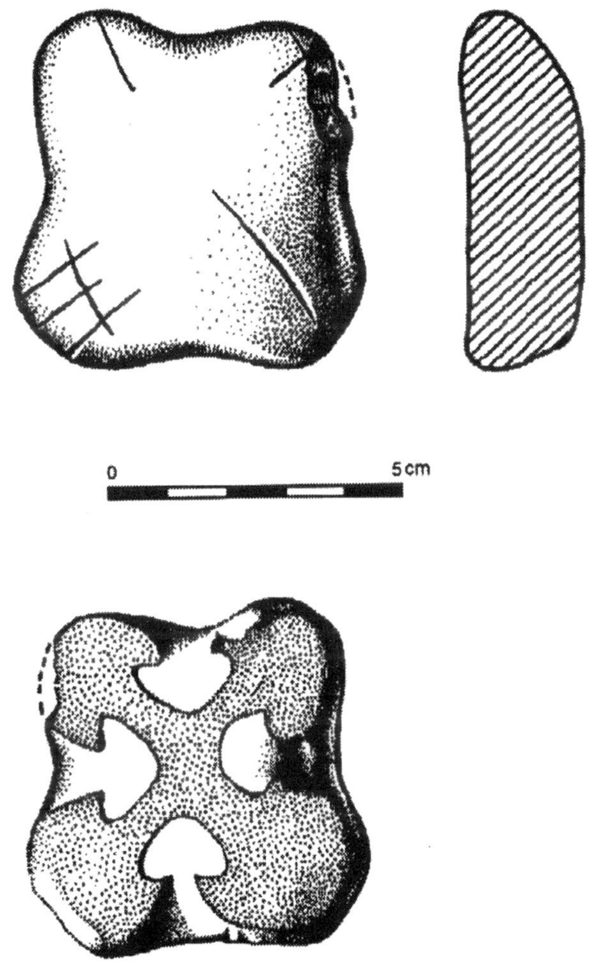

Figure 17. Ceramic bi-valve (?) casting mould from Nil Kham Haeng for the sequential production of small cordiform implements.

Bi-valve mould casting of artifacts

Besides smelting there are indications that artifact casting was also undertaken using bi-valve moulds bearing external linear incisions as at Non Pa Wai - although not found in comparable numbers. On one mould (Figure 17), the corner was incised with a geometric pattern. Such incised decoration was common at Non Pa Wai. What is particularly interesting is that this mould was for casting four small cordiform implements. A cordiform implement which actually fits the mould impression was excavated in a burial at Nil Kham Haeng. In order to cast four cordiform implements in rapid succession, the mould pair, somehow wrapped together (in green bamboo twine?) could have been stood on its edge and then the first pour made. It would then need to have been turned and then the casting made three more times. A pair of green bamboo lengths could have been used to hold and turn the hot mould for each casting - perhaps like tweezers.

Cordiform implements
The cordiform implements are blunt ended and very thin - about 2-3 mm thick (Figure 19). The copper implement blades bend very easily. Their fragile construction and the frequency of miscast cordiform implements suggest that these implements may not have been produced to perform a specific mechanical task. It is suggested that these implements may constitute a type of ingot - a standardized unit, easily recognized, counted, transported and melted - that was very useful in the exchange networks in which the sites in the Khao Wong Prachan Valley participated. Moreover, the high incidence of these curiously shaped implements in burials suggests their symbolic value as a commodity to those burying their dead. The potential exchange role of these implement/ingots may have contributed to their value.

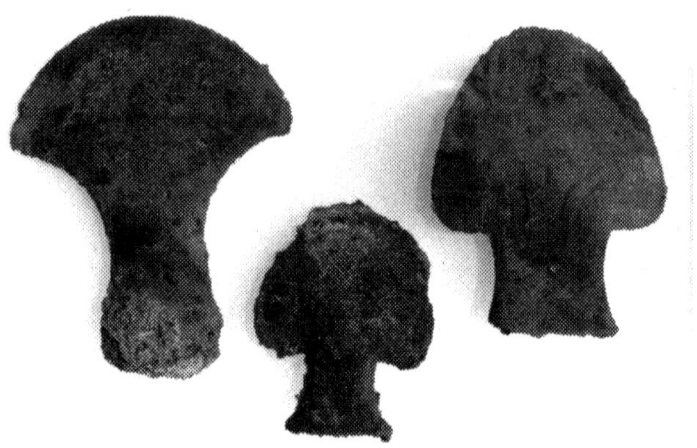

Figure 18. Implements from a burial cluster from Nil Kham Haeng. The large axe on the left has no socket and is tin-bronze. It was probably imported from a bronze producing area which may also have been importing copper from the Khao Wong Prachan Valley. The other two are socketed cordiform implements typical of production at Nil Kham Haeng.

Exchange
The presence in two burials at Nil Kham Haeng of two typologically exotic artifacts (an axe and a spear point), which were shown to be bronze by PIXE analysis, suggests that a far flung system of exchange existed in which the Khao Wong Prachan Valley sites participated. The axe is a solid casting, not socketed, and very heavy (see Figure 18). It could easily have served either as a tool or weapon. The spear point is socketed but is a decidedly rare artifact in any context in prehistoric Thailand. These two artifacts are the first indication of the presence of bronze at sites in this valley. Tin to make bronze could have been brought to the valley from the substantial deposits of western Thailand; however, we have no evidence as of yet to indicate that bronze was being produced in the valley.

A perhaps more likely scenario can be suggested. The bronze artifacts themselves may have been traded to central Thailand from locales where bronze was being made. The Khorat Plateau of northeast Thailand is one such region where sites such as Ban Chiang and Ban Na Di have yielded numerous bronze artifacts. Metalworkers at these sites could have obtained their tin from nearby Laos. One could go so far as to suggest that copper ingots from central Thailand were being traded to the bronze-making villages in northeast Thailand and in return some of that bronze may have been traded back to central Thailand sites like Nil Kham Haeng.

Concluding observations: some inferences on the nature of production in the valley
While copper production in the valley was 'industrial' in scale there is little to suggest differentiated task activity in a workshop-type organization (i.e., these were not specialists attached to, and working for, an elite hierarchy). Both Non Pa Wai and Nil Kham Haeng are industrial production sites which are remarkably homogeneous in terms of the nature of the cultural deposits. Non Pa Wai appears not to have been occupied during the industrial phase whereas at Nil Kham Haeng industrial and domestic contexts were intermingled.

Based on arguments made elsewhere (White and Pigott 1996) and given the deposit homogeneity, we ascribe production at Non Pa Wai and Nil Kham Haeng to small, independent, kin-based groups - families perhaps - who may have practised smelting at any spot on-site. Production increased with the investment of labour and time. The producers themselves had control over the results of their labour.

The primary product of the industry at both Non Pa Wai and Nil Kham Haeng appears to have been small, portable ingots distinctive in shape (e.g., see Figures 11 & 15). Their small and consistent size could have facilitated the creation of bronze by recipe with an ingot serving as a 'unit of measure'. While bronze was not being produced at these central Thailand sites, the ingots most probably were being exported to bronze-producing villages. The making of a bronze with 'standardized' ingot shapes would have not required sophisticated knowledge of bronze production. Producing a bronze may well have been relatively easy once the metals were obtained. Such a recipe approach would have been far simpler and more predictable when compared to smelting a mixture of copper ore and cassiterite to obtain bronze.

Production may have occurred only during the dry season (White and Pigott 1996:159). It is during the dry season in the valley today that strong winds prevail which could easily have driven multiple smelting operations like those which must have taken place at Non Pa Wai and Nil Kham Haeng. Thousands upon thousands of smelts had to have occurred to have helped form these two sites tens of thousands of square meters in size. No evidence for tuyeres was found at these sites and our reconstruction suggests the use of wind driven smelting contained in a furnace chimney with either a crucible or bowl furnace to collect the product. Moreover, the use of wind driven smelting as opposed to bellows driven smelting would have meant many less hours of human effort being devoted to the production process.

Why not year round production? Thailand is visited by the monsoons beginning in the spring and the rains continue for many months. Torrential rains would not only make mining and metal smelting difficult, but the presence of water vapour could have a deleterious effect on metal cooled from the molten state. In preparation for the dry season production activity, crucible and mould manufacturing might have been accomplished during the monsoon if fires sufficient to bake them could be maintained. Agriculture and animal husbandry combined with hunting and gathering would have demanded considerable time especially given the sizeable concentration of people suggested by the size of the sites. Thus copper production might have been limited to a seasonal, part-time activity.

At both Non Pa Wai and Nil Kham Haeng production centered around mostly smaller scale items namely ingots, implements and some ornaments. Clearly local consumption went on as some metal was being made into tools/implements, weapons, and personal ornaments. However, given the enormous volume of the production debris, the resulting substantial yield of copper - perhaps as ingots - was moving elsewhere as suggested above. Given the likelihood that the copper was moving off-site along the pathways of a network of exchange - the question concerning what was coming to the Khao Wong Prachan Valley in return needs to be asked. Is it possible that food comprised a portion of such exchange commodities? Less time devoted to agriculture, animal husbandry, hunting and gathering would have meant more time to be devoted to making metal.

We are entertaining the notion that the existence of two principal production centres - Non Pa Wai and Nil Kham Haeng - may correlate with a shift in ore source exploitation from the nearby easily accessible deposit at Khao Tab Kwai (near to Non Pa Wai), as it was slowly exhausted, to a more difficult deposit high up in on the mountain known as Phu Ka (near to Nil Kham Haeng). Moreover, the dramatic change in stratigraphic composition of the two sites may relate to a major change in processing technology in association with exploitation at Nil Kham Haeng of Phu Ka ores which were potentially of a lower grade and were increasingly rich in sulfur. The considerable human effort which went into the crushing of hostrock, copper ore and slag apparent at Nil Kham Haeng can be explained in such terms.

In closing, the current evidence from this central Thai valley suggests basically peaceful co-existing settlements whose inhabitants mined the massive deposits which over time, based on resource driven technological change, pursued somewhat different procurement strategies. The development of copper production within a strong community-based mode of organization prior to political centralization helped to establish this commodity in a decentralized economic realm even while society was becoming increasingly complex.

Notes

1. This paper comprises an overview with additional commentary of data to appear in Pigott *et al.*: 1998.

2. A study is currently underway by TAP researcher Judy Volker, a Ph.D candidate in the Dept. of Anthropology at State University of New York at Buffalo. She is conducting a petrographic and SEM study of the clays used in BVM manufacture.

3. Without entering into a discussion of the chronology of Non Pa Wai, there appears to have been considerable disturbance of Period 2A burials by Period 2 inhabitants. Period 2A is the era of the metalworker's burials mentioned in the text. It is also the era of the production of big axes using the moulds as seen in Figure 5. We believe that many, if not most, of the big axe mould fragments found in the Period 2 industrial deposit may come from 2A burials. Intense stratigraphic disturbance characterized the industrial phase deposit and is due primarily to pit digging for crucible smelting operations. In addition, the natural activities of burrowing animals, insects as well as deep penetrating roots systems of vegetation have added to the disturbance. Remember Non Pa Wai can't be plowed due to the amounts of industrial debris which characterize its matrix so when unoccupied it has always grown wild.

References

ARMSTRONG, L.A. 1994. Interacting Technologies: The Analysis of Ingot Moulds and Associated Industrial Ceramics from Non Pa Wai, a Prehistoric Smelting Site in Central Thailand. M.S. Thesis (Dept. of Anthropology), University of Pennsylvania.

CHERNYKH, E. N. 1980. Metallurgical Provinces of the 5th-2nd Millennia in Eastern Europe in Relation to the Process of Indo-Europeanization. Journal of Indo-European Studies 8, 3/4: 317-335.

1992. Ancient Metallurgy of the U.S.S.R.: The Early Metal Age. Cambridge: Cambridge University Press.

COSTIN, L.C. 1991. Craft Specialization: Issues in Defining, Documenting, and Explaining the Organization of Production. In Archaeological Method and Theory, Vol. 3, M.B. Schiffer, (ed.) Pp. 1-56. Tucson, AZ: University of Arizona Press.

HOFMAN, H.O. 1914. Metallurgy of Copper. New York: McGraw Hill

KROEBER, A. L. 1948. Anthropology, (New, Revised Edition). New York: Harcourt, Brace and Company.

PIGOTT, V.C., WEISS, A. & S. NATAPINTU. in press. The Archaeology of Copper Production: Excavations in the Khao Wong Prachan Valley, Central Thailand. To be published among the proceedings of the "Fifth Conference of the European Association of Southeast Asian Archaeologists," Rome, Italy, 1993.

ROSTOKER W.; PIGOTT, V.C.; & DVORAK, J.R. 1989. Direct Reduction to Copper Metal by Oxide/Sulfide Mineral Interaction. Archeomaterials 3:69-87.

NATAPINTU, S. 1988. Current Research on Ancient Copper-base Metallurgy in Thailand. In Prehistoric Studies: The Stone and Metal Ages in Thailand. P. Charoenwongsa and B. Bronson, eds., Pp. 107-124, Papers in Thai Antiquity 1. Bangkok: Thai Antiquity Working Group.

PIGOTT, V.C.; WEISS, A.D; & NATAPINTU, S. 1998. The Archaeology of Copper Production: Excavations in the Khao Wong Prachan Valley, Central Thailand. Publication forthcoming among the proceedings of the "Fifth Conference of the European Association of Southeast Asian Archaeologists," Rome, Italy, 1993. To be edited by Roberto Ciarla and Fiorella Rispoli, Is.IAO, Rome.

WHITE, J.C. 1988. Early East Asian Metallurgy: The Southern Tradition. In The Beginning of the Use of Metals and Alloys, R. Maddin, ed., pp. 175-181. Cambridge, MA: MIT Press.

WHITE, J.C. & V.C. PIGOTT. 1996. From Community Craft to Regional Specialization: Intensification of Copper Production in Pre-State Thailand. In Craft Specialization and Social Evolution: In Memory of V. Gordon Childe, B. Wailes, ed., Pp. 151-175. University Museum Monograph 93. Philadelphia, PA: University Museum Publications.

3

Envaluing metal: theorizing the Eneolithic 'hiatus'

TIMOTHY TAYLOR

Department of Archaeological Sciences, University of Bradford

This paper focuses on the Early Metal Age archaeological record for central- and southeast Europe, the Ukraine and European Russia, and - to a lesser extent - the Alps, in order to advance an alternative to the standard model of metallurgical development. I challenge the idea of a production 'hiatus' or 'recession' (Rückgang) *in the later Eneolithic. By approaching the archaeological record critically, developing arguments concerning skeuomorphism and grave-robbing, I suggest that there are grounds for believing that there was actually an* intensification *in the use of metal during this period, leading directly to the more general occurrence of harder and more function-specific copper-based metals in the earlier Bronze Age.*

Introduction

When prehistoric communities in Eurasia first began to name and to use metals, to make and remake metal objects, they started down a path of development that was unique. In order to conceptualize it for ourselves today however, as archaeologists or archaeometallurgists, we necessarily begin with the vast handicap that metals are familiar to us, that we are on the receiving end of the sequence of development upon whose origins we seek to shine light. All of our preconceptions about metal, no matter whether they come directly from our modern industrial and scientific knowledge, or from our ethnographic knowledge of communities first brought into contact with fully-fledged metals as part of the expansion of mercantile capitalism, are simply that – preconceptions. To throw away our knowledge does not seem a good way to begin to understand the inception of metallurgy. Yet, the people long ago started with none. Why should we not seek to think our way back to their position? This paper is part thought-experiment. But it attempts to be more than simply speculative or provocative. By thinking our way back into an unfamiliar possible world, we may be able to establish some rudimentary starting points for a middle-range theory of metals.

The formation of the Earlier Metal Age (EMA) archaeological record in southeast Europe – the Copper and Early Bronze Ages of the Carpatho-Balkan region and its steppe and forest steppe extensions – has too often been considered unproblematic. Metal was made and metal artefacts were deposited; metal ran short and few or no metal artefacts were deposited; new supplies of metal came on stream and metal was deposited *en masse* again. I do not believe in this account, for several reasons. To demonstrate that the record really is problematic, I use terminology from Binford and from Schiffer, whose empirical base is far removed in space and time from the Eurasian EMA. Although it has largely been lost sight of in the proliferation of self-reflexive readings of prehistory, middle-range theory still represents a critical step in improving interpretations and is worth applying explicitly to any body of archaeological observations. Taking a new look, focussing particularly on the nature of the evidence for a 'hiatus' in archaeological metal putatively occurring around the mid-fourth millennium BC, allows a tentative new formulation for a question that lies continually on the edge of the frame: how did copper become bronze?

When Lewis Binford coined the term Middle Range Theory (MRT or, better now, middle-range theory), he was attempting to formulate a set of procedures by which one could work back with some certainty from the 'statics' of the archaeological record to the 'dynamics' of past cultural systems (see Binford 1983). At the same time, Michael Schiffer proposed a more rigorous focus on what he termed 'site formation processes' (e.g., Schiffer 1976). I am aware that Binford and Schiffer have never accepted each other's terminology, and do not agree about whether and/or how the archaeological record either 'reflects' or 'distorts' past reality; however, their broad intentions are similar enough; here I mix and match their terms to the degree that seems heuristically useful for present purposes and, following Trigger (1995), I will used the term middle-range theory in a general sense. It is worth briefly outlining examples of Binford's, and of Schiffer's, reasoning:-

Binford on the basis of his work among the Nunamuit of Alaska and other North American indigenous hunting, fishing and foraging groups, explicitly formulated the idea of a society in which *curation* was central. That is, artefacts were part of a survival tool kit which required periodic replacement when they became worn out or were lost. The idea of loss suggested to Binford that the archaeological context of curated artefacts (their modern *static* situation, as found by the archaeologist) would often bear no clear spatial relationship to their original context of use (their ancient *dynamic* situation, as used by their maker).

By contrast Schiffer focussed on Pueblo villages, looking at ways in which material such as pottery was used and discarded and then sometimes reused (for example, as a temper for pisé wall construction), cycling through a continuous series of cultural and natural transformations (C-Transforms and N-Transforms); Schiffer developed a *discard algebra* in order to make estimates of the percentage relationship between the surviving or *archaeological population* of artefacts and the original or *target population* of artefacts in use in a variety of forms at any one point in the past.

These ideas are very useful, at the very least, in that they make explicit what many archaeologists already know but which in practical circumstances they tend to forget. For example, no-one has attempted to apply the principle of Schiffer's discard algebra to data from Copper Age Europe. However, it is true that both Binford and Schiffer have focussed on behavioural phenomena which they saw as effectively universal or cross-cultural, and which were based on an appreciation and assumption of familiar attitudes to materials in terms of costs and benefits. Both assume the operation of rational economy. This may be one reason why middle-range thinking has not been incorporated into reasoning about the detailed culture sequences and ritual behaviours of the communities of the Old World EMA. As Bruce Trigger has lucidly put it: "Middle-range theory ... is widely regarded as a useful means by which archaeologists can reconstruct human behaviour from a materialist and rationalist perspective" [but] "As a result of the widespread acceptance of Julian Steward's unjustified strictures against anthropologists studying the culturally specific, the kinds of middle-range theory that would permit archaeologists to deal systematically with these sorts of phenomena remain grossly undeveloped." (Trigger 1995: 450f and 453).

Clearly, Old World archaeologists and archaeometallurgists have accepted that there have been particular mechanisms by which material typically entered the archaeological record, and that these mechanisms give rise to different types of records for different past societies. For example, Balkan Neolithic and EMA communities produced massive settlement tells. These were created over time and through successive destructions and rebuildings, whether intentional (hygiene-based and/or ritual), or unintentional (accidental and/or conflict-related). In contrast, the Later Bronze Age Scandinavian communities of Montelius Stage IV made themselves archaeologically most visible through the deposition of metalwork in anaerobic peat bogs. However, these differences in the constitution of the archaeological record have become so familiar through 150 years of European prehistoric scholarship, that they are infrequently addressed explicitly in terms of what precise aspects they are likely to preserve of past dynamics, and which potential aspects they are likely not to inform on, or perhaps even disinform on, if not interpreted with care (Hodder 1986).

The nature of the Eneolithic metal sequence

For the Carpatho-Balkan Eneolithic, there has been an assumption that the inception of metal use was coterminous with both the beginnings of scientific analytical thinking and marked a significant step in the development of rational economy (e.g., Chernykh 1982a; cf. Budd & Taylor 1995). Yet the material on which this putative economic rationality is inferred, is predominantly from burials - from metal artefacts ritually returned to the earth alongside the bodies of the dead. As funeral rites do not constitute a strictly rational economic activity, variations in the pattern of artefact deposition over time, while they may well correlate in some way with developments in a broadly economic sphere, will not necessarily do so in the direct way hitherto assumed for this region and period. To develop this argument further, we need first to be clear about the general chronological and spatial frameworks.

It is a fact that the development of carbon-14 dating has had a massive impact on the history of scholarship concerning Old World metallurgy, starting with Renfrew's seminal paper postulating the autonomy of the southeast European Copper Age (Renfrew 1969). But the way that the dates are used in general, and the shape of the calibration curve during the Eneolithic period specifically, leaves broad and continuing scope for confusion and disagreement about both absolute and relative sequences (cf. Makkay 1976 and Vulpe 1976 - both retaining late or 'conventional' dating - Lichardus 1991; Forenbaher 1993; Rassamakin 1993; Sherratt 1997), including synchronization with the Near East (Sherratt 1997: caption to Fig.18.2). The large number of Carbon-14 date assays, now available for Cucuteni-Tripolye sites in particular, means that there is increasing certainty about the form and extent of the general framework, but one must proceed with care[1].

For present purposes, we can view the relevant section of the culture sequence in southeast Europe as being divided into roughly five phases, from the final Neolithic/initial Eneolithic to the early Bronze Age (based on literature including Bognár-Kutzián 1963; Jovanović 1971; Novotna 1976; Todorova 1978a, 1978b, 1981; Masson & Merpert 1982; Roman 1983; Demoule & Lichardus-Itten 1989; Muhly 1989; Mohen 1990; Chernykh 1991, 1992; Lichardus 1991; Pingel 1989; Rassamakin 1993: Todorova & Georgieva 1993; Sherratt 1993, 1997; Pernicka 1997); some comparisons are made with central, Alpine and northwestern Europe, along a trajectory that eventually reaches Britain. The periodization terminology used to label the phases 1-5 below is meant as a rough guide only; it cannot rectify alternative regional terminologies:-

(1) *The late Neolithic, from around 5000 BC. Copper both native and ore-derived has been in use from before 7000 BC in Anatolia and the Near East and now begins to appear rarely in Europe*: Earlier Karanovo culture in Bulgaria, Pre-Cucuteni in Moldavia and western Ukraine, Tisza in Hungary, contemporary with late LBK and early Stichbandkeramik in parts of central-western Europe and the late Mesolithic in northwest Europe.

(2) *The earlier Eneolithic, from around 4700 BC, and especially after 4500BC. The deposition of copper artefacts increases sharply*: Kirilovo and Karanovo V-Gumelniţa, Maritsa, Boian, early Cucuteni-Tripolye, Vinca-Plocnik, Tiszapolgár; early Skelya (a.k.a. Skelyansk) on steppe and forest steppe, Mariupol-Stredny Stog I. These cultures are broadly contemporary with the

middle Neolithic Lengyel culture in parts of central Europe.

(3) *The middle Eneolithic, from around 4200 BC. Widespread and occasionally spectacular appearance of gold, copper, and copper alloy in the record (Chernykh's 'boom'); well-documented interregional contacts, but metalwork poorer in the steppe and forest steppe than in the metal-source regions*: Kodzhaderman-Karanovo VI-Gumelniţa and Varna-Durankulak in Bulgaria; Cernavoda I in Romania and, moving east, Cucuteni B-Tripolye B2-C1, later Skelya, Sredny Stog II, and Kvalynsk in the Volga region; Bodrogkeresztúr I and II in Hungary; and in the Alps, the later Neolithic Michelsberg and Pfyn, then Cortaillod, cultures; TRB to the north, with the Chasseen in France. Bocca Lorenza (Vicenza) axes in Italy may belong to the Neolithic Square-Mouthed Pottery Culture.

(4) *The final Eneolithic-Bulgarian proto-Bronze Age, from around 3700 BC. Metal much reduced or absent from the record in the Carpatho-Balkans, especially after 3500BC (the 'hiatus'), while the Tripolye C2 culture zone breaks down after*: Karanovo VII-early Ezero, Cernavoda III and Coţofeni in the lower Danube basin and - moving east through Moldavia to the lower Don - Gordineşti, Usatovo, Kasperovo, Sofievka-Krasny Khutor, Zhivotilovo-Volchanskoe, and Repin in the Volga region; Novosvobodnaya and Maikop in TransCaucasus. In the northwestern Balkans and Hungary the Boleráz-Baden culture takes shape. Farther west, late Neolithic type economies continue, with occasional metal - later Cortaillod, Mondsee, Horgen and SOM cultures; early Remedello in N.Italy, with some metal, as now in northeast Iberia.

(5) *The early Bronze Age, from around 3100 BC. More metal and increasing alloying in archaeological contexts*: Later Ezero culture in the northern Balkans; Yamnaya (Pit Grave) on the steppe. Kura-Araks culture in the Caucasus. Baden and early tumuli in Hungary; the emergent Corded Ware extending west across the north European plain; after 2500 BC in central Germany and Bohemia *Fahlerz* and tin are utilized in the Únětice, while the Bell-Beaker phenomenon finally brings metallurgy to the Atlantic west - coterminous with increased organization of lithic distribution in the Saône-Rhône culture and British terminal Neolithic.

Conceptualizations of the development of the earliest metallurgy have, until recently, been strongly influenced by what may be termed the 'industrial orthodoxy' of Theodore Wertime, which is based on the earlier ideas of a rational economy of metals proposed by Gordon Childe (Childe 1944; Wertime 1964, 1973; cf. Budd *et al.* 1994, Budd & Taylor 1995: 136ff). In these terms, the early development of Eurasian metallurgy has standardly been viewed roughly as follows (divided into four phases here for ease of reference): (i) discovery of native copper and gold and, subsequently, attractive and easily converted oxides of copper, led to copper sheet-working and simple casting for decorative and, to some extent, functional purposes; (ii) early enthusiasm, (the 'boom') in areas such as the Balkans, created an export trade, with copper objects appearing in the steppe and forest steppes; (iii) a crisis - the easily exploited oxides run low or out; this period lasts for several hundred years (the 'hiatus') until, (iv), production restarts in earnest in the Carpathians and the Alps using the less tractable sulfide ores, especially chalcopyrite, and arsenical alloys are deliberately created in order to make tools more functionally useful.

The third phase in this development is clearly observable in the archaeological record as a shifting will-o'-the-wisp of presence-absence-presence in deposition, not always in phase across Eurasia, but remarkably widespread as a phenomenon. Most noticeable is the 'crash' which occurs between Karanovo VI and VII in Bulgaria (Chernykh 1978a), equally dramatically mirrored in Romania and Serbia (Vulpe 1976; Chernykh 1992), with knock-on effects in the break-up of Tripolye C2. While Chernykh (1982a and 1992: 52) views this as a reflection of "the growth of irrational elements in the general culture complex", Jovanović (1989: 16) speaks of *Rückgang* ('recession'), and others postulate the beginning of a sequence of invasions of warlike 'kurgan cultures' from the East (eg Danilenko 1974; Gimbutas 1977; Merpert 1991). The technologically-centred explanation, due to *oxide zone exhaustion*, is part of the industrial orthodoxy and may be described as the *production hiatus model.*

Versions of this model can be found in recent papers by Andrew Sherratt and by Stephen Shennan. Thus Sherratt writes: "Copper-Age production, in particular, had the character of 'boom and bust', since the accessible and easily smeltable ores were fairly rapidly exhausted; although new and more advanced techniques of mining and smelting might subsequently revitalize certain areas, as apparently in Transylvania in the later part of the early Bronze Age" (Sherratt 1993: 11f). The terminology of boom and bust invites us to envisage an economic cycle of a disembedded modern type; on the other hand, there is a paradox - when seen in strictly economic terms - of suggesting that, although the copper economy was in a 'bust' phase, there was a concerted investment in developing 'advanced techniques' in order to refloat copper. In a similar vein, Shennan warns that the "importance of the exhaustion of metal supplies should not be underestimated." He also draws a parallel with more recent times in terms of the documented silver shortage in early medieval Europe (Shennan 1993: 61), but concludes that "it is unlikely that Copper-Age societies and economies were articulated in the kind of system where a decline in supply, and the no doubt consequent increase in 'price' which occurred, would lead to large-scale new prospecting and technological change; the metal simply dropped out of use" (Shennan 1993:61).

Shennan writes (1993) of the patchy nature of the Copper Age, and the phenomenon of imitation of copper in flint, as at Le Grand-Pressigny (Petrequin & Petrequin 1988;

Mallet 1992), "Given the desirability of the material, the lack of metal in circulation is best seen as a result of the lack of easily worked copper ores". Shennan's view is partly based on a version of the *skeuomorphism* argument, which has been advanced in a similar sense for pottery, for example by Andrew Sherratt and me (Sherratt & Taylor 1997), and left similarly vague by us in its specifics. I return to this below.

It does, indeed, seem clear that there is an incremental shift from lowland to highland ore sources - a prospection-driven expansion prompted by the increasing effort needed to extract Balkan ores[2]. This is indicated in general terms by the fact that the Alpine region seems to 'come on line' later than the Balkans, where available ores are lower lying (Sherratt 1997 and note Fig.4.14). But this need not mean that there was an actual cessation in metal supply at any time. It happens that the actual chemistry of metal in geological deposits around the world is far more complicated than has been imagined in most cases, both at the macro level and also more importantly here, at the micro-level of the constitution of the ore bodies themselves. Secondary native copper is found cheek-by-jowl with sulfidic deposits, formed by the degradation of oxide-zone minerals towards the junction with the secondary enrichment zone (Craddock 1995: 94), and in practice clear, layered zonation is not present. Because of this complexity, there are many potential ways in which varying grades of usable metal may be produced, wittingly or not, from whatever mix of material happens to be dug out (O'Neill 1996; Budd pers. comm.).

Recent studies of pre-Columbian Andean metallurgy show that oxide ores containing a significant sulfide component were utilized from the outset, perhaps by deliberately mixing highland sulfide ores containing arsenic with lowland oxide deposits (Lechtman 1979, 1980), but more likely because ores obtained from a particular location might well be a mix of sulfides and oxides; such a mixture eliminates the need for the progressive roasting of the sulphide in the matte process (Epstein 1995: 142). Co-smelting of sulfide and oxide does not depend on exact relative proportions (Rostoker, Pigott and Dvorak 1989), while sulfide-rich ores can be smelted in small crucibles of Neolithic type (Zwicker *et al.* 1985). There is direct evidence from slag that sulfidic ores were indeed exploited as early the mid-fourth millennium BC in the southeast Europe (Glumac & Todd 1991).

A new picture is beginning to emerge for southeast Europe. It had previously been thought that the absence of slag in the initial phase of the Eneolithic provided evidence for the use of native copper. The recognition of small fragments of slag, indicative of crucible smelting, from Vinca culture sites, as well as from a woman's grave in the Lengyel culture cemetery at Zengövárkony in southwest Hungary, is indicative of a crucible technology and the ability to deal with sulfide-rich ores at an early time (Glumac & Todd 1991; Pernicka *et al.*. 1993)[3]. Copper artefacts with significant arsenic also occur early too, so that the best available evidence seems to support a picture of alloying initially emerging, not as part of any cognized 'scientific' thinking about metallic elements, but as an apparently random variation predicated on the particular mix in the available ore body (Budd 1992).

Curation, cycling, and the depositional hiatus

I want, now, to approach the Carpatho-Balkan hiatus in the spirit of middle-range theory, thinking explicitly and quantitatively about curation and cycling. Initially, we need some first-approximation quantification. There has been considerable speculation about the social and economic role of copper in Eneolithic societies (Ottaway & Strahm 1975; Chernykh 1982a; Ottaway 1981, 1989; Demoule & Lichardus-Itten 1989; Jovanović 1989; Shennan 1993; Sherratt 1993; Todorova & Georgieva 1993; Bernbeck & Müller 1994; Budd *et al.*. 1994; Budd & Taylor 1995; O'Neill 1996; Sherratt 1997), both on the basis of the patterning of the surviving archaeological record and in the light of potential ethnographic parallels. What has not been done is the sort of middle-range theorizing which would attempt to build clear bridges between general anthropologically-observable behaviours (*pace* Boyer 1994) and what might be expected of the Eneolithic archaeological record. A specific weakness has been the tacit assumption that the burial record will have remained a more or less constant, proportionally-representational index of the amount and type of metal in circulation at any one time, rather than being an integral part of a dynamic social response to an economically and symbolically revolutionary new material.

Chernykh has calculated Eneolithic copper production for the AiBunar mine in the Morozov (Bratan) metallogenic zone north of Stara Zagora, Bulgaria, at between 500 and 1000 tonnes (Chernykh 1978b, 1982b: 14). Pernicka *et al.* (1997: 42) estimate something of the order of 50 tonnes of finished metal from Rudna Glava. In the absence of further quantification for mines like Bor, Majdanpek, and Rudnik in Serbia (Jovanović 1989: 13), or those in the Romanian Carpathians (see below) total estimates from all mine sites currently known or likely to have been utilized in this period in the Carpatho-Balkans as a whole might reasonably multiply Chernykh's lower limit by ten, giving an estimated 5000 metric tonnes (five million kg) of copper produced. It nevertheless remains uncertain where the majority of Eneolithic metal actually came from, as the isotopic and other signatures of the artefacts do not closely match anayses from those ancient mine sites so far identified; according to Pernicka *et al.* (1997: 143) Rudna Glava metal, for example, is undetectable archaeologically: "we again have not a single artefact that would fit the lead isotope and trace element fingerprints as they are predicted, and measured, for metal smelted from Rudna Glava ores. However seminal and important Rudna Glava may have been in the development of early extractive copper metallurgy in the Balkans, quantitatively it apparently was of no significance."

It would be useful to be able to make a direct comparison between estimates of copper metal mined and extracted and the actual total weight of Eneolithic artefacts archaeologically recovered from the region, but only first-order approximation is possible. Pernicka *et al.*. (1997: 41) consider the total weight of copper artefacts in south-eastern Europe to be around 4,700 kg. Although they do not reference this estimate or give details of its calculation, it should be noted that the figure, while excluding the steppe, significantly includes the Eneolithic 'boom' in Hungary where individual late Eneolithic Mezőkeresztes hammer axes can each weigh 4kg. In 1982 Chernykh estimated that there were 900 known objects for the Carpatho-Balkan Eneolithic, excluding Hungary (Chernykh 1982a); nearly twenty years of subsequent excavations have significantly added to this tally, so that a figure in excess of 2000 now seems uncontroversial. This estimate includes small awls and pins, as well as larger objects such as axes. Unfortunately, too few site reports give the weights of copper artefacts recovered. For present purposes, it will be useful to be able to estimate a mean-average Eneolithic artefact weight. The best data are available for axes, but these can vary greatly in size; smaller sheet- and hammered copper artefacts, like pendants, hooks, knives and awls obviously weigh less. Based on a selection of published measurements[4], I have chosen a 250gm per artefact average weight, giving a total weight of archaeologically recovered copper from the Carpatho-Balkan region of 500 kg, which is one tenth of the figure proposed, including Hungary, by Pernicka *et al.* (1997).

Using my figures, we arrive at a hypothesized ratio between the target and archaeological artefact populations of 1:10,000. That is to say, what is known archaeologically represents a mere 0.01% of original production. On the reckoning of Pernicka *et al.*. 1997, archaeological representativity would be a factor of ten better, but still only 0.1%. Such ratios should not be surprising, since the 2,000 or so archaeologically-recovered artefacts are spread over a timescale of nearly a millennium and a wide region, for which an average rate of production of two items per year would be unrealistic. The production of c.20,000 artefacts per year is more plausible, and can be usefully looked at in terms of, for example, 1000 village-scale local communities scattered throughout the entire Carpatho-Balkans producing artefacts, great and small, at the average rate of less than two artefacts per month.

Clearly, whatever the precise amounts in use were, the estimate for the amount of mined copper provides a more plausible production picture than does the count of surviving artefacts. Viewed against this background, one implication of believing in the hiatus as a production phenomenon, is that we must accept the assumption that the 0.1-0.01% sample is directly representative of production at the 100% level. It is not clear that such an assumption is justifiable. Another implication of conceptualizing a 'crisis' in metal availability following immediately on from a period where 5000 tonnes of metal has been produced, is that the 5000 tonnes were irrevocably 'used up' and was no longer available in an economic sense.

In stark terms, if the figures cited here are anywhere near correct, then a vast amount of produced metal is missing from the archaeological record. What proportion of it was ever deposited, and of that, how much stayed deposited? Answers to these questions will not be found without explicitly theorizing modes and mechanisms by which metal was moved into and out of potentially preserving contexts ('artefact traps'). Two forms of cycling are of particular interest, viz. (1) legitimate cyling: most of the metal was never deposited in the record, but was recycled and reused in a continuous chain stretching into at least classical times; and (2) illegitimate cycling: metal that was initially deposited with permanent intent, for example in the form of grave goods, was later 'mined' as a raw material by other cultural groups or later generations.[5]

Mechanism 1 is accepted as a reality to some degree by most scholars (Lichardus & Lichardus-Itten 1995; Chernykh 1992, Rassamakin 1993; Pernicka *et al.*. 1993, 1997), but from what may be an evidentially weak position. Thus Chernykh writes "Along with the demonstrable utilization of metal in the Carpatho-Balkan metallurgical centres, with its characteristic 'signature' of impurities, we can also postulate supplies of copper coming [to the Pontic steppes] from the southwest and west in the form of ingots (as opposed to finished articles). The typological and technical originality of Tripolye metalworking provides clear evidence of this"; Pernicka *et al.* (1997) postulate that copper may reach the steppes predominantly from the Medni Rid, sources, but the massive, so far under-researched archaeologically, copper resources of the Carpathians and Apuseni mountain must also be considered. But while it is clear that metal was in 'circulation', it is not as clear whether this really was directional trade (metallic production » casting as ingots » transportation » working from ingots into artefacts in local styles » deposition in graves) or some other form of exchange and reworking.

Two observations are in order here. First, there is currently no evidence for the production of ingots in this early phase. Second, while copper may well travel from west to east, the implication that no copper travels from east to west is not necessarily supported by the elemental analyses, which seek to find a certain match between ore-bodies no longer in existence and artefacts produced through processes we cannot directly observe.

It seems at least possible, that west Ukrainian copper sources, previously noted but dismissed by scholars in the FSU, could have provided the basis for Cucuteni-Tripolye metalwork yet still remain undetectable in the chemical signature of surviving metalwork. On any realistic calculation, because such a tiny percentage of metalwork now survives, it is conceivable that the dilution of west Ukrainian sources into a predominantly Carpatho-Balkan mix would not be analytically obvious. What is significant here is not any denial that Carpatho-Balkan copper found its way out beyond the immediate communities on whose territories the primary sources were situated, but that other areas, such as western Ukraine had a small, yet significant, parcel of indigenous production that became its passport into the grand game. I use the word 'game' in a loose sense here, but I am envisaging something more like the anthropologically-studied Melanesian *kula* exchange ring (cf. Sherratt 1997: Fig 4.1) than directional trade with standard units (which is a concept more appropriate - although even then perhaps not fully - to the Bronze Age proper, with its *Ösenhalsringe* facilitating standardized but hand-to-hand exchange, and ox-hide ingots designed for bulk shipment). The fact that Cucuteni-Tripolye production is highly distinctive is not because it was worked from hypothetical blank ingots, but because it expressed a technology that the community was locally proud of, notwithstanding the fact that a predominantly high percentage of metal was ultimately derived from artefacts traded in from outside. Much cycling may have been continuous, never resulting in any final deposition back in the ground.

Mechanism 2 - illegitimate cycling - is just as important, but is not clearly acknowledged. In terms of dynamics this could cover several modes of operation: theft and booty-taking, the discovery and appropriation of the hoards of others, and grave robbing. In terms of statics, the mode most recognized (and recognizable) is grave robbing. It appears in the archaeological record not as an occasional aberration but as an organized mass phenomenon (Jankuhn 1978; Taylor 2000 forthcoming). For example, of the 258 graves in the early Bronze Age (Reinecke A1) Cemetery F at Gemeinlebarn, Lower Austria, only 15 had not been robbed in antiquity (Neugebauer 1991: 115). This was determined by using an excavation strategy sensitive enough to identify recutting and which integrated physical anthropological analysis with artefact and soil contexts on-site; a key indicator of mass systematic robbing in this flat cemetery was the presence of skeletal elements from one grave in an adjacent grave, which must therefore have been standing open at the same time (Neugebauer 1991: Abb.30).

According to the excavation technique and the division of specialist skill, it is clear that grave robbing is more or less easy to recognize on the basis of the extent of the disturbance and the form of the original burial. From personal experience, I consider it no accident that grave robbing has been recognized in southeast Europe and the steppe regions more in cist and barrow (*kurgan* or *mogila*) than in flat cemetery excavations: a combination of baking continental summers and the lack of intimate integration of physical anthropologists with excavation teams militates against the recognition of disturbance in flat cemeteries. The relatively frequent recording of disarticulated skeletons during the excavation of Eneolithic cemeteries in the Carpatho-Balkan and steppe regions is at least suggestive.

The motive for grave robbing in Eurasia, from its earliest recognized manifestation in the Lengyel cemetery at Freibritz (Neugebauer 1991: Anm.697) through to the Merovingian period and beyond, seems typically to be to obtain items of value, usually metallic (this could, in the light of Zengövárkony, and *contra* Neugebauer, even have been true in the Freibritz case). This may sometimes have a ritual-symbolic dimension, where, for example, obtaining the sword of the deceased is conceived a trial of nerve - Nordic "haugbrot" - but at sites like Gemeinlebarn, as at Pazyryk (Rudenko 1960), the clear-out of all reusable metal seems to have been the primary aim.

In the light of the foregoing, it may be suggested that, in the Eneolithic metal artefacts were initially seen as symbolically powerful as grave goods - having magical and transformative qualities - and that they rapidly became a liability; as metal came to play an ever greater economic role and ore sources, although never running out, became more controlled, so the placing of metal in graves came to be seen as carrying certain risks. For it is clear, that robbing is seen as desecration by the relatives of the deceased, even when that is not the primary intention of the robbers. It is thus entirely consistent that, after an early and enthusiastic inception, metal should suddenly become more elusive in the archaeological record. Not only was it being dug out from cemeteries as the highest quality, pre-smelted, raw material, but communities were taking the decision to remove it themselves, either before burial or between the initial and final funerary rites (*primary* and *secondary* burial in social anthropological terms (Metcalf & Huntington 1991 Taylor 2000 forthcoming). With flat cemeteries such as Gemeinlebarn, the robbing must have been near-contemporary, occurring while visible grave markers were still in place.

In these terms, it is no surprise that the funerary record of the late Eneolithic to early Bronze Age transition is characterized by the development of what must be considered 'defended' burial, with the first burial tumuli in Hungary and the emergence of the Yamnaya on the steppe (e.g., Srejović & Tasić 1987; Merpert 1991). That flat cemetries such as Varna survived at all, being guarded until the grave markers rotted, and then passing out of memory until their rediscovery in the twentieth century, is an extraordinary fluke. But once burial contexts were made more secure, metal was again more frequently placed in them. Of course, this did not stop grave robbing, but it marked an end of a distinct phase in the cat-and-mouse game of honouring one's own ancestors, while attempting to dishonour and score off one's neighbours. Barrows stay visible for a longer time, so the trade-in may be that grave robbing becomes both more arduous and more hit and miss (according to the mound size) and a potentially more delayed phenomenon, taking place several generations after the intitial interment. This kind of game yeilds two, polarized, solutions, viz. the exclusion of precious items from burials altogether, the choice in the Ancient Greek world as it entered a money and credit economy (Vickers

& Gill 1994) or the increasing fortification of grave contexts, as in Iron Age Scythia and in modern American 'car' burial (Taylor 1994, 1999 forthcoming).

Whether legitimately cycled above-ground, through curation, inheritance, and prestige exchange, or illegitimately liberated from below-ground funerary contexts and hoards, the lateral cycling of artefacts is probably the principal reason that only 0.1-0.01% or so of the copper produced in the Eneolithic has been archaeologically recovered. Direct evidence for this in Alpine Europe is provided by the discovery of the so-called Iceman who died accidently in the Ötzal Alps in the middle of the 'hiatus' period, and whose body and clothing and equipment were preserved by the Similaun glacier. Ottaway (1982) defined three horizons of pre-Bronze Age metal use, with heavy artefacts such as axes only occurring in the first and third; only a few small artefacts belong to the second horizon. The 2-Sigma interval for the combined 14 Radiocarbon-14 date determinations from the Iceman's soft tissue and bone is from 3350 to 3120 BC, with a death-date close to 3325 BC considered most likely (Barfield 1994: 20, Fig 9; Tab.1; cf. Spindler 1993: 31f; Tab.1 showing the rather wider 2-Sigma range for the associated botanical finds: 3,492-3,049 BC); the paucity of direct parallels for the axe he was carrying when he died could suggest that he belonged to Ottaway's second horizon. This artefact is of cast and hammered copper (Cu 99.7%, 0.22% As, and 0.08% Ag) with traces of edge wear (Egg 1993: 56f, Section C.I.4), and still in its hafting. The axe is close to the EBA Neyruz type (Egg 1993: 83ff; Mayer 1977), ancestral to *Randleistenbeile*, the earliest known example of which comes from the Corded Ware lacustrine settlement of Vinelz in Switzerland, with a dendrochronological date of 2745-2695 BC. De Marinis terms it the *Similaun* type (de Marinis 1992) and he and Marcus Egg draw close parallels with axes in the Remedello Sopra hoard in northern Italy.

Although such axes seem functionally unappealing to the modern eye (Egg speaks of it in relation to later alloyed ones, as "eine sehr viel schlechtere Bearbeitbarkeit"), the Iceman also carried a wood-handled flint knife – presumably with a better cutting edge than could be achieve in copper. Barfield, nevertheless, observes that the axe "was a working tool; hafted as it was, only 2.6 cm of the metal blade would have been visible anyway, and in this way could hardly have functioned as a display item. We are probably dealing here with some form of competitive, 'big man' type of social organization in which relative status was measured in stock rather than artefacts" (Barfield 1994: 20). What was the copper axe good for? It is rather oddly hafted in yew, which is springy, as Egg remarks, drawing on the experimental observations of a contemporary yew bowmaker (Egg 1993: 93 and Anm.125). One may wonder if there was some special use for this combination of softish metal and springy hafting - perhaps (could it be?) as a nicely balanced and forgiving ice-axe? What is clear is that this was a *curated* object in Binford's terms, very uncommon in burials of the period (given a choice the Iceman would - literally speaking - not have been seen dead with it), but demonstrating typological continuity and development through the 'hiatus' period.

The period of the Carpatho-Balkan hiatus, as well as the mid-Eneolithic downturn in the Alpine region, can thus be conceptualized as a period of increased circulation (Schiffer's *lateral cycling*) and increased liberation of metal, if not from ore bodies then from dead human bodies (although we can currently quantify neither activity). It was this lateral cycling that led, I believe, to the sort of widespread admixing of different metals that was the precondition for the explicitly cognized alloying identifiable in the Bronze Age proper. And, despite the lack of sufficient data for grave-robbing in the Carpatho-Balkan Eneolithic (for technical reasons which I have already alluded to), there is another way of inferring such past dynamics.

The argument from skeuomorphism

Skeuomorphism has already been touched on in terms of the emulation of metal forms in Le Grand-Pressigny flint production. This phenomenon has been recognized in Scandinavia, as well as France, a region where another distinct hiatus has been observed: Carpathian copper reaches Scandinavia early and then vanishes from the archaeological record until the first bronzes appear (Randsborg 1979). Skeuomorphism is marked throughout the neolithic, becoming very elaborate at the finish (as Nancy Sandars describes it "the now almost superseded flint imitates the natural splay of the bronze haft, and polished stone axes have the dorsal casting-seam of a metal prototype." 1968: 182). Was copper "swamped" out by the development of the local flint exchange economy (Sherratt 1993: 12), or was it simply not archaeologically trapped? The answer turns on how skeuomorphism is best understood in a particular case - does it express a yearning for something now totally unavailable, or is it a down-market or funerary version of something too precious not to be fiercely curated? For pottery at this time Andrew Sherratt has reached the latter conclusion, namely that the fictile mimicry implies the circulation of precious metal vessels which are absent in any direct form from the record (Sherratt & Taylor 1997).

Skeuomorphism in pottery is one of the most striking features of ceramic production in the period 3500 to 3000 BC. This is most noticeable in the Baden and Ezero culture areas, which are defined in part on the basis of distinctive strap-handled pottery forms clearly copied from the precious metal prototypes designed as part of the drinking rituals which typify the status expression of Secondary Products Complex economies (Sherratt & Taylor 1997: Figs 17.1, 17.2, 17.4;) in the Carpatho-Balkan region (but which are also seen, for example, as far east as Koisug in the Zhivotilovo-Volchan group on the lower Don). There are also distinctive outliers in central and northern Europe, such as the Lengfeld-Alkofen cup from Bavaria (mimicking an Aegean Vapheio handle in pottery) and the

Oldendorf cup from Saxony. Dating to c.3200 BC, the latter is a strap-handled drinking cup which has both an omphalos base and a skeuomorphic rivet (Sherratt & Taylor 1997: 436; Sherratt 1997: Fig.7.3). For the Baden and Ezero material we argued that "it is unlikely that metal prototypes would have been locally manufactured: probably rare examples circulated from farther south. This is also likely to be the explanation for the very rare metal-derived shapes in pottery from TRB and related contexts farther north" (Sherratt & Taylor 1997: 435f).

I feel less sure of this now. For the northern regions beyond the Carpatho-Balkans it seems fair enough, but in the environs of Varna itself, it seems hard to imagine that the skills and metallic wherewithal to make fine metal objects were in total abeyance. Such widespread skeuomorphism, as Baden and Ezero display, is, I would now argue, evidence of direct metal activity in the cultures themselves - which we know are producing and depositing some copper-alloy artefacts. The argument from skeuomorphism is that we should take the mimicry of forms in alternative, cheaper material, occurring in the static, archaeological context as *prima facie* evidence for the existence of more expensive curated materials in the dynamic, target context. This should be as true for flint as for pottery.

Conclusion: envaluing metal

The inception of metals among Eurasian communities was revolutionary, both economically and symbolically. Speaking of the hiatus period in Bulgaria (their 'proto-Bronze Age') Pernicka *et al.* (1997: 137) write that the "general appearance is as if a variety of new ores were being experimented with. Perhaps this was by necessity, because the old sources were exhausted. But it is possible as well that. . .a change in cultural orientation occurred". Taking a cue from Stephen Shennan, who rightly observes that "Trajectories of social and economic change are strongly affected by the nature of the categorization processes at work within society, which are themselves in turn affected by them" (Shennan 1994: 59), the process of *envaluation*[6]can be envisaged, as follows:-

Copper was soft; yet, for all that, it changed everything, allowing a multitude of tasks to be accomplished in a different manner (Jovanović 1989); even a soft-edged axe might have its uses. It is the first truly laterally cyclable artefactual product, which could be unmade and remade at will virtually *ad infinitum* without any necessary loss of basic material value. I believe that it is to be expected that there will be dramatic shifts in the depositional pattern of such a revolutionary material through time, and especially during the period of its inception. Such shifts would be underscored by the fact that the new material was also 'good for thinking.' It was not treated in a disembedded, secular manner: the act of making it and the remains of making it were as significant as the 'product' itself (Budd & Taylor 1994). Even slag was treasured. As Eugenia Herbert has shown for sub-Saharan Africa, slag is almost universally thought of as numinous, either positively or negatively, and is variously tabooed, used in healing, or considered as a menstrual or as a faecal product; it is not just left around (Herbert 1993). The putative female-associated slag at Zengövárkony may not so much give the lie to Childe's original postulate that Old World metal-making was a male specialism (Childe 1936: 9; cf. Budd & Taylor 1994: 137f), as indicate that metal was, from the outset, seen as good for gender-thinking, as it also was at Varna (Katincarov, Mohen, *et al.* 1989, No.311; Taylor 1996: 52-71, Ch. 7, and Fig 7.6: discussion of the famous sheet-gold 'penis piece').

Students of the EMA have hitherto drawn a sharp distinction between *raw materials* and *finished products*. For gold and copper, this is profoundly wrongheaded, because - unlike flint, pottery and wood, whose reuse and recycling is strictly limited and involves more or less rapid depletion - finished metal products are just as good as the raw material ore. Far better in fact, as all the energy involved in the conversion to metal is now gratis. For metal, the idea of a *discard* rate is strictly inappropriate; unlike Pueblo pottery, or an incrementally reduced lithic, it can go on and on; all one can attempt to calculate are *deposition* and *retrieval* rates. Often, all that lies between demand and supply are a few feet of graveyard earth and the sorry spirits of someone else's ancestors. Thus, although the deposition rate for Eneolithic copper artefacts could easily have averaged 50% of the 25,000 per-year produced (as extrapolated above) it was counterbalanced by a synchronous retrieval and recycling rate that could have been 90% of this half[7]. That finished metal artefacts required defending from the outset is indicated by the early advent of hoarding behaviour, as seen at Plo_nik (Stalio 1964; Jovanović 1989) or in the Karbuna hoard (Chernykh 1992), notwithstanding their clear difference from later and more massive Bronze Age hoards[8]. The sharp increase in defended settlement in the final Eneolithic - spanning the time when so little metal survives in the record - is best seen as an attempt to more effectively safeguard wealth[9]; of what metal has been recovered, Pernicka *et al.* (1997: 136) note the sudden "peculiar" dominance of daggers. This is not a 'crisis' or *Rückgang* or 'hiatus' - the relative absence of metal is a sign of its developing worth, and its growing association with - in some as yet admittedly ill-defined sense - funding and prosecuting warfare (as is clearly the case later on for Classical Greece: cf. Vickers & Gill 1994).

The near-three thousand years taken for archaeologically-recognizable metal use to move from southeast Europe to Britain and Scandinavia is not because the terrain between was unsettled, untraversable or otherwise impenetrable. What it does seem to imply (*pace* Budd & Taylor 1995) is that metal use developed within communities in an embedded way, not as a secular, economic add-on. Following this implication, it can be thought that ritualization of production allowed knowledge of how to produce metal from ore to remain esoteric, and, as it was *de facto* a monopoly, metal was economically beyond the general reach of Atlantic communities at a time when it was widespread in the Carpatho-Balkans. The other side of this coin is that it was also expensive for certain communities within the Carpatho-Balkans. What is desirable and what cannot be obtained consensually or legitimately will always be obtained by those prepared to hazard forceful or illegitimate means. Thus grave-robbing (motivated by a desire to disembed metal) became, by, I would suggest, the later Eneolithic, the easiest way for some communities to obtain 'raw material.'[10]

There is no precise ethnographic parallel for what happened in the Eneolithic. The process of the initial envaluation of metal in the Old World was unique. One can, of course, look at the adoption of different kinds of metal and metal artefacts in Australia or Canada in historic times as a correlate for the EMA in Eurasia, but the time-scale is very much shorter, the contact more intense and the contacters were at the end of a very long historical tradition of metal use, involving developed and relatively slow-changing conceptualizations of *metal*. We need then, not simply to back-project modern ideas of what metals mean today, nor strictly apply particular ethnographic comparisons. Rather, we need to bridge between, on the one hand, our developing anthropologically-informed knowledge of the behaviour of people in general and the modalities of particular kinds of living societies and, on the other hand, the patterning of the archaeological record. Doing so, we can attempt to envisage what particular and unique coincidence of materials-related behaviours could have occurred among Eneolithic communities, and what the testable archaeological correlates of such behaviours might be. Whether these species of behaviour are conceptualized as primarily economic or primarily symbolic is beside the point, as they are complementary aspects of what Wittgenstein would term our 'natural history'.

Acknowledgements

I would particularly like to thank the following, who have helped me to frame the ideas in this paper although they may not agree with all (or any) of it: Paul Budd, Peter O'Neill, Ernst Pernicka, Neville Steed, Andrew Sherratt, Gerhard Trnka, Suzanne Young, Michael Vickers and, finally and most importantly, Sarah Wright, my wife, whose translation of Chernykh's *Ancient Metallurgy in the USSR* (1992) and other recent works by Russian scholars has proved a constant intellectual provocation.

References

BARFIELD, L. 1994. The iceman reviewed, *Antiquity* 68: 10-26.

BERNBECK, R. & J.MÜLLER, 1996. Prestige und Prestigegüter aus kulturanthropologischer und archäologischer Sicht. In R.Bernbeck & J.Müller (ed.), *Prestige-Prestigegüter-Sozialstrukturen (Beispiele aus dem europeischen und vorderasiatischen Neolithikum)* Archäologische Berichte 6: 1-27. Bonn: Holos.

BINFORD, L.R. 1983. *In pursuit of the past: decoding the archaeological record*. London: Thames & Hudson.

BOGNÁR-KUTZIÁN, I. 1963. *The Copper Age cemetery of Tiszapolgár-Basatanya* (Archaeologia Hungarica N.S. 42; transl. G.P.Dienes). Budapest: Academiai Kiado.

BOYER, P. 1994. *The naturalness of religious ideas: a cognitive theory of religion*. New York: Cambridge University Press.

BUDD, P. 1992. Alloying and metalworking in the Copper Age of central Europe, *Bulletin of the Metals Museum of the Japan Institute of Metals* 17: 3-14.

BUDD, P., D.GALE, A.M.POLLARD, R.G.THOMAS & P.A.WILLIAMS, 1992. The early development of metallurgy in the British Isles, *Antiquity* 66: 677-86.

BUDD, P., B.SCAIFE, T.TAYLOR, & R.G.THOMAS, 1994. Untangling the web: some new views on the origins of prehistoric metallurgy, *Journal of the Historical Metallurgy Society* 28(2): 98-102.

BUDD, P. & T.TAYLOR, 1995. The faerie smith meets the bronze industry: magic versus science in the interpretation of prehistoric metal-making, *World Archaeology* 27(1): 133-43.

CHERNYKH, E.N. 1978a. *Gornoe delo i metallurgiya v drevneishei Bolgarii*. Sofia: Bulgarian Academy of Sciences.

1978b. [=Cernych] AiBunar - a Balkan copper mine of the fourth millennium BC, *Proceedings of the Prehistoric Society* 44: 203-17.

1982a. Otrazhenie ratsionalnogo i irratsionalnogo v arkheologicheskoi kulture, *Sovetskaya arkheologiya* 1982(4): 8-20.

1982b. Die ältesten Bergleute und Metallurgen Europas, *Das Altertum* 28(1): 5-15.

1991. Frühestes Kupfer in den Steppen- und Waldsteppenkulturen Osteuropas. In J.Lichardus (ed.), *Die Kupferzeit als historische Epoche* 2: 581-592 (Saarbrücker Beitrage zur Altertumskunde 55). Bonn: Habelt.

1992. *Ancient metallurgy in the USSR: the Early Metal Age* (transl. S.Wright). Cambridge: Cambridge University Press.

CHILDE, V.G. 1936. *Man makes himself*. London: Watts.

1944. Archaeological ages as technological stages, *Journal of the Royal Anthropological Society* 74: 7-24.

CRADDOCK, P.T. 1995. *Early metal mining and metallurgy*. Edinburgh: Edinburgh University Press.

DANILENKO, V.N. 1974. *Eneolit Ukrainy*. Kiev: Naukova dumka.

DE MARINIS 1992. Metallurgia. In F.Höpfel, W.Platzer & K.Spindler (ed.) *Der Mann im Eis* 1: *Bericht über das Internationale Symposium 1992 in Innsbruck* (*Veröffentlichungen der Universität Innsbruck* 187): 389-418. Innsbruck: Universität Innsbruck.

DEMOULE, J.-P. & M.LICHARDUS-ITTEN, 1989. Invention du métal et premières inégalités: la nécropole de Varna dans son contexte européen. In R.Katincarov,

J.-P.MOHEN, *et al.* (ed.) *Le premier or de l'humanité en Bulgarie, 5e millénaire*: 38-44. Paris: Editions de la Réunion des musées nationaux.

EGG, M.1993. C. Die Ausrüstungen des Toten. In M.Egg, R.Goedecker-Ciolek, W. Groenmann-van Waateringe & K.Spindler, *Die Gletschermumie vom Ende der Steinzeit aus den Ötztaler Alpen* (separately bound 'Sonderdruck' of the *Jahrbuch des Römisch-Germanischen Zentralmuseums* 39 (1992)): 35-100. Mainz: RGZM.

EPSTEIN, S.M. 1995. *Cultural choice and technological consequences: constraint of innovation in the late prehistoric copper smelting industry of Cerro Huaringa, Peru*. Philadelphia: University of Pennsylvania unpublished PhD dissertation.

FORENBAHER, S. 1993. Radiocarbon dates and absolute chronology of the central European Early Bronze Age, *Antiquity* 67: 218-56.

GIMBUTAS, M. 1977. The first wave of Eurasian steppe pastoralists into Copper Age Europe, *Journal of Indo-European Studies* 5: 277-339.

GLUMAC, P.D. & J.A.TODD, 1991. Eneolithic copper smelting slags from the middle Danube basin. In E. Pernicka & G.A.Wagner (ed.), *Archaeometry '90*: 155-64. Basel: Birkhauser Verlag.

HAUPTMANN, A., F.BEGEMMANN, E.HEITKEMPER, E.PERNICKA, & S.SCHMITT-STRECKER, 1992. Early copper produced at Feinan, Wadi Araba, Jordan: the composition of ores and copper, *Archaeomaterials* 6:1-33.

HERBERT, E.W. 1993. *Iron, gender, and power: rituals of transformation in African societies*. Bloomington: Indiana University Press.

HODDER, I. 1986. *Reading the past: current approaches to interpretation in archaeology*. Cambridge: Cambridge University Press.

JANKUHN, H. (ed.), 1978. *Zum Grabfrevel in vor- und frühgeschichtlicher Zeit, Untersuchungen zu Grabraub und "haugbrot" in Mittel- und Nordeuropa* (Abhandlungen der Akademie der Wissenschaften in Göttingen, Phil.-hist. Klasse 3. Folge 113).

JOVANOVIĆ B. 1971. *Metallurgy of the Eneolithic period in Yugoslavia* (transl. V.Poznanovic). Belgrade: Archaeological Institute.

1989. Die Rolle des Kupfers in der Ökonomie der frühäneolithischen Gruppen des Balkans. In A.Hauptmann, E.Pernicka, & G.A.Wagner (ed.) *Old World Archaeometallurgy* (Proceedings of the International Symposium 'Old World Archaeometallurgy', Heidelberg 1987; *Der Anschnitt*, Beiheft 7): 13-17. Bochum: Deutschen Bergbau-Museum.

KATINCAROV, R. & J.-P.MOHEN, *et al.* (ed.), 1989. *Le premier or de l'humanité en Bulgarie, 5e millénaire*. Paris: Editions de la Réunion des musées nationaux.

LECHTMAN, H. 1979. Issues in Andean metallurgy. In E.P.Benson (ed.) *Pre-columbian metallurgy of South America*: 1-40. Washington: Dumbarton Oaks.

1980. The central Andes: metallurgy without iron. In T.A.Wertime & J.D.Muhly (ed.) *The coming of the age of iron*: 267-334. New Haven: Yale

University Press.
LICHARDUS, J. (ed.), 1991. *Die Kupferzeit als historische Epoche* 1: 35-46. Saarbrücker Beitrage zur Altertumskunde 55. Bonn.
LICHARDUS, J. & M. LICHARDUS-ITTEN, 1995. Nordpontische Beziehungen während der frühen Vorgeschichte Bulgariens, *Thracia* 11:31-62.
MALLET, N. 1992. *Le Grand-Pressigny. Ses relations avec la civilisation Saône-Rhône* (2 vols). Supplément au bulletin de la Société des Amis du Musée du Grand-Pressigny: Indre et Loire, Ministère de la Culture.
MAKKAY, J. 1976. Problems concerning Copper Age chronology in the Carpathian basin. Copper Age gold pendants and gold discs in central and south-east Europe, *Acta Archaeologica Academiae Scientiarum Hungaricae* 28: 251-300.
MASSON, V.M. & N.YA. MERPERT (ed.), 1982. *Eneolit SSSR*. Moscow: Nauka.
MERPERT, N.YA. 1991. Die neolithisch-äneolithischen Denkmäler der pontisch-kaspischen Steppen und der Formirungsprozeß der frühen Grubengrabkultur. In J.Lichardus (ed.), *Die Kupferzeit als historische Epoche* 1: 35-46 (Saarbrücker Beitrage zur Altertumskunde 55). Bonn: Habelt.
MAYER, E.F. 1977. *Die Äxte und Beile in Österreich*, (Prähistorische Bronzefunde IX.9). Munich: C.H.Beck.
METCALF, P. & R.HUNTINGTON 1991. *Celebrations of death: the anthropology of mortuary ritual* (2nd edn). Cambridge: Cambridge University Press.
MOHEN, J.-P. 1990. *Metallurgie prehistorique*. Paris: Masson.
MUHLY, J.D. 1989. Çayönü Tepesi and the beginnings of metallurgy in the ancient world. In A.Hauptmann, E.Pernicka, & G.A.Wagner (ed.) *Old World Archaeometallurgy* (Proceedings of the International Symposium 'Old World Archaeometallurgy', Heidelberg 1987; *Der Anschnitt*, Beiheft 7): 13-17. Bochum: Deutschen Bergbau-Museum.
NEUGEBAUER, J.-W. 1991. *Die Nekropole F von Gemeinlebarn, Niederösterreich: Untersuchungen zu den Bestattungssitten und zum Grabraub in der ausgehenden Frühbronzezeit in Niederösterreich südlich der Donau zwischen Enns und Wienerwald*. Mainz: Phillip von Zabern.
NOVOTNA, M. 1976. Beginn der Metallverwendung und Verarbeitung im östlichen Mitteleuropa. In H.Müller-Karpe (ed.) *Les débuts de la metallurgie*: 118-33. Nice: Université de Nice.
O'NEILL, P.J. 1996. *An investigation into the European later Eneolithic metalworking 'hiatus' in the light of archaeometallurgical, archaeological, and ethnographic data*. Bradford: University of Bradford unpublished UG dissertation.
OTTAWAY, B.S. 1981. Modelle des Kupferhandels im Äneolithikum Europas, *Slovenska Archeologia* 29(1): 139-148.
1982. *Earliest copper artefacts of the Northalpine region: their analysis and evaluation*. (Schriften des Seminars für Urgeschichte der Universität Bern 7). Bern: University of Bern.
1989. Interactions of some of the earliest copper-using cultures in Central Europe. In A.Hauptmann, E.Pernicka, & G.A.Wagner (ed.) *Old World Archaeometallurgy* (Proceedings of the International Symposium 'Old World Archaeometallurgy', Heidelberg 1987; *Der Anschnitt*, Beiheft 7):19-23. Bochum: Deutschen Bergbau-Museum.
OTTAWAY, B.S. & C.STRAHM, 1975. Swiss Neolithic copper beads: currency, ornament or prestige items? *World Archaeology* 7: 307-20.
PERNICKA, E., F.BEGEMANN, S.SCHMITT-STRECKER, & G.A.WAGNER, 1993. Eneolithic and early Bronze Age copper artefacts from the Balkans and their relation to Serbian copper ores, *Praehistorishe Zeitschrift* 68(1): 1-54.
PERNICKA, E., F.BEGEMANN, S. SCHMITT-STRECKER, H. TODOROVA & I.KULEFF, 1997. Prehistoric copper in Bulgaria: its composition and provenance, *Eurasia Antiqua* 3: 41-180.
PETREQUIN, A.M. & P.PETREQUIN, 1988. *Le Néolithique des Lacs*. Paris: Errance.
PINGEL, V. 1989. Zum Beginn der Goldmetallurgie auf der Iberischen Halbinsel, In A.Hauptmann, E.Pernicka, & G.A.Wagner (ed.) *Old World Archaeometallurgy* (Proceedings of the International Symposium 'Old World Archaeometallurgy', Heidelberg 1987; *Der Anschnitt*, Beiheft 7):45-50. Bochum: Deutschen Bergbau-Museum.
RANDSBORG, K. 1979. Resource distribution and the function of copper in early Neolithic Denmark. In M.Ryan (ed.) *The origins of metallurgy in Atlantic Europe* (proceedings of the 5th Atlantic Colloquium, Dublin, 1978): 303-18. Dublin: Government Stationery Office.
RASSAMAKIN, Y. 1993. Eneolit stepnogo Prichernomorya i Priazovya (po pogrebalnym pamyatnikam). In P.Georgieva (ed.) *The Fourth Millennium BC*: 5-28. Sofia: New Bulgarian University.
RENFREW, A.C. 1969. The autonomy of the south east European Copper Age, *Proceedings of the Prehistoric Society* 35: 12-47.
ROMAN, P. 1983. Der Übergang vom Äneolithikum zur Bronzezeit auf dem Gebiet Rumäniens, *Godi_njak* 21: 115-134.
ROSTOKER, W., V.C.PIGOTT, & J.R.DVORAK, 1989. Direct reduction to copper metal by oxide-sulphide mineral interaction, *Archaeomaterials* 3: 69-87.
RUDENKO, S.I. 1960. *Kultura naseleniya Tsentralnogo Altaya v skifskoe vremya*. Moscow: Nauka.
SANDARS, N.K. 1968. *Prehistoric art in Europe* (The Pelican History of Art). Harmondsworth: Penguin.
SCHIFFER, M.B. 1976. *Behavioral archaeology*. New York: Academic Press.
SHENNAN, S. 1993. Commodities, transaction, and growth in the central-European early Bronze Age, *Journal of European Archaeology* 1(2):59-72.
SHERRATT, A.G. 1993. What would a Bronze Age world system look like? Relations between temperate Europe and the Mediterranean in later prehistory, *Journal of European Archaeology* 1(2): 1-57.
1997. *Economy and society in prehistoric Europe: changing perspectives*. (Selected and revised papers 1972-1995). Edinburgh: Edinburgh University Press.
SHERRATT, A.G. & T.TAYLOR, 1997. Metal vessels in Bronze Age Europe and the context of Vulchetrun, in A.G. Sherratt, *Economy and society in prehistoric Europe: changing perspectives*: 431-56. Edinburgh: Edinburgh University Press.
SPINDLER, K. 1993. B. Die Mumie. In M.Egg, R.Goedecker-Ciolek, W. Groenmann-van Waateringe & K.Spindler, *Die Gletschermumie vom Ende der Steinzeit aus den Ötztaler Alpen (Sonderdruck aus Jahrbuch des Römisch-Germanischen Zentralmuseums* 39 (1992)): 27-35. Mainz: RGZM.
SREJOVIĆ D. & N.TASIĆ(ed.) 1987. *Hügelbestattung in der Karparten-Donau-Balkan-Zone während der äneolithischen Periode*. Belgrade: Archaeological Institute.
STALIO, B. 1964. Novi metalni nalaz iz Plocnika kod Prokuplja (Dépôt d'objets métalliques nouvellement mis au jour à Plocnik prés de Prokuplje), *Zbornik Narodnog muzeja Beograd* 4: 35-41.
TAYLOR, T. 1989. Iron and Iron Age in the Carpatho-Balkan region: aspects of social and technological change. In M.-L.S.Sørensen & R.Thomas (ed.) *The Bronze Age-Iron Age transition in Europe* (BAR International Series S.483): 68-92. Oxford: British Archaeological Reports.
1994. Thracians, Scythians, and Dacians, 800 BC-AD 300. In B.W.Cunliffe (ed.) *The Oxford illustrated prehistory of Europe*. Oxford: Oxford University Press.
1996. *The prehistory of sex: four million years of human sexual culture*. London: Fourth Estate.
2000. forthcoming. *The invention of death: archaeology, evolution, and the soul*. London: Fourth Estate.
TODOROVA, H. 1978a. Die Nekropole bei Varna, *Zeitschrift für Archäologie* 12: 87-97.
1978b. *The Eneolithic period in Bulgaria in the fifth millennium B.C.* (BAR International Series, S.49). Oxford: British Archaeological Reports.
1981. *Die kupferzeitlichen Äxte und Beile in Bulgarien* (Prähistorische Bronzefunde IX.14). Munich: C.H.Beck.
TODOROVA, H. & P. GEORGIEVA, 1993. Der ethnokulturelle Inhalt des Übergangs vom Äneolithikum zur Frühbronzezeit Bulgariens, *Pulpudeva* 6: 165-177.
TRIGGER, B. 1995. Expanding middle-range theory, *Antiquity* 69: 449-58.
VICKERS, M. & D.GILL, 1994. *Artful crafts: ancient Greek silverware and pottery*. Oxford: Clarendon Press.
VULPE, A. 1976. Zu den Anfängen der Kupfer- und Bronzemetallurgie in Rümanien. In H. Müller-Karpe *Les débuts de la metallurgie*: 134-75. Nice: Université de Nice.
WERTIME, T.A. 1964. Man's first encounters with metallurgy, *Science* 146: 1257-67.
1973. The beginnings of metallurgy: a new look, *Science* 182: 1257-67.
ZWICKER, U., H,GREINER, K.-H.HOFMAN, & M.REITHINGER, 1985. Smelting, refining and alloying of copper alloys in crucible furnaces during prehistoric up to Roman times. In P.T.Craddock & M.J.Hughes (ed.) *Furnaces and smelting technology in antiquity* (British Museum Occasional Paper 48): 103-16. London: British Museum Publications.

Notes

[1] There has been a tendency to quote only 1-Sigma confidence intervals (68.2% likelihood), and to omit lab-error multipliers (Blockley pers. comm.). The reality for dates around 4700 BP, whose probability range centres at 3400 BC cal (i.e., at the apparent nadir of Carpatho-Balkan late Eneolithic metal production: Chernykh, 1992: 51 and Fig.16), is that the 2-Sigma interval (95.4% likelihood) will typically span 700 years (i.e., from 3750 to 3050 BC), and sometimes more.

[2] It is the eventual utilization of the glacially-truncated gossan bodies of the high Tatra - and Caucasus - that produces the first Bronze Age iron, in Reinecke A2: Taylor 1987).

[3] Although Pernicka *et al.* (1997: 125) cast some doubt of the reliability of the Zengövárkony evidence, they do note (1997: 127) the presence of crucible fragments at Durankulak dating to 4300 BC.
[4] Pernicka *et al.* (1997: 57ff) give weights for a large number of Bulgarian Eneolithic artefacts - from finger rings and borers weighing a gram or two, via bracelets in the 30-50gm range, to axes and axe-adzes weighing anything from 50gms to over 1kg; it is useful to compare the weights given for a selection of flat axes of the EBA Neyruz type which are close to the sort the Iceman, discussed below, carried: Mayer's (1977) figures range from around 225-275 gm (Nos 214, 215, 219, 226) down to 19gm (No225) with a number in the range 100-140gm (Nos 223, 224, 220, 222).
[5] I have deliberately avoided the more usual term *recycling* here, as this implies secondary use as salvage or afterthought; cycling, on the other hand, *pace* Schiffer, implies a continuous series of artefactual manifestations of the same material.
[6] The neologism *envaluing* is proposed here as I can find no established term that catches the sense of a new and strange material becoming valued as *metal*, with a specific semantic load and valency which is now broadly cross-culturally recognized. By *envaluing* I mean something like 'investing metal with value', but that phrase suggests something done to a particular pre-existent material category or object, and also suggests the possibility of the reverse process of disinvesting it.
[7] Of course this would still leave a deficit between what, at the end of the Eneolithic, was left untouched in the ground and what is now in museums. Such a deficit could be made up by reference to subsequent C- and N-Transforms: robbing out in all subsequent periods, including modern metal-detector use (which plagues eastern Europe), and erosion into riverine and marine systems where dispersal and restratification take place.
[8] The scale and composition of these discrete depositions may be as informed by ritual as by any more secular urge to safeguard 'property'; Sherratt considers that Karbuna may represent part of some kind of shamanic kit (Sherratt pers. comm.).
[9] Thus Jovanović (1989: 15): "Die Burgen sind daher die ersten Arsenale, gut gesicherte und befestigte Orte, wo nicht nur die Aufbewahrung, sondern auch die Erzeugung von Metallprodukten garantiert wurde."
[10] Reflecting on my thought-experiment in this paper, Andrew Sherratt neatly indicates potential compatibility with the Wertime/Sherratt/Shennan description in the following terms: "the exhaustion curve is time-transgressive; the envaluation curve is flatter, because it is more of a mental than material process. But I agree that (fluctuating) levels of above ground use go on continuously during periods of depositional cessation." (Sherratt pers.comm.).

4

Arsenical Copper in Early Irish Metallurgy

WILLIAM O'BRIEN

National University of Ireland, Dept of Archaeology, Galway, Ireland

Arsenical copper is a well known, if poorly understood, phenomenon in early metal-working traditions in Europe, Eurasia and the Near East. This is particularly true of the European Copper Age where many regional traditions can be identified as part of what, on first examination, appears to be a remarkably widespread metallurgical horizon spanning the fourth and third millennia BC. First identified over a century ago, this metal composition has provoked much debate on its significance for the early development of metallurgy (see Selimkhanov 1982 for historical review). Many researchers argue that arsenical copper can be regarded as the first deliberately produced 'alloy', representing a definite stage of technological innovation in the Bronze Age.

Several aspects of this phenomenon have engendered debate, from the functional and aesthetic value of arsenical metal compositions to the technology and mineral sources used in their manufacture. It is generally unclear what understanding early metalworkers had of this 'innovation' and, indeed, whether they regarded this metal as such. Much discussion has proceeded on the diffusionist assumption that arsenical copper was produced as part of a unified technology marking a particular stage in the development of metallurgy. However, numerous regional studies testify to considerable variation in the quality and use of this metal. This reflects both different perceptions of its value, as well as technological adaptation to a diversity of suitable mineral sources over a wide geographical area.

Wertime (1973) referred to a 'period of experimental alloying' in the 4th and 3rd millennia BC when metalworkers gradually learned to add such impurities as arsenic, antimony and bismuth to copper, to improve castings and metal hardness. He suggested that the origins of this alloying lay in the polymetallism of complex ore bodies and the use of fluxes, at a time of intense human curiosity about the properties of different earth minerals. In contrast, Jovanovic (1992) argued that the initial experimentation with arsenical copper ores came about as a result of a depletion of oxide and carbonate ores in the Balkan Copper Age. He proposed a gradual transition from an older technology of smelting oxide and carbonate ores to the smelting of arsenic and sulfide ores, the latter leading to increased metal availability in the late Copper Age and Early Bronze Age. Muhly (1988) says that a similar sequence may be identified in the Near East where arsenical copper first appears in the late fifth/early fourth millennia BC. Both clearly see the use of arsenical copper as a transition phase between the early exploitation of oxidized ores and the subsequent use of sulfides.

The study of arsenical copper has long centred on the technology of production, the raw materials used and the mechanical properties of the resulting metal artefacts. There is still divergence of opinion about whether early arsenical copper represents deliberate alloy practise, involving the addition of arsenic or arsenical substances to raw copper, or derives from the deliberate or unconscious use of naturally arsenical copper ores. Much of this uncertainty stems from a general lack of mining and metal-working contexts, leaving scientific analysis of finished artefacts as the principal avenue of investigation. Unfortunately, the scientific data gained from such studies rarely provides a direct path to the technology and mineral sources used and the literature is littered with speculation on these points. It is now possible to address these questions with the recent discovery of a Copper Age mine site in south-west Ireland where arsenical copper was made as part of a vigorous metallurgical tradition in the late third millennium BC.

Technological Innovation?

During the fourth and third millennia BC in Europe, immediately prior to the widespread adoption of tin-bronze, arsenic was a commonly occurring impurity in copper metalwork, ranging from less than 0.25% wt. at the lower end, typically up to 3-4%, with higher compositions reaching 10-12%. In some traditions, for example the Cycladic Early Bronze Age (Charles 1967), there is a sharp cut-off in arsenic content with the adoption of tin-bronze, suggesting a deliberate alloy practise. Elsewhere in Europe, it is common to find a considerable period of overlap between arsenic and tin contents, with the former more indicative of natural impurity levels in the ore used. Most researchers now draw a distinction between metals containing in excess of 5% arsenic and those with low arsenic compositions believed to derive from polymetallic ore processing. Deliberate alloying practise is particularly difficult to identify among the lower arsenic compositions which should probably be termed 'arsenicated' rather than 'arsenical' copper (Delibes *et al.* 1991).

As Northover (1989, 111) points out, defining early arsenical coppers as deliberately contrived alloys is too restrictive. There are several routes by which such compositions may have been achieved: some deliberate; some more or less accidental and a consequence of the ores being used. Of concern here is whether these copper-arsenic alloys offered any advantages over pure copper; were these metalworkers aware of these properties; did they seek to exploit them by selecting particular compositions for different tasks?

The benefits of a small arsenic content in early copper metal are well known (Charles 1967). These range from its role as a deoxidant leading to better castings, spectacularly seen in the famous Nahal Mishmar hoard from Palestine (Bar-Adon 1980), to increased workability, strength and hardness in the cold hammered state. Other potential benefits include corrosion resistance and decorative finish through surface segregation effects.

Finally, there is also the effect of arsenic content on lowering the melting temperature of raw metal and its less understood role in smelting procedures. Thus, we have several possible explanations, from initial smelting through to final casting, fabrication and use properties, as to why early metalworkers may have chosen to make arsenical copper.

With no evidence that arsenic was ever used as a metal, how did early metalworkers discover the properties of this substance when added to copper? Zwicker (1991, 331-2) lists the many ways in which arsenical copper can be produced, using native arsenic, arsenic sulfides and arsenic-bearing copper ores in different ways. Much attention has focused on the deliberate addition of arsenic or arsenical minerals to copper metal, in a similar fashion to the use of cassiterite in tin alloying. While native arsenic is very rare, several authors have cited the two principal arsenic ores, realgar (AsS) and orpiment (As2S3), as potential alloying agents. Craddock (1979) argued for deliberate arsenic alloying using these minerals in Iberia, while Pazukhin (1964) also favoured their use. Nevertheless, the scarcity of these minerals must be a major factor mitigating their widespread use in the production of arsenical copper.

The deliberate selection and smelting of arsenic-bearing copper ores is another way of producing arsenical copper metal. Researchers differ on how this may have been achieved and, specifically, on whether primary sulfarsenide ores or secondary, supergene minerals were used. Proponents of the former argue that direct roasting and smelting of copper-iron-arsenic sulfides can yield arsenical copper of the type prevalent in Copper Age Europe (e.g., Witter 1936; Eaton & McKerrell 1976). However, this has been disputed on the grounds that the necessary roasting of these ores prior to smelting would remove most of the arsenic content (Tylecote 1980). Instead, it is suggested that early arsenical coppers were produced by low temperature smelting of relatively pure oxide ores and that it was only in later Bronze Age times that the ability to smelt primary sulfide ores was developed. Several researchers have pointed to the use of secondary copper arsenate minerals from the surface zone of oxidation as the most likely source here (Charles 1967, 1979; Rapp 1988; Budd *et al.* 1992).

The diversity of mineral sources used to make arsenical copper is evident in the variability of final metal compositions. There is evidence for the deliberate manipulation of arsenic contents, most clearly seen in the use of this metal for high quality castings in Israel (Shalev 1991). Northover (1989) cites several examples where the mechanical properties of arsenical copper appear to have been deliberately exploited, for example, in the case of early copper daggers in second millennium BC Palestine and in Copper Age industries in Italy. Opposed to this is an increasing body of compositional and metallographic data which questions whether early metalworkers ever exploited the physical properties of this arsenical metal. Hook *et al.* (1990), in their examination of arsenical copper objects from Los Millares in south-east Spain, discovered that a surprisingly high number of objects were left in a soft annealed state. They concluded that either these objects were unfinished or that Copper Age metalworkers did not appreciate the benefits of work-hardening. These results are supported by metallographic analyses of early Irish metalwork which also indicate a failure to exploit the full potential of this metal (Allen *et al.* 1970; Northover 1989). Similar findings in a recent study of Copper Age metalwork from Austria led to the conclusion that the appearance of this arsenical metal "...is related to the exploitation of polymetallic secondary minerals and to developments in the smelting technology used to process these rather than to sudden recognition of the mechanical properties of alloys" (Budd 1991, 167).

Metallographic evidence should not be used solely to link arsenic content to perceived mechanical effects on finished metal artefacts. It is well known that the physical properties of this metal are not solely related to arsenic content, but to the degree of cold-working in the finished artefact. In the case of copper axeheads, a failure to exploit arsenic content by work-hardening does not necessarily imply total ignorance of the potential benefits of this metal. This knowledge is evident in contemporary blade fabrication, and the treatment of axeheads may instead reflect their non-utilitarian role in these early metal-using societies. Allen *et al.* (1970) concluded that while early British-Irish smiths probably did realise the advantages of arsenical compositions as seen in the work-hardening of halberd blades, in the case of many axeheads they deliberately chose not to exploit this property, possibly for fear of producing cracks in the working edge during prolonged work-hardening. Identifying selective use of metal compositions is further complicated when we consider the effect of casting methods on arsenic content, relevant in any comparison made between simple flat axes cast in open moulds and blades made in bivalve moulds.

Thus, scientific artefact studies are often ambiguous on the questions of whether arsenical copper was deliberately produced, the manner in which this was achieved and the use to which the resulting metal was put. Also unclear are the reasons behind the eventual replacement of arsenical copper by tin bronze which occurred over a wide area at approximately the same time. Charles (1967; 1979) argued that tin was introduced as a more easily controlled alternative to arsenic, the health implications of which would have become apparent to early metalworkers over time. Tin is a better alloying agent which, unlike arsenic, does not require extensive post-cast working to achieve improved hardness levels. A further factor here may have been the growing scarcity of arsenical minerals and ores. However, the health implications of early arsenical copper production can be questioned (McKerrell & Tylecote 1972), while for many parts of Europe, tin did not represent a particularly plentiful resource in comparison to occurrences of arsenical copper ore. This is particularly true of Ireland where the demise of a vigorous arsenical copper tradition in the Early Bronze Age must be explained against the known scarcity of tin on that island.

Early Irish Metallurgy

The earliest appearance of metal in Ireland is linked to the production of copper axeheads, daggers and halberds (as well as gold sheetwork) over a brief period prior to the adoption of tin-bronze metallurgy (Harbison 1969; Burgess 1979). Available evidence points to a Beaker-hosted introduction of metallurgy, most probably in the third quarter of the third millennium BC (see Sheridan 1983 and O'Brien 1995 for discussion). Copper axeheads, typologically connected to this Irish production and sharing the same distinctive composition, are among the earliest metal artefacts to be found in Britain. Along with daggers and halberds of the same metal type, these Irish 'imports' are believed by many to have contributed to the inception of metallurgy on that island (Case 1966; Northover 1982).

This early phase of metallurgy in Ireland is marked by two distinctive features, namely the prolific output of axeheads, daggers and halberds over a 300-400 year period and the consistency of the compositions used. Some 700 copper artefacts survive from this period, 95% of which are marked by a distinctive arsenical composition also containing other trace element impurities (Northover in press). The various flat axe forms typically contain <1-3% wt. arsenic, with some higher compositions for daggers and halberds in this period. This arsenical copper continued in circulation after the appearance of the first tin bronzes, c.2200-2100 BC. Thus, as elsewhere in Europe, early metal use in Ireland is marked by a short-lived 'Copper Age', generally regarded as the introductory phase of the Early Bronze Age.

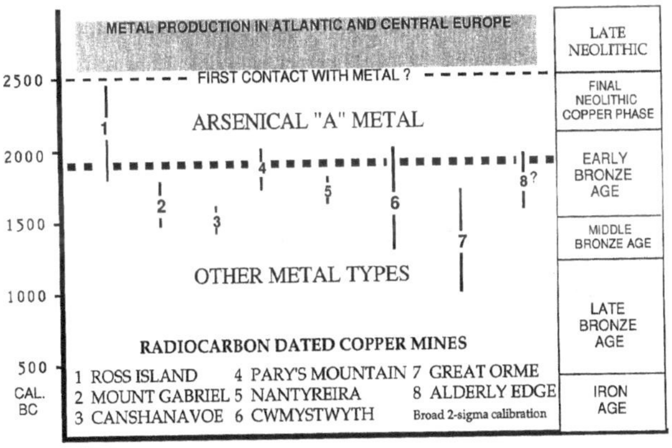

Figure 1. Arsenical copper and mine chronology. Ireland Britain.

Northover argued that small quantities of arsenic (up to 2%) offer very little improvement over pure copper and that it is only when arsenic levels exceed 4% that mechanical properties approaching tin bronze can be achieved (Northover 1989, 113, Figure. 13.3). Analysis of some two dozen copper flat axes from Ireland revealed arsenic contents typically in the 1-3% range. Metallographic analysis of these axes indicates that in most cases very little work-hardening was carried out and thus the potential of the arsenical composition is not being exploited. He contrasts this with Irish halberds which "...represent an appreciably greater level of technical sophistication in their manufacture" (Northover 1989, 114). While there may be some selective use of higher arsenic compositions in the casting of halberds, in general "...there was probably no deliberate creation of arsenic copper alloys" (Northover 1989, 117). Northover further speculates that this might also be the case for Spain where, in contrast to the Italian Copper Age material, there is a similar broad spread of compositions which do not break down according to object type.

The continued use of low arsenic copper in the earliest tin bronzes in Ireland supports the idea that arsenic levels were not intentional, but rather the product of a particular ore resource and/or production technology. Despite the prolific output of finished metalwork, this early Irish production was until recently marked by a total absence of site or artefact evidence for mining, smelting and fabrication. In the absence of mines and metallurgical evidence in the form of furnaces, slags, crucibles, and moulds, the source tracing and production of this metalwork had to be inferred indirectly from the scientific examination of finished artefacts. Several important studies carried out over the past forty years provide the basis for modern research.

Source Tracing

The first steps to source tracing this early Irish metal were taken in the 1950s as part of research conducted by the Ancient Mining and Metallurgy Committee of the Royal Anthropological Institute and the British Association. This research was made possible by the application of spectrographic techniques to metal analysis, as part of a source tracing approach pioneered on the Continent in the post-war period. By 1957, led by Herbert Coghlan and Humphrey Case, this team identified a distinctive metal composition which they associated with the earliest phase of mining and metal production in Ireland. This so-called Group 1 copper, found in axeheads, daggers and halberds, has an impurity pattern marked by high levels of arsenic, antimony and silver. This finding was confirmed by subsequent analytical programmes, most notably, those carried out by the Stuttgart (Junghans *et al* 1968), British Museum (Craddock 1979) and recent Oxford (Allen *et al.* 1970; Northover 1980 *et seq*) programmes.

Ruling out deliberate alloying practice, Coghlan and his colleagues argued that this metal derived entirely from the smelting of arsenic-bearing ores. They analyzed a range of Irish and British ore mineral samples in an attempt to correlate this metal with specific ore deposits (Biek 1957; Coghlan & Case 1957; Coghlan *et al.* 1963). The conclusion reached was that this early Irish copper was smelted from fahlerz ('grey') copper ores, most probably sourced in the Cork-Kerry region where a large number of early copper axes have been found. Here, they were undoubtedly influenced by current thinking on the continent, in particular the research of Otto & Witter (1952) which identified 'fahlerzmetalle' as the major source of early arsenical copper in Germany.

Coghlan & Case (1957, 96-8) argued that these early metalworkers in Ireland may have recognized the benefits of arsenic content and used it selectively in the production of axeheads, daggers and halberds. The arsenic content of this Group 1 metal was thus contrived, partly by selection of ore and partly by operating the smelting process. The exploitation of fahlerz ore was felt to be a high temperature process, involving a roasting procedure followed by furnace smelting and crucible refining (Coghlan & Case 1957, 93-4). They concluded that the ability to process sulfide ore at such an early date in Ireland reflected the introduction of metallurgy at an advanced technological level, by Beaker people from the continent (Case 1966, 168).

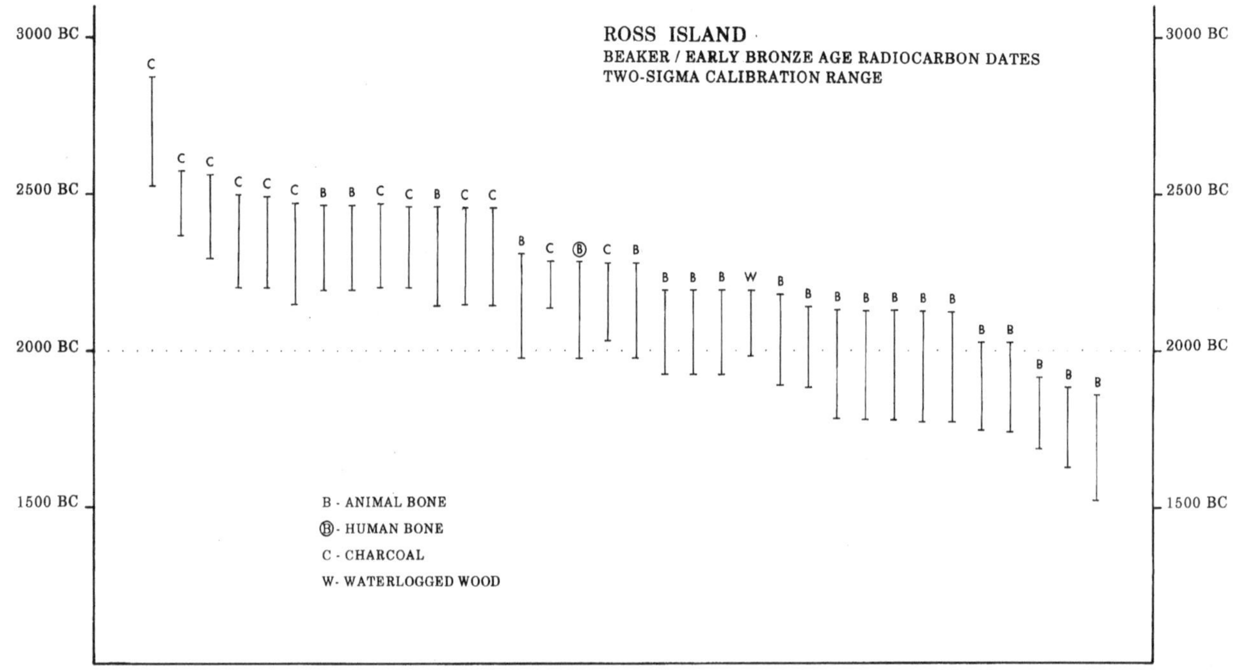

Figure 2. Radiocarbon chronology. Ross Isalnd copper mine.

These conclusions were taken a stage further by Peter Northover who identified some major trends in Bronze Age metal circulation across Britain and Ireland by combining compositional data with typological groupings (Northover 1980; 1982). This revealed that early Irish metal, marked by a distinctive arsenic-antimony-silver impurity pattern, was the first metal to circulate in western and northern Britain, also reaching some of the Beaker graves of southern England. Northover (in press) estimates that as many as 80% of the early copper axes in Highland Zone Britain are made from this Irish metal, the circulation of which continued into the early tin-bronze using period. Shortly after 2000 BC, the supply of this Irish metal was abruptly replaced by new compositions which reflect the establishment of regional mining and metalworking centres across Britain.

Thus, a model of a primary Irish 'industry', which dominated the earliest phase of insular metallurgy and achieved prolific output through the use of sulfarsenide ores mined in the Cork-Kerry region in presented. Opposed to this, is a recent suggestion by a University of Bradford team that the earliest copper metallurgy in Ireland and Britain was based on a simple, low-temperature treatment of secondary copper minerals (Pollard *et al.* 1990; Budd *et al.* 1992; Budd 1993). Central to their argument was a belief that arsenic-bearing primary copper sulfides could not be smelted directly to yield arsenical copper metal in the Bronze Age. The smelting of this ore would require a prior roasting treatment which Tylecote (1980) had argued would result in almost total loss of arsenic. They also argued that one-step smelting of fahlerz ore above 1300° C would yield excessive arsenic content in any resulting metal. With limited availability of native arsenic and other arsenic minerals, they concluded that the only feasible way to make arsenical copper in the insular Bronze Age was through the smelting of secondary copper minerals in a low temperature, non-slagging process.

Following earlier suggestions (Charles 1967; Rapp 1988; Tylecote 1980), the Bradford team suggested that this early arsenical copper was linked to the use of secondary copper arsenate minerals and not to sulfarsenide ores of the type envisaged by Coghlan & Case (1957). They proposed that arsenate minerals like olivenite (Cu2As04OH) were reduced to form low-arsenic compositions, by diffusion of arsenic into solid copper at temperatures as low as 700° C. This accounts for the low arsenic content of the resulting metal and also explains an apparent absence of slag associated with this early smelting technology. They concluded that the main advantage of making metal in this way was its lower melting point which allowed remelting in an open fire with a crucible and a blow-pipe (Budd 1993).

These two models of early insular metallurgy mirror a wider debate on the use of arsenate versus arsenide ores in the production of arsenical copper (see above). Recent research on early copper mining allows the examination of this question from the perspective of mine sources and available mineralization. A recent review confirms that, with one exception, none of the 30 or so known Bronze Age copper mines in Britain and Ireland have ore mineralization which could yield arsenical metal (Ixer & Budd in press). The exception is a recently discovered site located in the Killarney area of south-west Ireland, where recent research has identified a major mine source for early arsenical copper.

Ross Island

The Ross Island copper mine is located on a small peninsula off the eastern shore of the Lower Lake, Killarney, Co. Kerry (Figure 1). Recent research undertaken by the University of Galway has identified a long history of mining for copper, and possibly other metals, at this site (O'Brien 1995). This mining began in final Neolithic/Early Bronze Age times and continued in the Early Christian period, ending with large-scale industrial operations during the 18th and 19th centuries. Of particular interest here is the evidence for copper mining in the period 2400-2000 BC which can be directly linked to the users of early Beaker pottery and hence to the origins of Irish metallurgy.

Figure 3. Aerial view of Western Mine sector, Ross Island.

A major programme of radiocarbon dating, involving some 50 determinations, was undertaken as part of the Ross Island project. These results confirm an early phase of copper mining, beginning c.2400-2200 BC and continuing to shortly after 2000 BC (Figure 1). This implies that the earliest mining at Ross Island pre-dates the inception of tin bronze, recently dated by Northover (in press) to c.2200-2100 BC. On chronological grounds alone, the earliest activity at Ross Island mine can be securely linked to the copper axe horizon which represents the earliest phase of Irish metallurgy (see above).

This is supported by the mineralization of this mine which is the subject of an ongoing study co-ordinated by Dr Robert Ixer, University of Birmingham. The here is rich in arsenical mineralization which was highly accessible to Bronze Age miners. There is little doubt that this mineralization constituted a major source of arsenical copper (Coghlan and Case Group 1/Northover Group A) in the earliest phase of insular metallurgy. It now remains to examine the nature of these copper ores and the technology used to exploit them.

Mineralization

The Ross Island mine is part of a discontinuous zone of mineralization hosted in Carboniferous limestone across the Lower Lake area in Killarney. Within this zone, stratabound copper-lead-zinc-arsenic plus minor silver and cobalt mineralization is present as sulfides, arsenides and sulfarsenides (Ixer & Pattrick in press). This mineralization is intermittently developed in up to three metre thick beds within a 20m thick, heavily recrystallized, dolomitized and partly silicified horizon of bioclastic limestones of Courceyan age, close to the present day land surface.

The Ross Island mine is marked by an abundance of high grade copper- and arsenic-rich mineralization exposed at outcrop. The richness of this orebody is attested to by the sale of 4153 tons of dressed copper ore to the Swansea smelters during mining operations between 1804-1829. With the mine effectively worked out today, that production represents the ore remaining in this mine after the operations conducted in the Bronze Age, early medieval and 18th-century mining periods. All of this ore was raised from a surface zone of mineralization extending to only 13-18m in depth. Access to this mineralization was only limited by lake flooding which partly accounts for the episodic pattern of mining.

The mineralization at Ross Island is epigenetic and displays a number of different styles and mineral associations (ores). These occur within two geologically distinct parts of this orebody, namely the Western Mine and the Blue Hole, both of which were worked for copper and possibly other metals in antiquity. The Western Mine is located on strata-bound mineralization, occurring within fine-grained bioclastic limestones marked by extensive calcite veining. The mineralization here is disseminated through the 'Blue Limestone' bed to a variable depth of between 13-18m from outcrop, at which point it is sharply cut off by a thrust plane. This Blue Limestone is marked by an extensive development of thin cross-cutting veinlets and disseminations of copper ore, consisting of finely mixed chalcopyrite and tennantite, with up to 0.3% wt. silver and trace antimony, nickelferous cobaltite, molybdenite and pyrite.

Nineteenth-century mining records confirm the presence of more massive ores in the Western Mine, comprising chalcopyrite, bornite, tennantite, spionkopite and molybdenite. These may represent a local, more extensive, development of the veinlet-style mineralization at depth. These are coarse-grained, non-banded massive sulfides which form a continuum from tennantite plus minor chalcopyrite and trace amounts of molybdenite, pyrite, cobaltite and stromeyerite ('grey sulfuret ore'), to chalcopyrite or chalcopyrite-arsenopyrite plus minor amounts of tennantite, galena and molybdenite ('yellow sulfuret ore'), to bornite-idaite-chalcopyrite plus minor amounts of molybdenite, pyrite and tennantite ('purple sulfuret ore')

In the eastern sector of this mine, the Blue Hole open-cut is associated with a different style of mineralization consisting of fine-grained massive sulfide ores infilling void spaces within karstic surfaces. Here, chalcopyrite-pyrite with minor galena and sphalerite ores pass upwards into very finely banded sphalerite-galena with less amounts of pyrite and chalcopyrite. Minor to trace amounts of arsenopyrite and tennantite are present in all these ores, however, these are only of minor importance.

Ixer and Pattrick's study confirms that the mineralization at Ross Island consists of unaltered sulfides/sulfosalts, with only minor oxidation zone minerals. This is supported by 19th-century mine records and by recent geological and archaeological sampling. These sources indicate that there is a negligible occurrence of secondary mineralization in this ore-body, visible today mostly on surface spoil material. This trace occurrence of malachite and other secondary minerals would not constitute a viable 'ore' in any past mining efforts. In the absence of supergene mineralization, it is clear that mining in all periods at Ross Island must have centred on the extraction of sulfide and sulfarsenide ore.

Figure 4. Copper/Early Bronze Age mine workings, Western Mine.

Mining and Metallurgy

Recent archaeological excavations at Ross Island reveal evidence of Bronze Age mining and metal production which can be directly linked to the treatment of primary copper-arsenic ore. Primitive workings in the Western Mine, radiocarbon dated to before 2000 BC, are associated with the extraction of chalcopyrite/tennantite copper ore. A large cave-like opening connected to extensive underground workings has been identified here (figure 4). Mining techniques included the use of fire-setting in conjunction with stone hammers, bone and probably wooden tools. Spoil deposits from this primitive mining have been identified, as has the location of an adjacent work camp. The latter provides a unique insight into the activities associated with the processing of arsenical copper ore to metal.

Figure 5. Excavation of large pit, Beaker work camp, Ross Island.

The work camp is located on a level escarpment adjacent to Bronze Age workings in the Western Mine sector. This site was the setting for ore concentration, metal production and habitation-related activities spanning the Copper/Early Bronze Age phase of activity in this mine, c.2400-1800 BC. The foundation traces of hut structures used by these miners have been identified, as have animal bone food residues, worked flint and pottery sherds. The latter derive from simple cord-, comb- and line-decorated vessels of the early Beaker pottery tradition in Ireland- pottery consistent with a 2400-2200 BC date range.

The main activity in this work camp was the processing of copper ore to metal. This began with the hand-cobbing of mineralized rock brought up from the adjacent mine workings where a preliminary sorting had taken place. This ore was then finely crushed using stone hammers and anvil slabs, and hand-sorted to provide a rich concentrate. The latter may have been further reduced using large grinding slabs found in the site, as part of the final preparation for smelting. The location of this work camp immediately adjacent to a lake raises the possibility of water separation as part of this concentration process, though this cannot be archaeologically confirmed. The entire Bronze Age shoreline was unfortunately destroyed during 19th-century mining and dam construction.

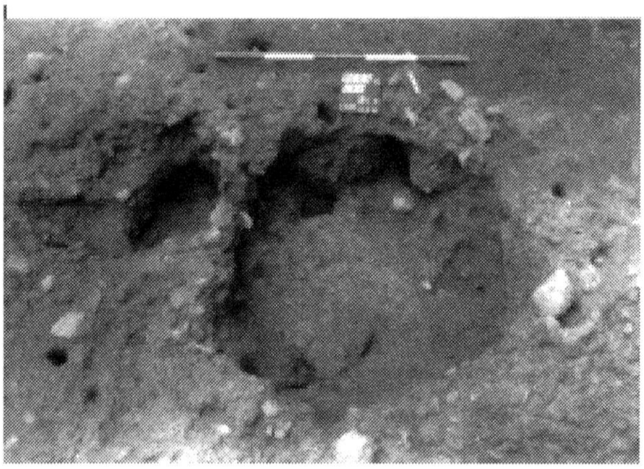

Figure 6. Large furnace pit, Beaker work camp, Ross Island.

Excavation of the Beaker work camp uncovered important evidence of on-site metallurgy, involving the reduction of ore concentrate to metal by smelting. This activity centred in an open-air work area adjacent to hut structures, where a series of pit features and surface sediments can be directly linked to the smelting of arsenical copper ore. Detailed sampling confirms that the miners here were processing tennantite-rich (fahlerz) and chalcopyrite-arsenopyrite mineralization, both of which were probably smelted to give an arsenical metal. The pit features are of particular importance, as they represent the first copper smelting furnaces to be discovered in the insular Bronze Age. They fall into two main types, namely large oval pits and smaller features marked by stone settings.

Six large oval pits with rounded profiles were discovered in close proximity in this metallurgical area. These pits were cut into underlying subsoil and measured between 0.8-1.2m in diameter and 0.4-0.8m in depth. They were surrounded in some cases by stakes, but otherwise had no apparent superstructure (figures 5-7). These pits are marked by traces of intense burning around the upper edges, however the sides and bases are not fire-reddened. These pits contained silty fill sediments with lenses of charcoal and fuel ash, however they had clearly been emptied of their original smelting products. The surrounding ground surface was marked by spreads of copper-rich, fine silty sediment, in one instance comprising a deep basin deposit. These appear to represent fuel ash residues derived from furnace operations in the adjacent pits. Spreads of pure quartz sand were also excavated in close proximity to these pits, as were numerous fragments of comminuted copper ore.

Figure 7. Large furnace pit, Beaker work camp, Ross Island.

Two smaller furnaces were also found in the same area, both marked by smaller subsoil depressions with stone settings. The first of these is an oval-shaped depression, measuring 0.3-0.4m in diameter and 0.24m in depth, bordered on one side by an arrangement of three stones, the latter possibly used to support a bellows (figure 7). When excavated, this pit contained a large mass of solid chalcopyrite-tennantite ore which the miners had apparently failed to smelt. A second small furnace consists of a carefully built, U-shaped arrangement of stones set in a shallow pit (figure 8).

Figure 8. Stone lined furnace, Beaker work camp, Ross Island.

The shape and depth of the large oval pits would appear to rule out their use in the roasting of sulfide/sulfarsenide ore, as would the size and stone settings of the two smaller examples. These features are more consistent with pit furnaces used to smelt ore concentrate to metal. This is supported by evidence of intense heat exposure in and around these pits and by the presence of spreads of pure quartz sand on the adjacent work surface. The latter was deliberately brought into this metallurgical area, possibly for use as a fluxing agent in smelting.

Any interpretation of these pits as smelting furnaces must consider the apparent absence of metallurgical slags, crucibles, tuyeres or other furnace residues from the excavation area. The absence of slag is difficult to explain if the aforementioned quartz sand was indeed used as a fluxing agent. It is possible that slag produced in these furnaces was subsequently removed to another location within the site, possibly to the adjacent lake shore, for crushing and water separation of copper prills. In that instance, it is surprising that small amounts of true slag do not survive in and around these pit features. This raises the possibility that the smelting of this ore involved a non-slagging process. Specifically, the question must be asked whether this was feasible for tennantite-only concentrates, for the more iron-rich chalcopyrite-arsenopyrite ore, and for mixed ore charges. Further consideration of this metallurgy must await smelting experiments and the analysis of a small number of altered ore and slag-like fragments from this area.

Conclusive evidence that metal was produced in this site is provided by the discovery of a metal droplet adjacent to one of the large oval pit furnaces. This droplet has recently been analyzed by Dr Peter Northover, University of Oxford, who has also examined a small copper slab from this site and a plano-convex copper ingot of Ross Island metal found in the Knockasarnet area of Killarney. Northover confirmed that the droplet consists of smelted raw copper adhering to copper sulfide or 'matte'. With a mean arsenic content of 7.35%, this droplet may represent an early stage in the production of copper here. It is possible that collection and subsequent melting of these droplets to form small ingots resulted in further oxidative loss of arsenic. The copper slab and the Knockasarnet ingot have a mean arsenic content of 1.0% and 0.67% respectively.

Discussion

Ross Island is the first copper mine discovered in Ireland or Britain which dates to the inception of insular metallurgy in the final Neolithic 'Copper Age'. On chronological grounds alone, this mine can almost certainly be linked to the production of copper axeheads, daggers and halberds made from a distinctive arsenical metal. This is supported by the discovery of early Beaker pottery and by the site evidence for extraction and smelting of arsenical copper ore. Some 30 Bronze Age copper mines are now known in Britain and Ireland, however, only Ross Island has mineralization which could produce metal with the Group 1 impurity range of the earliest copper metalwork.

The discovery of this Beaker copper mine offers a rare opportunity to examine early arsenical copper in a production site context with direct evidence of the ore minerals being used. Investigation of the Ross Island ore body confirms that the ability to process unaltered sulfide and sulfarsenide ores existed from a very early stage in Irish metallurgy. It is likely that low-arsenic copper was produced here, by direct roasting and smelting of pure tennantite and probably also chalcopyrite-arsenopyrite. Contrary to some earlier assertions (Tylecote 1980; Budd *et al.* 1992), it now seems likely that enough arsenic remained from the smelting of this sulfarsenide ore to account for the levels found in early metalwork. The question of whether this residual arsenic was of any significance for the subsequent casting and mechanical properties of the resulting metal must now be addressed.

It has been seen how several researchers question whether the low arsenic levels in early Irish metalwork are indicative of deliberately contrived alloys. This is supported by metallographic evidence which reveals a failure to take advantage of the potential benefits offered by arsenic content (Northover 1989). This alone must question the degree of knowledge which early Irish metallurgists had regarding the properties of arsenical copper. For Ross Island, it must now be asked to what extent the selection of arsenical copper ores was based on known properties of the resulting metal or was instead a function of the smelting process.

The fact that these early metallurgists were not exploiting the mechanical properties of this metal does not necessarily imply that the arsenic content of this metal is entirely adventitious. An important factor here may have been the lower melting point and castability of the resulting metal. Also relevant may be the effect of arsenic content on smelting temperature. Of particular interest here is the rich tennantite ore at Ross Island, with its low iron content and the possibility that it was reduced at comparatively low temperatures in these pit furnaces. A primitive non-slagging technology using sulfarsenide ore may be consistent with the low iron content of contemporary metalwork.
The arsenic content derived from this mine may be incidental to the final metal use and, instead, reflect an ability to process a particular type of copper ore using expertise gained from a similar technological environment on the continent.

McKerrell & Tylecote (1972) suggested that the arsenic content of early copper metalwork was down to correct ore source tracing and smelting procedure. They argued that the ore selection process was critical here, but also suggested that the arsenic content may have been controlled by smelting and hot-working under oxidizing conditions. In their experiments, they demonstrated that control of arsenic content through oxidative loss was a feasible technique to yield the low-arsenic copper typical of the insular Early Bronze Age. Zwicker (1991, 331-2) points out that arsenic content decreases in steps from the ore to final metal. The low arsenic content of early Irish metal may not be due to deliberate oxidative working, but may derive from incremental loss during the transition from roasting to smelting, to droplet melting and ingot formation, to casting and recycling. This is supported by the recent analysis of raw copper from Killarney (see above) and by evidence of arsenic variation in large and small copper axeheads in Ireland (Northover 1989). In this way, arsenic content is a factor in the selection of orebodies like Ross Island, but for reasons more connected to metal yield and ease of smelting and casting than to the mechanical properties of the resulting artefacts.

How can the consistency of metal circulation in the earliest phase of Irish metallurgy be explained? Is it possible that the arsenic content of early Irish-British copper reflects the exploitation of a particular mine source, namely Ross Island? Could this one source have supplied the bulk of metal produced in Ireland in the period 2500-2000 BC, gradually being surpassed by the emergence of other copper mines in Ireland and Britain in the early second millennium BC? It is significant that the radiocarbon time-scale of mining at Ross Island does correlate closely with the currency of arsenical metal in Ireland and Britain. The cessation of mining here coincides with the appearance of different metal compositions from new mine sources (Figure 2).

The prevalence of early arsenical metal may thus reflect the importance of one specific source, namely Ross Island, which dominated copper production in Ireland for a period of several centuries. Certainly, the ore occurrence here was rich enough to sustain the level of metal supply apparent in this first 500 years of insular metallurgy. On the question of a single mine source dominating production, parallels can be drawn with stone axe production in late Neolithic Ireland which featured very strong production and supply networks based around the porcellanite sources of north-east Antrim (Sheridan 1986). The distribution record from stone axe quarries and flint mines in Britain also supports the notion of strong output from individual early copper mines, at a time when the knowledge of metallurgy may have been restricted to certain groups, a control which gradually broke down after 2000 BC.

Whereas the earlier use of arsenical copper was largely a function of ore selection and metal production, the widespread adoption of tin alloying in Ireland after 2000 BC is probably linked to a desire to improve the mechanical properties of finished metalworks which are now increasingly entering the utilitarian sphere. The motivation to practice this metallurgy here must have been considerable given the known scarcity of tin sources in Ireland, at a time when there was apparently a plentiful supply of arsenical copper. When more serviceable work tools and weapons became important, tin alloying offered tangible benefits which could be easily controlled as opposed to the less discernible benefits of arsenical content. This factor, rather than any health implications of arsenical copper production, explains the rapid adoption of tin alloying. The latter had little to do with the eventual decline of arsenical copper use in Ireland c.1900-1800 BC, which can instead be attributed to the cessation of mining at Ross Island and possibly other centres in the Cork-Kerry region. The background to this shift in metal supply is not understood at present, however in the case of Ross Island, site-specific factors, such as the impact of lake-flooding, may have been involved.

Finally, while there has been considerable criticism of scientific approaches to source tracing metal in the Bronze Age, it seems likely that some considerable measure of success was achieved in an investigation of the earliest Irish metallurgy conducted some four decades ago. The research of Coghlan and Case combined ore and metal analyses with typological and distributional information to identify the major source region for arsenical copper in Ireland. This can now be confirmed by the recent discovery of Ross Island, one of the major mine sources of that tradition, if not the single most important supplier of arsenical copper in Ireland. Further support also comes from recent lead isotopes analyses for early Irish-British metalwork from the Oxford Isotrace laboratory. These point to a tight cluster of lead isotope values for Copper Age metalwork, which mirror the compositional data and may also correlate directly with the Ross Island orebody (Rohl unpublished).

Conclusions

The Beaker copper mine of Ross Island, Co. Kerry can be directly linked to the earliest phase of Irish metallurgy, c.2400-2000 BC. The mining and smelting of sulfarsenide ore here supplied much of the low arsenic copper which dominated metal circulation in that period. The currency of this metal correlates closely with the radiocarbon timescale for exploitation of the Ross Island orebody. While the precise details of smelting technology remain to be clarified, recent research confirms that the production of arsenical copper at Ross Island was based exclusively on the smelting of primary sulfarsenide ore. This evidence lends strong support to the model of a primary Irish 'industry' exploiting sulfide ore resources of the Cork-Kerry region (Coghlan & Case 1957; Northover 1982) and effectively dismisses recent suggestions to the contrary.

It now appears that the consistency of metal composition in the earliest phase of insular metallurgy reflects a limited number of mine sources and possibly also the dominant position which mines like Ross Island achieved over a period of several centuries. The utilization of arsenical copper is thus linked to the exploitation of a particular ore resource and not to any deliberate 'alloying' practise. Instead, it is suggested that incremental oxidative loss of arsenic through the production process may explain the consistent low arsenic contents of contemporary metalwork. This arsenic content is a function of a particular resource base which involved deliberate selection of arsenical copper ores, for reasons not solely connected to the quality of finished metal, but more importantly to ease of extraction and subsequent casting. The eventual decline of this primary Irish metal is more likely to be connected to a shift in ore supply through the demise of mining centres like Ross Island than to the widespread adoption of tin-bronze metallurgy.

The conclusions presented here are specifically relevant to the earliest phase of copper metallurgy in Ireland and do not necessarily extend to other arsenical copper traditions in Europe and beyond, for whom alternative production and resource models may be feasible. With the discovery of Ross Island, at least a certain clarity in the consideration of the resource base used by the first Irish metallurgists has been achieved. More importantly, 2400 BC or thereabouts has been established as an important marker date for the practise of sulfide metallurgy in this part of western Europe.

References

ALLEN, I.M., D. BRITTON & H. COGHLAN. 1970. *Metallurgical reports on British and Irish Bronze Age Implements and Weapons in the Pitt-Rivers Museum.* Oxford: Occasional Papers on Technology 10.

BAR-ADON, P. 1980. *The Cave of the Treasure.* Jerusalem: The Israel Explorations Society.

BIEK, L. 1957. The examination of some copper ores: a report of the Ancient Mining and Metallurgy Committee. *Man* 84: 72-76.

BUDD, P. 1991. Eneolithic arsenical copper: heat treatment and the metallographic interpretation of manufacturing processes, in E. Pernicka & G.A. Wagner (ed.), *Archaeometry'93.* Heidelberg, 35-44.

1993. Recasting the Bronze Age. *New Scientist*, 23 October 1993, 33-37.

BUDD, P., D. GALE, A.M. POLLARD, R.G. THOMAS & P.W. WILLIAMS. 1992. The early development of metallurgy in the British Isles. *Antiquity* 66: 677-86.

BURGESS, C. 1979. The Background of Early Metalworking in Ireland and Britain, in M. Ryan (ed.), *The Origins of Metallurgy in Atlantic Europe*, 207- 47. Dublin: Proceedings of the Fifth Atlantic Colloquium.

CASE, H.J. 1966. Were Beaker-people the first metallurgists in Ireland? *Palaeohistoria* 12: 141-77.

CHARLES, J.A. 1967. Early arsenical bronzes: a metallurgical view. *American Journal of Archaeology* 71: 21-6.

1979. From copper to iron: the origin of metallic materials. *Journal of Metals* 30: 8-13.

COGHLAN, H.H. & H.J. CASE. 1957. Early metallurgy of copper in Britain and Ireland. *Proceedings of the Prehistoric Society* 23: 91-123.

COGHLAN, H.H. 1963. A note on Irish copper ores and metals, in *Ores and Metals*, 1-33. A Report of the Ancient Mining and Metalurgy Committee, Royal Anthropological Institute.

CRADDOCK, P.T. 1979. Deliberate alloying in the Atlantic Bronze Age, in M. Ryan (ed.), *The Origins of Metallurgy in Atlantic Europe*, 369-85. Dublin: Proceedings of the Fifth Atlantic Colloquium.

DELIBES, G., M. FERNANDEZ MIRANDA, M.D. FERNANDEZ POSSE, C. MARTIN, I. MONTERO & S. ROVIRA. 1991. Almizaraque, Almeria, Spain, in J.P. Mohen & C. Eluere (ed.), *Decouverte du Metal*, 303-16. Picard: Amis du Musee des Antiquites Nationales.

EATON, E.R. & H. MCKERRELL. 1976. Near Eastern alloying and some textual evidence for the early use of arsenical copper. *World Archaeology* 8: 169-91.

HARBISON, P. 1969. *The Axes of the Early Bronze Age in Ireland.* Munich: Prahistorische Bronzefunde 9.1.

HOOK, D.R., I.C. FREESTONE, N.D. MEEKS, P.T. CRADDOCK & A. MORENO. 1990. The early production of copper-alloys in south-east Spain, in E. Pernicka & G.A. Wagner (ed.), *Archaeometry'93.* Heidelberg, 65-76.

IXER, R.A. & R.A.D. PATTRICK. In press. Copper-arsenic ores and Bronze Age mining and metallurgy with special reference to the British Isles, in P.T. Craddock & J. Lang (ed.), *Aspects of Early Mining and Extractive Metallurgy.* London: British Museum Press.

JOVANOVIC, B. 1992. Smelting of copper in the Eneolithic period of the Balkans, in P.T. Craddock & M.J. Hughes (ed.), *Furnaces and Smelting Technology in Antiquity*, 117-22. London: British Museum Occasional Paper 48.

JUNGHANS, S., E. Sangmeister & M. Schroder. 1968. *Kupfer und bronze in der fruhen Metallzeit Europas.* Berlin: Studien zu den Anfangen der Metallurgie 2.1-3..

MCKERRELL, H. & R.F. TYLECOTE. 1972. The working of copper-arsenic alloys in the Early Bronze Age and the effect on the determination of provenance. *Proceedings of the Prehistoric Society* 38: 209-218.

MUHLY, J.D. 1988. The beginnings of metallurgy in the Old World, in R. Maddin (ed.), *The Beginning of the Use of Metals and Alloys*, 2-20. Proceedings of the Zheng-Zhou colloquium, October 1986. Cambridge (MA): Harvard University Press

NORTHOVER, J.P. 1980. Bronze in the British Bronze Age, in W.A. Oddy (ed.), *Aspects of Early Metallurgy.* London: British Museum Occasional Paper 17: 63-70.

1982. The exploration and long distance movement of bronze in Bronze Age and Early Iron Age Europe. *Bulletin of the Institute of Archaeology* 19: 45-72.

1989. Properties and use of arsenic-copper alloys, in A.Hauptmann, E. Pernicka & G.A. Wagner (ed.), *Proceedings of the International Symposium on Old World Metallurgy, Heidelberg*: 111-118. Bochum: Deutschen Bergbau-Museums.

In press. The earliest metalworking in southern Britain. A. Hauptmann (ed.), *The Beginnings of Metallurgy.* Bochum: Proceedings of the April 1995 International Symposium.

O'BRIEN, W. 1995. Ross Island and the origins of Irish-British metallurgy, in J. Waddell & E. SheeTwohig(ed.),*Ireland in the Bronze Age*, 38-48. Dublin: Stationary Office.

OTTO, H. & W. WITTER. 1952. *Handbuch der altesten vorgeschichtlichen Metallurgie in Mitteleuropa.* Leipzig.

PAZUKHIN, V.A. 1964. The origin of arsenical copper in antiquity. *Izvestiya Akademii Nauk S.S.S.R* 1: 151-65.

POLLARD, A.M., R.G. THOMAS & P.A. WILLIAMS. 1990. Experimental smelting of arsenical copper ores: implications for Early Bronze Age copper production, in P. Crew & S. Crew (ed.), *Early Mining in the British Isles.* Snowdonia, 72-74.

RAPP, G. 1988. On the origins of copper and bronze alloying, in R. Maddin (ed.), *The Beginning of the Use of Metals and Alloys*, 21-27. Proceedings of the Zheng-Zhou colloquium, October 1986. Cambridge (MA): Harvard University Press.

SELIMKHANOV, I.R. 1982. Arsenical Copper Age in H.H. Coghlan's works and investigations in the Laboratory of the Institute of History of the Academy of Sciences of Azerbaijan. *Historical Metallurgy* 16(2): 50-7.

SHALEV, S. 1991. Two different copper industries in the chalcolithic cultures of Israel, in J.P. Mohen & C. Eluere (ed.), *Decouverte du Metal*, 413-24. Picard: Amis du Musee des Antiquites Nationales.
SHERIDAN, A. 1983. A reconsideration of the origins of Irish metallurgy. *The Journal of Irish Archaeology* 1: 11-19.
1986. Porcellanite artefacts: a new survey. *Ulster Journal of Archaeology* 49: 19-32.
TYLECOTE, R. F. 1980. Summary of results of experimental work on early copper smelting, in W.A. Oddy (ed.), *Aspects of Early Metallurgy*. London: British Museum Occasional Paper 17: 5-12.
WERTIME, T.A. 1973. The beginnings of metallurgy: a new look. *Science* 182: 875-887.
WITTER, W. 1936. *Die technische Verwendung von Kupfer-Arsenlegierungen im Altertum, Metall und Erz*. 33 Jahrg, Heft 6.
ZWICKER, U. 1991. Natural copper-arsenic alloys and smelted arsenic bronzes in early metal production. *Decouverte du Metal* 331-40. Picard: Amis du Musee des Antiquites Nationales.

5

The Role of Ore Geology and Ores in the Archaeological Provenancing of Metals

R.A.IXER

School of Earth Sciences, University of Birmingham, Edgbaston, UK

The degree of processing needed for the manufacture of metals means that matching them to their raw material sources is difficult. Attempts to do this based on comparisons between the major and minor element geochemistry of metal and potential ores and more recently on their stable isotope geochemistry have met with limited success. A major reason for this is that ore sampling strategies have either been neglected or based upon an incomplete understanding of the mineralogical and geochemical complexities inherent in any metalliferous ore deposit.

All ore deposits are spatially zoned and each zone has its own set of mineral assemblages and its own distinct geochemical characteristics. In addition, deposits are zoned in time and many are the result of multiple and often genetically unrelated mineralization events. Mineralogical changes at mine sites continue after mining, forming a new zone named the supragossan. Here, the removal and gain of elements leads to new mineral associations whilst the original composition of the mined ores is radically altered. In particular an understanding of the origin and behaviour of lead in the supragossan zone is an essential first step in stable lead isotope provenancing studies at mine sites.

The provenancing of arsenical copper and gold nicely illustrates these issues. Only by comprehensive sampling of a wide range of potential copper ores, plus stable isotope data, is it possible to suggest that Ross Island, Ireland, was a source for copper arsenic bronzes (type A metal) in the Chalcolithic of the British Isles. Primary gold grains from a small gold deposit at Calliachar, Scotland show a wide variation in chemistry between the main zones, demonstrating the need for extensive sampling of multiple grains. Gold in placer deposits potentially comprises three components, proximal and distal detrital grains and in situ gold and, as their relative amounts will vary throughout the deposit, bulk analyses rather than detailed analyses of individual gold grains would appear to be a better method of provenancing prehistoric gold.

Keywords: Ore Deposit Geology; Provenancing Studies; Gold; Lead; Ross Island.

Introduction

Provenance studies, in the sense of matching stone to its geological origin or, in the case of a processed artefact matching its component raw materials to their respective geological origins, potentially remains a very powerful tool in the armoury of archaeologists. The techniques for 'scientific provenancing' (as opposed to 'stylistic provenancing') for lithics and ceramics based on petrography or geochemistry are routine, well established and often successful. Indeed, recently it has been shown that a combination of 'total petrography' (the use of transmitted and reflected light techniques so that all the mineralogical and textural data are collected) alongside standard major and minor element geochemistry, can provide very precise provenancing of lithics (Ixer 1994). There is some suggestion that artefacts can be matched to individual outcrops or even within an outcrop down to a scale of tens of metres (Ixer 1994; Ixer 1996; Ixer 1997).

The provenancing of ceramics is harder than lithics because they comprise a mixture of clay and temper. Both components may be differently sourced and will have undergone different degrees of processing. However, the final ceramic remains a mechanical mixture of lithic fragments (still largely unaffected by any processing and so petrographically recognisable) within thermally altered clays.

With metal artefacts, provenancing becomes yet more difficult as the raw materials are homogenized into a chemical whole – the metal or metal alloy. For metalwork, therefore, provenancing studies have been almost exclusively geochemical; initially, comparing the major or trace element chemistry of 'ores' and metal and, most recently, by studies based on their stable lead isotope signatures. The results of these endeavours have been subject to much criticism, notably by Budd *et al.* (1994; 1996) who question the validity of such studies. It is because of these doubts and because the degree of processing that is required to produce a metal obscures much about the raw materials used in their manufacture, that a complete understanding of the origins of the raw materials, especially the ores, is needed. This falls into the branch of geology called mineral deposit studies that deals with the recognition, classification and the genesis of mineral deposits.

Introduction to Mineral Deposit Geology

Exploitable metalliferous mineral deposits are rare, geochemical aberrations often with complex mineralogies; there are classes of mineral deposit that are represented by a single example and many have less than ten, for example Carlin-type gold deposits or the Zambian Copperbelt-type copper deposits. Even commonplace mineral deposits, those with hundreds of examples worldwide, are geochemical anomalies. This is because most geological/geochemical processes work towards the gradual homogenization of the Earth's crustal composition but, ore deposits, almost by definition, comprise aberrant concentrations of metals, having 10^2 to 10^4 times the average concentration of those metals in the crust. These abnormal concentrations rarely form as the result of a single process but normally require multiple events. This is because a sequence of different concentration mechanisms are needed to up-grade non-ore rocks through a series of proto-ores finally into an exploitable deposit. These mechanisms may be related to each other but equally the final deposit may be the result of the superimposition of more than one genetically unrelated mineralization event. This means that, for example, a deposit may comprise copper mineralization that was differently sourced and of a different age from any lead-zinc mineralization that lies next to it. In addition, for most mineralization events the ore fluids are an evolving system with mineral precipitation taking place in response to changes in physico-chemical conditions, especially drop in temperature.

Fig 1a

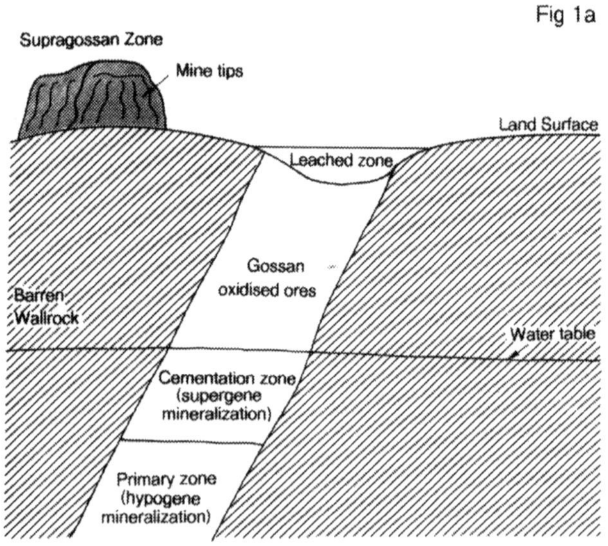

Fig 1b

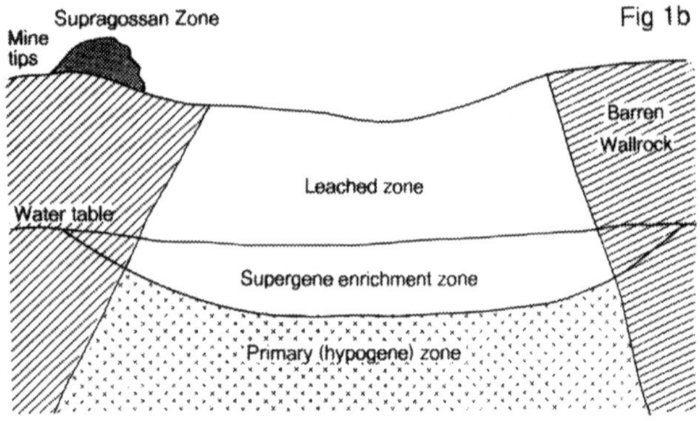

Figure 1. 1a. Ore zoning present in a base-metal sulfide vein deposit.
1b. Ore zoning associated with a disseminated sulfide orebody.

The majority of ore deposits are geochemically and therefore mineralogically variable; this variation occurs in space and time and is known as spatial zoning and paragenesis, respectively. Spatial zoning can be investigated and described in the field by plotting the distribution of ore types onto a map, section or plan but, whilst some aspects of temporal zoning can be determined in the field by collecting structural data, it is mainly studied in the laboratory by interpreting the textural relationships between minerals under the title of paragenetic studies.

Zoning and paragenesis are two aspects of the same phenomenon and hence can be cogenetic (Park & MacDiarmid 1970). It is vital for archaeologists who wish to match ore deposits and their ores to metal, using trace element geochemistry or stable isotope methods, to recognise the complexities of both these aspects of mineralization. This will ensure that only those mineral associations/ores that have archaeological significance are sampled and, equally, that none are overlooked.

Spatial Zoning within an Orebody

FIGURE 1a shows a representation of the spatial zoning in a base metal sulfide vein whilst FIGURE 1b shows zoning in a massive stockwerk deposit where the primary ore/proto-ore is finely disseminated within its host rock and the exploitable ore is present within a zone of supergene enrichment. As the figures show, the mineral associations/ores fall into a fourfold framework. It should be kept in mind that, within each of these four divisions, there might be further sub-divisions. These associations are: -

1. Host rock/wallrock assemblages – these are pre-mineralization assemblages often comprising simple sulfides or oxides. They are not ore but some wallrocks may be used as fluxes.
2. Primary (hypogene) mineralization assemblages. Commonly there is more than one primary mineralization event and the ores resulting from any one event may be zoned. Normally assemblages formed at high temperatures are separated in time and space from those that formed at low temperatures. The assemblages comprise many minerals including sulfides, arsenides, sulfosalts and oxides intergrown with 'unwanted' minerals known as gangue and often in complex textural relationships.
3. Secondary (supergene) mineralization assemblages. These are formed by the alteration of primary ores below the water table. They are especially important in copper and silver deposits, where they form very rich and hence very exploitable ores. The assemblages are simpler than the primary ores; they are dominated by 'simple' sulfides and replacement textures are characteristic.
4. Oxidation/weathering assemblages. Ores within the oxidation zones are often known as gossans (iron-hat) and are present on the surface. Brightly coloured carbonates, sulfates, arsenates, native metals etc are present within brown to black iron oxides and abundant hydrated iron oxides – the latter are collectively known as 'limonite'.

In should be noted that each of the four zones can carry very different ore types which in turn may need different mining and beneficiation techniques to exploit them. At its very simplest, friable carbonate-oxide ores need to be treated differently from massive sulfide ores. Hence with time and the continued exploitation of an orebody the ores will change from those found in gossans down through the supergene enrichment zone into the primary ores. Potentially each change in ore-type will be accompanied by a change in the composition of any metal produced from that ore and there may be cases where early and late metal from the same mine will have very different geochemical signatures and are not recognized as having the same source.

Temporal Zoning within an Orebody

Although FIGURE 1 shows the spatial relationships of the various mineral assemblages and ores within an orebody it cannot show their temporal variation, known as their paragenesis. A paragenetic diagram is the usual way of showing the evolution of a mineral assemblage with time and one is shown in FIGURE 2. Here the paragenesis for the copper and lead mineralization at the Bronze Age Great Orme Mine, partially based on Ixer & Davis (1996), is shown. The diagram employs the usual conventions so that relative time moves from left to right; the minerals are listed in order of their first appearance and the width and thickness of the bars indicate the length of time of deposition and relative amount of mineral deposited.

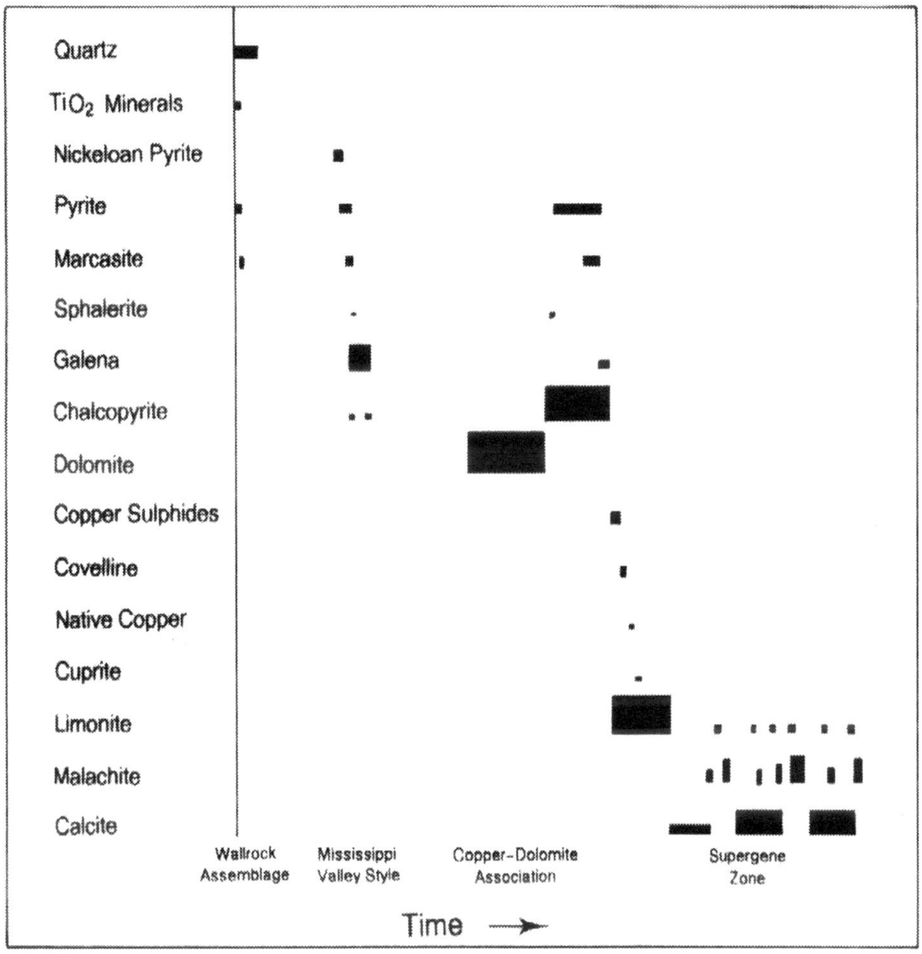

Figure 2. A paragenetic diagram for the lead and copper ores from the Great Orme Copper Mine, North Wales.

From the diagram it can be seen that at the Great Orme the limestone-dolostone wallrocks only carry a little pyrite and TiO_2 minerals, that there are two separate mineralization events, an earlier lead-rich and a later copper-rich one, which are followed by extensive supergene and oxidation zone alteration dominated by carbonates and limonite. Although the diagram separates the two primary mineralization events, it cannot by itself indicate the length of time between them in absolute terms, nor can it suggest if the events were cogenetic or genetically unrelated. In the Great Orme example it is other evidence that shows the lead mineralization is Mississippi Valley-style in origin and probably Permo-Triassic in age, whereas the copper mineralization is an example of the copper-dolomite association and Jurassic-Cretaceous in age (Ixer & Davies 1996; Ixer & Stanley 1996; Ixer & Budd 1998). Hence, at the Great Orme Mine, an unrelated episode of copper mineralization is superimposed upon earlier lead mineralization so that the mine carries two independent sets of ore. The archaeological implications of this, in particular on sampling procedures and their role in the use of stable lead isotopes to provenance copper metal, are discussed at length by Ixer (in press).

Once the ore geology and its complexities have been understood it becomes possible to collect, for archaeological purposes, *in situ* representative samples of all the potential ores alongside the other mineral associations. This is best done together with an economic or mine geologist and as much care should be taken in selecting the mineralized material as in excavating any archaeological artefacts (Ixer 1995). Whilst this is feasible for mines that are currently being exploited or abandoned mines with good access to the mined ore, it is often not possible for historical and prehistoric mine sites where the ore has been removed.

Material collected from abandoned mines and mine sites should be treated with extra caution. It is useful to consider the mine and mine dumps as belonging to a new and discrete geochemical/mineralogical zone. This zone, christened the supragossan, is the result of post- mining geochemical and hydrogeological processes that take place both underground and on the surface, namely in the spoil, mine tailings and slags. This zone shares many aspects with, and analogies to, the post-depositional changes found in artefacts recovered from excavations.

The Supragossan Zone

One of the main consequences of mineral exploitation (mining) is to change the local hydrological conditions by altering the chemistry and flow rates of the groundwaters. Underground, mining features like drives, shafts and stopes can provide pre-eminent pathways for groundwater flow especially if the country rocks/host rocks are impermeable (Younger 1995). The chemistry of this groundwater is radically altered as it passes through these open systems especially if mining has exposed pyrite (and other iron sulfides) to oxidation, which then acidify that groundwater. This acidification (lowering of the pH to between 2 – 4) and the passing into solution of base metal ions (iron, copper, lead and zinc) plus other elements like arsenic is the reason that many mine waters are both grossly polluted and highly reactive. These groundwaters not only remove metals but can also reprecipitate them, sometimes in new combinations as new mineral species, so that the surface geochemistry/mineralogy of the accessible underground 'ore' is not the same as that originally mined. The extent of this alteration zone depends on the porosity and permeability of the rocks but in many cases it will not be very thick.

On the surface, spoil heaps, rock waste (and slagheaps) react with rainwater and groundwaters in a similar way. Allard *et al.* (1989) have shown that the release of metals from mine waste and slags is promoted by lowering the pH of waters penetrating them but more especially by an increase in the redox potential when the material is exposed to air in the presence of water. Crushing increases by many orders of magnitude the surface area of mineralized rocks and if spoil heaps are left unreclaimed then oxidation, drainage and surface outflow occur in rapid succession (Younger 1995). Extreme geochemical conditions are created that lead to the formation of unusual suites of alteration products in the spoil. Similar processes occur underground as described by Jenkins & Johnson (1993). Many spoil heaps, especially those from base metal mines, have suffered rapid, intense and extreme degrees of alteration by geochemically aggressive fluids under pH and E_h conditions that are very different from those associated with the primary (and most of the supergene) mineralization. Although many of the mineral assemblages found in spoil heaps will be similar to those found in the gossan zone of the exploited ore (if it had one) there will be very important differences that can distort archaeological interpretations. For example, the recent proposal that copper arsenates, which are commonly present in spoil material (and some gossans) but rare in primary ores, were the original ores for the production of arsenical bronzes (Pollard *et al.* 1991; Budd *et al.* 1992), has been challenged on these grounds by Ixer & Budd (1998) and by Ixer & Pattrick (in press).

It should be remembered that the spoil has been anthropogenically selected. Richer and run-of-the-mill grade ores will be missing, and the original metal ratios of the deposit will be distorted. Cwmystwyth in central Wales is a good example, for here, at this Bronze Age copper mine, the excavated spoil is galena-rich and copper-poor (Timberlake 1995), quite different from the ore that was exploited.

Bachmann (1982) has similar concerns with regard to slags stating that 'A slag may undergo more or less extensive weathering depending on composition, climate and duration of deposition – leaching changes the overall composition of slag significantly. Chemical analyses therefore must be interpreted with care'.

These problems can be overcome. On many mine sites paragenetic studies of the available material will distinguish between the mineralogy of the supragossan zone and the original ore assemblages, or at least will allow those earlier assemblages to be mineralogically reconstructed. Restoring the original geochemistry of the ores and especially the relative importance of their minor elements is far more difficult.

Lead in the supragossan zone

The popularity of stable lead isotope ratios as a provenance tool means that an understanding of the geochemistry of lead within the post-mining, supragossan zone of a mine site is essential. Normally lead moves slowly in the environment and most groundwaters carry very little lead in solution but at mine sites it, alongside other heavy metals, is easily leached, although often it does not move very far before being reprecipitated within secondary lead-bearing minerals.

Gee *et al.* (1997) investigating this phenomenon compared Roman, Medieval and Post-Industrial Revolution lead smelting slags. They were able to show that the major controls on secondary lead precipitation are pH and the availability of Fe, Mn, P and CO_2, with lead being adsorbed onto iron or manganese oxides ('wad' and 'limonite' respectively), or incorporated into members of the phosphate-bearing, pyromorphite group of minerals or the carbonate minerals cerussite and hydrocerussite.

Poorly crystalline iron and manganese oxides/hydroxides are present at nearly all mine sites (from the oxidation of iron-bearing sulfides etc) whilst sources of phosphorus (bone, wood ash and other organic material) are commonplace at prehistoric mines, as are sources of carbonate. Therefore, within the supragossan zone there is an increased potential for the precipitation of secondary lead minerals from circulating groundwaters onto the spoil, slag or metal. It is vital that the origin of this lead is known. If lead remobilization is localized then the isotopic ratios of the lead in the secondary minerals remain the same as the initial ones from the ore (Faure 1986) and the movement has few archaeological consequences. By contrast, if the original ores are lead-poor, as are many copper or tin ores, and if the local groundwaters are lead-bearing, having come into contact with a nearby lead source, then it is the isotopic signature of the groundwaters that will be measured, not that of the ores (assuming some precipitation of the lead onto the ore). Hence it is important to know the isotopic signature of the local groundwater at the site and doubly so if one or more of the ore, flux, slag or metal are foreign to the excavation site, as their true signature may be swamped by that of the local, post-mining groundwater. Lead-shot, which is quite often found in modern gold placers, certainly in the highlands of Scotland, is an extreme example of this problem. Similarly if the deposit carries lead-rich and lead-poor ore assemblages then the isotopic signature of the lead-rich ores will dominate the groundwater and subsequently the supragossan material, leading to the potentially erroneous conclusion that the lead-rich ores alone were exploited.

Ross Island

The need for understanding the ore geology, for archaeologically dedicated collecting of ore samples and for having a detailed knowledge of the mineralogy can be clearly demonstrated by the studies at Ross Island, Killarney, Ireland. Preliminary descriptions of the geology and ores, and discussion of the importance of this Chalcolithic copper mines site for our understanding of the origins of metallurgy in the British Isles, are given in Ixer & Pattrick (1995; in press), Ixer & Budd (1998) and O'Brien (1995; this volume). It is the earliest known copper mine in the British Isles and has a very wide range of potential ores (in terms of their metal contents and grain-size). Its main importance, however, lies in being the only known Bronze Age copper mine in the British Isles with arsenic-rich ores (tennantite and/or arsenopyrite) capable of producing arsenical copper, namely Chalcolithic type A metal, as defined by Northover 1982, (this is equivalent to Coghlan and Case (1957) type 1 metal).

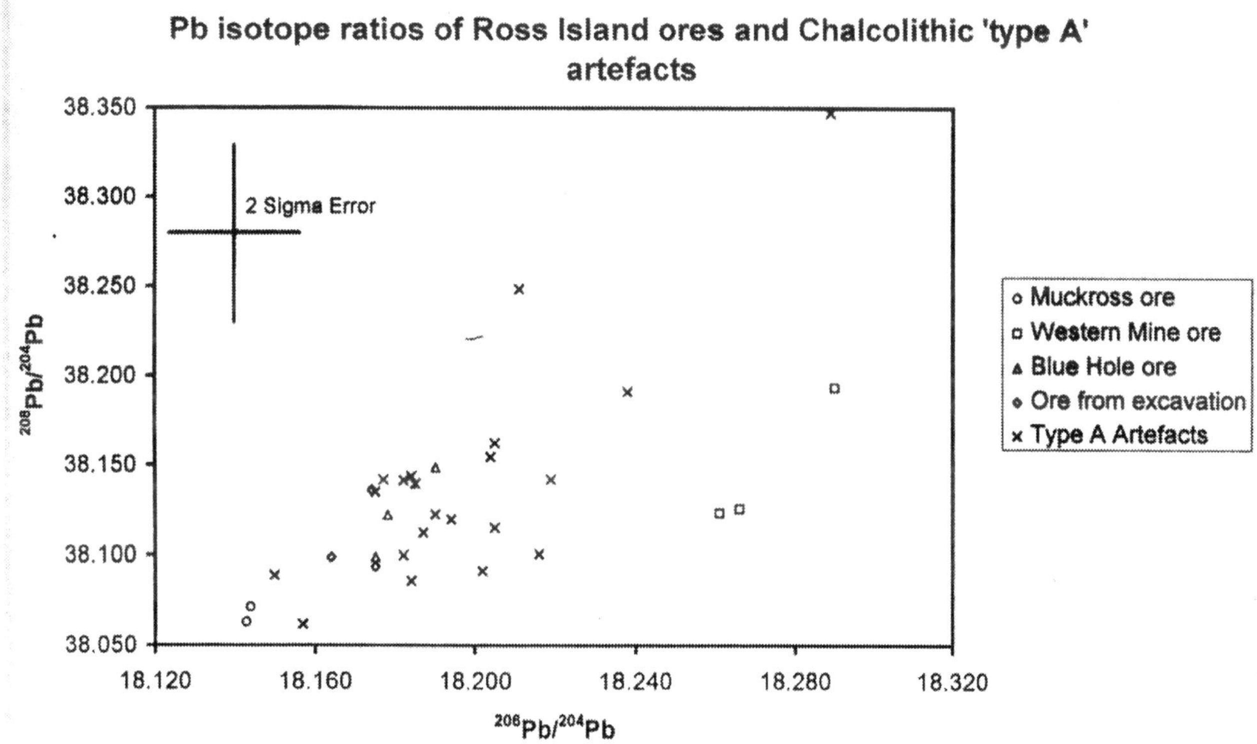

Figure 3. Lead isotope ratios of Ross Island ores and Chalcolithic 'type A' artefacts.

Although some lead isotope data have been reported by Ixer & Pattrick (1995) FIGURE 3 shows all the data for the mineralized specimens from Ross Island together with plots of Chalcolithic type A artefacts (Rohl 1995). Isotrace Laboratories and Dr. Rohl are thanked for making the Ross Island measurements and the wider archaeological implications of these data are discussed in Rohl & Needham (in press).

Plotted onto FIGURE 3 are three galena-poor, disseminated, mixed chalcopyrite-tennantite ores hosted within limestones (Western Mine ore); three massive-pyrite-chalcopyrite plus minor amounts of galena and sphalerite ores taken from two sulfide caches excavated from the Bronze Age work camp (Ore from Excavation) and three massive galena-sphalerite plus some pyrite and chalcopyrite ores taken from the adjacent site of Blue Hole, probably also exploited in the Bronze Age (Blue Hole ore). The 'Muckross Ores' are from similar massive sulfide mineralization to Blue Hole but a couple of kilometres away; they are not believed to be potential Bronze Age ores.

The figure shows an overlap between the massive sulfide ores from Ross Island and type A metal. Hence, at first sight the ore mineralogy (the copper ores are tennantite and/or arsenopyrite-rich), the dating of the mine and adjacent work camp (Beaker pottery and the C^{14} dates) (O'Brien 1995) and lead isotope data all concur with the suggestion that Ross Island supplied metal for type A artefacts. However, the data suggest that it was the massive ores rather than the chalcopyrite-tennantite, carbonate-hosted ores that were smelted, since the latter plot away from the artefacts, being more uranogenic than either the massive sulfide ores or the artefacts.

It should be noted in passing that the isotopic differences between the massive sulfide ores (with 1 wt% galena or more) and the limestone-hosted copper ores (with lead in the ppm range) provide a fine example of the isotopic divergence between lead-rich and lead-poor ores, that occurs in geologically old deposits, even when the ores began with the same initial lead ratios. This phenomenon, which is described in more detail by Macfarlane (this volume), is of extreme importance in provenance studies. The high Pb^{206}/Pb^{204} ratios, showing that the limestone-hosted copper ores are uranogenic, are also to be expected, as much of the mineralization is associated with carbonaceous matter and detrital heavy mineral grains along stylolitic junctions.

The archaeological interpretation of FIGURE 3 is more difficult than the mineralogical explanation of the differences shown by the ores. If it is assumed that the differences between the lead isotopes of the ores is real, then the simplest interpretation is that only the massive ores were exploited. However, this runs counter to the archaeological evidence, for it is the copper-tennantite ores that are present within the Bronze Age mine workings and make up the majority of the 'ore' found with the excavated work camp (O'Brien 1995). It seems more likely that the *in situ* chalcopyrite-tennantite ores are not representative of the Bronze Age copper ores that were taken from Western Mine and that these (richer) ores had sufficient galena to give the same ratios as those seen in the type A metal. It is also worth noting that, if only the disseminated, chalcopyrite-tennantite ores had been sampled in line with the excavation evidence, there would have been no isotopic match between Ross Island ores and Chalcolithic arsenical copper metal. In order to try to constrain the problem better, a far larger suite of samples, including richer chalcopyrite-tennantite ores, is currently being analyzed in co-operation with the British Geological Survey.

Although it is best to use slag or worked ore for provenancing rather than ores, there is no slag from secure Bronze Age contexts at Ross Island. Very rare samples of burned ore are present within the Bronze Age stratigraphy (for example specimen 1242) (Ixer & Pattrick in press). However, their void spaces are infilled with post-Bronze Age lead carbonate (cerussite) so that any lead signature from this material would be seriously compromized and so lead determinations on this material will not be attempted.

Ross Island shows that mineral deposits are zoned in space and time (the zoning continuing after mining) and comprise many different mineral assemblages some of which are/were ore. In terms of provenancing (the matching of the correct ores to metals) there is probably no such thing as a simple ore deposit nor is there a simple ore collecting strategy.

These truths will be illustrated again by using gold and the 'simplest' problem, namely the provenancing of unalloyed gold artefacts.

Introduction to the provenancing of gold

As stated in the introduction the provenancing of metals and alloys is inherently difficult. This is because the degree of processing that goes into the manufacture of an artefact is inversely related to the chances of a successful match between that object and its raw materials. This being so, the simplest metal to provenance should be unalloyed, single-sourced (obtained from a single mine site) gold.

Large numbers of quantitative and semi-quantitative major and minor element analyses for European gold artefacts have been published by Hartmann (1970; 1982). Taylor and her co-workers have had some success in using these analyses to match and classify these artefacts (Taylor 1980; Warner 1993) but there has been little progress in identifying their geological sources. Indeed recently Taylor *et al.* (1997) have acknowledged that 'major constituents and major trace analyses were inadequate for characterizing specific ore sources' and have proposed that qualitative, multi-elemental, association patterns will prove to be better at pin-pointing ore sources. The use of increasingly more sophisticated analytical techniques will certainly provide better data on individual gold grains, but this in itself can add little to further provenance studies as long as an incomplete understanding of the complexities of gold mineralization remains the main reason for poor or inadequate sampling.

Hard rock sources

Gold is a rare, but widely dispersed, metal occurring in many different mineral associations throughout a variety of geological environments (Boyle 1979; 1984). Although the majority of present day, economic gold deposits are primary, hard rock deposits associated with acid, intermediate or basic igneous rocks or found within a wide range of metamorphic terrains, much of the gold mined in prehistory was won from sedimentary placers in recent river gravels. These secondary deposits are formed by the mechanical concentration of gold and other minerals that survive physical and chemical weathering of a primary deposit and are then transported by water or ice to a new location and deposited.

Each major class of primary gold deposit has its own mineralogical and geochemical characteristics and exploration geologists spend much time in learning to recognise and identify these in order to find new sources of the yellow metal. The mineralogical and geochemical differences between classes of primary gold deposit not only affect their ore associations but also are manifested in a number of different ways within individual gold grains. Firstly, native gold grains can enclose other mineral phases, as small, 1-10μm diameter, inclusions; amongst these are a wide range of commonly occurring sulfides, sulfosalts, oxides and silicates, plus more unusual selenides and tellurides. Usually these minerals are part of the main mineral assemblage but some phases are only found within gold grains. However, in either case the minerals are of genetic significance and Leake *et al.* (1997), who list over fifty opaque phases from gold grains in Scotland, suggest that inclusion identification may be the most diagnostic tool in distinguishing between gold deposits. Secondly, the gold may itself vary, most notably in its fineness, chemical composition and homogeneity. Fineness is the numerical measure of the parts per thousand of gold in any gold-bearing alloy, so that 800 fine gold is 800 parts gold to 200 parts other metals: any naturally occurring gold-silver alloy that is 800 fine or less is called electrum. Silver and copper are the most common metal alloys but in a few gold occurrences palladium or mercury are important. Finally, in many deposits single gold grains are chemically inhomogenous with silver-rich areas separated from more gold-rich areas either in a random fashion or zonally but with a recognisable pattern (Leake *et al.* 1993).

These mineralogical phenomena are universal; they are part of the primary mineralization and are largely preserved during the weathering and erosion of the primary deposit and transport of the gold grains to their placer (Leake *et al.* 1993; Leake & Chapman 1996). Hence, for a number of decades it has been a standard and successful exploration technique to use placer gold grain characterization to predict the parent bedrock gold mineralization– a form of genetic provenancing.

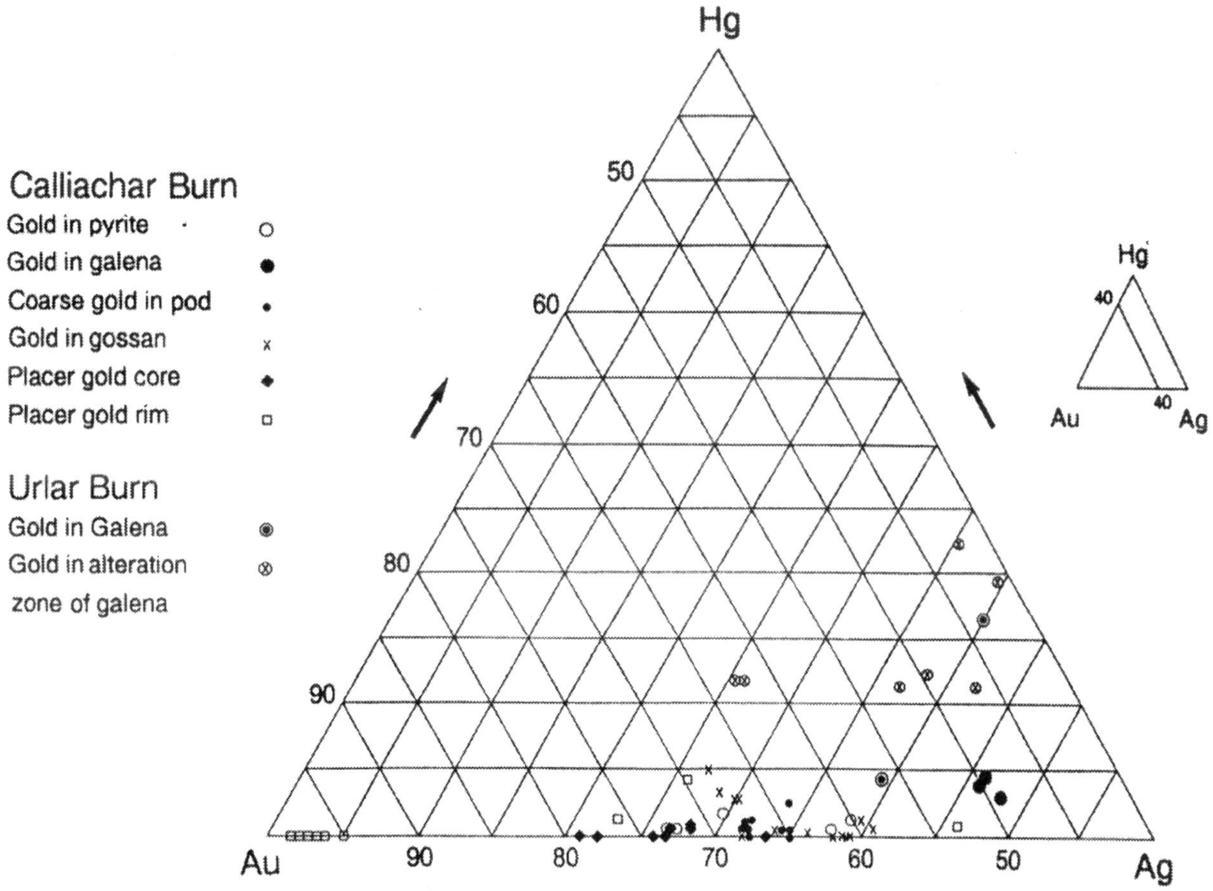

Figure 4. A Partial ternary plot for Au,Ag and Hg in wt% illustrating the compositional variation in gold and electrum grains from different mineral assemblages at Calliachar-Urlar Burn.

Potentially therefore, single gold grains show sufficient mineralogical and geochemical variation that they uniquely reflect the type of gold mineralization or even an individual gold deposit. However, gold in primary deposits is a complex mineral and as Leake & Chapman (1996) state 'there may be significant variation in the composition of gold derived from the different parts of the same mineralization'. Few studies have attempted to examine and quantify the extent of these primary gold variations within a small deposit and ever fewer have extended the research to include variations between primary gold mineralization and any associated placer. A gold prospect at Calliachar-Urlar Burns, Scotland was chosen to attempt such a study as the prospect is pristine and small enough that comprehensive and representative sampling from drill-core, surface outcrops and stream sediments was practicable. In addition, the richness of the gossan (100-400ppm gold has been reported) and the presence of gold grains up to 0.9 grams in weight in nearby streams (Chapman 1994) suggest that the prospect could have been mined successfully in prehistory had it been discovered.

Calliachar-Urlar Burns

Calliachar Burn and Urlar Burn lie approximately 1km from each other near Aberfeldy in the Highlands of Scotland. Here, fourteen, thin, quartz-galena-pyrite ± base metal sulfide ± gold veins and their overlying gossans lie upstream of a small gold placer. The mineralogy of all the mineral assemblages, including analyses of their associated gold grains, is reviewed and described in Ixer *et al.* (1997). Gold at Calliachar shows all the expected mineralogical features so that although many gold grains are inclusion-free, others enclose pyrite, galena and sphalerite and a few have 'unusual antimony sulfosalts'; the gold is alloyed with silver and mercury; and the primary and supergene gold grains are compositionally inhomogeneous with internal fineness variations.

Native gold (strictly speaking electrum together with a few gold alloys) from the different mineral associations from Calliachar and Urlar Burns are plotted in terms of their major metals, gold, silver and mercury in FIGURE 4. Each mineral association is listed in its paragenetic order from the oldest at the top to the youngest at the bottom. The figure demonstrates a very large variation in gold compositions between these associations so confirming the views of Leake & Chapman (1996). Some of this variation is routine, namely the overlap between the core composition of some of the placer grains and *in situ* hypogene and supergene (gossan) gold. The increase in fineness with time from hypogene electrum grains to supergene electrum grains to high fineness gold rims surrounding placer grains is also commonplace. Less expected are the marked differences between the gold compositions found at Calliachar and those of Urlar Burn approximately one kilometre away and in particular the differences in mercury content. These variations clearly demonstrate: that even in a small (easily mineable), primary gold deposit very different ores are present in different parts of the orebody and so potentially would yield metals with quite different minor and trace element geochemistries; as a consequence all parts of the orebody need to be sampled and many analyses are required, not just one or two, to show these complexities in their entirety. Indeed, the results from Calliachar strongly indicate that the twenty samples per ore deposit that are believed to be enough for lead isotope provenance studies (Reedy & Reedy 1998; Sayre *et al.* 1992) would be an insufficient number for a comprehensive geochemical characterization of all the potential metal producing associations. If the amount of geochemical variation in gold between Calliachar and Urlar Burns is a common phenomenon (and it may be), then Budd & Haggerty (1995) are correct in challenging the concept of a single regional (or even quite local) chemical, gold signature as suggested by Taylor (in Thwaite 1995).

Placer Gold

The requirement for extensive sampling is not restricted to primary gold deposits but is equally valid for placer deposits, as they too are not homogeneous in terms of the grade and composition of their ores. Some placer gold deposits are the result of the weathering of a single, primary bedrock gold source but even here the placer will carry gold grains with significant fineness variations, that will reflect the differences in composition between the hypogene and supergene zones and the degree of complexity of the original source. Most placers are more complex as they are multi-sourced with gold coming from nearby hardrock sources (proximal grains) as well as distant ones (distal grains) plus perhaps gold grains already modified by the sedimentary cycle and entering their second or third generation placer. In addition most placers have an *in situ* (authigenic) gold component in the form of gold-rich rims about the detrital grains and/or forming discrete nuggets; this gold is not detrital but is formed by the precipitation of gold from groundwaters.

Proximal, distal and authigenic gold grains not only vary in their chemistry but also in their size and shape and Bowles (1988) in an accessible review, summarises the chemical and morphological changes that take place in alluvial gold grains during their passage from bedrock to placer. In broad terms proximal grains, those close to the hardrock source, are angular, have bright, scratch-free surfaces, still have minerals attached to them or show sharp scars where they have been lost and have thin or missing gold-rich rims. During transport the surface of the grains becomes heavily scratched and matte and the grains are hammered, folded and kneaded until after travelling for approximately 100 kilometres they are finally deformed into one of two shapes, rods /cylinders or flakes/plates/sheets. At the same time the core composition of the grain increases in fineness and there is an increase in the thickness of the gold-rich rim. Most gold placers therefore comprise three different sorts of gold; angular, equidimensional, low to moderate fineness, proximal grains; cylindrical or flake-like, high fineness, distal grains; and very fine gold as rims enclosing detrital grains and/or present as discrete nuggets.

The absolute and relative amounts of the three components will vary throughout the placer, since all three classes of gold differ in their hydrodynamic properties and in grain size and because the precipitation of authigenic gold depends on very local geochemical conditions. Hence, authigenic nuggets may be the initial gold taken from the surface of a placer, whereas the bulk of the gold might be detrital and mined from concentrations to be found in irregularities at its base (Bowles 1988). Therefore as the ratios of proximal: distal: authigenic gold vary throughout the placer, with each component carrying its own independent and distinctive geochemical signature, so the composition of the gold metal produced from that part of the placer will change. Indeed, since the detrital and non-detrital components have very different geneses and are unlikely to be geochemically (including isotopically) alike, artefacts made from them may not be recognized as coming from the same place but assigned to different sources.

Finally, there is the question of what constitutes gold ore in a placer. This needs to be addressed before a placer can be correctly sampled. In gold placers ore cannot be represented by individual gold grains (unless they are large) but comprises collections of them. Perhaps as in forensic work, bulk gold, namely one to ten grammes collected from a single part of the placer, needs to be geochemically analyzed and compared to an artefact. Where this is not possible the number of gold grains needed to adequately represent all of the gold components must be established before extensive analyses are warranted. For some single-sourced, proximal deposits this may only require a few grains but there will be others where the complexities of gold their components may be such that the deposit cannot be included in any meaningful provenancing programme. There may even be a case where, if the artefact were made exclusively from authigenic (nugget) gold, none of which remains, trace element or stable isotope analyses of groundwater might prove a better provenancing tool than detrital gold grains.

For the minerals exploration industry, detailed mineralogical and geochemical data from single, placer gold grains are of great value as a genetic provenancing tool (indicating that the gold is associated with a porphyry copper deposit or has ophiolitic affinities) as exemplified by Leake and co-workers (1993) rather than being geographically specific (namely that the mother-lode crops out fifteen kilometres upstream), whilst the bulk chemical analyses of the same gold grains have little diagnostic worth. This contrasts with archaeometallurgy where the situation is reversed. Here artefacts are more likely to be successfully provenanced (given a specific geographical origin) if their chemistries are compared with those of bulked gold samples rather than with individual gold grains.

So it seems that although the provenancing of prehistoric gold does not have some of the problems associated with other metals, especially alloys like arsenic- or tin-bronzes, it remains difficult. It, like all other metal provenancing, depends upon recognizing and sampling the correct ores. This in turn requires a full understanding of the ore geology of the deposit and this must never be thought of as simple, however tempting that may be.

Conclusions

Trace element and isotopic provenancing of metals can only be as good as the ore sampling that it is based upon. This in turn depends on a proper understanding of the ore deposit geology, natural geochemical processes within the supra-gossan zone and the distinction between mineral samples and exploited ore. Once all this has been achieved then geochemical studies have much to contribute, both in terms of provenancing ores/mines to artefact but also temporally within the mine itself. Once the mineral zoning has been established in three dimensions, then for mines with centuries or millennia of exploitation it may be possible to determine their mining history, rather than just provenancing metal from their ores. Combining detailed and precisely located ores from within the mine and associated spoil and slag heaps with trace element and isotopic fingerprinting and if all this can be matched to datable metal artefacts it may lead to a proper understanding of the exploitation history of a mine.

Acknowledgements
Drs R.A.D.Pattrick and C.J.Stanley are thanked for their mineralogical prowess and for the many high quality gold analyses from Calliachar-Urlar Burn. In a similar vein Dr J. Naden provided numerous analyses of gold placer grains from throughout the world and introduced me to the complexities of genetic gold provenancing. Germane criticism by Drs. J. Tellam and R. Chapman helped to modify the text, as did the incisive referees' comments. C.R.Ixer provided the computer skills for this paper. The help of the University College Galway, Ross Island Research Project and its officers and very generous financial help from the Irish Government were, and are, welcome.

References

ALLARD, B., I. ARSENIE, K. HAKANSSON & S. KARLSSON. 1989. In J.P. Vernet (ed.): *Proceedings of an International Conference on Heavy Metals in the Environment*: 221-224. Edinburgh: CEP Consultants.

BACHMANN H. 1982. *The identification of slags from archaeological sites.* Occasional Publication 6. London: Institute of Archaeology.

BOWLES, J.F.W. 1988. Mechanical and chemical modification of alluvial gold. *The Australian Institution of Mining and Metallurgy Bulletin and Proceedings* 293: 9-11.

BOYLE, R.W. 1979. The geochemistry of gold and its deposits. *Bulletin of the Geological Survey of Canada* 280.

1984. Gold deposits: their geology, geochemistry and origin, in R.P. Foster (ed.): *Gold '82. Geological Society of Zimbabwe Special Publication* 1: 183-189. Rotterdam: A.A. Balkema.

BUDD, P., D.GALE, A.M.POLLARD, R.G. THOMAS & P.A.WILLIAMS. 1992. The early development of metallurgy in the British Isles. *Antiquity* 66: 677-686.

BUDD, P. & R. HAGGERTY. 1995. Old Gold. *New Scientist* 147: 50.

BUDD, P., R. HAGGERTY, A.M. POLLARD, B. SCAIFE & R.G. THOMAS. 1996. Rethinking the quest for provenance. *Antiquity* 70: 168-174.

BUDD, P., B.SCAIFE, T.TAYLOR & R.G.THOMAS. 1994. Untangling the web: some new views on the origins of prehistoric metallurgy. *Historical Metallurgy* 28: 98-102.

CHAPMAN, R. 1994. The geological origins and occurrences of gold in Britain and Ireland. Department of Adult and Continuing Education Course Notes. Leeds University.

COGHLAN, H.H. & CASE, H.J. 1957. Early metallurgy of copper in Britain and Ireland. *Proceedings of the Prehistoric Society:* 91-123.

GEE, C., M.H.RAMSEY, J. MASKALL & I. THORNTON. 1997. Mineralogy and weathering process in historical smelting slags and their effect on the mobilisation of lead. *Journal of Geochemical Exploration* 58: 249 – 257.

HARTMANN, A. 1970. Prähistorische Goldfunde aus Europa – Spektralanalytische Untersuchungen und der Auswertung – *Studien zu den Anf<ngen der Metallurgie 3.* Berlin: Gebr der Mann Verlag.

1982 Prähistorische Goldfunde aus Europa II – Spektralanalytische Untersuchungen und deren Auswertung – *Studien zu den Anf<ngen der Metallurgie* 5. Berlin: Gebr der Mann Verlag.

IXER, R.A. 1994. Does ore petrography have a practical role in the finger-printing of rocks? in N.Ashton and A. David (ed.): *Stories in Stone.* Lithic Studies Occasional Paper 4: 10-23. Oxford: Lithic Studies Society.

1995. Ancient Mining: Modern Collections, Abstract , *Mineralogy in the Field 1995* . A one-day symposium organised by The Russell Society in association with the University Museum at Keble College Oxford.

1996. Ore petrography and archaeological provenance. *Mineralogical Society Bulletin* 113: 17-19.

1997. Detailed provenancing of the Stonehenge Dolerites using reflected light petrography - a return to the light, in A. Sinclair, E. Slater and J.Gowlett (ed.): *Archaeological Sciences 1995.* Oxbow Monograph 64: 11-17.

In press. Potential and realizable ores from the Great Orme Mine. *Great Orme Mines Occasional Publication* 1. D. Jenkins (ed.). Llandudno.

IXER, R.A. & P.BUDD. 1998. The mineralogy of Bronze Age copper ores from the British Isles: implications for the composition of early metalwork. *Oxford Journal of Archaeology,* 17: 15-41

IXER, R.A. & J. DAVIES. 1996. Mineralization at the Great Orme Copper Mines, Llandudno, North Wales. *UK Journal of Mines and Minerals* 17: 7-14.

IXER, R.A.F. & R.A.D. PATTRICK. 1995. Cu-Pb-Zn-Co-Ag mineralization at Ross Island – Muckross Mine, Killarney, Ireland. Abstract. *Mineral Deposits Study Group meeting. Manchester.*

In press. Copper-arsenic ores and Bronze Age mining and metallurgy with special reference to the British Isles, in P.T.Craddock and J.Lang (ed.): *The Prehistory of Mining and Smelting.* London: British Museum Press.

IXER, R.A.F., R.A.D. PATTRICK & C.J. STANLEY. 1997. The geology, mineralogy and genesis of gold mineralization at Calliachar-Urlar Burn, Scotland. *Transactions of the Institution of Mining and Metallurgy. (Section B: Applied Earth Science)* 106: B98-B107.

IXER, R.A. & C.J. STANLEY. 1996. Siegenite-bearing assemblages found at the Great Orme Mine, Llandudno, North Wales, *Mineralogical Magazine* 60: 978-982.

JENKINS, D.A. & D.B. JOHNSON. 1993. Abandoned metal mines: a unique mineralogical and microbiological resource. *Journal of the Russell Society* 5: 40 – 44.

LEAKE, R.C., D.J.BLAND & C.COOPER. 1993. Source characterization of alluvial gold from mineral inclusions and internal compositional variation. *Transactions of the Institution of Mining and Metallurgy. (Section B: Applied Earth Science)* 102: B65-B82.

LEAKE R.C. & R.J. CHAPMAN. 1996. Microchemical characterization of alluvial gold from Scotland. Abstracts Volume. *Mineralisation in the Caledonides.* Institution of Mining and Metallurgy Meeting. Edinburgh.

LEAKE R.C., R.J. CHAPMAN, D.J. BLAND, E.CONCLIFFE & M.T. STYLES. 1997. Microchemical characterization of alluvial gold from Scotland. *Transactions of the Institution of Mining and Metallurgy (Section B: Applied Earth Science)* 106: B85 – B98.

NORTHOVER, J.P. 1982. The exploration and long distance movement of bronze in Bronze Age and Early Iron Age Europe. *Bulletin of the Institute of Archaeology* 19: 45-72.

O'BRIEN, W. 1995. Ross Island and the origins of Irish-British metallurgy in J. Waddell and E. Shee Twohig (ed): Ireland in the Bronze Age: 38-48. Dublin: Stationary Office.

PARK, C.F. & R.A.MACDIARMID. 1970. *Ore Deposits.* 2nd Edition. San Francisco: W.H.Freeman and Co.

POLLARD, M., R.G.THOMAS, D.P.WARE & P.A. WILLIAMS. 1991. Experimental smelting of secondary copper minerals: implications for Early Bronze Age metallurgy in Britain, in E. Pernicka and G.A.Wagner (ed.): *Archaeometry '90*: 127-136. Basel.

REEDY, C.L. & T.S.REEDY. 1988. Lead isotope analyses for provenance studies in the Aegean region: a re-evaluation, in E.V. Sayre *et al* (ed.): *Materials issues in art and archaeology*: 65-70. Pittsburgh: Materials Research Society.

ROHL, B.M. 1995. Application of lead isotope analysis to Bronze Age metal work from England and Wales. Unpublished D. Phil thesis. Oxford University.

ROHL, B.M. & S.P.NEEDHAM In press. The Circulation of Metal in the British Bronze Age: the application of lead isotope analysis. British Museum Occasional Paper 103.

SAYRE, E.V., K.A.YENER, E.C.JOEL & J.L. BARNES. 1992. Statistical evaluation of the presently accumulated lead isotope data from Anatolia and surrounding regions. *Archaeometry* 34: 73-106

TAYLOR, J.J. 1980. *Bronze Age Goldwork of the British Isles.* Cambridge. Cambridge University Press

TAYLOR, J.J., R.J. WATLING, C.A. SHELL, R.A.IXER, R.J. CHAPMAN, R.B. WARNER & M. CAHILL. 1997. From gold ores to artefacts in the British Isles: A preliminary study of a new LA-ICP-MS analytical approach, in A.Sinclair, E. Slater and J.Gowlett (ed.): *Archaeological Sciences '95.* Oxbow Monograph 64: 144-154.

TIMBERLAKE, S. 1995. 'Copa Hill' and 'Llancynfelin Mine'. *Archaeology in Wales* 35: 40-43

THWAITES, T. 1995. Gold fingerprints spot the fakes. *New Scientist* 147 (1985): 24.

WARNER, R.B. 1993. Irish Prehistoric goldwork: a provisional analysis. *Archaeomaterials* 7: 101– 113.

YOUNGER, P. 1995. Polluted groundwater in abandoned coalfields: some hydrogeological and geochemical idiosyncrasies. *Fourth annual conference on groundwater pollution.* 11pp. London, March 1995.

6

The Application of Laser Ablation Inductively Coupled Plasma Mass Spectrometry (LA-ICP-MS) for Establishing the Provenance of Gold Ores and Artefacts

R. JOHN WATLING[1], JOAN J. TAYLOR[2], COLIN A. SHELL[3], ROBERT J. CHAPMAN[4], RICHARD B. WARNER[5], MARY CAHILL[6], & ROBERT C LEAKE[7]

[1]*Department of Applied Chemistry, Curtin University, GPO Box 1987, Perth, Western Australia 6845*
[2]*School of Archaeology, Classics and Oriental Studies, University of Liverpool, William Hartley Building, Brownlow Street, Liverpool, L69 3BX, UK*
[3]*Department of Archaeology, University of Cambridge, Downing Street, Cambridge, CB2 3DZ, UK*
[4]*Department of Adult Continuing Education, University of Leeds, LS2 9JT, UK*
[5]*Department of Archaeology and Ethnography, Ulster Museum, Botanic Gardens, Belfast, BT9 5AB, UK*
[6]*Irish Antiquities Division, National Museum of Ireland, Kildaire Street, Dublin, Eire 2, Ireland*
[7]*Consultant, School of Earth Sciences, University of Leeds, LS2 9JT, UK*

Laser Ablation-Inductively Coupled Plasma-Mass Spectrometry (LA-ICP-MS) has been used successfully to identify elemental association patterns (fingerprints) in gold from the United Kingdom and Ireland. By comparison of these elemental associations it has been possible to relate alluvial gold to specific mineralising events and placer deposits and potentially to link these sources of gold to fabricated artefacts. The methodology facilitates identification of the uniqueness of gold, thereby providing definitive scientific evidence of provenance of alluvial gold and artefacts for archaeologists and metallurgists. "Gold" samples for analysis span a wide range of materials, from nuggets, alluvial and placer material to fabricated artefacts, with many requiring extensive prior separation and treatment to present the contained gold in a chemically "unmodified" state, and in a form suitable for LA-ICP-MS fingerprinting. The quality of the fingerprint, and hence its value, is directly related to the rigors of prior sample preparation and evaluation, to an appreciation of the history of the gold under investigation and to the skill of the analyst. At this stage of the development of the analytical protocols, and with an obviously limited database of artefacts and deposits, it has not been possible to unequivocally relate the fingerprint of an artefact to that of a named deposit. However, it must be considered only a matter of time before definitive associations can be established.

Keywords: Laser Ablation Inductively Coupled Plasma Mass Spectrometry; Forensic Archaeology; Gold Fingerprinting.

Gold "fingerprinting" is based on the premise that gold, associated with a specific mineralising event or extractive process, will inherit a trace/ultratrace element signature unique to that event or process (Taylor *et. al.* 1997). Because mineralising events are themselves unique, reflecting as they do the geochemical dynamics of ore fluid generation, migration, interaction and ore deposition, the trace/ultratrace element signature of gold, derived from an ore by either mechanical or "chemical" treatment processes, will be specific to that ore. However, the fingerprint of modern chemically-derived gold may be significantly different to that of natural unprocessed gold obtained from the same ore. In the past, however, extractive metallurgical processes were not as complicated or chemically aggressive as today and consequently the potential for associating a gold artefact with its source material is increased. Most of the starting materials require exhaustive prior separation and treatment to present the contained gold in a chemically "unmodified" state, and in a form, suitable for Laser Ablation-Inductively Coupled Plasma-Mass Spectrometry (LA-ICP-MS) fingerprinting. The concept of trace element signatures is not new and has formed the basis of petrogenetic assessments in geosciences for decades (Herbert 1987).

LA-ICP-MS has now been used successfully to obtain the inter-element association pattern, "fingerprint" of gold in forensic investigations internationally (Watling *et al.* 1994, Herbert & Watling 1996, Watling & Herbert 1995 a, and b) and for fingerprinting scene of crime evidence (Watling *et al.* 1997), diamonds (Watling *et al.* 1995a and Watling, 1996). The technique has also been successfully used to establish trace element patterns in oxide, silicate and, more importantly, sulfide minerals which are often associated with, and can be included in, gold (Perkins *et al.* 1992, Crain & Gallimore 1992, Perkins *et al.* 1991, Pearce *et al.* 1992, Perkins *et al.* 1993, Broadhead *et al.* 1990, van de Weijer *et al.* 1992 and Watling *et al.* 1995b). These minerals can act as an additional provenance-sourcing tool. Elemental fingerprinting relies on the fact that LA-ICP-MS can identify trace and ultratrace element assemblages and isotope abundance shifts at analyte concentrations in and below the μg/kg region. Because the detection limits are so low, a large number of inter-element associations can be employed to develop unique trace/ultratrace element associations in gold. Commonly, elemental data are obtained for over fifty elements. However, these inter-element associations are valid for inter-comparison of samples only in so far as they reflect the signature derived exclusively from the analyzed gold. Most naturally occurring gold is characterized by coeval inclusions of other mineral matter (FIGURE 1C) (Leake *et. al.* 1993, Leake *et. al.* 1996, Leake *et. al.* 1997, Leake *et. al.* in press), while many processed golds, especially those produced centuries ago, contain eutectic intergrowths of non-gold phases developed during smelting and artefact manufacture (FIGURES 1A and B). If these "inclusions" go undetected, or unrecognized, during ablation, a spurious signature, representing the combined signal from the gold and the included phase, may result and lead to incorrect interpretation. A detailed physical examination of the material, using electron optical techniques, is required prior to LA-ICP-MS analysis, to be certain that the material being analyzed does not contain inclusions, incorporated exotic debris and areas of anomalous exsolution phases associated with cooling of metallic melts.

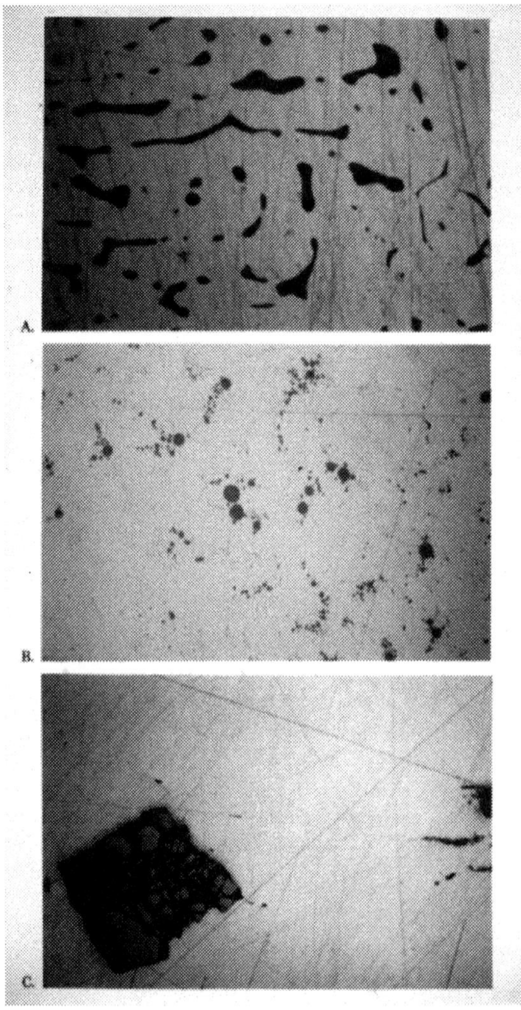

Figure 1: Optical photomicrographs illustrating: (A) smelt eutectic of lead sulfide, iron monosulfide, silver sulfide and silver telluride in gold; (B) eutectic of selenium in gold bullion bar; and (C) telluride grains in natural unprocessed gold.

Because the detection limits associated with the technique are so low, the number of elements that can usefully be employed to provide data to establish the fingerprint is much higher than any method previously available. This aspect increases the number of useful inter-element combinations and associations and thereby increases the probability of providing a unique fingerprint. Each ablation event consumes only a fraction of a microgram of sample and it is usually possible to retain a significant proportion of the sample for subsequent further forensic examination and corroborative analyses. This aspect has considerable appeal to conservators, charged with protecting the overall integrity of artefacts in their care.

In the case of most gold ores and artefacts, because insignificant modification and breakdown of the material has taken place in historical time, matching elemental association patterns in ores with those obtained from the artefact can form the basis of identifying early metallurgical and mining sites so providing information on early trade and cultural links. Comparison of the fingerprint in gold grave goods, from sites within a specific region, has identified both sources of raw materials and common "refined" gold being used to fabricate a number of separate articles. It has also been possible to associate broken and dispersed material to a single artefact, thus aiding reconstruction and to indicate the possibility of an article being a more modern forgery.

Experimental

Instrumentation and technique

A VG Turbo Plus Inductively Coupled Plasma-Mass Spectrometer with VG LaserLab accessory is central to the "fingerprinting" facility. The LaserLab is built around a 500mJ pulsed Nd-YAG laser operating at a wavelength of 1064nm. Essential ancillary equipment includes a range of specialized sample preparation, optical microscopic and scanning electron microscopic/energy dispersive X-ray analytical facilities.

The laser system is used in free running mode to allow the smallest possible laser crater to be formed on the surface of the sample. This significantly diminishes damage to any sample and facilitates the use of the smallest samples possible, so insuring minimal sample damage. Indeed, often only extremely small samples are available for analysis. The ICP-MS system is fitted with a high sensitivity upgrade and an additional, fine focus attachment to the laser facilitates the analysis of samples as small as 30μm in diameter. The microlaser produces craters of between 15 and 20μm in diameter, under the ablation protocol used; consequently the number of repeat laser shots per sample is dependent on the sample size. The size of the sample also limits the availability of inclusion- free sample sites. While the mass of the ablated sample being delivered to the plasma, after ablation with the microlaser, is reduced from that produced when a conventional IR laser is used, the analytical sensitivity of the system is not significantly degraded. Throughput of ions into the plasma is increased by the incorporation of a high sensitivity interface upgrade into the basic mass spectrometer design. This upgrade, consisting of an additional high volume pump, reduces the vacuum in the interface zone between the sample and skimmer cones. This improves ionic throughput into the lens stack and quadrupole and increases sensitivity by up to an order of magnitude.

The new system, with its vastly enhanced spatial resolution capability, facilitates the precise location of each analytical site. This greatly reduces the accidental incorporation of debris derived from associated non-gold minerals.

Because of the common occurrence of mineral inclusions and/or exsolved eutectic phases in the gold matrices submitted for fingerprint analysis, it is essential that a rigorous prior examination of the sample be undertaken, in order to both facilitate interpretation of the fingerprint after analysis and to identify the most appropriate areas on the sample surface for analysis. Normally, this involves using binocular and polarized light microscopy in association with Scanning Electron Microscopy/Energy Dispersive X-ray Analysis (SEM-EDXRA). A Cambridge Instruments Scanning Electron Microscope, fitted with an Energy Dispersive X-ray Analyzer, was used to map the nature and elemental composition of the sample under investigation. Elemental maps are then used in conjunction with the LA-ICP-MS protocols to identify areas for ablation. Throughout the matrix, all these phenomena will occur as significantly reduced size inclusions. Obviously, there is no way in which the distribution in depth of these phases can be determined from surface scans. However, best possible site location, together with SEM-EDXRA analysis of the visible inclusions and areas of exsolution, facilitates interpretation of the data from LA-ICP-MS analysis. In the authors' opinion these precautions ensure that the best possible data are used for future inter-comparisons, especially when the inter-element association pattern data are referenced to databases of non-gold minerals.

Laser Optimization

The two variables that significantly control analyte release from any sample are laser power and shot frequency. Increasing the frequency of laser shot repetition rate causes progressive ablation of an increasingly liquid pool of gold and a corresponding increase in the size of particles being removed. Removal of these large particles increases the loci for adsorption and attachment of smaller particles that would otherwise be carried into the plasma. This recombination phenomenon occurs inside the laser cell and in the interface tubing between the cell and the plasma. This not only reduces the amount of material available for ionization but also significantly increases the potential for inter-sample contamination. However, while trying to achieve small crater size it must be remembered that the lower the power and smaller the crater size, the more likely it is that selective volatilization of the more labile elements will occur. Although increasing laser power accentuates ablation of samples, it also increases the size of laser craters. Consequently, there is a compromise between laser power and shot repetition rate to ensure that representative volatilization of analytes, together with the smallest possible laser crater, will result.

Examination of numerous samples also demonstrates that there is a marked physical variation in the morphology of different gold samples (Watling *et al.* 1994). Often, this difference is exhibited more in natural gold than in fabricated artefacts. These variations give rise to significant differences in laser coupling efficiency and, consequently, produce varying amounts of debris from equivalent laser pulses. Because of this, quantitative analysis of the wide variety of gold types used in gold fingerprinting, is not possible using a conventional IR laser system. However, fingerprinting of gold does not require quantitative data to be used but identifies the provenance of samples by reference to the association, or lack of association, of specific suites of analytes. Therefore, it is this association of elements, and not their absolute concentrations, that is used to establish the fingerprint.

It is possible to normalize data from laser ablation ICP-MS, by employing a ratio of data to the magnitude of the gold argide peak. Theoretically, this improves analytical quantitation. However, detailed studies have established that selective volatilization of elements frequently occurs during ablation (Watling *et al.* 1994 and Chenery *et al.* 1992). Consequently, even if the mass of material removed under a uniform set of ablation conditions is the same for different gold samples, the magnitude of the gold argide peak will vary significantly and thereby invalidates its application. Therefore, the use of this peak is only acceptable within well-recognized, uniform, single matrix sample types, such as gold coins or refined gold bullion.

Nevertheless, it is necessary to establish an optimum series of parameters for laser ablation, which ensure that a robust inter-comparison mechanism is achievable for the different samples of gold. When establishing these conditions it is essential that the variations in relative response are as small as possible between samples. This ensures that fingerprint variations are as small as possible when analyses are performed using different instruments, an essential aspect when considerations for the establishment of international databases are made. Laser crater size should be as small as possible, while still providing enough representative material for analysis. In this way, small samples can be ablated under exactly the same conditions as large ones, maintaining the relevance of inter-sample comparison of the fingerprint. Consequently, low laser power and low repetition rate are preferred. Results from numerous experiments, using a number of instruments, indicate that an experimental protocol of two laser shots per second per sample site and a laser power setting of approximately 900V is optimum for most analytes. Where possible, samples should be ablated at a number of sites to ensure that the fingerprint is representative. However, it may only be possible to use a single site when dealing with extremely small samples, or those particularly rich in included or exsolved material.

Acquisition time for each data set is kept at approximately 120 seconds. In this way it is possible to use the multichannel analyzer, to visually inspect the accumulation of data for each fingerprint and to move the ablation site around the sample with reference to the SEM/EDXRA map. Extremely rapid increases for individual or a group of element isotopes can be easily recorded. These increases usually indicate the ablation of included material, probably in depth as the laser crater penetrates the sample. Under these circumstances it may be considered appropriate to terminate the ablation and restart the analysis. This is impossible for some small samples and under these circumstances a written record of the incident will assist interpretation of the final fingerprint data. In general, ablation of ten sites per sample is believed to be appropriate in order to obtain a suitable data set for interpretation.

Instrument Response Optimization

National Institute of Standards and Technology Standards (NIST) (Formerly National Bureau of Standards - NBS) Glasses, SRM610, SRM612 and SRM614 are used to optimize the instrument operating parameters. In addition, the use of these standards, at the beginning and end of a complete analytical protocol, allow quantification of changes in mass response with time to be made. This procedure also facilitates inter-comparison and normalization of data on a day to day basis. The choice of these samples and not gold is governed by the fact that optimization requires a steady state signal. Continuous ablation of gold would produce a significant amount of debris, that would coat the inside of the laser cell and deposit material on the interface tubing between the sample cell and the plasma, thus contaminating the system for the analysis of subsequent samples. That glass and gold are dissimilar sample types, and that the plasma loading will be different for the two matrices, is not relevant, as the interpretational protocol for fingerprinting is based on the inter-element association pattern of samples and not on absolute analyte quantitation. The use of NIST standards has proved to be reliable and extremely convenient. Once optimization has been achieved, three in-house gold reference samples are analyzed to provide a cross-reference to previously obtained data. In practice, these reference samples are analyzed either every twenty study samples or a minimum of three times during any specific sample batch. The in-house gold standards were prepared by combining residual gold from a number of case studies. The molten gold was quenched in chilled water to ensure minimal analyte segregation. The element mixture and concentration are immaterial, as the standards are used simply to ensure correspondence of day to day analyses and a minimum of 25g each was prepared to ensure their continued use over a number of years. Instrument calibration usually takes approximately thirty minutes.

Sample preparation

Each sample type requires preparation prior to analysis and the fundamental requirement is that the sample is presented to the laser in a suitable form for analysis. The principal restraints are that the sample preparation process must avoid or minimize contamination of the gold, the possible elimination or reduction of analyte concentrations and the homogenization of otherwise heterogeneous material. Specialized techniques are employed to ensure that these conditions are met. Final samples are usually mounted on quick setting plastic using cyanoacrylate glue. In addition, polished samples and thin sections are also used, but, if diamond paste is used in the final grinding, then a representative sample of the paste should also be analyzed with the samples, to ensure that relevant blank corrections can be made. However, some samples are so small or have shapes that preclude active surface cleaning that conventional sample preparation is impossible. For these samples a pre-ablation protocol is established. Namely the site to be analyzed is ablated by employing a de-tuned laser operating in Q-switched mode. Careful observation of the ablation site reveals when the surface contamination has been removed and the sample is ready for analysis. Most artefact material, submitted by museums, has one cut edge. In these circumstances, as an appropriate sample surface has already been provided, it may be necessary to pre-ablate this surface for only a few seconds to ensure that there is no contamination from the sample cutter.

Spectral interpretation

Interpretation of the mass spectral data, obtained using the protocols described, may be subdivided into four basic areas:

1) Direct comparison of the raw spectral data
2) Comparison of the raw count data for selected analytes
3) Comparison of ternary inter-element association plots
4) Comparison of data to the entire gold database, followed by re-comparison using protocols 1 to 3 above.

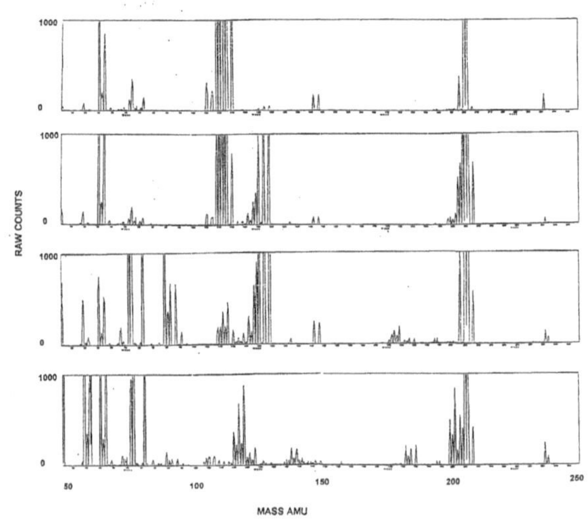

Figure 2: Comparative mass spectra for gold from four different sites in the United Kingdom and Ireland. 1) East Wicklow: 2) Co. Antrim: 3) Croag Patrick: 4) Suisgill Burn.

Direct Comparison of the Raw Spectral Data

Comparison of the inter-element association pattern (fingerprint) is the initial screening test to establish if there are general similarities, or differences, between the individual spectra under investigation. Spectra from alluvial gold, panned from four different sites in Ireland and the United Kingdom, are compared in FIGURE 2. Here, fundamental differences can be seen in the relative abundances of ^{59}Co, $^{58, 60, 61, 62, 64}$Ni, $^{63, 65}$Cu, $^{64, 66, 67, 68, 70}$Zn, $^{74, 76, 77, 78, 82}$Se, $^{106, 108, 110, 111, 112, 113, 114, 116}$Cd, $^{112, 114, 115, 116, 117, 118, 119, 120, 122, 124}$Sn, $^{174, 176, 177, 178, 179, 180}$Hf, $^{182, 183, 184, 186}$W, $^{196, 198, 199, 200, 201, 202, 204}$Hg, $^{203, 205}$Tl and $^{235, 238}$U. The four deposits, represented in this figure, are derived from four different sites of gold mineralization and the comparative plots of the spectra are simply designed to indicate the kind of differences that are expected with this type of analysis. These spectra should not be seen in isolation as being able to identify individual deposits. However, their inter-element association patterns would be expected to be different, as the gold that they represent has resulted from four different mineralising events in different geographical locations. Consequently, the confirmation that their fingerprints are different may be considered a geologically identifiable observation. The vertical scale on these spectra, which is an approximation of the relative concentration of the abundance of each isotope in the sample, is given on the ordinate. Providing that the spectra are graphically presented on comparable abscissa and ordinate scales, and that the laser ablation conditions are identical, visual comparison is a valid method of distinguishing between both similar and significantly different gold samples and mineralising events.

Whereas the inter-element association patterns of the gold from the four different sites are quite different, many gold samples, particularly if related to the same mineralising event, or from an artefact where the same gold has been used for manufacture, are markedly similar in gross terms and more detailed investigation is required to confirm the robustness of spectral comparability. For example, spectra from the Bracks Farm Torc (FIGURE 3) indicate almost exact inter-comparability and it is extremely difficult to distinguish between the spectra from different parts of the torc. When spectra from a separate gold flange, found associated with the torc, are compared with those from the torc itself, there is virtually no difference. This inter-comparability is again confirmed by reference to the spectra from two artefacts, a sleeve ornament and a dress ornament, found in County Antrim, Northern Ireland (Taylor 1980), indicating that the gold used in their fabrication was from the same original source material. This material is dealt with in detail in the following section.

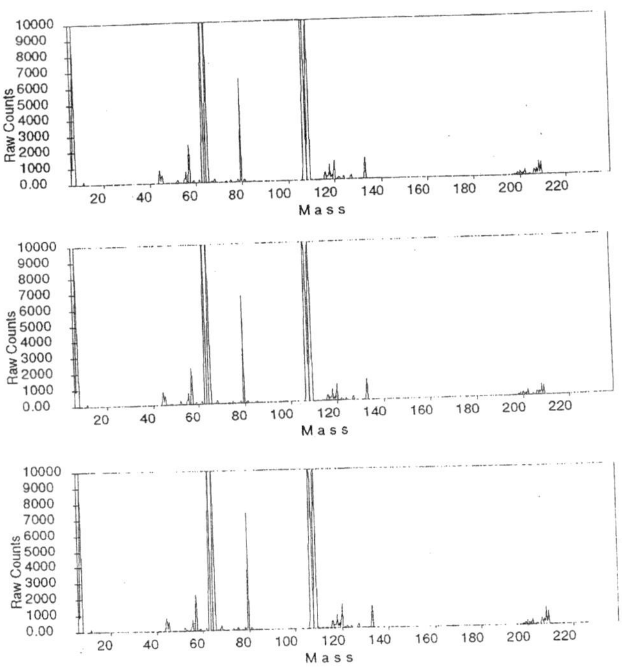

Figure 3: Comparative mass spectra for three samples taken from different sites on the Bracks Farm Torc.

Comparison of Raw Count Data for Selected Analytes

Visual comparison of spectral data is frequently demanding, with similarities and differences between the spectra being difficult to discern and quantify for both trained, and untrained, personnel. Consequently, comparison of raw count data for selected analytes is often preferred. Using this method, graphical plots of the relative concentration of the analyte isotopes are produced and coincidence between these elemental patterns are used as a direct comparison technique (Taylor *et. al.* 1997). In this way close relationship between samples of gold from the same site, and hence from potentially the same mineralising event, can be clearly demonstrated, while significant differences between samples from different sites are also easily seen.

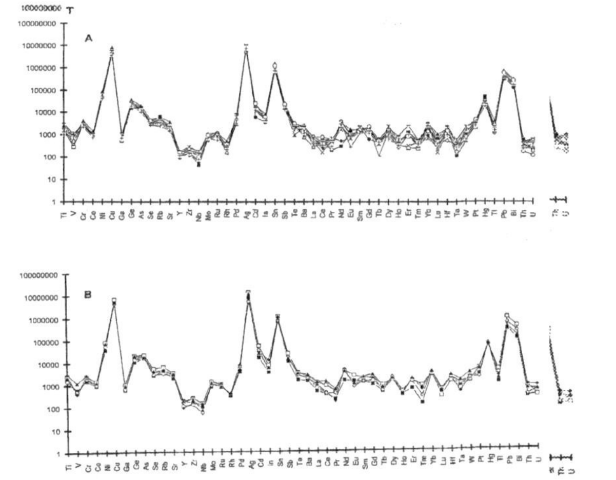

Figure 4: Comparative raw count data plots for thirteen samples of gold from the Bracks Farm Torc (A), and five gold samples from the separate Torc flange (B).

Comparison of the spectral pattern obtained from the Bracks Farm Torc (Taylor 1980), (FIGURES 4A and B) and from the sleeve and dress ornament from County Antrim (FIGURES 5A and B) demonstrates the remarkably close association of spectral patterns of the fingerprint within a specific artefact. These data would logically point to a single source of the gold for the Bracks Farm Torc and a single, although different, source of gold for the two County Antrim artefacts. Comparison of spectral patterns from these three different artefacts (FIGURES 4 and 5) confirms that, when different sources of gold are used for fabrication, then different spectral patterns will result. This observation provides a significant pointer to the potential ability of the technique of gold fingerprinting to relate the artefact gold back to a specific source/gold batch.

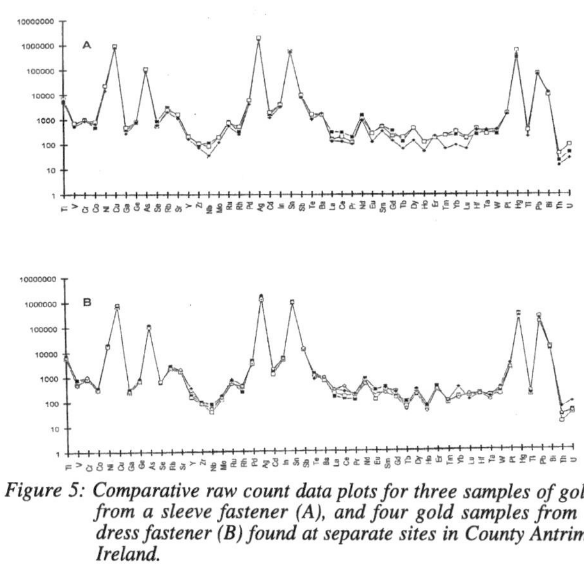

Figure 5: Comparative raw count data plots for three samples of gold from a sleeve fastener (A), and four gold samples from a dress fastener (B) found at separate sites in County Antrim, Ireland.

Certain elements represent inclusions of non-solid solution material. In particular significant rare earth element patterns would strongly indicate the presence of lithogenic material. This material is likely to be incorporated as discrete mineral grains. The relative concentrations of these elements can form the basis of identification of rock types associated with either the area of production or area of ore body emplacement. In either event the presence of such suites of elements as the rare earths can prove to be an added bonus in the discrimination or association of material.

Comparison of Ternary Inter-Element Association Plots
Ternary plots represent the direct comparison of the relationship between three components in a system. Positioning a data point anywhere within the ternary plot indicates the relative percentile inter-association of the three components defined by the diagram.

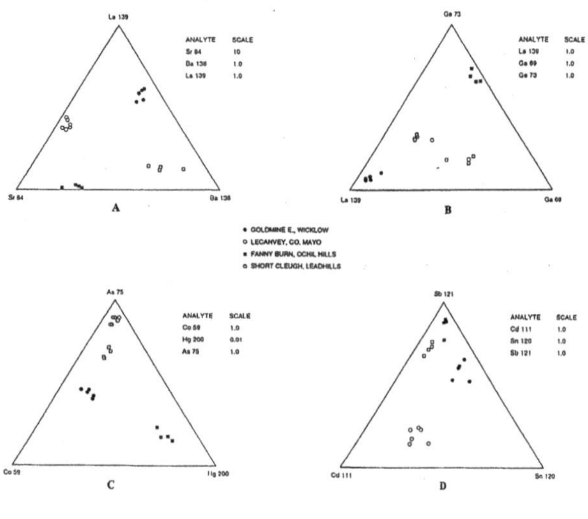

Figure 6: Comparative ternary plots for alluvial gold samples from four different sites in the United Kingdom and Ireland.

It is also possible to increase or decrease the value of a component by a specific binary amount. Within the analyte scale box on the diagram, unity indicates that the data have not been manipulated, while any other value indicates the degree of manipulation of the raw count data. Hence it becomes possible to plot ternary associations of analytes that differ. Lanthanum-Sr-Ba associations (FIGURE 6A) probably indicate variations in the relative trace element assemblage of specific gangue minerals, held as inclusions within the gold. These inclusions could come from either the site of production of an artefact or, in the case of natural placer gold, from the site of ore emplacement. Providing these inclusions have been observed and recorded during the sample preparation stage, it is possible to use these sorts of element associations for discrimination purposes. Ge-La-Ga associations (FIGURE 6B), represent a contribution of sulfide mineral phase inclusion with gangue minerals, while As-Co-Hg associations (Fig. 6C), and associations between Sb-Cd-Sn (FIGURE 6D) represent inter-metallic compositional changes in the gold itself and are extremely useful in provenance identification. The close groupings of data within these plots indicate the strong inter-element associations exhibited within gold from a single mineralising event, while facilitating identification and isolation of gold from specific sites.

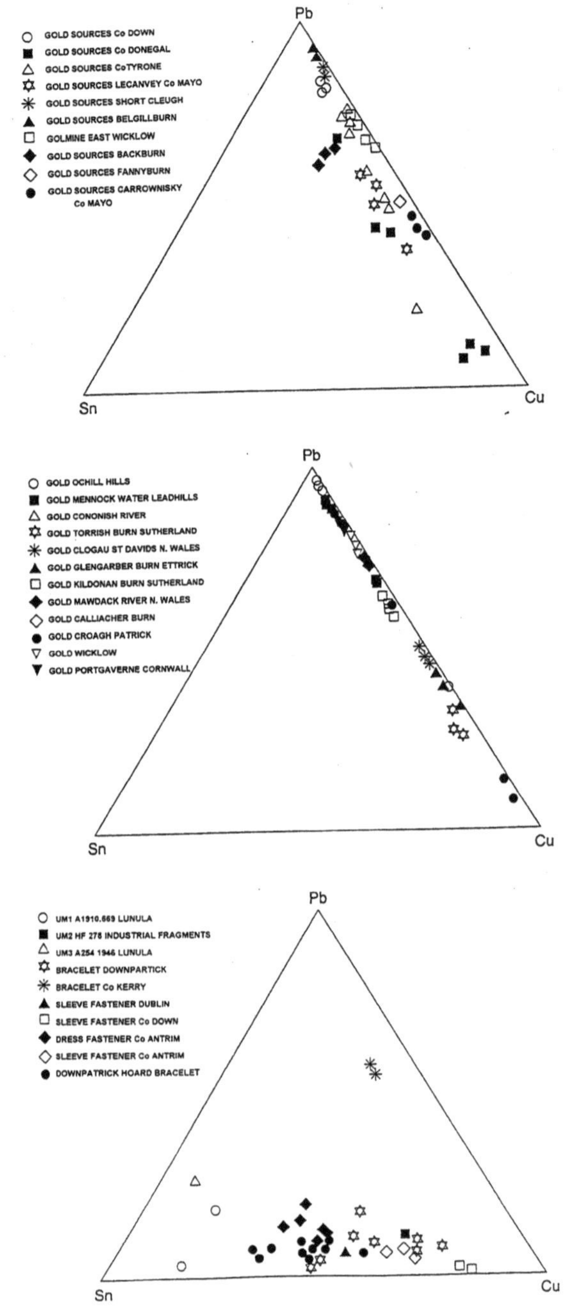

Figure 7: Comparative ternary plots (Pb-Sn-Cu) for samples from twenty-two alluvial gold sites and ten artefacts from the United Kingdom and Ireland.

It is relatively easy to discern differences between various gold samples and gold from different areas when only a small number of data points are used. However, when comparing data from a larger database results can be difficult to interpret. Data from twenty-two alluvial gold sites in the United Kingdom and Ireland and ten artefacts are shown in FIGURES 7-9. Replicate analyses, plotted with respect to the association of Pb-Sn-Cu (FIGURE 7 top and middle), indicate that there is a reasonable spread of data from the different areas with respect to these elements. However, clustering of data for a specific deposit is also apparent and relatively coherent, although some spread of data is present. This is to be expected, as all three elements are common in some gold deposits. Ternary plots from the ten artefacts are shown in FIGURE 7 bottom. Here, data are vectored away from the lead apex, indicating either an increase in both tin and copper with respect to lead or a decrease in lead with respect to tin and copper in the artefact material. Reference to the raw count data confirms the former of these two possibilities and points to a possible addition of the two elements during manufacture. A notable exception is the County Kerry bracelet, which appears to have a significantly higher lead concentration, with respect to tin and copper, than any of the other artefacts. However, it is difficult to reconcile the artefact to an area of gold production with reference to this data inter-relationship, especially if the addition of two elemental components of the ternary plot is a possibility.

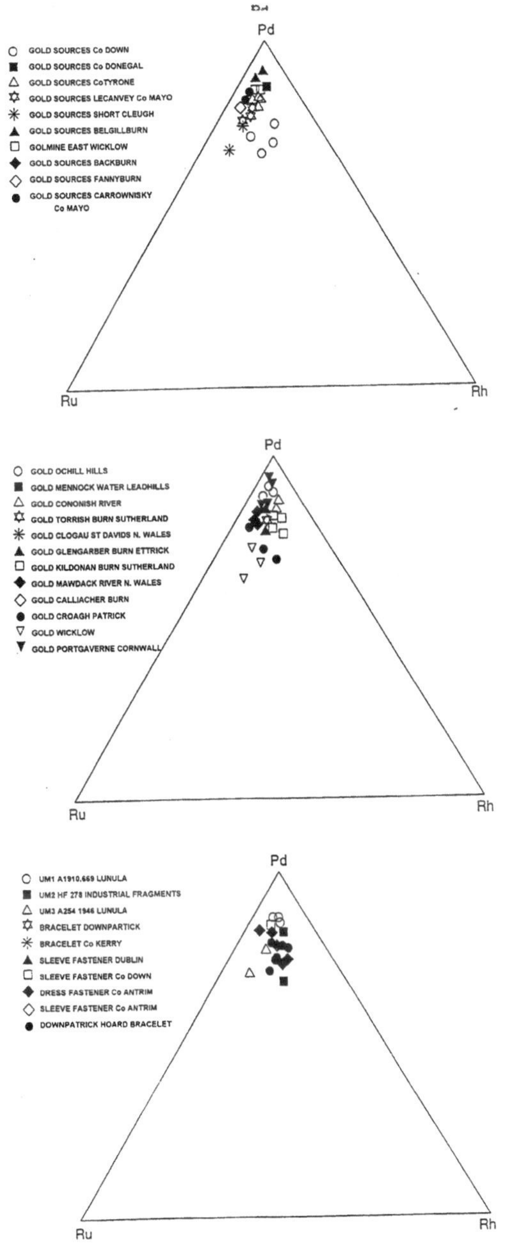

Figure 8: Comparative ternary plots (Pd-Pt-Sn) for samples from twenty-two alluvial gold sites and ten artefacts from the United Kingdom and Ireland.

If tin is plotted against platinum and palladium (FIGURE 8), a significant trend is observed. Specific deposits tend to vector away from each other and form sub-groups, possibly giving an indication of the style of mineralization and of the type of gold deposit. Very obvious in this respect are the Co. Donegal and Carrownisky, Co. Mayo, golds. Carrownisky, Co. Mayo, may be relatively important, as it does seem to have elevated tin levels with respect to the other two elements. However, there appears to be a relatively uniform association between palladium and platinum which is common to Scottish, Welsh and Irish gold but with the relatively higher tin golds occurring predominantly in the Scottish and Welsh sources. An equivalent ternary plot for the artefact material indicates that the association between palladium and platinum is still apparent. This is to be expected, as it would be unlikely that selective refining of these two elements could have been an option at this time. However, the relative increase in tin in the artefact material is very clearly shown and the anomalous position of the Kerry bracelet again confirmed.

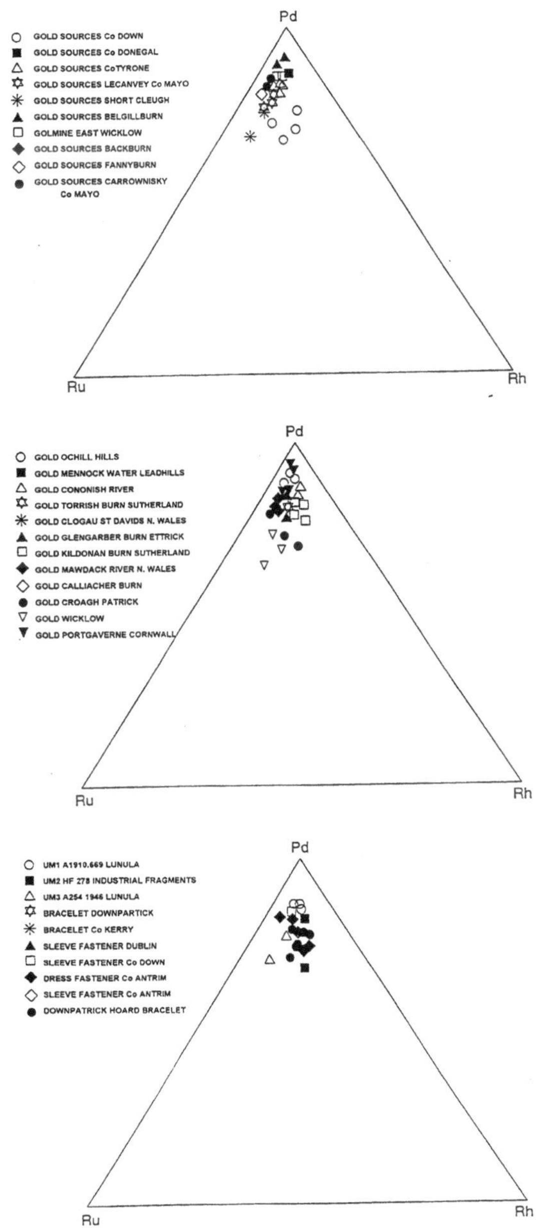

Figure 9: Comparative ternary plots (Pd-Ru-Rh) for samples from twenty-two alluvial gold sites and ten artefacts from the United Kingdom and Ireland

The inter-relationship between Pd-Rh-Ru in samples of alluvial gold (FIGURE 9) exhibits a relatively close association of all data points and even a commonality of data for artefact material when compared to natural gold. The specific addition of any of these elements during the manufacture of an artefact from this period is not feasible, for a variety of both mineralogical and metallurgical reasons. These elements all form solid solution products with gold and will not be volatilized or lost during manufacture of an artefact. Because of this, it would be expected that their ternary association, in both the ore and in the artefact resulting from processing the ore, should be identical. Reference to FIGURE 9 bottom confirms this, with all artefact material plotting in a relatively confined area towards the palladium apex of the ternary diagram. An interesting aspect is that close association with the palladium apex is not achieved by any of the artefact material. A fact possibly implying that the artefact material may not be derived from such relatively high palladium golds as those of Belgillburn, Ochill Hills or Portgaverne. In addition, the relatively high rhodium golds of the County Down source would also seem to be precluded from being source material for the artefacts.

From the use of ternary plots, it is apparent that associations and lack of associations between elements can be identified and that these associations can form the basis of a provenance identification protocol. However, the absolute number of associations, required to begin to identify a specific source for an artefact, is extremely large and the time taken to short-list potential sources very significant. Because of this, it is necessary to have some form of screening to identify quickly potential sources.

Database Search/Match Comparison

Establishment of a database search/match protocol has centred around quickly identifying differences between samples, using the raw count data for the isotopes rather than establishing difference between the absolute concentration of the elements. The pre-screening software, developed for this project, facilitates summing the differences between a search reference sample and all other samples in the database. The search reference sample data are also part of the database, as the search and match criteria used must achieve 100% fit to this sample's original database entry and, thereby, establish a datum point for relative comparison of all other comparability indices. The sum of the differences between raw count data for the individual isotopes of all samples from the database and the reference sample is then rationalized with respect to the number of isotopes used. A comparability index value, as a percentage, is then assigned to the relative comparison of each sample in the database to the reference sample. The percentile figure is not absolute and is based only on an assumption that the best comparability must be 100% (this is the comparability of the search reference sample to its original entry in the database) and 0% for the worse comparability of a sample to the reference. There are some five hundred samples in the database. While the comparability indices are not absolute, as a screening technique this method works extremely well. Raw count data of the highest ten comparable samples are then subjected to additional comparison using the first three spectral interpretation protocols described previously. After this examination, a subjective decision is made as to the acceptability of potential comparability between samples in the database. A plot of the database search/match fit, with respect to percentile comparability, also indicates where slope breaks in the relationship occur (FIGURE 10). At these points, significant degradation in potential inter-relationship occurs and below these points it is unlikely that a reference sample will find a match in the database. Application of the same search/match protocol to artefact material produces similar results, with slope breaks occurring at positions beyond which it is unlikely to find an acceptable fit of the reference sample to any in the database (FIGURE 11).

Conclusions

Preliminary investigations into the use of LA-ICP-MS for the analysis of gold ores and artefacts from Scotland, Wales and Ireland indicate that the technique provides reproducible data with respect to relative elemental isotopic distribution in samples. The data are not quantitative; however, the means of interpreting data relies not on the absolute value of the element concentration in the sample but on the inter-element association pattern of the sample. An advantage of the technique is that it can determine the trace element distribution in samples as small as 30μm in diameter, thus minimising apparent damage of precious artefacts and facilitating the use of extremely small samples of ore material, which in many cases is all that is available.

Comparison of inter-element association patterns in alluvial gold and artefact material are seen as the basis of eventually establishing artefact provenance and initial investigations have given credibility to this approach. A pre-screening computer program has been developed, to provide a first pass comparability index between a search reference sample and all other samples in the database. This approach facilitates screening of the database, to identify the most comparable samples to the search reference material under investigation. In this way, problems associated with comparing data with an ever-increasing database have been largely overcome. This has its major application in the comparison of similar materials with each other (artefact with artefact; natural gold with natural gold). However, it can also be used to obtain the best fit of artefact to natural gold, as a precursor to more detailed ternary plot interpretation based on selected relevant analytes.

While absolute provenance identification for a specific artefact has not yet been achieved, the basis for this procedure has been identified. Future work will concentrate on establishing inter-element association patterns that will facilitate ore field identification. Artefact provenancing to an ore field with ultimate identification of specific ore bodies within the field, potentially relating at least early unmodified gold objects back to a specific ore body and mineralising event, should then follow.

SAMPLE	% COMPARABILITY	SAMPLE	% COMPARABILITY
1/11/96 KILDONAN BURN SUTHERLAND	100.00	1/11/96 OCHILL HILLS	92.76
1/11/96 KILDONAN BURN SUTHERLAND	100.00	16/6/95 SAMPLE 9	92.64
1/11/96 KILDONAN BURN SUTHERLAND	98.17	16/6/95 SAMPLE 10	92.64
1/11/96 KILDONAN BURN SUTHERLAND	97.89	4/4/97 MINE WEST WICKLOW	92.41
1/11/96 KILDONAN BURN SUTHERLAND	97.71	4/4/97 RIVER TWEED TWEEDSMUIR	92.28
1/11/96 KILDONAN BURN SUTHERLAND	97.43	1/11/96 GLENGABER BURN ETTRICK	92.20
1/11/96 KILDONAN BURN SUTHERLAND	97.18	8/5/97 ATTAGH H2	91.98
1/11/96 CALLIACHER BURN SCOTLAND	95.78	27/10/95 SAMPLE BLOCK 3 4 2 UM 12	91.87
1/11/96 MENNOCK WATER LEADHILLS	95.53	1/11/96 OCHILL HILLS	91.74
1/11/96 CALLIACHER BURN SCOTLAND	95.34	27/10/95 SAMPLE UM 21 2	91.73
4/4/97 CARROWNISKY RIVER	95.27	4/4/97 SUISGILL BURN 1	91.71
1/11/96 CALLIACHER BURN SCOTLAND	95.17	27/10/95 SAMPLE BLOCK 3 4 1 UM 12	91.48
4/4/97 BELGHILL BURN	94.98	16/6/95 SAMPLE BELGILLBURN 38	91.40
4/4/97 UPPER MENNOCK LEADHILLS	94.84	1/11/96 TORRISH BURN SUTHERLAND	91.28
1/11/96 GLENGABER BURN ETTRICK	94.70	WICKLOW	91.25
1/11/96 GLENGABER BURN ETTRICK	94.64	4/4/97 CROAGH PATRICK HARD ROCK	91.19
1/11/96 MENNOCK WATER LEADHILLS	94.28	WICKLOW	91.07
1/11/96 TORRISH BURN SUTHERLAND	94.25	1/11/96 TORRISH BURN SUTHERLAND	90.99
4/4/97 ALLUVIAL LECHANVEY	94.24	16/6/95 SHORT CLEUGH 16	90.86
1/11/96 MENNOCK WATER LEADHILLS	94.05	16/6/95 SAMPLE FANNY BURN 56	90.85
4/4/97 BACK BURN	93.95	16/6/95 SAMPLE FANNY BURN 55	90.80
1/11/96 CONONISH RIVER	93.93	27/10/95 SAMPLE UM 21 1	90.77
1/11/96 CONONISH RIVER	93.55	16/6/95 UM3 A254.1946	90.75
16/6/95 SAMPLE CARROWNISKY 59	93.42	WICKLOW	90.61
1/11/96 GLENGABER BURN ETTRICK	93.40	8/5/97 ATTAGH H6	90.55
1/11/96 OCHILL HILLS	93.32	1/11/96 WINGATE BURN LEADHILLS	90.52
16/6/95 SAMPLE CARROWNISKY 61	93.02	27/10/95 SAMPLE UM 21 3	90.49
4/4/97 RIVER COTLI TOM TAYLOR?	92.96	16/6/95 SAMPLE FANNY BURN 57	90.37
16/6/95 LECKANVEY MAYO (8) (L SERIES)	92.93	WICKLOW	90.27
1/11/96 HAWKSHAW BURN TWEEDSMUIR	92.87	WICKLOW	90.24
16/6/95 SAMPLE CARROWNISKY 60	92.79	27/10/95 SAMPLE UM 18 1	90.21

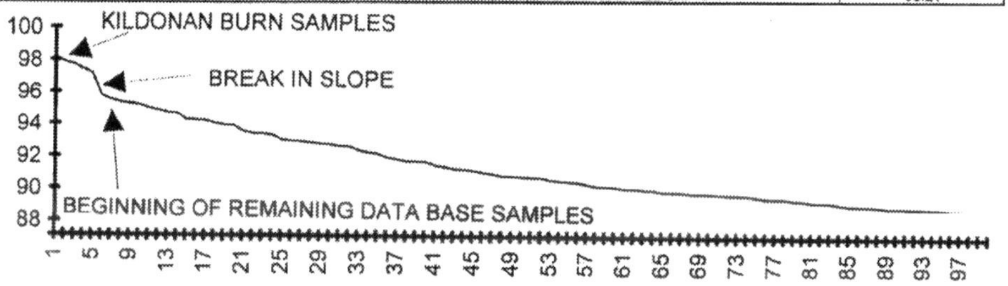

Figure 10: Database search/match comparability fits for Kildonan Burn (alluvial) gold samples showing percent comparability to the best match site samples in the database and relative position of the break in slope.

SAMPLE	% COMPARABILITY	SAMPLE	% COMPARABILITY
16/6/95 BRACKS FARM TORC 18	100.00	13/2/96 GOLD SAMPLE UM 39	92.91
16/6/95 BRACKS FARM TORC 18	100.00	13/2/96 GOLD SAMPLE UM 33	92.77
16/6/95 BRACKS FARM TORC 20	98.61	13/2/96 GOLD SAMPLE UM 34	92.40
16/6/95 BRACKS FARM TORC 21	98.29	27/10/95 GOLD SAMPLE UM 26 3	92.33
16/6/95 BRACKS FARM TORC 19	98.22	27/10/95 GOLD SAMPLE UM 24 1	92.32
16/6/95 BRACKS FARM TORC 23	98.03	27/10/95 GOLD SAMPLE UM 26 1	92.20
16/6/95 BRACKS FARM TORC 26	98.02	27/10/95 GOLD SAMPLE UM 31 2	92.19
16/6/95 BRACKS FARM TORC 28	97.94	LUNULA BANEMORE Co KERRY	92.17
16/6/95 BRACKS FARM FLANGE 4	97.87	27/10/95 GOLD SAMPLE UM 31 1	91.77
16/6/95 BRACKS FARM TORC 24	97.85	LUNULA ROSGARRON Co DERRY	91.77
16/6/95 BRACKS FARM FLANGE 5	97.84	27/10/95 GOLD SAMPLE UM 26 2	91.70
16/6/95 BRACKS FARM TORC 25	97.82	27/10/95 GOLD SAMPLE UM 25 1	91.66
16/6/95 BRACKS FARM FLANGE 2	97.61	13/2/96 GOLD SAMPLE UM 37	91.64
16/6/95 BRACKS FARM TORC 29	97.48	16/6/95 SAMPLE VIKING RING 39	91.41
16/6/95 BRACKS FARM TORC 30	97.46	13/2/96 GOLD SAMPLE UM 38	91.25
16/6/95 BRACKS FARM TORC 27	97.43	27/10/95 GOLD SAMPLE UM 30 3	91.04
16/6/95 BRACKS FARM FLANGE 3	97.14	13/2/96 GOLD SAMPLE UM 36	90.85
16/6/95 BRACKS FARM TORC 22	97.10	27/10/95 GOLD SAMPLE UM 29 1	90.67
16/6/95 BRACKS FARM FLANGE 6	96.76	27/10/95 GOLD SAMPLE UM 24 3	90.66
16/6/95 SAMPLE VIKING RING 44	94.92	13/2/96 GOLD SAMPLE UM 35	90.49
16/6/95 SAMPLE VIKING RING 40	93.95	LUNULA DUNFIERTH Co KILDAIRE	90.36
16/6/95 SAMPLE VIKING RING 41	93.93	27/10/95 GOLD SAMPLE UM 28 2	89.98
16/6/95 SAMPLE VIKING RING 45	93.80	27/10/95 GOLD SAMPLE UM 30 4	89.90
16/6/95 SAMPLE VIKING RING 42	93.63	27/10/95 GOLD SAMPLE UM 28 1	89.86
13/2/96 GOLD SAMPLE UM 43	93.45	27/10/95 GOLD SAMPLE UM 30 1	89.82
16/6/95 SAMPLE VIKING RING 43	93.44	13/2/96 GOLD SAMPLE UM 32	89.70
13/2/96 GOLD SAMPLE UM 41	93.37	27/10/95 GOLD SAMPLE UM 29 2	89.44
13/2/96 GOLD SAMPLE UM 42	93.13	GOLD SAMPLE BLOCK 3 3 UM 11	89.40
13/2/96 GOLD SAMPLE UM 40	93.07	27/10/95 GOLD SAMPLE UM 22 3	89.30

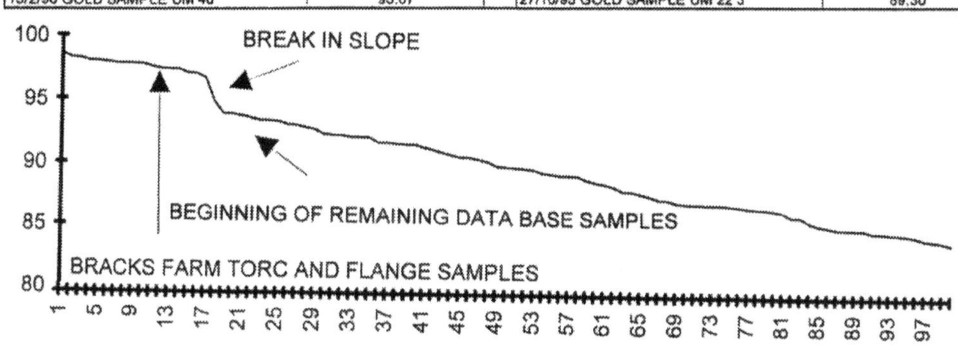

Figure 11: Database search/match comparability fits for Bracks Farm Torc (artefact) gold samples showing percent comparability to the best match sixty samples in the database and relative position of the break in slope.

References

BROADHEAD, M., R. BROADHEAD & J.W.HAGER. 1990. Laser sampling ICP-MS: semiquantitative determination of sixty-six elements in geological samples. *Atomic Spectroscopy* 11: 205- 209.

CHENERY, S., A. HUNT & M. THOMPSON. 1992. Laser ablation of minerals and chemical differentiation of the ejecta. *Journal of Analytical Atomic Spectrometry* 7: 647 - 652.

CRAIN, J. S. & D.L. GALLIMORE. 1992. Determination of trace impurities in uranium oxides by laser ablation inductively coupled plasma mass spectrometry. *Journal of Analytical Atomic Spectrometry* 7: 605 - 610.

HERBERT, H.K. 1987. Minor element compositions of sphalerite and pyrite as petrogenic indicators. Pacific Rim Congress 87, *Australian Institute of Mining and Metallurgy* 831- 841.

HERBERT, H.K. & R.J. WATLING. 1996. Gold fingerprinting and analysis. *Australian Criminal Intelligence Digest,* January 1996, 39-42.

LEAKE, R.C., D.J.BLAND & C.COOPER. 1993. Source characterisation of alluvial gold from mineral inclusions and internal composition variation. *Transactions of the Institute of Mining and Metallurgy, 102B:* 65-82.

LEAKE, R.C. & M.T. STYLES. 1996. Trace metals in native gold. Advances in gold processing technology. Institution of Mining and Metallurgy Conference. London.

LEAKE, R.C., R.J.CHAPMAN, D.J. BLAND, E.CONDLIFFE & M.T.STYLES. 1997. Microchemical characterisation of alluvial gold from Scotland. *Transactions of the Institute of Mining and Metallurgy, 106B:* 85-98.

LEAKE, R.C., R.J.CHAPMAN, D.J. BLAND, P. STONE, D.G. CAMERON & M.T. STYLES. In press. The origin of alluvial gold from the Leadhills region of Southern Scotland: evidence from internal characteristics. *Journal of Geochemical Exploration.*

PEARCE, N.J.G., W.T.PERKINS, I. ABELL, G.A.T. DULLER & R. FUGE. 1992. Mineral microanalysis by laser ablation inductively coupled plasma mass spectrometry. *Journal of Analytical Atomic Spectrometry* 7: 53.

PERKINS, W.T., R.FUGE & N.J.G.PEARCE. 1991. Quantitative analysis of trace elements in carbonates using laser ablation inductively coupled plasma mass spectrometry. *Journal of Analytical Atomic Spectrometry* 6: 445 - 449.

PERKINS, W.T., N.J.G.PEARCE & R. FUGE. 1992. The analysis of zircon by laser ablation and solution inductively coupled plasma mass spectrometry. *Journal of Analytical Atomic Spectrometry* 7: 611 - 616.

PERKINS, W.T., N.J.G.PEARCE, & T.E. JEFFRIES. 1993. Laser ablation ICP-MS: a new technique for the determination of trace and ultra-trace elements in silicates. *Geochimica et Cosmochimica Acta* 57: 475 - 482.

TAYLOR, J.J. 1980. *Bronze Age goldwork of the British Isles.* Cambridge University Press.

TAYLOR, J.J., R.J.WATLING, C.A.SHELL, R.A.F.IXER, R.J.CHAPMAN, R.B.WARNER & M.CAHILL. 1997. From gold ores to artefacts in the British Isles: a preliminary study of a new LA-ICP-MS approach. In A. J. Sinclair, E. A. Slater & J. Gowett (ed): *Archaeological Sciences 1995: Proceedings of a conference on the application of scientific techniques to the study of archaeology.* Oxbow Monograph 64, Liverpool. 144-154.

VAN DE WEIJER, P., L.M.BAETEN, M.H.J. BEKKERS & P.J.M. VULLINGS. 1992. Fast semiquantitative survey analysis of solids by laser ablation inductively coupled plasma mass spectrometry. *Journal of Analytical Atomic Spectrometry* 7: 599 - 603.

WATLING, R..J., H.K. HERBERT, D.DELEV & I.ABELL. 1994. Gold fingerprinting by Laser Ablation Inductively Coupled Plasma Mass Spectrometry. *Spectrochimica Acta* 9B: 205-219.

WATLING, R. J. & H.K.HERBERT. 1995a. Forensic gold analysis by laser ablation. Inductively Coupled Plasma Mass Spectrometry. *Randol Gold Forum, 1995.* Perth, Australia, 179-190.

WATLING, R. J. & H.K. HERBERT. 1995b. Gold fingerprinting: Tracing the origins of gold samples and artefacts. *Australian Science Teachers Journal* 41: 11-19.

WATLING, R. J., H.K.HERBERT, I.S.BARROW & A.G. THOMAS. 1995a. Analysis of Diamonds and Indicator Minerals for Diamond Exploration by Laser Ablation-Inductively Coupled Plasma Mass Spectrometry. *Analyst* 120: 1357-1364.

WATLING, R. J., H.K.HERBERT & I.D. ABELL. 1995b. The application of Laser Ablation-Inductively Coupled Plasma-Mass Spectrometry (LA-ICP-MS) to the analysis of selected sulphide minerals. *Chemical Geology* 124: 67-81.

WATLING, R. J. 1996. Novel Applications of Laser Ablation Inductively Coupled Plasma Mass Spectrometry in Inorganic Analytical Chemistry. *Rapid Communications in Mass Spectrometry* 10: 130-137.

WATLING, R. J., B.LYNCH & D. HERRING. 1997. Use of Laser Ablation-Inductively Coupled Plasma-Mass Spectrometry (LA-ICP-MS) for fingerprinting Scene of Crime Evidence. *Journal of Analytical Atomic Spectrometry* 12: 199-203.

7

The early development of metallurgy in the north-west of the Iberian peninsula

BEATRIZ COMENDADOR REY

Area de Prehistoria. Departamento de Historia, Facultad de Geografía e Historia. Universidad de Compostela, Pza. da Universidade s/n. 15703. Santiago de Compostela. Spain

This paper discusses the metallurgy of the North-west Iberian Peninsula (NIP) (Galicia and Asturias in Spain and the north of Portugal) during the long interval from the end of the 4th millennium until mid-way through the 2nd millennium BC, the Chalcolithic and Early through Middle Bronze Age in conventional terminology. The lack of defined contexts for early metalwork and of absolute dates is such that only a partial understanding of these, first phases, of metallurgy has been possible in the past. This has generally been based on the analysis of typology and metal manufacturing techniques, largely reliant on finds from funerary contexts. The traditional interpretation proposes a linear model of evolution with successive phases of metallurgical development, articulated around the Beaker Phenomenon, which is considered to be responsible for, or the promoter of, the widespread use of metal. This investigation, 'The Early Development of metallurgy in the North-west of the Iberian Peninsula' (Comendador 1998) allows a substantial revision of this interpretation on the basis of a wider review of the available literature. The different metallurgical phases argued for in the traditional model cannot be supported, at least at present, by technological or archaeological evidence. A new interpretative model is therefore suggested.

Keywords: Europe; Iberian Peninsula; Metalwork; Chalcolithic; Early Bronze Age; Beaker Phenomenon.

Introduction: The traditional interpretation of the beginnings of metallurgy in the NIP

Typological study of metal artefacts was fundamental to the development of the chronology of the Bronze Age in the NIP because the region lacked finds from clear contexts with reliable absolute dates. Thus an interpretative model of the first metallurgical production in the NIP was developed, the *traditional interpretation*, based on the typology of metalwork from isolated finds and funerary contexts. This, together with the lack of study of manufacturing processes, has resulted in an incomplete interpretation with an over emphasis on grave goods, especially those connected to the 'Beaker phenomenon'. The Beaker phenomenon has been considered as promoting the widespread use of metal and it is unsurprising therefore that a theoretical chronological sequence has been articulated around it. The development of early metallurgy is defined as a process of long duration beginning with a pre-Beaker phase around the end of the fourth millennium and early third millennium BC. According to the model, metallurgy was rare in this pre-Beaker phase, but took off in the late Beaker period and Early Bronze Age. Metal production is thought to have reached a peak of development in the post-Beaker phase persisting into the Middle Bronze Age c.1900-1600 BC.

In relation to copper technology, the introduction of the Beaker phenomenon is considered to herald a general improvement in the process of elaboration, including the adoption of innovations such as the alloying of copper and arsenic, as well as a short term increase of production. These characteristics, together with the new artifact types or forms, characterize the so-called *Beaker Metallurgical Tradition*. This sequence supposes technological improvement to follow a linear evolutionary model, which has its culmination in the introduction of the alloy of copper and tin in the Middle Bronze Age. The scheme is also applied to the production of gold, which is considered to have evolved from simple cold-working in the Chalcolithic to melting and casting at the end of the Early Bronze Age. The scarce xamples of silver work are considered to be imports from the south-east of the peninsula (Almagro 1996).

In typological terms, rudimentary metallic objects of small dimensions are usually attributed to the first phase. The Beaker phenomenon is associated with characteristic metal types such as tanged daggers and Palmela-type points which were long-lived and widely adopted. Only on the brink of the Middle Bronze Age are new types like riveted daggers introduced. An example which illustrates this is the hoard of Roufeiro (Nocelo da Pena, Sarriaus, Ourense), which Harbison (1967) interpreted as characteristic of the juxtaposition of two different metallurgical traditions at the end of the Early Bronze Age: *the Atlantic Beaker tradition*, represented by the tanged dagger, and the *Argaric southern tradition*, represented by various riveted daggers. The introduction of bronze, the alloy of copper and tin, is attributed to the Middle Bronze Age, with the gradual appearance of relatively low-tin alloys in certain new types of axes (Ruiz-Galvez 1984). Their appearance is considered to have been stimulated from the south-east of the Iberian Peninsula.

Revision of the traditional interpretation of the early development of metallurgy in the NIP.

The last few years have seen a considerable revision of data relating to the earliest metal production in the NIP (Galicia and Asturias-Spain-, and north of Portugal) (Comendador 1998) in which particular attention has been paid, not only to the finished objects, but to the remains of the process of manufacture found in domestic contexts. This has involved an analytical study (XRF and metallography) integrated within the program *Archeometallurgy of the Iberian*

Peninsula (D.G.I.C.Y T.PB92, N°0351). To date, 91 objects of the 378 collected in this study (Rovira *et al.*1997) have been analyzed. The greater part of the production was of copper, although gold objects were also a high proportion of the total; in Galicia they constitute 32% of the total and in the north of Portugal 18.5%.

In relation to pre-historic mining in the NIP, Asturias and León (Spain) are important mining centres, within which the most striking examples are the Asturian copper mines of the Aramo (Cangas de Onís) and the Milagro (Riosa). Both have absolute dates within the 3d millennium BC (Blas 1996). These dates parallel those of the earliest phases of mining at Ross Island (Co. Kerry, Ireland), one of the most ancient mines in north-western Europe, itself connected with Bell Beaker pottery (O'Brien 1995). No indications of pre-historic copper mining in Galicia or the north of Portugal have been found, but copper resources are located in areas 80-100km from settlements where metallurgical activity has been documented (Comendador 1998). Gold mining is considered to have involved the exploitation of placers and unlikely to have resulted in recognisable archaeological features.

Data relating to the process of manufacture and production within domestic contexts are very scarce. In the later levels of Buraco da Pala (Mirandela, Trás-os-Montes) a blow pipe and some metallic objects have been recovered. These are associated with four radiocarbon determinations that date the site in the second half of the 3d millennium BC (Sanches 1997). Two other Galician settlement sites have produced fragments of crucibles, although their dates, within the 3d millennium BC, are problematic due to their isolation and relatively large errors: O Fixón (Cangas, Pontevedra) GaK 11188: 3930±130 BP and Lavapés (Cangas, Pontevedra) GaK-11188: 3930±120 BP (Comendador 1998).

In relation to metal manufacture, the production of copper objects appears to have been strongly conservative, limited to a repertoire of approximately twelve or thirteen generic forms. With the Beaker phenomenon two new types appear: the tanged dagger and the Palmela-type point. These are introduced into funerary contexts without a substantial variation of the other, pre-existing, forms. Regional differences can be perceived in the production and circulation of metal objects; for instance, the introduction of certain elements typical of Bell Beaker grave goods is more pronounced in Galicia (except eastern Galicia) and in the north-west of Portugal than elsewhere.

One of the most interesting results of the investigation is the technological interpretation of the copper production process. This generally consisted of smelting and melting in poorly controlled conditions and was characterized by an absence of alloys and incomplete finishing treatments, when these were applied at all. The traditional interpretation, of a chronological framework articulated around the Beaker phenomenon, appears to be unproven as it is impossible to argue for distinct phases based on the manufacturing processes of objects. In general terms it is not possible to discern a metallurgical Beaker tradition at all, and in fact, it is difficult technologically to distinguish the early metalwork of the NIP from that of other areas of the Peninsula as the metal work presents very homogeneous characteristics throughout.

This conservatism in copper production contrasts with the diversity of forms of jewellery. As early as the mid-3rd millennium BC two systems for working in gold appear to have co-existed. This is shown by the settlement site of Buraco da Pala (Mirandela, Portugal) where gold sheets and several solid gold beads obtained by melting were found (Cavaheiro and Sanches 1996). The exceptional case of the hoard known as the Treasure of Caldas 1 (Pontevedra), which is attributed to the final phase of the early Bronze Age (Ruiz-Galvez 1978) and contains gold objects produced with different techniques (in sheets or melted), is more uncertain. The latest revisions (Ambruster 1996), show that the vases belonging to the hoard were elaborated using the lost wax technique. This is not documented in other areas of the Peninsula till the Late Bronze Age. This suggests either a reconsideration of the chronology of the hoard, or of the dating for the first use of the technique in the NIP.

Another issue of great interest is that of silver. It is rarely used in European settlement sites of the Beaker type. It begins to be documented with relative frequency at the start of the Bronze Age, in the Iberian Peninsula. In the graves of the Argar Culture, in the south-east of the Iberian Peninsula, over 300 silver objects were found, but no objects have been found which could be dated with security to the Chalcolithic. The oldest date belongs to a grave in Herrerías (Cuevas de Almanzora, Almería, Spain) CSIC-3670±70 BP (2270-1790 BC), that contemporary with the Bronze Age of the Armorican mounds (Montero *et al.* 1995: 98). In Europe there are known objects of imprecise chronology but related to the Beaker phenomenon, that seem to indicate the use of this metal at earlier dates than in the Peninsula. They are stylistically superior to those of the Peninsular, therefore could have come from very different metallurgical traditions (Montero *et al.* 1995:98).

In the NIP seven silver springs have been documented. These were found in funerary contexts as well as a group (with a weight less than 400g) from an isolated findspot. The Galician graves where they have appeared have no absolute dates, though in Atios (Porriño, Pontevedra), two springs appeared associated with two tanged daggers and two gold sheet cylinders inside a cist (Comendador 1998). In Portugal springs have been found in two monumental mounds of dates that are hard to relate to these objects: in Outeiro de Gregos 1 (Baião, Porto) they would mark a limit *ante quem* between 2140 and 1520 BC, and in Meninas do Crasto 4 (Baião, Porto) a limit *post quem* between 3308 and 2046 BC (Jorge et al 1988). One of the springs could be connected with the Bell Beaker pottery in the mound of Châo do Brinco 1 (Viseu, Beira Alta) (Silva 1995). In the female burial at Santioste (Villafáfila, Zamora), in the Spanish north plateau the grave-goods included a bracelet made of bone beads, a necklace of the same material with three small semispheric silver capsules, a button perforated in a V shape and possibly a wristguard, dated *post quem* by two dates between 2460 and 1920 BC (Beta-50709: 3780±80 BP, Beta-50710: 3750±80 BP) (Delibes 1993). These dates make it possible consider silver production in the NIP as being of a similar, if not earlier, date than that in the south-east of the peninsula.

This revision also indicates that tin bronze appeared before the Middle Bronze Age. Two awls found in the island settlement of Guidoiro Areoso (Illa de Arousa, Pontevedra) are high-tin bronzes containing more than 17% tin. Their fabrication technology is unusual for objects of similar chronology in that they were annealed to eliminate brittleness within the metal and obtain a homogenous structure. The dates for associated objects place them in the mid-3rd millennium BC making them exceptionally early bronze objects in the Iberian Peninsula (Comendador 1998). Although the site is still in the process of being studied, the materials show the presence of the bronze alloy from an early date. In the neighbouring coastal area several bronze drops were collected from the Beaker settlement site of O Fixon-A Costa da Seixeira, dated chrono-typologically to 1700-1500 BC (Suárez 1995:60).

Deposition mechanisms

Most early metal objects in the NIP are isolated finds, the majority of the remainder are from funerary contexts. The few hoards contain a greater number of objects than the grave sites. The preserved part of the hoard of Caldas 1 for instance contains 14.9kg of gold, while the gold found in all of the burial sites in the whole NIP amounts to only 2.75kg. This observation holds true even if isolated finds and other finds with badly defined contexts, that could also be isolated finds, are included. The contexts that have provided the least material are those of domestic type, where both remains of the production process and finished objects have been found.

The correlation between certain metal artefact forms and particular contexts suggest a pattern of selective deposition. Flat axes, for example, are often found as isolated finds as well as in hoards, often with equally large numbers of tanged daggers. In funerary contexts however, metal objects tend to appear together with non-metallic materials (pottery, lithics etc.). In the hoards the only non-metals that occasionally appear are ceramic vases used to contain the metalwork.

Harrison (1974a, 1974b) proposed a post/epi-Beaker horizon called *Montelavar Group*, in western Iberia that has been associated with the beginning of the Early Bronze Age in the NIP. He defines the group in terms of an association of two or more Palmela-type points with a tanged dagger in burial sites that are apparently individual and without bell beaker pottery. Given the absence of absolute dates in these contexts, a chrono-typological date between 1750 and 1500 BC has been proposed.

Further analysis of the documentation suggests that basing a chronological phase on the recurrent association of certain objects is unsound. Firstly, the concept has as a starting point the absence of Bell Beaker pottery, even though metalwork is connected with this pottery in a limited number of contexts. Furthermore, although some burials within the NIP (cists and mounds) contain large numbers of tanged daggers/swords, Palmela-type points, and silver and gold ornaments (these are the only metal types which appear to be associated with one another up to this date), grave contexts with multiple metal objects are rare. The great majority of graves contain only isolated pieces of metalwork sometimes accompanied by artifacts made from other materials. Therefore, Montelavar Group graves as defined by Harrison are exceptional in the north-west. In any case, associations of grave goods may be as a results of socio-cultural rather than temporal factors. Finally, although the concept of Montelavar Group has been defined in relation to funerary practices, this horizon has also been related to metal types, like halberds, that have never appeared in burial sites, but are found in hoards. These occur even more frequently in representations of the Galaic Group of cave paintings, which Bradley (1998) has recently related to the hoards.

For all these reasons, the concept of the Montelavar Group requires revision. It is necessary to define the beginning of the Bronze Age in the NIP on more complete information than that provided by funerary contexts alone.

Conclusions

The earliest metallurgical activity to be dated in the NIP took place within several settlement sites in the north of Portugal. These are Vinha da Soutilha (Chaves, Tras -Os-Montes) (Jorge 1986) and Buraco da Pala (Mirandela, Tras-os-Montes) (Sanches 1997), in the north and Castelho Velho (Beira Alta) on the Duero river (Jorge 1993). These appear towards the end of the 4th and beginning of the 3d millennium BC.

Local and independent invention of metallurgy in the region seems improbable. The earliest NIP dates are approximately coeval with the nearest metallurgical centres on the north Sub-Plateau of Spain (Delibes *et al.* 1996) and the horizon represented by Vila Nova de São Pedro-Zambujal in the centre and south of Portugal. The similarity of the earliest metallurgical production in all these areas makes it difficult to establish their relative chronologies. At present the earliest traces of metallurgical activity in the Iberian Peninsula are found in the South-east (BETA-90884: 5660±80 BP: 4700-4300 BC, BETA-90885: 5920±70 BP: 4940-4620 BC) (Montero & Ruiz 1996). In relation to other areas of western Europe, Ross Island has the earliest dates for copper mining in Ireland between 2500-2200 BC (O'Brien). In Britain, Northover (1996) and Needham (1996) suggest that the earliest use of copper occurs between 2500-2100 BC overlapping with the appearance of tin bronze around 2200-2100 BC. It seems likely therefore that copper metallurgy arrived in the NIP from the south.

There is no evidence for specialized metallurgical units within settlements, however the presence of metal is notable at two sites which do not appear to be strictly domestic. In the first, Buraco da Pala (Mirandela, Trás-os-Montes), the metalwork is associated with an accumulation of great quantities of cereal (wheat and barley), nuts and beans in a context interpreted as a storage silo (Sanches 1997). The deposit contains two concentrations of calaite (variscite) ore. The second site, the fort of Castelo Velho (Freixo de Numao, Beira Alta), has been interpreted as a village specialized in the production of flour and textiles, although fireplaces suggest some habitation as well (Jorge 1993). In these sites the metallurgy appears to be related to artisan activities, although perhaps as a minor component of a wider economy connected with food production and storage. In none of the settlements does it seem that metallurgy was on a large scale or dominant in the domestic economy, nor were such sites located to take advantage of copper ore resources.

Given that some of the metal artifacts recovered from these settlements are considered to be tools (awls, chisels, knives, axes), it may be that metallurgy appeared in the domestic context with metal artifacts having some functional value. However, the rarity and limited effectiveness of such tools, as well as the presence of metal ornaments, suggests that metallurgy may have fulfilled more of a social than functional role.

Although it has been suggested that metallurgy predated the Beaker phenomenon, at least in the north of Portugal, it is not clear if this is true of all areas. This study shows that it is not appropriate to single out different traditions within the Chalcolithic and Early Bronze Age. It is not therefore possible to define a *Beaker metallurgy* with particular characteristics, different from those of metal production in the periods immediately preceding of succeeding it, with the exception of the adoption of certain metal forms in funerary contexts. The difference that exists between the copper forms produced in the *pre-Beaker* and *Beaker/post-Beaker phases* was probably not as drastic as has been thought. In great measure this image is influenced by the differing natures of the contexts in which objects are found. The subjective and inconsistent nature of the term *Beaker phenomenon*, is reinforced by the results of the technological study presented here, making it a wholly inappropriate in relation to discussion of the chronological of early metallurgy.

The process by which new metal types and Bell Beaker pottery were introduced in funerary contexts should probably be seen in the light of developing social intensification, interaction and increasing complexity without necessarily implying a major discontinuity in development; even if such developments were geographically asymmetric. Developing funerary traditions may have made use of new elements (metallic, ceramic or lithic) as symbols to express growing social complexity. In this sense the NIP is integrated in a production pattern covering a large area of western Europe. The hoards indicate that certain quantities of metal, belonging to certain specific artifact forms, are withdrawn from the system of circulation, showing a greater availability of raw materials and an interest in the accumulation of metal.

In this context the appearance of bronze in the north-west from an early date can be explained, not as a linear process, but rather emerging in certain areas, while others continued to use copper alloys. The presence of awls in Guidoiro can be interpreted as a result of importation or of sporadic exchange. It is unlikely that tin-bronze originated in south-east Iberia as bronze appears only at an advanced stage of the Argar culture. In the Peninsula the first evidence for bronze appears in the north-east, from La Bauma del Serrat Pont (Gerdna, Spain) (4020±20 BP) (Alcalde *et al.* 1997) and Monte Aguila de las Bardenas Reales (Navarrs, Spain) associated with a date placing it between 1890 and 1750 BC (GrN-19671: 3510±20 BP) (Sesma & García 1994). Fernández-Miranda *et al.* (1995) suggest that bronze technology spread through the Pyrenees, slowly penetrating the Peninsula, and reaching the south-east at a late date. This can be seen in relation to the introduction of bronze in French Atlantic regions towards the end of the 3d millennium BC (Briard 1979).

In the NIP there is evidence for the early use of bronze in a coastal area with a greater diversity of mining resources and better communications. Given that metallurgy seems to have been limited in scale, it is advisable to be prudent in suggesting exchange of minerals by long-distance journeys, as between the north-east of the Peninsula and the British Isles or French Brittany for example. Nevertheless, the appearance of silver and bronze can probably be related to a greater interaction of social groups and more widespread circulation of materials.

By the end of the 3d millennium and early 2nd millennium BC the archaeological record indicates a change in habitation and funerary practices in the NIP. Although some finds with uncertain contexts might correspond to the new funerary practices and there is evidence for the persistence of mound structures, metal objects cease to be present in burials, appearing mostly as either isolated finds or in hoards. A few recovered objects show that the production of bronze was characterized by a different technology in the Middle Bronze Age. At present, it is unclear if this was innovative in relation to the presence of Palmela-type points and other types throughout the 2nd millennium BC. This is particularly difficult given the funerary character of the earlier artefacts and the period, during which, for the whole of the NIP, there were changes in relation to both funerary and domestic expressions.

In conclusion, the first development of metallurgy in the NIP does not correspond to the traditional interpretation of a linear model of consecutive phases of growing technological complexity which are distinguishable from one another. A new model is proposed here involving production in cycles which are not distinguishable on technologicl grounds. The development of metallurgy in the NIP was not linear, but differed in pace on an intraregional scale, even becoming regressive in some areas at times. The cultural development of each area played a role in the adoption of particular artifact forms which were only occasionally understood as technological innovations.

Acknowledgements
The study is integrated in the Program: Archaeometallurgy of the Iberian Peninsula: Technology and Cultural Change during Bronze Age (D.G.I.C. y T.PB92, Nº0351).

References

ALCALDE, G., MOLIST, M., SAÑA, M. & TOLEDO, A. 1997. Procés d' ocupació de la Bauma del Serrat del Pont (La Garrotxa) entre el 2900 y el 1450 cal AC. *Publicaciones Eventuals d' Arqueología de la Garrotxa,* 2. Museo Comarcal de la Garrotxa.
ALMAGRO-GORBEA, M. 1996. L'Eta del Bronzo in Iberia: Cronología Relativa. *Acta Archaeológica,* 67: 153-163.
ARMBRUSTER, B. 1996. Zu den technologischen aspekten der goldfunde aus dem bronzezeitlichen scahtz von Caldas de Reyes (Prov. Pontevedra). *Madrider Mitteilungen,* 37: 60-73. Instituto Arqueológico Alemán. Madrid.
BLAS, M. A. DE 1996. La primera minería metálica del N. Peninsular: las indicaciones del C-14 y la cronología prehistórica de las explotaciones cupríferas del Aramo y el Milagro. *Complutum Extra* (Homenaje a Manuel Fernández-Miranda), 6-Y: 217-266.
BRADLEY, R. 1998. Invisible warriors- Galician weapon carvings in their Iberian context. *A Idade do Bronce en Galicia: Novas Perspectivas.* R. Fábregas (De.), Cadernos do Seminario de Sargadelos, 77. Sada: 243-258.
BRIARD, J. 1979. Problemes Metallurgiques du Bronze Armoricain: Etain, Plomb et Argent. *Proceedings of the fifth Atlantic Colloquium.* M. Ryan (Ed.). Dublín: 81-96.
CAVALHEIRO, J. & SANCHES, M.J. 1996. Um caso de metalurgia primitiva de ouro na 1ª metade do IIIº milénio A.C.: abrigo de Buraco da Pala-Mirandela. *Actas del 1º Congreso Peninsular de Arqueología* (Porto, 1993), VIII. *Trabalhos de Antropología e Etnología,* 35 (4): 167-188. S.P.A.E. Porto.
COMENDADOR, B. 1998. *Los Inicios de la Metalurgia en el Noroeste de la Península Ibérica.* CD-ROM. Universidade de Santiago de Compostela.
DELIBES, G. 1993. Sal y jefaturas: una reflexión sobre el yacimiento del Bronce Antiguo de Santioste, en Villafáfila (Zamora). *Brigecio,* 3: 33-46.
DELIBES, G., FABIAN, J., FERNANDEZ-MANZANO, J., HERRAN, J., SANTIAGO, J. DE & VAL, J. DEL. 1996. Los más antiguos testimonios del uso y producción de metal en el suroeste de Submeseta Norte: consideraciones, tecnológicas y contextuales. *Humanitas. Estudios en Homenaje al Prof. C. Alonso.* Universidad de Santiago de Compostela.
FERNANDEZ-MIRANDA, M., MONTERO, I. & ROVIRA. S. 1995. Los primeros objetos de bronce en el occidente de Europa. *Trabajos de Prehistoria,* 52, 1: 57-69. Consejo Superior de Investigaciones Científicas. Madrid.
HARBISON, P. 1967. Mediterranean and atlantic elements in Early Bronze Age of Northern Portugal and Galicia. *Madrider Mitteilungen,* 8: 100-122. Instituto Arqueológico Almemán. Madrid.
HARRISON, R.J. 1974a. Ireland and Spain in the Early Bronze Age. Fresh evidences for Irish and British Contacts with Proto-Atlantic Bronze Age Spain in the second millennium BC *Journal of the Royal Society of Antiquaries of Ireland,* 109: 58-60.
1974b. A reconsideration of the Iberian Background to beaker metallurgy. *Palaeohistoria,* XVI: 63-105.
JORGE, S. 1986. *Povoados da Prehistória Redcente da Regiâo de Chaves-Vª. Pª. de Aguiar.* Instituto de Arqueología da Faculdade de Letras de Porto. Braga.
1993. O povoado de Castelo Velho (Freixo de Numâo, Vila Nova de Foz Côa) no contexto da Pré-história recente do Norte de Portugal. *Actas del 1º Congreso Peninsular de Arqueología* (Porto, 1993), VIII. *Trabalhos de Antropología e Etnología,* 33 (1-2): 179-216. S.P.A.E. Porto.
JORGE, V., ALONSO, F. & DELIBRIAS, G. 1998. Novas datas de Carbono 14 para Mamoas da Serra da Aboboreira. *Arqueología,* 18: 95-98. Grupos de Estudos Arqueológicos do Porto.
NEEDHAM, S. 1996. Chronology and periodisation in the British Bronze Age. *Acta Archaeológica,* 67: 121-140.
NORTHOVER, P. 1996. The Earliest Metalwork in Southern Britain. *Proceedings of the International Symposium The Beginnings of metallurgy,* Bochum, April, 1995.
MONTERO, I., ROVIRA, S.& GOMEZ, P. 1995. Plata Argárica. *Boletín de la Asociación Española de Amigos de la Arqueología,* 35: 97-106.
MONTERO, I. & RUIZ, A. 1996. Enterramiento colectiva y metalurgia en el yacimiento neolítico de Cerro Virtud (Cuevas de Almanzora, Almería). *Trabajos de Prehistoria,* 53 (2): 55-75.
O' BRIEN, W. 1995. Ross Island and the origins of the Irish-British metallurgy. *Proceedings of the Conference Ireland in the Bronze Age* (Dublín, 1995). J. Waddell, E. S. Towing. Dublín: 38-48.
ROVIRA, S., MONTERO, Y. & CONSUEGRA, S. 1997. *Las primeras etapas metalúrgicas en la Península Ibérica. Y. Análisis de los Materiales.* Fundación Ortega Gasset. Madrid.
RUIZ-GALVEZ, M. 1978. El Tesoro de Caldas de Reyes. *Trabajos de Prehistoria,* 35: 173-192. Consejo Superior de Investigaciones Científicas. Madrid.
1984: *La Península Ibérica y sus relaciones con el Círculo Cultural Atlántico.* Universidad Complutense, Madrid.
SANCHES, M.J. 1997. *Pré-história Recente de Trás-os-Montes e Alto Douro.* Sociedade Portuguesa de Antropología e Etnología. Braga.
SESMA, J. & GARCIA, M.L. 1994. La ocupación desde el Bronce Antiguo a la Edad Media en las Bárdenas Reales de Navarra. *Cuadernos de Arqueología de la Universidad de Navarra, 2:* 89-218.
SILVA, E.J. 1995. Megalitismo da Bacia do Douro (margem sul). *Actas Iº Congreso Peninsular de Arqueología (Porto, 1993), V. Trabalhos de Antropología y Etnología,* 35 (1): 35-46.
SUAREZ, J. 1995. O Fixón: Una nueva perspectiva del Bronce Inicial en Galicia. *Actas del XXII Congreso Nacional de Arqueología* (Vigo, 1993), I: 57-67. Xunta, Concello. Vigo.

8

Project Au For The Study Of Goldwork Technology And The Concept Of *Technological Domain Systems*

Alicia Perea

Centro de Estudios Históricos, C.S.I.C., Serrano 13, 28001 Madrid, Spain

Project **Au** *is a research program that aims at studying technological change taking as a paradigm gold metallurgy. We believe that technology is neither an ensemble of techniques, nor an operational sequence, but a complex cultural phenomenon in which social, economic and symbolic aspects take place. It is useless to look for technology's beginning in aleatory or evolutionary phenomena; more interesting clues can be found in human deliberate choice among alternatives in a socioeconomic context. Technology, then, has a meaning and can be explained.*

To deal with archaeological data, a theoretical construction called "technological domain system" is used. An example from the Early Iron Age of the Iberian Peninsula is given.

Keywords: Technology; Gold-metallurgy; Theory; Iberian Peninsula.

Project Au

The ability of an archaeological artefact to convey information varies considerably, depending on the question asked by the archaeologist and the methodology used. Coherence between these two variables is the minimum condition for research. This coherence can be described as a dialectic process which changes and takes on new forms with the passage of time.

In Europe, particularly in the Iberian Peninsula, archaeological research on gold work has generally been approached in two traditional ways. The first was based on the typological method, and answered questions related to time; the second was based on comparative methods used in the History of Art and answered questions related to space. Typology and style, time and space, were integrated into a positive theoretical framework which became weakened in the 1970s and was discarded in the '80s.

The Würtembergischen Landesmuseum Laboratories in Stuttgart adopted this positive theoretical framework for the ambitious research project it undertook to answer questions related to technology and trade in prehistoric Europe. The vast amount of spectrographic data on the chemical composition of metal artefacts handled, and the very wide geographical area covered, created expectations which were not realized as the results were published (Hartmann 1970, 1982). The verdict on the project did not take long to come: the methodology used could not answer the questions asked (Tylecote 1970; Harbison 1971; Coles 1982).

The search for alternatives was soon under way and archeao-metallurgy adopted new methods and strategies, becoming an academically well-established speciality, extending its field of study to issues which were, until recently, considered impossible to examine.

The idea of *Project Au* emerged in this context, aspiring to renew both the theoretical and the methodological approach to research into gold metallurgy of the Iberian Peninsula, where there is one of the richest sequences of this material in European archaeology, from the Chalcolithic to the Roman conquest. The first questions of How? and Why? were compared in a dialectic method with those of the previous stage - When? and Where?. Out of these, other questions emerged.

It is a long-term research project based on the study of the process of transforming precious metals, in order to determine the mechanisms which govern persistence, change and technological transmission, and ultimately social change. We think gold is an archaeological material with an enormous capacity for providing not just technological, but also ideological information, because of its ascribed symbolic content. This symbolic significance transcends its strictly economic value and is related to aspects of communication, political or religious legitimization, belonging, exclusion or identification within and between groups (Hays 1993). Thus this project covers a long period of time, the diachronic and dynamic process of change from the beginning of metallurgy until the Romanization of the Iberian Peninsula.

We shall now look at the methodology we have used to obtain the data.

Obtaining the Data

The methodology used is based on the association of the analytical data with its corresponding structural image, and both data and image, to a specific technological context which is provided by the archaeological information obtained from the artefact being studied.

This methodology is developed, therefore, in two phases: one topographical and the other microanalytical.

The topographical phase consists of the observation, identification and graphic documentation of the working traces and superficial microstructures using scanning electron microscopy (SEM) or other instruments of optical observation. Working traces are the result of the plastic deformation which the tools used or the process of mechanical manufacture leave on the surface of the metal;

while the superficial microstructures usually appear as a result of the application of heat during manufacture or chemical processes in site conditions.

The microanalytical phase is concerned with the chemical characterization of the alloys used and with the effects and transformations produced in them by the passage of time. The best tool for this purpose is the electron microprobe provided with an energy dispersive spectrometer (EDS). The most representative example is microsoldering in technical processes such as filigree and granulation.

Finally, the object is studied from a contextual point of view: within the site, within its archaeological structure and within its technological environment.

This information has been used to create the *Au Database*, which stores three types of data:
a) microanalytical data: from the results obtained from analyzing the soldering and other elements of interest
b) macro- and microstructural data: from the images obtained from the topographical examination;
c) contextual data: from information of archaeological interest.

The Theory

Data are only a succession of mute measurements and circumstances without meaning. It is of no use to accumulate data unless these are read in the light of a theory which gives them a voice and a research design which makes understanding possible.

We do not consider a theoretical proposal to be something separate from reality, but something which is produced by an on-going dialectic with *praxis* (Hodder 1995: 253); we could say that theory responds to the need of solving a crisis in the progress of any research (Perea 1996: 62). And this crisis occurs when it is time to connect the data with a possible reality, or model, which will offer an explanation.

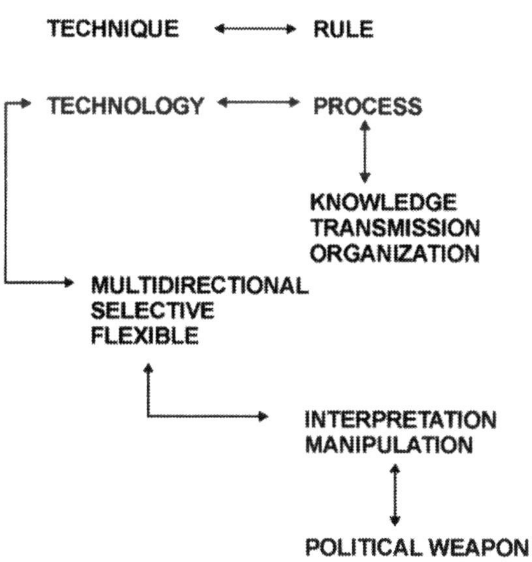

Figure 1. Interrelations in a model for the study of technology in prehistoric societies.

As in any model (Figure 1), the starting point is a series of assumptions; in this case, we start from those that have been developed by the still young Sociology of Technology within its program on the social construction of Technology (Pinch, Bijker 1994). This program characterizes technology as:
-multidirectional: in contrast with the linear models which emerged from the evolutionist framework;
-selective: by the individuals, groups or institutions which make decisions or exercize power;
-flexible: there are many ways of resolving an artefact or technical problem, depending on the social and economic situation.

If they are multidirectional, selective and flexible, the techniques or technological processes do not respond to random or evolutionary phenomena but to man's capacity for making choices and decisions; and this implies intentionality and meaning. In short, technology can be interpreted and manipulated; it thus becomes a potential political weapon (Perea 1996, 1997).

Starting from these premises, I have developed the concept of "technological domain system" (Perea 1995: 71-72), a method which acts as a kind of middle range theory. It is based on the concept of system which Hughes (1994) applies to the analysis of present-day technological systems, and on Pfaffenberger's interpretation (1992) for pre-industrial societies.

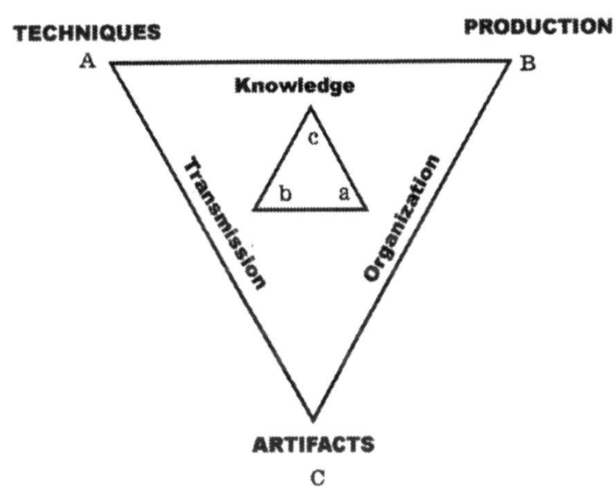

Figure 2. Technological domain system.

I understand a "technological domain system" to consist of a system delimited by the following variables (Figure 2):
A- **Techniques:** a specific set of manufacturing rules.
B- **Production:** a system of social relations of production.
C- **Artefact:** a product.

The variables which delineate this domain system are related to each other by various other forms of communication:
a- *organization*: which connects artefacts to production;
b- *transmission*: which connects technology to artefacts, and
c- *knowledge*: which links technology with production.

Within a technological domain system there is a series of mechanisms which react to the needs for internal change or to interactions with other domain systems. These mechanisms are: 1) differentiation, 2) cohabitation and 3) integration. Some are generative, which means that new "technological domain systems" arise from them, and others are sterile, which means that the contact is casual and without consequences.

An example, although necessarily simplified, is worth more than a thousand explanations.

An Example

During the transition from the Late Bronze Age to the Iron Age, 8th-7th century BC, the arrival of Phoenician colonists in the southern part of the Iberian Peninsula resulted in the reorganization of the centres of power, with a change in the social manifestation of that power and in the economic bases which supported it.

Figure 3. Pectorals and plaques from the Carambolo goldhoard, Seville, Spain.

The Carambolo gold hoard (Figure 3) (Perea & Armbruster 1997) is one of these manifestations and comprises 16 rectangular gold plaques, 2 pectorals and 2 bracelets. All of them contain apparently exact technical-ornamental elements, with all the characteristics of Phoenician gold work: filigree work, soldering of multiple elements and oriental iconography.

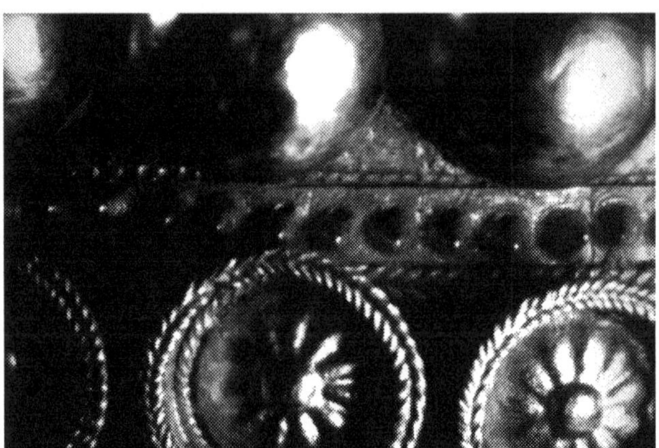

Figure 4. Macrophotograph of the ornamentation of the Carambolo plaques; in the center lost wax casting solid points.

However, there is one element which comes from Late Bronze Age goldwork- the row of points inserted between the rows of rosettes (Figure 4). They are solid points, produced by lost wax casting, and which originated in a well-defined "technological domain system" on the Atlantic seaboard (Armbruster & Perea 1994).

Figure 5. Carambolo bracelets.

Project Au (Perea & Armbruster in press) studied this hoard in detail and discovered that the points in the bracelets (Figure 5) were not solid, but hollow, and had been stamped from gold plate. From a superficial examination, it is impossible to distinguish between the points made using one technique or the other, and it is only because there are occasionally small holes (Figure 6) and deformities (Figure 7) in these hollow elements that it was possible to detect the difference without having to take the pieces apart.

Figure 6. Macrophotgraph of one bracelet showing small holes in the row of stamped points.

Figure 7. Macrophotograph of the same bracelet showing deformaties in the hollow points.

The Carambolo example is typical of a direct contact between two "technological domain systems", one represented by the Phoenician colonists, and the other characteristic of the indigenous inhabitants of the Peninsula. The interaction between the two resulted in a typological transmission, that of the ornamental element of points, without the technology of lost wax casting being transferred. The Phoenician craftsman interpreted the native form using a technique from his own technological domain.

The case study is more complex than we can relate here, but the consequence is a generative integration which results in the appearance of a "new technological domain system".

References

ARMBRUSTER, B. & A. PEREA. 1994. Tecnología de herramientas rotativas durante el Bronce Final atlántico. El depósito de Villena. *Trabajos de Prehistoria* 51 (2): 69-87.

COLES, J. 1982. The Bronze Age in Nothwestern Europe: problems and advances. *Advances in World Archaeology* I: 287-288.

HARBISON, P. 1971. Hartmann's gold analysis: a comment. *Journal of the Royal Society of Antiquaries Ireland* 101: 159-160.

HARTMANN, A. 1970. *Prähistorische Goldfunde aus Europa I*. Studien zu den Anfängen der Metallurgie, Band 3. Berlin.

1982. *Prähistorische Goldfunde aus Europa II*. Studien zu den Anfängen der Metallurgie, Band 5. Berlin.

HAYS, K.A. 1993. When is a symbol archaeologically meaningful?: meaning, function, and prehistoric visual arts, in: N. Yofee & A. Sherratt (ed.) *Archaeological Theory: who sets the Agenda?* Cambridge University Press: 81-92.

HODDER, I. 1995. Towards a contextual methodology, in: M. Kuna & N. Venclová (ed.) *Whither Archaeology? Papers in honour of E. Neustupny*. Prague: 249-254.

HUGHES, T.P. 1994 (1st 1987). The evolution of large technological systems, in: Bijker, W.E., Hughes, T.P. & Pinch, T.J. (ed.) *The Social Construction of Technological Systems*. MIT Press. Cambridge: 51-82.

PEREA, A. 1995. La metalurgia del oro en la fachada atlántica peninsular durante el Bronce Final: interacciones tecnológicas, in: M. Ruíz-Gálvez (ed.) *Ritos de Paso y Puntos de Paso. La Ría de Huelva en el mundo del Bronce Final europeo. Complutum*, extra 5:69-78.

1996. Propuesta teórica para una aproximaciÛn global a la imágen ibérica: el ejemplo del cambio y la transmisión iconogr·fica en metalistería, in: R. Olmos (ed.) *Al otro lado del Espejo*. Colección Lynx. Madrid: 61-71.

PEREA, A. & B. ARMBRUSTER. 1997. El Carambolo visto desde el Proyecto Au. *Revista de Arqueologìa*, 195 (Julio): 20-29.

In press. Cambio tecnológico y contacto entre Atlántico y Mediterráneo: el depósito de El Carambolo, Sevilla. *Trabajos de Prehistoria*.

PFAFFENBERGER, B. 1992. Social Anthropology of Technology. *Annual Review of Anthropology*, 21: 491-516.

PINCH, T.J. & W.E. BIJKER. 1994 (1st 1989). The social construction of facts and artifacts: or how the sociology of science an the sociology of technology might benefit each other, in: Bijker, W.E., Hughes, T.P. & Pinch, T.J. (ed.) *The Social Construction of Technological Systems*. MIT Press. Cambridge: 17-50.

TYLECOTE, R.F. 1970. The composition of metal artifacts: aguide to provenance?. *Antiquity*, XLIV: 19-23.

9

Copper production in the Eastern Alps during the Bronze Age: Technological change and the unintended consequences of social reorganization

ROGER C.P. DOONAN

Ancient Monuments Laboratory, English Heritage, 23 Savile Row, London W1X 1AB, UK

Introduction

The last two decades have been a period in which the practice of archaeology has changed dramatically. The 'New Archaeology' of the sixties and seventies which emphasized general processes and systems (i.e., Binford 1967) has been largely rejected in favour of a social archaeology, which, although defying categorization, may be loosely characterized by its reflectivity, context specificity and emphasis on human agency (see for instance Hodder 1990, 1989, Shanks & Tilley 1987, Barrett 1994). However, this major revision of how archaeology is practised has had little effect on 'archaeometallurgy' or archaeological science in general. This is an important observation because if 'archaeometallurgy' is not aligned with its parent discipline, it is in danger of becoming irrelevant to wider studies whilst reducing itself to little more than the metallurgy of old things. Despite this point having been noted by scholars within archaeometallurgy (Ehrenreich 1991) and archaeological science (e.g., Dunnell 1993), few archaeometallurgical studies have addressed current archaeological problems, or even archaeometallurgical problems from current archaeological perspectives. Typically archaeometallurgical studies seem only to be concerned with techno-chemical descriptions which emphasize the technical evolution of metallurgy, specifically the increasing efficiency of a process (e.g., Hauptmann *et al.* 1992), or the provenance of metal artefacts and ingots (e.g., Stos-Gale 1989).

With the advent of 'New Archaeology' in the 1960's, 'archaeometallurgy' underwent dramatic changes in the both the methodology employed and the type of material studied (Goodway 1991). As 'New Archaeology' replaced 'Cultural Archaeology', which had emphasized the diffusion of peoples and the description of artefacts, 'archaeometallurgy', the new phrase coined about this time for the study of metals within archaeology, shifted its emphasis from the analysis of finished artefacts to the analysis of all kinds of production debris from diverse metallurgical technologies (*ibid.*). This shift was no doubt stimulated by the importance that 'New Archaeology' attributed to socio-economic factors and exchange sub-systems. The analysis of production debris was seen as important, because the identification of production sites was a key issue in understanding resource procurement and production within exchange systems. Despite the broadening of 'archaeometallurgy' to include all kinds of production debris, the deeper underlying aims and practices of the discipline remained unchanged. These were, and to all extent still are, the continued accumulation of data to allow technological description/classification and for comparative studies to determine provenance. These aims were underpinned by the tacit assumption that 'Man's' relationship with metallurgy, as with all technology, was essentially one obsessed with increasing the efficiency of the process (Pfaffenberger 1992). The technological description and classification of archaeological material thus became a way of demonstrating a Darwinian evolutionary trajectory for societies where apparent technological sophistication equated with social complexity (Pfaffenberger 1992). Often these ideas led straight to environmental determinism, since technology was restricted by the availability of local resources. Nowhere in this world of techno-environmental determinism were issues of human choice or selection addressed; human existence was seen as the logical response to environmental factors facilitated by available technology. In such a deterministic world, humans could be easily removed from archaeological discourses, and hence the archaeology of metal production became nothing more than the documentation of technical factors and chemical pathways. These ideas, underpinned with an obsession for technical efficiency, probably tell us more about the mind-set of scholars and researchers working within a Western Capitalist society than it does about ancient human concepts of technology.

Whilst 'New Archaeology' aligned itself with the archaeological sciences, many current scholars of archaeology question the relevance of overly scientific studies to the discipline (see Thomas 1990). Such is the force of these questions within archaeology, that unless archaeometallurgy can adapt to this changing archaeological agenda, it is in danger of marginalizing itself and simultaneously being deemed an illegitimate discipline. The key reasons for archaeometallurgy being deemed irrelevant are complex, although not difficult to understand. Since today archaeology emphasizes the importance of social factors and human agency in explaining archaeological material, it should not be surprising that such studies are conducted at the human scale, and have focused mainly on landscape, monumental architecture and other spatial studies. Such studies, when conducted at the human scale force one to confront landscapes and architecture at a scale directly relevant to everyday human experience. When working within such a paradigm, it is not easy to remove the idea of human agency, existence even, from any archaeological discourse.

The world of archaeometallurgy, especially its scientific aspect, is one of indirect experience. It is a world inhabited by signals and readouts from analytical equipment. In order to inhabit such a world in a meaningful way, archaeo-metallurgists immerse themselves in an atomistic paradigm. In this paradigm, knowledge is constructed in a way which is invisible to everyday human experience, often counter-intuitive, quite literally beyond our senses. Data which has been collected within such a paradigm is not easy to interpret in terms of human experience. However it is not impossible. The following case study attempts to break away from the tradition of mere technical description/classification and its implications for understanding of past societies by adopting an embodied approach to furnace architecture. Its aim is to demonstrate how chemical studies, when combined with spatial analysis, become useful for understanding the range of possible technological choices- how concepts of human memory and experience are important elements when understanding 'objective' chemical data and how power over and control of resources could have been negotiated during the smelting of copper minerals in the Alpine Bronze Age.

Copper smelting in the Alpine region during the Bronze Age: *The History of the Problem*

The archaeology of prehistoric copper mining and smelting in the Eastern Alps has probably received more attention than any other prehistoric metallurgical remains in the world. By the second half of the 19th century, researchers such as Much (1878) had noted and begun to record evidence of 'the old men's workings.' Since then, there has been an almost uninterrupted tradition of archaeological research in this region. The more notable works are those of Zschocke and Preuschen (1932), Pittioni (1948, 1952, 1957, 1958a, 1958b, 1968, 1971, 1972) and more recently, that of Professor C. Eibner in conjunction with Sperl and Presslinger (Presslinger 1987, Presslinger *et al.* 1980, 1982, 1985, 1986, 1988,1992).

Despite the copious amount of archaeological and scientific work, no coherent picture of the past emerges from this mound of published literature. Any scholar choosing to read about this subject is faced with a plethora of papers, some apparently contradicting others, sometimes even when dealing with the same collection of material (e.g., Czedik-Eysenberg 1958, Presslinger *et al.* 1987, Moesta 1991, Doonan *et al.* 1996). This state of disparity has made some scholars declare that this area requires yet more study (Tylecote 1987:130), while others have simply chosen to ignore this important strand of European prehistory when writing general archaeological accounts (note absences in Hawkes 1939, Phillips 1980, Champion *et al.* 1984, Renfrew & Bahn 1996). To illustrate the present state of confusion in Alpine archaeometallurgy, it is worthwhile to review the various process models suggested by different authors. The disparity that exists in this area of study is unlike that in other areas of archaeology, where arguments tend to centre on issues of interpretation, such as the organization of production, degrees of specialization or how the control of resources empowered elite groups. The arguments in Alpine archaeometallurgy centre on what could be argued is 'objective' scientific data and the technical interpretation of scientific analytical data. How then can the present state of disagreement be understood?

The issue around which disagreement centres for 'The Eastern Alpine metallurgical process' is whether ***'the'*** process was reducing or oxidizing, and whether it took place in open hearths or shaft furnaces. Other important issues are the mineralogy of the smelted ore, whether the ore was partially roasted or dead roasted, and whether the product of smelting was copper metal, matte or both and whether matte, if produced, was re-roasted and smelted or simply converted under oxidizing conditions.

Moesta (1990) has concluded that the 'Eastern Alpine metallurgical process' was a single step process, whereby sulfidic ores were smelted in an oxidizing environment in an open hearth, similar to the Japanese Mabukki style process. This model is based mainly on evidence from Furnace F1, Hockheilhaus, Mitterberg which has been dated to the Early Bronze Age. The mineral assemblages from the walls of other furnaces do not argue as strongly for an oxidizing process (Moesta *et al.* 1990: 10-12). Presslinger *et al.* (1988) favour a very different model, in which the ore is first roasted, then smelted in a shaft furnace under reducing conditions to form a matte which is then reroasted and smelted until finally refined copper is produced. This model closely resembles that documented by Agricola for sixteenth century copper smelting in the Alps (Hoover & Hoover 1950:388).

It should be emphasized that these apparently contradictory models, although both based on scientific results, are based on archaeological evidence that is separated greatly in time and space. It is argued here, that it is this spatial and temporal separation of the metallurgical remains, which is the reason for confusion. All studies seem to aim toward identifying *'the'* process that was used in the Eastern Alps. In truth, there is no reason to suspect that a single technique ever existed at any one point in time, and even less reason to suspect that any process remained unchanged for almost a millennium. The need to explain the 'Eastern Alpine metallurgical process' in terms of a single tradition has more to do with scholars' assumptions about cultural and ethnic affinities during the period in question, than it does on any empirical archaeological evidence. Rather than extrapolating evidence from a single site or region to account for a pan-Alpine process, it may be wiser for individual scholars to study remains at a regional level and then to compare the outcomes of these different studies, recognizing differences where they exist, and hence explaining temporal and regional variations in copper smelting technology.

Why the archaeology of Alpine copper smelting has continued for so long without an explicit defined research strategy is most likely due to the way in which excavations and research have been managed. The remains of copper smelting are unapproachable by traditional archaeological methods, because there are few pottery finds, slag finds are difficult to categorize and interpret without some form of analysis, and archaeologists generally have a poor understanding of technical processes. This has meant that these sites have not attracted the attention of many 'thorough-bred' archaeologists (although there are very notable exceptions), and archaeologists have been content to allow these sites to be dug, examined, and reported by miners, metallurgists and engineers keen to understand the origins of their professions. This has had great benefit, as the skills they have brought to the investigations have been indispensable and often resulted in novel techniques of survey and excavation and excellent technical insight. Nevertheless, since it has been the objective of these investigators to explore the origins of 'their' profession, the final reporting of these sites has tended to concentrate on technical aspects with a resulting lack of archaeological valency. This should not be seen as a criticism of such studies or of those involved in them, rather it is a call for the management of such investigations, presumably by archaeologists, to acknowledge and direct research investigations accordingly whilst harnessing these resources.

A reassessment of Alpine copper smelting in the Bronze Age

The nature of the mineralization exploited is perhaps the least researched area of Alpine copper smelting. Nevertheless, it is worth mentioning what is known; the archaeological remains of copper smelting in the Eastern Alps are invariably located in the vicinity of copper mineralization. The main mineralized zone is the East-West oriented Grauwacken zone, south of the Northern Calcareous Alps (Oxburgh1968). The absence of 'exploitable' oxide minerals in the Alpine region appears widely accepted (Oxburgh 1968) and it is, no doubt, for this reason, in conjunction with archaeological evidence, that copper smelting in the Alps invariably exploited primary sulfidic ores, namely, chalcopyrite (Presslinger *et al.* 1992, Doonan *et al.*1996) and/or the Fahlerz ores (Czedik-Eysenberg 1958). Although the grade of these ores is frequently quoted (Presslinger *et al.* 1988, Bachmann 1982), it is perhaps unwise to do so, as it is now widely accepted that we cannot judge the composition of a mineral vein as it was in antiquity by the remnants that survive. Presumably these survive for the very reason that they were not exploited.

In proposing an 'archaeometallurgical' model, the choices that were made by the individuals who converted these minerals into metals are defined. Because of the multiple chemical pathways by which sulfide minerals may be converted to copper metal, the number of choices possible for sulfide smelting is far greater than in the case for oxide mineral smelting. In terms of the *chaine operatoire* (Edmonds 1990), sulfide smelting must be seen as more complex than oxide smelting, and hence the potential for regional and temporal variation greater. (For a review of the alternative chemical pathways in sulfide smelting see Doonan 1996, Biswas & Davenport 1980, and Yawaza 1980)

Against the background of the various 'chemistry only' models for Alpine copper smelting stands the work of Eibner (Presslinger *et al.* 1988). Although he, with other scholars, has proposed a technical model for the production of copper, his more important work, the creation of a furnace typological series, has been overlooked (see reference; Figure 1 omitted – no copyright permission granted). The true importance of this typology is that it is one of the few examples of scholastic work which actually suggests that Alpine copper smelting technology changes through time. Unfortunately, the implication of these changes is neither further discussed nor developed. It is argued here that these changes are central to understanding Alpine copper smelting in the Bronze Age.

Eibner's furnace typology clearly demonstrates that furnace dimensions appear to decrease through time while retaining the basic morphological unit and the general technique of construction, namely schist blocks luted with clay. In parallel with the dimensional change of the furnaces, features often interpreted as 'roasting beds' (Doonan 1994, Presslinger *et al.* 1988, Zschocke and Preuschen 1932) are seen to decrease in size from 18m to 2m in length. Despite these huge changes in 'architectural' form, the significance of these changes has, until now, never been discussed.

The concept of furnace architecture is an important one. Too often, furnace remains are referred to as 'merely' features. The use of the term 'feature' to describe architecture with a technological function effectively relegates it to a position of lesser importance than other architecture. Acknowledging buildings with a technological function as architecture permits thinking in terms of the way people moved around the buildings, interacted with them, grouped around them and how they employed them as a resource in recreating the act of copper smelting. It opens these buildings up for architectural and spatial analysis in much the way Barrett has treated prehistoric monumental architecture in Wessex (Barrett :1994). Admittedly, categorizing furnaces as architecture challenges our conceptions of architecture. Furnaces are odd architecture, people cannot move through them in the normal sense but nevertheless, they are buildings that facilitate certain movements and constrain others; they enclose an environment which human beings control and manipulate, drawing on their knowledge, expectations, and memories of smelting, to transform brittle stones to malleable shiny metal. When engaged in the excavation of these furnaces or in experimental reconstructions, the opportunity to inhabit the same kind of space as people did in the pastis presented. This gives the chance to explore possible encounters with the furnace in relation to human bodies; such analysis allows us to consider aspects of the smelting act on a scale directly relevant to human experience. When furnaces are considered- that is taking quite literally, an embodied approach to architectural analysis- one is confronted with a working space vastly different at different periods in time. For early furnaces, I refer to the one excavated at Hochkeilhaus, Mitterberg by Eibner and the subject of chemical analysis by Moesta *et al.* (1991); for a later example, to those in the Ramsau valley, Steiermark excavated by Klemm (Klemm 1993, Klemm1994, Doonan *et al.* 1996). According to Eibner's typology (Presslinger *et al.*1988), earlier furnaces have an anterior working platform approximately one metre wide, while in the later furnaces, it may be as little as 0.4m. The variation in size of the anterior platform has implications for the possible ways in which people may be grouped around and operate the furnace with bellows. For the earlier larger furnaces, it is apparent that one can move with ease in the space immediately in front of the furnace. There is ample room for three or possibly four adult people on the anterior working platform. When one is situated in this space, it is obvious that the furnace is out of reach of one pair of hands; it is not possible to envisage how one can operate such a furnace without the assistance of several others. This impression is in direct contrast to the experience of moving/sitting/kneeling round a furnace from the later Bronze Age or Urnfield period. In this case, one becomes aware of the lack of choice of where to orient oneself; it is as though the furnace itself forces you to engage with it. In relation to earlier furnaces, one's movement is very much more constrained: it is obvious than two people in this space would create a rather squashed crowd.

I have chosen to interpret the significance of these architectural changes in terms of the number of adult people who could have worked at the front of the furnace. It could be that changes in architectural dimension reflect something else; a change in bellow design, for instance. Of course, this remains possible but changes in furnace dimensions and the implication for smelting group size is also reflected in the structures termed roasting beds. These change from huge structures, 18m in length, which must have been managed by more individuals than roasting beds of 2m. Further, bellow design must be seen as being more conservative than furnace design, for simple reasons. Firstly, whereas a furnace only lasts one to three smelts without needing significant repairs or rebuilding, bellows may last for several generations if cared for correctly (Bealer 1995: 61). This means that the opportunity for innovation or change in form is many times higher for furnaces than that for bellows.

Based on evidence from architectural and spatial analysis, there is a change in the organization of labour in smelting- from one where many individuals worked at the furnace to one where, if any more than two worked at the furnace, the anterior region would be overly crowded. If the changes in the size of roasting beds are also considered in terms of embodied experience, then there is evidence to support the idea that metallurgy began as a large group activity, which by the advent of the Urnfield period, had changed to a smaller group activity.

The 'old' school of archaeometallurgy may well prefer to see changes in furnace architecture as the 'logical' outcome of 'Man's' desire to improve the efficiency of the process (see Pfaffenberger 1992 for a deconstruction of such a view). Indeed, such influences should not be overlooked, but in this case the evidence does not support such a contention. Had the copper-smelters been overly concerned with improving the efficiency of the process, then evidence for a rapid change in furnace design representing various attempts to modify the process should be found. This is opposite to what is found in the archaeological record, where only one morphological form of furnace, which gradually shrinks in size over a millennium, is found. The time frame within which these changes occur is important, as it determines whether these changes were perceived by individuals and incorporated in their memories. Memory is an important concept as it is through individuals' memory of past smelts that experience is made relevant to later smelts. Explanations contingent on perception and memory allow technical processes to be understood in terms of how people drew upon and recreated traditions, as they engaged in social reproduction. The change in size of Alpine smelting furnaces seems to have been slow and gradual, meaning that the degree of change within an individual's lifetime would not have been dramatic. Periodically, as the organization and availability of labour changed, certain smelters, maybe the elder or the more astute, may have drawn on their memories of previous smelts and 'felt' that the act was not as comfortable as before, work was harder, or that failure was more common with smaller groups. In order to counter such feelings, the rebuilding of furnaces may have been done so that the 'experience' or 'feel' of working a furnace during a smelt was similar to that remembered when working with larger groups and smelting may have worked well. This model gives a rather simplified impression of the furnace gradually shrinking, whereas the idea of a less clear transition is perhaps more realistic. Sometimes smelters may have built furnaces that worked well and other times not so well; they may have worked in larger groups and sometimes smaller groups. However, I would argue that the general trend was toward smaller groups and smaller furnaces. In truth, it could be argued that these architectural changes were made to make the act of smelting more effective or successful. However, this should not be confused with the hypothesis that the over-arching principle stimulating change was a desire to increase efficiency. Rather change comes about as the desire to maintain smelting operability in relation to the changes in the availability and organization of labour.

Such changes in the organization of labour are not inconsistent with other changes seen to occur during the European Bronze Age. From a very generalist perspective, Bradley (1990) summarizes these changes in the Bronze Age as a change from a society which emphasized the past and the community to one which emphasized individuals in the form of separate burials, the accumulation of a surplus and hence a preoccupation with the present and craft production. Invariably, these changes are inferred by evidence from burials, the common ground for archaeologists, since artefacts from these contexts lend themselves to traditional archaeological categorization. So far, this paper has attempted to demonstrate that the changes in metallurgical ritual architecture are direct evidence that the social organization of labour changed, suggesting that metal production became increasingly specialized, changing from a large group to a smaller more intensive group activity. This is important, as it shows that the study of technological debris can provide direct evidence for change in production strategy, and does not rely on indirect evidence or assertions from burials where elite classes and specialization are inferred from the nature of grave goods (cf. Bradley 1990:102, Barrett 1994:63 and Chpt 4). It is perhaps worth restating, that this change in smelting group size is not a response to changes in geological phenomena or any other technical factor, but rather an outcome of the complex social changes in the second millennium BC.

These technological changes are not the only ones to be associated with the re-organization of copper smelting. Accompanying the change to a smaller group activity, is a change in the distribution and number of smelting sites (Eibner, pers comm). Preliminary impressions are that the number of later sites dramatically outnumber earlier sites. In the later periods, sites are considerably smaller, although the distribution of these sites increases. Changes in site distribution and number suggest that, associated with the changes in labour organization, the location of production changes from a single, large, central site to multiple sites dispersed around the valleys. This is certainly at odds with the general accounts for the characteristic changes in the Bronze Age. It appears as though copper production undergoes decentralization, while the norm would be toward centralization of production, storage, and distribution. For those in power, centralization of production is the obvious strategy as it offers definite advantages for control and access of resources. The apparent decentralization, then, of Alpine copper production must therefore be considered odd, especially as it is the control of metals which has so often been argued as being central in the maintenance of hierarchies and elite dominance (Kristiansen 1987, Bradley 1990). However, since so many copper artefacts are found in contexts suggesting that their deposition constituted a social practice that was responsible for reinforcing the role of elite groups, (Bradley 1990) then it must be assumed that somehow control was exercised over this resource. The question then arises: If it was not through the centralization of production, how was the production of metals controlled in the later Alpine Bronze Age?

The control of copper smelting and copper resources in the Alpine Bronze Age.

It is argued here, that central to understanding the changes in the control and organization of production are the changes in process chemistry which were the unintended consequences of changes in the organization of labour within society. As shown above, changes in the organization of labour, specifically the move from larger to smaller groups, was a non-technical decision which affected the architectural dimensions of the smelting furnaces. It is argued here, that the changes in furnace design to accommodate the smaller smelting group size affected the chemistry of smelting, which was, in turn, taken advantage of by the empowered elite as a means by which they could control production from afar by consequence of the changes in the chaine operatoire.

Moesta *et al.* (1991) suggests that copper smelting in the Eastern Alps was an oxidizing process conducted in large open hearths (see above). Although Moesta has been criticized for his model reconstruction of the furnaces (Doonan 1996), the results of his Mossbauer studies on slag must still be largely respected. It is unequivocal from these results that the slag from the Early Bronze Age was produced in an oxidizing environment. It is apparent that at Hochkeil in the Mitterberg smelting took place on large sites where the process chemistry was oxidizing, chemically similar to the Japanese Mabukki process. Such a process would permit the production of copper metal in a single step, especially if the ore had been partially roasted first.

Studies by Doonan *et al.* (1996) and Presslinger *et al.*(1988) have dealt mainly with material from later sites. Although the results from the studies differ in many ways, they all tend to postulate a process that is considerably more reducing. Chemical and microstructural analysis has shown that the most common kind of metallic inclusion is matte, a copper iron sulfide intermediate, rather than refined copper. It is an enigma that evidence appears to be lacking for exactly how copper was produced at these sites, as evidence suggests that the main product was, in fact, copper matte. One of the obvious technical models suggested (Presslinger *et al.* 1988) and based on Agricola's accounts (Hoover *et al.* 1950) involves the repeated roasting and reduction smelting of matte until finally black copper is produced. Although theoretically possible, it is not supported by the archaeological evidence. In such a model, one would expect to find slags containing inclusions of varying grades of matte. Chemical and microstructural analysis of a large representative sample of slag from Ramsau S1 (Klemm 1993, 1994, Doonan *et al* 1996, Doonan 1996) shows that matte inclusion compositions are tightly clustered, suggesting they come from a single process utilizing similar resources.

To summarize this argument so far, it is suggested that the unintended consequences of the reorganization of labour led to a decrease in furnace size, with the effect that the smelting process changed from a single or two step oxidizing process to a reducing process which produced a functionally useless, yet inherently valuable intermediate, copper matte.

Earlier, it was mentioned that later sites lack evidence for the production of copper especially evidence for the conversion of matte to copper. Generally this is true. However, one important piece of evidence not yet mentioned is worthy of consideration. Zwicker *et al.* (1979) suggests that like Doonan (1996) the predominant product of these smaller later furnaces was matte although this was often accompanied by a little copper metal. Based on evidence from smelting site No 27 in the Jochberger Ache, Zwicker (*ibid*) suggests that the matte produced at smelting sites was transported to this site, where it was converted under oxidizing conditions to copper metal. Evidence for this comes in the form of magnetite saturated slags very different in appearance to those found on smelting sites. Disappointingly, the implications for resource control are not discussed.

The location of the site is interesting, as it is lower than all the other primary smelting sites so far identified in that valley (Zwicker & Goudarzloo 1979). Any matte brought down the valley would have been channelled towards this site where it would, presumably, be converted in to copper. This final conversion stage may well have been conducted in secret or have been associated with taboos, so that the empowered elite, which depended upon the control of copper production for social status, was not challenged by others carrying out illicit conversions elsewhere. It seems that the apparent decentralization of primary production was therefore possible without any loss of resource control, as access to refined useful copper was only possible with social position and the control of knowledge and resources necessary to bring about the final stage of conversion. As this was carried out at a single central site, direct control need only to have been exercized at this location.

In conclusion, it is worthwhile to summarize some of the points that this paper has intended to address. Firstly, this case clearly demonstrates how decisions made in non-technical spheres of society can have affects on purely technical factors such as smelting chemistry. This emphasizes the sociological basis of technology and shows that technological studies must be viewed as valid subjects of anthropological study. Such understandings of technology are possible when technological features are studied in combination with architectural, spatial, chemical and theoretical aspects. Such practice is essential, if atomistic data is to be reworked or rather explained at the human scale.

Hopefully, this paper has offered a meaningful archaeological explanation of the Bronze Age copper smelting remains found in the Alpine region. Unfortunately, this paper has also failed in many ways; although, hopefully, it has produced a discussion which will focus future research agendas to answering specific questions, rather than the continued documentation of the record. Although perhaps its biggest failing is criticizing the belief in a single alpine metallurgical process through time and space, it has itself only partially succeeded in dismissing the notion. While it has acknowledged changes through time, it has failed to do so through space, this should be seen as a reason for more detailed regional studies as there is still a definite lack of well excavated and dated sites in all the mineralized regions. Nevertheless, now that the importance of such studies has been identified, it is hoped that these criticisms and weaknesses will be acknowledged in future projects. Other critical questions which arise from this discussion are where did these copper smelters live, how did they feed themselves and what happened when they died. With these thoughts in mind, it seems certain that the tradition of research into Alpine copper smelting and life in general is certain to continue for at least another century

References

BACHMANN H. 1982. *The identification of slags from archaeological sites*. The Institute of Archaeology. Occ Paper No 6 London.

BARRETT, J.C. 1994. *Fragments from antiquity; An archaeology of social life in Britain, 2900-1200BC*. Cambridge. Blackwell.

BEALER A.W. 1995. *The art of Blacksmithing*. Castle books. London.

BISWAS A. & DAVENPORT W.G. 1980. *Extractive metallurgy of Copper*. Pergamon press Oxford. 2nd. Edition.

BRADLEY R. 1990. *The Passage of Arms: An archaeological analysis of prehistoric hoards and votive deposits*. Cambridge University Press. Cambridge.

CHAMPION T., GAMBLE C., SHENNAN S., & WHITTLE A. 1984. *Prehistoric Europe*. Academic Press.London.

COLES J., HARDING A. 1979. *The Bronze Age in Europe* . Methuen. London.

CZEDIK-EYSENBERG, F. 1958. Beiträge zur Metallurgie des Kupfers in der Urzeit. *Archaeologica Austriaca Beiheft* 3: 1-18.

DOONAN R.C.P. 1996. Old Flames, Slags and Society: Copper smelting technology in The Ramsau valley, Austria, during the Bronze Age. Ph.D Thesis. Unpublished. University of Sheffield.

DOONAN R.C.P, KLEMM S, OTTAWAY B, SPERL G, & WEINEK H. 1996. The East Alpine Bronze Age Copper Smelting Process: Evidence from the Ramsau Valley, Eisenerz, Austria. *Archaeometry 94*: The Proceedings of 29th international symposium of Archaeometry.(Ed) Demerci S, Ozer A., Summers G. Ankara. pp 17-22.

DUNNELL, R.C. 1993. Why archaeologists don't care about archaeometry. *Archaeomaterials* 7: 161-165.

EDMONDS M. 1990. Description, Understanding and the châine opêratoire. *Archaeological review from Cambridge* 9(1): 55-70.

EHRENREICH R. 1991: Metals in society: Theory beyond analysis: Introduction. in *MASCA research papers in science and archaeology* Vol 8, PtII .

EIBNER C. 1993. Fund- und Forschungsberict Mitterberg/Mühlbach am Hochkönig. *Archaeologia Österreichs*, 4(2): 34-36.

GOODWAY, M. 1991. Archaeometallurgy, evidence of a paradigm shift? P. B. Vandiver, J Druzik, and G. Wheeler (eds.) *Materials Issues in Art and Archaeology II, Materials Research Society Symposium Proceedings* Volume 185.pp. 705-712 Pittsburgh.

HAUPTMANN, A., BEGEMANN, F., HEITKEMPER, E., PERNICKA, E., & SCHMITT-STRECKER, S. 1992. Early Copper Production at Feinan, Wadi Arabah, Jordan, *Archaeomaterials* 6: 1-33.

AWKES C. 1954. Archaeological Theory and Method: some sugestions from the Old World *American Anthropologist* 56:155-68.

HODDER I. 1989. This is not an article about material culture as text. *Journal of AnthropologicalArchaeology* 8: 250-264.

1990. The Domestication of Europe: Structure and contingency in Neolithic societies. Blackwell. Oxford.

HOOVER H.C & HOVER L.H (trans) 1950. *Agricola: De re Metallica.*Dover publications. New York.

HUGHES T. 1988. The seamless web: Technology, Science, Et Cetera, Et Cetera in Elliott B (ed.), *Technology and Social Process*. Edinburgh University press.

KLEMM S. 1993. *Fundberichte Österreichs* 32:703-704.

1994. Ausgrabungen in der Eisenerzer Ramsau. *Archäologie Österreichs* 5(2): 28-29.

KRISTIANSEN K. 1987. From stone to bronze- the evolution of social complexity in Northern Europe, 2300-1200BC in Brumfiel E. & T. Earle (ed.). *Specialization, Exchange and ComplexSocieties* 1987 pp30-51.

MALINOWSKI B. 1935. *Coral Gardens and Their Magic*. London. Routledge.

MOESTA H. & SCHLICK G. 1991. The furnace of the Mitterberg. An oxidising Bronze Age process. *The Bulletin of the metals Museum.*Vol 14: 5-16.

MUCH M. 1878. Das vorgeschichtliche Kupferbergwerk auf dem Mitterberg (Salzburg). Mitt.Zentrum Kommission ppCXLVI-CLII.

OXBURGH E.R 1968. *The Geology of the Eastern Alps*. The Geologists Association London.

PFAFFENBERGER B. 1992. Social Anthropology of Technology. *Annual. Review of. Anthropology* pp491-516.

PHILLIPS P. 1980. *The Prehistory of Europe.*Chaucer Press. London.

PITTIONI R. 1948. Recent researches on Copper mining in Austria. *MAN* 48: 120.

1952. Prehistoric Copper-mining in Austria: Problems and facts. *Inst. of Archaeology (UCL) Annual report* 10. London pp16-43.

1957. Urzeitlicher Bergbau auf Kupfererz und Spurenanalyse. *Archäologica Austrica*, Beiheft 1, pp1-76, Abb1-20.

1958a. Der Schmelzplätz 13 des Bergbaubietes Jochberg bei Kitzbühel, Tirol. Beiheft 3: 19-40.

1958b. Neue Schmelzplätze im Bergbaugebiet Jochberg bei Kitzbühel, Tirol. Beiheft 3: 41-45.

1968. Stadtbuch Kitzbühel, Band II, Vorgeschichte und Bergbau, pp32-102.

1971. Spektralanalytische Miscellen. *Archäologica Austrica* 49: 48-62.

1972. Der gegenwärtige Stand der Bergbau-Urgeschichtsforschung in Österreich *Berg-u Hüttenmännische Monatshefte* 117, Heft 6. Springer Verlag. Wien-New York.

PRESSLINGER H. 1987. Montanarchäologie im Paltental-Untersuchungergebnisse von bronzezeitlichen Bergbau und bronzezeitlicher Verhüttung. Da Shau Herr, 1. January. Liezen.

PRESSLINGER H, EIBNER C, WALACH G, & SPERL G. 1980. Ergenis der Erforschung urnenfelderzeitlicher Kupfermetallurgie im Paltental. *Berg-u. Hüttenmännische Monatshefte*125: 131-142.

PRESSLINGER H, SPERL G, EIBNER C, & WALACH G. 1982. Kupferverhüttung in Österreich vor 3000 Jahren. Österreichisher Kalender fur Berghütte & Energie Montan-Verlag, Wien.

PRE LINGER H, GRUBER A, & PASCHEN P. 1985. Die Verhüttung sulfidischer Erze im Schachtofen in der Bronzezeit. *Metallwissenschaft und Technik* 39. heft 5. Mai pp 423-425.

PRESSLINGER H, WALACH G, EIBNER C, & GRUBER A. 1986. Geophysikalische Montanarchäologische u. metallurgische Untersuchungsergebnisse von bronzezeitlichen Ostalpinen Kupferhütten. *Berg-und hüttenmännische Monatshefte* 131: 225-230.

PRESSLINGER H, WALACH G, & EIBNER C. 1988. Bronzezeitl. Verhüttungsanlagen zur Kupferzeugung im den Ostalpen. *Berg-und hüttenmännische Monatshefte* 133: 338-344.

PRESSLINGER H. WALACH G. EIBNER C. & PROCHASKA W, 1992. Montanarchäologische Untersuchungsergebnisse eines urnenfelderzeitlichen Kupfererz-Verhüttungsplatze bei Mautern/ Steiermark. *Berg-und hüttenmännische Monatshefte* 137, Heft 1 pp31-37.

RENFREW C. & BAHN P. 1996. *Archaeology: Theories, methods and practice*. 2nd Edition. London Thames and Hudson.

SHANKS M. & TILLEY C. 1987. *Social Theory and Archaeology*. Albuquerque: Univ. New Mexico Press.

THOMAS J. 1990. Silent Running: the ills of environmental archaeology. *Scottish ArchaeologyReview* 7: 2-7.

STOS-GALE Z.A. 1989. Lead isotope studies of metals and the metal trade in the Bronze Age Mediterranean. in *Scientific Analysis in Archaeology*, J. Henderson (ed.), OUCA monograph 19, Oxbow, Oxford. Pp. 274-302.

TYLECOTE R. 1987. The early history of metallurgy in Europe. Longman. London.

YAZAWA A. 1980. Distribution of various elements between copper, matte and slag. *Erzmetall* 33: Nr 7/8 377-382.

ZSCHOCKE K & PREUSCHEN E. 1932. Das urzeitliche Bergbaugebeit von Mühlbach Bischofshofen. Materialien zur Urgeschichte Österreichs 6. Wein.

ZWICKER U. & GOUDARZLOO F 1979. Investigation on the distribution of metallic elements in copper slag, copper matte and copper and comparison with samples from prehistoric smelting places. *Archaeeophysika* 10. Bonn pp 360-37.

10

Questions in the Analysis of Ancient Tin

J.P. NORTHOVER[1] & C. GILLIS[2]

[1] Department of Materials, University of Oxford, Parks Road, Oxford, OX1 3PH, UK
[2] Dept. Of Classical Studies, University of Lund, Sulveg 2, S-22362 Lund, Sweden

The sources of tin and the distribution of the metal in the ancient world have been a matter of contention since the importance of bronze was first realized. The geochemistry and metallurgy of tin conspire to frustrate those attempting to explore this question and we need to use the few clues that are offered as fully as possible. This paper explores those aspects of tin which are accessible by chemical and metallurgical analysis using primarily inductively coupled plasma mass spectrometry (ICP-MS) and proton induced X-ray emission (PIXE). First, though, the current state of analysis of tin is reviewed including the results available from smelting slags. Following that the applications of ICP-MS to the bulk trace element analysis of Bronze Age tin objects from Europe and analysis by PIXE of intermetallic inclusions in tin ingots and tin foil decoration on ceramics are described and reviewed. It is concluded that where there the population of iron-tin and copper-tin intermetallics is sufficient enough characteristic trace elements segregate to them to offer a possibility of exploring whether a group of tin objects had a common history.

Keywords: Tin; Mass Spectrometry; Microprobe Analysis; Intermetallics; Trace Elements.

Introduction

Tin holds a special fascination for those concerned with the evolution of metallurgy from the earliest times. Here is a metal whose ores have a very restricted geographical distribution but which had a very widespread impact on the development of copper alloys as practical and utilitarian. Further, we must not underestimate the importance of other alloys of tin used in ancient and historical periods, especially pewters and solders, as well as recognizing a growing number of tin ingots and tin objects. Recent research has laid a particular emphasis on identifying sources of tin that contributed to the emergence of a "Bronze Age" and connecting tin ingots and artefacts with those sources (e.g., Franklin *et al.* 1978). This is a task of peculiar difficulty. The commonest tin ore, cassiterite (SnO_2), is itself often of very high purity and potentially diagnostic minerals and elements associated with it partition to the slag on smelting (e.g., Rapp 1978). It might be thought that tin with its ten naturally occurring isotopes could offer possibilities for tracing the movement and processing at tin but it is now clear that the isotopic fractionation of tin is so minimal in either natural or anthropogenic processes that it is barely possible to detect any effect at all (Gale 1997). Another approach is the isotopic analysis of lead impurities in tin which has some potential with tin ingots but has its dangers since lead is such a frequent alloying addition to tin. There may also be a further complication in the not infrequent cases (like Cornwall) where there are Pb-U associations which lead to highly anomalous Pb-isotope ratios. Therefore, given these obvious difficulties with isotopic analyses this paper examines a return to elemental analysis to refine our understanding of exactly what can be obtained from the compositional study of tin and tin artefacts. The material concerned comes mainly from the Bronze Age in Europe and includes both sound and heavily corroded metal.

Existing Analyses

Analyses of tin objects from prehistoric contexts have been reported in the archaeological literature for over four decades. Usually the analyses have been associated with programmes of research into copper and copper alloys and the same element sets have been sought. As examples there are the four later Bronze Age objects from Ireland analyzed by Coghlan and Case (1957) where the detected impurities are copper, lead and arsenic, all elements which need not have originated with the tin. From the extreme far eastern borders of Europe the analyses of tin objects have been presented by Selimkhanov (1978), again with an element set more usually associated with the analysis of copper and bronze. More usually the analysis of tin is less quantitative than that serving simply to identify an object as tin (Primas 1984).

In the last few years some attempts have been made to overcome these deficiencies. Analysis of drilled samples from Bronze Age tin objects from the beginning of the first millennium BC from Flag Fen, Cambridgeshire (Northover and Rohl 1995) by electron probe microanalysis (EPMA) failed to find any diagnostic impurities either qualitatively or quantitatively at typical microprobe detection limits of 100-200ppm. They found that the elements most regularly occurring were iron, copper, lead and bismuth, most of which could have been picked up as easily in the workshop as from the ore. Analyses by optical emission spectroscopy (OES) were carried out by Bollingberg (1995) on samples from Cornish tin ingots of various origins to examine the contribution that impurities in tin might make to the composition of Roman copper alloys. Again the results served to emphasize the purity of the tin although the author believed that a small number of impurities might have a diagnostic use. Analysis of other significant finds of tin ingots, such as those from the Cape Gelidonya and other Mediterranean shipwrecks, and those from the coasts of Israel and SW England is only now proceeding at a level that will add to our knowledge. More sensitive analytical techniques, notably neutron activation analysis (NAA), have been available to archaeology for a long time. NAA was used over twenty years ago by Rapp in his programme of analysis of cassiterite and experimentally smelted tin (1978). Recently Grant (forthcoming) has used NAA supported by other techniques to examine the relationship between Iron Age tin objects from the Rooiberg and other sites in South Africa to southern African tin ores in general and the Rooiberg in particular. He found that ratios of specific element pairs, for example arsenic/antimony and scandium/tantalum, were diagnostic in correlating Rooiberg tin with Rooiberg cassiterite. A much more recent addition to the range of analytical techniques being applied to tin is inductively coupled plasma mass spectrometry (ICP-MS), although in its solution based version it faces the perennial problem of tin analysis, that of keeping tin in solution and preventing it forming colloidal precipitates. An example of one experiment in ICP-MS analysis of tin artefacts is included in this paper.

A complication in the examination of archaeological tin and tin alloy artefacts is that they are often deeply corroded and may now largely comprise hydrated tin oxides, hydroxides and chlorides (see Tylecote 1978) with the possibility that trace elements have been contributed to the overall composition by the environment. This difficulty is further compounded by the fact that it is possible for bronze to corrode in such a way that virtually all the copper has been removed with the residue being practically indistinguishable from corroded tin. Two striking examples of daggers corroded in this way have been recovered from graves at Caerloggas Down, Cornwall, England, and Crug-yr-afan, Glamorgan, Wales, dating to the end of the Early Bronze Age (Biek 1978). A corroded tray from a Roman grave on the site of London Stansted International Airport yielded a similar analysis but did contain sufficient lead to suggest that it was made of pewter rather than tin (Northover *et al.* 1991). A sample could be taken for metallographic study which revealed the presence of many uncorroded inclusions of copper-tin and iron-tin intermetallics. The corrosion resistance of these intermetallic compounds was also evident in cross-sections of the tin foil decoration applied to Mycenaean pottery (cf. Gillis, this volume). Their significance for tin analysis is explored below.

Inductively-Coupled Mass Spectrometry of Tin Artefacts

The EPMA analysis of tin objects from Flag Fen by Northover and Rohl (1995) referred to above did little more than identify those objects which contained alloy levels of lead, and should therefore be described as pewters or solders, and suggest that copper and iron were common among the remaining impurities. To determine what impurities might be present in tin at levels below the100-200ppm limit set by the way in which the electron microprobe was used, and which might convey information as to the ore or its processing a series of ICP-MS analyses was arranged on selected Bronze Age tin artefacts.

The analyses addressed a specific archaeological question. The excavations at Flag Fen, near Peterborough, England produced some sixteen tin and tin alloy objects, all small and mostly beads and ornaments, some possibly votive. The main period of activity at Flag Fen, including all the timber constructions date to what in English terms is the end of the Middle Bronze Age and the beginning of the Late Bronze Age, the eleventh and tenth centuries BC (Pryor 1991). There are no parallels at all for these types of tin object in Britain, the nearest being found at exactly contemporary sites around some of the lakes in north-west Switzerland, notably Lake Neuchâtel. Therefore samples were taken from three objects (two four -lobed beads and a biconical bead) from Flag Fen (FF) and four (a wheel pendant, a pendant, and two lumps) from the site of Hauterive-Champréveyres (HR), a large settlement discovered during highway construction and yielding over 5000 metal artefacts (Rychner-Faraggi 1991, 45-46). Given the closeness of some parallels it was reasonable to ask whether the tin in the two cases might have had the same history and leave the same traces in the composition.

TABLE 1: ICP-MS ANALYSIS OF TIN

Sanple No.		Sn blnk	Tin G	Tin H	Tin I	Tin J	Tin K	Tin L	Tin M	Tin N	Tin O
Sample Name			FF 4-lobed bead (6)	FF 4-lobed bead (3)	FF biconic bead (4)	HR.18132 wheel	HR.6679 pendant	HR.19284 1/8	HR.19284 1/8	HR.19249 52/49	HR 19249 52/49
Dilution Factor			28954.178	17123.287	24220.023	18850.141	18264.840	18993.352	17969.451	38095.234	20387.359
Element	Isotope	ppm	ppm	ppm	ppm	ppm	ppm	ppm	ppm	ppm	ppm
Chromium	Cr 52	0.008	BBL	BBL	BBL	BBL	BBL	BBL	BBL	BBL	BBL
Manganese	Mn 55	0.001	3.446	BBL	22.358	BBL	BBL	BBL	BBL	BBL	BBL
Iron	Fe 57	0.099	76.575	BBL	2077.006	231.762	BBL	BBL	BBL	BBL	BBL
Cobalt	Co 59	0.000	12.491	11.024	10.859	12.325	11.965	10.521	11.783	11.358	12.261
Nickel	Ni 60	0.001	72.417	62.031	49.397	71.794	70.287	64.349	75.116	65.403	73.401
Copper	Cu 63	0.002	186.091	18.490	22.107	10.803	BBL	BBL	BBL	104.616	BBL
Zinc	Zn 64	0.006	BBL	BBL	5.852	1.424	BBL	BBL	8.469	BBL	BBL
Arsenic	As 75	0.040	BBL	BBL	BBL	BBL	BBL	BBL	BBL	BBL	BBL
Silver	Ag 107	0.001	BBL	BBL	BBL	BBL	BBL	BBL	BBL	BBL	BBL
Cadmium	Cd 111	0.000	30.321	23.395	25.784	25.329	26.382	28.980	25.298	31.720	29.937
Indium	In 115	0.000	342.159	173.689	275.316	189.470	200.719	198.885	177.194	327.960	237.953
Antimony	Sb 123	0.000	138.111	116.645	87.444	125.437	142.758	113.412	124.621	192.601	212.155
Tungsten	W 184	0.001	BBL	BBL	BBL	BBL	BBL	BBL	BBL	BBL	BBL
Iridium	Ir 191	0.000	BBL	BBL	BBL	BBL	BBL	BBL	BBL	BBL	BBL
Gold	Au 197	0.001	BBL	BBL	BBL	BBL	BBL	BBL	BBL	BBL	BBL
Lead	Pb 206	0.000	35.312	26.316	18.038	1382.341	13274.105	26.750	23.203	14047.421	12813.523
Thorium	Th 232	0.000	BBL	BBL	BBL	BBL	BBL	BBL	BBL	BBL	BBL
Uranium	U 238	0.000	BBL	BBL	BBL	BBL	BBL	BBL	BBL	BBL	BBL

Table 1. ICP-MS analyses of tin artefacts.

The analyses are given in Table 1, including repeat analyses of two of the objects from Hauterive-Champréveyres. Five elements were found to be present in all cases, cobalt, nickel, cadmium, indium, antimony and lead. These elements, with the exception of cadmium, were all reported by Grant (*op. cit.*). It was found, however, that the cadmium content was almost exactly the same in all cases, and also there was always more cadmium than zinc. The problem arises from an isobaric interference between ^{111}Cd and ^{112}Sn and the data were therefore excluded from Table 1 and the discussion. That aside, there are visible differences between the Swiss and the British objects.

Lead is present as a small trace in the Flag Fen pieces but is at percentage concentrations in two of the four Swiss examples. The two known tin-lead alloy objects at Flag Fen were deliberately avoided but those from Switzerland were selected solely on the basis of their suitability for sampling and so are a rather more random selection. It may, then, be more than accidental that half the items selected contained more than 1% lead. Another significant variable is copper, found in the three British objects analyzed here but only intermittently in the Swiss ones. Also more common in the Flag Fen artefacts is iron and where iron was identified at Flag Fen manganese is also present, but it is absent in those from Hauterive-Champréveyres. The last impurity to be considered, zinc, occurs too infrequently to offer any pattern.

On the basis of these data it is reasonable to conclude that there are groups with different metallurgical histories. One division could be between the two Flag Fen and one Hauterive-Champréveyres with iron, and those samples without, while another could be between those objects that had picked up copper and those which had not. However, it is still plausible that the tin has a common geological origin.

According to published alloy phase diagrams (Massalski 1990) the principal impurities can be divided into those where solid solubility in tin is negligible and often below 100ppm (manganese, iron, cobalt, nickel and copper) and those where there is a more significant solid solubility of 1% or more (lead and zinc). Where the solid solubility is limited intermetallic compounds form very readily; with the impurities listed they will most commonly be based on copper and iron and some at least of the other impurities will segregate to them. These intermetallic compounds will tend to float when the tin is first melted and will tend to be lost in subsequent re-meltings and pourings of the same metal. Thus, there is nothing in the data given in Table 1 to enable us to say categorically that the metal in the two groups did or did not have the same origin. If they did have the same origin the metal from Hauterive-Champréveyres has been melted and re-melted more often than that at Flag Fen. By the same token the biconical bead from Flag Fen is, perhaps, nearest to being newly smelted tin. There is nothing yet to tell us where that tin might have been smelted but the proportions of iron and manganese may ultimately assist us in reaching a conclusion.

Microprobe Analysis of Tin Slags

Compared with the abundance of tin sources the amount of tin smelting slags that has been recovered archaeologically is almost invisible. A serious consequence is that we do not know what to expect the evidence of tin smelting to look like, save that it may not bear much resemblance to the more abundant remains of historical and early modern industrial sites. A very strong possibility is that early tin smelting could very easily have been a crucible process (Salter 1997). Particles of metallic tin are attached to crucibles in the much debated material from the Early Bronze Age sites of Kestel/Göltepe in Anatolia (Yener and Vandiver 1993). More satisfactory for our purposes is the analysis of the slag and the tin prills it contained from a late Early Bronze Age grave from Caerloggas Down, Cornwall dating to about the 16th century BC (Salter *op. cit.*) . The data given for Caerloggas and for slag from the 16th-17th century blowing-house at Merrivale, Dartmoor, Devon show how slags from two sites may differ substantially in the abundance of specific elements. These data, the mean of twenty analyses over areas 30x40m for Caerloggas and twenty seven for Merrivale are set out in Table 2.

TABLE 2: TIN SLAG COMPOSITIONS

Oxide	Caerloggas	Merrivale
Na2O	1.09	1.55
MgO	2.64	3.41
Al2O3	16.12	14.24
SiO2	29.11	38.46
W2O5	1.68	1.90
ZrO2	0.09	1.60
K2O	0.91	1.44
SnO2	41.05	17.87
CaO	0.53	0.87
TiO2	1.49	6.47
MnO	0.42	1.01
FeO	4.53	10.42
CoO	0.01	0.02
NiO	0.04	0.02
CuO	0.05	0.02
ZnO	0.05	0.03
As2O3	0.18	0.17

Compositions in wt%

(After Salter 1997)

Table 2. Tin slag compositions.

The most important difference between the two slags and the processes they represent is not the most relevant to this discussion and that is that the Early Bronze Age process was very inefficient and left a very large amount of tin oxide unreduced. The other differences between the two slags will depend on the ores used, the methods used for their recovery and beneficiation, and on the conditions under which the ores were reduced. Some elements are always going to be confined to the slag phase because they will not be reduced in the prevailing reducing conditions; these include the obvious slag forming elements such as aluminium, silicon, calcium and the alkali elements as well as zirconium and tungsten. Other elements could be reduced, at least in part, and could be partitioned either to the slag or the metal or both. Iron, nickel, copper and zinc are examples of these.

The story does not end there because, as described above, many of these elements have extremely low solid solubilities in tin and will appear instead in intermetallic compounds Iron-rich intermetallics are a major contaminant in raw tin and are known as hardhead and are now a subject of study in tin ingots (Biek *et al.* 1994; Grant pers. comm.) although no detailed analyses are yet in print. The iron-tin and copper-tin intermetallics are likely to be the most abundant in the tin because of the concentration of iron and copper in the ores and each has its own capacity to substitute other elements depending on the specific intermetallic phase. For example, analysis of tinned copper alloy objects has shown that the ? phase can contain high levels of antimony and nickel while the ? phase has very much lower solubilities for these elements (Jones 1992).

Microprobe Analysis of Intermetallic Compounds

The above discussion has suggested that the analysis of intermetallic inclusions could be a useful approach to understanding the history of tin artefacts. Their corrosion resistance is an important factor in this because it gives them the ability to provide information even from otherwise wholly corroded objects (Northover *et al.* 1991; Biek *et al.* 1994). However there is the problem already identified that the elements concerned tend to be lost on repeated remelting, and in heavily mechanically worked tin, e,g, foil, the inclusions are going to be very small. In these circumstances microprobe analyses, whether by electron microprobe or scanning proton microprobe (SPM), offer the most effective first approach to their analysis.

This paper presents the results of experiments on both instruments with, first, an experimental tin ingot from South Africa and, second, a cross-section through the foil coating applied to a Mycenaean pottery vessel.

Electron microprobe analysis of the the intermetallic inclusions in the South African ingots was made using the same element set as was used on bronze simply to see whether any partition between them and the tin matrix was detectable at this simple level (Table 3). This proved to be the case with an enrichment in cobalt in particular. A next stage will be to carry out the same experiment with a more appropriate element set including elements in the first transition series.

TABLE 3: ANALYSIS OF TIN FOIL DECORATION

EPMA			SPM		
Element	1	2	Element	Sn-rich	Fe-rich
Fe	*0.28*	*0.05*	Ti	0.00	0.30
Co	*0.00*	*0.01*	V	0.07	0.52
Ni	*0.00*	*0.04*	Cr	0.00	0.11
Cu	*0.01*	*0.43*	Mn	0.09	0.51
Zn	*0.05*	*0.13*	Fe	0.43	5.39
As	*0.35*	*0.00*	Cu	0.42	2.27
Sb	*0.00*	*0.00*	Zn	0.34	2.78
Sn	*64.84*	*62.80*	Sn	98.59	87.66
Ag	*0.00*	*0.00*	Pb	0.05	0.46
Bi	*0.04*	*0.04*			
Pb	*0.03*	*0.07*			
Au	*0.00*	*0.02*			
S	*0.05*	*0.02*			

Figures in italics indicate analysis of corroded material

Table 3. EPMA of the the intermetallic inclusions in the South African ingots.

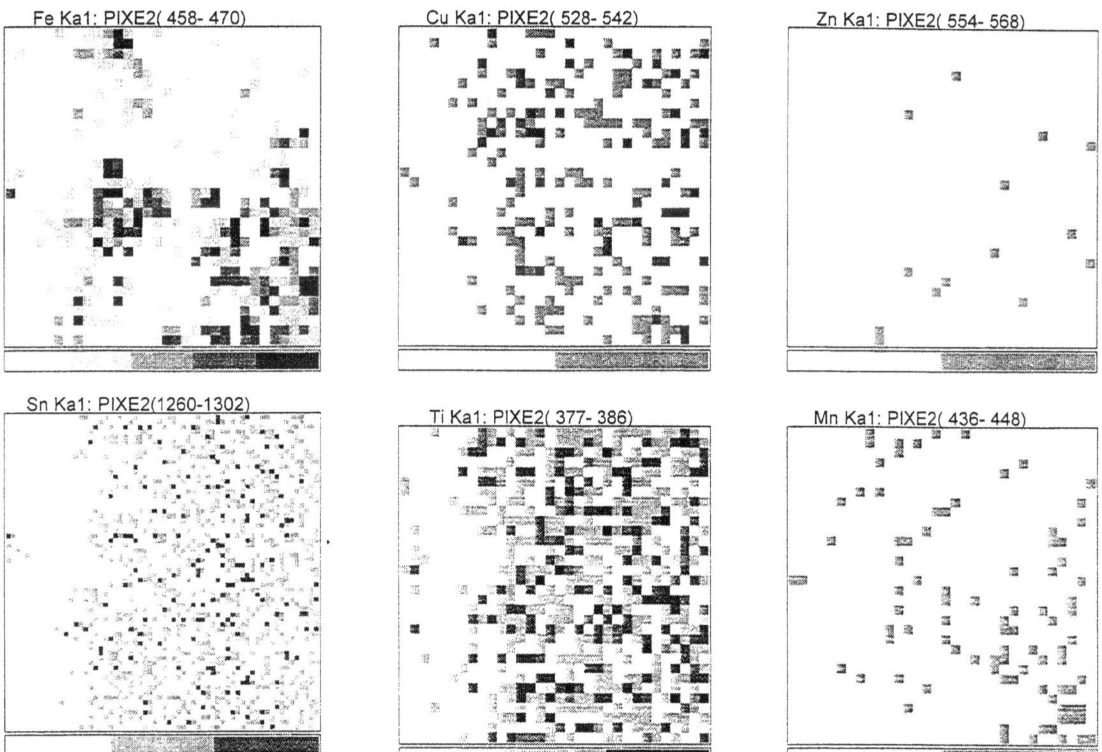

Figure 1. X-ray distribution maps for the sample from a tin ingot; area 1500μm square.

The scanning proton microprobe uses a proton beam in the same way as the electron beam is scanned in the electron microprobe and can provide elemental, and even topographic maps of the area scanned. With the SPM analyses the proton induced X-ray emission (PIXE) signal was first used to provide elemental maps for a number of elements of interest. The configuration of the operating system of the instrument in the Department of Materials in the University of Oxford is such that up to a total of 64 windows can be set on the X-ray, γ-ray, and Rutherford back-scattering spectra so that all elements on interest can be mapped simultaneously. The area mapped can vary from 25μm to 2500μm square with a maximum spatial resolution of better than 2μm. Points for analysis are selected from a cursor on the maps so that it was a simple matter to identify and analyse iron-rich areas and particles. Typical maps at a low resolution are given in Figure 1. As the iron-rich areas may contain more than one intermetallic compound, e.g., FeSn and $FeSn_2$ (see also Biek *et al. op. cit*) high resolution maps could both identify the different phases and map the partition of other impurities between them. Even with the very basic data exhibited in Figure 1 it would appear that the manganese is not homogeneously distributed within the iron[1].

Analysis of the corroded tin-foil decoration on the Mycenaean ceramic proved difficult in the electron microprobe because the $FeSn_2$ intermetallics differed too little in contrast in the back-scattered electron image from the corroded tin matrix. As a result only a few small area analyses were collected, again to determine whether there was a potential segregation of some elements to the intermetallics. The results were suggestive with evidence of both copper-tin and iron-tin compounds and cobalt and nickel as trace elements. appearing in the analysis. SPM, as will be seen, proved a much more informative technique. The combined analytical data are given in Table 4 and typical maps are presented in Figure 2 at low resolution and in Figure 3 at high resolution.

These figures show very well how the SPM data can be used to control the analysis process and to aid in interpretation. At low resolution, over a 500μm square, the maps of silicon, aluminium, calcium and titanium illustrate the structure of the ceramic, while that for tin Lα both confirms the presence of a tin coating and reveals something of its structure. The tin Kα line was also mapped but with a lower intensity than the Lα line a larger pixel size was required to integrate sufficient counts. With the non-metals the chlorine map records the chlorine in the epoxy mount together with the presence of chlorides at the interface between, while there is a sulfur rich layer on the tin surface. It was also possible to map some of the transition metals of interest to us. One of these, titanium, points up one of the hazards of X-ray spectrometry, that of interferences between lines of different elements. In the titanium Kα map the Ti-rich points within the ceramic matrix are undoubtedly from Ti-containing minerals, but at the surface the concentration is most probably of barium, the barium Lα line largely overlapping the titanium Kα. This problem can be dealt with in the software used for processing the spectrum for quantitative analysis but it is not possible to do this with the maps. Manganese and copper were identified within the tin layer but, surprisingly, iron was not at this resolution.

These maps were then used to select an area of the corroded tin coating for mapping at higher resolution. Clearly the tin coating as it now exists in corroded form has a layered structure. Chlorine has penetrated the outer layer from the environment, and below this is a thin band richer in some clay elements, certainly silicon and aluminium, but also calcium. There are strong peaks also in this band for titanium and sulfur, while the innermost layer of the tin coating is much freer of these elements. Copper, iron and manganese are irregularly distributed throughout the tin coating. The cursor was then used to select points both rich and depleted in iron for analysis. The quantification of the data is still in the course of development, so, as a semi-quantitative guide to the partition of elements between intermetallics and the matrix, the metallic elements are given in Table 4 normalized to 100% with tin. That there is a partition in these elements is plain enough, and to a sufficient degree for microprobe analyses to offer a very positive way forward for analysing archaeological tin objects.

TABLE 3: ANALYSIS OF TIN FOIL DECORATION

EPMA

Element	1	2
Fe	*0.28*	*0.05*
Co	*0.00*	*0.01*
Ni	*0.00*	*0.04*
Cu	*0.01*	*0.43*
Zn	*0.05*	*0.13*
As	*0.35*	*0.00*
Sb	*0.00*	*0.00*
Sn	*64.84*	*62.80*
Ag	*0.00*	*0.00*
Bi	*0.04*	*0.04*
Pb	*0.03*	*0.07*
Au	*0.00*	*0.02*
S	*0.05*	*0.02*

SPM

Element	Sn-rich	Fe-rich
Ti	0.00	0.30
V	0.07	0.52
Cr	0.00	0.11
Mn	0.09	0.51
Fe	0.43	5.39
Cu	0.42	2.27
Zn	0.34	2.78
Sn	98.59	87.66
Pb	0.05	0.46

Figures in italics indicate analysis of corroded material

Table 3. Analyses of the tin foil decoration.

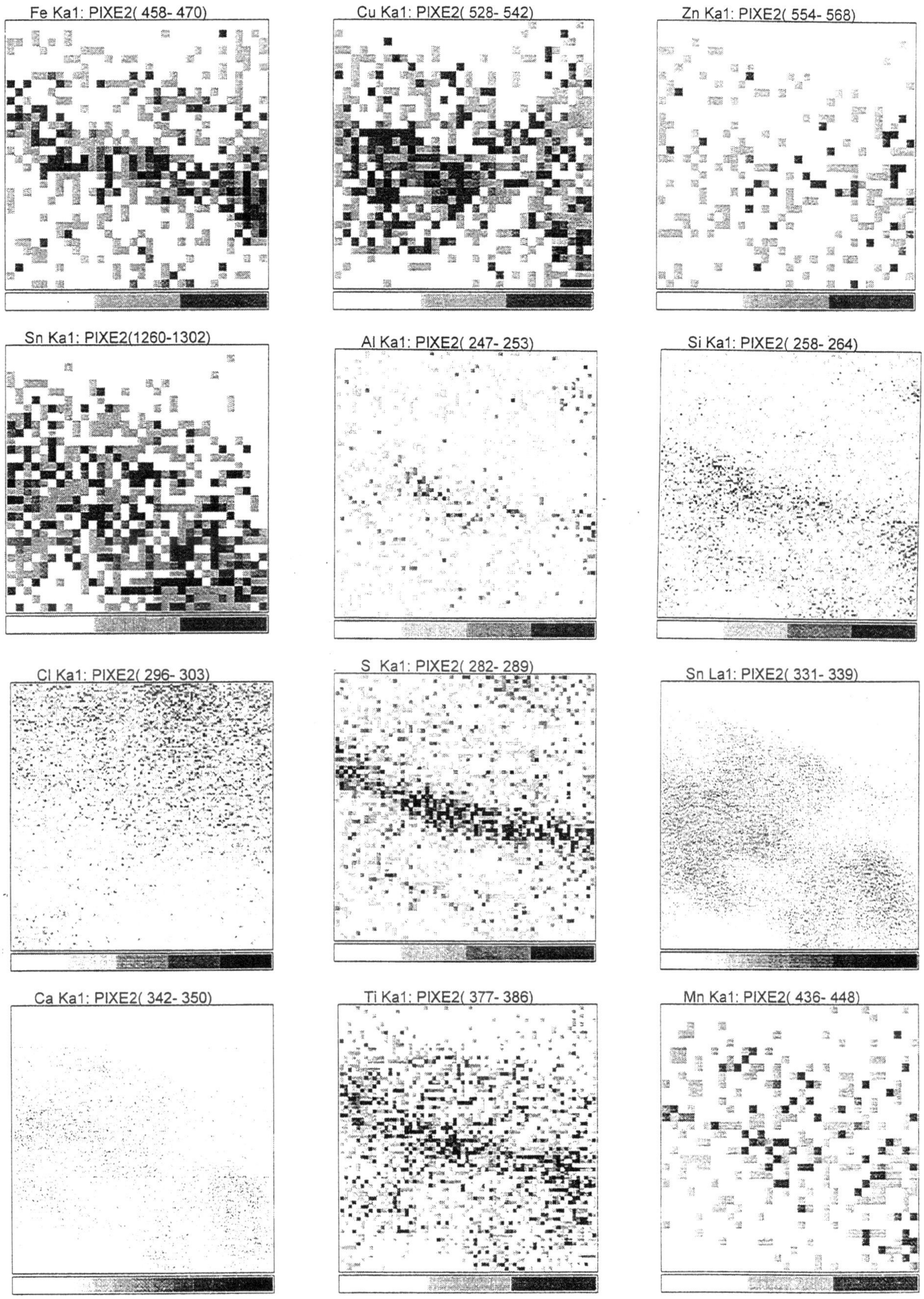

Figure 2. Low resolution X-ray distribution map for tin on ceramic; area 500µm square.

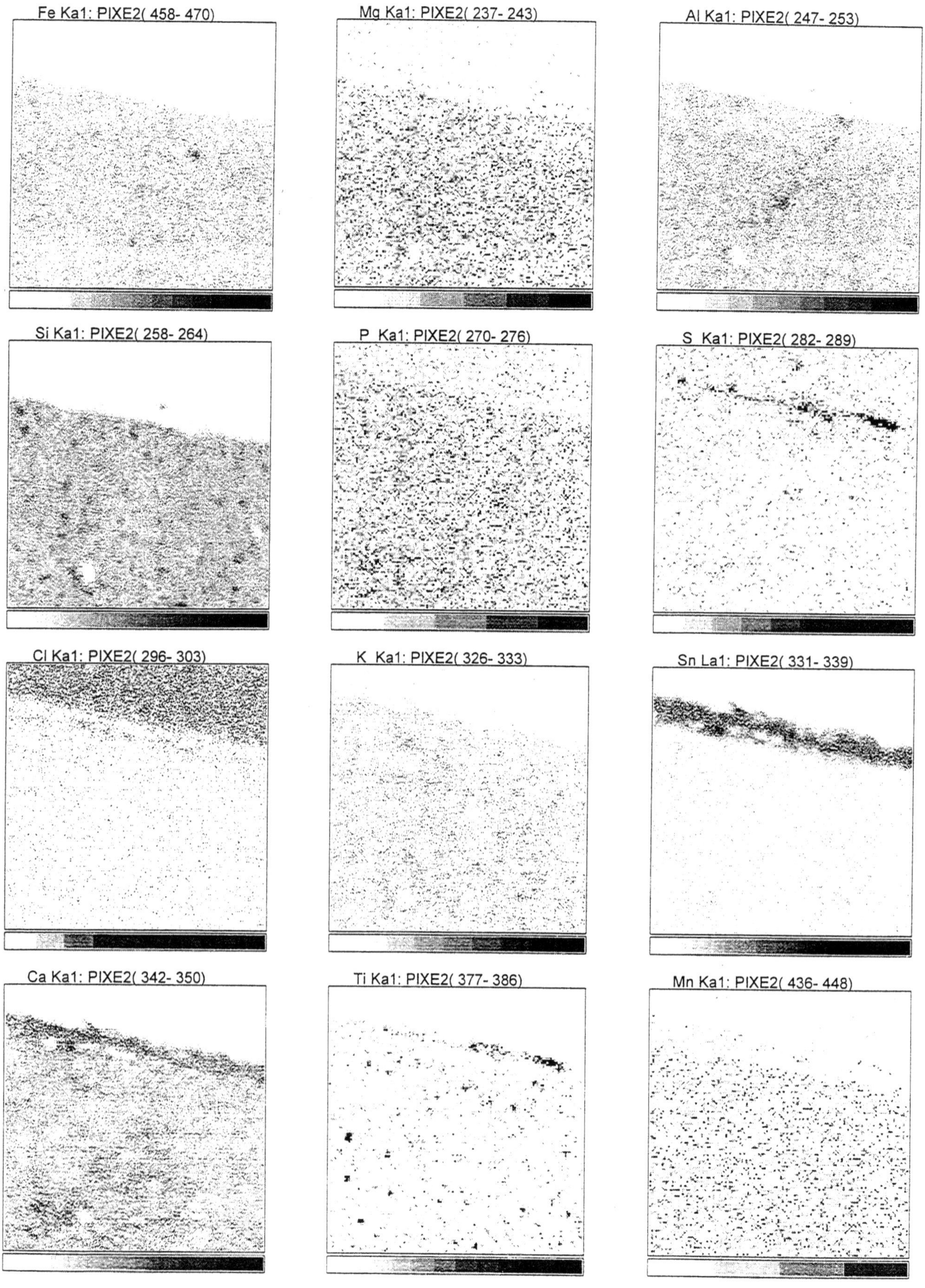

Figure 3. High resolution X-ray distribution map for tin on ceramic; area 50μm square

Conclusions

In any individual problem involving metals in antiquity no single analytical technique is sufficient to explore all its aspects. In this paper we have examined two basic approaches to understanding tin objects, trace element analysis by inductively-coupled mass spectrometry, and microprobe analysis of intermetallic inclusions. Others, notably optical and scanning electron metallography and neutron activation analysis have a great deal to offer while research continues on aspects stable isotope analysis in connection with tin. The work described here has shown that there is useful information to be gained from both the methods studied, and results would be enhanced by combining the two. However, the approach taken will undoubtedly be dictated by both curatorial concerns and the nature of the material and for corroded tin and tin alloys microprobe techniques offer the only simple way forward. For the microprobe to be successful the presence of intermetallic compounds is required and if the tin has become too refined or the particles too small that way is blocked. In these circumstances, too, trace element levels may become too low for mass spectrometry to be practical. The experiments carried out for this paper have been based on raw or unalloyed tin but within reason it should be possible to extend the techniques to some pewters and solders, the key being the continued presence of iron-tin intermetallic compounds. Suitable samples are available and this research is included in the continuing development of archaeological applications of the scanning proton microprobe.

Note

1. Unfortunately it could not be fully quantified because of a small error introduced by the then physical condition of the X-ray detector but a simple calibration curve suggests that the actual concentration was in the range 0.2-0.3%. In this example the other transition elements were undetectable. The experiment will be repeated with a repaired detector.

Acknowledgments

The writers must express their thanks to Mr C.J. Salter for the data from and discussions about the Caerloggas and Merrivale slags; to Valentin Rychner and Dr Francis Pryor for the samples for ICP-MS analysis and to Dr. S.M. Young for those analyses, and to Dr G.W. Grime for assistance with the operation of the scanning proton microprobe.

References

BIEK, L. 1979. The archaeological iron and tin cycles, in I. Scollar, (ed.), Proceedings of the 18th International Symposium on Archaeometry and Archaeological Prospection, Bonn, March 1978, *Arkaeophysika 10:* 75-81.

BIEK, L., CLARK, A.J., DOUCH, H.L., HARRIS, P.G., JONES, L.J., KAY, P.J., OTLET, R.L., PENHALLURICK, R.D.,SCOTT, B.J., 1994. Tin ingots found at Praa Sands, Breage, in 1974, *Cornish Archaeology 33:* 57-70.

BOLLINGBERG, H.J., 1995. Copper-alloyed artefacts from Roman imports in Scandinavia: elemental analysis, in P.B. Vandiver, J.R. Druzik, J.L. Galvan Madrid, I.C. Freestone & G.S. Wheeler, (eds.), *Materials Issues in Art and Archaeology IV* (Pittsburgh: Materials Research Society Symposium Proceedings, *352*): 621-630.

COGHLAN, H.H. & CASE, H.J., 1957. Early metallurgy of copper in Britain and Ireland, *Proceedings of the Prehistoric Society 23:* 91-123.

FRANKLIN, A.D., OLIN, J.S. & WERTIME, T.A. (eds.), 1978. *The search for ancient tin,* (Washington D.C.: The Smithsonian Institution).

GALE, N.H., 1997. The isotopic composition of tin in some ancient metals and the recycling problem in metal provenancing, *Archaeometry 39(1):* 71-82.

GRANT, M., forthcoming. The sourcing of South African Tin artefacts, paper presented at the International Symposium on Archaeometry, Urbana-Champaign, May 1996, to appear in *Journal of Archaeological Science* .

JONES, O.B.C., 1992. Tin plating, unpublished thesis for Part II of BA degree in Metallurgy and the Science of Materials, Department of Materials, University of Oxford.

MASSALSKI, T.B., 1990. *Binary alloy phase diagrams,* 2nd. Ed., 3 vols., (Metals Park, OH: American Society for Metals).

NORTHOVER, J.P., BROOKS, H., LISTER, J. & LLOYD-MORGAN, G. 1991. Fine metalwork from the Roman cemetery at Stansted Airport, Essex: tinning and high tin alloys, in P. Vandiver, J. Druzik, G. Williams & I.C. Freestone, (eds.), *Materials issues in art and archaeology II,* (Pittsburgh: Materials Research Society Symposium Proceedings, *185*): 719-724.

NORTHOVER, J.P. & ROHL, B.M. 1995. The metalwork from Flag Fen, *Historical Metallurgy 28(2),* (1994): 104-112.

PRIMAS, M. 1984. Bronzezeitlicher Schmuck aus Zinn, *Helvetia Archaeologia, 15, No. 57/60:* 33-42.

PRYOR, F. 1991. *Flag Fen: prehistoric Fenland centre,* (London: Batsford/English Heritage).

RAPP, G. 1978. Trace elements as a guide to the geographical source of tin ore, in Franklin *et al.* 1978: 59-63.

RYCHNER-FARAGGI, A.-M. 1991. *Hauterive-Champréveyres, 9, Métal et parure au Bronze final,* (Neuchâtel: Musée cantonal d'archéologie; Archéologie Neuchâteloise 17).

SALTER, C.J. 1997. A note on the tin slags from Caerloggas Down, Cornwall and the Upper Merrivale blowing-house, Devon, in P. Budd & D. Gale, (eds.), 1997: *Prehistoric extractive metallurgy in Cornwall* (Truro: Cornwall Archaeological Unit): 45-50.

SELIMKHANOV, I.R. 1978. Ancient tin objects of the Caucasus and the results of their analyses, in Franklin *et al.* 1978: 53-58.

TYLECOTE, R.F. 1978. Early tin ingots and tinstone from the western Mediterranean, in Franklin *et al.* 1978: 49-52.

YENER, K.A. & VANDIVER, P. 1993. Tin processing at Göltepe, an Early Bronze Age site in Anatolia, *American Journal of Archaeology 97:* 207-38.

11

Smelting and Sacrifice: Comparative Analysis of Greek and Near Eastern Cult Sites from the Late Bronze through the Classical Periods

SANDRA BLAKELY WESTOVER

American School of Classical Studies,54 Souidias St, 106 76 Athens, Greece

Metallurgical production was carried out in cult sites in both Greece and the Near East through the Late Bronze and Classical Periods. This paper examines the potential for shared religious sensibilities behind these occurrences. While there is a tradition in the scholarly literature of an assumption that the Near Eastern sites may cast light on the Greek material, closer examination suggests more distinctions than similarities; it does, however, also yield a typology of different ways to invoke the metallurgical procedure in a ritual setting, by the juxtaposition of ritual and smelting areas, appeals to divinities who are given either iconographic or titular association with the metallurgical process, or votives that recall the process itself. Religious activities also suggest two distinct purposes -- the exercise of elite control over an economically vital process, or apotropaic appeals by the metallurgists to the divinities they felt capable of offering protection in a dangerous procedure. Greek sites show less carefully articulated sensibilities, but seem largely a convenient juxtaposition of market and production, the larger historical context of Late Bronze and Iron Age Greece suggests reasons for this distinction in the intractability of the metals industry and less dependence upon metals in the economy of the areas in which metallurgy took place at the sanctuaries. The myths of the Greek daimones, however, offer an indication of more fundamental similarities between the two systems: the Greek myths show patterns parallel to those seen in the Near Eastern sites of a distinction between magical and managerial applications of religion. The magical would be much more likely to move with itinerant metallurgists who travelled between East and West -- these metallurgists, however, were not the typical smith in the Greek cultural sphere. Thus the decision to focus on the magical and foreign as the type of the divine smith suggests the creation of a religious image of alterity rather than identification; the daimonic myths are not designed to answer the needs of the metallurgists themselves, but of Greek society at large responding to the social, political and economic impact of technological advance.

Keywords: Cult; Semiotics; Daimones.

Among the most intriguing contexts for metallurgical production in the ancient Mediterranean are cult sites, temples and high places. Metallurgical installations are a part of a number of cult sites in both Greece and the Near East from the Late Bronze through the Classical periods. Two factors about this juxtaposition are basic in setting out on any kind of analysis. The setting is a ritual one; a ritual act, according to Walter Burkert, is one that is redirected for communicative purposes (1983:23). The metallurgical processes, moreover, are notably dangerous and noisy, as well as, inherently fascinating. This leads us naturally to ask what communicative force this juxtaposition could have. Two assumptions recur in the scholarship: first, that the Greek and Eastern sites are comparable in their semiotic employment of the metallurgical act, e.g. Kition on Cyprus offers a type that helps explain the Greek phenomena (Karageorghis 1973: 108; Hiller 1975:89; Leukart, 1975:184; Matthaus 1986:129-191; Risberg 1990:39; Hagg 1990:29) ; second, that the divinities of the Cypriot shrines could logically be expected to represent a metallurgical divinity, e.g. Hephaistos or one of his Eastern counterparts. (Catling 1971:29-30; Karageorghis 1973: 108-9; Robertson 1978:202-205)

With respect to the first hypothesis, comparative analysis of Greek and Eastern sites, as will be seen, reveals very different contexts and cultural dynamics. These comparisons can help us identify useful questions about the data we have, but certainly caution us against assuming, for example, Late Bronze Cypriote sensibilities at work in the Argolid.

The second hypothesis addresses the relationship between mythic ideas and archaeological evidence. These physical remains were embedded in cultural contexts whose icons of power used notions of the sacred that were expressed in a remarkably labile mythological corpus. The Greek Hephaistos, however, does little to explain temple-centred metallurgical activities, nor does Kothar wa Hasis clarify the Eastern exempla. There are, however, traces in the Greek world of a sophisticated mythological conundrum, the daimones known as Daktyloi, Telchines, Kabeiroi, Kouretes and Korybantes, whose association with smithing offers a finely nuanced look into the Greek cultural construction of metallurgy and its practitioners -- one, moreover, with strong connections to the East. The myths tell us that these daimones were magicians who came into Greece from the East, attended the Great Anatolian mother goddess, and invented iron, bronze, and the casting of statues; among them, the Telchines offer the clearest narrative of itinerancy, though the enduring confusion over which daimones truly belonged where, was often answered with a hard-stretched tale of migrations from Crete, the Troad, Rhodes, Samothrace, Chalcis and Euboia (Strabo 14.2.7-8; 10.3.19; Diodorus Siculus 5.70.2, 5.64.4). The myths of the daimones tend to cluster around areas whose economy was particularly metallurgically dependent -- thus Bakhuizen notes that it was in iron-rich Boiotia that Hesiod created the epic on the Idaian Daktyloi (cf. Schachter 1979:161), the metallic resources of Crete that, according to Faure, made the Cretan daimones a virtually inevitable mythological discussion (1966: 45, 78), and Morris notes the coincidence of rich ore deposits and legends of miraculous metallurgists in Rhodes, Euboia and Crete, (1992: 133, 157). Gernet was the first to propose that they represented the mythic articulation of itinerant metallurgical specialists who entered the Greek world from the East (1932: 79); his proposal has been echoed by a number of scholars (Burkert 1985: 281; Morris 1992: 142; Marinatos 1962: 93), but, not to date systematically examined, against such intriguing archaeological evidence as the 14th century Kas and 12th century Cape Gelidonya shipwrecks. However, they hardly offer a neatly legendized vision of metallurgical history – their record is notoriously confusing, evoking curses from such mythographic luminaries as Strabo in the 1st century and Lobeck in the 19th; the latter, in fact, concluded of them 'there is a certain art and science to knowing nothing' (1829: 1110)

The complexity of the daimonic corpus recommends a structural analysis; the outcome of this analysis, which concludes the first half of my dissertation research, reveals patterns in the daimonic myths that show them, as a group, articulating the stresses and strains of a society in the process of metallurgical development, even anticipating the models of Childe, Eliade and others of the itinerant technological specialist (Childe, 1944: 7-24; Eliade, 1962: 102; Renfrew, 1972: 17; Wertime 1964: 1258, 1265; cf. Budd and Taylor 1995: 140; Sandars 1971: 26, Rowlands 1971: 214). Intriguingly, despite the strong distinctions among these phenomena in the two distinct areas, the essential dynamics of the model the daimones offer seem to be legible in the juxtaposition of metallurgy and cult in both east and west. The vigorous and, it now seems, relatively uninterrupted commerce in metals across the Mediterranean from the Late Bronze through the Dark Ages would seem to suggest that cultural contact played some role in the coincidences between the two systems; the distinctions between them point us to the complexities of the cultural translation of ideas. Considering the archaeological evidence against this strange mythic corpus, I will argue that the daimones seem to be not, as has been proposed in the past, merely the religion of metallurgical practitioners, the importation of Eastern demons intact into the Greek cultural sphere. They do, however, seem to represent a Greek discussion of ongoing technological borrowings and artefact exchanges. Thus the genesis for similarities would not be a prototypical divine metal smith, but shared participation in the international Aegean metallurgical trade. The essential characteristics of this trade -- long-distance exchange, high-value items and their specialist crafters, and elites in search of an inalienable iconography -- form an essential commonality between such distinct cultural milieu as Ancient Near Eastern palace economies and the festal contexts of the emerging Greek poleis.

This paper thus draws from three kinds of data -- Near Eastern archaeology, Greek archaeology, and Greek myth. I will first present the results of my analysis of the Near Eastern sites, examining the evidence for any archaeologically legible koine as comparanda for the Greek material; I will then present the Greek data, studied first for any recurring patterns among the sites, then in comparison with the Near Eastern material. The final section of this paper will consider the social dynamics these data suggest against the Greek mythological corpus.

Turning first to the Eastern sites, an unofficial 'canon' has emerged as scholars publishing on one site refer to the others as comparanda. These sites are naturally divided into three cultural groups -- Egyptian, Levantine, and Cypriot, the latter with strong Aegean influence. Cypriot sites include Athienou , Kition , Enkomi, Kalopsidha and Tamassos; Egyptian are Serabit el Khadem, and Timna, which includes the temple of Hathor, a Midianite shrine, a bamah, and a funerary shrine. Semitic sites include the Middle Bronze Nahariya, the Late Bronze Megiddo, Ras Ibn Hani and Kamid el Loz, and Dan and Taanach from the Iron Age. I have considered three broad categories in the analysis of the sites: environment and economy, the type of metallurgical processes and artefacts found, and religion, e.g. the nature of the cult and metallurgy link, votives and cult equipment, divinities worshipped and iconography.

Comparative analysis of these Eastern sites is as instructive for what it does not reveal as for what it does suggest. We do not see Kabeiric sanctuaries that support the ancient and modern arguments for Near Eastern antecedents for the Greek divinities; we do not even see the same god worshipped in different places, or any kind of international koine in offerings, iconography, and cult paraphernalia from one site to the other. We have, instead, the local tendencies applied in each case, e.g. Bucrania in Cyprus, Hathor in Egypt, massebot and offering benches in Semitic shrines.

What do emerge, however, are some models of different ways to invoke metallurgy in a religious context -- using either the site itself, the god, or the votives. Thus, for sites we may have the introduction of metalworking into a sacred area -- a forge in the Vatican, as it were. The most famous example of this is Kition, where metallurgical production in direct communication with the holy of holies is attested from the 13th century through the Phoenician levels (Karageorghis 1976: 54-55, 62-63, 88-89, 112; 1973: 105); others include the 12th century bamah at Timna, which shows signs of metallurgical production and debris, as well as, normal votive offerings (Rothenberg 1972: 115-16), the Middle Bronze site of Nahariya, (Dothan 1981: 76-77) where a mould for the casting of a female figurine was found directly at the bamah, and at Taanach, where evidence of 10th century metal production in the cultic building suggests smithing under sacred auspices (Stech-Wheeler, Muhly, Maxwell-Hyslop and Maddin 1981: 249-51) . Other sites offer the reverse pattern, that is, the introduction of a sacred area into a primarily industrial context -- a shrine in Pittsburgh, as it were. Examples include Kamid el Loz, where offerings in workshop pits and the ritual slaughter of sheep and goats suggest cultic activities and magical performances (Frisch, Mansfeld, & Thiele 1985: 181-82) , and Dan, where a snake house and chalices were found in metal workshops (Biran 1994: 145-57).

A second vehicle for invoking metallurgical processes is the divinity itself. Enkomi has given us, in the god on the ingot, an iconographic association of the divinity with metallurgical processes (Courtois 1982: 156; Masson 1992:155-56) That the god in question is of the widely attested Near Eastern smiting god type suggests the application of metallurgical associations to an already familiar form, rather than the unique iconography of a specifically metallurgical divine. Timna and Serabit-el-Khadem provide examples of an analogous operation in appeals to Hathor as Lady of Turquoise or Malachite, giving a non-metallurgical divinity a titular association with mining (Beit-Arieh 1985: 115-116; Rothenberg 1972: 172; Muhly 1984: 291). An appeal to traditional metallurgical divinities by metallurgical workers -- which could perhaps be expected in these communities so uniquely dedicated to metal production -- is not at all in evidence.

Votives constitute a third method for associating religion with production. These can take the form of dedications of metallurgical waste material -- thus at Athienou, a centre for smelting, production and shipping, votive pits held offerings of metal scrap, and a number of the approximately 10,000 votive juglets in the courtyard held copper nodules. (Dothan and Ben-Tor 1983: 95) Dedications in the 14c Hathor temple at Timna included copper prills, fragments, spills and trimmings from the various stages of metal production, including copper waste pieces in the shape of animals that may represent primitive castings (Rothenberg 1972: 172-177). The offering of votive tools and other metalworking paraphernalia, seen at Kalopsidha, Athienou and Timna, similarly suggests the presentation of metallurgic concerns to the god; votive versions of the metallurgical product -- miniature ingots – attested at Enkomi offer a third type (Masson 1971: 449-55)

The possible functions proposed for these juxtapositions also vary from one site to the other; the distinctions cannot be ascribed merely to the interpretative predispositions of the scholars, but must show convincing support in the evidence itself. Thus, Timna and Nahariya support the model of on-site production of votives for purchase and dedication; Kition and Enkomi have been interpreted by Knapp as the elite manipulation of sacred symbols to ensure control over the metallurgical industry (1986: 4). Athienou and Timna, Dan and Kamid el Loz suggest sacred activities carried out by the metallurgists on their own behalf -- the dedication of the 'gleanings' of the metallurgical processes, employing the economically insignificant residue that would be available to the limited economic circumstances of the workers, suggests the private sensibilities of those workers much more than marketing interests or the flexing of elite muscle. The fact that these two religious sensibilities are not mutually exclusive, but can be seen in a single site such as Timna, we are cautioned against the assumption of a single meaning for sacra in a metallurgical context. There seems, moreover, to be a rough division between the cultural groups studied and these two sensibilities: the Levantine sites favour private dedications to unnamed and aniconic divinities; Cypriot sites suggest public and state sensibilities for metallurgy and cult in tandem; and Timna offers both, in the form of an official cult of Hathor in whose precinct metallurgical operations were carried out; and Midianite honours to an aniconic divinity. The activities of the latter were so contrary to the Hathor cult that the Midianites defaced the temple materials before reusing them in the construction of their Semitic shrine (Rothenberg 1972: 150-151).

Turning to Greek archaeological material, cult centres are associated with metallurgical production in both the Late Bronze and the Iron Age. Late Bronze evidence consists of three types: the juxtaposition of metalworking and cult areas in the Cretan palaces, metallurgical finds in caves at Psychro, Arkalochori and Ida, and written evidence from the linear B archives at Pylos. The evidence is less compelling, in all three categories, than the Near Eastern material. The palace workshops are approximately, but not emphatically, located near the shrines -- the shrine at Malia has not been found, but is assumed to have been on the upper floor, and the juxtaposition at Zakros is argued from the predominantly religious function of the west wing in which the workshops were found (Hagg 1992: 29-32) None of the caves have yielded evidence of actual kilning of the material, but the richness of the finds - huge numbers of votives, swords in perhaps half-finished form, bun-ingots - suggested to Marinatos the priestly execution of metallurgical processes as part of initiatory ritual (Marinatos 1962: 93-94). The much-argued Pylos tablets do not suggest any awareness of such functions, but seem to indicate a special association of the smiths with Potnia, (Hiller 1979: 190; Leukart 1975: 183-88; Hurst 1968: 94; Smith 1992:184; Chadwick 1973: 354) a particularly high status for bronze workers , (Gillis 1997:513; Chadwick 1973:292-3) or the requisitioning of what is ultimately a small portion of their production for the palace (Gillis 1997: 511). The data take very different forms beginning in the mid-8th century, when evidence of metallurgical production first appears in the form of a smithy at the shrine of Apollo at Eretrie; bronze and iron manufacturing at the sanctuaries of Artemis and Apollo at Kalapodi, Athena Alea at Tegea, Athena Itonia at Philia, and the panhellenic sanctuary at Olympia; other sites include the Heraion on Samos, Delphi, and Delos (Risberg 1992: 33; Treister 1996: 76; Morgan 1993: 23). Not only votives, but such pragmatic pieces as door hinges, fibulae, swords and armour were manufactured -- reminding us that, however numinous the setting, the sanctuaries were clearly also a lucrative market (Treister 1996: 105).

The distinctions between the Greek LB and IA examples are significant. The IA consist largely of itinerant workers in a festal context, producing votives for purchase and commission (Treister 1996: 77, 122-3; Heilmeyer 1969: 2, 21; 1979: 52; Morgan 1993:23). They occurred against a historical background of a significant increase in the amount and economic importance of metals in the Greek economy (Snodgrass 1980: 49-54) ; metal goods, moreover, as the sanctuaries themselves, played a significant role in the articulation of rivalry between new born states (Treister 1996: 20). Thus, with respect to the interest and energies dedicated to metals, the Greek Iron Age seems more analogous to the Near East Late Bronze metallurgical communities than the Minoan and Mycenaean palaces were.

Comparing the Greek with the Near Eastern material, we see significant distinctions in economic context, industry control, and religious articulation. The Greek context, even in the metal-enthusiastic Geometric period, is never as dependent on metals as the Near Eastern counterpart. Perhaps in accordance with this, there is less evidence of successful elite control of the industry in either the Late Bronze or the Iron ages: archaeological evidence from LB

Crete suggests the existence of extra-palatial workshops on the island's northern shore, and workshops in the palaces seem to have been limited to the production of sacral goods. Studies of 11-9 century burials, and analyses of the wider context of Iron Age metallurgical production, have suggested that we see in Greece not a lack of interest in controlling the metals industry, but more a failure to do so (Morris 1989: 508; Morgan 1993: 18-44). The religious articulations of metallurgy are also quite distinct between East and West: we see no attempt, in Greece, to make the divinity, at whose precinct the metallurgy was executed, particularly metallurgical in its aspect: we have no Apollo on the ingot or Artemis of the copper. The metallurgical act is aniconic and non-titular in those instances when it is juxtaposed to the sacred areas. We see neither metallurgists offering bits of slag, interesting looking spills or casting runoff as votives, nor special concern dedicating smithing tools. For Greek divinities that do have mythic connections to metallurgy, the daimones in particular, the occasional appearance of hammers in Kabeiric iconography, the possible identification of a lodestone at Samothrace and the wearing of iron rings by initiates is as close as we come to an invocation of the metallurgical act (Cole 1984: 30) Thus what is invoked ritually - through site configuration, iconography, and dedications -- in the Near East is approximated only mythically and metaphorically in Greece: none of the communicative potential of the ritual setting is brought to bear on what must then seem, in the Greek context, a merely convenient juxtaposition of marketplace and manufacturing.

The archaeological data thus seem to give a damning review to the possibility of any shared metallurgical ideologies between East and West. If we consider the corpus of daimonic myths, however, we may see three points in the relationship between myth and archaeological data that suggest a deeper similarity between the two systems. The first is the evidence for two kinds of religious sensibilities, one serving the managerial ambitions of an elite, the other serving the personal, magical and apotropaic sensibilities of the workers themselves. Both are legible in the archaeological data from the Near East; in the Greek sphere, however, they are much more visible in the daimonic myths. Structural analysis of the five daimonic types reveals an intriguing correspondence between the strength of the metallurgical association and the relationship to established powers. The two daimones, Daktyloi and Telchines, are the most clearly metallurgical (Diodorus Siculus 5.55; Pliny *Natural History* 7.57; Clement of Alexandria *Stromata* 1.362; Scholia Laurentiana to Apollonius Rhodius *Argonautica* I 1126-1131b, Strabo 10.473; Statius *Thebaiad* II 269-276; *Silvae* IV 6, 44-49), and are also the most magical, resistant to divine protocol, prone to physical deformation, and destructive. They are fratricides (Clement of Alexandria, *Protrepticus* II 19, 1.3.4; Nauck, *Tragicorum Graecorum Fragmenta* 337), infanticides (Eusebius, Praeparatio Evangelicum 4.16), proverbial for jealousy (Suetonius, peri dusfhmwn lecewn; Hesychius, TelcineV: Scholia to Statius *Thebaid* II 274 = Mythogr. Vat. ed. Bode II 185; Nikolaos of Damaskos fr. 114 Jacoby; Plutarch, *On Whether Virtue Can Be Taught* II. 387), antithetical to the life-giving forces of agriculture and the Great Mother (Plutarch, Alexandrian Proverbs 1 p.; Pherekydes of Athens, Istoriai: FGH 3 F47; Sophokles in Nauck, *Tragicorum Graecorum Fragmenta* 337; Zenobius IV.80; Kallimachus *Aitia* fr. 9 p 64 ff; Diodorus Siculus 5.55; Strabo 14.2.7-8 ; Ovid *Metamorphoses* 7.365-7) and even credited, in the sixth century AD, with the collapse of the dome of the Hagia Sophia (Paulus Silentiarius, *Description of the Hagia Sophia*, 193-199). The two daimones most loosely linked to the forge, the Kouretes and Korybantes, articulate, in contrast, enthusiastic support of kings, young men in military training, and of physical beauty: they protect the infant Zeus from his father's appetite (Sallust fr. 3.67; Lucretius II.600; Vergil *Georgics* 4.149-152; Strabo 10.3.7, 11; Martial 9.16; Hyginus *Fabulae* 139.3), instruct youths in war dances, invent laws and a host of other useful social practices (Diodorus Siculus 5.65-66). The Kabeiroi, a fifth type, form an intermediate category sharing the characteristics of both. This division in the mythic types of magical, secretive powers on the one hand, and normative public rituals on the other, echoes the model of two religious types in the Near Eastern evidence.

A second point addresses the issue of the metallurgist as magician. We know of the importance of Near Eastern metallurgists in the Greek milieu, and understand that they brought ideas, as well as, artefacts with them. Operating in the Iron Age as free agents rather than as servants of palace systems, it seems unlikely that they would have had much use for religious sensibilities that served the goals of a managerial elite. The need for magical, apotropaic forces, however, would remain as constant as the technological processes themselves, as they answer the Malinowskian expectation of appeals to the divine in the face of uncertain and dangerous technological procedures. The likely bearers, then, of magical sensibilities could well be expected to be marked as magicians in the cultural milieu they entered – and in the Daktyloi and Telchines, they are.

A final point returns to Gernet's injunction to consider the smiths themselves. Not all Iron Age smiths were Eastern itinerants; not all Greek Iron Age metals production took place at sanctuaries. The construction of the smith in the Greek mythic corpus, however, focuses precisely on the notion of a foreign itinerant inventing and producing substances (iron) and objects (statues, armour and weapons) of decisive economic and social impact. The historical referent for the itinerant foreign smith is the master, much more than the part-time worker, who is presumably an agriculturist able to meet his own simple metallurgical needs; the festal context, geared to the production of status and display items, is the most likely market for the itinerant master. Thus the 8th century emergence of festal metallurgical production -- semiotically marked and liminal in time, space and personnel -- may have provided the referent for the type used for the mythic discussion of metallurgy. Why, though, would the myths seize upon a nontypical, rarefied type instead of the more familiar phenomenon? The answer may lie in the fact that the Greek sphere does not provide any LB or IA indicators of ritual or religion by, or on behalf of, metallurgists. Based on my analysis of the daimonic myths, they do not represent a smith's view of smithing, but rather the perspective of society at large on the operation of metallurgists and the impact of their products. The smith in the Greek mythic context is thus an icon of alterity, rather than a point of divine identification; the embodiment of liminality, he enables discussion of the problems and tensions associated with a society in the processes of increasing social complexity, international contact, contested hierarchies, and need for self-identification -- the anthropological type for the emergence of new technologies. Condensing the ideas of alterity, expertise and disturbingly valuable power, the myths constitute a more useful category than merely metallurgists' religion, and one that helps suggest the remarkable longevity of the polymorphic Greek daimones of the forge.

References

BEIT-ARIEH, H. 1985. Serabit el-Khadim: new metallurgical and chronological aspects. *Levant* 17:89-116.

BIRAN, A. 1994. *Biblical Dan.* Jerusalem: Israel Exploration Society.

BUDD, P. & T. TAYLOR. 1995. The faerie smith meets the bronze industry: magic versus science in the interpretation of prehistoric metal-making. *World Archaeology* 27 (1): 133-143.

BURKERT, W. 1983. *Homo Necans: The Anthropology of Ancient Greek Sacrificial Ritula and Myth.* Berkeley: University of California Press.

CATLING, H.W. 1971. A Cypriot bronze statue in the Bomford collection. in *Alasia* Tome IV, Premiere Serie. ed. C.F.A. Schaeffer, 15-32. Paris: Klincksieck.

CHADWICK, J. 1973. *Documents in Mycenaean Greek.* Cambridge: Cambridge University Press.

CHILDE, V.G. 1944. Archaeological Ages as Technological Stages. *Journal of the Royal Anthropological Institute of Great Britain and Ireland* 74: 7-24.

COLE, S. 1984. *Theoi Megaloi: The Cult of the Great Gods of Samothrace.* Leiden: Brill.

COURTOIS, J.C. 1982. L'activite metallurgique et les bronzes d'Enkomi au Bronze Recent (1650-1000 avant J.C.) in J.D. Muhly, R. Maddin & V. Karageorghis, (ed.), *Early Metallurgy in Cyprus.* Nicosia: Pierides Foundation.

DOTHAN, T. & A. BEN-TOR. 1983. Excavations at Athienou, Cyprus 1971-1972. *Qedem* 16. Jerusalem: Monographs of the Institute of Archaeology, The Hebrew University.

DOTHAN, M. 1981. Sanctuaries along the Coast of Canaan in the Middle Bronze Period: Nahariyah in A. Biran (ed.), *Temples and High Places in Biblical Times.* Jerusalem: Nelson Glueck School of Biblical Archaeology of Hebrew Union College, Jewish Institute of Religion

ELIADE, M. 1962. *The Forge and the Crucible.* tr. Stephen Corin. New York: Harper.

FAURE, P. 1966. Les Minerais de la Crete Antique. *Revue Archaeologique* I: 45-78.

FRISCH, B., G. MANSFELD & W.-R. THIELE, (ed.) 1985. *Kamid el-Loz 6: Die Werkstatten.* Band 33: Saarbrucker Beitrage zur Altertumskunde, hrsg. Rolf Hachmann und Walter Schmitthenner Bonn: Dr. Rudolf Habelt GMBH.

GILLIS, C. 1997. The Smith in the Late Bronze Age - State Employee, Independent Artisan, or Both? in R. Laffineur & P. Betancourt, (ed.), *Techne: Craftsmen, Craftswomen and Craftsmanship in the Aegean Bronze Age.* Universite de Liege: 505-513.

HAGG, R. 1992. Sanctuaries and Workshops in the Bronze Age Aegean, in T. Linders & B. Alroth (ed.): 29-32.

HEILMEYER, W-D. 1969. Giessereibetriebe in Olympia. *Jahrbuch des Deutschen Archaologischen Instituts* 84:1-28.

1979: Fruhe olympische Bronzefiguren: Die Tiervotive. *Olympische Forschungen* 12. Berlin: De Gruyter.

HILLER, S. 1979. Ka-ko na-wi-jo. Notes on Interdependences of temple and bronze in the Aegean Bronze Age. in Risch & Muhlestein (ed.): 189-95.

HURST, A. 1968. Apopos des forgerons de Pylos. *Studi Micenei ed Egeo-Anatolici,* fascicolo quinto. Incunabula Graeca vol. 26. Roma: Edizioni dell'Ateneo. 92-96.

KARAGEORGHIS, V. 1973. Contributions to the religion of Cyprus in the 13th and 12th centuries B.C. in *Acts of the International Archaeological Symposium, 'The Mycenaeans in the Eastern Mediterranean'.* Nicosia: Deptartment of Antiquities, Cyprus and Chr. Nicolaou and Sons Ltd.

1976. View from the Bronze Age: Mycenaean and Phoenician Discoveries at Kition. New York: Dutton & Co.

KNAPP, A.B. 1986. Copper Production and Divine Protection: Archaeology, Ideology and Social Complexity on Bronze Age Cyprus. Goteborg: Paul Astroms Forlag.

LEUKART, A. 1979. Autour de ka-ko na-wi-jo: quelques criteres. in Risch & Muhlestein (ed.): 183-88.

LOBECK, C. 1961 (1829) Aglaophamus; sive, De theologiae mysticae Graecorum causis libri tres. Darmstadt: Wissenschaftliche Buchgesellschaft.

MASSON, O. 1971. Deux petits lingots de cuivre inscrits d'Enkomi in C.F.A. Schaeffer (ed.), *Alasia,* Tome IV. Paris: Klincksieck.

MARINATOS, S. 1962. Zur Frage der Grotte von Arkalochori. *Kadmos* I: 93.

MORGAN, C. 1993. The Origins of pan-Hellenism in N. Marinatos & R. Hagg (ed.), *Greek Sanctuaries: New Approeaches.* London: Routledge.

MORRIS, I. 1989. Circulation, Deposition, and the Formation of the Greek Iron Age. *Man* n.s. 24: 502-519.

MORRIS, S. 1992. *Daidalos and the Origins of Greek Art.* Princeton: Princeton University Press.

RENFREW, C. 1972. *The Emergence of Civilisation: The Cyclades and the Aegean in the Third Millennium B.C.* London: Methuen.

RISBERG, C. 1992. Metal Working in Greek Sanctuaries. in T. Linders & B. Alroth (ed.): 33-40.

ROBERTSON, N. 1978. The Goddess on the Ingot in Greco-Roman Times, *Report of the Department of Antiquities of Cyprus:* 202-205.

ROTHENBERG, B. 1972. *Timna: Valley of the Biblical Copper Mines.* Great Britain: Thames and Hudson.

ROWLANDS, M.J. 1971. The archaeological interpretation of prehistoric metalworking. World Archaeology 3: 210-224.

SANDARS, N.K. 1971. From Bronze Age to Iron Age: a sequel to a sequel. in J. Boardman, M. Brown & T.G.E. Powell, (ed.), *The European Community in Later Preshistory, Studies in honour of C.F.C. Hawkes.* London: Routledge & Kegan Paul.

SCHACHTER, A. 1979. The Boiotian Herakles. *Teiresias* Supplement 2: Proceedings of the Second International Conference on Boiotian Antiquities.

SMITH, J. 1992. The Pylos Jn Series. *Minos*: 27-28: 167-259.

SNODGRASS, A. 1980. *Archaic Greece.* London: J.M. Dent & Sons.

STECH-WHEELER, T., J.D. MUHLY, K.R. MAXWELL-HYSLOP & R. MADDIN. 1981. Iron at Taanach and Early Iron Metallurgy in the Eastern Mediterranean. *American Journal of Archaeology* 85 (3): 245-268.

TREISTER, M.Y. 1996. *The Role of Metals in Ancient Greek History.* Leiden: E.J. Brill

WERTIME, T. 1964. Man's first encounters with metallurgy, *Science* 146 (3649): 1257-67.

12

Bronze Age copper smelting technology in Cyprus - The evidence from Politico Phorades

VASILIKI KASSIANIDOU

Archaeological Research Unit, University of Cyprus, P.O. Box 537, CY 1678 Nicosia, Cyprus

During the Late Bronze Age, Cyprus was one of the main producers and exporters of copper in the Eastern Mediterranean. Research on the smelting technology of this period, however, has been based mainly on archaeometallurgical finds from urban centres located on the coast. These, presumably, represent secondary processes. During the 1996 season of the Sydney Cyprus Survey Project, a copper smelting workshop was located on the bank of an ancient creek bed at the site of Politico-Phorades. A joint team from the Universities of Glasgow and Cyprus excavated the site in August 1997. Pottery recovered during the excavation, as well as the radiocarbon measurement of a single sample of charcoal, suggest that the workshop dates to the Late Bronze Age. An early date is also indicated by the archaeometallurgical finds. The slags are plano-concave in shape, a morphology previously unknown from Cyprus, and preliminary analysis has shown that they have a rather high metal content. The slags, together with some well preserved furnace lining and tuyére fragments, provide a good insight into this, presumably early, smelting technology. This paper presents the findings of the first season of excavation at Politico Phorades, as well as, some of the preliminary analytical results and conclusions regarding the archaeometallurgical finds. These are compared and contrasted to those from other LBA Cypriot sites which, until now, have been considered representative of the island's Bronze Age smelting technology.

Keywords: Copper; Primary Smelting Workshop; Cyprus; Late Bronze Age.

Introduction

In the proceedings of the "Early Metallurgy in Cyprus" conference which were published over 15 years ago Tamara Stech introduced her paper on "Urban Metallurgy in Late Bronze Age Cyprus" with the following paragraph (1982: 105): "*(Yet) in spite of the archaeological and historical significance of the Cypriot copper industry, basic questions about its nature are unanswered: what were the smelting techniques? Where did smelting take place? What were the mechanisms through which copper - in various forms was distributed? How were copper production and trade administered?*".

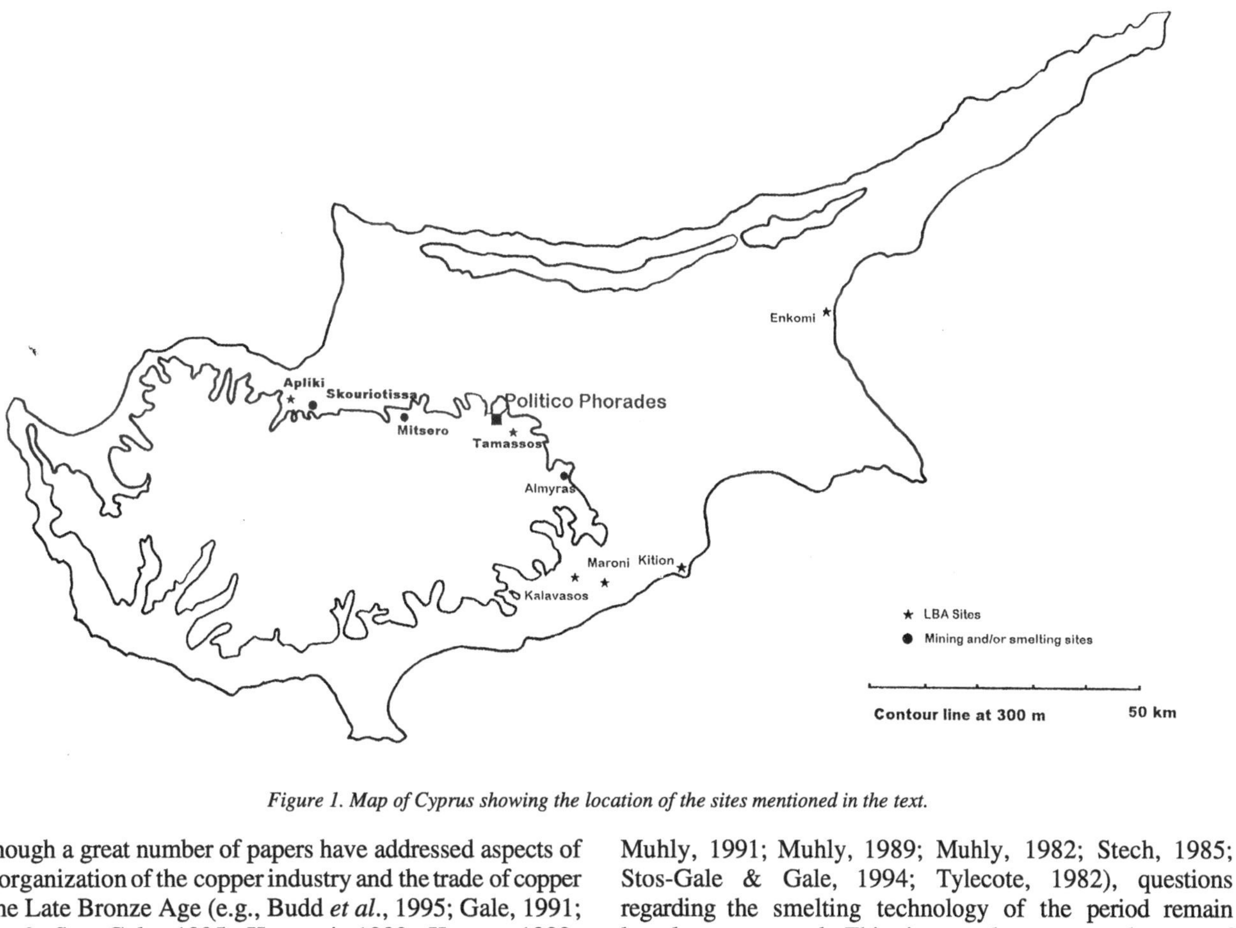

Figure 1. Map of Cyprus showing the location of the sites mentioned in the text.

Although a great number of papers have addressed aspects of the organization of the copper industry and the trade of copper in the Late Bronze Age (e.g., Budd *et al.*, 1995; Gale, 1991; Gale & Stos-Gale, 1995; Keswani, 1993; Knapp, 1988; Muhly, 1991; Muhly, 1989; Muhly, 1982; Stech, 1985; Stos-Gale & Gale, 1994; Tylecote, 1982), questions regarding the smelting technology of the period remain largely unanswered. This is not due to an absence of archaeometallurgical finds from excavated, Late Bronze Age sites. In fact, there are a plethora of such finds. The problem lies in the fact that all of the excavated sites are either urban and/or religious centres located along the coast or in the agricultural hinterland (Figure 1). It has, therefore, been argued that archaeometallurgical debris found there is unlikely to be connected with smelting, but only with secondary processes, such as the refining of black copper (the rather impure, iron and sulfur rich, metallic copper produced by the primary smelting of roasted sulfide ores), alloying and casting was produced somewhere near the mining districts, and then transported to the urban centres. There it was re-smelted in small furnaces to form a small pool of copper at the bottom of the furnace with a chunk of slag above. Copper and slag separation is presumed to have been poor leaving the slag rich in metal prills, especially within its lower and outer portions. Stech *et al.* proposed that the copper was recovered largely through the crushing of the slag, and the removal of the copper prills trapped within it. These were then thought to have been melted together to form a copper ingot. (Muhly, 1989: 302).

However, furnace conglomerate and 'primary' smelting slags, presumably the products of roasting and primary smelting processes, have also been found in the urban centres. In order to explain their presence, Stech *et al.* (1985: 399) have put forward a two-stage model of production. In this model the finds represent a primitive technology where a partially smelted furnace conglomerate

The reconstruction of Stech *et al.* although plausible was largely theoretical, and could only be verified by the excavation a primary smelting workshop and by comparing the processes and archaeometallurgical finds represented there with those found in the urban centres. Although there are large slag heaps scattered over the whole of the Troodos periphery, which clearly represent primary smelting workshops, no smelting furnace was known in Cyprus at the time of the conference in which the model was put forward.

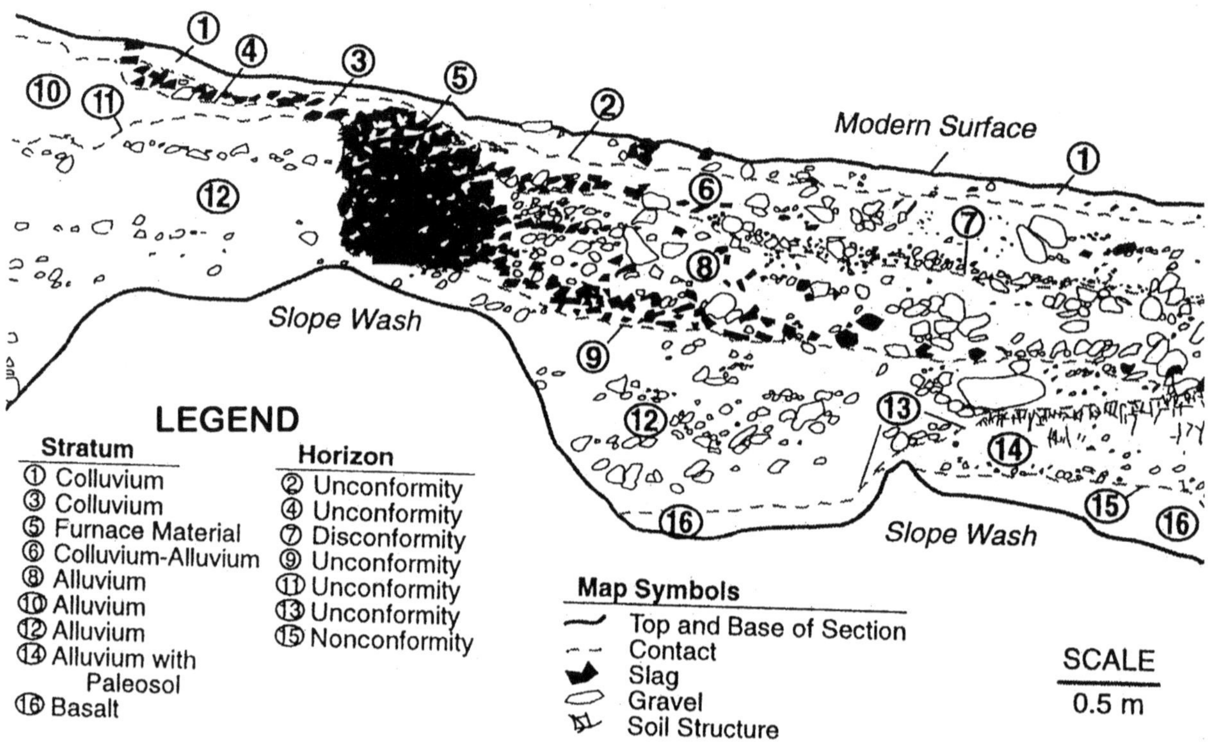

Figure 2. Drawing of the section of the site at Politico Phorades exposed by the creek.

Thus Weisgerber (1982: 29) concluded his paper at the same conference with the following words: *"To get detailed information of mining and smelting antiquities, these first have to be studied in small, one-period sites. I am rather sure that these either are known or still can be found on Cyprus by extensive survey and also by looking at river banks."*

Weisgerber was proven right when, in 1982 during a survey of the area around the MBA settlement of Alambra *Mouttes*, Walter Fasnacht located the site of Ayia Varvara-*Almyras*, the first primary copper smelting workshop to be discovered in Cyprus (Fasnacht, 1983: 89). However, the earliest of the three periods of occupation at Almyras dates one thousand years after the beginning of the Late Bronze Age (Fasnacht *et al.*, 1991: 102) and, therefore, does not help us answer the questions regarding Bronze Age smelting technology. Fortunately Weisgerber has now been proven right for a second time with the discovery of another primary smelting workshop, on the bank of a creek which, based on all the presently available evidence, dates to the Late Bronze Age. This site, Politiko *Phorades*, is the subject of the present paper.

Discovery of the Site

The workshop was discovered in 1996 during the fourth season of fieldwork of the Sydney Cyprus Survey Project in Special Interest Area 7, which lies at the south-eastern extreme of the area investigated (Figure1). The presence of a smelting workshop was indicated by a large quantity of slag and refractories projecting from the 4 meter high stream bank of a tributary creek of the Kouphos river.

After collecting all the slag and furnace lining from the surface of the slope, and cleaning back the section in order to expose the stratigraphy of the deposit, a series of strata were identified (Figure 2). Stratum 2 consists mainly of furnace lining, which was thought, possibly, to represent a collapsed furnace, as well as slag and an intact section of tuyére. Stratum 3, situated on the edge of the ancient creek channel, appears to be an intentional dump of the ancient smelters; in section it consists of an oblong-shaped deposit of slag, furnace lining, and tuyére fragments. Strata 7 and 9 contain detrital slag and furnace refractories from stratum 3 that were reworked into the channel, as well as, a single potsherd.

The section, therefore, was interpreted as a smelting workshop laying on the bank of an ancient creek into which slag was dumped. Subsequent to the smelting activities, the creek eroded the lower slag heap and incorporated this in alluvial deposits that eventually buried much of the smelting workshop. Subsequent migration of the Kouphos meanders, combined with river incision, have lowered the stream to its present position.

Preliminary Date

The single piece of pottery recovered in the 1996 season was of the White Slip 1 type which dates to the 16th century BC. This find provided only a tenuous indicator of the possible date of smelting activity, since these strata were formed as a result of erosion of the ancient creek bed. The White Slip sherd, therefore, may be contemporary with or predate the smelting activity.

Confirmation of the Late Bronze Age date was sought through the help of radiocarbon dating. Three samples of charcoal collected from three different fragments of furnace lining were sent to Oxford for AMS radiocarbon dating. Two of the samples were too small to give a reading while the third one (OxA-7013: 2865 ± 75 BP) renders a calibrated date in the 12th or 10th century BC (68.2% confidence 1210 BC-1180 BC or 1150 BC to 900 BC, 95.4% confidence 1260 BC-830 BC, calibration courtesy of Dr Sturt Manning, Department of Archaeology, University of Reading). The date range is considerably later than that suggested by the sherd. Nevertheless the earliest possible date (at 1 sigma confidence level) falls well within the LBA while the later possible date (again at 1 sigma confidence level) corresponds with the end of the LBA and the transition to the Iron Age

The early date of the site, the unique archaeometallurgical finds recovered from it and the fact that the workshop could soon be completely destroyed by erosion, led to the decision that the site be excavated. The excavation was undertaken by a joint team from the Universities of Glasgow and Cyprus under the direction of Bernard Knapp and the Field Direction of Michael Donnelly and the author.

Geomagnetic Survey

Before the excavation, a geomagnetic survey of the site was undertaken with the help of Dr Sturt Manning and Mr David Sewell of the University of Reading. In this case the methodology was problematic as both the slags and the underlying bedrock are ferromagnetic making it difficult to identify sub-surface archaeological and/or archaeometallurgical features. Nevertheless an area of 20 x 30m was surveyed with the help of a fluxgate gradiometer. Some anomalies were detected in the surveyed area which were taken into consideration when deciding on the location of excavation trenches.

The excavation

Three trenches were opened. Trench A, the largest, measuring 14 x 8m, was designed to include the edge of the creek's bank. Two other trenches, B and C were opened up in areas where the geomagnetic survey detected an anomaly. In both trenches the anomalies were due to geological rather than archaeological features. Trench A was divided into squares measuring 2 x 2m and a few of these squares were selected for excavation (Figure 3). Particular attention was paid to squares E and D, which represent the slag heap and layers of slag spilling on to the channel bed, Squares K and L, in order to detect the extent of the slag heap. Squares F, G, M, N were also excavated as were T and U, which in theory correspond to the ancient creek's bank on which the smelting activities may have been taking place. Finally Squares I and H, where an anomaly was detected by the geomagnetic survey, were also excavated.

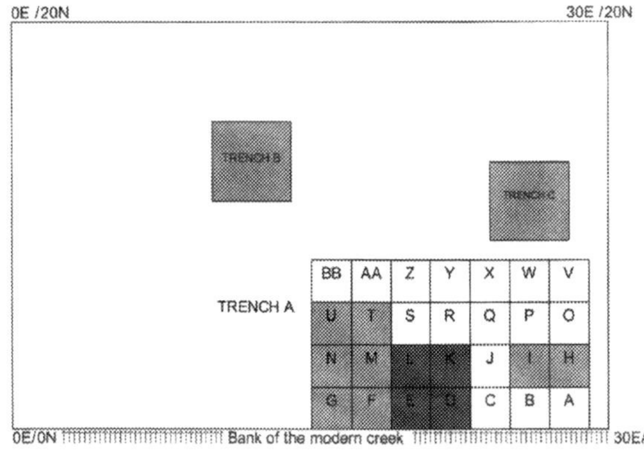

Figure 3. Schematic drawing of Trench A at Politico Phorades. Not drawn to scale.

As the excavation progressed the edge of the slag heap was defined. This turned out to be smaller than originally expected. Strata 7 and 9 were exposed in section. These correspond to layers of slag which slipped from the slag heap to line the bed of the channel. Large quantities of slag were recovered from these strata as well as more pottery sherds. Towards the eastern edge of the deposit a large concentration of very large fragments of furnace lining was found. The size and number of these seems to indicate that furnace lining was deliberately dumped there and soon covered before the stream had the chance to carry it much further down. Squares L and K produced the largest number of almost complete tuyéres from the excavation, as well as, some further pottery sherds found in close association with them.

Unfortunately, the squares corresponding to the creek's banks were fairly barren of archaeometallurgical or other finds. As the excavation of these squares progressed a set of plough marks was discovered which may explain the dearth of finds. When the layer containing the plough marks was removed in one of the squares, a second set of plough marks, perpendicular to the first was discovered. This set is at a considerable depth so that the chances of discovering features *in situ* are greatly diminished. However, further excavation at greater depth is planned in the next field season to investigate the possible survival of archaeological features.

Squares I and H, in which an anomaly was detected by the geomagnetic survey, produced no significant archaeometallurgical remains but a rather intriguing find. This is a large conical boulder set on a very hard layer of clay. A small deposit of bone was found at the base of the rock. The type of stone, diabase, is exotic to the area and would have had to be carried from a distance, either from the Kouphos river or from high up on the Troodos mountains. Although this is the type of rock used to make the stone tools needed to crush the ore, it bears no wear marks typical of such tools. Its conical shape, large size and exotic character, as well as the discovery of the bone deposit, may indicate a different, non-functional, use of this find (for similar examples see Karageorghis, 1992: 212). Its relationship with the archaeometallurgical remains will have to be ascertained during the next season of excavation.

Finally, the hill above Trench A, where roasting or smelting furnaces may have been placed, was examined. This hill is now practically stripped of any topsoil and the bedrock is exposed. Any features, therefore, that may have been located on this area are by now completely eroded away. In summary, although the excavation confirmed last year's observation of the presence of a slag heap within an old channel, no other features or installations have yet been found *in situ*.

Description of the Finds

Slags

The slags recovered from the site are all of a type previously unknown on Cyprus (see for example Koucky & Steinberg's, 1982 classification of Cypriot slags). Of the 2000 kilos of slag collected this year, only two small fragments preserve a flow structure on their surface characteristicof 'tap slags'. All the rest are from large slag cakes of plano-concave shape, the top surface of which is flat with an estimated diameter of 44 cm. The edge angles inward toward the bottom which is concave in shape (Figures 4 and 5). In section, the slag cakes nearly always exhibit a well defined, large gas pocket, approximately 1 cm thick, which extends horizontally throughout the length of the cake a third of the way between the top and bottom surfaces. The surface of the concave bottom generally exhibits many small gas vesicles and a well defined rim formed parallel to the outer edge. They are homogeneous, very dense, and magnetic, indicating that they are rich in iron or iron oxides. Based upon measurements of some of the larger pieces, the estimated weight of a complete slag cake is about 20 kg. There is only one large slag cake of a different type which is plano-convex in shape.

Figure 4. Section of the plano-concave slag cake, showing the flat top surface, the sharp edge of the cake, the large gas pocket and the concavity.

Preliminary analysis of slags from Phorades with the assistance of Yannis Bassiakos (NCSR Demokritos, Athens) has shown them to contain a large amount of metallic copper sulfides, finely dispersed among the fayalite crystals. Bulk X-ray fluorescence analysis indicated a copper content typically of ~2%. X-ray diffraction analysis of one of the samples detected fayalite and chalcopyrite, typical of matte, as well as wüstite.

Furnace Lining

Just as important and unique to the site is the large collection of furnace lining. A number of different types of rims, wall and base fragments have been collected. The furnace lining is made of coarse refractory clay mixed with a large amount of organic material. In some cases small pieces of

Figure 5. Large fragment of a plano-concave slag cake from Phorades. View of the bottom side showing the concavity. Some of the small gas vesicles are clearly seen on the lower left corner of the object.

gossan have been mixed with the clay, presumably to render it more refractory. Some of the furnace lining pieces are slagged while others are large enough to enable us to estimate the dimensions and form of the furnace. The average outer diameter is 42-44cm. The thickness of the walls varies and there are at least two examples were multiple layers of clay are evident.

One of the most important finds is a corner piece of the furnace bottom with adjoining furnace wall. The wall departs from the base at an angle of approximately 90°. A number of pieces have been found with finger impressions on the outer surface. These help reconstruct the way the furnaces were constructed. It seems the furnace bottom was formed first and the walls built around it. The finger prints indicate that the clay was pushed down and against the outer surface of the base.

The evidence suggests that the furnaces were cylindrical structures with a flat bottom and well defined rims. At the moment the full height of the furnaces is not known. It is also difficult to say whether the furnaces were free standing or set in a pit, although whenever the outer surface is preserved it is smooth and well shaped, which would argue for a free standing furnace or at least a furnace which was constructed in the round and perhaps later set in a pit. The position of the tuyéres is also problematic as no pieces of furnace wall have been found with any clear evidence as to where the tuyéres would have been placed. These, therefore, could either have been positioned at the top or in side walls of the furnaces.

Tuyéres

20 almost complete tuyéres as well as hundreds of fragments were recovered from Trench A. An excavated area of $8m^2$ in Phorades has produced a collection of tuyéres comparable in size and quality to the collections from Enkomi and Apliki. The tuyére fragments fall into four categories: cylindrical tuyéres with slagged tips, unslagged cylindrical tuyéres (possibly because both ends are broken, and therefore the tips are missing), double tuyéres, where one tuyére is set within another, and the fourth type is a tuyére which has a fine tip but is not slagged. This is either a tuyére which had not been used or a tuyére which acted as the connection to the bellows.

The two double tuyéres are intriguing. In both cases the outer tuyére seems to have been built around the first one, while in one case the inner tuyére is slagged which may indicate re-use.

Some of the slagged examples were fired to such an extent that they had started to melt. One broken tip was sealed altogether with slag. The slagging extends to an average 10cm in length which shows that the tips were set quite deeply into the furnace, whether this was from above or from the side.

Metal

As expected we found hardly any traces of the final product of the smelt. Besides the large concretions of copper carbonates observed on a number of the slags, the remains of the final product consist mainly of a few samples of slag which still preserve a layer of oxidized copper and some furnace lining which also preserves some metal. At this point we do not know whether any of this is still metallic. The only possible example of the product is a small flat piece with a thickness of approximately 0.5 cm which is covered with green copper carbonates. This is heavy and may still preserve a metallic core.

Archaeological finds - Chronology

Besides the archaeometallurgicalfinds the excavation produced a small number of artefacts. These were mostly ceramics and some lithics. As with the other finds the investigation of our ceramics is still in the preliminary stages. The surface layers of the site produced ceramics which date to the Iron Age. The layers, however, where the archaeometallurgical finds were found in abundance, produced Black Slip and White slip Pottery which is one of the most characteristic types of the Cypriot LBA. White slip I appears at the beginning of the Late Bronze Age which would date the activities to the 16th-15th centuries BC.

The problem of dating the metallurgical activities in Phorades, is, however, not quite solved, because, although the pottery was found in direct association with the archaeometallurgical finds, both come from layers which correspond to the river channel and the possibility still exists that they were carried there by the water. However, none of the pieces seems to be worn which would have been the case had they travelled a long distance in the creek's course. Charcoal samples have once again been collected and hopefully will enable more accurate dating the site.

Discussion and Preliminary Conclusions

Some preliminary conclusions may be drawn on the smelting technology practiced at the site as indicated by the archaeometallurgical finds. It should be emphasized that this was the first season of excavation and that the following discussion, which is based mainly on visual observation of the finds rather than a thorough scientific investigation, is extremely preliminary.

The nature of the finds, as well as the site and its location, point to a primary smelting workshop. The gossan of Kokkinorotsos is clearly visible from Phorades and presumably this is where the ore was mined. It lies 500 meters to the north, less than 30 minutes walk from the site. No crushing or grinding tools have been found at *Phorades*, nor have any ore roasting layers or roasting conglomerates. Unless these are discovered at a later stage of excavation this suggests that ore beneficiation and roasting were taking place either at the part of the site which has eroded away or somewhere else perhaps near the mine. Indeed remains from such processes have been located in the area of the Kokkinorotsos, but they are associated with slags and pottery of a later period.

The slags from Politiko Phorades are of a singular nature as no other such examples are known from Cyprus. The manner in which the cakes are shaped, in particular their concave bottom, suggests that the liquid slag may have cooled on top of a layer of liquid copper. Preliminary analysis, however, has detected copper sulfides and matte, but not copper metal. Thus the slag may in fact have been sitting on top a layer of matte rather than copper. This will be confirmed by the analysis of a significant set of samples as it may be coincidental that those analysed to date have not contained metallic copper. If no slags containing copper metal are found it may indicate that the process of secondary smelting and refining of copper was not taking place at this workshop but elsewhere. This would fit the pattern of finds from other excavated Bronze Age sites and the conclusions of the previous archaeometallurgical studies discussed above. However, the technology represented by these slags seems to be more sophisticated than the one suggested by Stech *et al.* (1985: 399). There is no evidence for the crushing of slag to remove copper prills: most of the pieces of slag are large to medium sized. Furthermore, wherever there is a fresh break there are no large copper prills visible to the naked eye trapped within the slag matrix. It is clear that the metallurgists active at the workshop had managed to produce a "good" slag which separated well from the smelted metal.

Long fayalite crystals show that the slag body was able to cool rather slowly which could mean that the slag and the metal were left to solidify in the furnace. However, there is evidence which argues against such an interpretation. First, if the slag cakes were left in the furnace, then the surface of the furnace lining would not have been slagged in the way that it is. Second, the angle of the slag cakes is obtuse while the angle of the furnace walls to the base is almost vertical. Finally, although in the majority of the slag cakes the edges are smooth, there is a considerable number of examples which have gravel embedded in their edges. Some of the gravel is quite large. Such gravel has not been detected in any of the pieces of furnaces lining examined to date and therefore it could only have been incorporated into the slag outside the furnace. The shape of the slags could have been produced by emptying the contents of the furnace, in other words both the slag and the metal, into a pit which was either dug into the gravel of the river bed or which was simply lined with gravel or clay. One of the most intriguing pieces of furnace lining is a bulky triangular shaped piece. Could this perhaps be the lining of one of these pits?

Among hundreds of fragments collected only two were found to have a surface which may indicate slag tapping. All the rest have an almost perfectly flat surface indicating that the slags were not very viscous when they left the furnace. This is further supported by the absence of large metallic prills within the slag matrix.

The slags from Phorades differ in shape from the other known LBA slags from Cyprus. In Apliki, for example, large tapped slag cakes were found (Muhly, 1989: 306), while in Kition some of the slags seem to be crucible shaped (i.e., plano convex) (Tylecote, 1982: 89). The closest parallel to the plano-concave cakes, was produced in smelting experiments by Bamberger and Wincierz (1990: 133). In one of their experiments the slag and the copper were tapped together in a sand mould outside the furnace. A complete separation of the two phases was achieved as the slag was highly liquid and it remained so up to 15 minutes after tapping. The slag cake formed exhibited the same flat surface, as well as, a rim where the two phases had been in contact.

Coming to the furnaces, one of the best known archaeometallurgical finds is a 'crucible' from Enkomi (Inv. 1640 in Dikaios, 1971: 644), which Tylecote (1982: 92) defined as a smelting furnace. This identification seems to be supported by qualitative analyses on a small slag sample from the bottom of the Enkomi crucible which detected the presence of small prills of copper sulfides (Zwicker, 1986: 82). The fabric and the general shape of the furnaces from Phorades, i.e., cylindrical with a flat base, are similar to the one from Enkomi, but they differ in size being rather larger. Could they be different in size because they were used in a different process? Primary smelting of roasted ore in theory would produce much larger quantity of slag and, therefore, would require a larger furnace, like the ones from Phorades, while secondary smelting and/or refining of black copper would require a smaller furnace, like the one from Enkomi. The estimated diameter of the Phorades furnace is similar to that of the furnaces from Ayia Varvara- Almyras (Fasnacht and Kassianidou 1992: 83), as well as, those from Timna which date to the Late Bronze Age (Rothenberg 1990: 9, 46), although their method of construction seems to be different.

Most published tuyéres from Cyprus come from Late Bronze Age sites such as Apliki (Du Plat Taylor, 1952), and Enkomi (Dikaios, 1971). Tylecote (1971: 55) identified three types: the straight cylindrical type, the straight 'D' shaped type, and the elbow tuyére. No elbow tuyéres have been found but this perhaps not surprising as elbow tuyéres appear to have been used most often for refining rather than smelting activities. The majority of tuyéres from Politiko are cylindrical, although there are also some other types previously unknown to Cyprus like the double tuyére.

Finally, it is important to consider the location of the site. This has ramifications for the location and study of other early smelting sites on Cyprus. Amongst the critical factors in deciding where to establish a smelting workshop in antiquity, the most important were the location of the ore and flux deposits and the availability of fuel, water, and refractory clay. Political, ideological or religious factors may also influence the location of the site, perhaps even more so at the time of occupation of Phorades. We know from other LBA sites that there are strong ties between copper production and religion (Knapp, 1986). Due to the lack of evidence, however, this aspect will not be considered for the moment. As the ore, the flux, and the fuel are equally available in the direct vicinity of the mines, it would not be unreasonable to suggest that the location for the workshop was chosen because of the creek, which would have provided the necessary raw materials, in other words, clay and water, for the construction of the furnaces. Furthermore, there is a fresh water spring in the immediate vicinity of the site which would have provided fresh drinking water for the workers even in the summer months.

The fact that the site is small (at least based on what has survived), leads to us to believe that this may be one of several similar workshops spread around the mining area of Kokkinorotsos. One of the teams of the Sydney Cyprus Survey Project will, therefore, Block survey the whole area around Phorades in an effort to locate other workshops, as well as, possible settlements from where the workers may have come.

As for Phorades, excavation will continue in 1998 in an effort to locate features and installations *in situ*. In the meantime archaeometallurgical finds from the primary smelting workshop will be subject to a thorough, and systematic scientific analysis which will hopefully enable us to address some of the largely unanswered, questions regarding Late Bronze smelting technology in Cyprus and eventually reconsider all of the previously proposed process models.

Acknowledgements
The excavation of Politico Phorades was funded by the University of Glasgow, the University of Cyprus and the Carnegie Trust for the Universities of Scotland. Michael Donnely (GUARD, University of Glasgow and Vasiliki Kassianidou (University of Cyprus) are the co-directors of the field excavations, and Bernard Knapp (University of Glasgow) is the overall project director (as part of the Sydney Cyprus Survey Project. The author is in charge of the study of the archaeometallurgical finds and takes full responsibility for their preliminary interpretation and any thoughts regarding the smelting technology. Jay Noller and Lisa Wells, the geomorphologists of the team, offered valuable advice throughout the excavation and interpretation of the site.

References

BAMBERGER, M., & WINCIERZ, P. 1990. Ancient smelting of oxide copper ore, Chapter III, in B. Rothenberg(ed.), *The ancient metallurgy of copper*: 123-157. London: Institute for Archaeometallurgical Studies, Institute of Archaeology, University College London.

BUDD, P., POLLARD, A.M., SCAIFE & R.G. THOMAS. 1995. Oxhide ingots, recycling and the Mediterranean metals trade, *Journal of Mediterranean Archaeology* 8(1): 1 - 32.

DIKAIOS, P. 1971. *Enkomi, Excavations 1948-1958. Volume II. Chronology, Summary and Conclusions, Catalogue, Appendices*. Mainz am Rhein: Verlag Philipp von Zabern.

DU PLAT TAYLOR, J. 1952. A Late Bronze Age Settlement at Apliki, Cyprus, *The Antiquaries Journal* XXXII: 133-67.

FASNACHT, W. 1983. Copper processing at Ayia Varvara -Almyras, *Reports of the Department of Antiquities of Cyprus* 89-91.

FASNACHT, W., MORRIS, R., & K. ZUBLER. 1991. Excavations at Ayia Almyras. Third preliminary report, *Reports of the Department of Antiquities of Cyprus* 97-108.

FASNACHT, W. & V. KASSIANIDOU. 1992. Copper at Almyras. A mining and smelting site in Cyprus, in A. Marangou & K. Psillides (ed.), *Cyprus Copper and the Sea*. 78 - 90. Nicosia: Government of Cyprus.

GALE, N.H. & Z.A. STOS-GALE. 1995. Comments on "Oxhide ingots, recycling and the Mediterranean metals trade" *Journal of Mediterranean Archaeology*. 8/1: 33-41.

GALE, N.H. 1991. Copper ox-hide ingots: their origin and their place in the Bronze Age metals trade, in N.H. Gale (ed.) *Bronze Age Trade in the Mediterranean*: 197- 239. SIMA vol. X. Jonsered: Paul Åströms Förlag.

KARAGEORGHIS, V. 1992. Miscellanea from Late Bronze Age Cyprus. 1. On 'baetyls' in Cyprus. *Levant* XXIV: 212.

KESWANI, P.S. 1993. Models of local exchange in Late Bronze Age Cyprus. *Bulletin of the American Schools of Oriental Research* 292: 73-83.

KNAPP, A.B. 1986. *Copper Production and Divine Protection: Archaeology, Ideology and Social Complexity on Bronze Age Cyprus*. Paul Åström's Vorlag: Gotenborg. Studies in Mediterranean Archaeology, Pocketbook 42.

1988. Ideology, archaeology and polity *Man* 23: 133-163.

KOUCKY, F.L. & A. STEINBERG. 1982. The ancient slags of Cyprus, in J.D Muhly, R. Maddin & V. Karageorghis (ed.) *Early Metallurgy in Cyprus 4000-500 BC*: 117-137 Nicosia: Pierides Foundation.

MUHLY, J.D. 1991. The development of copper metallurgy in Late Bronze Age Cyprus, in N.H. Gale (ed.) *Bronze Age Trade in the Mediterranean: 180-196*. Jonsered: Paul Åström's Förlag. Studies in Mediterranean Archaeology X.

1989. The organisation of the copper industry in Late Bronze Age Cyprus, in E. Peltenburg(ed.) *Early Society in Cyprus*. 298- 314. Edinburgh University Press in association with The National Museums of Scotland and the A.G. Leventis Foundation.

1982. The nature of trade in the Late Bronze Age Eastern Mediterranean: The organization of the metals trade and the role of Cyprus, in J.D Muhly, R. Maddin & V. Karageorghis *(ed.) Early Metallurgy in Cyprus 4000-500 BC*: 251 - 269. Nicosia: Pierides Foundation.

ROTHENBERG, B. 1990. Copper smelting furnaces, tuyéres, slags, ingot-moulds and ingots in the Arabah: the archaeological data, in B. Rothenberg (ed.) *The ancient metallurgy of copper*: 1-74. London: Institute for Archaeo-metallurgical Studies, Institute of Archaeology, University College London.

STECH, T. 1985. Copper and society in Late Bronze Age Cyprus, in A.B. Knapp & T. Stech (ed.) *Prehistoric Production and Exchange: the Aegean and the Eastern Mediterranean*: 100-105. Los Angeles: UCLA Institute of Archaeology Monographs 25.

1982. Urban metallurgy in LBA Cyprus, in J.D Muhly, R. Maddin & V. Karageorghis (ed.) *Early Metallurgy in Cyprus 4000-500 BC*: 105 -115. Nicosia: Pierides Foundation.

STECH, T. R. MADDIN & J.D. MUHLY. 1985 Production at Kition in the Late Bronze Age, in V. Karageorghis & M. Demas (ed.) *Excavations at Kition VI: The Pre-Phoenician Levels, Areas I and II, Part I*: 388 - 402. Nicosia: Department of Antiquities.

STOS-GALE, Z. & N. GALE. 1994. Metals, in Knapp, A.B. & J.F. Cherry (ed.) *Provenience studies and Bronze Age Cyprus: Production, exchange and politico - economic change*: 92 - 121. Madison: Prehistory Press. Monographs in World Archaeology 21.

TYLECOTE, R.F. 1982. The Late Bronze Age: copper and bronze metallurgy at Enkomi and Kition, in J.D. Muhly, R. Maddin, & V. Karageorghis(ed) *Early Metallurgy of Cyprus, 4000-500 BC*: 81-100 Nicosia: Pierides Foundation.

1971. Observations on cypriot copper smelting. *Reports of the Department of Antiquities of Cyprus*. 53-58

WEISGERBER, G. 1982. Towards a History of Copper Mining in Cyprus and the Near East: Possibilities of Mining Archaeology, in J.D. Muhly, R. Maddin & V. Karageorghis (ed.) *Early Metallurgy in Cyprus 4000 - 500 BC*: 25-32. Nicosia: Pierides Foundation.

ZWICKER, U. 1986 Ancient Metallurgical Methods for Copper Production in Cyprus- Part 1 -Natural Copper-, Oxide-, Sulphate and Silicate Ore. *Bulletin of the Cyprus Association of Geologists and Mining Engineers*. 3: 79-91.

13

The Archaeology of Mining: Fieldwork Perspectives from the Sydney Cyprus Survey Project (SCSP)

A. BERNARD KNAPP

Department of Archaeology, University of Glasgow, Glasgow G12 8QQ, UK

Ethnographic and historical evidence demonstrates that mining is a significant factor in the relationship between people and their landscape: the huge spoil heap at Mitsero Kokkinopezoula *mine is a monumental example of how human modification of the landscape helped to define a village, gave its inhabitants a sense of identity, and provided the economic wherewithal to survive in the landscape. One of the higher goals of the Sydney Cyprus Survey Project (SCSP) is to unravel some of the factors that have influenced such social and historical processes on the island of Cyprus over the past 5000 years. This paper presents the background to the SCSP and its commitment to an archaeology of mining, reports on the way we have attempted to link our fieldwork to this commitment, and presents the results of four seasons of fieldwork from this perspective.*

Introduction

The Sydney Cyprus Survey Project (SCSP) is undertaking an intensive archaeological survey in the north-central foothills of the Troodos Mountains, within a 65-square km area around the modern villages of Politiko and Mitsero (Figure 1). The Troodos foothills have always been an important zone for mineral and agricultural resources, not least because it includes the copper sulfide ore deposits of Cyprus' Lower Pillow Lavas. SCSP is looking at possible ways to understand better the relationship between metallurgical resources, agricultural production, and trade, on the one hand, and on the other the emergence, development and changing configurations of complex social systems. To address these issues, SCSP was designed in part to examine site location and settlement patterns through time, in relation to both metallurgical and agricultural resources. As such, SCSP represents the first co-ordinated attempt anywhere in the Troodos to locate industrial sites and agricultural villages; to consider and redefine the existence of a site hierarchy (mining sites, agricultural villages, urban centres); and to reconstruct early industrial landscapes. 'Industrial sites' refers to installations such as smelting areas, furnaces, mines, slag heaps, or anything else related to the extraction and production of copper from Cyprus' sulfide ore bodies (Kassianidou n.d.; Knapp and Seretis 1996:11-12).

This paper presents first some general thoughts on an archaeology of mining, and then discusses SCSP's theoretical and methodological program in terms of regional survey strategy. The previous seasons' fieldwork is then summarized and the aims and results of the two most recent (1996-1997) seasons are discussed, including the move from a site less to a site-oriented survey. Finally, the paper discusses the way in which SCSP's interdisciplinary approach may contribute to an archaeology of mining, concluding with a brief overview of the Bronze Age smelting site of Politiko *Phorades* (within the SCSP area).

The archaeology of mining and miners

Around the northern, eastern and south-eastern rims of the Troodos Mountains that occupy much of central and western Cyprus, copper-bearing ore deposits have been worked intermittently for over four thousand years. In economic terms, the most productive copper-fields are spread in an arc along the northern rim, from Ambelikou in the north-west to Mathiati and Sha in the east. Slags, ore and other debris from Kalavasos and Mousoulos mark the south-eastern limit, but other sites along the southern rim of the Troodos provide some indicators of prehistoric exploitation (e.g., Ayios Mamas). There are significant ore deposits at Limni and Kinousa in the west, and at the small Troulli outlier in the east, which is almost equidistant from the Bronze Age sites of Athienou and Kition (Catling 1969: 81-82 and Figure. 2;Gale 1989: 249, Figure. 29.3).

Even so, the archaeometallurgy of Bronze Age Cyprus may still be likened to that of some yet undiscovered El Dorado. Despite the amount of geophysical, technological and analytical research conducted, little survey work aimed specifically at mining and metallurgy has been undertaken by archaeologists. Over the past century, several excavation reports have appeared (e.g., Murray *et al.* 1900; Bruce 1937; Du Plat Taylor 1952; Koucky & Steinberg 1974; Maddin *et al.*1983; Merrillees 1984; various papers in Muhly *et al.*1982; Stech *et al.* 1985; Fasnacht 1991; Fasnacht *et al.* 1992, 1996; Knapp *et al.* 1992, 1994), but only Gerd Weisgerber, an archaeometallurgist whose fieldwork focuses on Oman, has been concerned with an archaeology of mining in Cyprus and the Near East (1982, 1989).

The history of pre-1970archaeometallurgical research in Cyprus was by no means limited (Balthazar1990: 4-21), but the quality and long-term significance of that work was compromised by the level of (modern) technology and the failure to orient research procedure around relevant cultural or social questions (Knapp & Cherry 1994). Work in modern, opencast mines revealed what were believed to

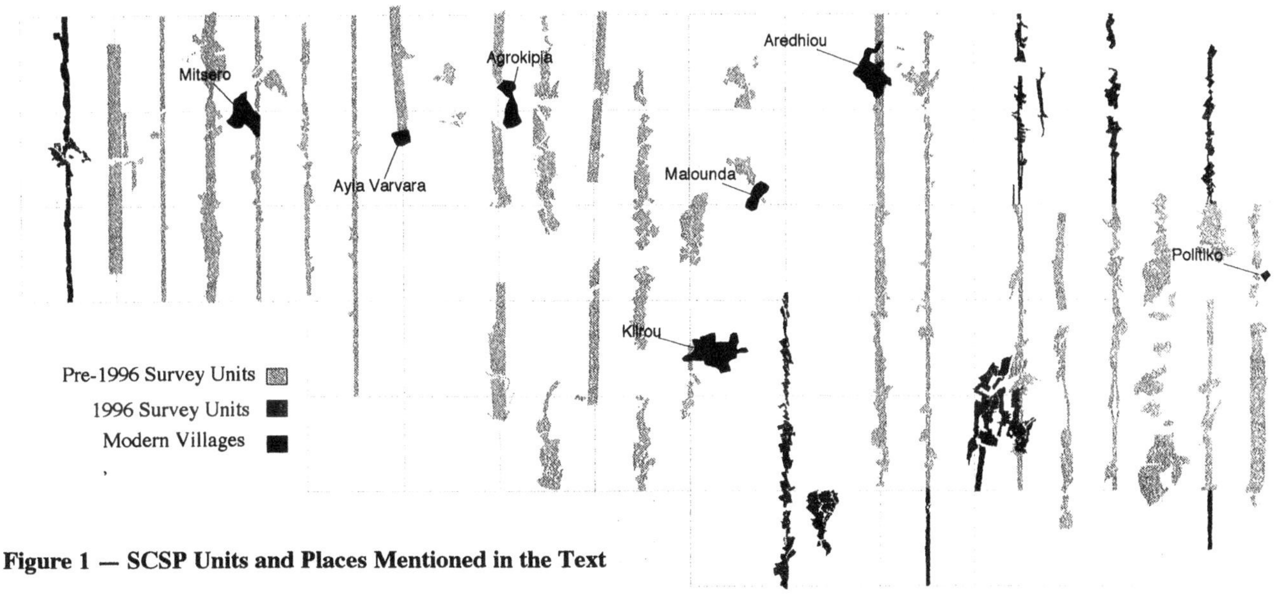

Figure 1. SCSP Survey Area.

be Hellenistic and Roman galleries and shafts (Bruce 1937),whilst huge quantities of slag were recorded and relegated to 'Phoenician' or 'Roman' activity, more on the basis of colour and chemical content than on archaeological context or definitive chronological-technological criteria (e.g., Zwicker *et al.*1972, 1977; Koucky & Steinberg 1974: 149-172; 1982a; 1982b: 120-121; Bachmann 1982a; Muhly *et al.* 1980: 86; Stech 1982: 105). In fact, the colour of Cypriot slags results from the degree of oxidation present in the iron compounds contained within all slags and has nothing to do with specific time periods (Catling 1969: 82; Muhly 1989: 306); recent ^{14}C analyses indeed now provide more reliable dates for some Cypriot slags and mining timbers (Weisgerber 1982: 26, platel; Zwicker 1986). The sheer amount of slag preserved - over two million tons estimated at Skouriotissa and Mavrovouni alone, with an equal amount scattered over the island at other smelting sites - is impressive and thus far unparalleled in eastern Mediterranean or Near Eastern mining regions (Constantinou 1982: 19; Wertime 1982; Weisgerber 1982: 25). Such quantities of slag suggest mass production for a commercial market (Levy & Shalev 1989: 362), and compliment a suite of other material that reflects Cyprus's links to an interregional, eastern Mediterranean economy (e.g., Catling 1969: 84-86; Artzy 1985; Portugali & Knapp 1985: 64-67; Knapp 1997; Sherratt & Sherratt 1993; Sherratt 1994; Rupp 1998).

The evidence for actual prehistoric mining activities on Cyprus is more limited and enigmatic, and equally difficult to date precisely. Once Cypriot copper mines had been reopened for exploration in the 1920s, it became evident that all the ore bodies being worked in modern times had been worked previously, and that the ancient or prehistoric miners must have had adequate experience with Cyprus's sulfide ores to know where to look for additional ore bodies (Catling 1969: 81-83; Constantinou 1982: 18).Modern mining exploration occasionally encountered caves or galleries completely filled with re-consolidated 'cores' of ancient mining waste. These rock-hard seams only became evident after the discovery within them of materials such as wood, charcoal and even tools, all of which would provide dating clues for the waste, but not necessarily for the galleries. Bruce (1937: 684) published what he described as 'wooden props' thought to be used as supports in galleries of Hellenistic or Roman date. Such wooden supports were well-preserved wherever they had been placed or wound up in water or under dripping water that bore copper sulfates; sometimes native copper even became solidified within the wood (Weisgerber 1982: 26). Amongst Cyprus's mountainous slap heaps, stones thought to have been used in crushing ores and lining furnaces have been recovered (Constantinou1982: 19). A series of roasting furnaces have been uncovered at the Iron Age-Hellenistic site of Ayia Varvara *Almyras*, one of which has been radiocarbon-dated to the 'Late Hellenistic' period (Fasnacht 1991: 50).

Dating evidence for prehistoric Cypriot mining is more limited. Weisgerber's (1982: 26 Figure. 1) 'Chalcolithic' dating of a grooved pebble hammer found at Mathiati (Bruce 1937: Figure. 374) is based on stylistic similarities with similar artefacts in Neolithic European flint mines, and gives some sense of the arguments being made and the quality of the data being used. Weisgerber (1982: 26 and plate 1, 28) provides a ^{14}C date of 420 BC for a pit prop from the Apliki mine and emphasizes the rarity of mining or smelting evidence. Charcoal found within a large slag-heap at Enkomi has been radiocarbon-dated to 1200±150 BC (Buchholz 1977: 299; Zwicker *et al.* 1977: 309). Zwicker (1986) likewise provides a series of ^{14}C dates for slag heaps and mines on the island.

The earliest irrefutable evidence for mining (and possibly smelting) activity stems from the Prehistoric Bronze Age (PreBA) 2 site of Ambelikou *Aletri* in the north west Troodos (Merrillees 1984: 6-11). Two radiocarbon dates from charcoal at Ambelikou, calibrated to 2044-1929 BC (Olsson 1987: 33), help to confirm the dating. Several Red Polished III sherds from *pithoi* were reported from trial shafts sunk by the Hellenic Mining Company; these shafts intersected ancient galleries of the pre-existing mine near Ambelikou in 1942 (Dikaios 1946: 245); stone axes and some rounded stones (for extracting ores?) were recovered in the same shafts. Given the limited amount of known evidence for copper smelting activity at or near the mines and ore bodies of Cyprus, it is worth mentioning that other material recovered at Ambelikou *Aletri* - lumps of ore, a terracotta tuyére, a coarse clay crucible, and a double-sided terracotta mould - could indicate a full range of industrial processes at the site (Zwicker 1982; Merrillees 1984: 7, 11); at this stage, suggestions of a workshop nearby for smelting copper remains unproved.

Strong arguments, nonetheless, have been made for production centres near the mines of Cyprus (Stech 1982: 111-112; 1985:103), a situation that holds true for most Old World metallurgical sites (Craddock & Giumlia-Mair 1985). The Protohistoric Bronze Age (ProBA) 2-3 site of Apliki *Karamallos*, with easy access to the rich copper deposits on the North Hill at Apliki, provided sound evidence for primary smelting activity (Du Plat Taylor 1952: 150; Muhly1989). Similar to the situation at Ambelikou, the site at Apliki was first discovered by modern-day miners during the course of prospecting operations (Du Plat Taylor 1952: 133). Although slag was found stratified on the floors of rooms filled with pottery of 13th-12th century BC date, the excavator regarded it - like much of the other metallurgical activity at the site (Muhly *et al.* 1982, Plates III and IV) - to be of Roman date (Du Plat Taylor 1952: 150). This material, however, as well as yet undiscovered mining galleries or shafts, must date to the Late Bronze Age (Stech 1985: 102; Muhly 1989: 309), and the slag recovered (together with sheet bronze, tuyéres, crucible fragments, and stone hammers) may have come from industrial rather than domestic contexts (cf. Du Plat Taylor 1952: 150-153; Muhly 1989: 307-309). The remnants of an industrial installation observed in a road scarp near Apliki are unlikely to be a roasting furnace (Koucky & Steinberg 1974: 150-151, n.4, Figure. 109; Muhly 1989: 309); the size and filling suggest instead that it many have been a roasting pit (Weisgerber 1982: 30 n.2). If the Apliki material - including great blocks of primary melting slag, massive tuyéres, etc. - is ever published in final form (Muhly 1989: 306; 1991), it should provide the best evidence yet for a Cypriot Bronze Age mining village. (The remnants of the site have now been obliterated by renewed mining activity within the Skouriotissa region.) With its furnaces, tuyéres, mining tools, evidence of ore crushing,slags and a possible mining pit nearby, the Iron Age production site of Ayia Varvara Almyras provides further key evidence for ancient Cypriot metallurgy (Fasnacht 1991, Fasnacht *et al.* 1991, 1992, 1996).

Primary smelting slags have also been found at the ProBA 1-2 coastal sites of Enkomi, Hala Sultan Tekke, Kition, and *Toumba tou Skourou.* Several ProBA urban centres, however, including Enkomi, Kalavasos *Ayios Dhimitrios*, Maroni *Vournes*, and Myrtou *Pigadhes*, were involved in the secondary re-smelting of large chunks of impure copper (termed 'furnace conglomerate' by Bachmann 1982b), most likely transported thence by road or river, not by sea, from the mining sites (Muhly 1989: 302). Based on a reconstruction of Steinberg & Koucky (1974: 176-177), it has been argued that these conglomerate chunks were (re)smelted in small furnaces (best exemplified at Enkomi - Stech *et al.* 1985: 388-391), which produced a copper-enriched, solidified chunk of slag, with small pools of copper, or prills of copper - later melted down to produce copper ingots - formed in the lower, outer portions of the chunk (Stech 1982: 106-107,111-112; Zwicker 1982: 64; Stech *et al.* 1985: 399). Muhly (1989: 304; 1991), however, now contends that extracting prills is an inefficient method that probably characterized Chalcolithic metallurgy, and that the production of 'conglomerates' at mining sites represents a 'very primitive smelting technology' that does not adequately explain the type of slags found at Apliki *Karamallos.* As was the case by the late 13th-early 12th century BC at Site 2 at Timna, in the Negev (Bamberger *et al.* 1986: 3 and Table 2), by the ProBA on Cyprus the technology should have developed to the extent that the metal could be separated from the slag. The latter would have been tapped off into a pit dug next to the furnace, whilst the metal itself is thought to have taken the form of an ingot in the bottom of the furnace (Muhly 1989: 303-304). The recovery of tapped slags would provide conclusive evidence of this technological procedure, but there is little evidence of such slags at Cyprus's ProBA 2 urban centres (Zwicker *et al.* 1977: 310; Muhly *et al.* 1980: 90); the slag from Apliki *Karamallos*is still under study (Muhly 1989: 306).

Cypriot copper deposits that could have been worked in antiquity are found both in Diabase and Pillow Lava formations, but the latter is what attracted the attention of ancient miners (Catling 1969: 82-83 and Figure. 2). It is usually agreed (Bachmann 1982c) that Cypriot copper production involved chiefly sulfidic ores (Constantinou 1982: 13), and that the ores mined, smelted and reduced - chalcopyrite, covellite, bornite, chalcocite, cuprite - derived from the secondary enrichment zone of the sulfide deposits within the Pillow Lavas (Catling 1969: 83; Stech1982: 111; Bachmann 1982c: 377). In order to 'slag' the iron out of the predominant, chalcopyrite ores, fluxes had to be added to the smelting charges. There is a great deal of variability in the four million plus tons of slag recorded from about forty different smelting sites, not least because several steps were necessary to convert sulfide ore to metal. What produced the majority of slag, of course, was the primary smelting that eliminated the many impurities, chiefly sulfur and iron, associated with the ore. Charcoal was the main fuel used in the smelting process, but it is still debated whether this was a 'one-step' process of direct smelting of the roasted sulfide ore(s) and flux to form copper, or a 'two-step' process that required the intermediate production of 'matte': a mixed copper and iron sulfide requiring further roasting and smelting to produce copper (see Catling 1969: 83; compare Stech 1982: 106-107, 111-112 with Bachman 1982a). With the intensified production that occurred on Cyprus by the ProBA, when the massive sulfide deposits hosted by the lower Pillow Lavas were exploited, Bachmann argues that matte smelting would have been mandatory (1982a:148-149; 1982c: 378); the massive quantities of slag found at settlement sites points to the same conclusion (Catling 1969: 83). Although arguments for hydro-metallurgical practices for producing copper have been presented (Koucky & Steinberg 1982a: 169-176; Merrillees 1984:11-12), archaeological evidence is wanting, and archaeometallurgical data seem to contradict the idea of such a process. What seems evident, from the variety of slags and array of smelting or melting equipment recovered from several sites, is that multiple technologies were being used to extract the maximum amount of metal from the ores. This phenomenon is a clear sign of the centrality of copper production to the Cypriot economy (Stech 1982: 112).

All of this activity and its material remains on Cyprus represent increasingly specialized production and exchange, as well as economic growth, and vindicate the call for an archaeology of mining and miners (Weisgerber 1982, 1989; Hardesty 1988). The literature treating the anthropology and history of mining is well represented (e.g., Paul 1963; Eliade 1978; Brown 1979; Taussig 1980; Godoy 1985), and a forthcoming publication (Knapp *et al.* 1998) seeks to integrate archaeological (historical and prehistoric) approaches with those of anthropology and social history. Such a task will be worthwhile only if the technology of mining, and the productive process overall, are examined in light of their social, economic and ideological dimensions.

The Sydney Cyprus Survey Project (SCSP)

Regional survey today is a multi-disciplinary endeavour involving a wide variety of specialist expertise; in that respect it broadly echoes developments within archaeological field excavation. The interdisciplinary approach of the SCSP facilitates a new and different perspective on the archaeology of mining, which is presented in more detail in the following section. First, however, it is necessary to pave the way by discussing briefly our field methodology.

Regional survey methodology usually incorporates eco-environmental approaches and settlement pattern analysis, both of which have been criticized as being overly reductionist. The SCSP has responded to these tensions over the role of the physical environment in the development of human culture by trying to understand how the disposition of the survey area's two great natural resources - metal-bearing ore and arable lands - conditioned settlement patterning. In reaction to a purely economic reconstruction of the past landscape in our survey area, SCSP seeks to understand the place of religion, ideology, community and politics in a number of ways, specifically by

(1) engaging textual analysis with an emphasis on perception, experience and symbolic aspects of the landscape (Tilley 1991, 1994);

(2)regarding the landscape as a dialectical relationship between socio-economic formations and the environment (Ruiz *et al.* 1992); and

(3) articulating a theoretical framework that insists on the irreducibility of the human-environment relationship, a process wherein the environment evolves as a phenomenon inseparable from human society (McGlade 1995).

One of the project's main goals has been to investigate the effect of metallurgical and agricultural resources on the patterning of settlement in early industrial societies. SCSP's geomorphological research aims to describe the human processes - e.g., the clearing of vegetation, terracing and filling, and the use of check dams - that have resulted in the present human-modified landscape. An integral part of this landscape is the construction and location of the built environment (resource extraction and processing locales, industrial facilities, agricultural communities, ceremonial sites, and burial plots). The spatial disposition of surface remains and the reconstruction of the human processes that affected the sedimentary, hydrological, fluoristic and biological cycles are key areas through which one may attempt to access the past.

Here I describe the evolving strategy in response to some of the concerns just outlined. SCSP's survey universe is a naturally defined area, the interface between the igneous ore bodies and the sedimentary soils. This area also has significance because it is suitable for investigating the articulation between agricultural and mining sites within early industrial landscapes.

SCSP's initial stages of survey work were designed to provide a systematic coverage of the survey universe and, at the same time, to generate knowledge which could be applied constructively to succeeding stages of our research. The goal was to develop a broad understanding of the cultural remains and recoverability potential across the survey universe before any stratification was undertaken. Accordingly, a multi-stage research design was adopted, and during the first stage of the survey intensified coverage as opposed to aerial extent was chosen.

Field walking is undertaken either in 50metre-transects (25x100m units) or in fields plots readily identifiable on aerial photographs. Three five-person field teams walk each transect or field in parallel lines at 5m spacing. For each survey unit, several environmental and locational factors (geomorphology, topography, surface character, sediment cover, erosional pattern, land use, slope, and visibility) are recorded. In each unit the teams collect 'diagnostic' cultural materials, including pottery, lithics, ground stone, metals, slag, ores and fluxes, glass and tiles. All non-diagnostic material is counted and left in place, and the information recorded daily in a relational database (*File Maker Pro* 3.0). Photo-maps produced by the project's GIS (see below) are coded with both the aerial extent of the survey unit and any special features (Knapp & Johnson 1994).

The study was initiated by completing a number of transects that cross the survey universe in a north to south direction, at every half-kilometre mark. A geomorphological profile of the survey universe was also created. Overall we gained a great deal of knowledge about the material culture and exposure conditions of the survey region(see Knapp *et al.* 1994; Knapp & Given 1996; Knapp & Johnson 1994). Toward the end of the 1995 season, we began the first field researches to be designed with a specific recovery goal in mind.

By the onset of the1996 season the main field strategy of the SCSP was:

1) To finalize a broad systematic sample of the survey area which would give an overview of the material data and soil conditions that typify the region's diverse topography and geology;
2) to determine which topographic, geological, geo-botanical and land use factors conditioned the occurrence of exposed cultural materials;
3) to conduct 'block' survey (a series of units generally following natural or agricultural boundaries, or else a series of units within transects of much greater width than the standard 50 meters) of certain areas known, based on previous experience or other knowledge, to have evidence of early mining, industrial or agricultural activities.

SCSP and the archaeology of mining

On a broader scale, the intensive stratified survey methodology allows the relationship between mining and the social and physical landscape in which it takes place to be explored. The most visible material remains of ancient mining and metalworking are the slag heaps scattered throughout the Pillow Lavas, and detailed investigations have carried out at two of these (Clough, *et al.*, in Knapp and Given 1996: 308-323). Here, the work at these sites is summarize as an illustration of SCSP's multi-faceted approach to the archaeology of mining.

Slag heaps

Several major slag heaps, as well as the slag scattered throughout the survey area, indicate extensive ancient metallurgical activity in the region. Over 500 units, or 25% of the area surveyed, contains slag, ores or flux material. Some of this results from secondary activity (e.g., the use of slag for road metalling in the early 20th century), but the prominence of this figure indicates widespread metallurgical activity in the area.

During the 1995 field season, cleaning and recording took place of the existing sections of two prominent slag heaps and metal-working areas. The first, Mitsero *Kouloupakhis*, consists of a slag heap, workshop floors and building foundations (Clough, in Knapp & Given 1996:308-316); the second is an ancient mineral dressing and roasting site at the modern mine of Agrokipia *Kriadis* (Kassianidou & Wright, in Knapp & Given 1996: 316-323). Mitsero *Kouloupakhis* revealed stratigraphic evidence of structures, workshop floors, furnace fragments, large chunks of charcoal, and other industrial remains, as well as enough pottery for a preliminary chronology. Before the original four slag heaps were quarried by the Hellenic Mining Company in the 1950s, they would have been almost 60m in diameter and could have been as high as 15-20m, suggesting production to the extent of approximately 50,000 tons of metallic copper.

At least one substantial building, constructed of mortared limestone with some finely cut blocks and walls up to 15m long, was situated within this slag heap. The bottom of the section reveals walls and floors buried without any slag, indicating that the building was constructed and used before copper was produced on the site. Adjoining the slag heap and extending some 100m to the north is an extensive Classical to Roman settlement which was surveyed in its entirety. During the Hellenistic or Early Roman period, copper production began in earnest and the building may have been refurbished at some point during this secondary usage. On the basis of the remnant pottery, we suggest that the main phase of copper production at *Kouloupakhis* took place in the middle and late Roman periods. The stratification revealed in section also shows debris that resulted from the successive stages of the smelting process. Section MK 1 shows nine successive workshop floors, whilst in section MK 2 the successive dumping episodes are clearly visible.

The other principal metal-working area investigated in the 1995 season consists of two sections on the southern edge of the modern mine of Agrokipia *Kriadis*. Exposed by modern mining operations, these sections were cleaned in order to expose more clearly the stratified layers of crushed and roasted ore and gossan, as well as slag and charcoal. Section 1A cuts through two ancient spoil heaps and consists of 43 stratified layers which rise to a height of some six metres; the bottom of the deposit has not yet been reached. Commonest are layers of crushed gossan, which sometimes alternate with layers of roasted ore and gossan mixed with ash and charcoal. There were only scattered pieces of tapped slag, suggesting that there was no major smelting operation nearby.

The nature of the debris at *Kriadis* suggests that the metallurgical activities carried out consisted of mineral dressing rather than smelting. The purpose of mineral dressing is to improve the ore before smelting by the mechanical removal of as much of the unwanted material as possible. This process - often done near the mine to avoid the unnecessary transport of heavy ore to the smelting site - usually involved selecting the richest fragments of ore, crushing the rock and then grinding it to a powder, and lastly washing this powder to remove the lighter gangue material. At *Kriadis*, this crushed and ground material appears in the section, suggesting that we are dealing with spoil heaps from the mineral dressing process. Some fragments of furnace conglomerate were also found in section 1A.

These preliminary investigations have revealed key information about the different stages of copper production in the Mitsero area. Mitsero *Kouloupakhis* is particularly important for its association with structures and its position adjoining a major settlement.

Mining and landscape

Seen through the eyes of a mining community, certain aspects come to the fore in the appropriation of the local landscape: natural resources; agriculture; sites for working and dwelling; and communications. Mining itself can be an important factor in the relationship between people and their landscape. In the late 1950s and 1960s the huge spoil heap at the Kokkinopezoulas mine overlooking the village of Mitsero gave the villagers not just a means of living but a sense of identity and purpose: this vast human modification of the landscape clearly defined it as a mining village.

The geological knowledge of ancient prospectors in Cyprus is well known. The bright reds and yellows of the gossans attracted the prospectors to the copper deposits which lay underneath them, and virtually all of these gossans were investigated in ancient times (Constantinou 1992: 57; Lavender 1962). Gossans, however, were not simply used to pinpoint deposits, but also as sources of flux for smelting. Fluxes, particularly iron and manganese oxides, are needed to lower the melting point and viscosity of the slag when smelting, thus facilitating the separation of silica-rich gangue from metal (Constantinou 1992: 64-67). Alternatively, silica may have to be added for the same purpose where the gangue is naturally iron-or manganese-rich. Haematite can be found at most of the ore sites in the form of the gossan cap, and silica is widely available in the area. For most ores, the best flux is manganese oxide, which in the Pillow Lavas occurs mainly in the form of umber and manganese concretions. These are particularly common in the south-eastern part of the survey area.

Because mining galleries in unstable ore deposits need pit-props, and smelting needs fuel, metal production needs forests. Timber preserved in mining galleries at Skouriotissa has shown that *Pinus brutia*, a species particularly suited to the Pillow Lavas, was used most commonly as pit props (Constantinou 1992: 61).Olive and hawthorn would have been the commonest hardwoods in the survey area in antiquity; because they produce a long, hot burn and can easily be coppiced, they would have been ideal as a source of fuel for roasting and charcoal for smelting.

The natural resources necessary for copper production must be supplemented by food resources to feed the miners and their families. Within the SCSP area, Aredhiou *Vouppes* is a likely agricultural support settlement of Late Bronze Age date, with an abundance of pithos sherds (Knapp *et al.* 1994: 337-338, 341). Just south east of the survey area is another Late Bronze Age site, Analiontas *Palioklichia*, which is also a candidate for an agricultural village that supported a mining community (Keswani 1993; Webb & Frankel 1994).

On the basis of current evidence, the peak of copper production in the SCSP area probably occurred during the Roman and Late Roman period. This peak of metalworking activity coincides with a considerable rise in the number of settlements, many of which are in good agricultural areas with easy access to the mining and smelting sites. Mitsero *Kouloupakhis*, for example, lies at the northern edge of a fertile agricultural plain that extends about 2km in each direction. Today this area produces much of the barley, fruit, nut and vegetable crops consumed by the villagers of Mitsero. Survey evidence indicates that this area was similarly well-used in the Roman-Late Roman periods. Extensive settlement evidence immediately adjoining the slag heaps at Mitsero *Kouloupakhis* may have housed workers and service industries, rather than being solely an agricultural village which could spare a few workers in off seasons.

When copper is produced and exported on a large scale, as was the case in the Hellenistic and Roman periods (Hill 1940: 178; Raber 1984: 125-126), a substantial communications system is required. Within the survey area, it is possible to extrapolate likely routes from the terrain and the positions of major sites. There is clearly a route running eastwards from the smelting sites and settlements round Mitsero to the plain of the Akaki river. This flat river valley provides a clear route to the north-east, joining the Roman town at Klirou to major sites at Malounda and Aredhiou. Another important axis of communication runs eastwards to the Archaic-period village of Tamassos. The most suitable route lies along the modern road from Aredhiouto Episkopio, beginning at the Roman settlement of Aredhiou *Panayia Odhiyitria*. This route would have linked the copper producing areas of Mitsero, Agrokipia and Klirou to the city of Tamassos, the most important site in the region during the Roman period.

People also relate to their landscape in ideological or symbolic terms, and there are indications that copper production was part of a symbolic system during the Archaic period. At the rural sanctuary of Mathiati *Mavrovouni*, one of several centring round the city-kingdom of Tamassos (Given 1991: 52; Figure. 14), terracottas were found in direct contact with a slag heap (*BCH* 1984: 964-5). In the village of Tamassos itself, copper workshops were found immediately adjacent to the principal Archaic sanctuary (Buchholz 1985: 240-2), with

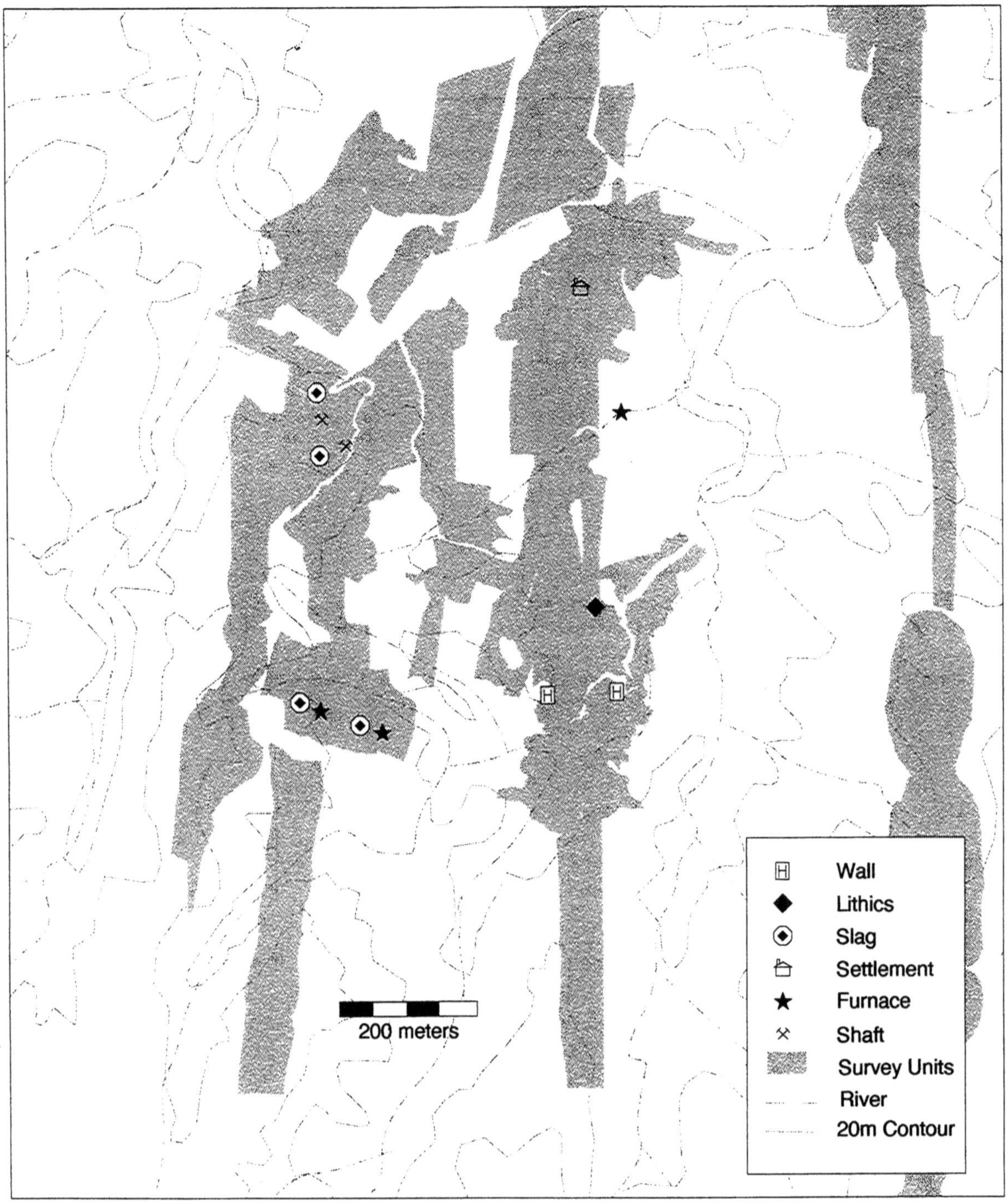

Figure — SIA7 POSIs with underlay of surveyed units

Figure 2. SIA 7 - Politiko Kokkinorotsos.

a considerable amount of slag spread throughout the neighbouring buildings and in the courtyard of the sanctuary itself. Many other products of the countryside are present in the sanctuary at Tamassos, including the bones of wild and domestic animals, olive pits, grain and, judging from the vessels, oil and wine. Copper is clearly one of several products that are won from the surrounding landscape and brought to the urban centre. Their dedication in the sanctuary is an outward expression of this process (Given 1991: 54-55).

Politiko *Phorades* 1: a Bronze Age smelting site?

In the Pillow Lavas approximately three kilometres south-west of Politiko, a long ridge of gossan runs southwards above a branch of the Kouphos river. Ancient slag is scattered across the surrounding fields and slopes, and exposed in road sections and stream beds. The upper Kouphos River valley is an area crucial to the investigation of the relationship between mining, agriculture and settlement. In SCSP terms, it is a Special Interest Area

(SIA 7) which has been named Politiko *Kokkinorotsos* ('red rock') after the locality name of the gossan. Amongst several places of special interest in this region, two have been examined in detail: the copper smelting workshop of Politiko *Phorades* 1 (SCY100); and the Roman settlement at Politiko *Phorades* 2 (SCY101).

The copper smelting workshop of Politiko *Phorades* 1 is situated about 500m east of the *Kokkinorotsos* gossan, the nearest ore deposit, in a bend of a tributary of the River Kouphos. Preliminary investigation revealed what appears to be a unique smelting technology, slag morphology and refractories (Kassianidou n.d.; this volume). Most of the riverine landscape that once surrounded this site has been lost to erosion subsequent to human activity at the smelting site.

The smelting operation at Politiko *Phorades* was much smaller than most others known on Cyprus, for example those at Skouriotissa or even Mitsero *Sykamies* (Kassianidou & Wright, in Knapp & Given 1996: 321-322). The projected amount of slag from the *Phorades* site is no more than a couple of tons, as opposed to several thousand or millions of tons at the other sites. Consequently its closest parallel is the Iron Age smelting site at Ayia Varvara *Almyras* (Fasnacht *et al.* 1991, 1992, 1996). However, the slags from Politiko *Phorades* 1 are completely different from the crucible-shaped slags described by Koucky & Steinberg (1982b: 118). The manner in which the cakes are shaped, in particular the concave recess in the bottom, suggests that the slag may have cooled on top of an ingot of copper, not unlike the process described by Muhly (1989: 303-304). The smooth flat top of the slag, and the adherence of small pebbles and sand to its edges, suggest that the slag and copper mix were tapped from the furnace into a shaped pit, where they would have cooled into the form described.

Several factors bear upon the date of ancient metallurgical activity at *Phorades.* One reliable indicator of age is a mid-Holocene, 3000-5000 year-old soil profile, which developed on deposits overlying the materials from the smelting area: this makes a Bronze Age or early Iron Age date for the smelting workshop plausible. AMS dating (University of Oxford's radiocarbon unit) of a single sample of charcoal taken from furnace material (2865±75 BP: OxA-7013) renders a calibrated date in the 12th-10th century BC (calibration courtesy of Sturt Manning, University of Reading). Survey work during the 1996 season showed that the two nearest survey units contained Cypro-Geometric III to Cypro-Archaic I pottery, which hints at the latest possible date for smelting activity at *Phorades*. During the 1997 excavations at *Phorades*, surface layers produced sherds datable to the Iron Age, but the layers containing archaeometallurgical material revealed several Black Slip II and White Slip I sherds, all of which suggest a date at the very end of the Middle Bronze or beginning of Late Bronze (16th century BC in most general terms). During the 1996 survey season, cleaning of the section at *Phorades* revealed one piece of White Slip 1 pottery. This does not mean, however, that the date of metallurgical activity at *Phorades* is fully resolved: despite the fact that the pottery was recovered in direct association with archaeometallurgical material, at this stage all finds still derive from layers that correspond to the river's channel, and thus it is still possible that they were carried there by water. Charcoal samples have once again been recovered and six more have been submitted to the University of Oxford's Radiocarbon Unit for AMS dating, which will be essential to confirm the chronological positioning of the site. *Prima facis*, however, Politiko *Phorades* appears to be the earliest primary smelting site yet discovered on Cyprus, by about 1000 years.

All other traces of ancient metallurgical activity in this region were found on the slopes of the gossan or close to its edge. Most of these consist of deposits of tapped slag, probably considerably later than the slag discovered at *Phorades* 1. One of these, Politiko *Kokkinorotsos* 1 (SCY116), just beyond the eastern side of the gossan, shows not only a large slag scatter but also an exposed section with stratified layers of roasting conglomerate, furnace lining and furnace floor. This may have been an ore roasting and smelting site. The 'ropy' nature of the tapped slag found at various locations nearby indicate a different mode of production than that seen at the smelting workshop of *Phorades* 1.

The landscape of Politiko *Kokkinorotsos* has been exploited and modified by human activity for at least 3500 years. Much of that activity clearly focused on the exploitation of the area's copper ore resources, which would also impact on the environment through collection of fuel, smelting, communications and the need to feed and house all those engaged in such activities. Geomorphological work indicates that this dramatic landscape changed considerably during the period of human occupation, and that much of the evidence has been obscured. Future research in this area will continue to elucidate the functional, chronological and social links between the locales of different human activities in the landscape.

Overview and conclusion

There is sound evidence for mining activities in the survey area from the Classical period at least until the end of the Roman Empire. It seems clear that the resources of the area were exploited as early as the Bronze Age and we believe we can demonstrate that copper-production continued on Cyprus after the fall of the Roman Empire, if not into the Medieval period: by itself, this contradicts all standard histories on the subject, and laboratory work (radiocarbon dating, slag analyses) seeks to establish firmer chronological and technological parameters. The Bronze Age smelting site of Politiko *Phorades* 1 reveals archaeometallurgical and technological developments that are unprecedented on an island that has been a centre for Mediterranean copper production over the past 4000 years. All of this mining activity and its material remains - in the survey area and beyond – represent increasingly specialized production and trade, as well as economic growth, and likewise vindicate the call for an archaeology of mining.

In a recent issue of *Archaeometry*, Stos-Gale *et al.* (1997) have argued for the predominance of the Skouriotissa region, and in particular of the site of Apliki *Karamallos*, in the production of all copper oxhide ingots thus far characterized by lead isotope analysis as being consistent with production from Cypriot (and especially Apliki) ores. This is, essentially, tantamount to claiming that a large proportion of Bronze Age mining and primary smelting activities were conducted at one site. Whilst Politiko *Phorades* by itself cannot (yet) necessarily refute their argument, it is here counter-suggested that sites like *Phorades* were ubiquitous throughout the pillow lavas that encircle the Troodos Mountains, and that the majority of such sites have been destroyed or obscured by twentieth-century mining activity. This is not idle speculation: SCSP can demonstrate, through work at Agrokipia *Kriadis,* that evidence of ancient beneficiation and ore preparation activity is buried within modern spoil heaps. Also, and most unfortunately, modern-day mining activities nearby have been entirely obliterated the Bronze Age site of Apliki and its surrounds.

One may offer as a contrast to the Stos-Gale *et al.* model a very preliminary reconstruction of an industrial landscape based on the work at Politiko *Phorades* and in the region that surrounds it, which has been surveyed intensively by the SCSP. *Prima facie*, it would appear that the ores exploited came from mines on the gossan ridge above the smelting site, whilst the ore preparation and roasting took place at an intermediate point, farther down the hill. Smelting took place at *Phorades* itself, some 500-800 metres distant, along the banks of a river. Geobotanical indices as reflected in satellite spectral images and geomorphological parameters which indicate a markedly different landscape around the smelting site, both help to substantiate the notion of a micro-regional settlement pattern, characterized by a multi-phase production process with specialized production units represented by the various 'sites' and installations. What we see here in micro-regional terms may represent a localized settlement pattern within a broader settlement hierarchy (see Knapp 1997), a self-contained metallurgical area with all the essential raw materials, with its own organization and communication network, and with the potential to emerge into the rapid technological revolution that is seen in the pyrotechnological material from *Phorades* and which we already know on other grounds to have taken place during the Late Bronze Age.[1]

The cultural and economic model proposed - namely that a single site and single ore deposit was producing all or most of the copper oxhide ingots found in the Bronze Age Mediterranean - whilst bold and provocative, and coherent in analytical terms, makes little sense in economic, spatial or even mineralogical terms, contradicts a model we can now propose on the basis of work at *Phorades*, and represents a marked change from earlier interpretations that proposed exploitation of multiple ore deposits round the Troodos. Even if Stos-Gale *et al.* were to be proven correct in their assumptions, their statement (1997: 115) that 'If indeed all of these oxhide ingots were smelted in the region of Apliki-Skouriotissa then the production of copper there must have been organized with professional efficiency' completely disregards and overlooks an extensive literature on the social and political organization of Late Bronze Age Cypriot society (for the most recent discussion with full references, see Knapp 1997). An archaeology of mining, and an awareness of the industrial landscape of prehistoric Cyprus, as gained from the kind of interdisciplinary survey, excavation, exploration and analytical work being undertaken by SCSP, will serve as a corrective to reconstructions based solely on problematic geological reconstructions and laboratory analyses. The author has long argued that there must be direct, long-term collaboration between archaeologists and science-based archaeologists if both fields are to move beyond their limited horizons and if the people involved are to make meaningful contributions to the interpretations of past human activity. I strongly reiterate that call for interdisciplinary work and co-operation.

The relationship between a society and its environment is a product of both human consciousness and material reality. The study of social and cultural landscapes considers the way people engage with their surroundings in terms of their own individual experience. A focus on human settlements, from the individuals and households that comprise them, to the social landscape that contextualizes them, is central to a social archaeology. People use the landscape that surrounds them and the settlements within it for everything from producing food, to expressing formal design, to making social statements. Ethnographic and historical evidence demonstrates that mining can be a very significant factor in this relationship: the huge spoil heap at Mitsero's *Kokkinopezoula* mine stands as a monumental example of how human modification of the landscape actually defines a village, gives its inhabitants a sense of identity, and provides the economic basis which makes it possible to survive in the landscape. Particular places or features in the landscape may be imbued with immense personal, ideological and economic significance; in this dynamic, two-way process, the landscape influences personal and cultural identity. One of SCSP's 'higher goals' is to examine and unravel some of those factors which have influenced social and historical processes on the island of Cyprus over the past 5000 years.

References

ARTZY, M. 1985. Supply and demand: a study of second millennium Cypriote pottery in the Levant. In A.B. Knapp & T. Stech (Eds), *Prehistoric Production and Exchange: The Aegean and East Mediterranean.* UCLA Inst. Archaeology, Monograph 25: 93-99. UCLA Institute of Archaeology: Los Angeles.

BACHMANN, H.-G. 1982a. *The Identification of Slags from Archaeological Sites.* Institute of Archaeology, Occasional Publication 6. Institute of Archaeology, University of London: London.

1982b. Copper smelting slags from Cyprus: review and classification of analytical data. In J.D. Muhly, R. Maddin, & V. Karageorghis (eds.), *Early Metallurgy in Cyprus, 4000-500 BC*, pp. 143-152. Pierides Foundation: Nicosia.

1982c. Review of papers on archaeo-metallurgical topics. In J.D. Muhly, R. Maddin, & V. Karageorghis (eds.), *Early Metallurgy in Cyprus, 4000-500 BC*, pp.377-380. Pierides Foundation: Nicosia.

BALTHAZAR, J.W. 1990. *Copper and Bronze Working in Early through Middle Bronze Age Cyprus.* Studies in Mediterranean Archaeology and Literature, Pocket-book 84. P. Åström's Förlag: Göteborg.

BAMBERGER,M., P. WINCIERZ, H.-G. BACHMANN, & B. ROTHENBERG. 1986. Ancient smelting of oxide copper ore. *Metall* 40: 1166-1174.

BROWN, R.C. 1979. *Hard-Rock Miners: The Intermountain West, 1860-1920.* Texas A and M University Press: College Station, TX.

BRUCE, J.L. 1937 Antiquities in the mines of Cyprus: 639-671. In E. Gjerstad, J. Lindros, E. Sjoqvist, & A. Westholm, *Swedish Cyprus Expedition* 3: 639-671. SCE: Stockholm.

BUCHHOLZ, H.-G. 1977. Bemerkungen zu einigen neuen C-14-AnalysenZyperns und Griechenlands. *Report of the Department of Antiquities, Cyprus*: 290-308.

1985 Der Beitrag der Ausgrabungen von Tamassos zur antiken Baugeschichte Zyperns, in V. Karageorghis (ed.), *Archaeology in Cyprus 1960-1985*, pp. 238-255. Nicosia: Leventis Foundation.

CATLING, H.W. 1969. The Cypriote copper industry. *Archaeologia Viva* 1(3): 81-88.

CONSTANTINOU, GEORGE. 1982. Geological Features and Ancient Exploitation of the Cupriferous Sulphide Orebodies of Cyprus. In J.D. Muhly, R. Maddin & V. Karageorghis (eds.), *Early Metallurgy in Cyprus, 4000-500 BC*, pp. 13-24. Nicosia: Pierides Foundation.

1992 Ancient copper mining in Cyprus. In A. Marangou & K.Psillides (eds.), *Cyprus, Copper and the Sea*, pp. 43-74.Government of Cyprus: Nicosia.

CRADDOCK, P.T., & A.R. GIUMLIA-MAIR. 1988. Problems and possibilities for provenancing bronzes by chemical composition. In J. Curtis (ed.), *Bronze working Centres of Western Asia c.1000-539 B.C.*, pp. 317-327. Kegan Paul: London.

DIKAIOS, P. 1946. Early Copper Age discoveries in Cyprus, 3rd millennium B.C. copper mining. *Illustrated London News* (2 March 1946): 244-245.

DU PLAT TAYLOR, J. 1952. A Late Bronze Age settlement at Apliki, Cyprus. *Antiquaries Journal* 32: 133-167.

ELIADE, M. 1978. *The Forge and the Crucible.* (2nd ed.) Chicago: University of Chicago Press

FASNACHT, W. 1991. Copper mining and smelting at Ayia Varvara-Almyras. In P. Åström (Ed.), *Acta Cypria: Acts of an International Congress on Cypriote Archaeology* (Göteborg,22-24 August 1991). Studies in Mediterranean Archaeology and Literature, Pocketbook 107: 49-61. P. Åström's Förlag: Jonsered.

FASNACHT, W. (with Robert Morris & Kurt Zubler). 1991. Excavation sat Ayia Vavara-Almyras. Third preliminary report. *Report of the Department of Antiquities, Cyprus*: 97-108.

FASNACHT, W. *et al.* 1992. Excavations at Ayia Varvara Almyras. Fourth preliminary report. Report of the Department of Antiquities, Cyprus: 59-74.

FASNACHT, W. (with Vasiliki Kassianidou, Anne Connoly, Carine Deslex, Johanna Kunz, Tunde Maradi, Kurt Zubler & Peter Roll). 1996. Excavations at Ayia Vavara Almyras: fifth preliminary report. *Report of the Department of Antiquities, Cyprus*:

GALE, N.H. 1989. Archaeometallurgical studies of Late Bronze Age copper oxhide ingots from the Mediterranean region. In A. Hauptmann, E.Pernicka, & G.A. Wagner (eds.), *Old World Archaeometallurgy.* Der Anschnitt 7:247-268. Deutschen Bergbau-Museums: Bochum.

GIVEN, MICHAEL. 1991. Symbols, Power and the Construction of Identity in the City-Kingdoms of Ancient Cyprus, c.750-312 B.C. Unpublished PhD thesis, University of Cambridge.

GODOY, R.1985. Mining: anthropological perspectives. *Annual Review of Anthropology* 14: 199-217.

HARDESTY, D.L. 1988. *The Archaeology of Mining and Miners: A View from the Silver State.* Society for Historical Archaeology, Special Publication 6.SHA: Pleasant Hill, CA.

HILL, GEORGE. 1940. *A History of Cyprus.* Volume 1. Cambridge: Cambridge University Press.

KASSIANIDOU, VASILIKI. n.d. Developments in the Archaeometallurgy of Cyprus (in Greek). In *The Third International Athens Conference on Archaeometry.* Athens: Hellenic Archaeometrical Society. (in press)

KESWANI, P.S. 1993. Models of local exchange in Late Bronze Age Cyprus. *Bulletin of American Schools of Oriental Research* 292: 73-83.

KNAPP, A. BERNARD. 1997. *The Archaeology of Late Bronze Age Cypriot Society: The Study of Settlement, Survey and Landscape.* University of Glasgow, Department of Archaeology, Occasional Paper 4. Glasgow: Department of Archaeology.

KNAPP, A.B., & J.F. CHERRY. 1994. *Provenance Studies and Bronze Age Cyprus: Production, Exchange, and Politico-Economic Change.* Monographs in World Archaeology 21. Prehistory Press: Madison, WI.

KNAPP, A. BERNARD, & MICHAEL GIVEN. 1996. A Report on the Sydney-Cyprus Survey Project (SCSP) - Third Season (1995). Report of the Department of Antiquities, Cyprus:295-366.

KNAPP, A. BERNARD, STEVE.O. HELD, IAN JOHNSON, & EBERHARD ZANGGER. 1992. The Sydney Cyprus Survey Project – First Preliminary Season (1992). *Report of the Department of Antquities, Cyprus*: 319-336.

KNAPP, A.BERNARD, STEVE. O. HELD, IAN JOHNSON, & PRISCILLA S. KESWANI. 1994. The Sydney Cyprus Survey Project. Second preliminary season (1993). *Report of the Department of Antquities, Cyprus*: 329-343.

KNAPP, A.B., V. PIGOTT &E. HERBERT. (eds.) 1998. *Social Approaches to an Industrial Past.: The Archaeology and Anthropology of Mining.*

Routledge: London.

KNAPP, A. BERNARD, & IAN JOHNSON. 1994. Quantifying Survey Data from Cyprus: The Use of Aerial Photos for Field Recording and GIS Input. In Ian Johnson (ed.), *Methods in the Mountains: Proceedings of the UISPP Committee IV Meeting.* Sydney University Archaeological Methods Series 2: 157-164. Archaeological Computing Laboratory, University of Sydney.

KNAPP, A.BERNARD, & KYLIE SERETIS. 1996. The Sydney-Cyprus Survey Project (SCSP), fourth season (1996): brief preliminary report. *Old World Archaeology Newsletter* 20(1): 9-13.

KOUCKY, FRANK L., & ARTHUR STEINBERG. 1974. Preliminary metallurgical research on the ancient Cypriot copper industry. In L.E. Stager, A. Walker & G.E.Wright (eds.), *The American Expedition to Idalion, Cyprus.* BASOR Supplementary Volume 18:148-178. American Schools of Oriental Research: Cambridge MA.

1982a. Ancient mining and mineral dressing on Cyprus. In T.E. & S.F. Wertime (eds.), *Early Pyrotechnology: The Evolution of the First Fire- Using Industries*, pp. 149-180. Smithsonian Institute Press: Washington, D.C.

1982b. The Ancient Slags of Cyprus. In J. D. Muhly, R. Maddin & V. Karageorghis (eds.), *Early Metallurgy in Cyprus, 4000-500 BC* , pp. 117-142 Nicosia: Pierides Foundation.

LAVENDER, D. 1962. *The Storyof Cyprus Mines Corporation.* Huntington Library: San Marino,CA.

LEVY, T.E., & S. SHALEV. 1989. Prehistoric metalworking in the southern Levant: archaeometallurgical and social perspectives. *World Archaeology* 20: 352-372.

MCGLADE, J. 1995Archaeology and the Ecodynamics of Human-Modified Landscapes. *Antiquity* 69: 113-132.

MADDIN, R., J.D. MUHLY, & T. STECH WHEELER. 1983. Metal working. In T. Dothan & A. Ben-Tor, *Excavations at Athienou, Cyprus, 1971-1972.* Qedem 16: 132-138.Institute of Archaeology, Hebrew University: Jerusalem.

MERRILLEES,R.S. 1984. Ambelikou-Aletri: a preliminary report. *Report of the Department ofAntiquities, Cyprus*: 1-13.

MUHLY, J.D. 1989. The organisation of the copper industry in Late Bronze Age Cyprus. In E.J. Peltenburg (ed.), *Early Society in Cyprus*, pp. 298-314. Edinburgh: Edinburgh University Press.

MUHLY, J.D., R. MADDIN, & V. KARAGEORGHIS. (eds.) 1982. *Acta of the International Archaeological Symposium: Early Metallurgy in Cyprus, 4000-500 B.C.* Pierides Foundation: Nicosia.

MULHY, J.D.,R. MADDIN, & T.S. WHEELER. 1980. The oxhide ingots from Enkomi and Mathati and Late Bronze Age copper smelting in Cyprus. Report of the Department of Antiquities, Cyprus: 84-99.

MURRAY, A.S., A.H. SMITH, & H.B. WALTERS. 1900. *Excavations in Cyprus.* British Museum Publications: London.

OLSSON, I. 1987. Carbon-14dating and the interpretation of the validity of some dates from the Bronze Age in the Aegean. In P. Astrom (ed.), High, Middle or Low. Part Studies in Mediterranean Archaeology. Pocketbook 57: 4-38. P. Astrom's Forlag:Gotheburg.

PAUL, R. 1963. *Mining Frontiers of the Far West, 1848-1880.* Rinehart and Winston: New York.

PORTUGALI,Y., & A. B. KNAPP. 1985. Cyprus and the Aegean: a spatial analysis of interaction in the 17th - 14th centuries B.C. In A. B. Knapp & T. Stech (eds.), *Prehistoric Production and Exchange, The Aegean and the eastern Mediterranean.* UCLA Institute of Archaeology, Monograph 25: 44-78. UCLA Institute of Archaeology: Los Angeles.

RABER, PAUL A. 1984. The Organization and Development of Early Copper Metallurgy in the Polis Region, Western Cyprus. Unpublished PhD dissertation: Pennsylvania State University.

RUIZ, MALTIDE, ROBERTO RISCH, PALOMA GONZALEZ MARCEN, PEDRO CASTRO, VICENTE LULL, & ROBERT CHAPMAN. 1992. Environmental Exploitation and Social Structure in Prehistoric Southeast Spain. *Journal of Mediterranean Archaeology* 5: 3-38.

RUPP,D.W. 1998. The seven kings of the Land of Ia', a district on Ia-ad-na-na: Achaean bluebloods, Cypriot parvenus, or both? In Kim J. Hartswick & Mary Sturgeon (eds.), Stefano§: Studies in Honor of Brunilde Sismondo Ridgway , pp.209-222. University Museum, University of Pennsylvania: Philadelphia.

SHERRATT, A.G., & S.SHERRATT. 1993. The growth of the Mediterranean economy in the early first millennium BC. *World Archaeology* 24: 361-378.

SHERRATT, S. 1994. Commerce, iron and ideology: metallurgical innovation in 12th 11th century Cyprus. In V. Karageorghis (ed*.), Proceedings of the International Symposium: Cyprus in the 11th Century B.C.* (30-31 October 1993), pp.59-106. Leventis Foundation: Nicosia.

STECH, T. 1982. Urban metallurgy in Late Bronze Age Cyprus. In J.D. Muhly, R. Maddin & V. Karageorghis (eds.),*Acts of the International Archaeologcial Symposium: Early Metallurgy in Cyprus*, pp. 105-115. Pierides Foundation: Nicosia.

1985. Copper and society in Late Bronze Age Cyprus. In A.B. Knapp & T. Stech (eds.), *Prehistoric Production and Exchange: The Aegean and East Mediterranean.* UCLA Institute of Archaeology, Monograph 25:100-105. UCLA Institute of Archaeology: Los Angeles.

STECH, T., R. MADDIN, & J.D. MUHLY. 1985. Copper production at Kition in the Late Bronze Age. In V. Karageorghis & M. Demas, *Excavations at Kition* V.1: *The Pre-Phoenician Levels*, pp. 388-402.Department of Antiquities: Nicosia.

STOS-GALE, S., G. MALIOTIS, N. GALE & N. ANNETTS. 1997. Lead isotope characteristics of the Cyprus copper ore deposits applied to provenance studies of copper oxhide ingots. *Archaeometry* 39: 83-123.

TAUSSIG, M. 1980. *The Devil and Commodity Fetishism in South America.* University of North Carolina Press: Chapel Hill.

TILLEY, CHRISTOPHER Y. 1991. *Material Culture and Text : The Art of Ambiguity.* London: Routledge.

1994. *A Phenomenology of Landscape.* Berg: Oxford.

WEBB, JENNIFER M., & DAVID FRANKEL. 1994. Making an impression: storage and surplus finance in Late Bronze Age Cyprus. *Journal of Mediterranean Archaeology* 7: 5-26.

WEISGERBER, GERD. 1982a. Towards a History of Copper Mining in Cyprus and the Near East: Possibilities of Mining Archaeology. In J.D. Muhly, R. Maddin & V. Karageorghis (eds.), *Early Metallurgy of Cyprus, 4000-500 BC*, pp. 25-32. Nicosia: Pierides Foundation.

1989. Recent research on prehistoric and ancient mining in Greece. In C. Domergue (ed.), M *ineria y Metalurgia en las Antiguas Civilizaciones*

Mediterraneas y Europeas. Colloque Internacional Asociadio Madrid 24-28 Octobre 1985), pp.197-203. Ministeriode Cultura, Dirrecion General de Belllas Artes y Archivos, Instituto deConservacion y Restauacion de Bienes Culturales: Madrid.

WERTIME, T.A.1982. Cypriot metallurgy against the backdrop of Mediterranean pyrotechnology: energy reconsidered. In J.D. Muhly, R. Maddin & V. Karageorghis (eds.), *Early Metallurgy in Cyprus 4000-500 BC*, pp. 351-362. Pierides Foundation: Nicosia.

ZWICKER, U. 1982. Bronze Age metallurgy at Ambelikou-Aletri and arsenical copper in acrucible from Episkopi-Phaneromeni. In J.D. Muhly, R. Maddin & V. Karageorghis (eds.), *Early Metallurgy in Cyprus, 4000-500 BC*, pp. 63-68. Pierides Foundation: Nicosia.

ZWICKER, U.1986. Ancient metallurgical methods for copper production in Cyprus: Part II: sulphide ore and copper-arsenic-alloy production. *Cyprus Association of Geologists and Mining Engineers Bulletin* 3: 92-111.

ZWICKER, U., E. GREMBLER & H. ROLLIG. 1977. Investigations on copper-slags from Cyprus (second report). *Report of the Department of Antiquities, Cyprus*: 309-316.

ZWICKER, U., H. ROLLIG, & U. SCHWARZ. 1972. Investigations on prehistoric and early historic copper-slag from Cyprus. *Report of the Department of Antiquities, Cyprus*: 34-46.

[1] I thank Sven van Lokeren (Institute of Archaeology, University College, London) for his critical input to these ideas and for his participation in the fieldwork at Politiko *Phorades* and with SCSP during the1997 season.

14

Lead Isotope Characterization of the Ore Deposits of Cyprus and Sardinia and its Application to the Discovery of the Sources of Copper for Late Bronze Age Oxhide Ingots

N H GALE

University of Oxford, Isotrace Laboratory, Nuclear Physics Building, Keble Road, Oxford, OX1 3RH, UK

The ore deposit geology of Sardinia and Cyprus is briefly described; new lead isotope data are presented for ores from both islands. The basic principles of lead isotope provenancing of metals are reviewed and illustrated by a comparison of the lead isotope data for copper oxhide ingots with those of Cypriot and Sardinian ore deposits. It is concluded that lead isotope provenancing of metals is valid. Further the evidence is that all the later (post-1250BC) oxhide ingots were made from copper smelted from ores coming from the Apliki region in Cyprus, and that the oxhide ingots found on Sardinia were also made of Cypriot copper. However, the Apliki region did not totally dominate production of copper in Late Bronze Age Cyprus as only about 15% of Late Bronze Age Cypriot tools and weapons were made of Apliki region copper. The lead isotope data indicate that other Cypriot artefacts originated from diverse copper sources around the Troodos Mountains. Lead isotope data for the earliest copper oxhide ingots from Crete show that they were not made of copper from ore deposits in Cyprus, Crete or the Mediterranean/Anatolian region as a whole. Although the oxhide ingots from the Cape Gelidonya wreck are made of Apliki region copper, those from the earlier Uluburun shipwreck are not, although they are probably made from Cypriot ores. Most bun ingots from the Gelidonya shipwreck have lead isotope compositions characteristic of the Apliki region ores. However, some bar and bun ingots from these two shipwrecks have lead isotope compositions overlapping with the ores from Lavrion in Attica and since no analyzed oxhide ingots have been found to have Lavrion compositions, it is suggested that Lavrion copper may have been "traded" in the form of bun or bar ingots.

Keywords: Lead Isotopes; Ores; Oxhide Ingots; Bronze Age; Cyprus; Sardinia; Lavrion; Metal Provenance.

Introduction

One of the stated themes of the symposium was a discussion of the geology of ore deposits thought to have been exploited by prehistoric miners and metallurgists, and the implications for their possible characterization by trace element and isotopic analyses. Another theme was the possible alteration (or otherwise) of the trace element or isotopic composition of metals during smelting or fire refining and the implications for metal provenancing or the use of those data in reconstructing ancient metallurgical technology. With these themes in mind the metalliferous ore deposits of Cyprus and Sardinia were taken as a basis for discussing the possibility of using lead isotope analyses for metal provenancing. The discussion is focussed on the possibility of solving the long-standing problem of establishing the copper ores that supplied the metal from which the Late Bronze Age (LBA) copper oxhide ingots were made. These ingots, weighing on average about 30 kilograms, were a principal form in which copper was traded around the Mediterranean region in the period roughly between 1600 BC and 1100 BC (Buchholz 1959; Bass 1967).

A principal aim of research at the Isotrace Laboratory is to establish Bronze Age trading patterns in the Mediterranean, especially for metals (Gale & Stos-Gale 1982; 1989, 1992a). The single site which has contributed most towards a general appreciation of the range of goods involved in this trade is that of the Uluburun shipwreck, which foundered off the coast of Turkey slightly before 1318 BC. The excavators of the Uluburun shipwreck summarized their ideas about Late Bronze Age trade in the Mediterranean in a well-illustrated article which included a map suggesting possible trade routes (Bass 1987). Much of this they knew to be speculative, especially for the sources of tin, copper, glass, ivory, etc. Their map, for example, implies that the two chief copper sources for the oxhide ingot trade were Cyprus and Sardinia; it will be shown that this is correct for Cyprus but wrong for Sardinia. This paper is chiefly about the metals trade, in the light of comparative lead isotope studies of copper oxhide ingots and ore deposits in Cyprus and Sardinia. Some quite new data will be presented, together with a discussion of the significance of older data.

Oxhide ingots

The largest group of copper oxhide ingots, over 300 in number and weighing some 10 tonnes, comes from the Uluburun shipwreck dated to about 1318 BC (Pulak 1988); this ship also carried about 1 tonne of tin metal ingots. Bass (1987) illustrates the find circumstances of these ingots on the seabed in a series of dramatic colour photographs. A ship which sank off Cape Gelidonya about a century later (Bass 1967) also carried many copper oxhide ingots, both whole and fragmentary. FIGURE 1 is a map showing the distribution of oxhide ingots and the sites where these two Bronze Age ships foundered.

FIGURE 1 also shows that copper oxhide ingots are found all over the Mediterranean, especially on the islands of Cyprus, Crete and Sardinia, but also in Turkey, Bulgaria, Greece, Syria and Sicily. However, only one oxhide ingot mould has been found so far, at Ras Shamra in coastal Syria (Lagarce *et al.* 1983). To establish the trading patterns of oxhide ingots, which may have been different in different phases of the Late Bronze Age, it is essential to recognize the origin of the copper in the ingots, namely to discover their copper ores. Normal archaeological methods, based on typological studies or studies of impressed signs, have failed to establish where any particular ingot originated. Chemical analysis and metallographic examination have, by themselves, also failed to contribute much to the provenance question, however much they may

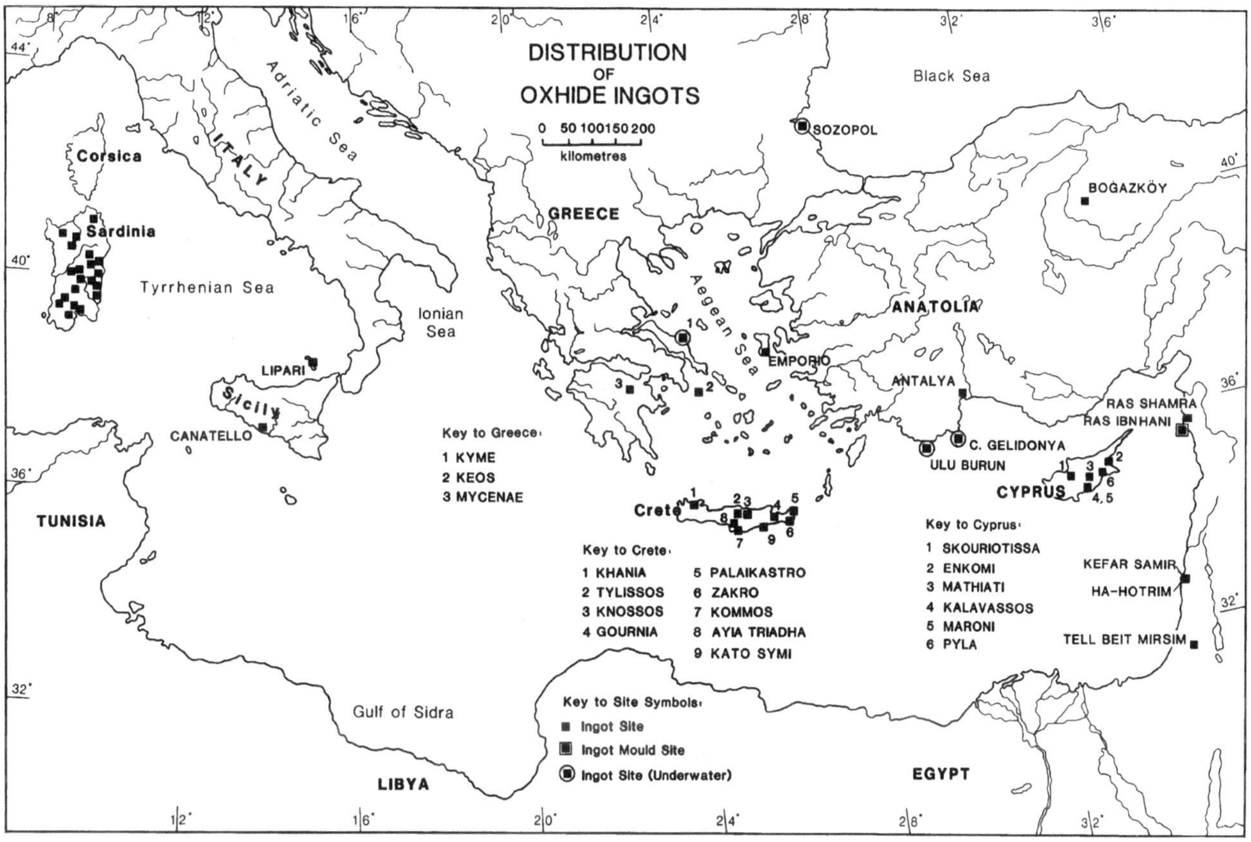

Figure 1: Map of the distribution of copper oxhide ingots and of the location of the two Bronze Age shipwrecks of Uluburun and Cape Gelidonya. Note that, although previously published as found underwater, it is now known that the ingot found in Bulgaria was found at a mainland coastal site near Sozopol and Burgas.

have contributed to an understanding of the technological aspects of these ingots.

Provenancing metals using lead isotope analyses

Indeed, thus far, only lead isotope provenance studies have contributed significantly to the solution of metal provenancing problems.

This approach requires that a number of lead isotope analyses of ores must be determined so as to define the fields characteristic of the ore deposits which could, in principle, have supplied the particular metal under consideration. All three possible lead isotope ratios must be used, conveniently depicted in two two-dimensional lead isotope diagrams. Only if <u>all</u> the lead isotope ratios of an artefact fall within the fields for an ore source in <u>both</u> diagrams could that ore source have provided the metal for the artefact.

The concept of a lead isotope field for a single deposit is useful for the majority of uranium-poor ore deposits in the Mediterranean region. It is less useful for the rare deposits (such as Timna in Israel and Feinan in Jordan) where present day U/Pb ratios are significant. Attempts at quantitative definition of lead isotope fields, or the assignment of the numerical probability that a particular artefact derives from a particular isotopic field, may present statistical difficulties, although this has yet to be demonstrated unequivocally (Baxter & Gale in press). The extension of the concept to an island, or a geographical region, is fraught with difficulties, as will be seen.

It is the coincidence of lead isotope compositions of particular artefacts with particular ore deposits that is required before it can be concluded that the metal of the artefact came from the named ore deposit. If some ore deposits have overlapping lead isotope compositions, then the conclusions for artefacts plotting in such regions must be less well defined.

FIGURE 2 sets out the quite simple basis of lead isotope analyses as used for provenancing the metal of an ancient artefact. This figure is based on lead isotope data for the ore deposits at Lavrion in Attica, and in the Cycladic islands of Seriphos, Keos and Thera (data in Stos-Gale *et al.* 1996; Gale 1998). Two examples are given in FIGURE 2, for imaginary lead artefacts having lead isotope compositions plotting at points A and B. Artefact A was made of lead metal which was not extracted from any of the four ore deposits whose lead isotope fields appear in FIGURE 2. Artefact B is consistent only with having been made from metal smelted from the Keos lead ore deposits, since it plots outside the Lavrion field in the lower part of the diagram.

It is largely a question of semantics to engage in arguments that lead isotope analyses can only exclude ore sources, and can at best establish only that an ore deposit is consistent with being the metal source for particular artefacts. If a set of copper artefacts have lead isotope compositions matching only one copper ore deposit, amongst the set of copper ore deposits which it is geographically/archaeologically reasonable to consider, the normal application of scientific method would lead to the conclusion that, in the present state of knowledge, those artefacts were made of copper from that deposit. Naturally such conclusions must rest on a comprehensive lead isotope investigation of all appropriate ore deposits.

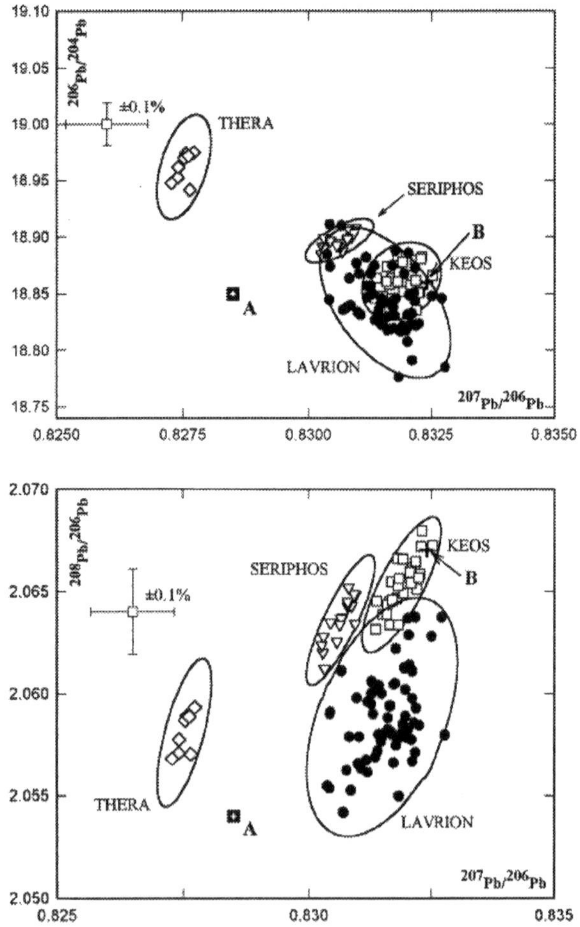

Figure 2: Simple illustration of the principles of lead isotope provenancing, based on the lead isotope fields for ore deposits of Lavrion and the Cycladic islands of Thera, Keos and Seriphos, in comparison with the lead isotope compositions of two hypothetical metal artefacts, A and B.

One must of course avoid the trap of believing that science provides infallible answers to our questions. As Feynman (1992) reminds us "science is uncertain" and he also points out that the universal *modus operandi* of science is to make guesses or predictions on the basis of laws only known exactly for limited regions.

Lead isotope analysis provenancing of metals does depend critically on there being no measurable alteration in the metal, of the isotopic composition of lead in the ores, by such processes as smelting, by subsequent fire refining of the metal, or processes such as cupellation to produce silver. That there are no measurable alterations of lead isotope composition due to such processes has now been proved (Gale & Stos-Gale 1996 and references therein).

In order to establish the provenance of the copper oxhide ingots, the lead isotope compositions of the larger copper deposits in Cyprus, Sardinia and Lavrion in Greece are of importance, particularly in view of the relatively large number of oxhide ingot fragments found in Sardinia and Cyprus

Geology and lead isotope compositions of the Sardinian ore deposits.

The metalliferous ore deposits and geology of Sardinia are summarized on the metallogenic map of Sardinia (Marcello *et al.* 1978) and discussed by Zuffardi (1980, 1989). Del Caldo *et al.* (1974) is a useful guide to the localities, and basic ore mineralogy, of the mines. The ore deposits are hosted in Cambrian to Tertiary rocks, which gives rise to a wide range of lead isotope compositions.

Most of the ore deposits, which are lead-zinc- rich together with a few copper deposits, are situated in SW Sardinia, in the Iglesiente-Sulcis region, and are hosted in Lower Palaeozoic rocks. Tertiary copper deposits occur in NW Sardinia, although most, with the notable exception of the Calabona deposit, are far too small to have supplied the relatively large amounts of copper needed for Late Bronze Age copper oxhide ingots. Pernicka (pers. comm.), who has independently surveyed the Sardinian ore deposits, has confirmed this view.

Sardinian ore deposits have varied and complex geneses including stratabound ores, intrusive-related epigenetic ores, contact metamorphic deposits associated with Hercynian granitic intrusions and palaeokarst deposits. Many deposits have suffered extensive remobilization, partly due to the Caledonian (510-470 Ma) but largely to the Hercynian (360-300 Ma) orogenies.

The Hercynian, is also associated with hydrothermal mineralization whereas the final Alpine orogeny (95-10 Ma) is only associated with normal faulting and partial rejuvenation of former fracture zones.

The geological age of ore deposits acts as a first-order constraint on their lead isotope compositions. Hence the ores are briefly described in stratigraphical order. The Cambrian (530-510Ma) crops out in southwestern Sardinia in the Iglesiente-Sulcis districts. Here most of the mineralization is hosted within limestones and dolomites and interbedded volcanics. This sequence is locally known as the 'Metalliferous', and is host to major lead, zinc, pyrite, baryte and fluorite deposits, including the large deposits at Monteponi, Campo Pisano and Masua. In the Sulcis region, Cambrian rocks host metamorphosed synsedimentary, polymetallic sulfide deposits such as Rosas/Sa Marchesa, Monte Tamara, and Truba Niedda. The karstic, oxidized, small copper deposit of Sa Duchessa occurs in the Oridda district.

Silurian, black, graphitic shales and minor limestones and interbedded volcanics host stratiform, polymetallic, sulfide deposits, among them Funtana Raminosa, Correboi, Pranu e' Sanguini and Perda s'Oliu. These ore deposits display sedimentary features although many have been remobilized.

Polymetallic vein deposits, including the extremely rich and extensive one at Montevecchio, are associated with a series of Hercynian acid to intermediate plutonic rocks. These deposits include the famous lead-zinc deposits of the Arburese district (Montevecchio-Ingurtosu-Gennamare); some of them (Gozzura, Sos Enattos, Montevecchio) also were copper producers. Other examples include the small chalcopyrite/pyrite deposit of Torpe-Canale Barisoni, and the Sn-Zn-Cu-Pb deposit of Monte Mannu-Canale Serci.

A large ophiolite complex was emplaced in Italy during the Jurassic-Palaeocene (200-50 Ma); in northwestern Sardinia this is seen as a small, poorly dated, dioritic mass (Calabona region) which contains a small, low-grade, porphyry copper deposit (Ridge 1990). Other Tertiary/Quaternary polymetallic, sulfide veins are present in northwestern Sardinia in a huge Tertiary, trachyte-andesite complex. Most deposits are in tuffaceous horizons, with metal grades of up to 6-8% combined copper, lead and zinc. Individual ore bodies are small; the largest cupriferous pyrite vein, at Capo Marargiu, is itself too small to have provided copper on the scale needed for significant oxhide ingot production.

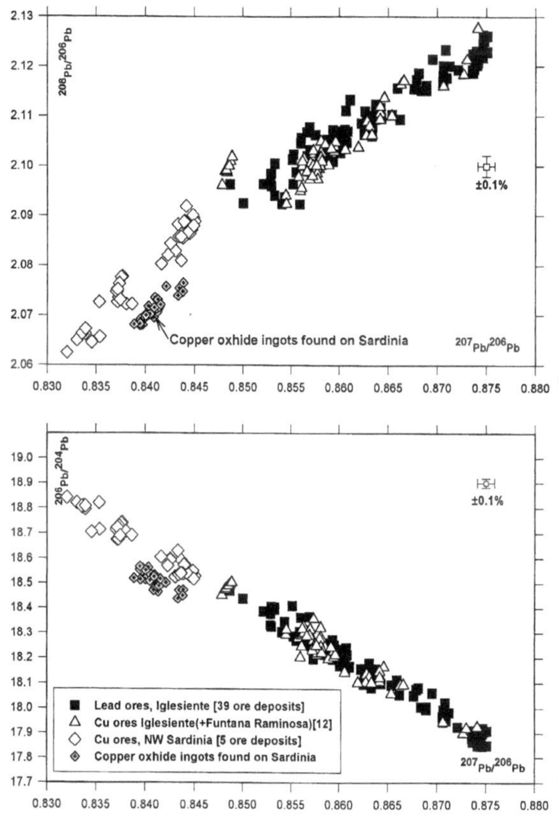

Figure 3: Lead isotope compositions for copper and Pb-Zn-(Cu) ores from Sardinia and for the copper oxhide ingots found on Sardinia.

Since the lead isotope composition of an ore deposit is governed to the first order by its age, the Cambrian (≈500 Ma) Sardinian copper deposits should be isotopically distinguishable from Late Cretaceous (≈80 Ma) Cypriot deposits. The lead isotopes for over 30 Sardinian copper and lead ore deposits have been measured, and this discrimination is, indeed, easy to make (Gale & Stos-Gale 1987, 1988, 1992b; Stos-Gale *et al.* 1995).

Lead isotope compositions

FIGURE 3 plots the lead isotope compositions of Sardinian ores: they exhibit a very large range, amounting to about 3%. Individual deposits, however, show a quite small within-deposit range.

The solid squares represent 39 individual copper-poor lead-zinc deposits from deposits in SW Sardinia, the Barbagia, and Argentiera areas in NW Sardinia. Data for the older copper deposits (open triangles), including Funtana Raminosa, fall within the range for the older lead-zinc deposits. Data for the ostensibly copper-poor deposits are included since, although relatively improbable in this case, it is claimed (Pernicka 1992) that such deposits formerly might have contained sufficient copper to have been exploited in the Bronze Age. Data for ore deposits of NW Sardinia fall in the lower region of the upper diagram in FIGURE 3.

To compare ore deposit data with those for copper-based artefacts it would be incorrect to try to represent the range of isotopic compositions of Sardinian ores by a single field drawn within an overall encompassing closed curve on these diagrams. Such a curve would enclose isotopic space not populated by Sardinian ores. Moreover individual deposits occupy quite limited parts of isotopic space, and other parts of the isotopic space within such a curve are occupied by copper-poor, lead-zinc deposits.

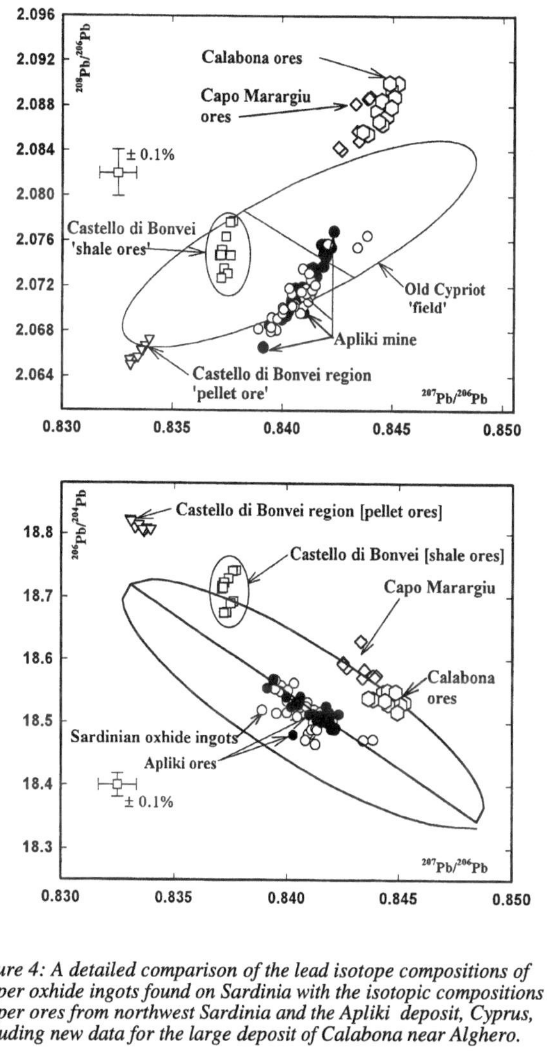

Figure 4: A detailed comparison of the lead isotope compositions of copper oxhide ingots found on Sardinia with the isotopic compositions of copper ores from northwest Sardinia and the Apliki deposit, Cyprus, including new data for the large deposit of Calabona near Alghero.

Data for the Tertiary copper deposits of Capo Marargiu and NW Sardinia, including new data for the large deposit of Calabona and for the minor copper occurrences of Castello di Bonvei are plotted, on an expanded scale, in FIGURE 4.

FIGURES 3 & 4 also show lead isotope data for copper oxhide ingots found on Sardinia. Although the ingots lie closest to the Sardinian Tertiary copper ores, there is no overlap between the lead isotope compositions of the oxhide ingots and any Sardinian ores. Note, however, that there is overlap between the "Sardinian" oxhide ingots and ores from the Cypriot copper mine of Apliki.

Geology of Cypriot ores

These ores are essentially pyritic with varying amounts of copper and are known as 'Cyprus-type' ores.

The Mineral Resources Map of Cyprus (Constantinou and Panayiotou 1982) shows that the metalliferous ore deposits lie in the Lower Pillow Lava Unit around the lower parts of the Troodos Mountains. The ore deposits were formed under the ocean about 90Ma ago, and were subsequently uplifted to their present positions approximately 25Ma ago. Another version of this map shows a great number of "ancient", mostly undated, copper slag heaps in Cyprus (see Stos-Gale, *et al.* in press for a detailed discussion). These range from the remains of the huge one at Skouriotissa (FIGURE 5) to a relatively small scatter at Petramoutti which has now been dated to about 410 BC.

Figure 5: Photograph of a section through part of the vast slag heap at Skouriotissa, Cyprus, which Lavender (1962) estimates to have originally contained a million tons of slag. Zwicker (1986) has dated slag from Skouriotissa by C-14 from 200 AD to 310 AD, and has dated timbers from an ancient gallery at Skouriotissa to 620±50 BC. For the scale of this slag heap, note the pylons and vehicles.

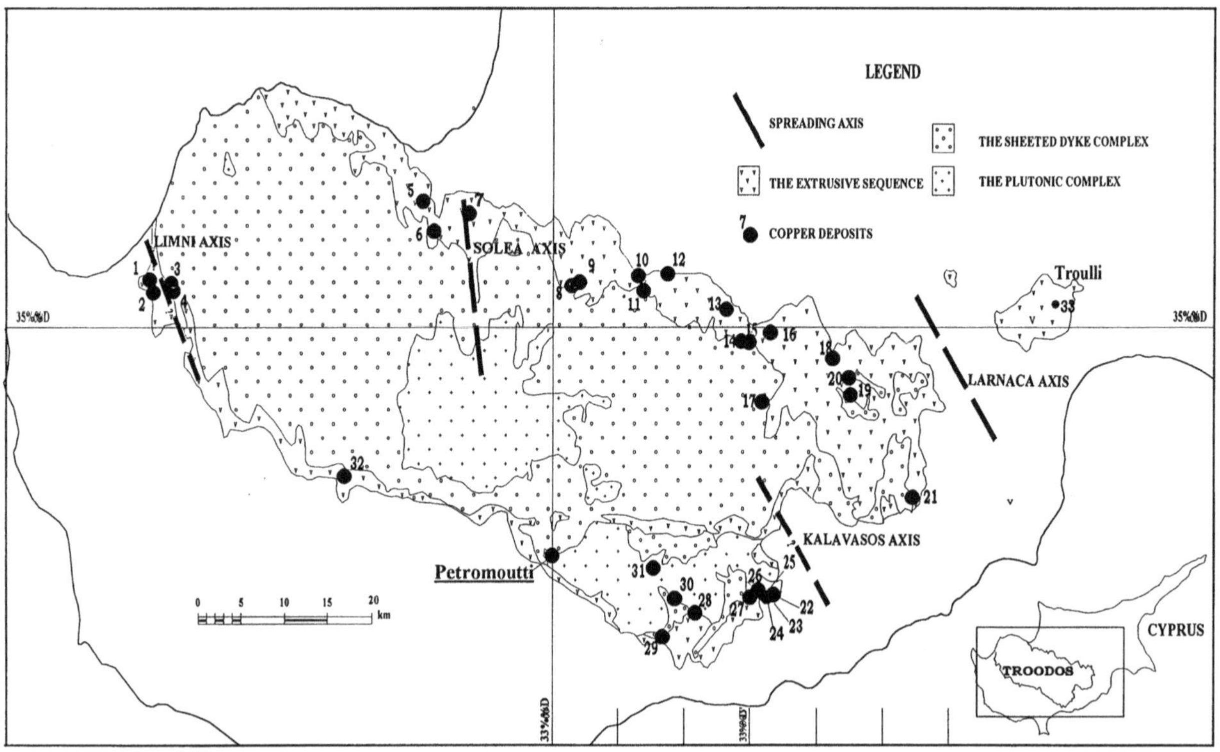

Figure 6: Sketch map of the Troodos ophiolite complex of Cyprus, showing the location of the major copper deposits and of the fossil sea floor spreading axes. ***Limni axis****: (1) Limni: (2) Evloymeni; (3) Kinousa; (4) Uncle Charles.* ***Solea axis****: (5) Mavrovouni; (6) Apliki; (7) Skouriotissa; (8) Alestos; (9) Memi.* ***Larnaca axis****: (10) Kokkinoyia; (11) Kokkinopezoula; (12) Agrokipia; (13) Klirou; (14) Pitharochoma; (15) Peristerka; (16) Kambia; (17) Kapedhes; (18) Mathiati; (19) Sha; (20) Strongyli; (21) Anglisidhes; (33) Troulli.* ***Kalavasos axis****: (22) Petra; (23) Mousoulos; (24) Mavridhia; (25) Landaria; (26) Mavri Sykia; (27) Platies; (28) Mazokambos; (29) Mangaleni.* ***Other occurrences:*** *(30) Pevkos; (31) Laxia tou Mavrou and Pevkos; (32) Perevasa.*

The metallogenesis of the Cypriot ore bodies has been the subject of a great deal of geological debate over the past decades. Geophysical and geological studies in Cyprus (Maliotis *et al.* 1994: Maliotis 1995); have suggested that the Troodos ophiolite may have evolved under the sea at several spreading axes. Stos-Gale *et al.* (1997) summarize in schematic form one current geological view of the formation of Cyprus-type ore deposits in association with submarine axial spreading axes (see figure 1 of Stos-Gale *et al.*1997). In this model faults associated with spreading axes act as conduits for hydrothermal, metal-bearing fluids. On emergence of the hot hydrothermal fluids into cold seawater, metals are precipitated out, mainly as sulfides, forming the ore deposits. In this model older ore deposits are further away from a fossil spreading axis than younger ones.

The ore deposits in Cyprus have been assigned to particular fossil spreading axes; from West to East these are the Limni, Solea, Kalavasos and Larnaca axes (FIGURE 6).

Since different ore deposits formed at slightly different times they could have tapped different ore source fluids from geographically different places themselves with different U,Th/Pb ratios. Consequently, somewhat different lead isotope compositions may have been frozen into the different copper-iron ore deposits in Cyprus.

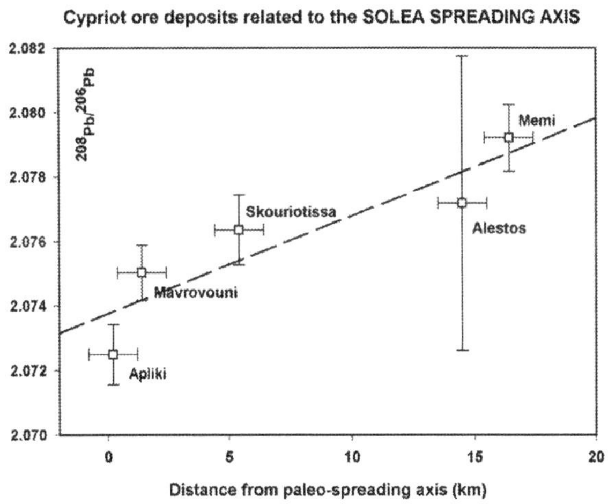

Figure 7: A plot of the mean $^{208}Pb/^{206}Pb$ data for the various mines associated with the Solea axis as a function of the distance of the respective mines away from the Solea fossil spreading axis.

FIGURE 7 shows that plotting the mean $^{208}Pb/^{206}Pb$ data for the various mines associated with the Solea axis is consistent with a monotonic, roughly linear, increase away from it. Similar findings apply to other ratios and for other fossil spreading axes. This evidence is consistent with the model of formation of Cypriot ores sketched above, but may not negate other models.

A number of other, non Cyprus-type, deposits are found in the upper levels of the lava sequence, and are believed to have resulted from later, off-axis, magmatic activity

FIGURE 8 shows that earlier ideas (based on about 15% of the data now available) that the Cypriot ores could be represented by a single lead isotope field (Gale & Stos-Gale 1989), were erroneous. Doubts were expressed about this by Gale & Stos-Gale (1992c) and by Sayre *et al.* (1992). With more data it can be seen that it is unhelpful to draw single closed curves enclosing all the Cypriot lead isotope data and referring to these as outlining a "Cypriot field", since such curves also enclose regions of lead isotope space not populated by any Cypriot ore.

FIGURE 8 also shows lead isotope data for deposits associated with the different spreading axes. The Sha mine demonstrably has a quite distinctive lead isotope composition, which may accord with Maliotis' view (pers.comm.) that it cannot be firmly linked with any fossil spreading axis. The lead isotope data for the Sha mine and Lavrion, Attica clearly do not overlap. Ore deposits associated with a particular fossil spreading axis do have similar lead isotope compositions, whilst the isotopic compositions of ores associated with different axes are, to differing degrees, separable. However, there is overlap between the Larnaca axis and the Limmasol Forest ores, and partial overlap between the Limni axis and the Solea axis ores (at the top end in the upper diagram of FIGURE 8, outside the region where the oxhide ingots

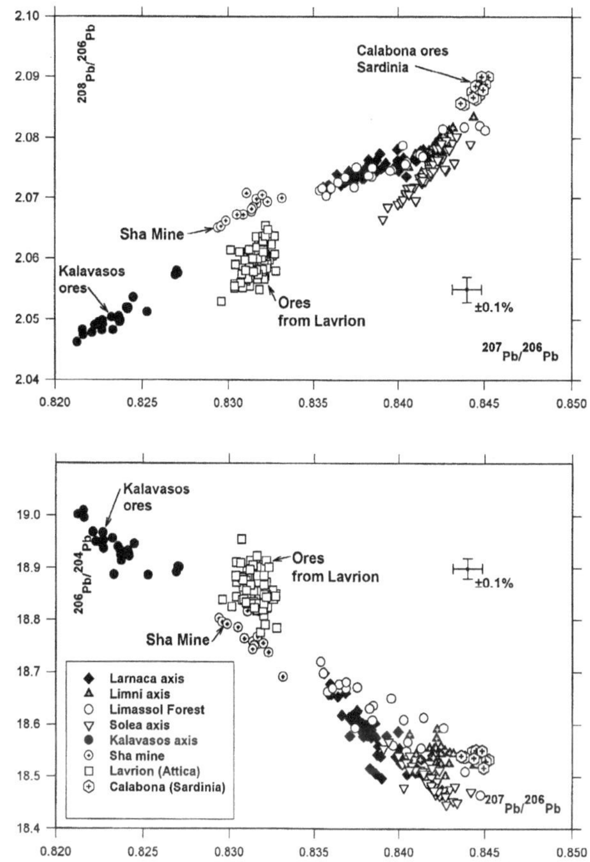

Figure 8: Lead isotope data for the Cypriot ores associated with the four fossil sea floor spreading axes; other Cypriot ores, and ores from Lavrion and the Calabona copper deposit in northwest Sardinia.

plot). Budd (pers.comm.) agrees that the complex, intergrown, Limmasol Forest ores, many of them containing little copper, were unlikely to have been recognized as copper ores in the Late Bronze Age and would have been very difficult, if not impossible, to smelt. Indeed, Zwicker (1986) failed in his attempts to smelt these ores, although he did succeed in transferring arsenic from Pevkos ore to molten copper using a relatively complicated process.

The lead isotope compositions of the seven mines of the Kalavasos axis are very, very different from those of any other Cypriot deposits. Surprisingly, no analyzed copper metal or slag from the nearby Late Cypriot IIC site of Kalavasos-Ayios Dhimitrios (South 1984) matches the adjacent Kalavasos axis ores. Similarly, Pernicka *et al.* (1993) have demonstrated that the well-known Eneolithic Serbian mine of Rudna Glava was not the source of copper for nearby contemporary sites. Clearly it cannot be assumed that the nearest metal source to an archaeological site was the source of metal for that site.

Manganiferous umbers are relatively common on Cyprus. It has long been known that some slags in Cyprus are rich in manganese, whilst others are manganese-free (Davies 1928/29). Bachmann (1982) discusses the technology and advantages of using manganese fluxes in copper smelting, namely to avoid the reduction of iron and its incorporation into the copper metal. He pointed, however, to the need to obtain manganese oxides, in Cyprus, from the umber layers overlying the pillow lavas. He further noted that it would take a conscious effort to seek these out when iron oxides were readily available from the gossans of the copper ore deposits near all the major smelting sites. He also opined that, although manganese oxide fluxing was used as early as the end of the second millennium in Timna (where manganese ores are, in contrast with Cyprus, abundant and close to the copper ores), this technique was adopted much later in Cyprus. Bachmann noted that slags containing manganese occurred at Skouriotissa, Lythrodonda, and Polis. Table 5 in Stos-Gale *et al.* (in press) shows that all of these slag heaps are much later in date than the Bronze Age. Bachmann also noted a progression in slag technology at Skouriotissa, with increasing manganese content; Zwicker's dates for Skouriotissa (Table 5 in Stos-Gale *et al.* 1998) suggest that these developments may all have taken place in the Roman period. It seems quite probable therefore that manganese oxide fluxing was, in Cyprus, a Roman introduction. Koucky and Steinberg (1982) suggest the same, and point to evidence that manganese-rich slags also are associated with the Byzantine and Mediaeval periods in Cyprus. What is certain is that Bachmann found the Bronze Age slags from Hala Sultan Tekke to classify as manganese-free slags (Bachmann 1982). Recent chemical analyses of Bronze Age slags from Hala Sultan Tekke, Ayios Dhimitrios-Kalavasos, Kition and Enkomi show them all to be manganese-free. On the other hand chemical analyses of Cypriot umbers show them to be relatively rich in manganese (Robertson & Hudson 1973; Constantinou 1982). On present evidence it seems that manganese oxide was not used as a flux in the Cypriot Bronze Age, and consequently that umbers were not introduced into the smelting charge. It was for this reason that the incorporation of lead isotope analyses of umbers into the lead isotope 'field' for Cyprus was rejected when dealing with the provenancing of Bronze Age copper based artefacts (Gale *et al.* 1997).

FIGURE 9 shows the lead isotope data for copper oxhide ingots found from the four separate sites of Ayios Dhimitrios, Maroni, Pyla Kokkinikremos, Enkomi, and the two findspots of Mathiati and Skouriotissa on Cyprus. The ingots from Maroni date to LCIIC, as do those from Pyla, Maa, Ayios Dhimitrios and probably those from Enkomi. The findspots of those ingots generally referred to as Skouriotissa and Mathiati are not well recorded, and cannot be dated except by typological analogy with other ingots. These oxhide ingots vary widely in findspot, from Enkomi on the East coast to Maroni on the central Southern coast. For clarity FIGURE 9 embraces a smaller part of isotopic space than FIGURE 8 and does not include the Sha or Kalavasos axis ore deposits, nor the Lavrion polymetallic deposit; none of these deposits have isotopic compositions overlapping any copper oxhide ingot. Clearly these particular Cypriot ingots have isotopic compositions which coincide with those associated with ores from the Solea axis in the NW of Cyprus, and in particular with the ores from the Apliki mine (Mine No 6 on FIGURE 6).

It will not be a surprise to Aegean and Cypriot archaeologists that the copper oxhide ingots found on Cyprus were indeed made from Cypriot copper deposits. It may surprise some that they were made of copper from such a restricted number of ore deposits in the northwest of the Troodos. Nevertheless the scientific evidence strongly supports that all the later copper oxhide ingots discussed here were made from copper smelted from ores coming

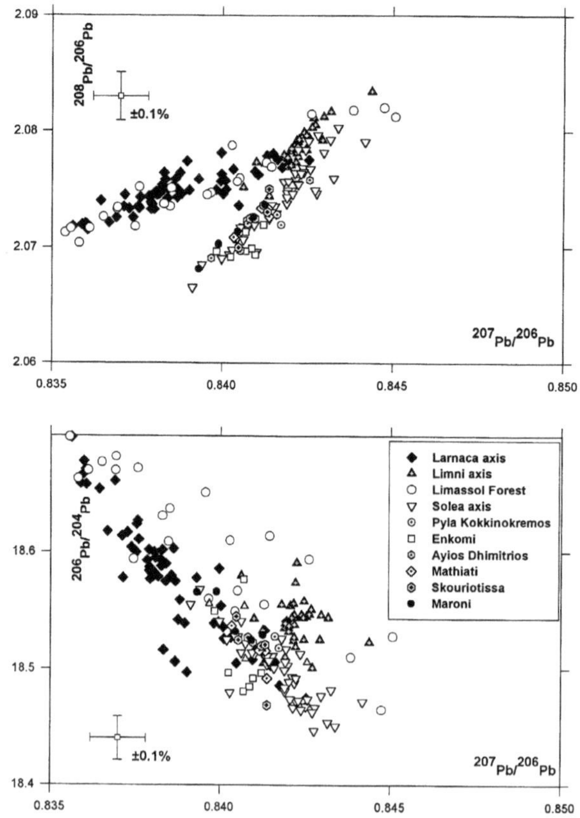

Figure 9: Plot of lead isotope analyses for copper oxhide ingots found on Cyprus in relation to the isotopic compositions of Cypriot ores. The numbers in brackets are the numbers of analyzed oxhide ingots from each site.

solely from the Apliki/Skouriotissa/Mavrovouni region in Cyprus. Later discussion shows that the situation is more complicated for oxhide ingots dating prior to 1250 BC.

Detailed examination of the lead isotope compositions of these 29 Cypriot copper oxhide ingots and the ores of the Solea axis reveals that there is no overlap with the ore deposits of Memi or Alestos. Indeed, within error limits, 27 of these ingots overlap only with the isotopic compositions of ores from Apliki. One oxhide ingot from Skouriotissa and one from Maroni overlap isotopically not only with Apliki ores, but fall also into a very limited area where (in the upper diagram of FIGURE 9) Apliki ores overlap slightly with the lower extension of the isotopic fields for Mavrovouni and Skouriotissa. Consequently it is not possible to exclude the possibility that these two ingots were made from copper smelted from ores from Skouriotissa or Mavrovouni, although they are also consistent with manufacture from copper from Apliki ores.

Since over 90% of these oxhide ingots have lead isotope compositions only overlapping with ores from the Apliki deposit, it is not unreasonable to conjecture that the two ingots from Maroni and Skouriotissa were also made from Apliki ores. The Apliki, Skouriotissa and Mavrovouni mines are separated from each other by less than 5km. Therefore, conservatively it can be said that the analyzed post-1250 BC copper oxhide ingots from six sites on Cyprus were made from copper smelted from Cypriot ores found in the Apliki region.

The restriction of the Apliki region as the ore source of copper for oxhide ingots on Cyprus seems to be a special case which cannot be extended to the source of copper for copper tools, weapons etc. There is clear archaeological evidence of Late Bronze Age copper production at numerous sites in Cyprus (primary and perhaps secondary slags, copper spills, moulds, etc.); sites both coastal and near to copper deposits, although there is no published archaeological evidence yet to prove from where a particular site was getting its copper metal or ore. It seems probable that copper ores from various parts of the Troodos were smelted at a number of different locations to provide copper for making tools and weapons, in contrast to a more centralized use of very large copper ore deposits for the bulk "trade" in massive oxhide ingots. A comparison of the new data for Cypriot ores (Gale *et al.* 1997) with the published lead isotope data for prehistoric Cypriot tools and weapons (Stos-Gale & Gale 1994) shows that, although some artefacts are indeed consistent with a derivation from the Apliki mine, most are consistent with the copper metal having come from many other Cypriot copper deposits around the Troodos. Support for this suggestion may come from the probable Bronze Age copper smelting site of Politiko Phorades (Knapp this volume).

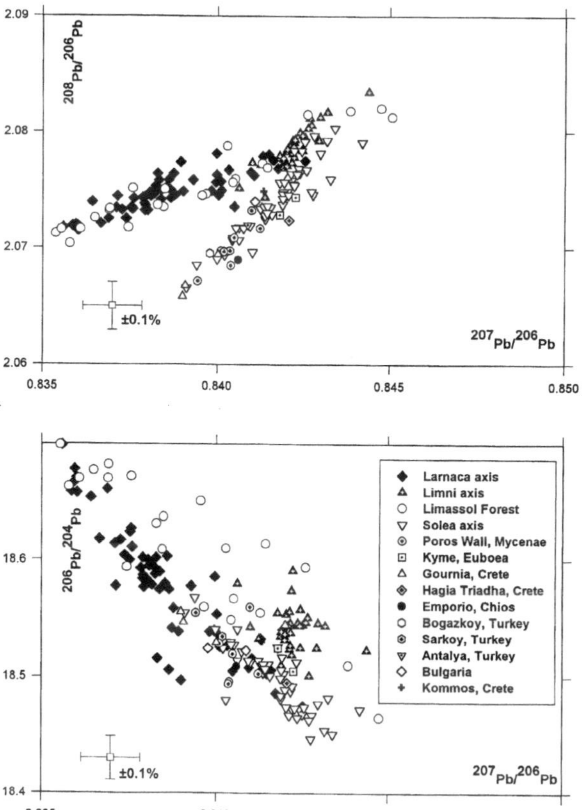

Figure 10: Lead isotope data for miscellaneous late copper oxhide ingots found or excavated on Crete, the Greek mainland, Chios and Turkey, compared with Cypriot copper ores.

Although surprisingly there is evidence for the breaking up of copper oxhide ingots within Cyprus (fragments have been found at Enkomi, Mathiati, Maroni, Ayios Dhimitrios and Pyla), this seems not to have been a common event. The oxhide ingots were not, within Cyprus, a normal source of copper for Late Bronze Age Cypriot people; rather these ingots were produced chiefly to supply a foreign "market". It would have been perverse indeed to go to the considerable trouble of casting large oxhide ingots, only to break them up again in order to make weapons and tools within Cyprus itself. Local needs for copper were probably usually supplied from a number of ore sources other than that devoted principally to the large-scale copper "trade". There are nevertheless some lead isotope data showing that the Apliki region source was sometimes used to satisfy local needs, since roughly 15% of Cypriot copper tools etc. are made of Apliki copper.

FIGURE 10 gives lead isotope data for miscellaneous copper oxhide ingots found in the larger 'Mediterranean' world. These include undated ingots found underwater from Kyme, Euboea, those from LHIIIB Mycenae, from the Black Sea coast of Bulgaria, from Bogazkoy and Sarkoy in Turkey, from Antalya off the Turkish coast, from the island of Chios, and from Kommos, Gournia and Hagia Triadha in Crete (FIGURE 1). These ingots have lead isotope compositions which group extremely well with the Apliki region ores. The 13th century BC ingot fragment from Quantir in the Nile Delta (Pusch in press) also plots with the Apliki region ores. A detailed comparison of the lead isotope compositions of these 21 oxhide ingots with the Solea axis ores reveals that five ingots, one each from Kyme, Gournia, Kommos, Bulgaria and Qantir, fall into the region of overlap with ores from Mavrovouni and Skouriotissa. The remaining 16 ingots match the Apliki ore deposit.

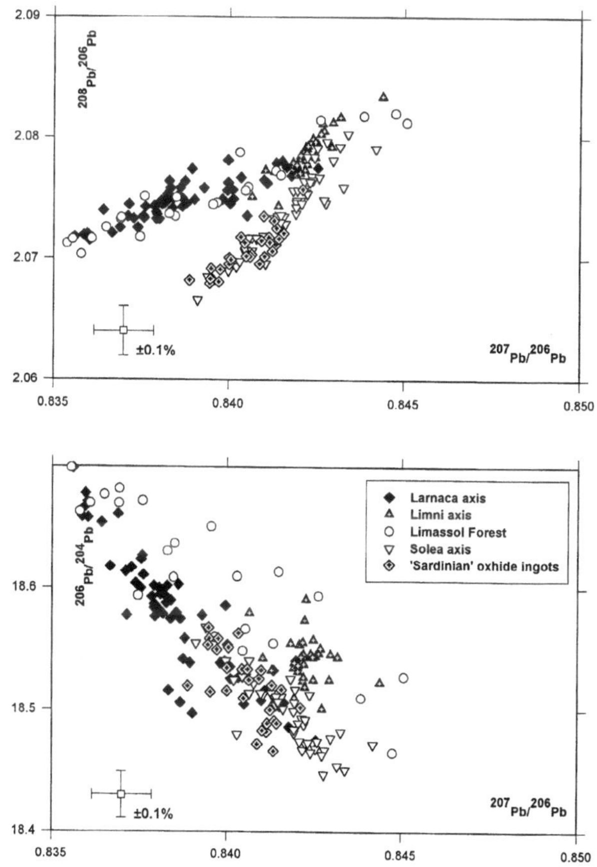

Figure 11: Lead isotope analyses of copper oxhide ingots found on Sardinia in relation to the isotopic compositions of Cypriot ores.

FIGURE 11 shows lead isotope data for 31 copper oxhide ingots and fragments found on Sardinia; unfortunately many of these do not have a clear archaeological time horizon (Lo Schiavo 1990). These oxhide ingots have lead isotope compositions that fit closely with the Apliki ores, with only two fragments plotting into the region of overlap between the Apliki and the Skouriotissa ores. It was demonstrated above that Sardinian Tertiary copper ores, including the relatively large Tertiary Calabona mine, have lead isotope compositions very different from those of the Solea axis ores. Therefore, it is difficult to avoid the conclusion that the oxhide ingots found on Sardinia were made of copper smelted from mines associated with the Solea axis; in particular overwhelmingly, or wholly, from the Apliki ore deposit. Perhaps these ingots were brought to Sardinia by a ship or ships similar to that wrecked off Cape Gelidonya.

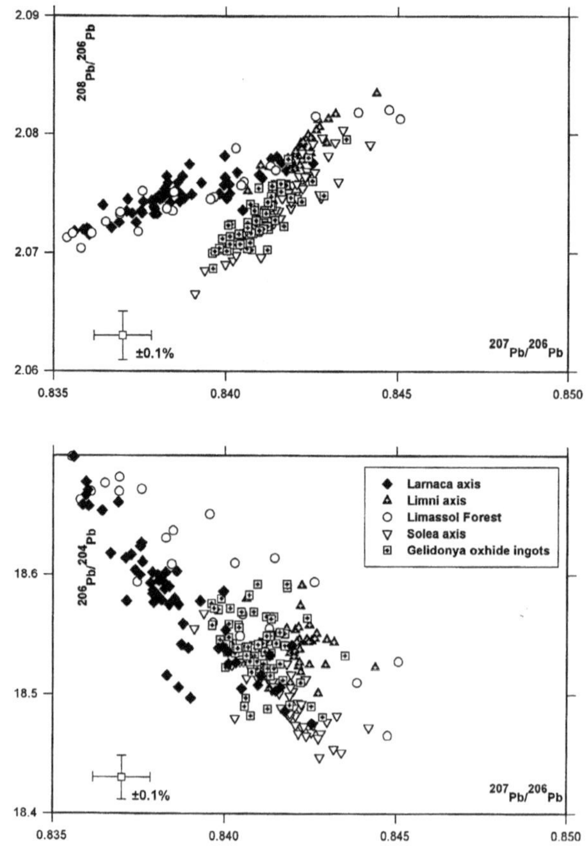

Figure 12: Lead isotope compositions of copper oxhide ingots from the Cape Gelidonya Bronze Age shipwreck compared with the lead isotope compositions of Cypriot ores.

That pre-supposes, of course, that the copper oxhide ingots carried by the Cape Gelidonya ship were indeed made of Cypriot copper. FIGURE 12 presents lead isotope data for 81 oxhide ingots and fragments from the Gelidonya shipwreck. Thirteen fragments fall in the area where the lead isotope compositions of ores from the Mavrovouni and Skouriotissa deposits overlap with those characteristic of the Apliki ores. The remaining 68 ingots have lead isotope compositions which overlap solely with Apliki ores. Since all 81 analyzed oxhide ingot fragments from the Cape Gelidonya wreck overlap with Apliki deposit ores it cannot be ruled out that this is the ore source for them all. Conservatively, it is true that this group of ingots derive from the Apliki region, namely the ore deposits of Mavrovouni, Skouriotissa and Apliki.

The copper metal cargoes of the Cape Gelidonya and Uluburun ships were not restricted to oxhide ingots; both also contained bun and slab ingots. FIGURE 13 shows that some of these differently shaped ingots are consistent with having been smelted from copper ores coming from the Lavrion mines in Attica, Greece. Five slab ingots from the Cape Gelidonya wreck and one bun ingot from the Uluburun wreck overlap very closely with the ores from Lavrion in both diagrams; these are the first Bronze Age

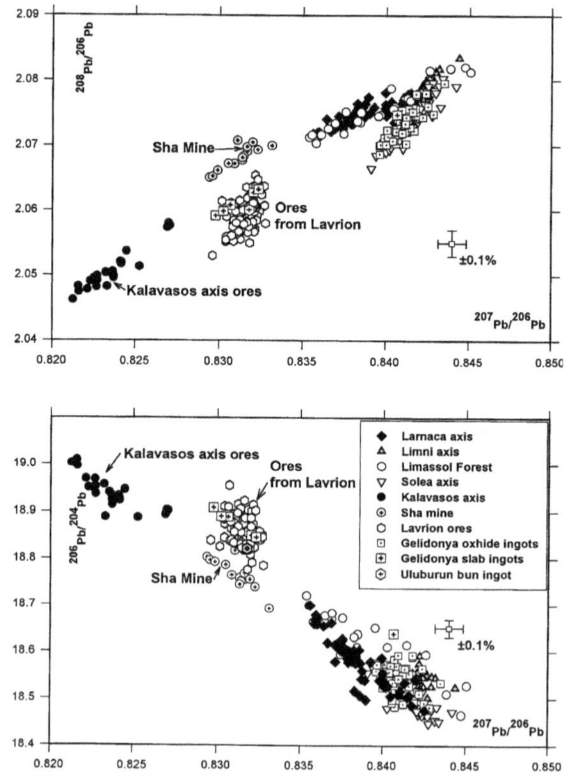

Figure 13: Lead isotope compositions for some of the bun and slab ingots from the Bronze Age shipwrecks of Uluburun and Cape Gelidonya. Also shown are the lead isotope compositions of copper ores from Lavrion and Cyprus.

copper ingots found to have this characteristic lead isotope composition. This finding strengthens the long-standing conclusion that Lavrion was an important copper source in the Bronze Age, Minoan and Mycenaean, Mediterranean world (Stos-Gale & Macdonald 1989; Stos-Gale 1993). Since no copper oxhide shape ingot has yet been found to have the lead isotope composition characteristic of the Lavrion ores; it seems that Lavrion copper may have been traded in the form of bun or slab ingots. FIGURE 13 shows that one analyzed slab ingot from the Cape Gelidonya wreck is consistent with a derivation from Cypriot ores, so some Cypriot copper was traded in that form, although the majority seems to have been traded as oxhide ingots.

FIGURE 14 shows rather surprising lead isotope data that are beginning to show that the Uluburun oxhide ingots form an isotopically different group from the Gelidonya group, with only a few overlapping the isotopic compositions of Apliki ores. The Uluburun ingots date before 1318 BC, and the Gelidonya ingots come from a shipwreck some 100 years later. Different sources of copper at the differing times of lading of these two ships are not unlikely. There are now preliminary lead isotope data for the Phoenix mine, located close to the Skouriotissa mine in Cyprus, that suggest that the Uluburun copper oxhide ingots may have been smelted from Phoenix copper ores. At present this remains a tentative idea.

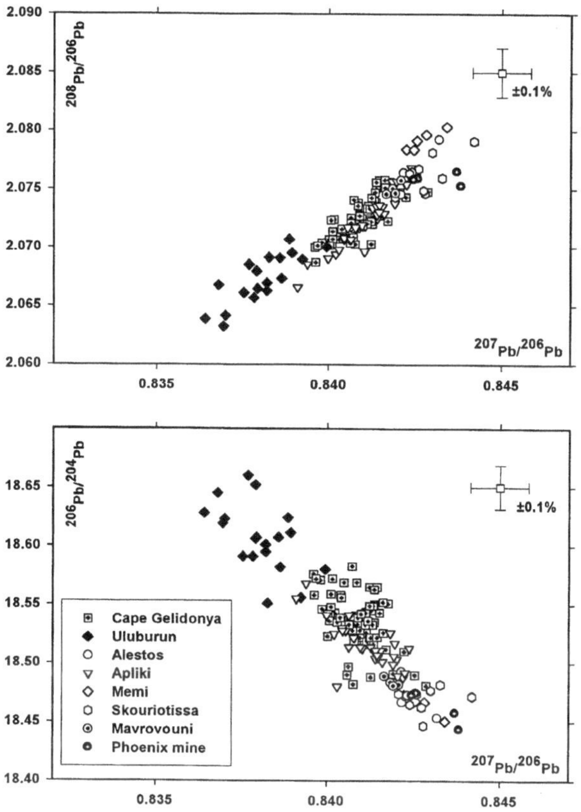

Figure 14: Lead isotope compositions for copper oxhide ingots from the Uluburun shipwreck compared with the isotopic compositions of the later oxhide ingots from the Cape Gelidonya wreck and Cypriot ores.

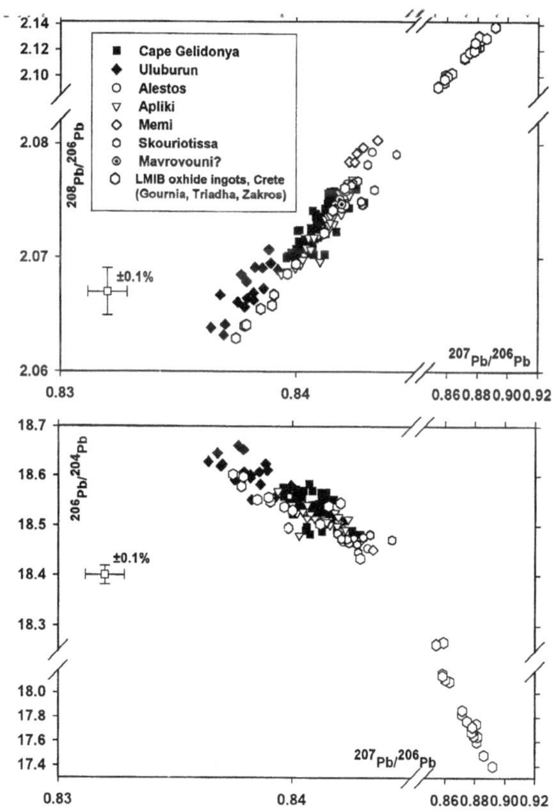

Figure 15: Lead isotope data for copper oxhide ingots from the Late Bronze Age shipwrecks of Uluburun and Cape Gelidonya and for certain LM IB Minoan copper oxhide ingots, compared with isotopic data for Solea axis ore deposits in Cyprus.

This suggestion that different copper ore sources may have been used at different times to supply the copper for oxhide ingots is supported by other data. The earliest copper oxhide ingots, excavated on Crete at Hagia Triadha and dating probably to LMI (Georgiou 1979) or about 1550-1450 BC, have a completely different lead isotope composition from all other oxide ingots (Gale & Stos-Gale 1986).

This is demonstrated in FIGURE 15, where the breaks in the axes are made in order to be able to plot the lead isotope compositions of these Minoan ingots on this graph since they are so different from those of all other oxhide ingots. On the other hand, some of the Late Minoan oxhide ingots, from Gournia, plot with the Uluburun ingots, consistent with the idea that a sister ship broadly contemporary with the Uluburun vessel managed to make a successful journey to Crete. The chronology of the oxhide fragments from Gournia is not certain, although Betancourt and Weinstein (1976) believe they date to LMIB; this date is some 200 years earlier than that of the Uluburun ship and about contemporary with the Hagia Triadha ingots. It is not impossible that different ships could have carried cargoes of copper from different sources at the same time as each other. The archaeological evidence may not be precise enough to fix the relative times of deposition of the Gournia and Hagia Triadha ingots to better than ± 100 years.

Conclusions

Normally it seems inadvisable or unhelpful to attempt to define overall Pb isotope fields for large geographical regions, or for islands as large as Sardinia or Crete, or even for islands as small as Kythnos, where ores of quite different geological history, and/or ores of different metals, are present. There may be very limited instances where the use of an overall field can be useful, such as its use in enclosing the lead isotope compositions of all Aegean ore deposits to illustrate the contrast with the exotic isotopic compositions of some bronzes found in EBII Troy (Gale in press). However, the value of such an application is outweighed by the confusion which the past use of such overall fields has brought to the lead isotope provenancing of metals.

Attention should be focused on particular mines or ore deposits and the ore geology and mineralogy should be properly taken into account (Ixer 1999, this volume). Simple errors can thus be avoided, such as looking for irrelevant overlaps between the isotopic compositions of Sardinian lead-zinc ores and Cypriot copper-iron sulfide ores.

It has to be recognized that much work has to be done, amassing large numbers of lead isotopic analyses of ores taken in comprehensive, wide-ranging, field programmes, before a comprehensive picture of an ore district can be assembled. An important region where the lead isotope characterization of the ore deposits has been left in an inchoate state is that of Anatolia, where exemplary German fieldwork (Wagner *et al.* 1989) was unfortunately followed by relatively small numbers of lead isotope analyses of the collected ores, so that the lead isotope fields for most ore deposits remain undefined.

Finally, comparative lead isotope analysis of ores and artefacts is a viable way of provenancing copper, and of unravelling facts about the Bronze Age Mediterranean trade in copper oxhide ingots, a problem which no other approach has yet been capable of solving. Using this approach it has been possible to demonstrate that all analyzed post-1250BC copper oxhide ingots were made from copper smelted from ores from the Apliki region in Cyprus, possibly just from Apliki itself. It has further been shown that the earliest copper oxhide ingots (dating to 1550-1450BC) were made from a completely different copper source, whilst those excavated from the Uluburun shipwreck (dating to about 1318BC) came from yet another copper source, probably in NW Troodos in Cyprus. These conclusions come purely from a comparison of lead isotope data for ores with that for the oxhide ingots; they are quite independent of any ore genesis model for the ore deposits involved. It should be emphasized that the dominance of the Apliki region copper ore deposits as a source for the post-1250BC copper oxhide ingots does not imply the same dominance as a source of copper for Cypriot weapons, tools, etc. Lead isotope analyses of such Late Bronze Age Cypriot artefacts (Stos-Gale and Gale 1994) show that they were made from copper coming from a number of other Cypriot ore deposits.

Finally, Stos-Gale *et al.* (1997) drew attention to the LCIIC - LCIIIA site of Apliki-Karamallos, situated in Cyprus near the Apliki deposits. This site produced a large quantity of copper smelting slag in good archaeological contexts and was characterized by Muhly (1989) as a Late Bronze Age miners' village. Stos-Gale *et al.* (1997) mentioned this site only as providing evidence that mining and copper smelting were proceeding at, or very near, the Apliki deposits at the period of the Late Bronze Age in which it is known copper oxhide ingots were produced. However, *contra* Knapp (this volume), they did not state or imply that Apliki-Karamallos was the only, or even a major, site producing copper for oxhide ingots. Instead the lead isotope evidence points to the ore deposits of the Apliki region, perhaps only the Apliki deposit itself, as the source of copper for the post-1250BC copper oxhide ingots. It is certainly not argued that a major proportion of Bronze Age mining and primary smelting activities was conducted at the one site of Apliki-Karamallos. Indeed, as discussed above, the Oxford lead isotope evidence for Bronze Age Cypriot tools and weapons points to mining and smelting activities at many other sites around the Troodos Mountains.

Acknowledgements

I would like to acknowledge the debt I have to many collaborators, so many as to defy listing fully here, but principally I must mention Zofia Stos-Gale and George Maliotis, also George Bass, Cemal Pulak and Fulvia Lo Schiavo. I should also like to acknowledge the late Professor Wolfgang Gentner of Heidelberg, and Lord Colin Renfrew of Cambridge, for inveigling me into this line of research many years ago. I thank R.A. Ixer and two anonymous referees for greatly improving this paper. Finally we thank the Institute for Aegean Prehistory, New York, without whose support this work would not have been possible.

Numerical lead isotope data for the ingots and ores can be obtained by application to the author.

References

BACHMANN, H- G. 1982. Copper smelting slags from Cyprus: Review and classification of analytical data. In J. Muhly, R. Maddin and V. Karageorghis. (ed.): *Early Metallurgy in Cyprus, 4000-500 BC*: 143-152. Nicosia: Pierides Foundation.

BASS, G.F. 1967. Cape Gelidonya: A Bronze Age Shipwreck. *Transactions of the American Philosophical Society* 57/8. Philadelphia.

1987. Oldest known shipwreck reveals Bronze Age splendors. *National Geographic* 172: 691-733.

BAXTER, M.J. & N.H. GALE. In press. Testing for multivariate normality via univariate tests: a case study using lead-isotope ratio data. *Journal of Applied Statistics.*

BETANCOURT, P. & G.A. WEINSTEIN. 1976. Carbon-14 and the beginning of the Late Bronze Age. *American Journal of Archaeology* 80: 329.

BUCHHOLZ, H. - G. 1959. Keftiubarren und Erzhandel im zweiten vorchristlichen Jahrtausend, *Prähistorische Zeitschrift*, 37: 1-40.

CONSTANTINOU, G. 1982. Geological features and ancient exploitation of the cupriferous sulphide orebodies of Cyprus. In J. Muhly, R. Maddin and V. Karageorghis. (ed.): *Early Metallurgy in Cyprus, 4000-500 BC*: 13-24. Nicosia: Pierides Foundation.

CONSTANTINOU, G. & A. PANAYIOTOU. 1982. *Mineral Resources Map of Cyprus*, Geological Survey Department, Nicosia, Cyprus.

DAVIES, O. 1928/29. The copper mines of Cyprus, *Annual British School at Athens* 30: 74-85.

DEL CALDO, A., C. MORO, C. MARIA GRAMACCIOLI & M. BOSCARDIN. 1974. *Guida ai Minerali*, Milan: Fratelli Fabri Editori.

FEYNMAN, R.P. 1992. *The Character of Physical Law.* London: Penguin Books.

GALE, N.H. In press. Metal sources for Early Bronze Age Troy and the Aegean. In M. Treister (ed.): *Proceedings of the International Conference: Troy and its Treasures,* November 1996, Moscow.

1998. The role of Kea in metal production and trade in the Late Bronze Age. In L.G. Mendoni & A. Mazarakis-Ainian (ed.): *Kea-Kythnos: History and Archaeology,* Meletimata 27, Athens: 1-17.

GALE, N.H. & Z.A. STOS-GALE. 1982. Bronze Age copper sources in the Mediterranean: a new approach. *Science* 216: 11-19.

1986. Oxhide copper ingots in Crete and Cyprus and the Bronze Age metals trade. *Annual of the British School at Athens* 81: 81-100.

1987. Oxhide ingots from Sardinia, Crete and Cyprus: new scientific evidence, in M.S. Balmuth (ed.): *Studies in Sardinian Archaeology III. Nuragic Sardinia and the Mediterranean World:* 135-178. Oxford: British Archaeological Reports S387.

1988. Recent evidence for a possible Bronze Age metal trade between Sardinia and the Aegean. In E.B. French & K.A. Wardle (ed.): *Problems in Greek Prehistory:* 349-84. Bristol: Bristol Classical Press.

1989. Bronze Age archaeology of the Mediterranean: the impact of lead isotope studies. In R.O. Allen (ed.): *Archaeological Chemistry IV:* 159-198. Washington: *Advances in Chemistry Series* 220.

1992a. Lead isotope studies in the Aegean (The British Academy Project). *Proceedings of the British Academy* 77: 63-108.

1992b. New light on the provenance of the copper oxhide ingots found on Sardinia. In R.H. Tykot & T.K. Andrews (ed.): *Sardinia in the Mediterranean: a Footprint in the Sea*: 317-46. Sheffield: Sheffield Academic Press.

1992c. Evaluating lead isotope data: Comments I. *Archaeometry* 34: 311-317.

1996. Lead isotope methodology: the possible fractionation of lead isotope compositions during metallurgical processes. In S. Demirci, A.M. Ozer
1997. &
G.D. Summers (ed.): *Archaeometry '94, The Proceedings of the 29th. International Symposium on Archaeometry, May 1994:* 287-299. Ankara: Tubitak.

GALE, N.H., Z.A. STOS-GALE, G. MALIOTIS & N. ANNETTS. 1997. Lead isotope data from the Isotrace Laboratory, Oxford: Archaeometry data base 4, Ores from Cyprus, *Archaeometry* 39: 237-46.

GEORGIOU, H. 1979. *The Late Minoan destruction of Crete: metal groups and stratigraphic considerations.* Los Angeles: University of California, *Monograph IX.*

IXER, R.A. 1999 (this volume). The role of ore geology and ores in the archaeological provenancing of metals. In S. Young, P. Budd, R. Ixer & M. Pollard (ed.): *Metals in Antiquity:* 00-00. Oxford: Archaeopress.

KNAPP, A.B. 1999 (this volume). The archaeology of mining: Fieldwork perspectives from the Sydney Cyprus Survey Project (SCSP). In S. Young, P. Budd, R. Ixer & M. Pollard (ed.): *Metals in Antiquity:* 00-00. Oxford: Archaeopress.

KOUCKY, F. & A.STEINBERG. 1982: The ancient slags of Cyprus. In J. Muhly, R. Maddin and V. Karageorghis. (ed.): *Early Metallurgy in Cyprus, 4000-500 BC*: 117-142. Nicosia: Pierides Foundation.

LAGARCE, J., E. LAGARCE, A. BOUNNI. & N. SALIBY. 1983. Les fouilles à Ras Ibn Hani en Syrie *Académies des Inscriptions et Belles Lettres: Comptes Rendus,* Avril-Juin: 249-290.

LAVENDER, D. 1962. *The story of Cyprus Mines Corporation,* The Huntingdon Library, San Marino, California.

LO SCHIAVO, F. 1990. Lingotti ox-hide e lingotti piano-convessi. In F. Lo Schiavo, R. Maddin, J. Merkel, J.D. Muhly & T. Stech. (ed.): *Analysi Metallurgiche e Statistiche sui Lingotti di Rame della Sardegna*: 16-40. *Quaderni,* 17. Ozieri: Sardinia.

MALIOTIS, G. 1995. The structural control of the mineralization of north-northeast Troodos, Cyprus, *XV Congress of the Carpatho-Balcan Geological Association, Special Publication No.4/2:* 767-770.

MALIOTIS, G, N.J.COOPER & M.A. KHAN. 1994. The use of geophysics in the search for sulphide deposits in the Troodos Ophiolite Complex, Cyprus, *Proceedings 2nd Congress of the Hellenic Geophysical Union:* 184-193.

MARCELLO A., S. PRETTI & I, SALVADORI. 1978. *Carta Metallogenica della Sardegna.* Firenze, Servizio Geologico d'Italia.

MUHLY, J. 1989. The organisation of the copper industry in Late Bronze Age Cyprus. In E. Peltenburg (ed.): *Early Society in Cyprus:* 298-314. Edinburgh: Edinburgh University Press.

PERNICKA, E. 1992. Evaluating lead isotope data: Comments...III. *Archaeometry,* 34: 322-26.

PERNICKA, E, F. BEGEMANN, S. SCHMITT-STRECKER & G.A. WAGNER. 1993. Eneolithic and Early Bronze Age copper artefacts from the Balkans and their relation to Serbian copper ores, *Prähistorische Zeitschrift,* 68: 1-54

PULAK, C. 1988. The Bronze Age Shipwreck at Ulu Burun, Turkey: 1985 Campaign, *American Journal of Archaeology* 92: 1-37.

PUSCH, E.B. In press. High temperature industries in the Late Bronze Age capital Piramesse Qantir. In K. Barakat (ed.): *Proceedings of the First International Conference on Ancient Egyptian Mining and Metallurgy,* Cairo, 1995.

RIDGE J.D. (1990). *Annotated Bibliographies of Mineral Deposits in Europe: Part 2:* 323-333. Oxford. Pergamon.

ROBERTSON, A.H.F. & J.D. HUDSON. 1973. Cyprus umbers: chemical precipitates on a Tethyan ocean ridge. *Earth and Planetary Science Letters* 28: 385-94.

SAYRE, E., K.A.YENER, E.JOEL & I.L.BARNES.1992. Statistical evaluation of the presently accumulated lead isotope data from Anatolia and surrounding regions. *Archaeometry* 34: 73-106.

SOUTH, A., 1984. Kalavasos-Ayios Dhimitrios 1983. *Report of the Department of Antiquities, Cyprus:* 14-41.

STOS-GALE. Z.A. 1993. Lead isotope provenance studies -do they really work? *Archaeologia Polona* 31: 149-180.

STOS-GALE, Z.A. & N.H. GALE. 1994. Metals. In A.B. Knapp and J.F. Cherry (ed.): *Provenience Studies and Bronze Age Cyprus:* 92-121. Madison: Prehistory Press.

STOS-GALE, Z.A & C.F. MACDONALD. 1989. Sources of metals and trade in the Bronze Age Aegean. In N.H. Gale (ed.): *Bronze Age Trade in the Mediterranean:* 249-288. Jonsered: Paul Åström's Förlag. *Studies in Mediterranean Archaeology,* 90.

STOS-GALE, Z.A., N.H. GALE, J. HOUGHTON & R. SPEAKMAN. 1995. Lead isotope data from the Isotrace Laboratory, Oxford: Archaeometry Data Base 1, Ores from the Western Mediterranean. *Archaeometry* 37: 407-415.

STOS-GALE, Z.A., N.H. GALE, & N. ANNETTS. 1996. Lead isotope data from the Isotrace Laboratory, Oxford: Archaeometry Data Base 3, Ores from the Aegean Part 1. *Archaeometry* 38: 381-390.

STOS-GALE, Z.A., G. MALIOTIS, N.H. GALE & N. ANNETTS. (1997). Lead isotope characteristics of the Cyprus copper ore deposits applied to provenance studies of copper oxhide ingots. *Archaeometry* 39: 83-124

STOS-GALE, Z.A., G. MALIOTIS & N.H. GALE. 1998 In press. A preliminary study of the Cypriot slag heaps and their contribution to the reconstruction of copper production on Cyprus. In Th. Rehren, A. Hauptmann & J.D. Muhly (ed.): *Metallurgica Antiqua, in honour of H.G. Bachmann and R. Maddin.* Bochum: Deutschen Bergbau-Museum. *Der Anschnitt,* Beiheft 8.

WAGNER, G., O. OZTUNALI & C. EIBNER. 1989. Early copper in Anatolia: archaeometallurgical fieldwork. In A. Hauptmann, E. Pernicka and G.A. Wagner (ed.): *Old World Archaeo-metallurgy*: 299-306. Bochum: Deutschen Bergbau Museum. *Der Anschnitt,* Beiheft 7.

ZUFFARDI, P. 1980: Copper Deposits of Italy: Metallogenesis and Evaluation. In S. Jankovic and R.H. Sillitoe (ed.): *European Copper Deposits.* Belgrade: Belgrade University Press

ZUFFARDI, P. 1989: Italy. In F.W. Dunning, P. Garrard, H.W. Haslam & R.A. Ixer (ed.): *Mineral Deposits of Europe Volume 4/5: Southwest and Eastern Europe, with Iceland:* 221-277. London: Institute of Mining and Metallurgy.

ZWICKER, U. 1986. Ancient metallurgical methods for copper production in Cyprus, Part II: sulphide ore and copper-arsenic-alloy production. *Bulletin 3, Cyprus Association of Geologists and Mining Engineers*: 92-111.

15

Lead Isotope Analysis, Oxhide Ingots and the Presentation of Scientific Data in Archaeology

B. SCAIFE[1], P. BUDD, J.G. MCDONNELL & A.M. POLLARD

Department of Archaeological Sciences, University of Bradford, Bradford BD7 1DP, UK, UK
[1]Present address: Sub-Unit for Medical Statistics, Leeds University, 32 Hyde Terrace, Leeds LS2 9LN, UK

Lead isotope analysis (LIA) seems at first sight to be an ideal technique for metal provenance studies. Measurements can be made very accurately and manufacturing processes do not measurably alter the isotope ratios. However, the use of LIA in archaeology over the last twenty-five years illustrates many of the successes and pitfalls of archaeological science as a whole. LIA is a technique which provides an extra dimension of information when discussing the ultimate source or sources of the metal used in an artefact but the numbers that result from the Thermal Ionization Mass Spectrometry (TIMS) analysis of archaeological artefacts and the diagrams used to illustrate them are not necessarily simply related to the presumed source orefield. This paper introduces an alternative method of visualizing lead isotope data based on Kernel Density Estimation, and speculates on some of the lessons which might be learned from the debates over the last twenty years.

Keywords: Lead Isotope Analysis; LIA, Oxhide Ingots; Provenance; Kernel Density Estimation; Mediterranean; Cyprus; Sardinia; LBA; Late Bronze Age.

Introduction

The debate over the origin of oxhide ingots and the implications for trade in the Late Bronze Age Mediterranean is one that brings into focus many of the disagreements and worries over the use of LIA in archaeology. The application of LIA to this question started in the early 1980s (Gale & Stos-Gale 1982) and the debate over the interpretation of the results and the methodology has intensified in subsequent publications, culminating in a debate in the *Journal of Mediterranean Archaeology* in 1995 (Budd *et al.* 1995a and following papers). Recently there has been a restatement of the Oxford Group's position, utilizing new ore data (Stos-Gale *et al.* 1997). Aspects of this debate will be used to illustrate the following discussion so a little background information on the oxhide ingots and the Cypriot orebodies will be summarized.

Oxhide ingots

Oxhide copper ingots are one of the most distinctive and discussed groups of artefacts from the Late Bronze Age Mediterranean (c. 13th-12th century BC). They have been found at sites from Syria to Sardinia (Knapp & Cherry 1991; Gale & Stos-Gale 1992a:90 and their fig. 19), including part of the cargoes of two shipwrecks off the Anatolian coast. The ingots are large (around 30 kg) and are composed of unalloyed copper without any major impurities (Maddin & Muhly 1974; Muhly *et al.* 1977; Muhly 1982; Muhly *et al.* 1980). This lack of impurities could suggest that the ingots are primary smelting products, unaffected by subsequent alloying or other processes. As such they would represent a good opportunity for various provenancing techniques such as trace element analysis or LIA to be applied. LIA obviously has the potential in this case for helping to answer specific important archaeological questions relating to LBA Mediterranean trade.

Of the various groups of oxhide ingots discussed in the literature, only a group from Ayia Triadha, Crete have been interpreted in a way that raises no real argument. These ingots have lead isotope compositions which are markedly different from those of all the other oxhide ingots. Their isotopic compositions are also different from those of ores from Cyprus, Sardinia, Laurion, and indeed from any Mediterranean ore source so-far characterized (Gale & Stos-Gale 1988; Gale & Stos-Gale 1992a). Thus we have a classic example of the exclusion principle operating; all researchers can agree that the source of the ore for these ingots is not any of the orefields studied so far.

The major debate referred to above largely focused on the contention of the Oxford Group that all the oxhide ingots, apart from the Ayia Triadha group, were manufactured from Cypriot ores. This caused disquiet amongst some researchers who found it hard to accept that the ingots would have been transported all the way to Sardinia, an island with abundant copper deposits. In the latest contribution (Gale *et al.* 1997) the discussion focuses on three groups of ingots: Cypriot ingots; Sardinian ingots; and a miscellaneous group comprising ingots from Bulgaria, Greece (three sites) and Turkey (three sites), dated variously between 1500 BC and 1200 BC In total seventy-nine ingots are analysed from all over the Mediterranean and all are found to have isotopic compositions in agreement with a small number of mines in the Apliki/Skouriotissa region of the Troodos ophiolite of Cyprus. Indeed, the agreement of most ingots' composition with the isotope field of the Apliki Mine lead Gale *et al.* to propose this mine as the predominant source for the copper for ingots that were shipped around the whole Mediterranean in the Late Bronze Age.

There are two orefields in Cyprus of interest to the discussion of oxhide ingots - those belonging to the Troodos ophiolite complex and those of the Western Limassol Forest complex. The data in Stos-Gale *et al.* 1997 (discussed in Gale *et al.* 1997) from the Troodos are presented as representing four different geological spreading axes, resulting from hydrothermal deposition and having distinctive sets of isotope ratios. Data from individual mine sites within these axis groups also form discrete groups. In general the groups for individual mines overlap in three dimensions but there are some cases where the data form differentiated groups. The Western Limassol Forest complex differs from the Troodos in that it has a far more complex tectonic and magmatic history. The process of formation did not involve hydrothermal deposition but rather magmatic segregation. The mineralization in the Limassol Forest is multi-metallic, generally characterized as copper-nickel-cobalt-iron; the copper content of the ores is very variable but averages 3.32% (Panayiotou 1980). *Chalcopyrite* ($CuFeS_2$) is abundant at Laxia tou Mavrou and common at Pevkos, the two Limassol Forest sites sampled by Gale *et al.*.

Sampling orefields and constructing isotope fields

Only geologically well-provenanced samples should be used to define an isotope field for an orefield. Not all samples, however well provenanced, will be applicable to all studies. For example, if Laurion is postulated as a source for silver for a group of artefacts then only the isotopic compositions of *galenas* (PbS) from Laurion should be used to define the Laurion isotope field, since argentiferous galena would have been the mineral used to produce silver. This criterion can be eased if it can be shown that other minerals must have undergone identical patterns of isotope evolution. The situation becomes more complex for copper production. Initially we should only take into account the minerals capable of being smelted by the technology of the time. In the Troodos ophiolite for Bronze Age production this would mean using samples of *chalcopyrite*. If, however, the production process involved the use of fluxing agents to aid slag formation and these fluxing agents contain lead then they have also to be used in the construction of the isotope field. It may even be the case that fluxing agents contain substantially greater concentrations of lead than the copper ores themselves and it may happen that in some cases the relevant isotope field for provenancing will be that constructed from the fluxes rather than the ores.

The necessity of using only the minerals that relate to the metals in artefacts being considered was noted by the Oxford Group some time ago (Gale 1989) but their decision on precisely which samples were needed to define the field was not purely geologically-based. It was stated variously that it required between thirty and fifty samples (Gale 1989) or possibly twenty geologically well selected analyses (Gale & Stos-Gale 1992a) to define the field. This statement was based on assumptions that the isotope field would conform to a pattern of a limited spread of approximately multivariate normal structure. The most recent data set for the Cypriot orefields is considerably different from earlier versions. The earliest data (Spooner & Gale 1982) appeared in a geologically-based paper in *Nature* and comprised 11 samples. The latest data set, which the Oxford Group now considers to be definitive consists of 210 samples from four spreading axes of the Troodos ophiolite and from the Limassol Forest complex (Gale *et al.* 1997). It is both expected and welcomed that the number of analyses has grown over fifteen years. The growth, however, has been fitful and has involved a number of changes which have been somewhat contentious. Sometimes samples have been excluded for reasons which may be justified archaeologically or technologically. For example, the 1997 compilation excludes umbers used in Cypriot field definitions between 1982 and 1994 on the grounds that work by other researchers carried out in 1973 and 1982 shows that they could not have been used as fluxes (Gale *et al.* 1997).

Following the publication of a compilation of Cypriot data in 1994 (Stos-Gale & Gale 1994) Budd *et al.* (1995a) listed a set of samples whose values had changed from earlier published values or had been dropped from the data set, and also noted that only thirty-three samples were published by the Oxford Group despite public claims of over a hundred analyses. These omissions were finally explained in 1997 by the Oxford Group (Gale *et al.* 1997). In the 1997 compilation six samples were omitted because they were considered bad analyses and there was no sample left or the sample did not contain sufficient copper mineral. Five samples were omitted because they did not have a good geological context. Umbers were now omitted as they were considered not to have been part of the production process.

The history of two particular samples from Mousoulos first analysed in 1982 (Spooner & Gale 1982) is interesting. These two samples plotted well away from all the other samples from Cyprus published before 1997 (see Budd *et al.* 1995a). These samples did not appear in any of the diagrams used by the Oxford Group prior to 1997. It was considered justified to exclude them on the grounds that there is no evidence for mining there (Gale & Stos-Gale 1995). This could be a reasonable position but it is not one which is applied consistently. It is acknowledged, for example, that there is no evidence of Bronze Age mining at any of the mines, including Apliki, which feature in the latest compilation of data (Stos-Gale *et al.* 1997). As explanation for the previous exclusion and current inclusion of the Mousoulos samples, Gale *et al.* (1997) state that the previous analyses were poor and needed to be repeated. One sample had been used up and thus there was only one to reanalyse, CY/75/100B. Unusually, the reanalysis agreed with original analysis to the last decimal place in all three measured ratios. It is difficult to escape the conclusion that the Mousoulos samples were initially as good as any of their contemporary analyses and that their exclusion from the field was for other reasons.

Stos-Gale *et al.* (1997) give two reasons as an alternative explanation for the exclusion of the Mousoulos samples. Firstly, no ancient metals show isotopic compositions overlapping those of these ores from Mousoulos and, secondly, that the Mousoulos deposit is too deep to have been exploited by ancient miners. The first reason is unconvincing: the full range of isotope values must be shown for any orefield that is a possible source for metals. If there are extreme values lying a long way from the majority of samples it simply means the isotope field is larger than anticipated and possibly discontinuous. The second reason is very acceptable if the methodology for provenancing with LIA is being strictly and uniformly applied. If a mineral cannot have been used in a particular period then it should not be included in an isotope field. The 1997 compilation of data, however, contains seven samples from Mousoulos. Their inclusion is justified on the grounds that there are other deposits on the Kalavasos axis, not deeply buried, which show the same isotopic composition. This is inconsistent - if the Mousoulos deposits could not have been considered a source of ancient copper in 1994 then they should not be a potential source in 1997.

The inclusion of the Mousoulos samples is justifiable, and is indeed essential, if the aim of is to give the maximum possible range of isotope values that Cypriot ores might show. The isotope values of deeply buried ores might be taken as representative of possible undiscovered workable deposits. This would then give us fields which are larger and more overlapping than those customarily used and would make provenancing *to* an orefield more tentative. It would, however, make provenacing *away* from the orefield more certain, using the more cautious exclusion principle. It seems odd that during the 1980s and early 1990s the Oxford Group deliberately did not encompass all the ores of the island, when they themselves proposed a single all-encompassing Cypriot field. As noted by Budd *et al.* (1995c), the 1994 field represented neither the full extent of the metallurgical remains on Cyprus nor the isotopic variation within the Cypriot orebodies. The 1997 paper represents a welcome attempt, largely successful, to demonstrate that the Cypriot field can be considered as a group of smaller fields representing individual spreading axes or even individual mines. If, however, the rationale is now to allow the identification of individual ancient mines then it would seem unnecessary to include deposits which, it is claimed, could never have been exploited by ancient miners.

This situation is indicative of two problems relating to the use LIA in archaeology. One is a general problem, relating to the exclusion principle. Whilst all researchers acknowledge that in principle all that can be said with certainty by LIA about an artefact is that it does *not* originate from a particular orebody, there is obviously a great desire to say that it *does* originate from somewhere specific. Thus, Stos-Gale *et al.* (1997) demonstrate that oxhide ingots from Turkey, Bulgaria and Greece do not originate from local ore sources. They also show that the composition of Sardinian ingots is different from all of the measured deposits on Sardinia, whilst noting that the Calabona deposits need to be analysed to ensure that there are no deposits on the island matching the ingots. The close similarity of the oxhide ingots with the Solea deposits of Cyprus and particularly with the Apliki Mine suggests very strongly that the ingots originated there, but it is clear that it does not prove it. The creation of a small, tight isotope ore fields which might be matched by a group of artefacts creates the impression in the minds of readers of provenance *to* an orefield. The exclusion of the Mousoulos samples prior to 1997 'tightened' the Cypriot field and created just this impression. As doubts grew (e.g., Budd *et al.* 1995a) about the relationship of the oxhide ingots' isotope field to the Cypriot field, particularly the way in which the ingots plotted to the extreme edge of the published confidence ellipse, the Oxford Group reconsidered the situation. They worked tentatively (Gale & Stos-Gale 1992b) towards a position of regarding individual mines as having limited, compact isotope ranges, which culminated in the 1997 compilation. Once this interpretation is accepted then it becomes useful to have another distinct mine, Mousoulos, reinforcing the impression gained from the other sites. It now becomes an asset in that it plots well away from the other samples and can be dismissed as a source for oxhide ingots on isotopic grounds (despite this dismissal being unnecessary because of its deep burial).

There are further problems associated with the treatment of Oxford LIA data. As noted above, there have been unexplained disappearances of data points between successive compilations, the majority of which were somewhat belatedly documented in the 1997 paper. The discussion above centred on the Mousoulos samples for which archaeological and geological reasons were given for their exclusion or inclusion, even though these are inconsistent. The treatment of two of the Sardinian oxhide ingots remains worrying. If Figure 1 is compared with the diagram from Stos-Gale *et al.* (1997:116 fig. 12) a difference is observed. The samples labelled A and B on Figure 1 do not appear on the published diagrams (A is sample SI C CA1, B is BS9). These represent extreme values, although BS9 may lie within the error limits of the Apliki field (since the correlation of the errors is not given in the original, only the 0.1% error bars reproduced here, the exact error is hard to determine). Of course, the Oxford group do not contend that Apliki is the sole potential source for these ingots - they accept that other Solea axis ores may have made a small contribution to the production. Figure 2 indicates that SI C CA1 probably lies outside the limits of the Solea field.

It is not contended here that these two omitted data points radically alter the interpretation given by Stos-Gale *et al.* (1997). Both lie well within the limits of other Cypriot ores. We would also expect, statistically, one or two analyses to be outside any error envelope if sufficient analyses were carried out. Also, it is possible that the true errors on low-lead samples in LIA are greater than is usually acknowledged and that both data points should probably lie within a (larger) error envelope of the Apliki field. The point is that data points which look contrary to the argument being presented seem to be omitted from the associated diagrams. This cannot be a deliberate attempt to deceive - if it were the relevant data would not have been included in tabular form. It is possible that preconceived notions of what the plots ought to look like have allowed the erroneous omission of outlying points to slip through unnoticed. This is both unfortunate and unnecessary, and undermines confidence, especially when similar lapses have occurred previously and have been comprehensively pointed out in print.

The final point of interest in the selection of ore samples for the Cypriot field is the rejection of the samples from the Limassol Forest when considering oxhide ingots or other Bronze Age metals (Stos-Gale *et al.* 1997). This cannot be done on purely isotopic grounds as some Limassol ores lie within error of some oxhide ingots. SI C

CA1 mentioned above actually matches well with the Limassol ores. Stos-Gale *et al.*, however, reject the Limassol Forest on the grounds that the chemical composition of the ores is unusual (and presumably would produce a distinctive trace element pattern), that they would have been difficult to smelt in the Bronze Age and

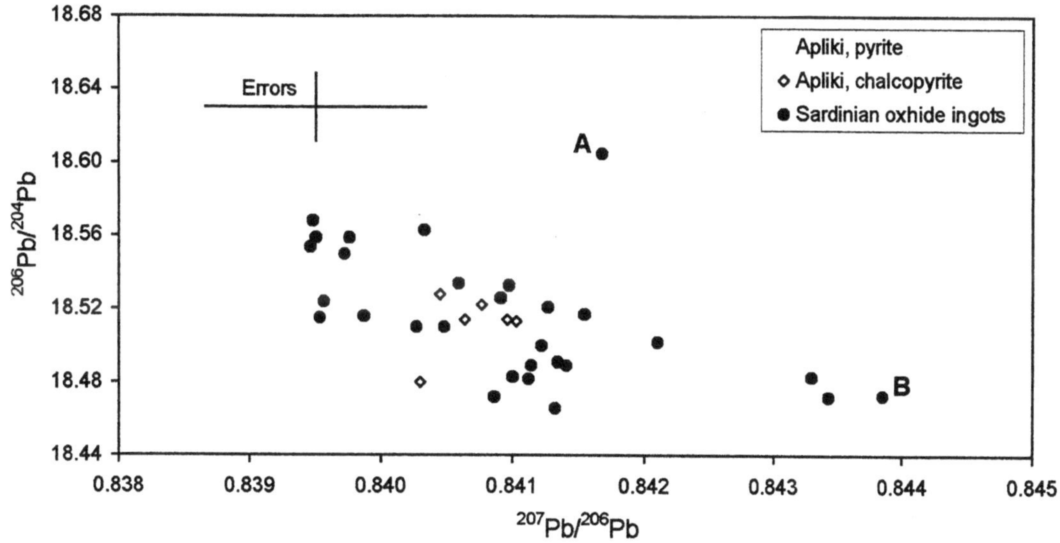

Fig. 1. Isotope ratios for ores from Apliki and Sardinian oxhide ingots. Data from Gale et al. *(1997) & Stos-Gale* et al. *(1997).*

that there is no evidence of any mining prior to modern times. It should be noted that whilst the ores' chemical composition is unusual it is also very variable. Whilst the ores may be difficult to smelt, they are easier to smelt for copper than the pyrite samples making up the bulk of the Cypriot samples and the geology of the area does not preclude the existence of outcrops of more easily smeltable ores. Finally, as observed above, the absence of evidence of ancient mining is not a criteria rigorously applied to the rest of the Cyprus.

The selection of samples to form a database of ores from Cyprus has improved considerably since the idea that a fixed number of ore samples, or ore *and* slag samples (Stos-Gale 1989), would suffice. The idea of using only well-provenanced ore samples collected from all areas that may contribute distinctive features to the isotope field will surely be welcomed by all past critics of the 'Cypriot Field'. It is commendable to see so much data in the 1997 compilation, particularly from the Limassol Forest, which is not specifically tied to the study in hand, and which will undoubtedly be of use to future researchers. It is possible that we may not yet be seeing the full range of Cypriot isotopic variability, or that of individual mines. The current

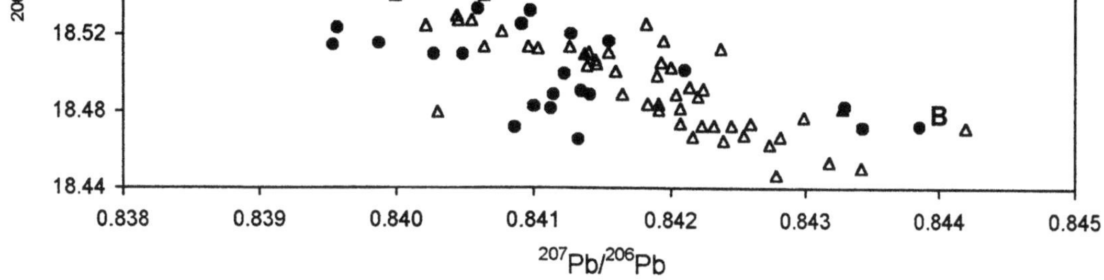

Fig. 2. Isotope ratios for ores from the Solea axis and Sardinian oxhide ingots. Data from Gale et al. *(1997) and Stos-Gale* et al. *(1997).*

compilation still contains errors and omissions of a similar nature to those pointed out in the past. It is recognized, however, that scientific research which spans two decades is almost certain to contain some inconsistencies, and the latest compilation must be recognized as the most authoritative and reliable corpus of LIA data yet produced on Cyprus.

The Presentation and Statistical Interpretation of LIA data

As noted above, the diagrams which present LIA data are of the utmost importance in both its interpretation and the impression they create in the minds of the reader. A group of artefact data points sitting inside a perfect ellipse which is claimed to encompass 95% of all knowngeological samples from an ore source tends to convince the reader that the metal originated at that ore source, even if the accompanying text is rightly somewhat more circumspect.

Archaeological LIA data is usually presented in the form of bivariate plots, usually of ratios with the form $*/^{206}Pb$. The use of the geologically more common $*/^{204}Pb$ diagrams would be preferable, since these allow for easier interpretation of the underlying geological processes. Whichever form of diagram is used, however, there is a problem that fields which are distinct in three dimensions may appear to overlap on these two-dimensional plots. The resolution of this has been one of the motivations for the use of more sophisticated statistical techniques applied to LIA data.

Both linear discriminant function analysis (LDFA) (e.g., Gale & Stos-Gale 1992a; 1995) and the multivariate normal approach of the Smithsonian Group (e.g., Sayre *et al.* 1992; 1995) have been used in relation to the Cypriot and other lead isotope fields. Both of these techniques rely on an assumption of multivariate normality in LIA data sets. This is an assumption which has never been demonstrated and, given the nature of the geological models used to describe lead isotope evolution, is extremely unlikely to ever be generally true. It is remarkable that these techniques could feature so strongly in such a contentious area without it ever being firmly established that they were valid for the data being discussed.

The technique used by the Smithsonian Group uses a multivariate normal technique to define isotope fields from sets of lead isotope data (Sayre *et al.* 1992). They "used the multivariate probability calculations of specimens matching groups as [their] primary statistical tool for evaluation because it is a very straightforward statistic that can be calculated for each specimen and because it is the only statistic that takes into account sample size." (Sayre *et al.* 1992:82). It must be noted that a great many statistical methods take sample size into account and that just because a statistic is straightforward does not mean it is appropriate. Examples of this approach can be seen in Yener *et al.* (1991), Joel *et al.* (1995) and in the debate over the origin of oxhide ingots (Sayre *et al.* 1995). Yener *et al.* (1991) provides one of the clearest illustrations of the inherent shortcomings of this method. In this study, fifty samples from the Taurus Mountains of Turkey were analysed and were then divided into four groups (Taurus 1A, 1B, 2A, 2B) of between eight and sixteen specimens each, along with seven "outliers", using multivariate normal techniques. The groups were constructed as post-hoc statistical entities. They were constructed with no reference to the underlying geology (or indeed even the geography) of the area. Some ore samples were considered outliers from the groups whilst archaeological artefacts were used to construct the groups.

There are many problems with this approach. Artefacts should not be used to form source groups for LIA work. An artefact will tend to be an average, to a lesser or greater extent, in its lead isotope content of the ores making it up, thus tending to make a group more compact. Furthermore, an ore sample can never be an outlier from its isotope field - the ore samples themselves by definition define the boundaries of isotope field. Almost any scatter of points will allow for the construction of groups if any chosen points can be dismissed as outliers. It is frequently possible to construct many sets of statistically valid groups from multivariate data, particularly if one is permitted to reject outliers and combine samples without regard to ore geology. It would seem reasonable to expect that samples in the same group should all come from a geologically well-defined ore deposit, whatever its geographical extent. Once a deposit has been identified then all samples collected from it must be used in the definition of its isotope field. Once the samples making up the field have been defined it can then be decided if any simpler model (such as multivariate normality) is applicable. The concept of samples being eliminated from an isotope field, despite being collected from the same ore deposit and having been measured accurately, on the basis of being statistically defined outliers, is wrong.

The most-used picture of Cypriot ores is the 'old confidence ellipse'. The Oxford Group point out that for the 1997 compilation "The great majority of Cypriot copper ores, taken together, still form a compact group within ±1% range of lead isotope ratios for all copper ore minerals. An ellipse drawn around this group is essentially the same as the ellipse which defined the old Cypriot field" (Gale *et al.* 1997:238). They do note that some ores fall outside this ellipse and justify its appearance in most of their diagrams as being a reference to earlier publications. This does not disguise the fact that this ellipse itself is meaningless, and has no part to play in any description or interpretation of current data sets. Its use perpetuates the feeling that there is some sense in which the ellipse represents the field, albeit imperfectly. The acceptance that the "overall distribution of lead isotope data for ore deposits from the [Troodos] axes is not bivariate normal in the mathematical sense" (Stos-Gale *et al.* 1997) is welcome but is somewhat negated by the continued use of ellipses.

The Oxford Group has tended to use Linear Discriminant Function Analysis (LDFA) to try to clarify which data points constitute separate isotope fields and how likely artefacts are to have originated from them. The use of LDFA in relation to LIA seems to have first been suggested by Reedy & Reedy (1988) as way of dealing with the sparse and apparently overlapping isotope fields of different parts of the Aegean. In addition to giving a clear graphical indication of the maximum possible separation between groups, LDFA can be used to give a quantitative assessment of the probability of an unknown sample originating from any of the groups under consideration (Gale & Stos-Gale 1992a). Although LDFA is attractive as a tool for manipulating LIA data, it can be used in an inappropriate manner.

LDFA is one of a range of techniques for reducing the number of dimensions required to describe a data set, and is useful when dealing with large numbers of variables and where there is independent evidence for prior groupings. In LIA, however, there are only three variables and hence LDFA is less useful. The full structure of the three-dimensional data set can be easily viewed from any angle using many widely available computer packages which can rotate three-dimensional data in real time. In essence, LDFA of a three-dimensional data set will produce the two-dimensional view which best shows separation between the defined groups. It is perfectly possible to get an equivalent view by three-dimensional rotation of the data set (Scaife *et al.* 1996). The Oxford Group have also made use of stepwise LDFA, (e.g., Gale & Stos-Gale 1992a; Stos-Gale & Gale 1992). In stepwise LDFA, the original variables are considered one at a time and an assessment is made as to whether they improve the discrimination. Obviously, starting with only three variables means that if one is not used the result is a simple bivariate plot which will be a rotation of one of the original isotopic ratio bivariate plots, and should have been obvious from initial inspection of the data. The stepwise inclusion of the third ratio returns us to a three dimensional plot, as discussed above.

The quantitative aspects of LDFA have been used to emphasize the separation between various orefields. For example, Gale (1991:208) carries out a LDFA of the Cyprus, Kythnos and Laurion orefields. Ore samples are quoted as belonging to their correct groups with probabilities of 98% or 99% or 81%, or similar values (although LDFA misclassified one specimen). These numbers seem impressive, but to be meaningful several criteria must be fulfilled. Firstly, but in practice unrealistically, *all* possible sources should be included in the analysis. Secondly, some statistical criteria cannot be avoided: each data set must have the same dispersion matrix, and all must be multivariate normal. Both criteria are rarely met. More importantly, however, the following question should be considered: what does it mean to say that a sample of ore collected from an orefield has only an 81% chance of belonging to that orefield? The statistics lead to confusion rather than clarity.

Kernel Density Estimation

It appears that the most commonly used statistical methods in LIA are inappropriate to the data. There is clearly a requirement to be able to picture three-dimensional data in two dimensions and also to be able to summarize the general distribution of large numbers of points in a simpler form. Rotated views of the three dimensional data can easily be produced with modern computer packages, and is a useful way of graphically demonstrating any separation of isotope fields. It does not, however, aid in-depth analysis of the data from a geological perspective and also makes comparison between different groups of data difficult. Thus we would agree with the Heidelberg-Mainz group (Pernicka *et al.* 1990) that the use of more conventional bivariate plots in addition to any other diagrams is still to be recommended. Simple plotting does not, however, lead to an estimate of the true extent of the parent isotope field. Kernel Density Estimation (KDE) offers the prospect of building up an estimate of the true shape and size of an isotope field, whilst making few extra assumption about the data. KDE is a relatively new technique in archaeological analysis (Baxter *et al.* 1997). We therefore briefly summarize KDE before discussing its possible role in archaeological LIA.

KDE, in one dimension, can be seen as an extension of the histogram, overcoming the problems of 'bin width' and starting point associated with drawing such histograms. In a KDE each individual data point is replaced by a 'kernel', which can be thought of as a bump on the axis. These bumps are superimposed and a curve is built up which is an estimate of the underlying probability density function of the data set from which the points were drawn. There is a wide range of choice for the kernel used (Silverman 1986:43 gives equations for five). There is little to chose between most kernels in terms of their ability to reproduce probability density functions (PDFs) accurately (Hand 1982; Silverman 1986; Wand & Jones 1995). Thus, as Silverman (1986:43) points out, the choice of kernel can be made on other grounds, such as the degree of computational effort required. The density estimations below have been carried out using an Epanechnikov kernel, which was chosen as a compromise between the smoothness of the final PDF curves and speed of computation. One crucial choice has to be made when using KDE: the width of the individual kernel (Hand 1982; Silverman 1986; Wand & Jones 1995). Thin, tall kernels produce jagged curves, whilst short, wide kernels produce flat, oversmoothed curves showing little detail. The smoothing parameter, h, which governs the width of the kernel can be determined by a variety of data-based techniques (see Silverman 1986 or Wand & Jones 1995 for an overview), or can be varied by the researcher to aid data exploration. Figure 3 demonstrates the effect of different values of h.

The technique can also be used to determine non-parametric confidence intervals. Rather than take the mean of a distribution and then include samples that lie within a certain number of standard deviations, the fact the density has been estimated at each point along the axis can be used. By taking the points that show the largest density in turn a confidence interval of any size required can be constructed. Figure 3 shows that this confidence interval does not have to be symmetric about the mean, nor does it have to be continuous. Figure 4 shows the final estimates of the density of $^{208}Pb/^{204}Pb$ for the Troodos Mountain ores. Now, it is already known that the Troodos Mountain ores come from four different spreading axes and so there is some justification for not treating them as part of the same statistical distribution. If this information were not available prior to statistical analysis there would still be some clues as to its underlying nature in the KDE plots. From inspection of the $^{208}Pb/^{204}Pb$ plot it would certainly be expected that there were three groups in the data which merited closer inspection. Comparison of the peaks in the distribution with the means of the spreading axes marked on the same diagram shows how this diagram tends to indicate the underlying groups. It is necessary to remember that the distribution could be distorted by the number of samples being collected from each spreading axis not being representative of their abundance in the orefield as a whole. However, given the limitations of the data set a remarkably good picture is obtained from these diagrams.

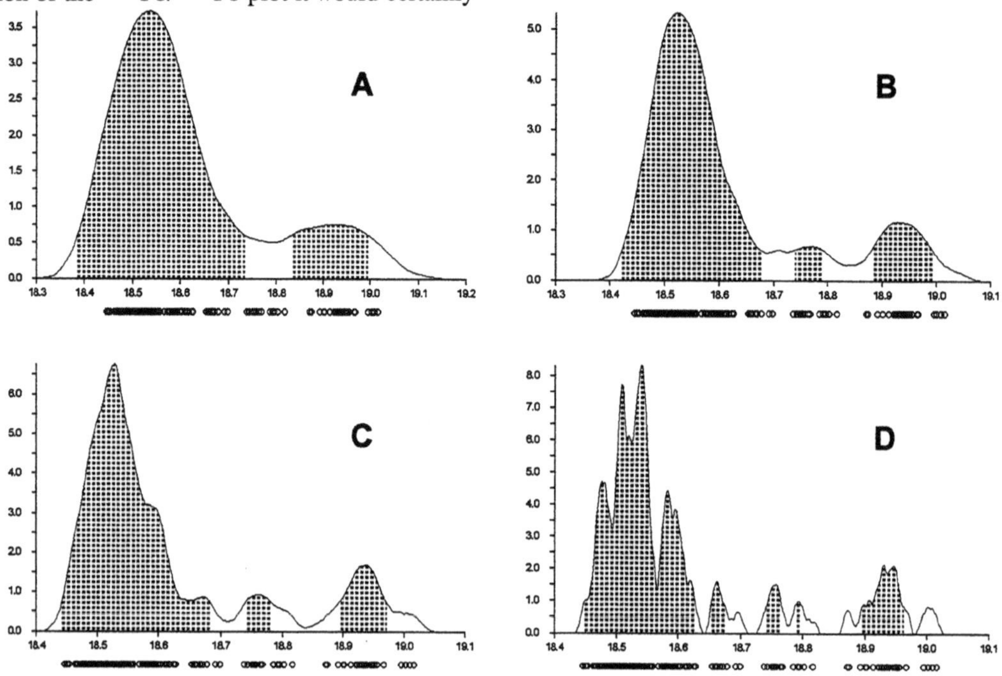

Fig. 3: Density estimation of distribution of $^{206}Pb/^{204}Pb$ values (horizontal axes) for ores from the Troodos Mountains, Cyprus (Gale et al. *1997). Epanechnikov kernel used in all estimations, with A h = 0.06; B h = 0.03; C h = 0.015; D: h = 0.005. 90% confidence interval shown shaded. Vertical axis in this and other univariate diagrams is a frequency axis. The individual data points are shown by points below the horizontal axis.*

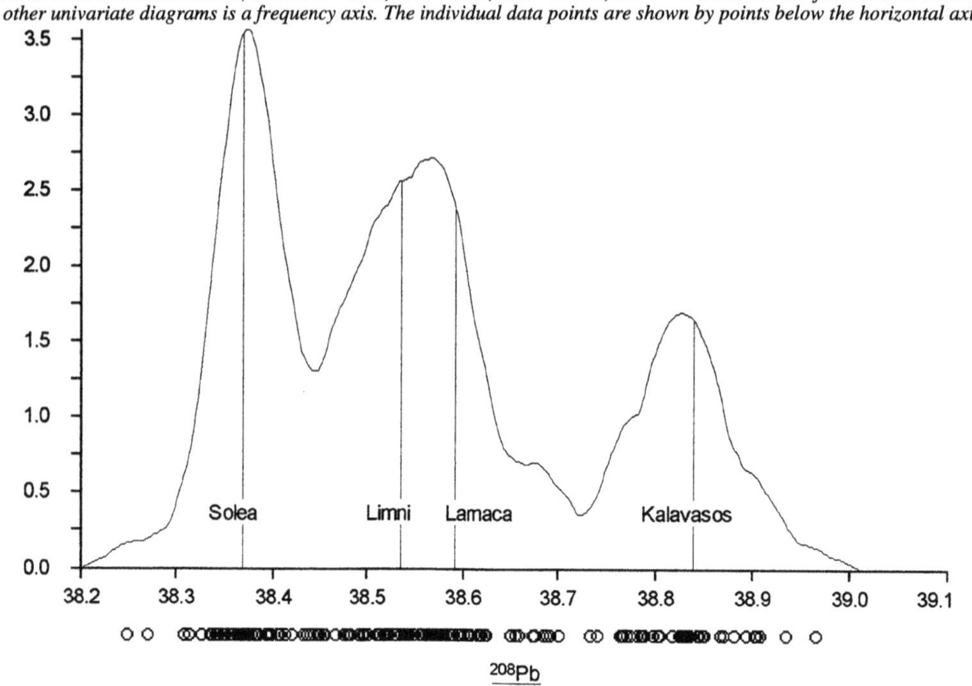

Fig. 4: Density estimation of distribution of $^{208}Pb/^{204}Pb$ values for ores from the Troodos Mountains, Cyprus (Gale et al. *1997). Epanechnikov kernel, h = 0.02. The means for the four different spreading axes are indicated.*

Although the technique of univariate KDE could be more widespread in archaeology as a way of investigating and presenting data, these diagrams are of limited use in the archaeological application of LIA.. In LIA consideration needs to be given to KDE in two, or even three dimensions. The extension to multiple dimensions is fairly trivial in description: at each point a kernel estimate of the density due to each member of the dataset in each direction is used. These densities are then multiplied together and scaled according to the smoothing parameters h_d used in each direction The diagrams which follow were produced by first making a density estimate on a 100x100 grid and then smoothing the resulting density contours. Figure 5 shows a bivariate KDE for the Troodos Mountain ores. Already, it is clear that the picture is more representative of the data than a simple bivariate ellipse. There are, however, some problems with this diagram. Firstly, the data have probably been oversmoothed somewhat and this has not only blurred the picture but has also emphasized the second problem. The kernels are aligned parallel to the plotting axes whereas the data tends to run at an angle to them, which leads to the density estimation being distorted in the direction of the plotting axes. In an attempt to deal with this the data were rotated onto new axes which ran

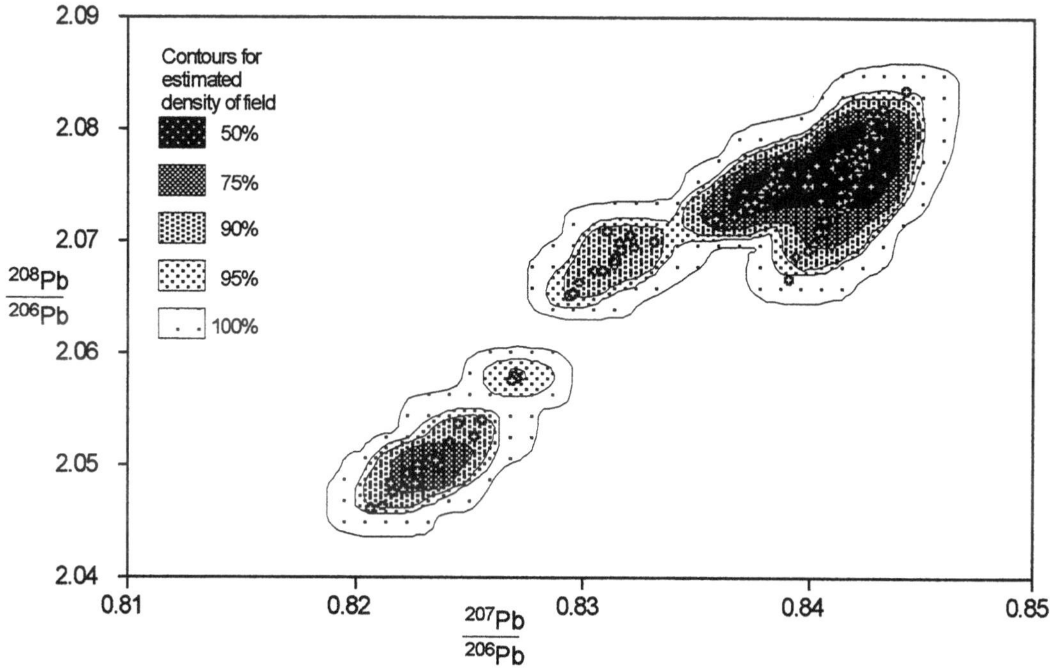

Fig. 5. Kernel density estimation of the Troodos orefield, Cyprus. Epanechnikov kernel used, smoothing factors: 0.00250 for $^{207}Pb/^{206}Pb$; 0.00277 for $^{208}Pb/^{206}Pb$ (data from Gale et al., *1997).*

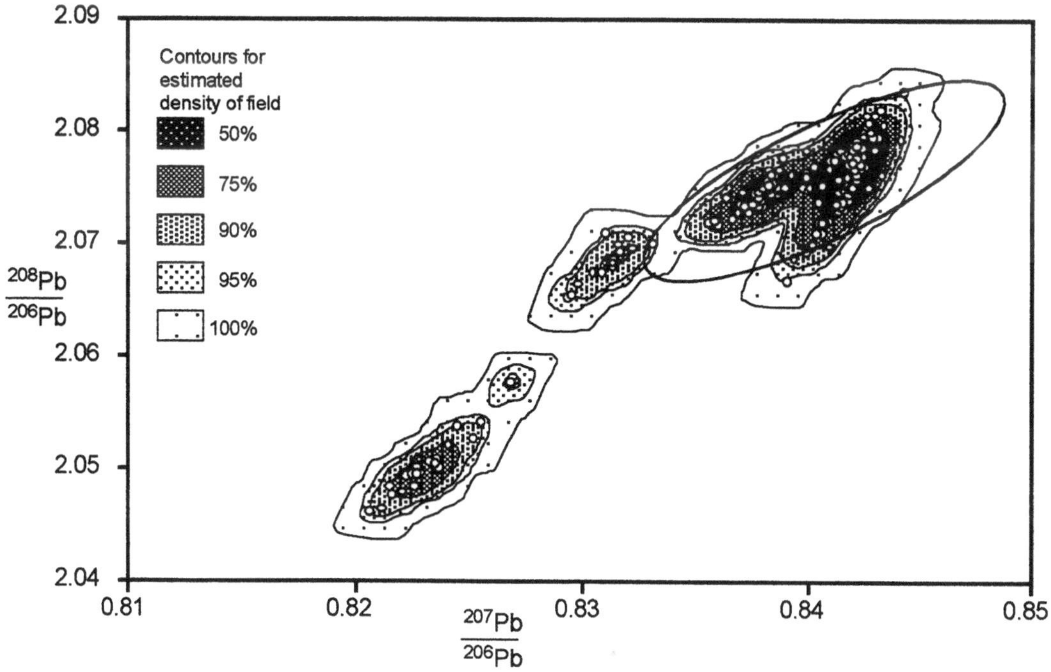

Fig. 6. Kernel density estimation of part of the Troodos orefield, Cyprus. Epanechnikov kernel used, along axes not parallel to the variable axes (data from Gale et al., *1997). The ellipse, shown for reference, is the one used to represent the Cyprus field in many publications of the Oxford group.*

parallel and orthogonal to the 121 Ma isochron. The 121 Ma isochron was chosen somewhat arbitrarily (it is the mean isochron of the Limni data) but the isochrons in the region of the data run almost parallel so it suffices for the present. This represents a reasonable approximation to the main axis of variation of the data, bearing in mind that the different subsets of the data run in different directions. A KDE was then carried out, using kernels parallel to these new axes and a density estimation calculated. The density estimation was then rotated back to the original isotope ratio axes, resulting in Figure 6. This is clearly a much better representation of the data than the 'Oxford ellipse', shown for comparison, and surely indicates that this technique could form the basis of the presentation of lead isotope fields in the future.

One further possibility is to reconstruct this diagram as four separate densities, one for each spreading axis. Whether this is desirable is, to some extent, dependent on the archaeological questions being asked. There are also other refinements which could be developed. These might include determining how the kernel size can be varied according to the local density in a way which does not obscure the possibility of detecting subsets of data and takes into account the specific ways in which LIA data tends to vary. A more subtle, and potentially more rewarding approach might be to use mathematical knowledge of the evolution of lead isotopes to suggest kernels which are not orthogonal to one another. Kernels could then be constructed in the direction of the isochrons (i.e., simulating variations in the initial composition of the source fluids) and combined with kernels in the direction of the growth curves (simulating variations in the model ages of the ores). This approach of basing kernels on known facts about the data surely offers a chance of making a better estimation of the underlying PDF. Such an approach would demand a large amount of computation, but this is now becoming available at the personal computer level. It would obviously also be preferable to produce KDE derived pictures in three dimensions. There is nothing, in principle to stop this. In practice, however, the computational effort is much greater.

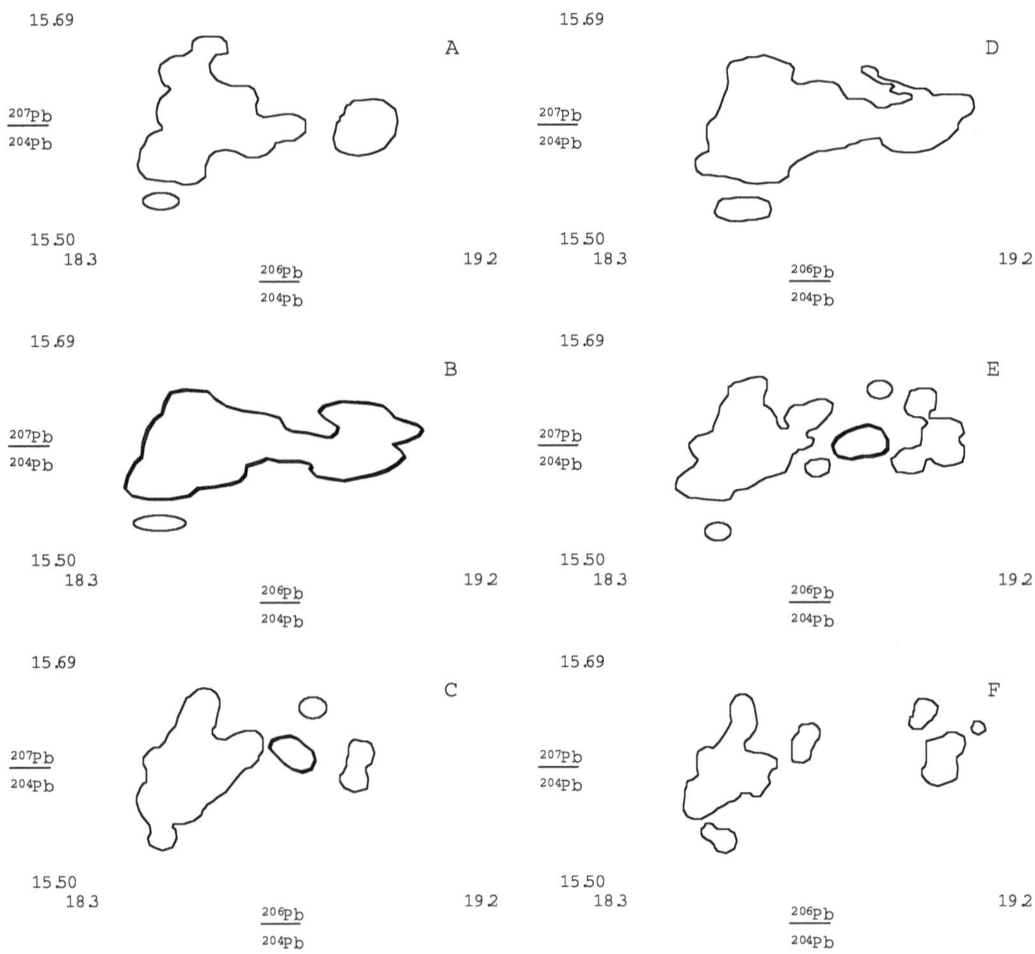

Fig. 7. 95% contours of kernel density estimations of subsets of the Troodos Mountain ore data to illustrate the effect of sample size. See text for details.

Although the pictures obtained from even this simple approach to bivariate KDE seem good there remain questions as to how representative of the underlying density they are unless it is certain that the sample size is large enough to minimize errors. Unfortunately there is little hard data to go on. Silverman (1986:94) indicates minimum sample sizes of 19 and 67 for two and three dimensions respectively. These figures were arrived at from estimating multivariate normal density with a normal kernel with a suitable smoothing parameter. They thus represent the absolute minima in the best possible case. LIA data will probably be less well behaved. Figure 7 shows the results of a preliminary investigation into the effect of sample size. Three of the density estimates (A, B and C) were constructed from random samples of 47 (25%) samples from the Troodos ores (Gale *et al.* 1997). D and E were estimated from samples of 94 ores. The diagrams show the 95% contour, chosen as a compromise between the full size of the field and the desire to show some detail, given the constraint of space. It is clear that the estimate has not settled down sufficiently at either of these sample sizes for us to be able to use the estimates to make any decisions. The 100% contours (not shown) for D and E are more similar, perhaps suggesting that the full extent of the field is easier to estimate than its inner detail. However, much more work needs to be done on the sample size problem before KDEs can be used to make statistical attributions of unknown data points to estimated fields.

Diagram F of Figure 7 is included to illustrate the effect of biased sampling on this type of data. It is also estimated from 94 ores but this time all 52 Solea axis specimens were included with the other 42 samples chosen at random from the other three axes. As expected this produces a distortion in the estimated field. Perhaps more clearly than any other diagram this one emphasizes the problems of sampling an orefield and describing it statistically. How do we decide on the number of samples from each mine or spreading axis or region? Simply picking up more samples from one location will bias a description of a field towards its composition. At the same time not all areas will have equal importance and there may be an argument for biasing the fields towards particular regions.

We would argue, therefore, that whilst KDE gives a much better picture of lead isotope fields compared to other statistical techniques, its real benefits probably lie with other types of data. It has been suggested (Scaife *et al.* 1996) that the best means of defining a sample's provenance is to restrict a possible attribution only to those samples which fall within the full three dimensional space defined by the ore samples. This is, in a sense, an unhelpful statement: what is required is a method which *defines* the three-dimensional space for the ore samples. It is necessary, however, to keep a focus on the primacy of the data: the data's structure must define the field, not the imposition of some arbitrary parameterized distribution. The primacy of the data makes the sampling procedure of great importance in LIA. The change in the perceived nature of the Cypriot field between 1994 and 1997 indicates this clearly. In 1994 the samples published were few in number and poorly provenanced; in 1997 there were more than four times as many samples, largely well-provenanced. In 1994 the Cyprus field was considered by many to be an ellipse; in 1997 even the Oxford Group had abandoned this description for the field as a whole. The implication is that if provenancing decisions based on the existing data are in doubt then the best way of improving them is by a more extensive and intensive ore-sampling strategies, not by resorting to potentially confusing and inappropriate statistics. It seems reasonable to agree with Pernicka (1992) that the multivariate methods used in the past can lead to over-analysis of the data. Indeed, a simple plot of the data, in some two-dimensional view, with associated errors has always been able to provide as compelling an argument as is possible.

Summary

The conclusion of Gale *et al.'s* 1997 paper in relation to the provenance of Sardinian oxhide ingots found on Sardinia is that they probably originated on Cyprus at the Apliki Mine. All the other ingots analysed are also provenanced to either Apliki or one of the nearby mines, but overwhelmingly to Apliki. It would seem difficult not to accept this thesis on the basis of present LIA data, with the possible exception of SI C CA1 which, however, still shows good agreement with other Cypriot ores. It seems likely that until many more analyses of oxhide ingots are published (with, perhaps, confirmatory results from another laboratory) there will remain a degree of disquiet amongst the archaeological community as to the overwhelming importance of Apliki in the production of oxhide ingots.

It is important to note that, despite the claims of Stos-Gale *et al.* (1997) that it is LIA which has given the "definite indications" of the provenance of the ingots, there is a more complicated interpretative process occurring here. For example, the dismissal of the Pevkos ores as a possible source for at least one ingot depends not on the isotopic analyses but on the absence of archaeological evidence of mining and on experimental work on smelting. To leave umbers out of the isotope field depends on trace element analysis of slags and theories about the techniques of LBA smelting technology. The whole question of recycling of metal in the ingots is one which is impossible to prove either way by LIA but relies on interpretation of vague scientific evidence and assumptions about the cultural use of metals in the LBA. LIA should be seen as one technique amongst many in this interpretative process, albeit in this case an extremely important one.

If, as seems likely at the moment, Apliki proves to be the source of LBA oxhide ingots there are a number of key questions which need to answered. There are obvious implications for the social structure of LBA Cyprus and its relationship with the wider Mediterranean. There are also problems about the use of the ingots themselves. An ingot mould found at Ras Ibn Hani, Syria had droplets of copper around it whose isotopic composition agreed with the Cypriot ores (Gale & Stos-Gale 1986). This seems a clear indication that at the very least ingots were re-moulded from Cypriot copper outside Cyprus. There are no analyses of LBA artefacts from Sardinia whose isotopic composition matches that of the ingots, i.e., no positive indication that they were being used in manufacturing processes. Perhaps the idea of a *koine* of metal, traded but not used greatly in manufacture, raised by Budd *et al.*

(1995b) is partially correct and there is a circulating pool of metal but it all originates on Cyprus. LIA has a role to play in answering all these questions but we may get more useful answers from other more traditional archaeological techniques.

The debate around LIA in archaeology has been characterized by a higher level of aggravation than would be expected in a discussion about a 'precise', 'scientific' technique. This is perhaps partly due to the pressures of being required to produce definitive answers in order to qualify for the next research grant (Muhly 1995). Given the limited number of research groups able to actively take part in the research, there is inevitably a degree of personalization. Pernicka (1995) stated that the Oxford group give the impression that LIA can only be performed and interpreted at Oxford. The inconsistencies documented in the presentation and re-presentation of the data do not help that contention. There is, perhaps, a more fundamental reason for irritation: it might seem galling to some scientists when they have a very expensive machine which can produce measurements accurate to five decimal places to have their conclusions challenged by a humanities-based researcher just because the archaeological conclusions do not seem to make sense. There is also the question of errors. Unavoidable measurement uncertainty is 'error', but avoidable mistakes are also 'error'. It can sometimes be unclear which of these is implied when one researcher questions the 'error' of another.

Although the LIA debate has been quite robust during the 1990s it has also been useful. The apparent clarity of the latest compilation of Cypriot data in relation to the oxhide ingot question is unlikely to have emerged without the vigorous criticisms directed at previous papers. Hindsight suggests that the development of LIA methodology in archaeology could have proceeded along more rational lines. Muhly (1995) notes that all the mistakes made in radiocarbon work in the 1950s and 1960s have been repeated in LIA in the 1970s and 1980s. It would have been expected that the limits of LIA would have been established early on in its history of use. The accuracy, both intra- and inter-lab, should have been established for various grades of minerals and ores. If the use of statistical techniques really was necessary then the nature of LIA data should have been investigated early on and techniques should have been selected that were appropriate to the data. The possible problems of anthropogenic alteration of isotope ratio should have been dealt with conclusively in the 1970s rather than being debated in the 1990s (Budd *et al.* 1995a; Gale & Stos-Gale 1996). Knapp & Cherry (1991) raised a number of questions in relation to the use of LIA seven years ago. Whilst some of these have been partially answered others have not. Thus, error bars are now generally seen on isotope ratio plots yet they are unlikely to represent the true errors in both size and direction. The plea for corrections based on inter-laboratory standards has gone completely unheeded.

The presentation of LIA data to the non-scientists in the archaeological community obviously raises problems, as is shown by the frustration of Chippindale's (1994) *Antiquity* editorial. The statistical interpretation of the data has been extensive and has never clarified an argument. The presentation of confidence ellipses and LDFA diagrams has given a spurious air of authority to what are essentially scatterplots. Provenancing decisions have always been based on whether or not artefact points lie within fields defined by ore data points. The use of these statistics further removes the interpretation from the non-technical reader and increases the mystification of the whole technique.

The bemusement of many archaeologists when confronted by the LIA debate reveals that LIA has been increasingly cut off from other areas of archaeology, and the technique needs to be re-established as an understandable, usable technique. The development of rapid plasma-based technologies for measuring lead isotope ratios with precision approaching that of TIMS should revolutionize the role of LIA, providing adequate understanding of the limitations are accepted. It should be a valuable part of the archaeologists' armoury for the analysis of artefacts. It is a technique which is relatively robust to anthropogenic processes and produces data sets of sufficient simplicity to be understood and visualized without sophisticated mathematical techniques. There is a good, and continually developing, background of geochemical models to aid interpretation of the data. Archaeological LIA data does not exist in splendid isolation, nor does it have some quality that automatically renders it as more important than other forms of data. Care needs to be taken when interpreting LIA data and the 'scientific' accuracy of five decimal places of measurement should not overwhelm any scepticism based on other equally valid criteria. Although LIA is a sound technique that can provide information not otherwise available we must remember that LIA is part of a whole - one argument amongst many and as open to question as any.

Acknowledgements
This research was supported in part by the award of a NERC studentship to one of the authors (BS), and a NERC Advanced Fellowship to PB. We are grateful for all the comments, helpful and otherwise, we have received during our reconsideration of the application of LIA in the Bronze Age Aegean.

References

BAXTER, M.J., C.C. BEARDAH & R.V.S. WRIGHT. 1997. Some archaeological applications of kernel density estimates, *Journal of Archaeological Science* 24: 347-354.

BUDD, P., A.M. POLLARD, B. SCAIFE & R.G. THOMAS. 1995a. Oxhide ingots, recycling and the Mediterranean metals trade, *Journal of Mediterranean Archaeology* 8: 1-32.

1995b. The possible fractionation of lead isotopes in ancient metallurgical processes, *Archaeometry* 37: 143-150.

1995c. Lead isotope analysis and oxhide ingots: a final comment, *Journal of Mediterranean Archaeology* 8: 70-75.

CHIPPINDALE, C. 1994. Editorial, *Antiquity*, 68: 1-9.

GALE, N.H. 1989. Lead isotope analysis applied to provenance studies - a brief review, in Y. Maniatis (ed.), *Archaeometry, Proceedings of the 25th International Symposium:* 469-502. Amsterdam: Elsevier.

1991. Copper oxhide ingots: their origin and their place in the Bronze Age metals trade in the Mediterranean, in N.H. Gale (ed.), *Bronze Age Trade in*

the Mediterranean: 197-239. Jonsered: P. Åströms Förlag.
GALE N.H. & Z.A. STOS-GALE. 1982. Bronze Age copper sources in the Mediterranean, *Science* 216: 11-19.
1986. Oxhide copper ingots in Crete and Cyprus and the Bronze Age metals trade, *Annual of the British School at Athens* 81: 81-100.
1988. Recent evidence for a possible Bronze Age metal trade between Sardinia and the Aegean, in E.B. French & K.A. Wardle (eds.), *Problems in Greek Prehistory*: 349-384. Bristol: Classical Press.
1992a. Lead isotope studies in the Aegean (The British Academy project), in A.M. Pollard (ed.), *New Developments in Archaeological Science:* 630-108. Oxford: Oxford University Press (Proceedings of the British Academy 77).
1992b. Comments on E.V. Sayre, K.A. Yener, E.C. Joel, and I.L. Barnes, 'Statistical evaluation of the presently accumulated lead isotope data from Anatolia and surrounding regions', Archaeometry 34(I) (1992) I, *Archaeometry* 34: 311-317.
1995. Comments on P. Budd,, A.M. Pollard, B. Scaife & R.G. Thomas, 'Oxhide ingots, recycling and the Mediterranean metals trade', *Journal of Mediterranean Archaeology* 8: 33-41.
1996. Lead isotope methodology: the possible fractionation of lead isotope compositions during metallurgical processes , in S. Demirci, A.M. Özer & G.D. Summers (eds.), *Archaeometry 94*: 287-299. Ankara: Tübitak.
GALE, N.H., Z.A. STOS-GALE, G. MALIOTIS & N. ANNETTS. 1997. Lead isotope data from the Isotrace Laboratory, Oxford: Archaeometry data base 4, ores from Cyprus, *Archaeometry* 39: 237-246.
HAND, D.J. 1982. *Kernel Discriminant Analysis*. Chichester: Research Studies Press.
JOEL, E.C., E.V. SAYRE, R.D. VOCKE & F. WILLETT. 1995. Stable lead isotope characterization of various copper alloys used in West Africa: an interim report, *Journal of the Historical Metallurgy Society* 29: 25-33.
KNAPP, A.B. & J.F. CHERRY. 1991. Archaeological science, statistics and cultural solutions: trade patterns in the Bronze Age Eastern Mediterranean, in E. Pernicka & G.A. Wagner (eds.), *Archeometry '90*: 183-198. Basel: Birkhäuser Verlag.
MADDIN, R. & J.D. MUHLY. 1974. Some notes on the copper trade in the ancient Mid-East, *Journal of Metals* 26: 24-30.
MUHLY, J.D. 1982. The nature of the trade in the Late Bronze Age Eastern Mediterranean: the organisation of the metals trade and the role of Cyprus, in J.D. Muhly, R. Maddin & V. Karageorghis (eds.), *Early Metallurgy in* Cyprus: 251-266. Nicosia: Pierdes Foundation.
1995. Lead isotope analysis and the archaeologist, *Journal of Mediterranean Archaeology* 8: 54-58.
MUHLY, J.D., T. STECH WHEELER & R. MADDIN. 1977. The Cape Gelidonya shipwreck and the Bronze Age metals trade in the eastern Mediterranean, *Journal of Field Archaeology* 4: 353-362.
MUHLY, J.D., R. MADDIN & T. STECH WHEELER. 1980. The copper oxhide ingots from Enkomi and Mathiati and Late Bronze Age copper smelting in Cyprus, *Annual Report of the Department of Antiquities, Cyprus, 1980*: 281-298. Nicosia: Department of Antiquities.
PANAYIOTOU, A. 1980. Cu-Ni-Co-Fe mineralization, Limassol Forest, Cyprus, in A. Panayiotou (ed.), *Ophiolites. Proceedings of the International Ophiolite Symposium, Cyprus* 1979: 102-116. Cyprus: Geological Survey Department.
PERNICKA, E. 1992. Comments on E.V. Sayre, K.A. Yener, E.C. Joel, and I.L. Barnes, 'Statistical evaluation of the presently accumulated lead isotope data from Anatolia and surrounding regions', Archaeometry 34(1) (1992) III, *Archaeometry* 34: 322-326.
1995. Crisis or catharsis in Lead Isotope Analysis?, *Journal of Mediterranean Archaeology* 8: 59-64.
PERNICKA, E., F. BEGEMANN, S. SCHMITT-STRECKER & A.P. GRIMANIS. 1990. On the composition and provenance of metal artefacts from Poliochni on Lemnos, *Oxford Journal of Archaeology* 9: 263-298.
REEDY, C.L. & T.J. REEDY. 1988. Lead isotope analysis for provenance studies in the Aegean region: a re-evaluation, in E.V. Sayre, P.B. Vandiver, J. Druzik & C. Stevenson (eds.), *Materials Issues in Art and Archaeology*: 65-70. Pittsburgh: Materials Research Society.
SAYRE, E.V., K.A. YENER, E.C. JOEL, & I.L. BARNES. 1992. Statistical evaluation of the presently accumulated lead isotope data from Anatolia and surrounding regions, *Archaeometry* 34: 73-105.
SAYRE, E.V., K.A. YENER & E.C. JOEL. 1995. Comments on P. Budd,, A.M. Pollard, B. Scaife & R.G. Thomas, 'Oxhide ingots, recycling and the Mediterranean metals trade', *Journal of Mediterranean Archaeology* 8: 35-53.
SCAIFE, B., P. BUDD, J.G. MCDONNELL, A.M. POLLARD & R.G. THOMAS. 1996. A reappraisal of statistical techniques used in lead isotope analysis, in S. Demirci, A.M. Özer & G.D. Summers (eds.), *Archaeometry 94*: 301-307. Ankara: Tübitak.
SILVERMAN, B.W. 1986. *Density Estimation for Statistics and Data Analysis*. London: Chapman & Hall.
SPOONER E.T.C. & N.H. GALE. 1982. Pb isotopic composition of ophiolitic volcanogenic sulphide deposits, Troodos Complex, Cyprus, *Nature* 296: 236-242.
STOS-GALE, Z.A. 1989. Lead isotope studies of metals and the metal trade in the Bronze Age Mediterranean, in J. Henderson (ed.), *Scientific Analysis in Archaeology and its interpretation*: 274-301. Oxford: Oxford University Committee for Archaeology.
STOS-GALE, Z.A. & N.H. GALE. 1992. New light on the provenience of the copper oxhide ingots found on Sardinia, in R.H. Tykot & T.K. Andrews (eds.), *Sardinia in the Mediterranean: A Footprint in the Sea*: 317-344. Sheffield: Sheffield Academic Press.
1994. Metals, in A.B. Knapp & J.F. Cherry (eds.), *Provenience Studies and Bronze Age Cyprus*: 92-121. Madison: Prehistory Press.
STOS-GALE, Z.A., G. MALIOTIS, N.H.. GALE & N. ANNETTS. 1997. Lead isotope characteristics of the Cyprus copper ore deposits applied to provenance studies of copper oxhide ingots, *Archaeometry* 39: 83-123.
WAND, M.P. & M.C. JONES. 1995. *Kernel Smoothing*. London: Chapman & Hall.
YENER, K.A., E.V. SAYRE, E.C. JOEL, H. ÖZBAL, I.L. BARNES & R.H. BRILL. 1991. Stable lead isotope studies of Central Taurus ore sources and related artifacts from Eastern Mediterranean Chalcolithic and Bronze Age sites, *Journal of Archaeological Science* 18: 541-57

16

An Investigation into the Fractionation of Copper Isotopes and its possible Application to Archaeometallurgy

A. P. WOODHEAD[1], N. H. GALE, & Z. A. STOS-GALE.

Isotrace Laboratory, Nuclear Physics Building, Keble Road, Oxford, OX1 3RH, UK
[1]also Department of Inorganic Chemistry, University of Oxford.

Using a low temperature, thermal ionization technique, copper isotopic compositions have been measured in both natural copper minerals from supergene/oxidation zones and in ancient metal artefacts. Significant variations in their isotopic composition were found in both types of material, suggesting that there may be considerable potential for copper isotope analyses in metal provenance studies, at the least as a supplement to lead isotope studies. This study also made an initial examination that suggests that fractionation of the isotopic composition of copper does not occur during any of the smelting processes thought to have been used in ancient times. In addition it has also shown that seven Late Bronze Age copper oxhide ingots have a range of copper isotope compositions.

Keywords: Copper; Isotopes; TIMS; Anomalies; Provenancing; Archaeometallurgy.

Introduction

Significant variations in the isotopic composition of copper were first reported by Walker *et al.* (1958), who examined the possibility of local variations, within mineral deposits, which they postulated might occur due to isotopic exchange or diffusion. The mass spectrometer source used in this work was a Knudsen effusion cell, measuring the ion beam from a charge of CuI. Unfortunately, as Shields *et al.* (1965) state '...an error in their calculations obscured their data'.

In 1964 Shields and co-workers at the National Bureau of Standards in the USA, set out to measure the absolute atomic weight for copper, from the accurately known masses and the measured isotopic ratio of $^{63}Cu/^{65}Cu$. These workers calibrated their thermal ionization mass spectrometer for instrumental fractionation effects in terms of a range of synthetic isotope standards, which they created from almost completely separated isotopes of ^{63}Cu and ^{65}Cu (>99.6% enrichment). Their method used a triple rhenium filament configuration, with ~200µg of copper electroplated onto two side filaments, and a central ionization filament heated to 2100°C.

In a companion study, Shields *et al.* (1965), examined the isotopic composition of 106 copper minerals, of various origins from around the world. They found a small variation in the copper isotope ratio, beyond the limit of experimental error, but no systematic geographical variation. The key variations that occurred were linked with the type of mineral studied, certain minerals showing a much greater range of ratios than others. Some secondary minerals appeared to show gross isotopic variation, the observed range being from 0.3% enrichment to 1.0% depletion of the isotope ^{63}Cu.

Even though two laboratories seemed to have independently discovered natural variations in the isotopic composition of copper, with Walker *et al.* (1958) suggesting possible applications to geological research, this topic appears to have lain fallow for over 30 years. This may, in part, have been due to scepticism on the part of isotope geochemists about the reality of the claimed isotopic variations, and in part due to the difficulty of obtaining reproducible data for the isotopic composition of copper using thermal ionization mass spectrometry. Shields *et al.* (1964; 1965) obtained $^{63}Cu/^{65}Cu$ ratio data having a 95% confidence limit for a single ratio of 0.15% of the ratio only by adhering to a very rigorous time/ion beam regime. They found that ratios 1% to 2% away from the correct value, thus obscuring natural variations, easily resulted by simply overheating the filament or changing the time/signal relationship. The difficulty in thermally ionizing copper on a hot filament is illustrated by the fact that, using their best high temperature triple filament technique, Shields *et al.* (1964; 1965) could achieve but a $5x10^{-13}$ A ion beam current for up to only 40 minutes. This undoubtedly set a limit to the precision of their data.

In 1995 the Isotrace Laboratory, Oxford, decided to investigate whether the sort of natural variations in the isotopic composition of copper reported by Walker *et al.* (1958) and by Shields *et al.* (1965) really exist. The motivation was two-fold. On one hand such variations might have applications to the sort of geological enquiries foreshadowed by Walker *et al.* (1958); on the other there might be applications for the direct provenancing of copper metal in archaeology, at the least as a supplement to lead isotope provenancing.

Methodology

The thermal ionization mass spectrometry (TIMS) procedure used by Shields and his colleagues at the U.S. National Bureau of Standards in 1965 could not produce data of the highest quality. To produce very high precision isotope ratio data using TIMS it is usual to correct out mass spectrometric fractionation effects by correcting all ratios in terms of an invariant isotope ratio in the sample, either naturally occurring (Sr or Nd, see Faure 1986) or by adding a 'double-spike' of a known isotopic ratio (Woodhead *et al.* 1997). This is not available for copper, for which only the two naturally occurring isotopes ^{63}Cu and ^{65}Cu are stable.

A survey of the analytical literature revealed that interest over the past decade in the mass spectrometry of copper had largely been limited to development of methods suitable for the determination of low concentrations of copper by mass spectrometric isotope dilution. Though some of these methods, using a silica gel activator (Moody *et al.* 1987) similar to that used in lead isotope analysis, produced large copper ion beams, the accuracy of isotope ratio measurement attained was only about ± 1%, adequate for isotope dilution but not for the investigation of isotopic variations in copper.

However, Hosoe *et al.* (1988) developed a new method using the copper sample in nitrate form mounted on a double filament, finding that a low temperature, 900°C, variant of their technique gave very slowly decaying copper ion beams of about 10-12 A for several hours, yielding precise measurements of the ratio $^{63}Cu/^{65}Cu$. For the certified NBS/NIST SRM 976 copper isotopic standard Hosoe *et al.* (1988) determined $^{63}Cu/^{65}Cu$ = 2.2448 ± 0.0007 as the average value for eight measurements (1σ); this is to be compared with the certified value (Shields *et al.* 1964) of 2.2440 ± 0.0021 (95% confidence interval).

This Japanese work appeared to have resulted in a low temperature double-filament TIMS technique for the isotopic analysis of copper which is superior to the high temperature triple filament technique used at NBS. They found that this method displayed very little variation with time of the measured isotopic ratio during an analytical run and needed no instrumental correction factor, since the directly measured isotopic ratio for the SRM 976 standard fell well within the experimental errors of the certified value. However Hosoe *et al.* (1988) did not investigate possible natural variations in the isotopic composition of copper.

The Isotrace Laboratory technique, and data for the NBS/NIST SRM976 standard

The Japanese work used a single collector Finnegan-MAT mass spectrometer. Recent advances in mass spectrometer design, particularly the use of multiple ion beam collectors, have increased the precision with which isotope ratio measurements may be made. Consequently it was decided, in 1995, to use a VG Isotech three-stage EBE computer-controlled multicollector thermal ionization mass spectrometer to develop a method for the isotopic analysis of copper, based on the Japanese technique but adapted to the different source design of the VG Isotech instrument.

A two-filament, low-temperature, static double collector TIMS procedure for copper (loaded in nitrate form) was developed, which considerably increases the precision and reduces the mass spectrometric fractionation, probably partly due to the lower temperature used (900°C instead of 2100°C in the NBS method), and partly due to evaporation of the copper as heavier $Cu(NO_3)_2$ molecules. Details of the new technique will be published elsewhere. Using this special TIMS technique, and collecting the isotopes ^{63}Cu and ^{65}Cu simultaneously in a static analysis, data for the NBS/NIST copper isotopic standard SRM 976 were obtained, as shown in FIGURE 1.

Within run precision of better than ± 0.004% standard error for a single sample analysis is relatively easy to obtain. However, even with this new TIMS technique, measurement of the isotopic composition of copper by thermal ionization mass spectrometry is not easy, and it is

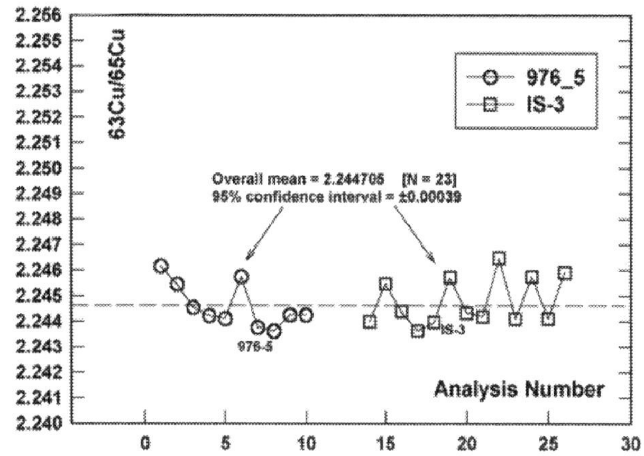

Figure 1: Copper isotope data for the NBS/NIST SRM 976 copper isotope standard, obtained using a low temperature TIMS method with the copper samples loaded onto a rhenium filament in nitrate form.

difficult to achieve run-to-run reproducibility. For this reason many repeat analyses were made, each involving a newly loaded sample on a new filament bead, for each sample, in this case the SRM976 standard, taking the mean of such sets of data. FIGURE 1 shows, for SRM976, two different sets of data, 976-5 and IS-3, taken several months apart in 1995. The overall mean is $^{63}Cu/^{65}Cu$ = is 2.2447 ± 0.0004, where the error is the 95% confidence interval computed in the standard way, and no correction for instrumental fractionation has been applied. This value is in complete agreement with the NBS/NIST certified value of 2.2440 ± 0.0021, and is to be compared with the value of 2.2448 ± 0.0007(1σ) obtained by Hosoe *et al.* (1988). Since the calculation of a 95% confidence interval is based on the normal distribution, which may or may not be an exact description of the distribution of the repeat analyses, the standard deviation of the 23 repeat analyses is also given, which is 0.000894. It is clear that these data for the isotopic standard SRM976 are closely comparable to the data obtained by Hosoe *et al.* (1988), and of better precision than that obtained by Shields *et al.* (1964; 1965). Moreover the technique used by Shields *et al.* (1964; 1965) produced a directly measured ratio $^{63}Cu/^{65}Cu$ of 2.2752, which had to be corrected for instrumental fractionation by measurements involving the nearly 100% separated isotopes ^{63}Cu and ^{65}Cu before the final corrected value of 2.2440 could be computed. One of the great advantages of the new Oxford technique is that the $^{63}Cu/^{65}Cu$ ratio is produced directly, without the need for instrumental fractionation corrections.

Natural variations in the isotopic composition of copper

Having established a reliable and precise, TIMS method for the isotopic analysis of copper, it was possible, in 1995, to investigate potential isotopic differences in a number of copper minerals. Minerals were chosen from mines for which Shields *et al.* (1965) had claimed fairly large isotopic anomalies (the term 'isotopic anomaly' is used to describe copper isotopic compositions that are different from that of the NIST SRM 976 copper isotope standard). The mineral samples were provided by F.B. Atkins, Keeper of the mineralogical collections of the University Museum, Oxford.

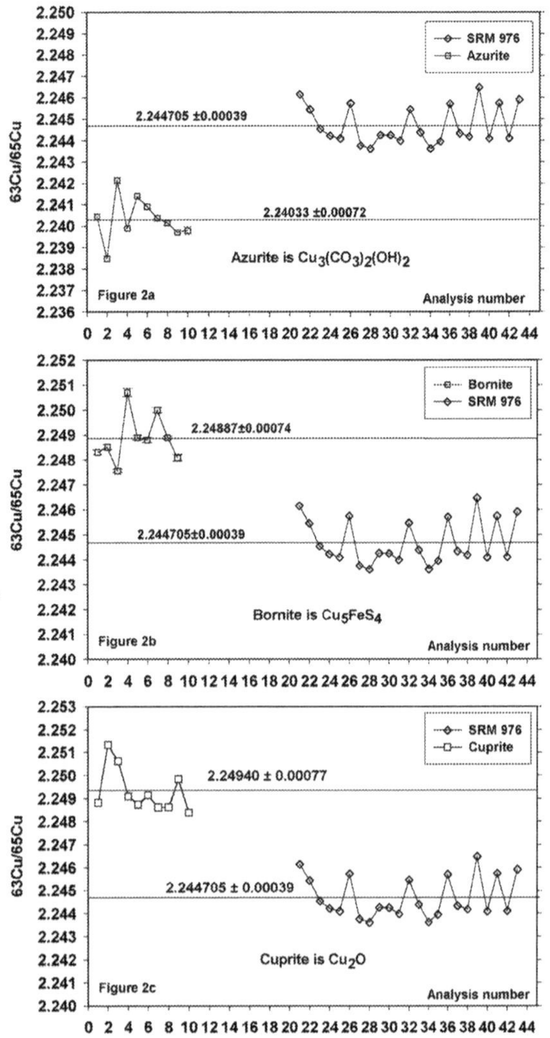

Figure 2: Copper isotope data, obtained using the low temperature TIMS technique, for Figure 2a azurite from the Chessy mine region, France; Figure 2b, for bornite from the Mufulira mine, Zambia; Figure 2c, for cuprite from the Morenci mine, Arizona.

FIGURE 2 gives data for a sample of azurite from the Chessy mine region near Lyon, France (Beyschlag *et al.* 1914), compared with isotopic data for the SRM976 isotopic standard. There is visually a clear difference in copper isotopic composition between the azurite sample and the SRM976 standard. A mean $^{63}Cu/^{65}Cu$ ratio of 2.24033 ± 0.00072 (95% confidence interval) for the azurite was obtained, 0.163% lower than the SRM976 standard. For a different sample of azurite from the same location Shields *et al.* (1965) obtained $^{63}Cu/^{65}Cu$ = 2.2407 ± 0.0033, a result closely similar to the new data. Consequently it may be the case that the copper isotopic composition does not vary much between azurite nodules found in the Keuper Sandstone of this mine.

FIGURE 2b depicts data for a sample of bornite from the Mufulira mine, Zambia, again compared with the SRM976 copper isotope standard; there is a clear difference in isotopic composition. These data give an isotopic composition for the bornite sample of $^{63}Cu/^{65}Cu$ = 2.24887 ± 0.00074 (95% confidence interval), or 0.217% higher than the SRM976 standard. Shields *et al.* (1965) measured a value of $^{63}Cu/^{65}Cu$ = 2.2513 ± 0.0033 for a different sample of bornite from the same mine, not very different from the new result. This suggests that although this mine is characterized by anomalous copper isotopic compositions, the isotopic composition of copper may be homogeneous within the mine.

FIGURE 2c plots data for a sample of cuprite from the Morenci mine, Arizona; again this mineral has a demonstrably different isotopic composition from the SRM976 standard. This sample of cuprite analysed at Oxford has $^{63}Cu/^{65}Cu$ = 2.24940 ± 0.00077, which is 0.241% higher than the SRM976 standard. Shields *et al.* (1965) measured $^{63}Cu/^{65}Cu$ = 2.2341 ± 0.0033 for another sample of cuprite from the same mine, which is very different. These results may suggest a greater variability in the copper isotope compositions within this mine, but this needs to be confirmed by many more analyses.

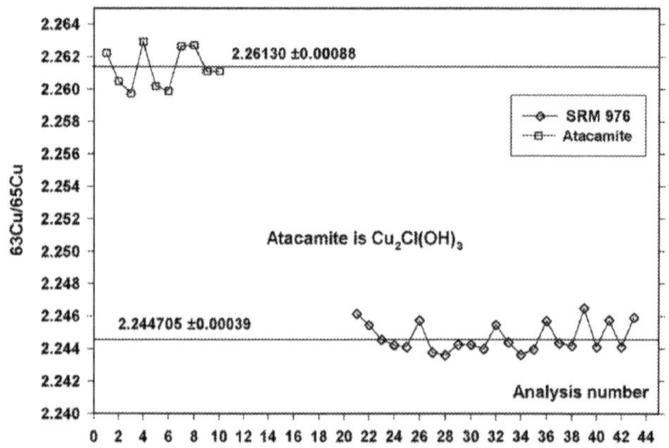

Figure 3: Low temperature TIMS data for the copper isotopic composition of atacamite from the Atacama mine, Chile.

FIGURE 3 shows copper isotope data for a sample of atacamite from its type location in the Atacama mine, Chile. With a mean $^{63}Cu/^{65}Cu$ = 2.26130 ± 0.00088 (95% confidence interval), this is by far the most extreme copper isotope composition yet measured by the Isotrace Laboratory, being 0.771% higher than the isotopic composition of the SRM976 standard. Here again Shields *et al.* (1965) report a quite different value, with their sample of atacamite, from the same mine, having $^{63}Cu/^{65}Cu$ = 2.2477 ± 0.0033. There is always doubt about the localities of mineral samples from any museum collection, so that these discrepancies must await resolution from analyses of mineral samples collected for this purpose from the mines involved.

What is no longer in doubt, following these measurements in Oxford, is that there are indeed substantial natural variations in the isotopic composition of copper; this clearly substantiates the earlier work of some 30 years ago. This immediately opens up the possibility of a new field of isotope geochemistry. It also suggests that variations of the isotopic composition of copper might be useful at least as an additional tool to lead isotope analyses for provenancing copper in ancient metal artefacts, and/or in exploring the possible mixing/recycling of copper alloys in the past. This would depend on there being no perturbations of the copper isotope composition by such anthropogenic operations as the smelting of copper ores to produce copper metal, or in subsequent operations such as the fire-refining of copper. It has been proved that lead isotope compositions are not measurably altered in metallurgical processes (Gale & Stos-Gale 1996 and references therein) but, due to copper isotopes being of much lower mass, there would be a greater possibility that metallurgical processes might alter isotopic compositions. Should this be the case, and depending on the magnitude of such possible effects, the use of copper isotope analyses for provenancing copper might be vitiated, though there then might be possible applications to archaeometallurgy of a technological nature. The simple application of thermodynamic theory, largely based on a model of evaporation into a vacuum, such as has been applied by the Bradford Archaeometallurgical Research Group to lead and tin (Budd *et al.* 1995), would predict measurable alteration of the isotopic composition of copper in smelting or refining if the losses of copper in such processes were appreciable. However such predictions have been shown not to be quantitatively valid in practise, by the Isotrace Laboratory for lead (Gale & Stos-Gale 1996) and tin (Gale 1997).

For copper the uncertainty remains, especially since the work of Fujii *et al.* (1986) has shown how easy it is to produce changes in the $^{63}Cu/^{65}Cu$ ratio. Consequently it was important to carry out direct experimental investigations of this matter using material from John Merkel who provided products from his experimental, full-scale, reconstructions of ancient copper smelting and fire-refining processes. Copper isotope data are presented here which suggest that these processes do not produce measurable alterations in the isotopic composition of copper, and that copper isotopes might therefore make a contribution to provenancing, although further investigation of these matters remains necessary.

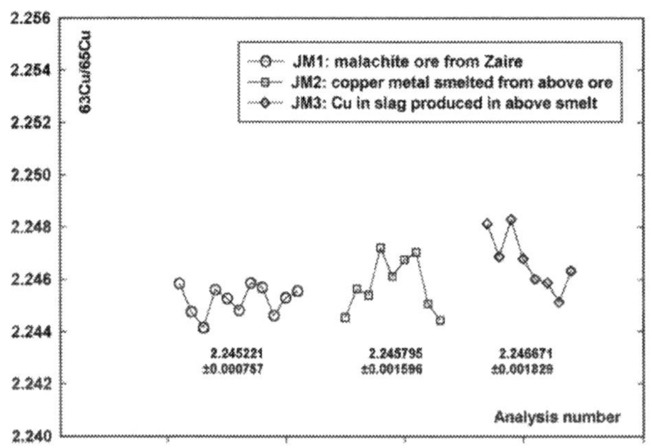

Figure 4: Copper isotopic compositions for malachite ore from Zaire, for copper metal smelted from this ore and for copper in the slag produced in the smelting experiment which produced the copper metal.

FIGURE 4 presents the results of copper isotope measurements on the source material and smelting products of a smelting experiment carried out at Abu Dhabi by Merkel using malachite from Zaire. The isotopic composition of copper was determined in the original malachite, in the copper metal smelted from this ore, and in copper extracted (by dissolution and ion exchange separation) from the slag produced in this smelting process. These data are visually very similar in FIGURE 6. Statistical tests were applied to the data to check for any differences in copper isotopic composition between the ore and the products of smelting. Paired Students t-tests, which assume the data are distributed according to the normal distribution, were performed, as were Mann-Whitney rank-sum tests that are independent of the distribution of the data. These tests showed that, in this experiment, there are no statistically significant differences in the $^{63}Cu/^{65}Cu$ ratios between the starting malachite ore, the copper smelted from it, or the traces of copper remaining in the slag. On the basis of this experiment there is, at present, no evidence that smelting changes the isotopic composition of copper in the smelted metal away from that in the original copper ore.

FIGURE 5 shows copper isotope data for fire- refined copper; the copper metal used for these experiments is

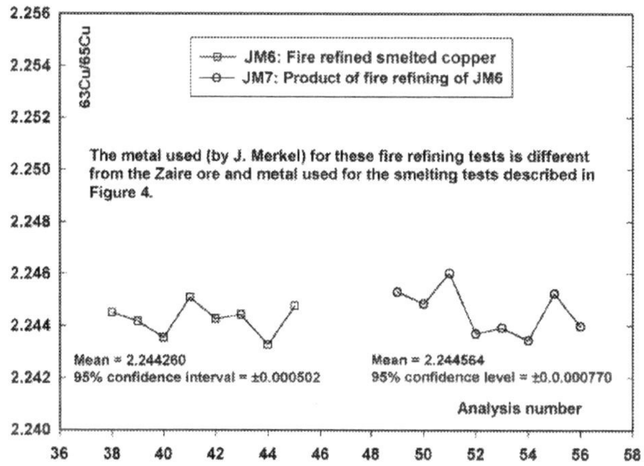

Figure 5: Copper isotopic compositions for batches of copper metal that have undergone one or two successive fire refining processes.

not related to the malachite ore from Zaire, nor the metal smelted from it, which are the subject of FIGURE 4. JM6 is a particular batch of copper metal which had been once fire- refined by Merkel; JM7 is a batch of copper metal which was the direct product of fire- refining batch JM6. Paired t-tests and the alternative Mann-Whitney rank-sum tests show that there is no statistically significant difference in the ratio $^{63}Cu/^{65}Cu$ between batch JM6 and JM7, so that fire-refining does not, on this evidence, alter the isotopic composition of copper.

Although further investigation of these matters is necessary, present evidence strongly suggests that smelting and fire-refining do not alter the $^{63}Cu/^{65}Cu$ ratio away from that in the smelted copper ore, and that copper isotopes might therefore make a contribution to provenancing archaeological objects made of copper-based alloys.

The question clearly arises: are copper isotopic differences to be found in archaeological copper objects? There seems to be no evidence in the literature that this has ever been investigated? Hence a few analyses to test this were made on copper oxhide ingots, found in Crete, Cyprus and Sardinia. Bass (1964) gives an archaeological discussion of this widely traded form of copper in the Bronze Age Mediterranean region.

FIGURE 6a presents copper isotope data for two oxhide ingots from Cyprus (from the find spots of Mathiati and Skouriotissa respectively) and one from the LM IIIA2-IIIB site of Kommos in Crete, in comparison with data for the SRM976 isotopic standard. The Mann-Whitney rank-sum test detects no statistically significant difference between the $^{63}Cu/^{65}Cu$ ratios of all three ingots and the SRM976 standard.

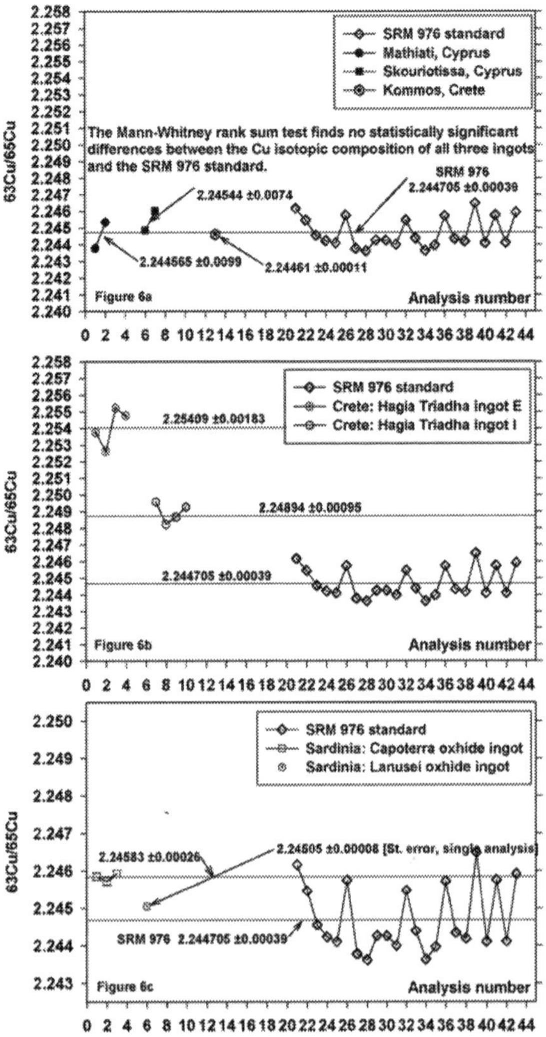

Figure 6: Copper isotopic compositions for Late Bronze Age copper oxhide ingots; Figure 6a: oxhide ingots from Cyprus and Crete; Figure 6b: earlier oxhide ingots from Hagia Triadha, Crete; Figure 6c: oxhide ingots found on Sardinia.

On the other hand, as shown in FIGURE 6b, two copper oxhide ingots from the LM IB site of Hagia Triadha in Crete clearly show different isotopic compositions from each other and from the SRM976 standard, differences which are confirmed by the Mann Whitney rank-sum test. These two Minoan ingots are typologically of an earlier form than the ingots considered in FIGURE 6a. They are from the group of oxhide ingots, that also have Pb isotope compositions which do not match the isotopic compositions of any copper deposits so far analysed in the Mediterranean region, including Cyprus and Turkey (Gale & Stos-Gale 1986).

FIGURE 6c gives copper isotope data for two copper oxhide ingots found in Sardinia, from the findspots of Capoterra and Lanusei respectively (Lo Schiavo 1990) which again show no anomalies outside experimental error. This may be due to their being made of Cypriot copper, that seems to show no anomalies judging by the evidence shown in FIGURE 6a. As Gale (this volume) suggests, the lead isotopic evidence is now overwhelmingly in favour of the copper oxhide ingots found in Sardinia having been made from copper smelted from the Apliki ore deposit region in Cyprus.

Conclusions

The 1965 work of Shields and co-workers has been verified; natural variations of the isotopic composition of copper do occur in copper minerals. The copper minerals studied were secondary minerals from the supergene/oxidation zone; modern work on primary copper minerals remains to be done. A new low temperature TIMS technique, using a multicollector and simultaneous measurement of both isotopes, has improved measurement precision.

Copper smelting or fire-refining do not, on present evidence, alter the isotopic composition of copper; consequently copper isotope analyses may be a valuable addition to the lead isotope provenancing of copper metal.

Copper isotope anomalies have been found in some LBA copper oxhide ingots, but not in others.

The present low temperature TIMS method, although an advance on former TIMS techniques, is difficult to control, as is clear from the data scatter which is much larger than the internal precision of individual runs (the internal precision is, on the diagrams, of about the size of the symbols). This means that the method is very time consuming and labour intensive, with at least five repeated mass spectrometric runs being necessary for the same sample to define a reasonable confidence level.

The way ahead is to use ICP-MS, with magnetic sector mass dispersion and a multiple set of faraday collectors to allow simultaneous collection and measurement of both isotopes. Copper is easily ionized in a plasma, which overcomes the difficulty of thermal ionization caused by the high ionization potential [7.726 eV] of copper. Moreover TIMS is bedevilled by time-dependent fractionation effect that cannot, for copper, be corrected out by a double-spike method. In contrast ICP magnetic sector multiple collector mass spectrometry has a fixed fractionation which is time-independent and independent of the chemical properties of the element or admixing of other elements (Lee & Halliday 1995). This allows mass fractionation to be corrected out, for copper, in terms of the fixed isotopic ratio, $^{68}Zn/^{70}Zn$, on addition of an in-house standard of zinc. Such measurements are underway at the Isotrace Laboratory.

Acknowledgements.
We are grateful to J. Merkel for supplying samples from his reconstructions of ancient smelting and metal refining processes, and to the Institute of Aegean Prehistory, New York, without whose financial support the work would not have been possible. The paper was improved by the editor and two anonymous referees.

References

BASS, G.F. 1964. Cape Gelidonya: A Bronze Age Shipwreck. *Transactions of the American Philosophical Society* 57/8, Philadelphia.

BEYSCHLAG, F., J.H.L. VOGT & P. KRUSCH. S.J. TRUSCOTT (Translator). 1914. *The Deposits of the Useful Minerals and Rocks*. London: Macmillan.

BUDD, P., R.HAGGERTY, M. POLLARD, B.SCAIFE & R.G.THOMAS. 1995. New heavy isotope studies in archaeology, *Israel Journal of Chemistry* 35: 125-130.

FAURE, G. 1986. *Principles of Isotope Geology*, 2nd. Edition. London: John Wiley & Sons.

FUJII, Y., M.HOSOE, & M.OKAMOTO. 1986. Copper isotope separation by ion exchange electromigration and copper isotope analysis by thermo-ionisation mass spectrometry, *Zeitschrift für Naturforschungen*, 41: 769-770.

GALE, N.H. 1997. The isotopic composition of tin in some ancient metals and the recycling problem in metal provenancing. *Archaeometry* 39: 71-82.

GALE, N.H. & Z.A. STOS-GALE. 1986. Oxhide Ingots in Crete and Cyprus and the Bronze Age Metals trade, *Annual of the British School at Athens* 81: 81-100.

1996. Lead isotope methodology: the possible fractionation of lead isotope compositions during metallurgical processes. In S. Demirci, A.M. Ozer & G.D. Summers (ed.): *Archaeometry '94* The Proceedings of the 29th. International Symposium on Archaeometry, 1994. Ankara, 287-299.

HOSOE, M.,Y. FUJII, & O. MAKOTO. 1988. Mass spectrometric measurements of copper isotopic ratios using low-temperature thermal ionisation. *Analytical Chemistry*, 60: 1812-1815.

LEE, DER-CHUEN & A. HALLIDAY. 1995. Precise determinations of the isotopic compositions and atomic weights of molybdenum, tellurium, tin and tungsten using ICP magnetic sector multiple collector mass spectrometry. *International Journal of Mass Spectrometry and Ion Processes*, 146/147: 35-46.

LO SCHIAVO, F. 1990. Lingotti ox-hide e lingotti piano-convessi. In F. Lo Schiavo, R. Maddin, J. Merkel, J.D. Muhly & T. Stech (ed.): *Analysi Metallurgiche e Statistiche sui Lingotti di Rame della Sardegna.* Quaderni 17: 16-40, Ozieri, Sardinia.

MOODY, J.R., BEARY, E.S., K.A.BRLETIC & P.J.PAULSEN. 1987. A comparison of isotope dilution mass spectrometric methods for assay of copper in copper ore reference materials. *Analyst* 112: 441-444.

SHIELDS, W.R., T.J. MURPHY & E.L. GARNER. 1964. Absolute isotopic abundance ratio and the atomic weight of a reference sample of copper. *Journal of Research of the National Bureau of Standards*, 68A: 589-592.

SHIELDS, W.R., E.L. GARNER, S.S. GOLDICH & T.J. MURPHY. 1965. Natural variations in the abundance ratio and atomic weight of copper. *Journal of Geophysical Research*, 70: 479-491.

WALKER, E.C., F. CUTTITTA, & F.E. SENFTLE. 1958. Some natural variations in the relative abundance of copper isotopes. *Geochimica et Cosmochimica Acta*, 15: 183-194.

J.D. WOODHEAD & J.M. HERGT. 1997. Application of the 'double spike' technique to Pb-isotope geochronology. *Chemical Geology* 138: 311-321.

17

The Economic Value and Colour Symbolism of Tin

CAROLE GILLIS

Dept. of Classical Archaeology and Ancient History, Box 200, S-405 30 Göteborg, Sweden

Introduction

The primary purpose of this article is to discuss the various roles and the enormous importance that tin played in the Aegean Bronze Age, especially in the latter part of the Late Bronze Age (c. 1400-c.1150 BC). A few remarks must be made first about the general importance of metal. In the first place, the need to use and/or acquire metals in order to make tools, weapons and status objects is not a universal. This can be seen for example in West Mexico, where despite rich natural metal ore resources and traditions of metal production and use in neighbouring civilizations, metal was not used at all until c. 600/800 AD. Even then, it was employed mainly for status and religious items rather than utility items (Hosler 1988, 1994 :12 and Ch. 3). In Mesoamerica, metal was virtually unknown until post-classical times, after 900 AD. In Teotihuacán, for example, the capital of the largest Mesoamerican Classical Period State (100-750 AD), metal was never used (Hosler 1994:11-12).[1] A slightly different example: in Peru, although a rich use of metal alloys was seen from c. 1500 BC, and despite the fact that highland Peru has some of the world's richest cassiterite deposits, the earliest tin bronze artefacts date to c. 500 AD (Lechtman 1977:8; Hosler 1997) — seemingly the superior utility of tin bronze was never a factor.. Thus, general statements about the importance of (certain) metals for a society and about societal development as a direct result of its need to create institutions and strategies to permit the use and/or import of metals cannot be made. Each society must be studied and evaluated on an individual basis.

Further, the relative status of different metals can vary: gold was not always the metal with the highest status. For nearly 600 years in western Mexico (c. 600/800 - 1200 AD), copper was almost exclusively the only metal in use and was employed only for religious and status objects, not utilitarian ones, although there are existing tin sources. Gold and silver were also available (Hosler 1994).

Thus, before one can start to discuss the importance and use of metals in general, and tin, in particular, in the Late Bronze Age Aegean society and its impact on the social, political, religious and economic development of that society, it is necessary to examine other factors: the existence and use of alternative non-metallic materials like obsidian, stone types, wood, bone, horn, ivory, etc.; metal sources; the level of social development of the particular society; the various contacts with metals and metallurgy which that society had and the social position it held among these other civilizations; and perhaps most important, the choices — aesthetic, economic, social and technological — that can be and are made by different members and levels in the society.

The Aegean and Metals

The Aegean area is geographically and perhaps psychologically[2] a water mass surrounded by land, with maritime communication and trade, perhaps as important for the Aegean economy, as agriculture, and always an integral part of the island and coastal societies.[3] As such, it is not surprising that, although the Aegean lies somewhat offside, contacts with the various areas in the Eastern Mediterranean and even central Europe were established in the 3rd M. if not before, as is well attested by imports such as stone vases, ivory, tin and even amber from a variety of areas. Thus, the Aegean was exposed to new impulses and technology, as well as, to non-local raw materials and objects in the Neolithic, Chalcolithic and Early Bronze Ages. Although flint, stone and obsidian are indigenous to the Aegean area and were used extensively in the Neolithic period for tools and weapons, and more or less still existed throughout the whole of the Bronze Age, metals quickly became more widely used from the early Early Bronze Age (e.g., Renfrew 1972:313). Their use was most likely due to direct influence from outside Greece.

The earliest indications of copper mining in Europe are from the 4th M. at Rudna Glava (late Vinca), Yugoslavia, and Ai Bunar, Bulgaria (Karanova IV) (e.g., Jovanovic 1979,[4]1998; Pernicka *et al.* 1998; Tylecote 1992:12). Evidence of smelting at these mining sites has been found, as well as moulds (Jovanovic 1979:109) and slags (McGeehan Liritzis 1996:146 and her references). Further evidence of advanced knowledge of smelting, specifically concerning smelting copper sulfides ores from the Eneolithic period in the Balkans from the mid-4th M, has been reported (Ryndina *et al.*1996), since the Aegean area lies more or less contiguous with the Balkans, the knowledge of metals, their mining, their casting, and ore smelting could well have filtered down to Greece in the course of the 4th M.

It is highly likely that Greece was also exposed to Anatolian metallurgy, which was well-advanced in the 4th M. in parts of Anatolia, through cultural contacts of various kinds (e.g., de Jesus, P. 1980 :135; Tylecote 1992:11-13; Muhly 1973 + 1976, ch. II and 91-104 in the 1976 Supplement; Renfrew 1972:311-312, Adriaens *et al.* 1998, and via the eastern Aegean area, Kaptan 1997).

In the Levant area, there are indications that mining and smelting were carried on in the Timna valley in present-day Israel in the late 5th M. (Rothenberg 1998). Advanced metalworking was carried on in northeast Syria where tin-copper alloys were common from the late 4th M. (Northover & Crane 1996)

There are many examples in Eastern Mediterranean cultures of using copper alloys as well as precious metals at an early stage (e.g., Muhly 1985; McGeehan Liritzis 1996:148-151 and table 4.3.1.4).

It is suggested that for weapons and tools, metal was chosen over alternative materials and was part of the assemblage of status goods in the Balkan and Eastern Mediterranean areas. It is not unreasonable to assume that in the late 4th and 3rd M., the rising elites in the Aegean would choose the same markers to express their social and political status as the civilizations they wished to impress and emulate, their trading partners *in spe* . Metal technology would have won acceptance through a kind of 'follow-the-leader' effect — observing a non-local phenomenon and following after, adopting or adapting it to the indigenous western Aegean cultures.

The Chalcolithic in the Aegean was fairly short-lived, lasting only a few hundred years. Various copper alloys, usually arsenical copper but also tin bronze mainly in the islands (McGeehan Liritzis 1996:table 4.3.1.4:352-353), were in use from Early Bronze Age I, generally considered to start around 3000 BC (Muhly 1985:255 and references in note 100; McGeehan Liritzis 1996:151, 1988). By the rise of the first palace economies on Crete, the walled town of Ayia Irini on Kea, the expansion of city II at Phylakopi on Melos, and the continuation of the fortified settlement Kolonna on Aegina (e.g., Forsén 1992, 114-117 and her references) that is, somewhere around 2000 BC, or shortly after, tin bronze was the leading (and perhaps only) copper alloy (e.g., Tylecote 1992:27; McGeehan Liritzis 1996:154).

On the mainland of Greece, an important source of copper, from at least the middle of the 3rd M., was Laurion in south Attica, but copper was probably mined and definitely smelted and processed already during the Early Bronze Age on Kythnos (Bassiakos and Hadjianastasiou, this volume). Possible sources on Kea (Caskey *et al.* 1988) and Thera (Bassiakos & Michailidou 1998) could have been exploited also as well as a variety of minor sources.[5] Gold is found in northern Greece (e.g., Tylecote 1992:5 and Table 2:4); silver and lead come from the Cyclades (e.g., Tylecote 1992 and McGeehan Liritzis 1996:154-156) and Laurion. The only vital metal, not indigenous to Greece is tin (Benton 1963), which had to be imported - mostly likely through long-distance trade networks (e.g., Muhly 1973 + 1976: chs. IV and V and 107-115 in 1976 supplement, 1985; Franklin *et al.* 1978, Penhallurick 1984; Gillis 1991). The need for the Aegean civilizations to acquire tin was obvious, given the proclivity to use tin bronze for the manufacture of tools and weapons. Furthermore, judging from the number of status metal objects found in rich tombs (for Late Bronze Age III Argolid, for example, Gillis 1996a), metals played a major role as elite markers and symbols of status and prestige.

Thus, in the Aegean, metals and metal products were important and were chosen to show social status, to maintain military and political freedom and independence from potential enemies both through creating alliances based on gift-giving/exchange and by being strong enough to avert attack by neighbours, and to ensure sufficient returns in agricultural and craft production to allow the populace to survive and to fill the palace coffers.

Tin

Utilitarian value

As mentioned above, tin is not indigenous to the Aegean area. As it was the preferred alloy for bronze, rather than arsenic or other alloying components from the Middle Bronze Age, it was obviously an important import for the production of tools, weapons, and other objects, and thus of great utilitarian value. While certainly not belittling this use of tin, it is, on the other hand, nothing that needs to be belaboured.

Socio-economic symbolism

As mentioned above, a look at some of the richer Late Bronze Age tombs in the Aegean shows the use of various metals, as well as other luxury items (usually acquired through long-distance trade) to show status. Can tin, which was primarily a more or less invisible, although vital, component in bronze, have had socio-economic or other status in its own right?

There are very few pure tin objects found in the Aegean Bronze Age: an Early Bronze bracelet or ring from Lesbos (Lamb 1936:39 - called 'lead')[6], two rings from the Late Bronze III sanctuary at Phylakopi on the island of Melos (Renfrew & Cherry 1985:311-313, fig. 8.6:813, 2802).[7] Modern discoveries of several shipwrecks off the coasts of Turkey and Israel, with tin ingots among their cargoes indicate the status of tin as a long-distance trade commodity (for Cape Gelidonya, Bass 1961, 1967, 1991; for the Ulu Burun (Kas) wreck, Bass 1986, Pulak 1988; for the shipwrecks off the coast of Haifa,[8] see Galili & Shmueli 1983, Galili *et al.* 1986, Wachsmann & Raveh 1981, 1984; and in general, Bass 1991.) It would appear that tin was little used except as an alloy of bronze, perhaps because it oxidizes easily and becomes powdery,[9] and can even be attacked by tin pest.[10]

Figure 1. A tin-covered conical cup, MM Acc. 1068, from Asine Chamber Tomb 1:1, now in the Medelhavsmuseum, Stockholm (photo M. Sjöblom).

Yet, tin is found in another form. An often unnoticed phenomenon from the end of the Late Bronze Age in the Aegean area (c. 1400-1200/1150 BC) is that of ceramic funerary vessels covered with a thin layer of tin foil (Immerwahr 1966; Gillis 1991, 1992, 1994, 1996a, 1996b). (*Fig. 1*) These vessels are found only in rich chamber and tholos tombs over the greater part of mainland Greece, on Crete, on Rhodes, and the coast of

Asia Minor, but never in settlement contexts (Gillis 1991 esp. appendix 1). While conducting replication experiments, it was discovered that the tin foil when heated to a temperature just below melting point (232°C.) oxidized to a colour, for all intents and purposes, identical to pure gold (Gillis & Bohm 1994). Various analyses with X-ray Photo-electron Spectroscopy (XPS) indicated that the ancient foil was most probably also 'gilded' before application (Gillis & Holmberg 1995; Gillis *et al.* 1995).

sample	vase i.d.	Si	C	Ca	N400	N407	Sn	O	S	Al
'gilded' tin			80.74				1.55	17.70		
silver tin (metallic tin foil)			44.04				17.49	38.04		
DARK SURF	39 KSDS	1.09	75.28	0.06	0.37	1.52	---	21.20	0.46	---
WHITE SURF	39 SWS	1.09	60.07	1.69	1.58	2.15	0.06	31.77	1.39	0.21
BROWN SURF	39 SBS	2.33	60.62	2.15	2.09	0.42	0.11	29.48	1.68	1.11
BLACK SPOT	39 FSBS	5.27	49.47	1.28	1.77	0.71	1.49	37.89	1.26	0.86
CLAY SURF	39 FSXS	2.28	60.14	1.48	2.68	0.43	0.08	29.99	1.79	1.14
BROWN SURF	39 FSRS	1.20	69.07	0.83	2.38	0.51	0.07	24.33	0.75	0.86
SHINY SPOT	39 FSGS	2.19	67.44	1.24	2.92	0.15	0.28	23.25	0.84	1.69
BROWN STRIP	41 SBS	0.87	60.92	0.61	0.97	4.40	---	31.21	0.42	0.56
LIGHT SURF	41 SLS	1.81	60.99	0.99	2.58	1.72	0.13	29.27	1.21	1.30
VERY DK SURF	44 ASDS	1.28	72.72	0.51	2.22	1.36	0.15	20.73	0.51	0.51
CLAY SURF	44 AUSS	1.49	67.48	0.76	3.11	0.70	0.08	24.57	0.79	1.02
BLACK SURF	44 ASNP	0.99	69.86	0.31	1.57	1.82	0.07	24.18	0.84	0.35
BLACK SURF	44 BSFR	0.70	77.31	0.42	1.11	1.31	---	18.26	0.87	---
BLACK SURF	44 BSNR	0.45	74.89	0.59	1.81	1.21	0.04	19.63	0.78	0.14
LIGHT SPOT	44 BSLS	0.73	66.19	0.73	1.69	1.12	0.05	27.02	0.91	1.16
DARK,SOME SH	49 ISDS	0.70	54.89	0.16	0.30	7.38	---	35.46	0.51	0.52
DARK,NO SHINE	49 ISS	0.47	59.73	0.14	0.32	5.48	0.05	33.44	0.38	---
GREY SURF	49 ISGS	2.01	53.81	1.01	0.87	5.05	0.16	35.54	1.55	---
CLAY SURF	49 ISCS	3.16	52.97	1.83	1.76	1.40	0.19	35.79	2.17	0.73
DARK SURF	49 TSDS	0.47	56.49	0.32	---	6.46	0.09	35.30	0.46	0.40
BROWNISH	49 TSBS	1.00	63.70	0.17	0.19	3.30	0.09	30.43	0.08	1.03
YELLOW AREA	68 SBS	0.35	75.38	0.22	0.67	0.21	0.03	23.09	0.03	0.02
YELLOW SHINY	68 SYS	1.03	73.95	0.25	0.36	0.35	0.01	23.95	---	0.10
GREY SURF	68 SGS	1.81	61.74	0.61	2.97	2.03	0.13	29.11	0.29	1.30
CLAY SURF	68 ISCS	2.05	65.45	0.51	1.87	2.17	0.07	26.67	0.66	0.54
WHITE SPOT	68 ISWS	2.92	64.19	0.65	3.43	1.21	0.48	26.55	0.20	0.87
SHINY SURF	70 SGS	1.35	58.33	0.50	0.33	5.35	0.10	33.77	0.17	0.11
YELLOW SPOT	70 ISYS	0.58	54.79	0.11	---	7.51	---	36.52	0.06	0.38
DARK SURF	70 ISDS	1.53	58.81	0.26	0.38	4.60	0.14	33.14	0.46	0.68
LIGHT SURF	70 ISLS	0.89	55.93	0.41	0.11	3.82	0.23	37.37	0.61	0.64
DARKRED SURF	73 SRS	2.69	49.75	2.71	1.71	1.38	0.14	37.65	3.01	0.96
BLACK SURF	73 SBS	5.36	50.93	0.93	1.74	1.99	0.21	36.25	0.89	1.70
RED SURF	73 ISRS	1.66	65.97	1.19	1.65	1.86	0.19	26.42	0.75	0.30
GREY SURF	73 ISGS	2.37	68.53	1.01	2.16	0.58	0.22	22.66	0.72	1.76

Figure 2. XPS analyses results of 44 samples taken from 7 vases from Asine ChTI:1, plus two modern tin analyses. N400 in the table ≈ 400 eV and N407 ≈ 407 eV. The values for C and O for the vases indicated by the boxes as compared with the modern 'gilded' tin results are considered probable indications that the tin on these vases was most probably 'gilded'.

Pieces of tin foil in geometric shapes have been used for decorative designs and patterns on ceramics [11] and tin even in solid form as studs inserted in wooden vessels from burials in Denmark in the 13th c. BC (Boye 1896). In only two cases to my knowledge, apart from the Aegean one, does the foil envelop the vessel completely. In the one, there are two cups and a bowl with seeming tin coating from 6th c. BC China;[12] and in the other, there are five ornate spouted or bridged vessels and two ordinary vases from the classical, or middle, period in the Sicán civilization in northern Peru dating to 900-1100 AD. These are also covered with foil, although in this case made of *tumbaga* , a copper(-silver)-gold alloy (Shimada 1997:69 and figs. 99, 101-103, pp. 196, 198 and figs. 1-3, p. 203, and 5, 204; Shimada in press:12-13, 29).[13]

What significance, regarding socio-economic values, can be drawn from this almost unique phenomenon in the Aegean of covering pottery completely with tin foil? It has been established above, that metals played an important social role for the Aegean. In the use of the tin, it is possible of course that the Mycenaeans thought they had made gold, but the evidence seems to refute this. First of all, the shapes of foil-covered vessels can be closed, as well as open, an occurrence never found in solid metal vessels. Secondly, the XPS analysis for nitrogen (N) on the vases indicated that the tin had flaked or powdered off the vessels, during the lifetime of the tomb (Gillis *et al.* 1995).[14] At successive burials, it could be plainly seen by looking at real gold vessels and then at these, that the latter were clay vases coated with foil which had darkened, dulled and partially fallen off: clearly not gold. It is not likely that someone (the producer? the consumer?) tried to fool others into thinking despite its form and its tendency to turn black and powdery and reveal a core of clay that the vessels were solid gold: gold is heavy. A solid gold kylix would weigh considerably more than a comparably sized clay kylix covered with 0.009 mm-thick foil: anyone holding a tin-covered vessel would easily have noticed the difference.[15]

But should these vases be regarded as a kind of ersatz gold or did tin have high status in its own right? Considering the great importance of tin for bronze making, its rare nature and the fact that it must be imported to the Aegean from as far away as perhaps Afghanistan (Cleuziou & Berthoud 1982) through international long-distance trading, one is led inevitably to the conclusion that the phenomenon of coating pottery with tin foil to deposit in graves must be a status statement *par excellence.* Only a few persons, assumedly those in leading positions in the society, would be in a position to have surplus tin. It is not unjustified to suggest that for the Aegean Late Bronze Age society, tin ranked as a status metal on a par with silver, and perhaps not far from gold. It was used *in its own right* in these prestigious burials, conferring status and prestige on the deceased and his/her family.

Colour symbolism

In New World archaeology, there is a phenomenon that appears in Mexico, Meso- and northwestern South America for almost two thousand years. That is the use of metals, not for their economic value, but for their religious and ideological symbolism and their connection with power and divinity.[16] In various cultures, these metals can be used as they are — gold is used for gold, silver for silver — but far more often there is a mixing of metals which produces a wide spectrum of colours from white to yellow (and reddish variants): the metallurgical technology is different from area to area, but the effect is the same.[17] The golden and silvery metal objects represent dichotomies yellow-white, sun-moon, birth-death, male potency-female fecundity (Hosler 1988:849, 1994:228-233; Lechtman 1977:7-9, 1994:9-19). Thus, wearing objects, such as tweezers or bells made of these metals, or carrying them into the grave is an undeniable symbol of power and connection with the divine. The use of metals and their colours (and even their sounds - e.g., Hosler 1994) is social, (establishing status, importance, closeness to divinity,) but is not socio-economic, (expressing status through worldly, material wealth). It is socio-symbolic, as it combines power with religion and the abstract world.

While this use of metals in the New World seems to have been a long-term and wide-spread occurrence, it also seems to have been unique for that part of the world. Could colour and colour symbolism have been an aspect in the Late Bronze Aegean (and in specific, the Mycenaean) culture as well, or at least in burial ritual and as a connection with power and status? To investigate this, the XPS analyses were reviewed. Above in Fig. 2, it was seen that three of the seven vases, 39[18] ('chalice', or kylix), 44 (kylix) and 68 (conical cup), had C, O, and C:O ratios that matched well with those of modern 'gilded' tin. Thus this means that the Mycenaeans knew how to oxidize tin to a colour like gold and did so. Yet, what about the rest of the seven vases. A review of the other results *(Fig. 3)* shows that 49 (kylix), 70 (conical cup) and 73 (conical cup) have C, O, and C:O ratios very similar to those of modern natural tin and quite different from the 'golden' ones. Instantly, parallels with the symbolic use of colours in metals in Mexico, Central and South American cultures leap to mind.

sample	vase i.d.	Si	C	Ca	N400	N407	Sn	O	S	Al
'gilded' tin			80.74				1.55	17.70		
silver tin (metallic tin foil)			[44.04]				17.49	[38.04]		
DARK SURF	39 KSDS	1.09	75.28	0.06	0.37	1.52	---	21.20	0.46	---
WHITE SURF	39 SWS	1.09	60.07	1.69	1.58	2.15	0.06	31.77	1.39	0.21
BROWN SURF	39 SBS	2.33	60.62	2.15	2.09	0.42	0.11	29.48	1.68	1.11
BLACK SPOT	39 FSBS	5.27	49.47	1.28	1.77	0.71	1.49	37.89	1.26	0.86
CLAY SURF	39 FSXS	2.28	60.14	1.48	2.68	0.43	0.08	29.99	1.79	1.14
BROWN SURF	39 FSRS	1.20	69.07	0.83	2.38	0.51	0.07	24.33	0.75	0.86
SHINY SPOT	39 FSGS	2.19	67.44	1.24	2.92	0.15	0.28	23.25	0.84	1.69
BROWN STRIP	41 SBS	0.87	60.92	0.61	0.97	4.40	---	31.21	0.42	0.56
LIGHT SURF	41 SLS	1.81	60.99	0.99	2.58	1.72	0.13	29.27	1.21	1.30
VERY DK SURF	44 ASDS	1.28	72.72	0.51	2.22	1.36	0.15	20.73	0.51	0.51
CLAY SURF	44 AUSS	1.49	67.48	0.76	3.11	0.70	0.08	24.57	0.79	1.02
BLACK SURF	44 ASNP	0.99	69.86	0.31	1.57	1.82	0.07	24.18	0.84	0.35
BLACK SURF	44 BSFR	0.70	77.31	0.42	1.11	1.31	---	18.26	0.87	---
BLACK SURF	44 BSNR	0.45	74.89	0.59	1.81	1.21	0.04	19.63	0.78	0.14
LIGHT SPOT	44 BSLS	0.73	66.19	0.73	1.69	1.12	0.05	27.02	0.91	1.16
DARK,SOME SH	[49] ISDS	0.70	[54.89]	0.16	0.30	7.38	---	[35.46]	0.51	0.52
DARK,NO SHINE	[49] ISS	0.47	[59.73]	0.14	0.32	5.48	0.05	[33.44]	0.38	---
GREY SURF	[49] ISGS	2.01	[53.81]	1.01	0.87	5.05	0.16	[35.54]	1.55	---
CLAY SURF	[49] ISCS	3.16	[52.97]	1.83	1.76	1.40	0.19	[35.79]	2.17	0.73
DARK SURF	[49] TSDS	0.47	[56.49]	0.32	---	6.46	0.09	[35.30]	0.46	0.40
BROWNISH	[49] TSBS	1.00	[63.70]	0.17	0.19	3.30	0.09	[30.43]	0.08	1.03
YELLOW AREA	68 SBS	0.35	75.38	0.22	0.67	0.21	0.03	23.09	0.03	0.02
YELLOW SHINY	68 SYS	1.03	73.95	0.25	0.36	0.35	0.01	23.95	---	0.10
GREY SURF	68 SGS	1.81	61.74	0.61	2.97	2.03	0.13	29.11	0.29	1.30
CLAY SURF	68 ISCS	2.05	65.45	0.51	1.87	2.17	0.07	26.67	0.66	0.54
WHITE SPOT	68 ISWS	2.92	64.19	0.65	3.43	1.21	0.48	26.55	0.20	0.87
SHINY SURF	[70] SGS	1.35	[58.33]	0.50	0.33	5.35	0.10	[33.77]	0.17	0.11
YELLOW SPOT	[70] ISYS	0.58	[54.79]	0.11	---	7.51	---	[36.52]	0.06	0.38
DARK SURF	[70] ISDS	1.53	[58.81]	0.26	0.38	4.60	0.14	[33.14]	0.46	0.68
LIGHT SURF	[70] ISLS	0.89	[55.93]	0.41	0.11	3.82	0.23	[37.37]	0.61	0.64
DARKRED SURF	[73] SRS	2.69	[49.75]	2.71	1.71	1.38	0.14	[37.65]	3.01	0.96
BLACK SURF	[73] SBS	5.36	[50.93]	0.93	1.74	1.99	0.21	[36.25]	0.89	1.70
RED SURF	[73] ISRS	1.66	[65.97]	1.19	1.65	1.86	0.19	[26.42]	0.75	0.30
GREY SURF	[73] ISGS	2.37	[68.53]	1.01	2.16	0.58	0.22	[22.66]	0.72	1.76

Figure 3. XPS analyses results of 44 samples taken from 7 vases from Asine ChTI:1, plus two modern tin analyses. N400 in the table ≈ 400 eV and N407 ≈ 407 eV. The values for C and O indicated by the boxes as compared with the modern white-colored tin results are considered probable indications that the tin on these vases was silver-colored also.

In using ethnographic, or ethnoarchaeological, examples, it is always tempting to draw great, sweeping parallels. In this case, there are many differences among the Mycenaean culture and the various New World ones: totally different parts of the world with no contact with each other; different times, separated in the most extreme case by more than 2500 years; totally different attitudes toward metals in many cases, and different ranking of them; different uses. However, there are also similarities: the various New World cultures were on a more or less equal level of social development with the Mycenaean one; their social and economic structures were quite similar; they had similar manifestations of prestige and status in the material culture. Thus it is quite possible, that some of the choices which were made could have been the same, despite the obvious, dissimilarities.

In view of the analysis results from the Mycenaean pottery discussed above and seen in Figs. 2 and 3, it seems clear that while almost half of the analyzed tin from the seven vases had in all probability been 'gilded', almost half had not.[19] There are several possible explanations for this. First of all, there is the possibility that the difference is one of time, that in one period all of them were one colour, and in another, the other colour. Another possibility is that certain forms were silver, certain ones 'gold'. However, a check of the seven analyzed vases shows that neither is the case — one 'gilded' (39) and one 'white' (49) kylix belong to an earlier period (Late Helladic IIIA1) while the third one, a 'gilded' one(44), belongs to Late Helladic IIIA2.[20] (For the cups, the shapes are quite conservative, and it is very difficult to divide them into sub periods. They are both golden and silvery (Gillis 1990).) Yet another possibility is that the use of gold and/or silver colours reflects individual choice, either on the part of the producer or of the consumer, representing differences in burial practices or tradition, even among the family or families who used a single chamber tomb. A larger study of the phenomenon will perhaps clarify this point.

Is it possible, however, to see colour symbolism here? Was the use of 'gilded' and 'natural' tin guided by symbolic consideration, that the persons having them in burials wished to manifest not only material, physical wealth or aesthetic preference, but also spiritual wealth and power? How would one go about demonstrating the supposed link between colour and spiritual power? And would the symbolism be different or the same as the documented examples?

It would further the argument, if there were clear indications of such a symbolism in other fields in the Mycenaean world. Lechtman (1977, 1994) discusses the link with textiles, with colours applied on each other creating an effect similar to surface depletion, and Shimade (in press) looks at cross-craft relationships as well; but nothing remains today of Mycenaean textiles, except their representations on frescoes. However, it is perhaps not merely by chance that the yellow saffron-producing crocus was sacred at least on Thera, as seen both by the portrayal of the flower itself in various generally religiously-interpreted contexts and by crocus decoration as embroidery on clothing and jewellery worn by persons considered as goddesses or priestesses.[21] The famous daggers with scenes in gold and silver inlaid in a black mass called niello[22] in the bronze blade could be an example of the symbolic use of colour, as well. Future study will perhaps yield more indications.

Thus, although there is no definite proof for now of the existence of colour symbolism in Mycenaean religious beliefs and even less of its interpretations, the use of colour variation in tin as shown above can be considered an indication, or at least a hypothesis, of the existence of colour symbolism. Even without the added dimension of colour symbolism, tin is a metal that has properties unlike any other - it can be both pure yellow and pure white, with all the inbetween variations, depending on the degree of oxidization. It can metamorphosize, seemingly by magic. In any case, it has magical characteristics that no other metal has. If colour symbolism — yellow and white — existed here, then this would only enhance the magical, symbolic value and status of tin, as well as, perhaps conferring religious prestige and power on the deceased and his/her family .

Conclusion

Metals, and even metallurgy, played an important part in the social, political, economic and religious spheres of the Aegean Bronze Age. Tin had three very important characteristics and filled functions in three different spheres of social life. As a utilitarian metal, it was vital as an alloy with copper to make bronze. Further, when used in foil form to wrap around vessels in rich, prestigious burials, it transmitted a message of socio-economic importance, as it was rare and imported from afar and imbued status on the person powerful enough to have a surplus. On a more abstract level, the colour permutations, that only tin can achieve, gave it a magical power that no other metal had. There is perhaps yet another dimension, that of symbolic meaning. It is possible that there was also a symbolic understanding in the colours of yellow and white, symbolizing perhaps the sun and moon, life and death. If this were the case, it would give the tin and its user a spiritual prestige and power beyond the socio-economic status. For these reasons, tin was of enormous importance in the Aegean Bronze Age, and especially, the latter part of the Late Bronze Age.

Tables 1 & 2 reprinted from "Aegean Bronze Age Tinned Vessels: Analyses and Social Implications" in *Ceramics Cultural Heritage* Vincenzia (ed.), 1995. Reproduced by permission of the author.

References

BASS, G.D.. 1961. The Cape Gelidonjy Wreck: Preliminary Report, *AJA* 65:267-276

1967. *Cape Gelidonya, a bronze age shipwreck* (TAPS57:8), Philadelphia.

1986. Bronze Age Shipwreck at Ulu Burun (Kas): 1984 Campaign, *AJA* 90:269-296.

1991. Evidence of trade from Bronze Age Shipwrecks, in N.H.Gale (ed.) Bronze Age Trade in the Mediterranean (SIMA 90), Paul Åström's Förlag, Jonsered: 69-82.

BENTON, S. 1964. No Tin from Kirrha in Phokis, *Antiquity* 38:138.

BOYE, VILHELM.1896 (reprinted 1986). *Fund af Egekister fra den Ældre Bronzealder.*

CASKEY, M. *et al.* 1988. Metals in Keos: a first approach, in P.G.Marinos & G.C.Koukis (ed.) *The engineering geology of anient works, monuments and historical sites*. Vol. 3. Rotterdam:1739-1745.

CLEUZIOU, S. & BERTHOUD, F. 1982. Early Tin in the Near East,*Expedition* 25/1:14-19.

DE JESUS, P. 1978. *The Development of Prehistoric Mining and Metallurgy in Anatolia* (B.A.R. 74 i, ii), Oxford.

DOUMAS, C. 1992. *The Wall -paintings of Thera.* Athens.

FORSÉN, J. 1992. *The Twilight of the Early Helladics* (SIMA-PB 116), Jonsered: Paul Åström's Förlag.

FRANKLIN, A, WERTIME, T. & OLIN, J. (ed.) 1978.*The Search for Ancient Tin*. Washington.

FRÖDIN, O. & PERSSON, A.W. 1938. *Asine. Results of the Swedish Excavations 1922-1930.* Stockholm.

GALILI & SHMUELI 1983. notes in *INJA* 12.2:178.

GALILI, SHMUELI & ARTZY 1986. Bronze Age ship's cargo of copper and tin, *INJA* 15.1:25-37.

GILLIS, C. 1990. *Minoan Conical Cups - Form, Function and Significance* (SIMA 89), Jonsered.

1991. Tin in the Aegean Bronze Age , *Hydra* 8, Athens: 1-30.

1992. All that Glitters is not Gold in *Medelhavsmuseets Bulletin* 26-27 for 1991-1992:24-32 .

1994. Binding Evidence, *OpAth* 20:57-61.

1996a. The Asine Chamber Tombs- Graves of Kings? in E. di Miro, L. Godart, A. Sacconi (eds.) *Atti e memorie del secondo congresso internazionale di Micenologia Roma-Napoli, 14-20 ottobre 1991* (Incunabula Graeca 98:3): 3, Rome,:1193-1203.

1996b. Tin at Asine, in R.Hägg, G.C. Nordquist & B. Wells (eds.) *ASINE III Supplementary Studies on the Swedish Excavations 1922-1930* fasc. 1 (SkrAth 4°, 45:1), Stockholm, 93-100.

GILLIS, C. & BOHM, C. 1994. Tin-Covered Vessels in the Aegean Bronze Age: Methods of Application in F. Burragato, O. Grubessi & L. Lazzarini(eds.)*1st European workshop on archaeological ceramics,* Rome: 26-34.

GILLIS, C. & HOLMBERG, B. 1995. Tin-coated Ceramic Vessels in the Aegean Late Bronze Age: Results and Implications from ToF SIMS and ESCA Surface Analyses in M.Vendrell-Saz, R. Pradell, J. Molera & M.Garcia (eds.) *Estudis Sobre Ceràmica Antiga: Acte del simposi sobre ceràmica antiga,* Barcelona:35-40.

GILLIS, C., HOLMBERG, B. & WIDELÖV, A. 1995. Aegean Bronze Age Tinned Vessels: Analyses and Social Implications, in P. Vincenzini (ed.) *The Cultural Ceramic Heritage*, (Monograph in Materials and Society 2), Faenza: 251-260

HÄGG, R., NORDQUIST, G.C. & WELLS B. (eds.)*ASINE III Supplementary Studies on the Swedish Excavations 1922-1930* fasc. 1 (SkrAth 4°, 45:1), Stockholm.

HOSLER, DOROTHY, 1988. Ancient West Mexican Metallurgy: South and Central American Origins and West Mexican Transformation, *American Anthropologist* 90:832-855.

1994. *The Sounds and Colors of Power*, Cambridge USA.

1997. E-mail communication on arch-metal discussion group. 14 February 1997.

IMMERWAHR, S. 1966.The Use of Tin on Mycenaean Vases, *Hesperia* 35:381-396.

LAFFINEUR, R. & BOSS, M. 1997. Mycenaean Metal Inlay: A Technique in Context, in *Tachne, Craftsmen, craftswomen and craftsmanship in the Aegean Bronze Age* (Aegaeum 16), Phildelphia: 191-197.

LAMB, W. 1936. Excavations at Thermi, section on metals by C.D., *BSA* 30, 1928-1930:1-52

LECHTMAN, H. 1977. Style in Technology — Some Early Thoughts, in H. Lechtman & R. Merrill (eds.), *Material culture : Styles, Organization, and Dynamics of Technology,* New York.

1994. The Materials Science of Material Culture: Exmples from the Andean Past, in David Scott (ed.), *Archaeometry of Pre.-Columbian Sites and Artifacts.* Getty Conservation Institute

MARINATOS, N. 1984. *Art and Religion in Thera*. Athens.

MOLTESSEN, M. 1988. A Group of Late-Etruscan Silver-imitating vases, in Melander *et al* (eds.) *Ancient Grrek and Related Pottery*, Copenhagen:435-444.

MOUNTJOY, P. 1996.Asine Chamber Tomb I:1: the pottery, in R.Hägg, G.C. Nordquist & B. Wells (eds.)*ASINE III Supplementary Studies on the Swedish Excavations 1922-1930* fasc. 1 (SkrAth 4°, 45:1), Stockholm, 47-67.

MUHLY, J.D. 1973 + 1976. *Copper and Tin* and *Copper and Tin supplement.* Hamden, Ct. USA

1985. Sources of Tin and the Beginnings of Bronze Metallurgy, *AJA* 89:275-291.

PENHALLURICK, R.D. 1984. *Tin in Antiquity*. London.

PHOTOS-JONES, EFFIE. in press.

PRIMAS, M. 1985. Tin Objects in Bronze Age Europe, in *Studi di Paletnologia in Onore di Salvatore M. Puglisi*: 555-562.

PULAK, C. 1988. The Bronze Age Shipwreck at Ulu Burun, Turkey: 1985 Campaign, *AJA* 92:1-37.

RENFREW, C. & CHERRY, J. 1985. The finds, *Archaeology of Cult.* London: 299-359.

SELLING, D. 1951. Problem kring Vikingatida keramikkanner, *fornvännen* 46:279-295

1955. *Wikingerzeitliche und frühmittelalterliche Keramik in Schweden.* Stockholm.

SHIMADA, I. *et al.* 1997. *Sicán - ein Fürstengrab in Alt-Peru* (catalogue from Museum Rietberg, Zurich)

in press. Sicán Metallurgy and Cross-craft Relationships, in*Boletín del Museo del Oro,* Bogotá

STJERNQUIST, B. 1958. Ornamentation métallique sur vases d'argile, *Meddelande från Lunds Universityt Historiska Museet,* Lund: 107-169.

TYLECOTE, R. F. 1992. *A History of Metallurgy* 2, London: Institute of Metals.

WACHSMANN, S. & RAVEH, 1981. An Underwater Salvage excavation near Kibbutz ha-Hotrim, Israel, *INJA* 10.2:160.

1984. Concerning a lead ingot fragment from ha-Hotrim, Israel, *INJA* 13.23:169-176.
WYSS, R. 1972.*Bronzezeitliches Töpferhandwerk* (=aus dem Schweizerische Landesmuseum 31), Switzerlnd.

---

[1] "But metal was unknown at Teotihuacán: not one of the varied subsistence endeavors, construction activities, military undertakings, or religious and ceremonial events at this great urban center was carried out using metal implements, metal weapons, or metal religious or ceremonial objects." (Hosler 1994:11). I wish to thank Georgia Nakou for informing me about this book by Hosler.

[2] As the rocky geography of the landscape made transportation much easier by water than by land, and as a consequence much of the Aegean exchange activity was maritime, to the various islands and farther abroad, it is natural that the Mediterranean Sea would unite rather than divide, and would naturally be the hub, the centre, with the various coastal regions the perimeter of the wheel. The word 'Mediterranean' and similar constructions in the Romance languages mean 'in the middle of the land, the earth'; in the Scandinavian languages and German 'Medelhav' or 'Mittelmeer' means 'the sea in the middle'.

[3]Excavations in the Franchthi Cave, in the Greek Peloponnese, indicate that people were paddling from the Greek mainland to the island of Melos to get obsidian in the 8th M. BC Although various routes for the spread of agriculture and animal husbandry heralding the 'neolithic revolution' have been proposed — north to south, northeast to southwest, or a general east to west, as most recently discussed by C. Doumas, M. Marthari and T. Marketou, and P. Sotirakopoulou, respectively at the 'Aegean in the Neolithic, Chalcolithic and Early Bronze Age' symposium held at Urla, Turkey, in Oct. 1997 — the common denominator is maritime communication.

[4] My thanks to the anonymous kind referee who provided me with this reference and gave so many constructive suggestions.

[5] See also Muhly 1973+ 1976: ch. II:4. Naturally the existence of a source does not mean it was exploited during the BA.

[6] Perhaps not a pure coincidence that finds of moulds and crucibles were made at the site (Lamb 1936:37). Nor is it without significance that in the same horde as the tin bracelet was also found a bracelet of iron (!) with copper wire wound around it (eadem :39, fig. on 38).

[7] There is also a mention of a tin pilgrim flask from or found in Egypt, but I lack a reference for this.

[8] It is not sure whether there was one wreck that jettisoned part of its cargo first or that was spread over a great distance, or two different wrecks. See a summary of the evidence in Gillis 1991: 13.

[9] In personal communication with Mr. Robert Heath at the Metals in Antiquity conference, he told me that he had made an object from tin that he had on his mantel piece. After about 2 years, it had dissolved into a pile of powder.

[10] A chemical change in the composition of the tin caused by exposure to temperatures under 13°c. for a relatively short period of time, perhaps two-three months. The result of this change is similar to that of oxidation, causing the metal to lose its lustre and become matte and flaky, or powdery

[11] There is a very wide range in both time and space of this occasional practice, from the 13th c. BC lake dwellers in Switzerland (Wyss 1972 and recent works by Callista Fischer), through La Tène, northern Italy (e.g., Moltessen, M. 1988), Hungary, up to 9th c. AD Birka in central Sweden (Selling 1951,1955), to name some. See also Stjernquist 1958; Primas 1985.

[12] My sincere thanks to Dr. Thomas Chase, Washington , D.C., for taking the time and interest to show these to me at the Sackler and Freer Galleries, Smithsonian. Although the coating has not been analysed as yet, the appearance is quite similar to those from the Aegean.

[13] Many thanks to Prof. Izumi Shimada for sharing his knowledge with me and for sending me his catalogue and manuscript, Shimada in press.

[14] N, released from decomposing bodies, was found on the outside of the tin and on the naked clay surface of the vase, but not underneath the tin, i.e., between the tin and the clay fabric. Thus the N could only have attached to the clay surface if the tin were already gone.

[15] I thank Niklas Hillbom, Dept. of Classical Studies, Lund, for pointing this out to me and for his comments on an earlier version of this article.

[16] "The interest in these colours [gold and silver], and the association of gold and silver metal with the sun and moon, were clearly metallurgical preoccupations that had become pan-American in scope." (Hosler 1988:849)

[17] In Western Mexico, the gold and silver colours were achieved by mixing tin or arsenic with copper (Hosler 1988:850) while *tumbaga,* an alloy of copper and gold, or copper, silver and gold was used often in Central and South American. The desired colour was achieved either through surface depletion or enhancement on sheet metal, as in Ecuador and Peru, or enrichment techniques on artefacts made of alloys such as Cu-Au, Cu-Au-Ag, or Cu-Ag as in lower Central American and Columbia through the surface depletion technique (e.g., Lechtman 1994:6-7; Hosler 1988:849). West Mexico was unique in using bronze alloys (CuSn or CuAs). My anonymous referee also noted that regarding colour transformations, ..."the great variety of colours that can be produced from CU-based alloys as well as by tin alone. The addition of arsenic, antimony, tin, with or without heat/surface treatment, [produces] colours from copper-red through to yellow, white, black and purple..." I thank him/her again.

[18] All numbers given here are prefixed in the Medelhavsmuseet with MM Acc. 10xx: e.g., MM Acc. 1039.

[19] All these vases come from the same chamber tomb, which had been in use for several hundred years. For the tomb, see Frödin & Persson 1938 and Hägg et.al. 1996.

[20] Mountjoy 1996.

[21] In specific the frescoes in Zesté 3. See e.g., Marinatos 1984 and Doumas 1992. On the fresco in the doorway between rooms 4 and 5, West House, a girl holding a censor is identified by many as a priestess because of her saffron yellow clothing (albeit circle reasoning).

[22] For a discussion about the so-called niello, see Laffineur & Boss 1997, and Jones *et al.* in press. I thank Prof. Eva Rystedt, Lund, for reminding me about these daggers and for reading and commenting on the ms. in an earlier stage.

18

Metals Technology versus Context in Late Minoan Burials

E. BABOULA[1] & P. NORTHOVER[2]

[1]*Lincoln College, Oxford OX1 3DR, UK*
[2]*Department of Materials, Parks Road, University of Oxford, Oxford OX1 3PH, UK*

Bronze artefacts from the Minoan Collection in the Ashmolean Museum, Oxford were sampled and examined metallographically. The production history of the samples amends the picture of the Minoans as skilful craftsmen with a good knowledge of their alloys. Alloys and manner of working were combined to suit the function of the produced artefact, whether a weapon or an object of every-day household use.

Introduction and archaeological background

This paper reports an analytical project involving samples of bronze artefacts belonging to the Minoan collection of the Ashmolean Museum in Oxford. Minoan metalwork has had its share in copper provenance studies (Gale, this volume), although physical examination has been intended mainly for determining issues of composition (e.g., Catling & Jones 1977). Metallurgical evidence relating to the production history of the artefacts has until now been sought in a very small number of studies. Such work has involved tools at the Minoan collection of the Ashmolean Museum (Northover & Evely 1995) and the examination of a dagger, a strip and a needle fragment from Gournia (Charles 1968; Betancourt *et al.* 1978).

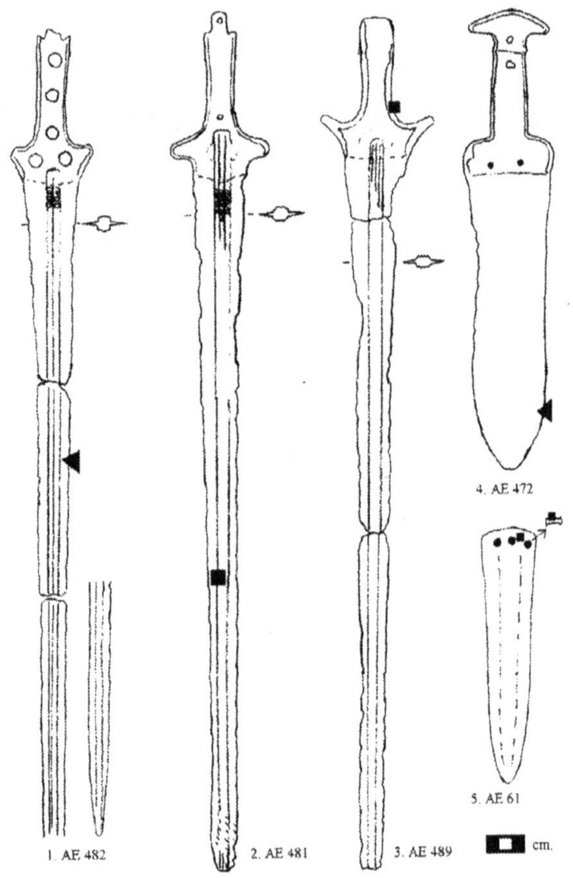

***Figurel. Swords** (Evans 1906; Evans 1921-1935, vol. IV:849; Sanders 1963; Kilian-Dirlmeier 1993).*

The artefacts sampled and presented here are largely derived from the Late Bronze Age cemetery of Zapher Papoura, one of the burial sites adjacent to the administrative centre of Knossos. The use of the cemetery began after the destruction and discontinuing of most of the Late Bronze Age I sites on the island, including the handful of palatial structures. Knossos survived the turmoil however, until LMIIIA1/2 (early 14th C. BC) or the end of LMIIIB (end of 13th C. BC) according to the on-going debate. Burials at Zapher Papoura continued until LMIIIB, although most of them were earlier. Characteristic of the site are single burials with rich goods accompanying them mostly in the earlier stages of use, whilst 'family' burials of two to four people with generally fewer goods are also present. Few of the artefacts here can be matched with specific graves.

Sampling

Cut samples and drillings for further metallographic and elemental analysis were taken from the less severely corroded (less severely conserved) objects. A jeweller's piercing saw with a blade of 32 teeth per cm. and a handheld modelmaker's electric drill with a 0.9mm or 0.7mm diameter bit were used for the cut samples and the drillings respectively. Wherever an artefact had been sampled before for composition and provenance analysis (Craddock 1976; Gale & Stos-Gale 1982:16) the drilling was made inside the previous cavities as far as possible. Cut samples are recorded as black triangles on the following drawings, drillings as black squares. What is presented here is based on the metallographic analysis of the cut samples in both the as polished and etched states, together with brief observations made on the surfaces of the objects through a binocular microscope.

Results

The metallographic data are discussed according to object typology. The numbers prefixed by AE are Ashmolean Museum catalogue numbers.

The cut sample from AE 482 (Fig. 1.1, Plate 1) showed thick corrosion crust largely composed of *cuprite* (Cu_2O); extensive intergranular corrosion with transgranular attack along slip traces and twin boundaries; a very few elongated sulfide inclusions; no coring; fully recrystallized equiaxed grain structure with a large grain size of 50-60 mm. Many grains show slip traces and there is some duplex slip.

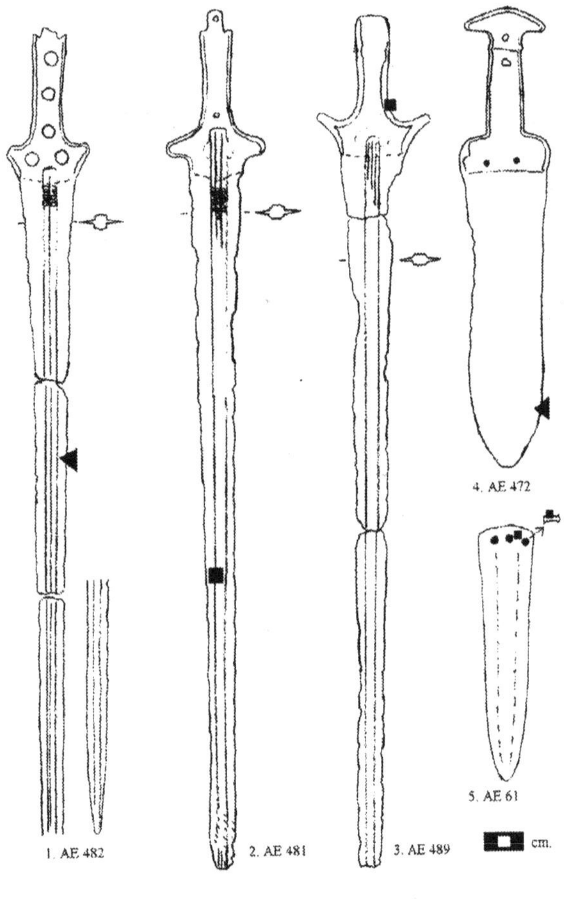

Plate1. Photomicrograph of sample from AE 482 showing fully recrystallised equiaxed grain structure; slip traces and some duplex slip. x 500.

This sword with horn-shaped hilt guard comes from Zapher Papoura shaft grave 44. The weapon is preserved in three pieces, not fitting perfectly together, the breaks possibly being the outcome of 'ritual killing'. Sampling involved cutting a small wedge from the surviving edge at the mid-length of the blade. The large grain size and lack of coring in the blade sample suggests that the sample area has been subjected to rather prolonged annealing and, probably, limited deformation. The extensive corrosion, reflected in the thick crust and deep penetration of the sample means that we do not know how close the surviving metal is to the originally prepared cutting-edge. It is likely that the sample site is a little distance from the original edge and has had less hammering than that edge. Hammering at that location would require a hammer with a small face and would also be limited by the shape of the midrib and its elaborate decorative pattern of two lines of spirals running both along the midrib and along the hilt flanges.

The combination of design and manufacture was to produce a good thrusting weapon and, although restricted by the thick mid-rib, the blade edge would also be effective for cutting. It is hoped that further metallographic samples will eventually be available to see how consistently the method of working this blade is applied to other weapons.

Sword AE 481 (Fig. 1.2) with a cruciform hilt-guard comes from Zapher Papoura Grave 55, a LMIIIA1 pit-cave. Three lines of running spirals, previously unobserved, were traced along the midrib, resembling AE 482. The decoration is very faint due to severe corrosion; however, it is better preserved on the shoulder and handle flanges. Organic remains at the end of the blade may derive from covering the weapon in a piece of cloth or sheath.

Horn-shaped sword AE 489 (Fig. 1.3) is one of the weapons from the Acropolis Tomb, a Knossian warrior grave. The other finds of the group comprise a fragment of a cruciform sword (AE 1564a,b), three spearheads (AE 490-92) and an alabastron. All of the bronzes from this tomb are in the Ashmolean collection. The sword is preserved in three pieces, the two lower ones bent, together forming an S-like shape. This is the result of 'ritual killing', probably through deliberate heavy blows with a hammer.

Dirks - Daggers (Benton 1952; Sandars 1963; Kilian-Dirlmeier 1993).
The cut sample from AE 472 (Fig. 1.4, from Zapher Papoura) proved to be completely corroded, mostly to *cuprite*. Sufficient ghost structure remained to suggest a homogeneous, fully worked and annealed blade. There are remains of the organic handle around the rivets, but the corrosion has obscured any visible traces of use. The shape of the blade suggests that the dirk is suitable for both cutting and thrusting, the use of bronze providing sufficient strength and a wrought and annealed blade sufficient toughness.

Lato-Goulas in eastern Crete is mentioned as the provenance of dagger blade AE 61 (Fig. 1.5). It seems to have been used for cutting on both sides but the wear and sharpening marks are more prominent on one face than on the other. AE 61 is of the same Neopalatial type as the longer dagger blade examined by Charles (1968).

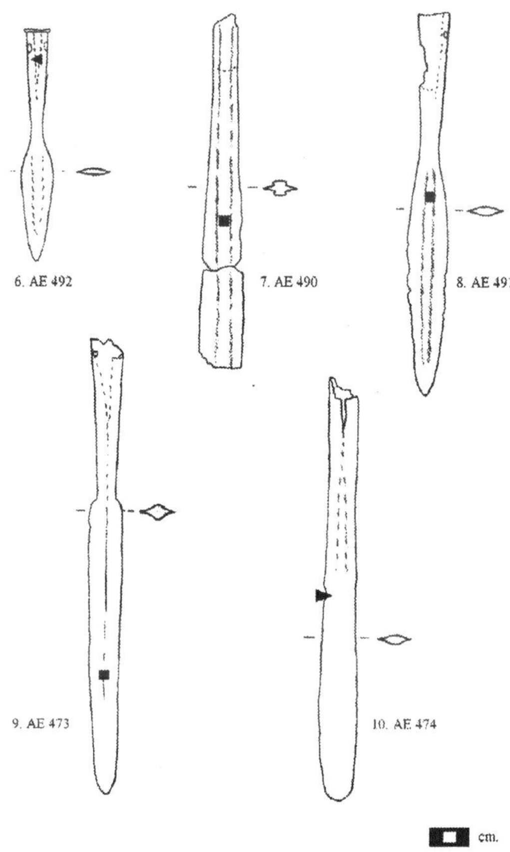

Figure 2. Spearheads *(Sandars 1963; Höckman 1980).*

The cut sample from AE 492 (Fig. 2.6, Plate 2) was taken from the edge of the split in the socket and so will reflect the manufacture of the socket rather than the blade. The surface is deeply pitted with some material loosely adhering which appears to be a mixture of soil and corrosion products. There is some sub-surface penetration by corrosion and scattered copper sulfide inclusions. The sample was difficult to etch but could be shown to have a

Plate 2. Photomicrograph of sample from AE 492 showing scattered copper sulphide inclusions; fully recrystallised grain structure with annealing twinn; no coring. x 500.

Plate 3. Photomicrograph of sample from AE 474 showing elongated sulphide inclusions; deformed fully recrystallised equiaxed grain structure with annealing twins and slip traces; no coring. x 500.

fully recrystallized grain structure with annealing twins; slip traces were not etched but some grain outlines were distorted implying that the socket was left cold worked after, presumably, final closure round the shaft. There was no coring so that annealing time and temperature had been sufficient for homogenization.

Two shiny spearheads in the collection, AE 492 and AE 474 (Fig. 2.10, Plate 3) owe their colour to conservation. Fragmentary AE 490 (Fig. 2.7) is preserved in two pieces, a possible result of ritual 'killing'. This weapon and AE 491 (Fig. 2.8) are heavily corroded 'one-piece spearheads', a type common at Knossian cemeteries of LMII-IIIA1 date. The size and shape of AE 473 (Fig. 2.9) would make it awkward for throwing but even with the rounded tip it could be used for thrusting. From an unknown grave of Zapher Papoura, AE 473 is close to Catling's type d (Catling 1964:119-20).

The cut sample from AE 474 presents a cross-section of the cutting-edge which cleaning has stripped of any surface corrosion products. There is some internal corrosion, both inter- and transgranular. Sulfide inclusions are strongly elongated parallel to the blade surface and point to a large (>50%) reduction in thickness from the original cast edge. Etching revealed a deformed fully recrystallized equiaxed grain structure with annealing twins; grain diameter = 20 - 25 mm; there is no coring. Most grains show slip traces which increase in density towards the surfaces and cutting-edge.

AE 474 comes from an unidentified grave at Zapher Papoura. With a length of c. 33 cm., the general shape would appear suitable for thrusting despite the rounded tip. The cutting-edge is well worked and annealed and left cold worked. In section the edge is now slightly rounded but the original profile has plainly been lost in cleaning. The style of working is very similar to that in knife AE 1831 (see below) and represents the norm for edged blades.

Although three spearheads in this collection comprise a closed group from the Acropolis Tomb it is noticeable that they are typologically different from each other, either because a fine line separates their serving different needs before interment in the tomb or more probably because spearheads, like swords, were highly personalized weapons.

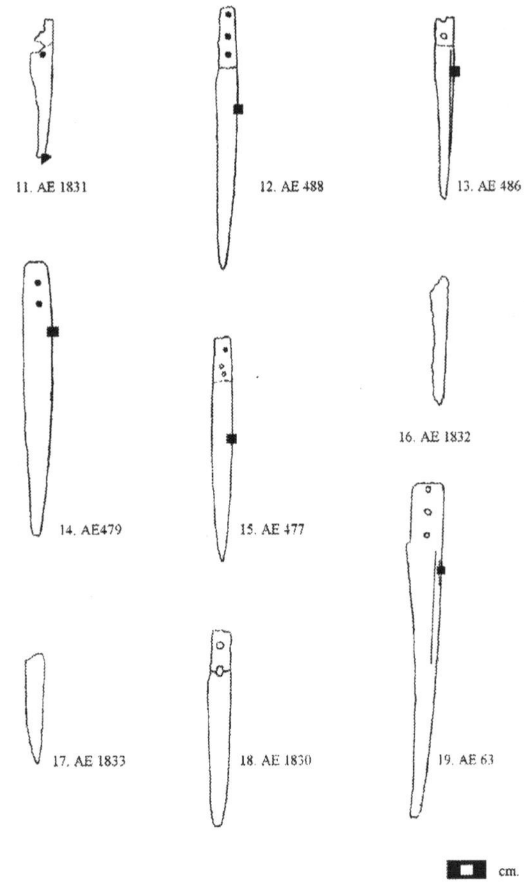

***Figure 3. Knives** (Evans 1897:356 no. 3; Evans 1906; Sandars 1955).*

The cut sample from AE 1831 (Fig. 3.11, Plate 4) presents a cross-section through the cutting-edge, at the tip of the preserved blade; the surface is heavily corroded with a band of inter- and transgranular corrosion merging into a solid layer of cuprite; there is some delamination. There are numerous sulfide inclusions heavily elongated parallel to the blade surface and marking the direction of working. Etching revealed a deformed, fully recrystallized equiaxed

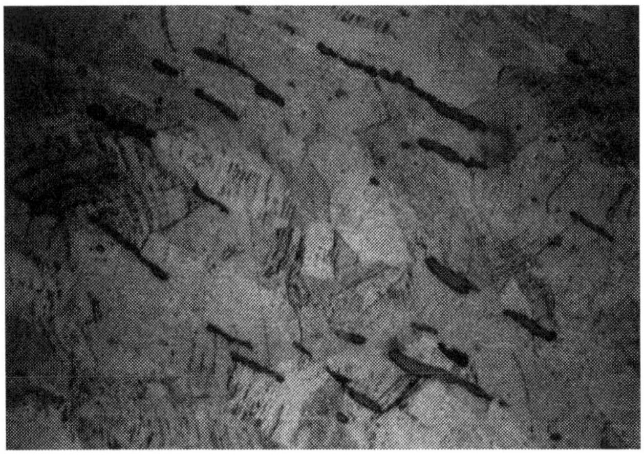

Plate 4. Photomicrograph of sample from AE 1831 showing elongated sulphide inclusions; deformed, fully recrystallised equiaxed grain structure with annealing twinn; some duplex slip. x 500.

Plate 5. Photomicrograph of sample from AE 1831 showing penetration by inter- and transgranular corrosion; fully recrystallised, equiaxed grain structure; no coring. x 500.

grain structure with annealing twins; grain diameter = 20 - 25 mm. The deformation is more intense close to the surface where deformed grains and duplex slip point to a cold reduction of 30 - 35%.

Knife AE 1831, probably from a Zapher Papoura burial, looks flimsy but, with a small grain size and not over-worked at the edge, would form a practical tool. There are marks of use on the blade. Visible black lines across the surface of the blade reflect the elongation of grains occurring during hammering or they may be an effect of the burial environment or the conservation procedure.

Knife AE 488 (Fig. 3.12) from Zapher Papoura pit-cave no. 64, was certainly a small utilitarian knife, since the edge of its blade has very visible wear signs. In places the marks from the conservation go under the corrosion. The use of copper could have been for the colour contrast, for a more easily formed and closed rivet or, was the result of a practical smith making rivets of whatever odds and ends of suitable metal were to hand.

Very clear wear and polishing signs are visible on most of the blades. AE 1832 (Fig. 3.16) and AE 1833 (Fig. 3.17) are partly preserved and completely mineralized. AE 1830 (Fig. 3.18) is severely corroded and is broken across the rivet hole which is closer to the blade. The break has resulted in two different colours and corrosion patterns on the two preserved parts, an outcome either of the micro-environment in burial of the two pieces separately or even of differences in conservation. All of the knives derive reportedly from Zapher Papoura except from AE 63 (Fig. 3.19) which comes from a tholos tomb at Kamares.

The cut sample from AE 1835 (Fig. 4.21, Plate 5) is very heavily corroded and almost all included areas of apparently sound metal are penetrated by inter- and transgranular corrosion which effectively etch a fully recrystallized, equiaxed grain structure with a medium grain size (30 - 35 mm) and a moderate amount of cold work. Etching confirmed this and also highlighted elongated sulfide inclusions. There was no coring although it is suspected that some residual segregation of impurities has an effect on the intensity of corrosion.

The sample was taken from the existing but perhaps not original edge. More care has been invested in its shape than in other razors, with grooved lines (four in all) parallel to the non-cutting edge. The metal appears to be folded at the edge of the razor and it is possible that a small area of metal was hammered back to the main body after casting. The metallography indicates the same good, basic standard of working edges as already noted for spearheads and knives.

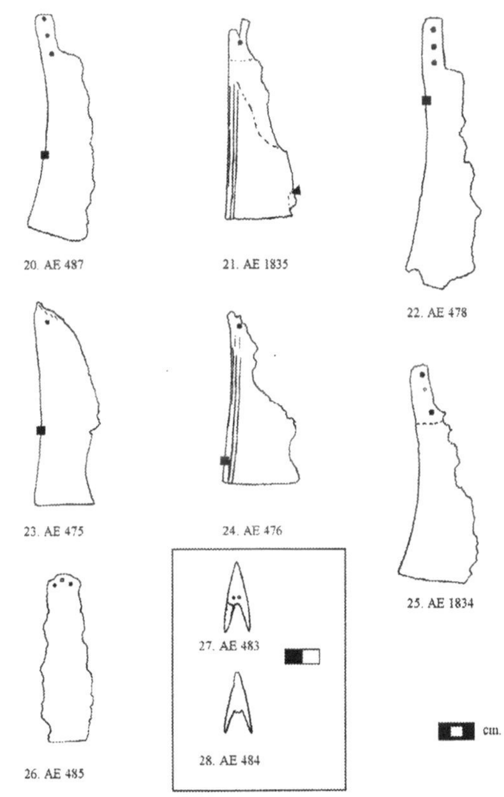

***Figure 5. Razors** (Evans 1906).*

Only razor AE 487 (Fig. 4.20) has a known context, Zapher Papoura grave no. 4, while the other seven razors from the series possibly come from the same cemetery but from unidentified burials. As in the case of AE 487 the rivets of AE 478 AE (Fig. 4.22) get slightly longer at the base of the handle, presumably to allow for some tapering of the handle plates in order to make the handle stronger and easier to hold. The edges of AE 478 and AE 475 (Fig. 4.23)are not preserved and they are heavily corroded, so it is not possible to discern any traces of use. Care for the shape of AE 476 (Fig. 4.24) has been taken as in AE 1835. There are marks of use very close to the end of where the cutting edge would be and it has the same finish as knives. AE 1834 (Fig. 4.25) is very corroded while AE 485 (Fig. 4.26) is a completely mineralized and heavily glued leaf-shaped razor.

There are two types of razors in the collection, corresponding to Sandars' two basic types (Hood *et al.* 1958-59:234-5; Hood 1956:96-97): those with a straight back and a generally straight base, and those with a convex-concave form of their cutting and non-cutting edges. Their function is uncertain although distinct uses may be implied by the different forms (Popham *et al.* 1974:245). In addition there may be a differentiation in use between 'leaf-razors' and razors since the two types are almost mutually exclusive in burials.

Among the flat blades the knives form a group of their own due to their excellent preservation in relation to the rest of the material and to the very clear marks of polishing and use. They do not seem to have had standard different compositions from the rest of the material, so their good state of preservation may be due to a modified micro-environment as compared to the rest of the types inside the grave or to a relatively lower tin content as compared to the bigger weapons. After the knives some of the razors are the next well preserved objects. Although their thin blades are more sensitive to corrosion than the other artefacts, they appear to have had smooth surfaces and certainly useable edges. The spearheads are either too much corroded or too much conserved while the swords and dagger are thoroughly corroded.

If a general division can be made about the conditions of burial of the different types of artefacts we have to note that at least one of the swords in this group preserves organic remains, either from the burial or from a sheath which covered them. Sheathing is not unusual in the Aegean or Crete specifically. If the more practical every-day objects such as small knives and razors did not require sheathing, it could help explain the general difference in the preservation and corrosion patterns.

Arrowheads (Evans 1906:32).
The two cast arrowheads (AE 483, Fig. 4.27; AE 484 Fig. 4.28) in our group, from the Hunter's Grave (no. 10) at Zapher Papoura, are both completely corroded. Their shapes are characteristic for the Late Bronze Age, one pierced with two holes and the other tanged.

Vessel (Evans 1906:89; Matthäus 1980:99)
A previous sampling of this heavily reconstructed pan (AE 494) has demonstrated a tin content of 9.1% (and 1.1% lead), a figure only indicative of a bronze composition however, since it was apparently measured on the corrosion products. Hammered out of a cast blank with repeated anneals and with cast handles attached to the body, it is a piece of standard technology. The vessel comes from Zapher Papoura chamber tomb 99, apparently a family burial rich in offerings and dated to LMIIIB.

Mirror (Evans 1906:90).
The bronze of the cut sample (AE 480, Plate 6) was completely corroded but contained sufficient ghost structure to show that it had been cold worked and annealed from a cast blank; there was insufficient evidence to indicate how close to the final form the blank was cast. The mirror plate was found at Zapher Papoura chamber tomb 99 (see above). The type is the typical Aegean circular plate, but bearing four holes instead of the common two for the attachment of the handle.

Plate 6. Photomicrograph of sample from AE 1835 showing completely corroded microstructure; ghost structure remaining. . x 500.

Fibulae (Boardman 1961:90, 92).
The final two objects (AE 122a-1891.665; AE 122b-1891.655) are both Geometric in date, come from a possible burial context (Kissamos), and both show the same technology in manufacture. Soft black marks lie over the decoration and patina; their nature has yet to be determined.

Discussion
The limited amount of metallographic investigation prior to this study has already demonstrated the levels of metallurgical skill available in Late Minoan Crete. The application of silver plating on the dagger from Gournia examined by Charles (1968) was described by him as a highly controlled procedure and the best way both of using the silver and for creating the bonding between copper and silver.

The two objects which Betancourt and colleagues examined (Betancourt *et al.* 1978), a strip and a needle, were impure copper with trace levels of tin. They were both cast, coldworked and then annealed, but with no indication of cold working after the final stage of annealing. The study made by Northover & Evely (1995) showed that all their sampled tools had been worked after casting by combining cold forging with annealing, so that their toughness was ensured.

Our own metallographic examination of the weapons and personal objects at the Ashmolean Museum demonstrated a competent standard of working for all of the samples. The metal smiths had a good understanding of the alloys they were using, making specific choices of alloy and production style to suit particular artefacts to particular functions, definitely differentiating weapons as objects in need of special skills. One difficulty in assessing the actual degree of differentiation is the condition of the objects and the uncertainty of comparing like with like when the original cutting-edge of an object no longer survives. The majority of objects examined by us, the spearheads and flat blades, required only simple casting skills to produce a near-shape blank. The forging skills on all objects with sound metal were of a similar order with the metal well homogenized, consistent annealing practice, and no over-hardening by cold work. The difference in microstructure in the sword blade (AE 482) with its larger grain size has a simple origin, either due to the location of the sample site, or equally to the habits of a different smith. The swordsmith's skills in forging were nonetheless of a high order because of the narrowness of the blade and the thickness of the mid-rib. In contrast the hammering of a simple flat blade was much simpler.

There is nothing here to contradict the suggestion that has been made that the swords, spearheads and knives could have been made in the same workshop (e.g., Macdonald 1987; cf. Gillis 1997). Certainly the swordsmith would have had all the skills and knowledge to make those simpler objects but the converse need not be true, that is the skill-set required to make a knife blade would not necessarily enable the smith to make a sword. The difference lies in the casting technology and the ability to make and fill a long, slender mould with a complex section rather than casting a simple flat blank. The applied decoration enhances the difference. It would be an additional advantage if the pattern and mould makers achieved a mould for the sword that was very close to the finished shape, reducing the need for forging to produce the necessary finish to both edge and surface.

Another way of differentiating between weapons, especially as symbols of prestige, and other blades is in the choice of alloy. We have already noted the use of higher tin bronze alloys in a number of weapons. The compositions with around 11% tin offer a good combination of hardenability and toughness while being attractive in colour. Further, their long freezing range would assist in casting. For the simple, forged shapes a lower tin content would still be practical although the lower hardenability would mean some increase in final cold work to achieve a given hardness. If tin, or imported higher tin bronzes, were in restricted supply this discrimination would natural enhance the perceived value of the weapons without interfering with the practicality of the smaller items.

Another purpose of our physical examination of the metal artefacts was to determine whether they had been used or had been interred with the dead bodies without ever having fulfilled a practical purpose. The weapons, although corroded and therefore not preserving signs of use, were worked so that they were effective. In this sense, they do reflect the militaristic values of their owners. The spearheads show the greatest variation in typology, while types like razors and knives conform to basic shapes, maybe confirming utilitarian, non-weaponlike functions. It is also clear that the knives and razors have been used: they are neither specially made for burial nor worn out and unfit for further use.

There is evidence for metalworking at several sites on Crete, e.g., Gournia, Kommos, Palaikastro, Knossos. However, there is little if any direct evidence for the relative proportions of new metal, whether brought to the site as copper and tin or already in the form of bronze and scrap supplied to these workshops. That there was a structure to the use of metal supplies in LMII-IIIA is marked by the concentration of higher tin alloys in specific items, notably weapons. It could be, then, that the alloying of copper and tin was reserved to certain workshops making these weapons, with the corollary that recycled metal was excluded from them. Equally, the tin contents could be maintained because these objects were excluded from recycling systems.

Careful control and selection from a scrap metal supply could also maintain the pattern of alloying so that like was always re-made into like and mixing, which is excluded by lead isotope evidence (Gale, this volume) was avoided. Not only could the content of tin vary between different types, it could vary between the different components of a single artefact. An example is the knife AE 488 where the body is bronze and the rivet copper with an impurity level of tin. The low level of tin implies a mixing of copper and bronze scrap as the tin content is too low to be useful. Similarly, the traces of tin in the Gournia samples suggested the use of scrap, some of which was tin bronze. However, there might also have been a distinction between places where metalworking was under the auspices of a palace or a strong local aristocracy and those which would work independently on metal for more practical uses, especially tools.

Considering the context of the sampled objects demands discussion of their social role and value. The issues that they are intricately connected to this include what the objects may be emblematic of in relation to their owners and manipulators, the burial process and its symbolism. It is important to remember that these very objects have as a common denominator the fact that they have been removed from circulation and everyday use, however ceremonial or utilitarian that use may have been. One connecting factor is their material and its 'financial' value. Bronze however, is a conspicuous part of the funerary assemblage but in no way the only one as a rule. Maybe what is more important is the effect of burying metal which means restriction of access to people outside and ultimately inside the burying group.

Northover & Evely (1995) stressed the need for

metallographic studies in the Aegean. We are confident that the physical examination of metal gravegoods in a combination of methods adds to our perspective of the industry as well as social issues connected with the burial customs of Late Bronze Age Crete.

Acknowledgements
We wish to thank Michael Vickers in the Ashmolean Museum for granting permission to study the material from the Arthur Evans collection and especially Mark Norman (Conservation Laboratory, Ashmolean Museum) for his valuable help.

References

BENTON, S. 1952. The Pelynt sword-hilt, *Proceedings of the Prehistoric Society* 18: 237-8.

BETANCOURT, P., T.S. WHEELER, R. MADDIN & J.D. MUHLY. 1978. Metallurgy at Gournia, *MASCA Journal* 1: 7-8.

BOARDMAN, J. 1961. *The Cretan collection in Oxford: the Dictaean Cave and Iron Age Crete*. Oxford: Clarendon Press.

CATLING, H.W. 1964. *Cypriot bronzework in the Mycenaen world*. Oxford: Clarendon Press.

CATLING, H.W. & R.E. JONES. 1977. Analyses of copper and bronze artefacts from the Unexplored Mansion, Knossos, *Archaeometry* 19: 57-66.

CHARLES, J.A. 1968. The first Sheffield plate, *Antiquity* 42: 278-85 (with Appendix by F.H. Stubbings: 284-5).

CRADDOCK, P.T. 1976. The composition of the copper alloys used by the Greek, Etruscan and Roman civilizations, 1: the Greeks before the archaic period, *Journal of Archaeological Science* 3: 93-113.

EVANS, A.J. 1897. Further discoveries of Cretan and Aegean script: with Libyan and Proto-Egyptian comparisons, *Journal of Hellenic Studies* 17: 327-95.

1906. *The Prehistoric Tombs of Knossos*. London: Quaritch (Offprint from *Archaeologia* LIX).

1921-1935. *The Palace of Minos at Knossos*, vols. I-IV. London: Macmillan & Co.

GALE, N.H. & Z.A. STOS-GALE. 1982. Bronze Age copper sources in the Mediterannean: a new approach, *Science* 216: 11-19.

GILLIS, C. 1997. The smith in the Late Bronze Age- state employee, independent artisan, or both?, in R. Laffineur & P.P. Betancourt (eds.), *AEGAEUM* 16. *TEXNH Craftsmen, craftswomen and craftsmanship in the Aegean Bronze Age: Proceedings of the 6th International Aegean Conference, Temple University, 18-21 April 1996*: 505-13. Liège: Université de Liège.

HÖCKMANN, O. 1980. Lanze und Speer im spätminoischen und mykenischen Griechenland, *Jahrbuch des Römisch-Germanischen Zentralmuseums Mainz* 27: 13-158.

HOOD, M.S.F. 1956. Another warrior-grave at Ayios Ioannis near Knossos, *Annual of the British School at Athens* 51: 81-99.

HOOD, S., G. HUXLEY & N. SANDARS. 1958-1959. A Minoan cemetery on Upper Gypsadhes, *Annual of the British School at Athens* 53-54: 194-262.

KILIAN-DIRLMEIER, I. 1993. *Die Schwerter in Griechenland (ausserhalb der Peloponnes), Bulgarien und Albanien*. Stuttgart: Franz Steiner. Prähistorische Bronzefunde IV(12).

MACDONALD, C. 1987. A Knossian weapon workshop in Late Minoan II and IIIA, in R. Hägg & N. Marinatos (eds.), *The function of the Minoan palaces: Proceedings of the fourth international symposium at the Swedish Institute in Athens, 10-16 June, 1984*: 293-5. Stockholm: Paul Åström.

MATTHÄUS, H. 1980. *Die Bronzegefässe der kretisch-mykenischen Kultur*. München: C.H. Beck Prähistorische Bronzefunde II(1).

NORTHOVER, P. & D. EVELY 1995. Towards an appreciation of Minoan metallurgical techniques: Information provided by copper alloy tools from the Ashmolean Museum, Oxford, *Annual of the British School at Athens* 90: 83-105.

POPHAM, M.R., E.A. CATLING & H.W. CATLING 1974. Sellopoulo tombs three and four: two LM graves at Knossos, *Annual of the British School at Athens* 69: 195-257.

SANDARS, N.K. 1955. The antiquity of the one-edged bronze knife in the Aegean, *Proceedings of the Prehistoric Society* 21: 174-97.

1963. Later Bronze Age swords, *American Journal of Archaeology* 67: 117-53.

19

Using Neutron Activation Analysis to Source Ancient Tin (Cassiterite)

GEORGE (RIP) RAPP, JR., RUSSELL ROTHE & ZHICHUN JING

Archaeometry Laboratory, University of Minnesota, Duluth MN 55812, USA

Although excavation archaeologists have not found any cassiterite (the main ore of tin) at eastern Mediterranean / Near Eastern Bronze Age sites, true smelting sites should contain sporadic pieces of cassiterite. Presented here are the first extensive trace-element analyses of eastern Mediterranean cassiterite from ten separate deposits; these illustrate, using discriminant analyses, the viability of this method to source cassiterite and smelted tin. The database includes 44 samples from four deposits in the Eastern Desert of Egypt and 87 samples from eight deposits in the Erzgebirge region of the Czech Republic. Cassiterite samples were irradiated by both short and long half-life INAA procedures. Twelve trace elements were used in the multivariate analyses. These elements are Fe, Hf, La, Sc, Se, U and W (long-life), and Al, In, Mn, Ti and V (short-life). Out of 131 samples only four were misclassified with regard to countries (Egypt vs. Czech Republic). For 71 samples from the five major Czech deposits the misclassification rate is only 10.8%. Of the 44 samples from the four Egyptian deposits only two (4.5%) were misclassified. Ten samples from each of the Egyptian deposits were smelted and the tin analyzed. These smelted tin samples showed close trace-element associations with their corresponding cassiterite sources, indicating the potential also for tracing smelted tin to specific sources.

Keywords: Provenance Studies; Cassiterite; Tin; Neutron Activation Analysis; Trace-Element Sourcing.

Introduction

At least since the seminal work of Muhly (1973) and Wertime (1973) and the volumes by Franklin *et al.* (1978) and Penhallurick (1986) the source of tin for the Bronze Age of the eastern Mediterranean area has been on the front burner for metals in antiquity. This paper will not review the extensive recent discussions of possible tin sources as for example McGeehan-Liritzis & Taylor (1987); Hall & Steadman (1991); Earl & Ozbal (1996); Pernicka *et al.* (1992) and Muhly *et al.* (1991) who focus on potential tin sources in Anatolia. Rather it reports on trace-element studies of cassiterite from deposits in the Eastern Desert of Egypt and the Erzgebirge of the Czech Republic. These areas are the first two to be investigated and many areas remain to be studied. However, the results are promising and it may be possible to differentiate, using trace elements, all possible sources for tin in the Bronze Age of the eastern Mediterranean. F.Begemann (pers.comm.) and Noel Gale (pers.comm.) separately are pursing tin isotope studies for cassiterite provenance. However, McNaughton & Rosman (1991) have reported that they could not detect any isotope differentiation on seven samples and possibly significant fractionation on one. It seems therefore unlikely that deposits will have a useful Sn isotope signature.

The Cassiterite Samples

Exploitation of the cassiterite (SnO_2) deposits in the Erzgebirge (ore mountains) of the Czech Republic has been underway since the twelfth century AD and perhaps as early as the Central European Bronze Age (Taylor 1983; Bourzek *et al.* 1989). The Erzgebirge extends in a northeast-southwest direction along the border between Saxony and Bohemia. Cassiterite deposits are always closely associated with granite.

The geological and metallogenetic contexts for these deposits are given in Tischendorf (1973) and Stemprok (1974, 1995). Although mining of lode deposits continued until recently, vast reserves were also present in stream placers. In 1990 the senior author, accompanied by the Czech expert, M. Stemprok, collected cassiterite widely in the Erzgebirge from placer and lode deposits. FIGURE 1 shows the location of the Czech Republic cassiterite deposits sampled for this study.

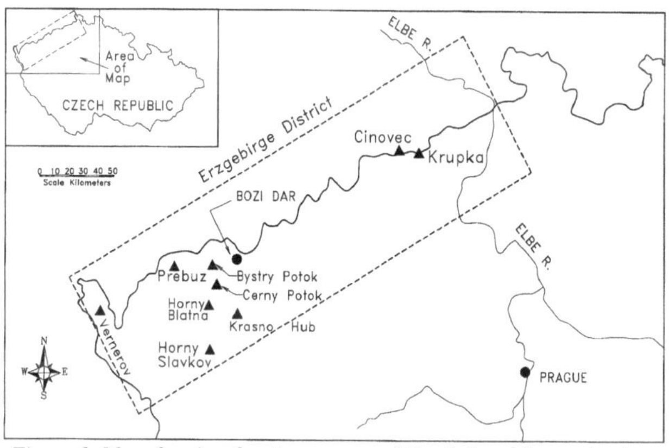

Figure 1. Map showing the nine sources of Czech Republic cassiterite.

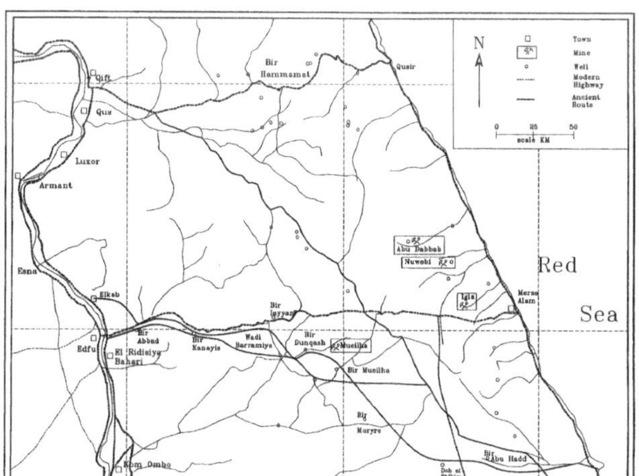

Figure 2. Map showing the four sources of cassiterite from the Egyptian Eastern Desert.

Cassiterite occurs in four deposits in the southern Eastern Desert of Egypt (FIGURE 2: Hume 1937; El-Ramley &

Al-Far 1955; Hussein & Sharkawi 1990). At all four locations cassiterite crystals occur at the contact zone between cross-cutting hydrothermal quartz veins and greisenized granite. Cassiterite is very resistant to both physical and chemical weathering so that when liberated from the lode, it washes into the wadi alluvium. At Mueilha, steeply dipping beds form natural riffles, trapping the dense cassiterite (FIGURE 3). All four locations have been worked in this century and one of them, Igla, is being worked currently.

Figure 3. Cassiterite concentrated in a natural riffle at Wadi Mueilha.

In conjunction with related fieldwork carried out over the past five years by the Archaeometry Laboratory, samples of cassiterite have been collected from the wadis draining the primary hardrock sources. In 1995, ten samples each from Igla, Nueiba, and Mueilha, and fourteen samples from Abu Dabbab were prepared and analyzed using INAA. Additionally, in 1996, ten samples from each of the four sites were smelted into tin and analyzed. TABLE 1 lists the sites and number of samples that form the basis for this study.

Smelting Studies

Cassiterite samples were prepared by pulverizing and homogenizing single cassiterite crystals. A 200 mg sample was taken from the powder and placed in an NAA vial which was sealed and numbered. When preparing the tin samples, one crystal per sample was fractured rather than pulverized and the entire sample was placed in a graphite crucible along with a small amount of laboratory grade powdered charcoal to act as a reductant. Based on previous experiments it was known that the amount of charcoal was not critical, as any extra reductant required would come from the graphite in the crucibles. All ten samples from each site were placed in a furnace at 1100° C for 40 minutes. The samples were removed and placed on a piece of polished gabbro and allowed to cool. After cooling, each sample was carefully cleaned using a dispersant in distilled water in an ultrasonic cleaner and a 200 mg sample removed and placed in an NAA vial.

Country	Site	Number of Samples
Egypt	Abu-Dabbab	10
	Igla	10
	Mueilha	10
	Nueiba	14
Czech	Brebuz	7
	Bystry Potok	15
	Cerny Potok	7
	Horni Blatna	9
	Horny Slavkov	26
	Krasno Hub	14
	Other *	9

* including Chaloupky, Cinovec, Krupka, and Venerov Deposits.

Table 1. Cassiterite Database by occurrence.

The 1100° C temperature was somewhat arbitrary since tin will smelt somewhat below that temperature although the process takes considerably longer. It has been shown that cassiterite can be smelted/alloyed in a single step directly into copper with the temperature remaining well below the melting point of copper (1083° C). For example bronze has been made from laboratory grade copper and Egyptian cassiterite at 1020° C. (Rothe 1997)

Analytical Method

Instrumental neutron activation analysis (INAA) was employed to determine the trace-element compositions of cassiterite and smelted tin. INAA is a physical method of determining trace-element concentrations with high precision and sensitivity. Many chemical elements can be detected at the low ppm and ppb levels and a wide range of elements can be measured simultaneously with no loss in precision.

Each cassiterite ore sample was crushed using an agate mortar and pestle; then cassiterite crystals were picked out under a microscope. The specimens were loaded in a plastic vial for neutron activation analysis at the University of Wisconsin reactor facility where both irradiation and counting were performed. Specimens in sets of eight were irradiated together with a 200 mg Canadian Standard Reference Material SO-4. Twenty-five isotopes having a short half-life (irradiated for 3 or 10 seconds with 18 minutes decay time) and forty-six isotopes with a long half-life (irradiated for 2 hours and a decay time of 7 to 8.5 days) were measured. For many samples, most elements had concentrations that were below the detection limits of the technique. Since the absence of elements from an appreciable number of samples will significantly affect the reliability of the statistical analyses, the analyzed elements were screened first on the basis of having a reasonable number of detectable values.

Source	Al	In	Mn	Ti	V	Fe
Czech	930.67±1709.73	27.03±31.34	640.76±1309.03	9001.67±21054.49	50.93± 86.31	11743.21 ± 25492.95
Brebuz	855.71± 1016.72	31.29± 20.99	5.27±7.74	3178.57± 2437.53	32.93± 36.02	1528.57±944.66
Bystry Potok	1905.33± 1881.01	26.36± 20.36	2273.40 ± 1867.95	7800.00± 8865.18	62.93± 58.07	41420.00 ± 43444.58
Cerny Potok	3800.00± 3321.65	21.93± 20.52	1915.71 ± 1275.30	71714.29±19172.15	258.57± 41.80	30857.14 ± 16354.70
Horni Blatna	516.67±570.07	101.33± 10.33	0.00±0.00	984.44±671.60	0.00±0.00	4088.89±416.67
Horny Slavkov	96.50±147.38	12.05± 10.41	63.69±254.06	1549.62± 1084.93	34.30± 73.76	325.38±599.83
Krasno Hub	304.50±477.18	8.20±4.62	55.38±171.62	837.86± 1057.04	6.86± 25.09	1622.14 ± 3368.85
Egypt	421.14±354.54	6.67± 7.94	154.02±148.20	5201.82±3306.80	96.50±118.14	2234.09 ±2654.96
Abu-Dabbab	388.00±252.80	0.22±0.26	234.80±185.31	2618.00±2281.45	29.10±19.03	910.00 ±1930.14
Igla	773.00±271.99	19.20±4.44	181.20±173.96	9670.00± 2200.53	282.80±118.03	4800.00 ± 2421.20
Mueilha	344.00±257.60	2.26±2.49	120.50±139.38	3230.00± 1702.97	63.20± 39.92	2750.00 ± 2114.63
Nueiba	248.57±376.36	5.48±4.60	100.86± 70.54	5264.29± 1928.17	35.36±9.69	978.57 ± 2294.87

Source	Hf	La	Sc	Se	U	W
Czech	14.36±42.37	8.07±25.94	20.15±10.02	12819.36±20489.89	13.24±21.21	1208.36 ±3085.39
Brebuz	0.00±0.00	1.08±0.99	18.71±6.50	1755.71± 1114.18	25.13± 12.53	1485.71 ±338.77
Bystry Potok	15.33± 59.39	37.12± 50.53	23.20±5.14	8070.67± 5687.06	5.83±8.07	0.00±0.00
Cerny Potok	88.43± 61.73	6.99±3.78	35.29±8.71	1362.86±741.84	10.46±2.48	0.00±0.00
Horni Blatna	0.00±0.00	1.02±1.21	22.33±2.06	2200.00±400.00	30.11±3.79	893.33 ±144.40
Horny Slavkov	6.92± 26.10	0.27±0.65	19.40± 11.16	13796.92±27955.71	16.01± 32.07	1720.46 ± 4612.69
Krasno Hub	6.50± 19.52	0.02±0.05	10.01±4.40	34178.57±14633.48	0.65±2.43	2220.00 ± 3345.83
Egypt	161.52±163.65	1.70±2.21	51.83±62.28	3043.41±2423.73	405.59±580.32	114.52 ±149.51
Abu-Dabbab	171.30±275.70	0.32±0.42	27.72± 33.22	3974.00± 2028.49	449.00±879.65	86.60 ± 77.80
Igla	160.70± 66.64	4.26±2.32	126.60± 56.51	586.00±602.87	211.50±178.55	327.00 ±161.39
Mueilha	170.80± 69.99	2.26±2.16	58.00± 68.95	2071.00± 1348.65	544.60±375.77	88.10 ± 54.78
Nueiba	148.49±170.79	0.46±0.86	11.25±3.85	4828.57± 2366.57	413.92±642.68	1.57 ±5.88

Table 2. Descriptive statistics for each cassiterite deposit.

In this study the selected elements are present in at least half of source samples. In spite of being present in most samples, Ta, Tm, and Lu, Mo and Na were not used in the statistical analyses due to their geochemical or instrumental uncertainties (such as strong peak interference from other elements). In INAA Niobium irradiates to very short or to very long half-life products that are not easy to detect so it was not used in this study. Tantalum, potentially a good diagnostic element, was also not used because of problems with analytical interference. Thus, twelve elements were included in the statistical analyses: Al, In, Mn, Ti and V (short half-life), and Fe, Hf, La, Sc, Se, U and W (long half-life). TABLE 2 is a summary of the descriptive statistics for the cassiterite deposits. All trace-element concentration data in the tables are in ppm.

Deposit	Classification Result 1	2	3	4	5	6	7	8	9	10	*I**	
Brebuz (1)	**42.86**	0.00	0.00	14.29	14.29	0.00	0.00	28.57	0.00	0.00	0.37	100.0
Bystry Potok (2)	0.00	**66.67**	0.00	0.00	0.00	0.00	6.67	6.67	13.33	6.67	0.63	100.0
Cerny Potok (3)	0.00	0.00	**100.00**	0.00	0.00	0.00	0.00	0.00	0.00	0.00	1.00	100.0
Horni Blatna (4)	0.00	0.00	0.00	**100.00**	0.00	0.00	0.00	0.00	0.00	0.00	1.00	100.0
Horny Slavkov (5)	7.69	0.00	0.00	0.00	**61.54**	23.08	0.00	0.00	3.85	3.85	0.57	100.0
Krasno Hub (6)	0.00	0.00	0.00	0.00	7.14	**85.71**	0.00	0.00	0.00	7.14	0.84	100.0
Abu-Dabbab (7)	0.00	0.00	0.00	0.00	0.00	0.00	**80.00**	0.00	20.00	0.00	0.78	100.0
Igla (8)	0.00	0.00	0.00	0.00	0.00	0.00	0.00	**100.00**	0.00	0.00	1.00	100.0
Mueilha (9)	0.00	0.00	0.00	0.00	0.00	0.00	10.00	0.00	**90.00**	0.00	0.89	100.0
Nueiba (10)	0.00	0.00	0.00	0.00	0.00	0.00	0.00	0.00	7.14	**92.86**	0.92	100.0
Classification												0.0
Hit Rate (%)	42.86	66.67	100.00	100.00	61.54	85.71	80.00	100.00	90.00	92.86		819.6
Total Hit Rate (%):		81.06		**Proportional Chance Hit Rate (%):**					0.1			
Overall Reduction-in-Error:											0.79	

Note: 7 elements included in analysis: Al, In, Mn, Ti, La, Se, W. Classification hit rate is calculated by leave-one-out(L-O-O) method. (based on the values with square-root transformation)
I* - reduction-in-error index

Table 3. Discriminant classification results of the ten separate cassiterite deposits.

The Chemistry of Cassiterite

Cassiterite is by far the most important ore of tin. It is typically found in veins closely associated with granite, in granite pegmatites, and in greisen (a hydrothermally altered granitic rock composed largely of quartz and white mica). Cassiterite departs extensively from its defined chemical formula, SnO_2, as it is non-stoichiometric. Columbite-tantalite, ixiolite, and rutile often exsolve from cassiterite during formation or are present as relict phases and so many published analyses may be on material from which minute inclusions were not removed. Also, cassiterite is chemically zoned. It would have been impractical, if not impossible, to X-ray all the samples (to determine any inclusions present) that were analyzed for trace-element concentrations. However, ancient metallurgists used the cassiterite as mined, so these mineralogical/chemical issues do not present a problem in this investigation.

Niobium and tantalum are almost universally present in cassiterite and ferrous/ferric iron, manganese, tungsten, titanium, and zirconium are common. There is an extensive literature on the composition of cassiterite including Little (1960); Dulski *et al.* (1982); Moller and Dulski (1983); Hassan (1986); Plimer *et al.* (1991); and Neiva (1996).

Discriminant Analyses

In this study discriminant analysis was considered as an appropriate multivariate technique for discriminating and characterizing the deposits. Both predictive and descriptive discriminant analyses were utilized. Predictive discriminant analysis involves deriving classification rules from two or more independent variables (elemental concentrations) that will discriminate best between the *a priori* defined groups (cassiterite deposits). The classification rules enable object group membership to be predicted (McLachlan 1992; Huberty 1994). Descriptive discriminant analysis uses the linear functions of the variables to describe or elucidate the differences between two or more groups. It has a focus on separation, and it is rather exploratory in nature. The goals of descriptive discriminant analysis include identifying the relative contribution of the variables to separation of the groups and finding the optimal plane on which the observations can be projected to best illustrate the configuration of the groups (Johnson & Wichern 1992; Rencher 1995; Huberty 1994). It needs to be emphasized that good classification requires significant separation but significant separation does not necessarily imply good classification. SAS for UNIX was employed to conduct discriminant analysis and other relevant statistical analyses.

Multivariate normality within groups (deposits) is an important underlying assumption in discriminant analysis. While real data are never exactly multivariate normal, the accuracy and precision of classification to a certain degree depend on how closely the true sample population resembles the multivariate normal form. Non-normal data may be made more normal by transformation of the data. It was found that the cassiterite data became more nearly normal with a square-root transformation. Better classification results were also achieved with square-root transformed data. Thus, square-root transformed data were used.

In predictive discriminant analysis, a subset of elements was always selected because fewer elements can yield greater classification accuracy. However, all the 12 elements were used in the descriptive discriminant analysis because fewer elements did not yield greater separation.

Considering the small sample size for each cassiterite deposit in the current database, the linear classification rule was selected (McLachlan 1992). In this study, the classification results were estimated by using the leave-one-out (L-O-O) method, an object and reliable hit-rate estimation. The degree of classification effectiveness was assessed by a reduction-in-error index $I = (H_0 - H_e) / (1 - H_e)$, suggested by Huberty (1994), where H_0 is the observed hit rate and H_e is the hit rate expected by chance.

Results of the Discriminant Analysis and Discussion

1. Classification and Separation for ten Separate Cassiterite Deposits

At the outset, linear classification was conducted for all ten major cassiterite deposits from both Egypt and the Czech Republic. TABLE 3 is the classification result, showing both overall and separate group hit rates. The overall hit rate is 81%, significantly better than the 10% expected by chance. A reduction-in-error of 0.79 is achieved, which means, by using the linear classification rule, about 79% fewer classification errors would be made than if the classification were done by chance. Separate group hit rates are consistently better than the chance hit rate. Nine out of ten deposits have a reduction-in-error larger than 0.5. All these suggest that the linear rule derived from the ten separate cassiterite deposits is highly effective in terms of classification accuracy. In this linear discriminant analysis, a subset of 7 elements (Al, In, Mn, Ti, La, Se and W) was used. Here, In and Ti are relatively more important than the rest of elements in terms of overall classification accuracy. Descriptive discriminant analysis was also undertaken to investigate the potential partitioning of cassiterite deposits. FIGURE 4 is a plot of group centroids projected onto the first two linear discriminant functions (LDFs) which describe 68% of the total variance in the data set. Visual inspection indicates that the first LDF accounts for the separation of the four Egyptian deposits from the Czech deposits that tend to form two clusters. The first LDF, which accounts for 43% of the total variance, is primarily defined by Ti and Fe (pooled within-group LDF-

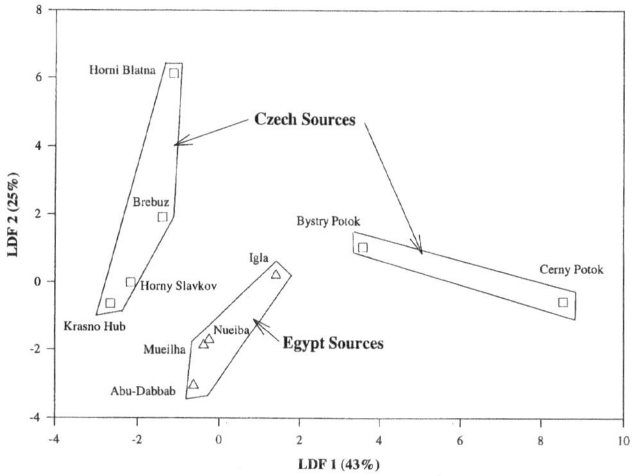

Figure 4. Plot of group centroids showing separation between Egyptian and the Czech Republic sources.

variable correlation *r* s=0.65 and 0.50, respectively; and standardized coefficients=0.56 and 0.71, respectively). The second LDF, which accounts for an additional 25% of the variance, is highly correlated with In (*r* s=0.75 and standardized coefficients=1.05); it is partially responsible for separation of Egyptian deposits from the Czech deposits. In terms of both classification accuracy and separation, Ti and In are relatively important with regard to all the 10 major deposits in the current database.

	Classification Result			
Deposit	**Czech**	**Egypt**	***I****	
Czech	**72** **(92.31%)**	6 (7.69%)	0.85	78.0
Egypt	**1** (2.27%)	43 **(97.73%)**	0.95	0.0
Total Hit Rate (%):	95.02			
Proportional Chance (%):	50.00			
Overall Reduction-in-Error:			0.90	

Note: 7 elements included in analysis: In, Mn, Ti, Fe, Hf, Se, W.
Classification hit rate is calculated by leave-one-out(L-O-O) method.
(based on the values with square-root transformation)
I - reduction-in-error index*

Table 4. Classification results of Egyptian vs. Czech cassiterite deposits.

2. Classification and Separation with Regard to Countries - Egypt vs. Czech Republic

Based on a seven element model (In, Mn, Ti, Fe, Hf, Se and W), a predictive discriminant analysis was performed to examine if a good discrimination could be achieved between the cassiterite deposits from Egypt and those from the Czech Republic. TABLE 4 is the classification result showing an overall hit rate of 95%. Only one out of 44

	Classificati on Result					
Deposit	**Abu-Dabbab**	**Igla**	**Mueilha**	**Nueiba**	***I****	
Abu-Dabbab	**9**	0	1	0	0.87	10.0
	(90%)	0	(10%)	0		0.0
Igla	0	**10**	0	0	1.00	10.0
		(100%)				0.0
Mueilha	0	0	**10**	0	1.00	10.0
			(100%)	0		0.0
Nueiba	0	0	0	**14**	1.00	14.0
			0	**(100%)**		0.0
Total Hit Rate (%):		97.50				
Proportional Chance (%):		25.00				
Overall Reduction-in-Error:					0.97	

Note: 4 elements included in analysis: In, Mn, La, W.
Classification hit rate is calculated by leave-one-out(L-O-O) method.
(based on the values with square-root transformation)
I - reduction-in-error index*

Table 5. Classification results for Egyptian cassiterite deposits.

samples from Egypt was misclassified into the Czech Republic; six out of 78 samples from the Czech Republic were not assigned back to their own group. The overall reduction-in-error indicates that the linear classification rule would yield 90% fewer classification errors than could be expected by chance.

Hafnium is the most important element in the classification with regard to countries, being responsible for the separation of cassiterite deposits from Egypt and the Czech Republic (*r* s=-0.67 and standardized coefficients=-0.84). Cassiterite from all four deposits in Egypt has a much higher Hf content than Czech cassiterite.

3. Classification and Separation of the four Egyptian Deposits

Using a subset of four elements (In, Mn, La and W), a predictive discriminant analysis was conducted on the data set containing the four cassiterite deposits from Egypt (TABLE 5). The classification results show an overall hit rate of 97.5%, significantly higher than the 25% derived by chance. Only one sample from Abu-Dabbab wasmisclassified. This suggests that the linear classification rule is extremely effective in discriminating between the four Egyptian deposits. In terms of classification accuracy, W and In are the most important elements.

FIGURE 5 is a plot of the first two LDFs for four Egyptian deposits. There is only a small overlap between the Abu-Dabbab and Mueilha deposits at a 90% probability level. The first and second LDFs accounts for 64% and 32% of the total variance, respectively. The first LDF is primarily defined by In, V, and W (*r* s=0.49, 0.5 and 0.44, respectively; standardized coefficients=1.04, -0.45 and 0.67, respectively); it accounts for the separation of Igla from the other three deposits. Igla cassiterite has the highest contents of In, V and W. The second LDF is responsible for the separation of Nueiba from Abu-Dabbab and Mueilha; it is mainly related to W (*r* s=0.44 and standardized coefficients=1.18). Based on these analyses, W and In are important to both the classification and separation of the four Egyptian cassiterite deposits.

4. Classification and Separation of the six Major Deposits from the Czech Republic

Both predictive and descriptive discriminant analyses were performed on the data set containing six major cassiterite deposits from the Erzgebirge region of the Czech Republic. The classification results, based on a nine element model (Al, In, Ti, Fe, Hf, La, Sc, U and W), are shown in TABLE 6. An overall hit rate of 86% with a reduction-in-error of 0.83 was obtained, suggesting the high effectiveness of the linear classification rule in distinguishing between cassiterite deposits from the Czech Republic. However, the classification rule for Czech cassiterite is not as powerful as that for the Egyptian deposits. In terms of classification accuracy, Ti and U are relatively more powerful in discriminating between the Czech deposits. By conducting descriptive discriminant analysis, a LDF plot was constructed to show the potential separation of these six deposits in terms of their trace-

	Classification Result							
Deposit	**Brebuz**	**Brstry Potok**	**Cerny Potok**	**Horni Blatna**	**Horny Slavkov**	**Krasno Hub**	*I**	
Brebuz	**5**	0	0	1	1	0	0.66	7.0
	(71.43%)	0	0	(14.29%)	(14.29%)	0		0.0
Brstry Potok	0	**14**	1	0	0	0	0.92	
		(93.33%)	(6.67%)	0				
Cerny Potok	0	**0**	**7**	0	0	0	1.00	
		0	**(100%)**	0				
Horni Blatna	0	**0**	0	**9**	0	0	1.00	9.0
		0	0	**(100%)**				0.0
Horny Slavkov	5	**0**	0	0	**15**	6	0.49	26.0
	(19.23)	**0**	0	0	**(57.69%)**	(23.08%)		0.0
Krasno Hub	0	**0**	0	0	1	**13**	0.91	14.0
		0	0	0	(7.14%)	**(92.86%)**		0.0
Total Hit Rate (%):				85.89				
Proportional Chance (%):				0.17				
Overall Reduction-in-Error:							0.83	

Note: 9 elements included in analysis: Al, In, Ti, Fe, Hf, La, Sc, U, W. Classification hit rate is calculated by leave-one-out(L-O-O) method. (based on the values with square-root transformation)
I - reduction-in-error index*

Table 6. Classification results of the cassiterite deposit from the Czech Republic

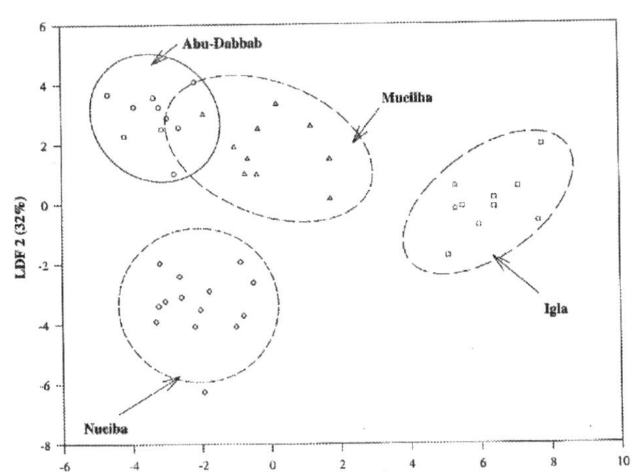

Figure 5. Plot of cassiterite deposits from Egypt showing separation of sources.

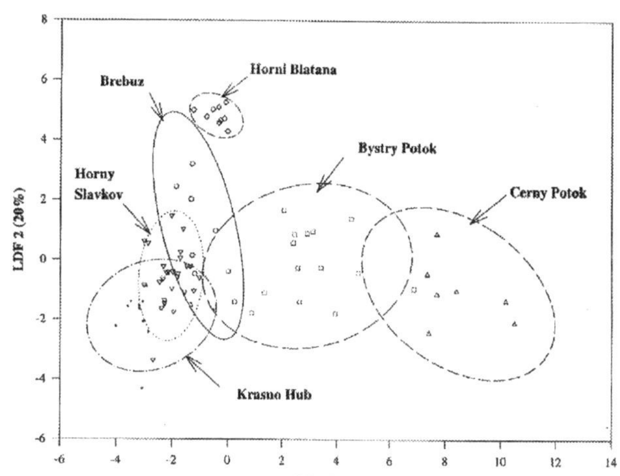

Figure 6. Plot of smelted tin from Egypt showing separation of sources.

		(90%)	(10%)			0.0
Mueilha	0	0 0	**10** **(100%)**	0 0	1.00	10.0 0.0
Nueiba	0	0	4 (40%)	**6** **(60%)**	0.47	10.0 0.0
Total Hit Rate (%):		79.17				
Proportional Chance (%):		25.00				
Overall Reduction-in-Error:					0.72	

Note: 5 elements included in analysis: Al, Mn, V, Hf, W.
Classification hit rate is calculated by leave-one-out(L-O-O) method.
(based on the values with square-root transformation)
I - reduction-in-error index*

Table 7. Classification results for Egyptian smelted tin.

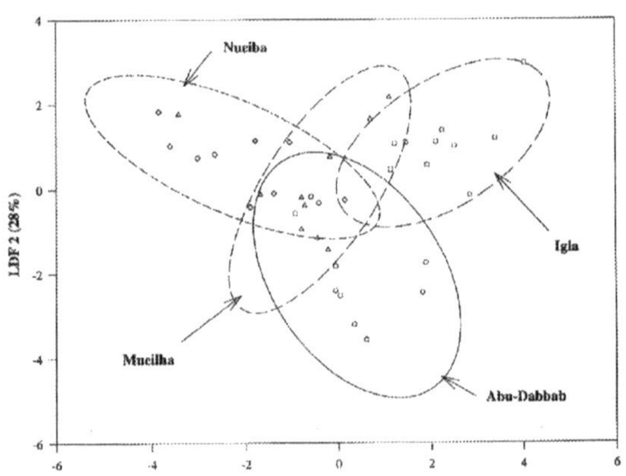

Figure 7. Plot of smelted tin from Egypt showing separation of sources.

element patterns (FIGURE 6). In spite of some overlaps, a general separation among the six cassiterite deposits is apparent. The first LDF, which accounts for 61%, is primarily related to Ti (*r* s=0.61 and standardized coefficients=0.63); it is mainly responsible for the separation of Cerny from Bystry Potok from the rest of the deposits. Cassiterites from Cerny Potok have the highest content of Ti within the whole database. The second LDF, which accounts for an additional 20% of the variance, is highly correlated with In (*r* s=0.68 and standardized coefficients=1.06); it accounts for the separation of Horni Blatana, characterized by relatively high contents of In, from the rest of the deposits. On the LDF plot, Horny Slavkov appears to be rather undifferentiated from both Krasno Hub and Brebuz.

5. *Classification and Separation for Smelted Tin*
Ten cassiterite samples from each of the four deposits in Egypt were smelted into tin. Predictive discriminant analyses were conducted to see if they could be effectively reassigned. One sample from Abu-Dabbab was deleted from the analysis as an outlier because it has extremely large values of Mn, Fe, La, and W. TABLE 7 is the classification result based on a subset of five elements (Al, Mn, V, Hf, and W). An overall hit rate of 79% was obtained; much higher than the 25% derived by chance. In comparison with the corresponding cassiterite classification, smelted tin yielded higher classification errors. Mn and W are relatively important elements in the classification of smelted tin specimens.

The separation of smelted tin specimens was evaluated with the aid of an LDF plot (FIGURE 7). There are certain overlaps among different sources of smelted tin, in particular the smelted tin from Mueilha has an appreciable overlap with both Abu-Dabbab and Nueiba. The separation of the Igla tin from the Nueiba tin can be attributed to the first LDF which describes 53% of the total variance. The first LDF is highly correlated with W (*r* s=0.51 and standardized coefficients=0.92). The smelted tin from Igla has the highest content of W, corresponding to the cassiterite deposit from Igla, which has a higher content of W than the other the cassiterite deposits in Egypt. The second LDF, which accounts for an additional 28% of the variance, is responsible for the separation of Abu-Dabbab on one hand versus Igla and Nueiba on the other. The second LDF is largely defined by In (*r* s=0.68 and standardized coefficients=0.50). Tin smelted from Abu-Dabbab cassiterite is characterized by substantially low contents of In.

6. *Impact of Tin Smelting on Trace-Element Patterns*
The underlying question is how the smelting process changes the original trace-element patterns in cassiterite. This question has been addressed by Rapp (1978) and is of obvious importance in the study of ancient tin sources. TABLE 8 shows the comparison of analyzed trace elements between smelted tin and the original cassiterite samples for each source in Egypt. For each, the smelted tin has different elemental concentrations than its corresponding cassiterite. Al, Mn, Ti, V, Fe, Sc and W are always lower in smelted tin than in the original cassiterite but In is enriched slightly. All other elements behave differently depending upon their original sources.

The change of trace-element patterns by smelting cassiterite to tin can also be observed in the LDF plots which show group centroids over the first two LDFs (FIGURE 8). Here, four cassiterite deposits and their corresponding smelted tin were treated as separate groups in descriptive discriminant analysis. The general separation of smelted tin versus the original cassiterite can be attributed to the first LDF that accounts for 49% of the variance. The first LDF is primarily defined by V and Ti (*r* s=0.70 and 0.53, respectively; standardized coefficients=0.55 and 0.80, respectively). The contents of both V and Ti decreased considerably during smelting.

Element	Abu-Dabbab	Abu-Dabbab (Tin)	Igla	Igla (Tin)
Al	388.00±252.80	124.22±169.93	773.00±271.99	111.00±239.19
In	0.22±0.26	0.83±1.05	19.20±4.44	23.40±5.80
Mn	234.80±185.31	131.07±137.15	181.20±173.96	31.86±52.01
Ti	2618.00±2281.45	577.78±1156.26	9670.00±2200.53	660.00±1419.86
V	29.10±19.03	2.48±4.33	282.80±118.03	16.50±27.06
Fe	910.00±1930.14	818.89±707.42	4800.00±2421.20	2706.00±1549.02
Hf	171.30±275.70	83.13±128.78	160.70±66.64	1.97±4.99
La	0.32±0.42	0.01±0.03	4.26±2.32	0.15±0.21
Sc	27.72±33.22	4.04±7.44	126.60±56.51	6.07±10.47
Se	3974.00±2028.49	94945.56±171077.87	586.00±602.87	1447.00±3469.72
U	449.00±879.65	3833.20±7482.17	211.50±178.55	4.10±9.34
W	86.60±77.80	13.33±22.29	327.00±161.39	30.30±34.89

Element	Mueilha	Mueilha (Tin)	Nueiba	Nueiba (Tin)
Al	344.00±257.60	0.00±0.00	248.57±376.36	95.00±137.62
In	2.26±2.49	11.83±22.68	5.48±4.60	7.27±4.85
Mn	120.50±139.38	16.07±13.87	100.86±70.54	70.00±85.71
Ti	3230.00±1702.97	0.00±0.00	5264.29±1928.17	1180.00±1594.99
V	63.20±39.92	0.11±0.35	35.36±9.69	7.94±11.11
Fe	2750.00±2114.63	722.00±720.41	978.57±2294.87	482.00±606.28
Hf	170.80±69.99	0.89±1.68	148.49±170.79	198.06±205.19
La	2.26±2.16	0.00±0.00	0.46±0.86	0.67±1.14
Sc	58.00±68.95	0.03±0.05	11.25±3.85	2.54±3.10
Se	2071.00±1348.65	478.00±1207.18	4828.57±2366.57	127714.00±151386.61
U	544.60±375.77	4.51±6.77	413.92±642.68	5421.53±6511.12
W	88.10±54.78	0.00±0.00	1.57±5.88	0.00±0.00

Table 8. Trace element comparisons between cassiterite and smelted tin.

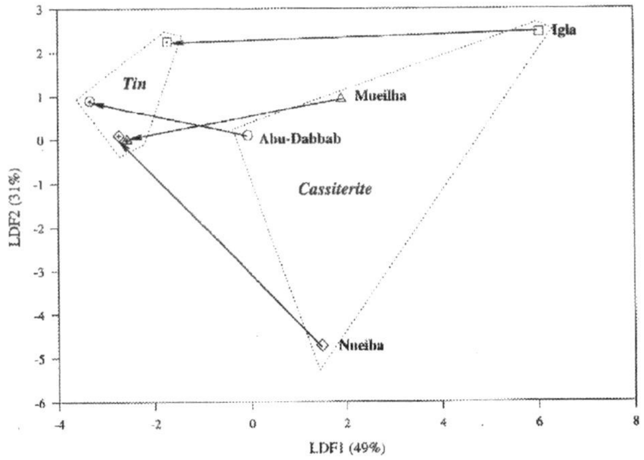

Figure 8. Plot of smelted Egyptian tin showing relationships to cassiterite sources.

Conclusion

The results presented in this paper indicate that if cassiterite is found at an ancient smelting site it may well be possible to trace its source deposit. The smelting studies reported here and in Rapp (1978) are exploratory. Although the results are very encouraging, much additional work remains to be done to fully understand the trace-element changes during cassiterite smelting.

Acknowledgements

Field work could not have been done without the assistance of many individuals, too numerous to name. This project relied extensively on the critical assistance of M. Stemprok of the Czech Geological Survey and M. El-Ramley and A.A. Hussein of the Egyptian Geological Survey. The neutron activation could not have been undertaken without the assistance of Richard Cashwell, director of the University of Wisconsin Reactor Facility. We also acknowledge Department of Energy grant DE-FG02-95NE38143 for its support of the Reactor Sharing Program. The comments of two anonymous reviewers were unfailingly helpful and significant.

References

BOUZEK, J., D. KOUTECKY & K. SIMON. 1989. Tin and Prehistoric Mining in the Erzgebirge (Ore Mountains): Some New Evidence, *Oxford Journal of Archaeology* 8: 203-212.

DULSKI, P., P. MOLLER, A. VILLAPANDO & H.-J.SCHNEIDER. 1982. Correlation of trace element fractionation in cassiterites with the genesis of the Bolivian Metallotect, in A.M. Evans (ed.) *Metallization Associated with Acid Magmatism*, New York: John Wiley & Sons, Ltd.

EARL, B. & H. OZBAL. 1996. Early Bronze Age Tin Processing at Kestel/Goltepe, Anatolia, *Archaeometry* 38: 289-303.

EL-RAMLEY, M.F. & D.M. AL-FAR. 1955. *Geology of El-Mueilha-Dunqash District*, Cairo: Les Editions Universitaires d'Egypte.

FRANKLIN, A. D., J.S. OLIN & T.A. WERTIME. 1978. *The Search for Ancient Tin*, Washington, D.C.: Smithsonian Institution.

HALL, M.E. & S.R. STEADMAN. 1991. Tin and Anatolia: Another Look, *Journal of Mediterranean Archaeology* 4: 217-234.

HASSAN, W.F. 1986, Aspects of geochemistry of Malaysian cassiterites, *Bulletin of the Geological Society of Malaysia* 1: 223-248.

HUBERTY, C.J. 1994. *Applied Discriminant Analysis*. New York: John Wiley & Sons, Inc.

HUME, W.F. 1937. *Geology of Egypt*, Cairo: Government Press.

HUSSEIN, A.A. & M.A. SHARKAWI. 1990. Mineral Deposits, in Rushdi Said (ed.), *The Geology of Egypt*, Rotterdam: A.A. Balkema.

JOHNSON, R.A. & D.W. WICHERN. 1992. *Applied Multivariate Statistical Analysis*. Englewood Cliffs, New Jersey: Prentice Hall.

LITTLE, W.M. 1960. Inclusions in cassiterite and associated minerals, *Economic Geology* 55: 485-509.

MCGEEHAN-LIRITZIS, V. & J. W. TAYLOR. 1987. Yugoslavian Tin Deposits and the Early Bronze Age Industries of the Aegean Region, *Oxford Journal of Archaeology* 6: 287-300.

MCLACHLAN, G.J. 1992. *Discriminant Analysis and Statistical Pattern Recognition*, New York: John Wiley & Sons, Inc.

MCNAUGHTON, N.J. & K.J.R. ROSMAN. 1991. Tin Isotope Fractionation in Terrestrial Cassiterites. *Geochemica et Cosmochimica Acta* 55: 499-504.

MOLLER, P. P. DULSKI. 1983. Fractionation of Zr and Hf in Cassiterite, *Chemical Geology* 40: 1-12.

MUHLY, J.D. 1973. *Copper and Tin: the distribution of mineral resources and the nature of the metals trade in the Bronze Age*. Hamden, Connecticut: Archon Books (reprint from *Transactions Connecticut Academy Arts & Sciences* 43: 155-535, with a *Supplement to Copper and Tin*, 1976, 46: 77-136.)

MUHLY, J.D., F. BEGEMANN, O. OZTUNALI, E. PERNICKA, S. SCHMITT-STRECKER, & G. A. WAGNER. 1991. The Bronze Metallurgy of Anatolia and the Question of Local Tin, in E. Pernicka & G. A. Wagner (ed.), *Archaeometry '90*, Boston: Birkhauser Verlag.

NEIVA, A.M.R. 1996. Geochemistry of cassiterite and its inclusions and exsolution products from tin and tungsten deposits in Portugal, *The Canadian Mineralogist* 34: 745-768.

PENHALLURICK, R.D. 1986. *Tin in Antiquity*. London: The Institute of Metals.

PERNICKA, E., G.A. WAGNER, J.D. MUHLY & O. OZTUNALI. 1992. Comments on the Discussion of Ancient Tin Sources in Anatolia, *Journal of Mediterranean Archaeology* 5: 91-98.

PLIMER, I.R., J. LU & J.D. KLEEMAN. 1991. Trace and rare earth elements in cassiterite - sources of components for the tin deposits of the Mole Granite, Australia. *Mineralium Deposita* 26: 267-274.

RAPP, G., JR. 1978. Trace elements as a guide to the geographical source of tin ore: smelting experiments, in A.D. Franklin, J.S.Olin & T.A.Wertime (ed.): *The Search for Ancient Tin* 59-63. Washington, D.C.: Smithsonian Institution.

RENCHER, A.C. 1995. *Methods of Multivariate Analysis*. New York: John Wiley & Sons, Inc.

ROTHE, R.D. 1997. Tin Smelting/Bronze Alloying Experiments Using Cassiterite from the Egyptian Desert, Paper presented to the annual meeting of the American Research Center in Egypt, Ann Arbor Michigan.

STEMPROK, M. 1974. Relation of tin and tungsten metallogeny to acid magmatism in the Krusne hory - Erzgebirge, in *Metallization Associated with Acid Magmatism*, 1, IGC Programme, Rospravy Ustredniho Ustavu Gelogickeho., Prague, 127-131.

1995. A Comparison of the Krusne hory-Erzgebirge (Czech Republic-Germany) and Cornish (UK) granites and their related mineralisation. *Proceedings of the Ussher Society* 8: 347-356.

TAYLOR, J.W. 1983. Erzgebirge Tin: A Closer Look, *Oxford Journal of Archaeology* 2: 295-300.

TISCHENDORF, G. 1973. The metallogenetic basis of tin exploration in the Erzgebirge, *Transactions of the Institution of Mining and Metallurgy (Section B: Applied Earth Science)* 82: B9-B24.

WERTIME, T. W. November 1973. The Beginning of Metallurgy: A New Look, *Science* 182: 875-87.

20

Trace Element Fingerprinting of Ancient Copper: A Guide to Technology or Provenance?

E. PERNICKA

Max-Planck-Institut für Kernphysik, Postbox 103980, D-69029 Heidelberg, Germany
Present address: Lehrstuhl für Archäometallurgie, TU Bergakademie Freiberg, Gustav-Zeuner-Str. 5, D-09596 Freiberg, Germany

The hypothesis that trace element concentrations should be a guide to the provenance of ancient metals has been around for more than a hundred years. Although very large analytical programs for ancient metal objects have been based on this concept, its validity is still in dispute with opinions that range from complete rejection to virtual acceptance. However, the apparent dichotomy expressed in the title is a false one. There are a number of elements whose concentrations are predominantly governed by ore mineralogy/composition and could thus, in principle, be used for provenance determinations. Other elements, found in ancient copper metal artefacts, are more closely related to the smelting process and so are of little use in provenance studies.

The relationships between copper ore and metal, as deduced from physico-chemical considerations and data; smelting experiments in the laboratory; and well-defined archaeological case studies such as the ancient copper production centres of Timna and Feinan in the Near East, are reviewed and principles governing these relationships are established. The usefulness of these principles and the conclusions drawn from them are shown in a large-scale investigation of Chalcolithic copper production and distribution in southeastern Europe; a study that included both trace element and lead isotope measurements. Long standing questions such as the possible use of native copper for the manufacture of heavy implements and the role in the early metallurgy of the Balkans of the, as yet earliest known, copper mines at Rudna Glava and Ai Bunar are discussed.

Keywords: Trace Element Fingerprinting; Provenance Studies; Chalcolithic Copper Production.

Introduction

The hypothesis that trace element concentrations can be a guide to the provenance of ancient metals was formulated more than hundred years ago. One early example is Göbel (1842) with an extended title that reads like an abstract. In (my own) translation it reads: "On the impact of chemistry on the tracing of prehistoric peoples, or results of the chemical investigations of ancient metal objects, especially of those from the Baltic region, to determine the peoples from whom they derive." He drew his conclusions from the geographical distribution of about 120 analysed objects and ascribed them to seemingly well-defined ethnic groups as was normal in those days. It should be remembered that the three period system had been proposed only a few years before, by Thomsen in 1836, and that an additional motivation for the analyses was the desire to date metal objects based on their compositions. It was discovered that minor elements were useful in determining the nature of the ore from which the metal came and perhaps even its geographical origin (von Fellenberg 1860-66; von Bibra 1869). Furthermore it was found that compositional differences between copper metal were to be expected when native copper, oxide or sulfide ores were used for smelting (Wibel 1865), an idea repeated almost exactly hundred years later by Friedman *et al.* (1966).

Very large analytical programs for ancient metal objects were performed along these lines but their results are still in dispute and the conclusions that can be drawn from them seem unclear. Opinions range from complete rejection to close to acceptance. The largest and most heavily criticized project was that undertaken by the Württembergisches Landesmuseum in Stuttgart (Junghans *et al.* 1960; 1968; 1974) with more than 20000 analyses of prehistoric metal objects from all over Europe. They were classified according to their chemical composition and the distribution of these metal groups was studied in time and space. Obviously, the classification scheme (based on a frequency analysis of the concentration of As, Sb, Ag, Ni, and Bi) was not understood, because the most strident critics of this work achieved almost the same result when they used a small subset of the data (Waterbolk & Butler 1965). Another point of concern was the choice of elements. These were based on an earlier work (Otto & Witter 1952) which assumed that, of all elements analysed, this set (As, Sb, Ag, Ni and Bi) would best characterize the original ore.

With the application of lead isotope analysis to copper-based alloys (Gale & Stos-Gale 1982) chemical analysis of ancient metal objects seems to have become obsolete. Indeed, it is constantly said that chemical analyses alone will not allow copper-alloy artefacts to be matched to their parent copper ores. Although this is often correct, there are cases where the trace element pattern may be more indicative of an ore source than lead isotope ratios. At Feinan, Jordan, for example, the ore deposit is chemically homogeneous but shows wide variations in its lead isotope ratios (Hauptmann *et al.* 1992). In addition, lead isotope ratios are strongly correlated so that only a small part of the theoretically possible, three-dimensional space is occupied resulting in a tendency for different ore deposits to overlap. In such a situation it is common sense that a combination of both sets of data - lead isotope ratios and trace element concentrations - will provide better discrimination between different sources. This approach was adopted by the Heidelberg/ Mainz group from the very outset (Pernicka *et al.* 1984).

The behaviour of trace elements

In the first half of this century geochemists divided the elements into three groups. Siderophile elements that are concentrated in the iron core, chalcophile elements in the outer core and mantle, and lithophile elements concentrated in the crust (FIGURE 1). There is also a group of atmophile elements that are gaseous under normal conditions. Since this concept was partially derived from observing the distribution of elements during metal smelting, it is certainly a useful one in the context of the analysis of ancient metals.

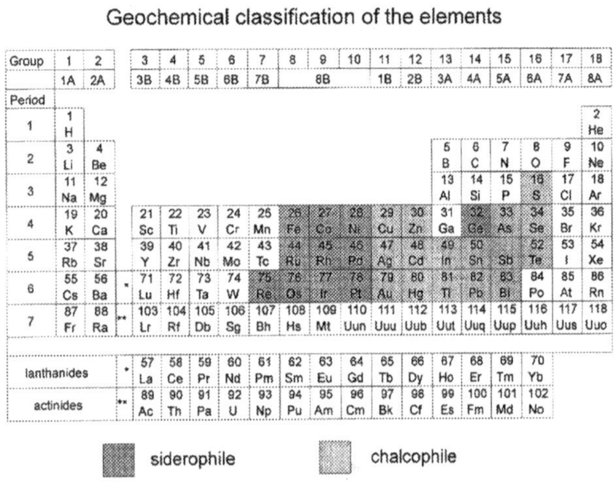

Figure 1: Geochemical classification of the elements. Elements not shaded are either lithophile or atmophile and of little use in provenance studies.

In provenance studies only those elements that follow copper during smelting are useful, which means that the element/copper ratio largely remains the same between the ore and the final smelted copper. Three major complications have to be considered at this point: Ores are generally inhomogeneous on all scales; they show zoning (Ixer this volume). The objection has been raised that this precludes any correlation between artefacts and ores but this need not be so as the prehistoric mining region of the Mitterberg in Salzburg, Austria shows. Here, trace element concentrations in chalcopyrite-rich ores vary over two orders of magnitude (FIGURE 2), yet this variation is not a random one. It is known, for example, that nickel occurs mainly in the form of Ni-As minerals such as gersdorffite (NiAsS) so that copper produced from Mitterberg ores is characterized by a combination of nickel and arsenic as major impurities with variable concentrations combined with relatively low amounts of antimony, silver and bismuth. In a recent study of some 800 Late Bronze Age copper artefacts from the adjacent area it was found that about 80% of the copper alloys conformed to this general pattern. Knowing that the peak production period at the Mitterberg was during the Late Bronze Age Urnfield Culture, it is only reasonable to assume that a correlation between ores and artefacts does exist.

It is essential that any ore deposit suspected of being a metal source in antiquity needs to be investigated, not only chemically but also geologically, to obtain as much information on its structure, mineral content and variability as is possible. The major disadvantage of this approach is that it needs a sizeable number of samples. In addition, it is often difficult to find equivalent samples to the rich ore that may have been used by ancient smelters because subsequent mining has removed all of the ores that were available in prehistory, as at Majdanpek, Serbia (Pernicka *et al.* 1993). Under such circumstances only very generalized conclusions can be drawn from the mineralization.

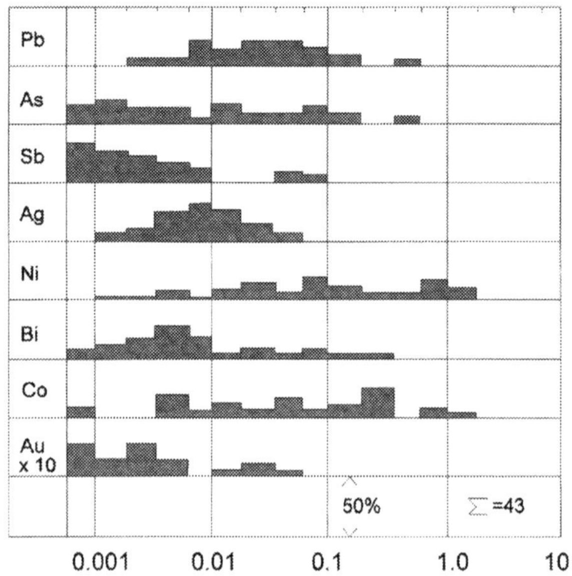

Figure 2: Variation of elements with possible reference to provenance studies in chalcopyrite concentrates from the Mitterburg area (after Pernicka 1987). The histogram is drawn in the style of Waterbolk & Butler (1965) for a better comparison with artefact analyses. The raw data have been recalculated so that the sum of copper and the plotted elements total 100%.

A second complication arises from the smelting process, as this will differ depending on the type of ore being used. Reduction of oxide ores is quite straightforward compared with the processing of sulfide ores. Although it cannot be expected that reduction happened under chemical equilibrium conditions, it is possible to use thermodynamic data for equilibrium reactions to predict the general behaviour of certain trace elements during smelting (Pernicka 1987). From these data it can be deduced that the concentration of some elements, notably nickel and arsenic, strongly depends on the smelting temperature and that copper, with or without nickel, could be produced from the same ore (Pollard *et al.* 1991; Budd *et al.* 1992). This is, however, very theoretical and needs the assumption that high and low impurity coppers were being produced intermittently due to different smelting temperatures. It is hard to imagine that ancient smelters did not know what they were doing and so sometimes smelted with low efficiency and without slag formation below the melting point of copper and at other times at very high temperatures above 1200°C. It is far more likely that people who were able to cast copper and thus achieve temperatures above 1100°C would strive to smelt at the maximum temperature obtainable with charcoal (between 1200 and 1300°C). Under these circumstances both slag and metal are liquid and the smelting process is easier to control and much more efficient. Consequently, it is most likely that the reduction smelting of ores containing nickel and arsenic would produce copper rich in both elements.

On the other hand, sulfide ores cannot be reduced directly but must be oxidized (roasted) before reduction. This is usually accomplished at around 700°C, well below the melting point of copper and most ore minerals. Nevertheless, it has been suggested that volatile elements like zinc, arsenic antimony and, presumably, selenium and tellurium, are partly lost during roasting (Tylecote *et al.* 1977). This may need more investigation, as was demonstrated by the experimental smelting of fahlerz from Cabri res in southern France. Here several Chalcolithic mines have been discovered (Ambert 1990/91; Ambert & Barge-Mahieu 1991) and copper mainly occurs in the form of tetrahedrite, a complex Cu-Sb sulfosalt. It is interesting to note that fahlerz-type copper dominates southern France during the late Chalcolithic. FIGURE 3 shows the concentrations of major impurity elements in the ore, the roasted ore and the raw copper metal. For comparison, all concentrations were recalculated so that the sum of the analysed elements and copper is 100%. This is equivalent to the assumption that all three elements (arsenic, antimony and silver) pass without loss from the ore into the metal. From FIGURE 3 it is evident that this assumption does not hold for antimony, whose concentration is reproducibly reduced by a factor of ten, while silver stays at the same level as expected. Interestingly, the arsenic content is far less reduced, although arsenic is more volatile than antimony. This is almost certainly due to the much lower concentrations of arsenic in the ore, for it seems that under these conditions volatilization of arsenic is reduced.

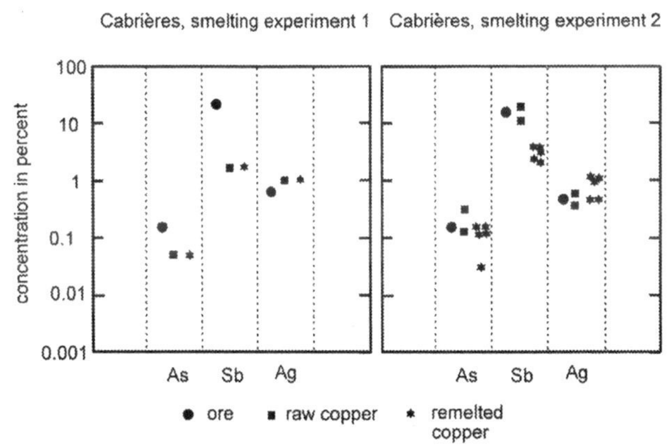

Figure 3: Results of smelting experiments of copper ores with high antimony concentrations (fahlerz) from Cabri res in southern France. For comparison the ore data have been recalculated according to the same method as in Fig.2.

The third complication arises from the treatment of the smelted copper to produce the finished object. Nowadays raw copper is generally refined to remove sulfur, iron and other impurities and this may also have been true in prehistoric times. This is an easy process because all it needs is to remelt the copper. Iron and other easily oxidized impurities are then removed as dross. It has been suggested that the refining of copper erases most of the characteristics that survived from the ore (Merkel 1983; 1990). However, using Merkel's experimental data it can be shown that this is not the case for elements like silver, nickel and antimony (FIGURE 4).

In a similar experiment to Merkel, Zwicker prepared a copper alloy with 2.02% Ni, covered it with charcoal and inserted it into a furnace at 600 °C. After 4 hours, when 1100°C was reached, the alloy began to melt; after 4.5 hours a temperature of 1150 °C was achieved and the alloy was completely molten but still covered with charcoal. Then the metal was partly poured out and chilled (sample 1). Charcoal was removed and the remaining metal left for further 15 minutes at 1150 °C in the furnace, then a further portion was poured out and chilled (sample 2). The remaining metal was left to cool in the crucible under a gentle airflow (sample 3). Chemical analysis showed 1.57% Ni for sample 1, 1.23% and 0.97% for samples 2 and 3, respectively. This demonstrates that, although some nickel is lost during simple re-melting, it is not so much as to make inferences impossible. It must be remembered that all processes, from ore formation to smelting and refining, depend on the distribution of the elements between various phases. Since the distribution laws are rational it is vital to compare ores and artefacts on a logarithmic scale, as shown in FIGURE 4. In such a diagram a loss of 25% nickel as in sample 1 does not appear dramatic, especially if considered alongside the range in concentrations seen in ores within a single deposit.

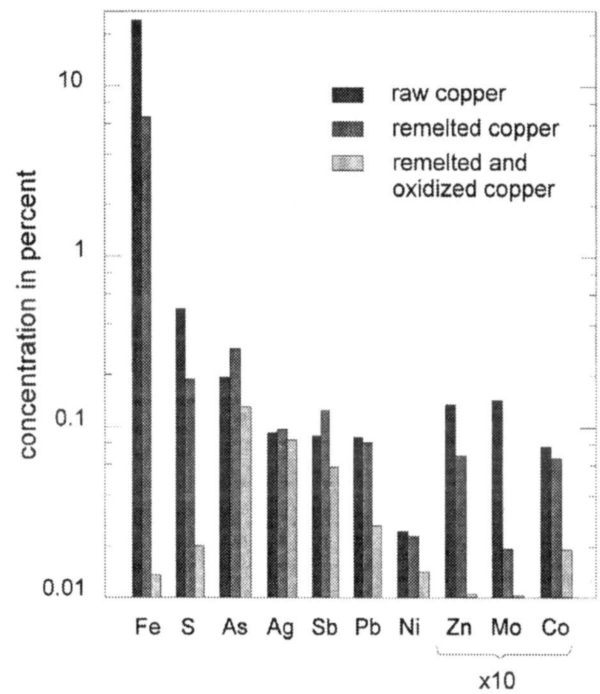

Figure 4: Summary of refining experiments performed by Merkel (1983;1990). It is evident that the concentrations of most elements relative to copper do not significantly change on simple remelting, except for Fe, S and Mo, all of which are irrelevant for provenance studies. It is unlikely that molten copper was regularly exposed to a blast of air for extended time as indicated in the third column (remelted and oxidized copper). But even then only Co, Ni and Pb are reduced by a factor of 2, which is of little significance when ores are compared with artefacts (after Pernicka 1987).

A final consideration is that some elements were deliberately alloyed with copper. In antiquity these were mainly tin, lead and - starting with the late Hellenistic period - zinc. Thus TABLE 1 gives a summary of the information provided by various elements that, in principle, can be found in ancient copper. There are few elements that are solely indicators of provenance. However, in earlier times when it can be assumed that deliberate alloying did not occur, a whole suite of elements are available that can be useful in helping to determine provenance (columns 2 and 3).

See Table + Caption at end of paper.

Trace elements and technology

Most copper ores contain iron and sulfur as major elements. Indeed, the principle task of smelting is the separation of copper from these two elements. Accordingly, their concentrations in copper alloys reflect how efficiently this separation was accomplished, hence they provide information about the smelting and purification processes. Friedman *et al.* (1966) suggested that an increasing iron content could distinguish copper produced from the three essentially different ore types (native copper, oxide, and sulfide ores). By contrast, Tylecote *et al.* (1977) found in smelting experiments that copper produced from oxide ores tended to contain higher concentrations of iron than from sulfide ores. On the other hand, Craddock & Meeks (1987) found that the iron contents of copper artefacts tend to increase from very early to later and technologically more advanced times. They suggest that, in the early periods, only very high grade copper ores with more than 60% copper were smelted at relatively low temperatures and under poor reducing conditions. They did not attempt to distinguish between oxide and sulfide ores.

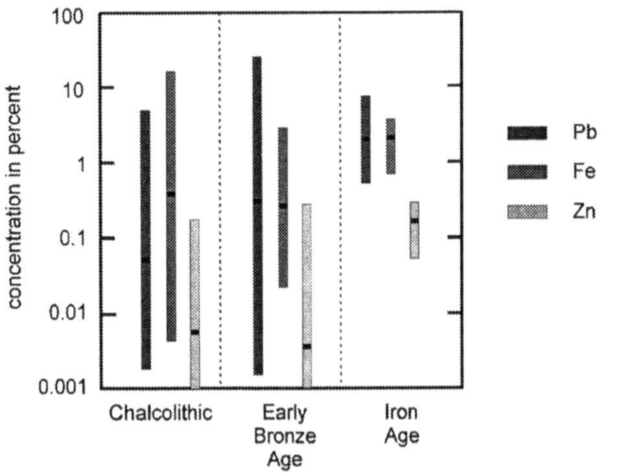

Figure 5: Variation of three elements that may be sensitive to redox conditions during smelting. The diagram shows that there was a major change in the technology of copper production between earlier periods and the Iron Age. In the latter period furnace conditions were presumably more reducing and better controlled as is indicated by the much smaller variations in Fe, Pb and Zn in copper prills extracted from slags (after Hauptmann et al. 1992).

The smelting of sulfide ores always includes the formation of matte; a mixture of copper and iron sulfides of variable composition that, together with the slag, takes up most of the iron. The matte can then be dead roasted and afterwards reduced like an oxide ore (an unfavourable process because of high fuel consumption) or it can be partly oxidized, in an open crucible, whereupon it yields quite pure copper. When oxide ores are reduced, then the iron content of the raw copper is bound to increase with increasing efficiency of the smelting conditions, namely with more reducing conditions. This is nicely demonstrated at Feinan (Hauptmann *et al.* 1992) where the copper ore is mainly present as oxides and where there is a clear increase in the iron content of copper metal between the Early Bronze Age and the Iron Age (FIGURE 5) by about an order of magnitude. This increase is accompanied by an increase in other elements that are sensitive to redox conditions in the furnace like lead, zinc, and cobalt (not shown in FIGURE 5). Although part of this change may be due to the exploitation of different ore charges, field evidence shows that considerable changes in the basic features of furnace construction occurred during this period. Presumably, these changes resulted in generally better controlled, more reducing furnace conditions, as is suggested by the decrease in variability in the contents of these elements in Iron Age copper. This conclusion is supported by the observation that if iron occurs in Chalcolithic copper prills at Feinan then it is always in an oxide form, magnetite or iron-rich slag, whilst in the Iron Age iron in copper occurs mainly as metallic iron or as carbon-containing iron phosphides (Hauptmann *et al.* 1992).

The Stuttgart analyses (Junghans *et al.* 1960, 1968; 1974) established that a major change in the metal composition occurred between the European Late Chalcolithic and the Early Bronze Age. The most abundant metal groups are chemically similar to the composition of fahlerz, so that it has been suggested that this may reflect the change from oxide to sulfide/sulfosalt ores. In this case rather low concentrations of iron would be expected, which is indeed observed in some 2500 Early Bronze Age copper-based artefacts from eastern Germany (FIGURE 6). Analyses were performed with an energy-dispersive X-ray spectrometer on drill samples (Lutz & Pernicka 1996) and in roughly half of the objects the iron content was below the detection limit of about 0.03%. Similar numbers of Early Bronze Age artefacts from Mesopotamia, analysed by the same method, exhibit a clearly different distribution of iron that was above the detection limit in almost all samples, ranging up to several percent (FIGURE 6). It is therefore suggested that most copper used in Mesopotamia in the third millennium BC derives from oxide ores. A possible explanation could be that in the semi-arid region of the Middle East the oxidized zones of the ore bodies are much deeper than in Europe, where most oxide ores have been eroded away during the Pleistocene so that present day oxide zones are far shallower.

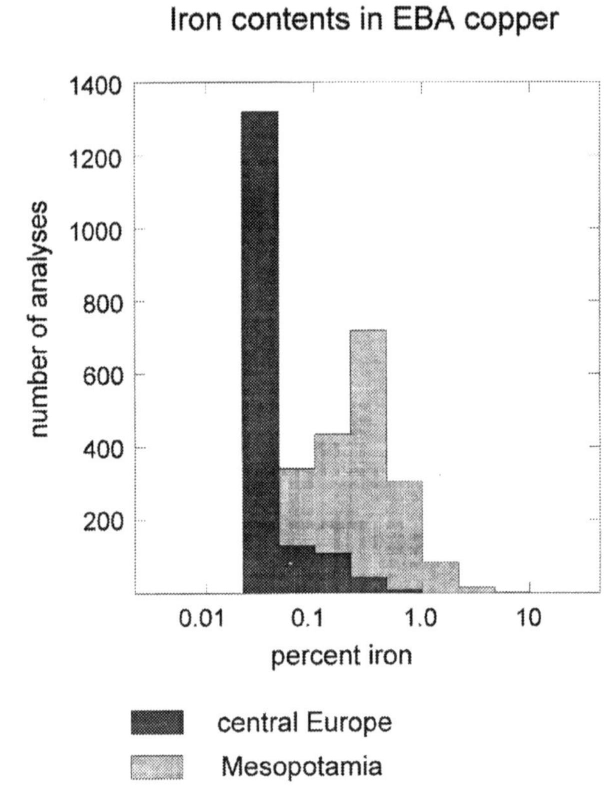

Figure 6: Iron contents in Early Bronze Age copper-based artefacts from eastern Germany and from Mesopotamia. Analyses were performed with energy-dispersive X-ray fluorescence on drill shavings.

Native copper is a special case. Its identification in artefacts is highly desirable, because it is generally held that the working of metals began with the use of native copper (Muhly 1989). The most straightforward method for its identification is investigation of its metallographic structure in polished section. This only works, however, if the native copper object has not been cast but been formed simply by hammering, and possibly tempering. Since it is more likely to find native copper in very early periods and because these early artefacts are usually very rare and small, e.g., beads or small tools, permission to take a large sample for metallographic study is extremely difficult to obtain. Often it is possible only to take a small sample by drilling or rubbing that is only suitable for chemical analysis. It would, therefore, be desirable to identify native copper by its chemical composition but this method is contentious. Native copper is usually very pure but it has been suggested that copper smelted from pure malachite or chalcopyrite could be just as free from impurities as native copper (Maddin *et al.* 1980).

Wibel (1865) proposed that high purity Neolithic and Chalcolithic copper artefacts from southeastern Europe were evidence, if not proof, that they were manufactured from native copper. Coghlan (1962) cast doubts on the use of trace element chemistry for the identification of native copper but conceded that it might provide at least an indication of it. Among the large number of prehistoric, copper-based artefacts investigated by Otto & Witter (1952) by optical emission spectroscopy, there is a group that stands out by its high purity having only silver as a major impurity. This metal composition was interpreted as indicative of native copper. Junghans *et al.* (1968), in their very comprehensive study of prehistoric metal objects from all over Europe, observed a comparatively high proportion of spectroscopically pure copper - their metal types N and E00 - in southeast Europe. They followed the suggestion of Otto & Witter (1952) that this was most probably native copper. Charles (1969) suggested that two Chalcolithic axe-adzes from Hungary and the former Yugoslavia were made of native copper, again on the basis of their purity.

One aspect seems to have been overlooked in this discussion, namely that the presence of volatile metals such as mercury could provide unequivocal chemical evidence for the presence of native copper. This was first suggested by Pernicka *et al.* (1993) after it had been found that many samples of native copper contain a few ppm of mercury. Contrary to common expectations mercury is not easily volatilized from copper, even at temperatures of several hundred degrees centigrade (FIGURE 7). Thus it could easily survive the tempering of native copper that seems to have been applied from the very earliest periods onwards (Maddin *et al.* 1991; Yalcin & Pernicka 1998).

On the other hand, smelting of copper ores removes mercury very effectively, so that this element might possibly serve as a useful indicator for native copper. A problem of this method is that laboratory tests have shown that, in principle, mercury can also be adsorbed on copper from ground water during burial so that, again, the identification is not unequivocal. This seems to have occurred with three Chalcolithic borers from Kacica, Bulgaria (Pernicka *et al.* 1997) dated to the late fourth millennium BC. Although they are quite different in composition (one contains 2.35% tin) and may even be of different ages they all contain between 6 and 9 ppm mercury. It is very unlikely that this indicates the use of native copper, rather it suggests that the mercury content is due to contamination from the environment either during burial or, more likely, at some later stage.

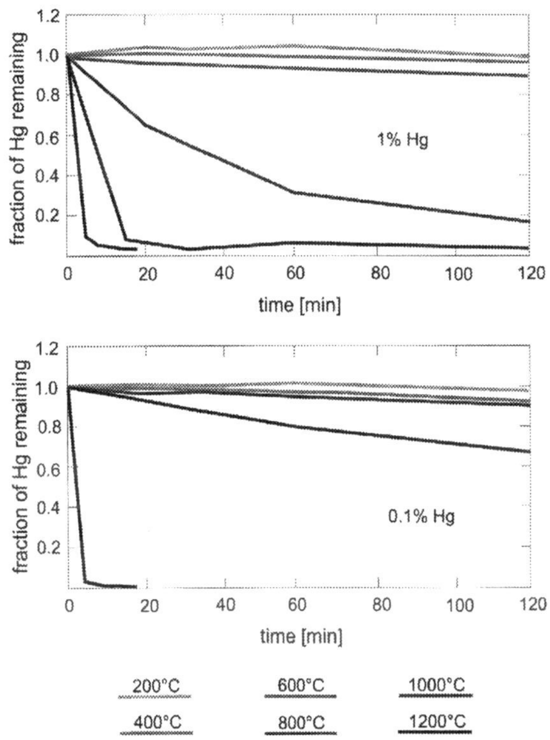

Figure 7: Results of desorption experiments of mercury from copper. Copper with 1% and with 0.1% mercury was prepared and heated in air over the time shown. Especially at lower concentrations (that are still high compared with those found in samples of native copper) mercury is only quantitatively removed above the melting point of copper.

Returning to the trace element pattern, it is true that native copper is usually a high-purity metal, with silver (and locally arsenic) as major impurities. Unfortunately, it is also true that very pure ores do exist that produce similarly pure copper on smelting, as represented by millimeter-sized copper prills found in Chalcolithic and Early Bronze Age slags from Feinan (Hauptmann *et al.* 1992). These demonstrate that silver, gold, and antimony in smelted copper can, for all practical purposes, be as low as those typically found in native copper; that arsenic shows a wide distribution in concentrations overlapping that observed in native copper, but that nickel and cobalt are almost invariably higher than in native copper (FIGURE 8). Most native copper is formed in the lower part of the oxidation zone, lying above primary ore deposits, by dissolution and re-precipitation of the metal. Geochemical considerations indicate that some elements are more mobile than others under these oxidizing conditions. This results in the separation of elemental pairs that behave similarly during smelting, such as arsenic/antimony and cobalt/nickel. It is well known that native copper can contain appreciable concentrations of arsenic but antimony is invariably very low, because antimony is immobile in the oxidation zone due to the formation of antimonates (Pernicka 1990). Similarly, cobalt is always low, because unlike nickel it is often precipitated as a carbonate. For the same reason gold

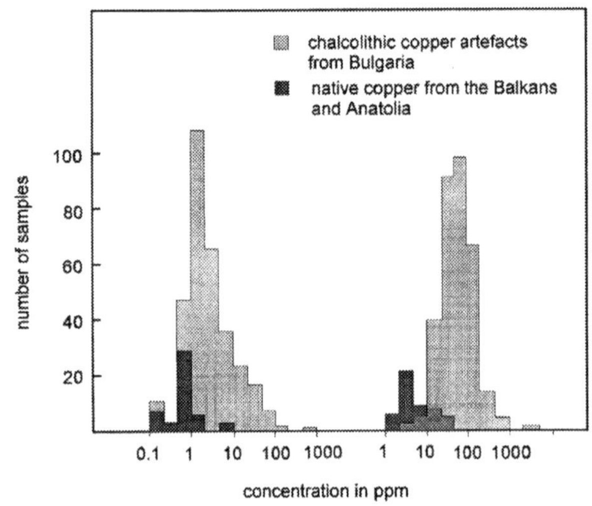

Figure 8: Nickel and cobalt concentrations in native copper samples from southeastern Europe and Anatolia. It is obvious that they do not have the same distributions although there is some overlap. Nevertheless, both elements seem to be useful in distinguishing native copper from smelted copper (after Pernicka et al. 1997).

should be low while silver concentrations can reach up to the percent level. Thus a pattern of low cobalt, antimony, and gold contents combined with high silver and/or arsenic contents can certainly be regarded as indicative of native copper. Unfortunately, not all native coppers show such a pattern. They can simply have low concentrations of all impurities and then it may be difficult to decide between native and smelted copper. Nevertheless, in a large sample suite certain trends should become apparent.

Such a suite has recently been published from material found in southeastern Europe (Pernicka *et al.* 1997). This is where most early, very pure, copper metal artefacts have been found (Junghans *et al.* 1968). They are generally interpreted as being made from native copper although many objects were undoubtedly cast. This interpretation has had a profound impact on our perception of the beginnings of metallurgy in southeastern Europe. It has been suggested that abundant sources of native copper may have given rise to an independent invention of metallurgy in this region (Renfrew 1969). However, new analyses with improved precision and detection limits provide evidence that, contrary to common belief, most Chalcolithic copper objects in Bulgaria were made out of smelted copper and it seems that metallurgy began in this area with smelting without a recognisable native copper phase (FIGURE 9). The iron contents have a bimodal distribution (FIGURE 10) that indicate that oxide ores were smelted along with either sulfide ones or, more likely, very pure ores with high copper concentrations. Since smelting of copper ores is a complex technique that requires considerable experience, it seems more likely that this knowledge was already available when metallurgy began in Bulgaria. Thus transfer of technology, most likely from Anatolia, seems to be a better model for the development of metallurgy in southeastern Europe than an indigenous re-invention. A similar model has recently been proposed for central Europe (Strahm 1994).

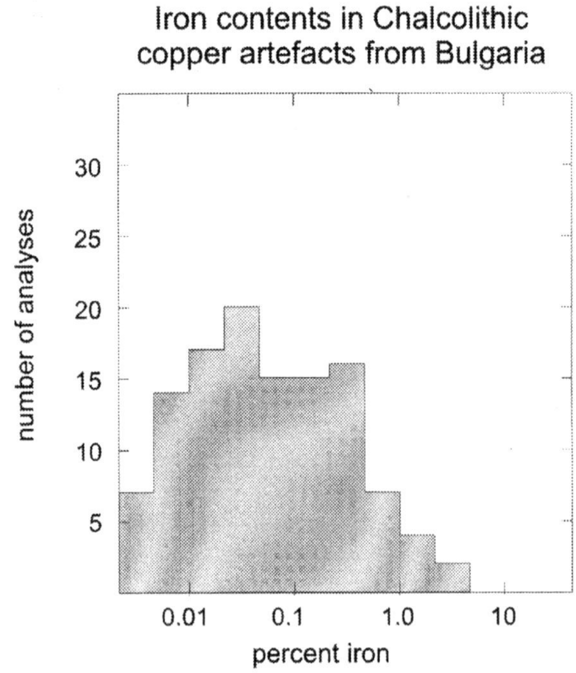

Figure 9: There is a clear difference in the distribution of S, Au and Se in native copper and Chalcolithic copper objects from Bulgaria suggesting that native copper did not play a major role in the early metallurgy of southeastern Europe.

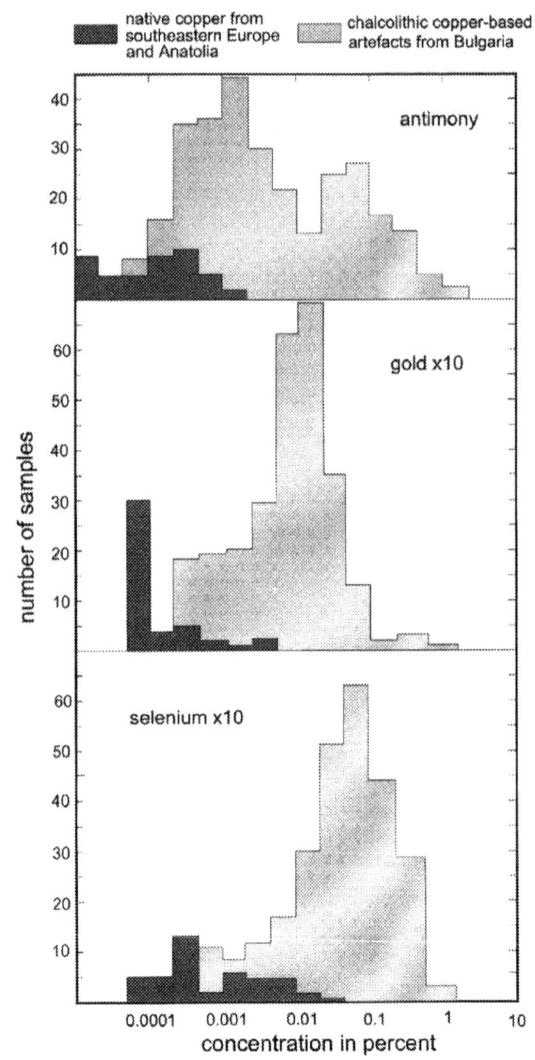

Figure 10: The concentration of iron in Chalcolithic copper objects from Bulgaria shows a bimodal distribution.

Trace elements and provenance

Trace element analysis of ancient metal objects was so discredited that for some time hardly anybody looked at the data that were available. Indeed, it is possible that the wealth of data was so overwhelming that it was actually difficult to digest. In any case, many of the arguments raised against metal analysis are no longer valid or were wrong from the start. It has been shown that the largest database namely that published by the Stuttgart group (Junghans *et al.* 1960; 1968; 1974) is quite accurate but that its precision is only moderate compared with modern analytical techniques (Pernicka 1984). The selection of elements used for grouping appears to be sensible. Tin was not used, because it was a deliberate addition in most cases and is, therefore, irrelevant to the source of copper. The same could apply to lead but in European Early Bronze Age artefacts lead concentrations are usually low and vary rather irregularly. Thus only five elements are left that can be determined in most samples, namely As, Sb, Ag, Ni, and Bi. It happens that all of them are indicative of ore sources, at least to a certain extent. Frequency analysis shows that groups of artefacts can be found with narrow quasi-normal distributions of all elements, which suggests that all of them were produced from the same ore applying similar processes. This is now increasingly being acknowledged by archaeologists (Krause 1988; Liversage 1993; Krause & Pernicka 1996) and metal composition is regarded a valuable tool for the grouping of artefacts.

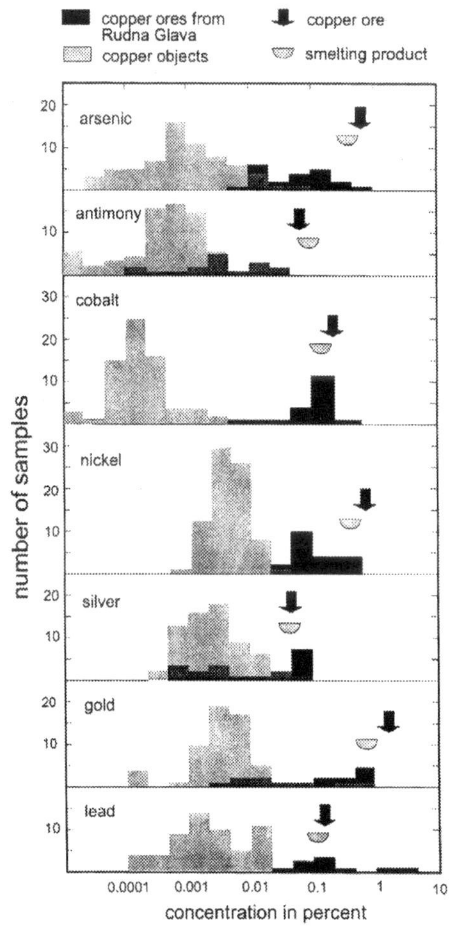

Figure 11: Concentration ranges of elements in Chalcolithic copper artefacts from Serbia and in copper ores from Rudna Glava. Ore analyses have been recalculated in the same manner as in Fig.2. Arrows indicate the concentrations in a prehistoric ore concentrate from Rudna Glava (Tylecote 1982) and bowls indicate copper smelted from the same ore (after Pernicka et al. 1993).

However, few cases have been reported where there is a convincing association between a group of artefacts and a specific ore deposit. This is mainly due to the lack of a comparable number of ore analyses and the similarity of many copper deposits, which makes it difficult to distinguish them unequivocally. Incidentally, this problem is shared with lead isotope analysis, for its real value lies in the exclusion of suspected ore sources. These are often situated in the geographical vicinity of the artefacts under study, and most archaeologists tend to assume a direct association if the distance between artefact and ore deposit is small.

A good example is the Chalcolithic copper mine at Rudna Glava in Serbia. Most archaeologists believed that the abundance of Chalcolithic copper finds in the central Balkans was directly related to this mine, as the earliest heavy implements such as those from Plocnik (Grbic 1929, Stalio 1964; 1973) appeared at roughly the same time that Rudna Glava was being exploited (Jovanovic 1982). In a study that set out to determine the role of Rudna Glava in the early metallurgy of southeastern Europe, some 90 Chalcolithic copper artefacts were analysed chemically and isotopically (Pernicka *et al.* 1993). The result was surprising, because Rudna Glava could be definitely ruled out as the source for all of the analysed objects. This was based on the combined trace element and lead isotope compositional data but even the trace element data alone would have been sufficient to show this (FIGURE 11). In addition it was discovered that the heavy implements from the Plocnik hoards, that are typologically very similar, consist of copper with at least three different trace element patterns, suggesting that the metal derives from different ore sources. However, these could not be identified (FIGURE 12).

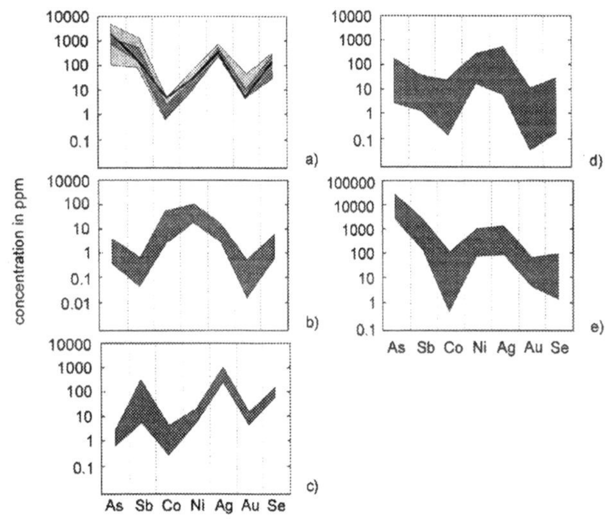

Figure 12: Trace element patterns of three different chemical groups found among heavy implements from the Plocnik hoard (a – c), in other Chalcolithic copper artefacts, and in final Chalcolithic copper-based objects from Serbia. Also included in panel a is a group of Chalcolithic copper artefacts from Bulgaria that are chemically indistinguishable from those from Plocnik lighter shaded area and one ore sample from the Chalcolithic mine at Ai Bunar near Stara Zagora, Bulgaria (solid line). The very similar trace element patterns, in shape as well as in concentrations, suggest a common origin of the objects and, in addition, Ai Bunar as the source of this copper. This conclusion is decisively strengthened by matching lead isotope ratios in the artefacts and in ores from Ai Bunar (Pernicka et al. 1997).

An extension of this study into present-day Bulgaria (Pernicka *et al.* 1997) revealed that more Chalcolithic copper objects with similar compositions occurred there and, moreover and most important, that one group chemically and isotopically resembled ores from another Chalcolithic copper mine, namely Ai Bunar near Stara Zagora. Leaving aside the isotopic evidence, the similarity of one ore sample from Ai Bunar with this group of objects is rather striking (FIGURE 12a). However, only one out of nine ore samples provided such a close match but it is by far the most relevant one with 53% copper whilst the other samples contained between 3 and 22% copper. These lower grade samples possibly represent rejected ore rather than that smelted. Again this confirms the major difficulty connected with the chemical characterization of ore deposits, for it is desirable to analyse many more samples than is considered necessary for lead isotope analysis. Yet it is often difficult to find enough ore that is representative of the material available in prehistoric times and left over by the miners. Fortunately, at Rudna Glava and Ai Bunar this was possible. However, at Majdanpek in Serbia modern opencast mining has completely removed the ore body and, in addition, covered a large area with waste rock so that finding any remains of ancient mining and the ore exploited in antiquity is very unlikely. Under such circumstances lead isotope ratios are virtually the only available tool that will allow inferences to be made about ancient exploitation (Pernicka *et al.* 1993).

In both studies it became clear that the two sets of data provide independent information on the provenance of ancient copper. These may or may not agree with each other. In the case of agreement between the data sets (still rare) this comes very close to a positive identification of an ore source, since it is extremely unlikely that several deposits have exactly the same chemical and isotopic patterns. On the other hand, some deposits turned out to be chemically homogeneous but not isotopically, such as Rudna Glava and others are isotopically homogeneous but show distinct chemical variation as is suggested by some groups of Bulgarian artefacts (Pernicka *et al.* 1997).

technology	provenance and/or technology	provenance
Al, B, Ba, *Be, Ca,* **Cr,** *Cs,* **Fe,** *Ga, Ge,* Hf, *K, Li, Mg, Mn,* Mo, *Na, Nb, P,* Rb, *S,* Sc, REE[b], *Si, Sr,* Ta, *Ti,* Th, U, *V, W, Y,* Zr	**As, Cd[a], Co,** In, **Hg[a],** Re, **Sb, Se, Te,** *Tl[a]*	**Au Ag, *Bi,* Ir, Ni,** Os, Pd, Pt, *Rh,* Ru
Sn > ca. 1% **Zn** > ca. 5% ***Pb*** > ca. 5%	**Sn** < ca. 1% **Zn** < ca. 5% ***Pb*** < ca. 5%	

[a] only applicable with native copper
[b] rare earth elements (La, Ce, *Pr,* Nd, Sm, Eu, *Gd,* Tb, *Dy, Ho, Er, Tm,* Yb, Lu)

Table 1: List of elements that have been reported in ancient and native copper and their potential uses in archaeometallurgy. Elements in bold type are routinely determined at the Max-Planck-Institut für Kernphysik, Heidelberg, by neutron activation (NAA) and atomic absorption analysis (bold italics). Elements in normal type can be determined by NAA, if desired (Kuleff & Pernicka 1995). Elements in italics are either difficult to determine in copper with NAA or cannot be determined.

References

AMBERT, P. 1990/1991. L'émergence de la métallurgie chalcolithique dans le Midi de la France. *Archéologie en Languedoc* 10/11: 51-58.
AMBERT, P. & H.BARGE-MAHIEU. 1991. Les mines préhistoriques de Cabrières (Herault). In: Mohen, J.-P., Éluère, C. (ed.): *Découverte du Métal,* Paris, 259-277.
BIBRA, E. VON 1869. Die Bronzen und Kupferlegierungen der alten und ältesten Völker, mit Rücksichtnahme auf jene der Neuzeit. Erlangen.
BUDD, P., D.GALE, A.M. POLLARD, R.G. THOMAS & P.A.WILLIAMS. 1992. The early development of metallurgy in the British Isles. *Antiquity* 66: 677-86.
CHARLES J.A. 1969. A metallurgical examination of south-east European copper axes. *Proceedings of the Prehistoric Society* 35: 40-42.
COGHLAN, H.H. 1962. A note upon native copper: its occurrence and properties. *Proceedings of the Prehistoric Society* 28: 58-67.
CRADDOCK, P.T. & N.D. MEEKS. 1987. Iron in ancient copper. *Archaeometry* 29, 187-204.
FELLENBERG, L.R. VON 1860-66: *Analysen antiker Bronzen, Mitteilungen der naturforschenden Gesellschaft* Bern 1860-1866.
FRIEDMAN, A.M., M. CONWAY, M. KASTNER, J. MILSTED, D. METTA, P.R. FIELDS & E. OLSEN. 1966. Copper artifacts: correlation with some types of copper ores. *Science* 152: 1504-1506.
GALE, N.H. & Z.A.STOS-GALE. 1982. Bronze Age copper sources in the Mediterranean, *Science,* 216: 11-19.
GÖBEL, F. 1842. Über den Einfluë der Chemie auf die Ermittlung der Völker der Vorzeit oder Resultate der hemischen Untersuchung metallischer Alterhümer, insbesondere der in den Ostsee-gouvernements vorkommenden, behufs der Ermittlung der Völker, von welchen sie abstammen Erlangen.
GRBIC, M. 1929. Plocnik. *Eine prähistorische Ansiedlung aus der Kupferzeit.* Nationalmuseum Beograd.
HAUPTMANN, A., F. BEGEMANN, E. HEITKEMPER, E. PERNICKA & S. SCHMITT-STRECKER 1992. Early Copper Produced at Feinan, Wadi Araba, Jordan. Part I: The Composition of Ores and Copper. *Archeomaterials* 6: 1-33.
IXER, R.A. 1999 (this volume). The role of ore geology and ores in the archaeological provenancing of metals. In S.Young, P.Budd, R.Ixer & M.Pollard (ed.): *Metals in Antiquity.* Oxford: Archaeopress.
JOVANOVIC, B. 1982. *Rudna Glava: Der älteste Kupferbergbau im Zentralbalkan.* Bor und Belgrad.
JUNGHANS, S., E. SANGMEISTER & M. SCHRÖDER. 1960. *Metallanalysen kupferzeitlicher und frühbronzezeitlicher Bodenfunde aus Europa,* Berlin
1968. *Kupfer und Bronze in der frühen Metallzeit Europas* 1-3, Berlin
1974. *Kupfer und Bronze in der frühen Metallzeit Europas* 4, Berlin
KRAUSE, R. 1988. Die endneolithischen und frühbronzezeitlichen Grabfunde auf der Nordstadtterrasse von Singen am Hohentwiel. *Forschungen und Berichte zur Vor- und Frühgeschichte in Baden-Württemberg,* Bd. 32.

KRAUSE, R. & E. PERNICKA. 1996. Das neue Stuttgarter Metallanalysenprojekt "SMAP". *Archäologisches Nachrichtenblatt* 1: 274-291.
LIVERSAGE, D. 1993. Impurity patterns and cultural history - an inquiry into compositional patterns in the Bronze Age. *Archaeology and Natural Science* 1: 73-89.
LUTZ, J. & E. PERNICKA. 1996. Energy dispersive X-ray fluorescence analysis of ancient copper alloys: empirical values for precision and accuracy. *Archaeometry* 38: 313-323.
MADDIN, R., T. STECH WHEELER & J.D. MUHLY. 1980. Distinguishing artifacts made of native copper. *Journal of Archaeological Science* 7: 211-225.
MADDIN, R., T. STECH & J.D. MUHLY. 1991. Çayönü Tepesi: The earliest archaeological artifacts. In: J-P. Mohen und C. Eluère (ed.): *Découverte du Métal*. Paris, Picard, 375-386.
MERKEL, J. 1983. Summary of experimental results for Late Bronze Age copper smelting and refining. *Museum Applied Science Center Journal* 2/6: 173-179.
1990. Experimental reconstruction of Bronze Age copper smelting based on archaeological evidence from Timna. In: B. Rothenberg (ed.): *The ancient metallurgy of copper*. London. Institute of Archaeo-Metallurgical Studies.
MUHLY, J.D. 1989. Çayönü Tepesi and the beginnings of metallurgy in the Ancient World. In: A. Hauptmann, E. Pernicka und G.A. Wagner (ed.): *Old World Archaeometallurgy*. Der Anschnitt, Beiheft 7: 1-11.
OTTO, H. & W. WITTER. 1952. *Handbuch der ältesten vorgeschichtlichen Metallurgie in Mitteleuropa*, Barth Verlag, Leipzig.
PERNICKA, E. 1984. Instrumentelle Multi-Elementanalyse archäologischer Kupfer- und Bronzeartefakte: Ein Methodenvergleich. *Jahrbuch des Römisch.-Germanischen Zentralmuseums*. 31: 517-531.
1987. Erzlagerstätten in der Ägäis und ihre Ausbeutung im Altertum: Geochemische Untersuchungen zur Herkunftsbestimmung archäologischer Metallobjekte. *Jahrbuch des Römisch-Germanischen Zentralmuseums* 34: 607-714.
1990. Entstehung und Ausbreitung der Metallurgie in prähistoricher Zeit. *Jahrbuch des Romisch-Germanischen Zentralmuseums,* 37: 21-129.
PERNICKA, E., T.C. SEELIGER, G.A.WAGNER, F. BEGEMANN, S. SCHMITT-STRECKER, C. EIBNER, Ö. ÖZTUNALI & I. BARANYI, 1984. Archäometallurgische Untersuchungen in Nordwestanatolien. *Jahrbuch des Römisch-Germanischen Zentralmuseums* 31: 533-599.
PERNICKA, E., F. BEGEMANN, S. SCHMITT-STRECKER & G.A.WAGNER. 1993. Eneolithic and Early Bronze Age copper artefacts from the Balkans and their relation to Serbian copper ores. *Prähistorische Zeitschrift* 68: 1-54.
PERNICKA, E., F.BEGEMANN, S. SCHMITT-STRECKER, H. TODOROVA & I.KULEFF. 1997. Prehistoric copper in Bulgaria: Its composition and provenance. *Eurasia Antiqua* 3: 41-180.
POLLARD, A.M., R.G.THOMAS, D.P. WARE & P.A.WILLIAMS. 1991. Experimental smelting of secondary copper minerals: Implications for Early Bronze Age metallurgy in Britain. In: E. Pernicka & G.A. Wagner (ed.): *Archaeometry '90*. Basel Birkhäuser Verlag, 127-136.
RENFREW, C., 1969: The autonomy of the south-east European Copper Age. *Proceedings of the. Prehistoric Society*. 35: 12-47.
STALIO, B. 1964. Novi metalni nalazi iz Plocnika kod Prokuplja. Zbornik Narodnog Muzeja 4: 35-41.
1973. Cetvrti nalaz bakarnog i kamenog orudja sa Plocnika kod Prokuplja. *Zbornik Narodnog Muzeja* 7: 157-161.
STRAHM, C. 1994. Die Anfänge der Metallurgie in Mitteleuropa. *Helvetica Archaeologica* 97: 1-39.
TYLECOTE, R.F., H.A.GHAZNAVI & P.J.BOYDELL. 1977. Partitioning of trace elements between the ores, fluxes, slags and metal during the smelting of copper. *Journal of Archaeological Science* 4: 305-333.
WATERBOLK, H.T. & J.J.BUTLER. 1965. Comments on the use of metallurgical analysis in prehistoric studies. *Helinium* 5: 227-251.
WIBEL, F. 1865. Die Cultur der Bronze-Zeit Nord- und Mittel-Europas. Chemisch-antiquarische Studien über unsere vorgeschichtliche Vergangenheit und deren Bergbau, Hüttenkunde, Technik und Handel. 26. *Berichte der Schleswig-Holsteinischen Lauenburger Gesellschaft zur Sammlung und Erhlatung vaterländischer Altertümer,* Kiel.
YALCIN, Ü. & E.PERNICKA. 1998. Frühneolithische Metallbearbeitung am Asikli Höyük, Türkei.In: A. Hauptmann, E. Pernicka, T. Rehren, Ü. Yalcin (ed.): *The Beginnings of Metallurgy*. Der Anschnitt, Beiheft 8, Deutsches Bergbau-Museum, Bochum.

21

Analysis of a Collection of Egyptian Tools and Weapons at the Los Angeles County Museum of Art using Typology, Metallurgy and ICP

M. ABRAHAM

Department of Materials, Oxford University, Oxford, UK and
Los Angeles County Museum of Art, Conservation Sciences, Los Angeles, CA, USA

The objects discussed here are part of a private collection of copper/bronze tools and weapons, assembled in Egypt, and now in the Los Angeles County Museum of Art, some of which have inscriptions of questionable authenticity. This study uses a wide variety of techniques to look at the objects, and the results are mixed. All of the objects are identified as ancient and most of the inscriptions are also thought to be ancient, but unfortunately without more study of the inscriptions, the objects will remain essentially unprovenanced. On the positive side, the objects can be added to the corpus of tools and weapons from ancient Egypt and, because of the inscriptions, may add to our understanding of the development of tool and weapon typologies as well as their uses. In the future, more work with the inscriptions and comparisons with other inscribed tools and weapons, which do have provenance, may lead to a better understanding of where these objects come from and what they were used for.

Keywords: Egypt; Metallography; ICP; Typlogy; Tools; Weapons; Copper; Bronze; Authenticity; Inscriptions.

Introduction

This study concerns a group of objects, from the Los Angeles County Museum of Art, which have no provenance. The objects are a set of copper/bronze tools and weapons from a much larger private collection, assembled in Egypt, whose inscriptions, where they exist, are of questionable authenticity. The original collector, Mr. Michailidis, favoured objects with inscriptions, leading to the possibility that the inscriptions were added in modern times, in order to enhance their desirability. Because of this, many of the objects from this collection have been somewhat discounted by Egyptologists and curators in the various museums into which the collection was dispersed. This is unfortunate, since the objects would be of particular interest to Egyptologists if the inscriptions were found to be authentic (Kaplony 1965). While the objects are from a wide temporal range (Old Kingdom through the New Kingdom, a period spanning two millennia) they are, in all likelihood, part of a long tradition of using inscribed tools and weapons as ritual objects and to indicate status. The inscriptions describe more then the royal personage named. Rather they indicate the concerns and aspirations of the government during the particular reign, be it warfare under Ramesses the Great or building programs during the early part of the New Kingdom. Inscribed weapons, foundation deposits, and burial goods were part of a state run propaganda system which influenced the daily life of ancient Egyptians. Thus it is important to study the typologies, inscriptions and forms of state controlled manufacture of bronze objects to further understand what the government of Egypt was trying to convey to the population at the time, and the relative importance of these messages.

After placing the objects in the context of known typologies, some standard scientific methods for identifying objects as ancient, using metallurgy, and for measuring trace elements in order to asses their agreement with the known metal compositions from the time periods involved, using Inductively Coupled Plasma Optical Emission Spectrometry (ICP), were employed. First the objects were carefully examined under a microscope. The corrosion products which have developed over years are a good indicator of antiquity. Furthermore, the remnants of corrosion products in the chased grooves of the inscription are an indication of the antiquity of the inscriptions. Small metallurgical samples were taken in order to give a clear indication of the antiquity of the copper object. If the sample contains deep intergranular, transgranular or interdendritic corrosion then the sample is most probably ancient. Unfortunately, this is no indication of how ancient since corrosion proceeds at widely varying rates depending on the conditions under which the object existed.

Trace element analysis is useful in a slightly different way. The development of metallurgy can be seen in the ever increasing complexity of the metal alloys. The earliest copper metals were usually native, or unsmelted, and typically contain very low amounts of trace elements. Smelted metals can often be identified by an increase in the iron content, which results from the smelting process. The addition of tin to make bronze is a still later invention. Thus the agreement of the major constituent elements as well as the trace element content with the technology available during the period that the object was thought to have been manufactured, based on typologies, is an added indication of the veracity of the object. Finally, in later periods, the inclusion of large amounts of tin or other elements may indicate something about the usefulness and importance of the object as a daily tool or as a ritual object.

Unfortunately, none of the above mentioned scientific techniques are able to do more than identify the objects as ancient and possibly to place them in a broad time period. Therefore, the study of un-provenanced objects is inherently limited. In the case of this group of objects the inscriptions add to the information available to scholars to understand the objects themselves. Having established the antiquity of the objects and of the inscriptions, they can be studied to help place them in time and in space.

Ancient Egyptian Metal Tools and Weapons

We can hope to understand the tools and weapons in this study better by understanding the value and uses of tools and weapons in ancient Egypt. The value, to the ancient Egyptians, of the objects themselves was inherent in the fact that they were made of copper. In fact, a measure of copper known as a *deben* (about 91 g) was the basic standard for trade in ancient Egypt. Since the value of tools was in large part equal to the value of the copper from which they were made, it can be assumed that it was the copper and not the tools which was the most important commodity to the Egyptians. Thus evolved the practice of weighing workman's tools at the workshop when they were issued in the morning and again when they were returned at the end of the day (Janssen 1975:314). Furthermore, the tools are likely to have been made, stored and apportioned by state administrative bodies in state run facilities. This explains the remarkable consistency in typology and inscriptions through out much of Ancient Egypt's history. The tools themselves were usually used by organized groups of craftsmen. The most common metal tools were the axes and adzes which were used by woodworkers and boat builders from the Pre-Dynastic Period onwards to cut and smooth wood. Other tools included saws, drills, leather punches, scissors, etc.

Like tools, weapons were commonly stored in state run stores such as the "broad hall" or armoury of the royal residence and distributed to soldiers in time of need. The Twelfth Dynasty tomb of Nakht at Asyut contained a model armoury among the grave goods. Whilst in the Old Kingdom the army and related armoury probably consisted of only a loosely developed hierarchy with an overseer in charge, by the New Kingdom things had developed considerably (Shaw 1991:27). The New kingdom army was a large and professional force with professional soldiers divided into military organizations. Each army consisted of about 5000 soldiers, including professionals, foreign mercenaries and conscripts. These soldiers used bows, spears, axes and maces.

The weapons, like the tools, were probably produced in large manufacturing areas run by the state. While we cannot exclude the possibility of independent craftsmen travelling to smaller sites in order to recast and repair bronze items, the Egyptian hierarchy must have controlled the majority of the raw materials and production facilities which used the large quantities of copper brought, first from the Sinai and Eastern Desert area, and later from Cyprus.

A further area of use for tools and weapons found in ancient Egypt was as ceremonial objects. Tools and weapons were commonly used in funerary contexts (Swain 1995) and as foundation deposits (Weinstein 1973). As far back as the Pre-Dynastic Period the use of model grave goods has been documented. Unlike the above-mentioned tools and weapons, those made for entombment were not intended to be used in the context of the living world and therefore often did not need to be made either as strong or as large. The objects were often smaller than life size and were sometimes made of materials other than that of their counterparts which were intended for use. A magical process was believed to be used by the dead to make these articles useful in the afterlife. A good example of a model battle axe is found in this collection (M80-203-43) and is inscribed with the cartouche of Seqenenre Tao.

The other common ceremonial use of tools was the foundation deposit. The foundation deposits were part of the ceremony used by the person who dedicated the building. There was a well established ritual which included the laying out of the building's foundations and the placing of ritual objects, for the purpose of maintaining the building, in pits at the corners and around the periphery of the building. These ceremonies are depicted on the walls of a number of tombs and temples in Egypt and were always part of a royal ceremony, even if Pharaoh sent a representative (Porter & Moss 1972: Sanctuary of Thutmose III at the Temple of Amun-Re at Karnak). Therefore, they must be viewed in the context of royal ritual. By the beginning of the New Kingdom, the form of the ritual objects was well established including the form of the inscriptions on the objects. The inscription on the adze in this study (M80-203-38) is of the form used in foundation deposits and would probably have been buried with other objects including brick moulds, grinders, sacrificed animals, and votive objects.

The Collection and Typology

The twenty copper alloy tools and weapons examined in this study can be broadly grouped into three subsets: five axes and adze; seven knives; and three daggers. Clearly these categories are somewhat arbitrary (small knives may have been used as razors for example), but since they are among those categories originally used by Sir Flinders Petrie they function as part of a standard in typologies of Egyptian tools and weapons.

The objects were examined in terms of typology and with regard to the hieroglyphic inscriptions which are on many of them. In addition each object was examined under a microscope to investigate the nature of the corrosion layer, and a small (approximately 20-30 mg) sample was taken for metallographic analysis. Finally, a 25 mg sample was taken for trace element analysis using inductively coupled plasma emission spectroscopy (ICP).

In most cases, objects very similar to the twenty aforementioned tools and weapons can be found in Petrie's publication (Petrie 1917) or in the British Museum publication by Davies (1987). Therefore, one can began to date the objects based strictly on stylistic grounds. In the subsequent discussion of the objects such a comparison is made.

Once one is convinced of their authenticity, the hieroglyphs on many of the objects are also a tremendous aid in the dating of the pieces. Although the area around some of the hieroglyphs has been cleaned at some point in the past, there is still a layer of corrosion in the chased-in grooves which make up the inscriptions on all but one of the objects. This corrosion is a clear indication that the inscriptions are ancient. Many of the inscriptions include royal names and therefore aid in dating the objects once they are determined to be authentic.

Metallurgical and ICP Analysis

The veracity of an object's ancient status can be difficult to ascertain in the case of metals. For bronzes, often the best indication of age is a deep corrosion pattern running into the object itself. Whereas a patina of corrosion and even crystals of *cuprite* (Cu_2O) may be applied to the exterior of an object, the pattern of corrosion running intergranularly, dendritically or interdendritically is virtually impossible to create in a short period of time. Therefore, by taking a small sample from the object, it is often possible to determine if it is old. Particularly in the case of the set of knives which have an odd paint applied to the hieroglyphs, there has been some concern as to whether, in fact, they are ancient. Although the corrosion proceeds, in a continuous layer, under the paint in the grooves of the hieroglyphs it was felt that metallographic sampling of the objects would conclusively indicate their antiquity. Metallography can also give important clues as to the methods used in the manufacture of an object (Scott 1991). In addition, one can often tell a little about the composition of a copper-based alloy from metallurgical samples.

As previously discussed, many of the objects can be dated based on style (particularly the knives and axes) as well as on the hieroglyphs. Those which cannot are ideal candidates for ICP analysis, to determine the elemental composition of the object. In Egypt, the addition of appreciable amounts of tin (greater than 1%) seldom occurred prior to the Middle Kingdom (2040 - 1640 BC) and did not become commonplace until the Second Intermediate Period (1640 - 1532 BC) or the 18th dynasty (1550 BC). Therefore, large amounts of tin in an object probably indicate that the object is not older than the Second Intermediate Period. Furthermore since tin was a valuable commodity, the high content of tin in an object may indicate that it was intended as a high status object.

In the case of ancient metals, the issue of when and what alloy elements were introduced into the metal matrix has always been of interest to archaeologists. The introduction of bronze metallurgy has often been seen as a marker used to indicate increasing social complexity and craft specialization. But the issue of when the ancient Egyptians first began to use bronze is complicated by the relative scarcity of tin in Egypt before the Second Intermediate Period, and by the fact that lead was also used for making copper alloy objects.

To better understand their intended use as well as to help place them temporally, trace element analysis was performed on many of the objects in this collection. The method used is a form of atomic emission spectroscopy, known as Inductively Coupled Plasma (ICP). The basic operation involves dissolving a small amount of sample in dilute acid. The liquid is then pumped into the ICP machine where a plasma is struck and the resulting light wavelengths and intensities are measured. The following are the results of metallurgical and ICP analysis (summarized in Table 1 at end of paper) for the objects sampled.

The Knives

This group is made up of seven small knives (M80-203-23, 24, 25, 28, 29, 30, 31) all inscribed with the same hieroglyphs (Fig. 1). These knives consist of varieties which would have been useful for many day to day activities such as shaving, carving and cutting fabrics. The majority of the knives have a double-convex edge (a leaf shape blade) and the blade and handle are cast in one piece. This type of knife was common throughout Egypt from the 12th dynasty through the New Kingdom (approximately throughout the second millennium BC) although the shape did vary a little with time.

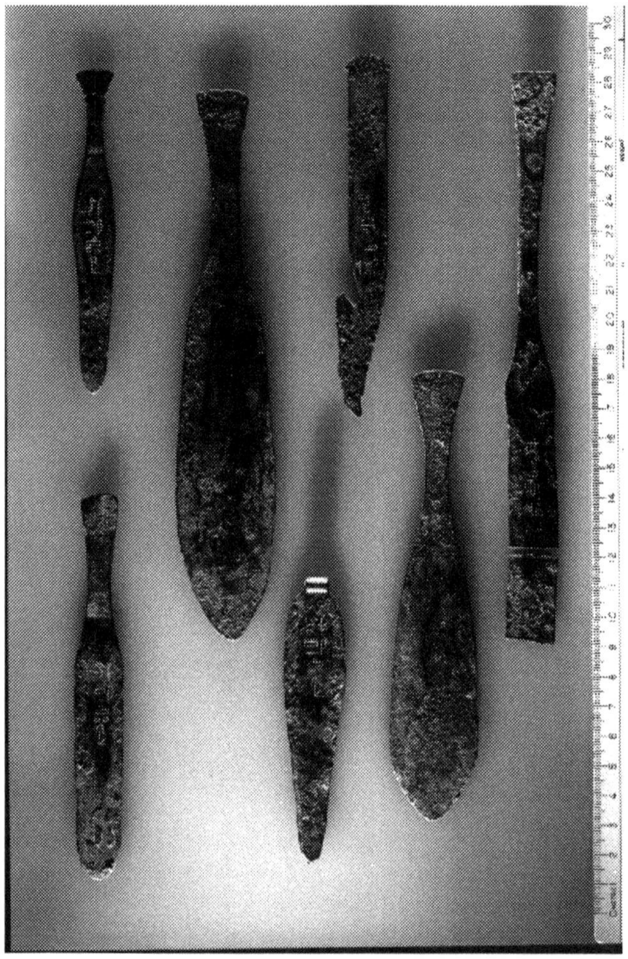

Figure 1. Knives M80-203-23, 24, 25, 28, 29, 30, 31

All of the knives studied could have been made by casting and then cold working the metal or by simply hammering a cold piece of copper until it was thin, although there seems to be some indication of casting dendrites in one of the samples. The additional step of annealing the thinned and work hardened metal would have substantially improved the plasticity of the knives so they would be more durable. In most cases it can be clearly seen in the samples that this additional step was undertaken. Although most of the knives were too corroded to yield good ICP samples, a qualitative XRF spectrum did indicate some tin in several of the samples.

The inscriptions on this set of knives (which read "son of the king Amenophis") were so badly enhanced in recent times, with both modern tools and paints, that they will always appear suspect. However, the incised lines on the knives do contain a corrosion product consistent with that covering the rest of the objects - an indication that the inscriptions are possibly original to the piece.

Based on the typology of the pieces and on the inscriptions as well as the fact that the knives contained tin and were well made, it is probable that they are from the New Kingdom period (c. 1500 to 1000 BC).

Axes/Adzes

The four axes and one adze in this collection are of a varied nature both typologically and chronologically. The collection consists of two axe heads from the Early Dynastic (beginning of the third millennium BC), as well as a model "battle axe" from the 17th dynasty (c. 1575 BC) and a small round axe. Finally there is an interesting adze which cannot be dated absolutely due to a lacuna in the hieroglyph, but which was probably part of a foundation deposit. Regardless of their temporal and spatial diversity, the pieces are of great interest due to the importance of the inscriptional evidence on the objects themselves.

Figure 2. Axe M80-203-36

Figure 3. Axe M80-203-42

The axes and adze were all cast and than cold worked at the edge. In some cases the tools were annealed prior to the final working of the blade. The two Early Dynastic axes (M80-203-36, 42; Figs. 2 and 3) are in pure copper and one of them is inscribed with a lovely example of representations of two Horus figures on Standards (possibly a reference to the "two lands" of Egypt). The model battle axe M80-203-43 (Fig. 4), as previously mentioned, is inscribed with the cartouche for Seqenenre, a 17th Dynasty Theban King (c. 1775 BC). It is also in copper and is typologically similar to weapons found from the 10th dynasty to the Ramiside (c. 1550 to 1350: Davies 1987:61). The object was heavily cold worked.

Figure 4. Model axe M80-203-43

A second "model round battle axe" (M80-203-90, Fig. 5) with butt lugs for hafting onto a shaft, is the one example of an inscription which was clearly added in the modern era. As previously described, these hieroglyphs were painted on an ancient object. Unfortunately the object was too corroded to sample for ICP, but the metallography revealed a heavily cold worked specimen with some lead inclusions and no evidence of destannification. Based on typology the object is probably from the Middle Kingdom (c. 2040 - 1640 BC).

Figure 5. Model axe with modern inscription M80-203-90

The last object in this group is the foundation deposit adze (M80-203-38; Fig. 6) of a necked variety. Again, remnants of corrosion products in the incised lines of the inscription lead the investigator to believe that the inscription is ancient. The trace element analysis revealed minor amounts of tin and iron. Metallographic studies revealed flow lines from cold working and some destannification (Fig. 7). The presence of twin planes indicates annealing. In addition, there is some directionality to the corrosion and lead, which is another indication of cold working. On one edge there is a thick layer of dense red corrosion.

Daggers

The following daggers are all from the periods from the 18th to the 19th Dynasties (early to mid second millennium BC). In addition, two of the inscriptions probably refer to the reign of particular rulers and therefore can aid dramatically in the dating of the objects. Again, all of the inscriptions appear to be genuine. In the case of the

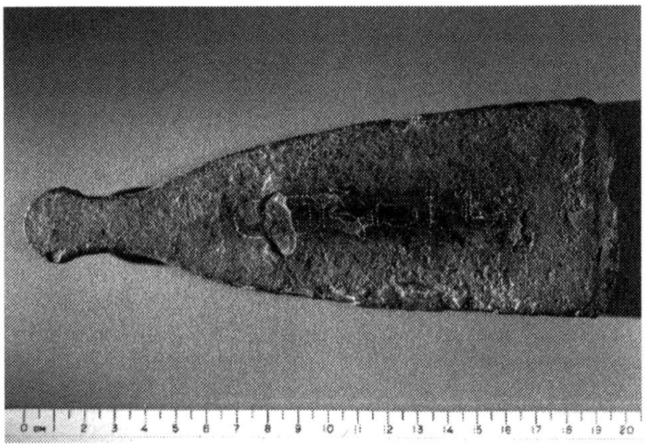

Figure 6. Adze M80-203-38

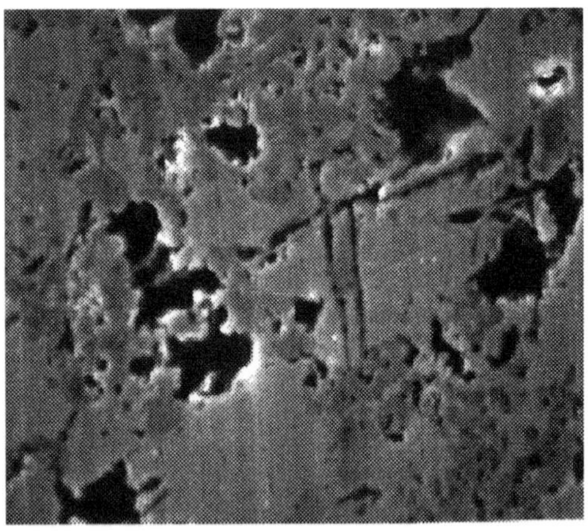

Figure 7. Adze M80-203-38, Scanning Electron Microscope image

first two objects this is determined by corrosion in the incised lines. In the case of the last, the inscription was cast into the object and must necessarily be original to the piece. The metallography of the daggers is very interesting. All of the daggers showed evidence of deep level corrosion. Further, the cast structure of the objects is very pronounced and allows for considerable speculation as to the processes used in the manufacture of the weapons.

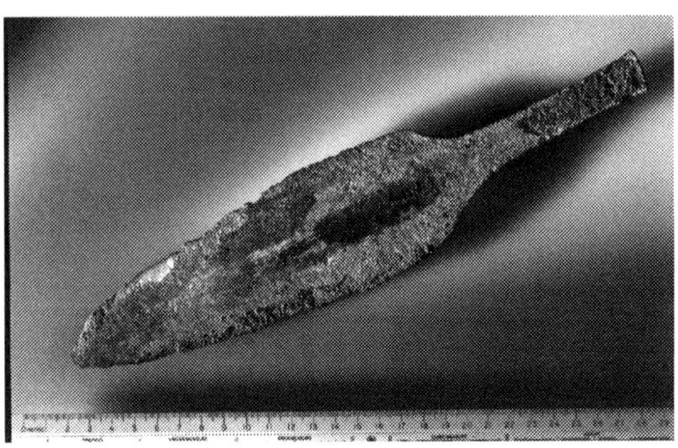

Figure 8. Dagger M80-203-39

The first dagger (M80-203-39; Fig. 8) is from the Middle Kingdom (2040 -1640 BC) and bears an inscription which reads "seal bearer of the King of Lower Egypt, the house of the ruler Meni". It contains traces of lead, tin and iron, and has an integral handle. Metallurgical sampling (Fig. 9) revealed an annealed cast structure. The sample had very large hexagonal grains, and many very distinct annealing twins. There is a variation in grain sizes and some lead which is elongated, indicating some cold working before annealing. There seems to be some redeposited copper in the corrosion layer.

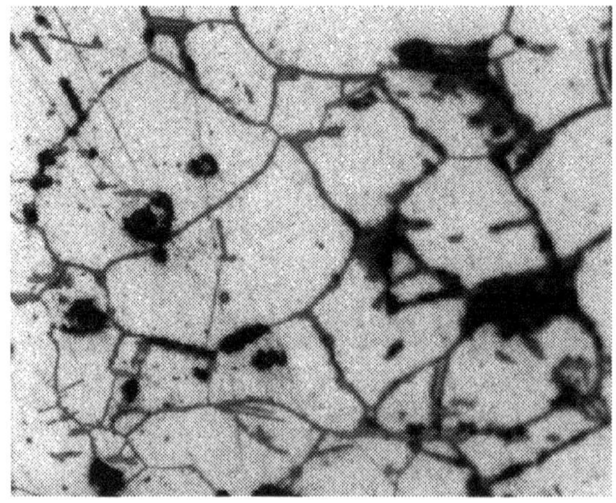

Figure 9. Dagger M80-203-39, micrograph x400

The second object (M80-203-48; Fig. 10) is a fragment of a dagger which had an inlaid handle. What we have now is part of the handle where it attached to the blade and the top section of the blade. The inlay is gone from the handle, leaving only the frame. ICP revealed significant amounts of tin which would have aided greatly in the casting process.

Figure 10. Dagger fragment M80-203-48

The blade has a cast set of hieroglyphs which reads "King of Upper and Lower Egypt" and then, in a cartouche, the name used by Ramesses II . This would indicate a date for the dagger of the 19th Dynasty (c. 1300 to 1200) during his reign, although the name was occasionally used in later times.

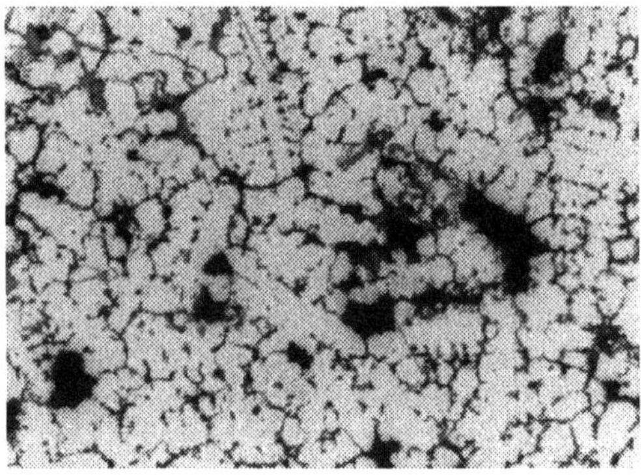

Figure 11. Dagger M80-203-48, micrograph x100

The metallurgical sample (Fig. 11) revealed the classic dendritic structure of a piece which was cast and not worked or annealed. The lack of working can be attributed to the fact that the sample was not taken from the blade edge (it is lost), where working to sharpen the dagger would have occurred. The dendrites show evidence of coring and there are inclusions in the interstitial spaces between the dendrites which may be oxides or a delta phase.

Figure 12. Dagger M80-203-46

The last dagger (M80-203-46; Fig. 12) is of a type not particular to a specific time in Egyptian culture although it is similar to one pictured in Petrie (Petrie Dagger D40:29). The hieroglyphs are New Kingdom and read "Son of the King, waj-msw" which is most probably a reference to the son of Tuthmosis I (c. 1528 to 1510 BC) from the New Kingdom. The results of ICP revealed significant amounts of tin in this bronze.

Metallurgical sampling (Fig. 13) showed a very fine grained, cold worked and annealed structure. There are some finely divided inclusions, which have also been elongated by cold working. In addition there is some destannification around the corroded areas, further indicating the presence of tin in the metal.

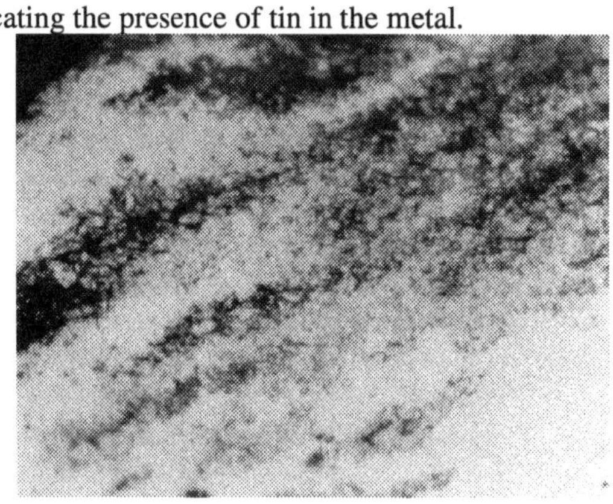

Figure 13. Dagger M80-203-46, micrograph x400

Conclusion

The purpose of this study was to investigate the antiquity of the objects, the authenticity of the inscriptions, and the objects' relative importance to the study of Egyptology. To that end the objects were examined typologically, microscopically, and metallographically. Further, the hieroglyphs were examined and compared with the typologies to insure that they were temporally aligned with the artifact typologies. Finally, trace element analysis was conducted and used as a further confirmation of the temporal sequence and to indicate function.

By beginning with the typologies, the assumption was made that the objects must be classified by known archaeological methods and that all the other means of analysis should agree with the original assessment. In most cases, the objects did compare favourably with objects listed in some of the standard references used for ancient Egyptian typologies. There was one exception, the rather impressive dagger M80-203-36 whose type is not listed in Petrie and for which there appears to be no direct parallel in other references.

Microscopic examination revealed a good corrosion layer on all of the objects which is an indication of antiquity. Many of the objects had developed large *cuprite* crystals as well as *malachite* ($Cu_2CO_3.(OH)_2$) patinas. All but one of the objects' hieroglyphs (M80-203-90) appear authentic although some have been augmented with white paint and/or cleaned in the area around the inscription. The best indicator of antiquity in the case of the inscriptions is that the corrosion layer is well developed down in the incised lines which form the hieroglyphs. Even on those objects which were cleaned, the cleaning was not so vigorous as to have removed the corrosion in the incised lines. In addition, the hieroglyphs are all well made and the dates indicated by the inscriptions are all in good agreement with the dates indicated by the typologies and other scientific examinations.

There is one exception to the general positive response to the inscriptions. In the case of M80-203-90, a model axe-head, the inscription is only painted on and the forms of the hieroglyphs themselves make this a highly suspect

inscription. It is also worth mentioning that on the Early Dynastic axe-head numbered M80-203-42 the white powder used to augment the Horus figures is probably from antiquity as the corrosion layer has grown over the white pigment.
Metallographic samples were taken from most of the objects. In all cases, the sample revealed deep inter-granular corrosion which is an excellent indication that all the objects are indeed ancient. In addition most of the objects were heavily worked and annealed. This is further confirmation of the ancient Egyptians skill in metal working and is important because the fine lined inscriptions, of the type found on many of these annealed objects, might have been cut using a work-hardened bronze chisel.

Further interesting information regarding the objects came from the trace element analysis. The trace element analysis agreed with the predicted results based on the typologies. That is, that the Pre-Dynastic knife and Early Dynastic axe heads were indeed copper and had no significant alloy elements. The small amount of iron found in these objects may have been the result of iron flux used to smelt the copper or may have been a constituent part of the original ore. Since the addition of tin to copper objects did not become commonplace until the Middle Kingdom or Second Intermediate Period, it would be unusual to find large amounts of tin in an Early Dynastic object.

In no case did the objects appear to be modern constructions. In fact the metallography indicates that all the objects originated in antiquity. Further, the inscriptions and trace element analysis confirm both the fact that the objects are ancient and that they fit well into known typologies. Finally, initial studies of the inscriptions indicate that many of the objects may have been for ritual or burial purposes and were designed to convey a social message. Further study of the inscriptions by philologists and Egyptologists is clearly warranted.

Acknowledgments
The author would like to gratefully acknowledge contributions to this work by the following: J Twilley, Dr. P. Meyers, Dr. N. Thomas (Los Angeles County Museum of Art); Dr. S. Morris, Dr. D. Polz, Dr. S. Downey, Dr. A. Lopreano, D. Winter (University of California Los Angeles); Dr. P. Northover (Oxford University).

References

DAVIES, W.V. 1987. *A catalogue of Egyptian Antiquities in the British Museum, VII Tools and Weapons: I axes.* London: British Museum Publications.
JANSSEN, J. 1975. *Commodity Prices from the Ramessid Period.* Leiden: E.J. Brill.
KAPLONY, P. 1965. Bemerkungen zu einigen mit archaischen konigsnamen, *Mitteilungen des Deutschen Archaeologischen Instituts Abteilung Kairo* 20: 1-47.
PETRIE, W.M.F. 1917. *Tools and Weapons.* London: British School of Archaeology in Egypt.
PORTER, B. & R.L.B. MOSS. 1972. *Topographical Bibliography of Ancient Egyptian Heiroglyphic Texts, Reliefs and Paintings, Vol II Theban Temples.* Oxford Clarendon Press (2nd edn.).
SCOTT, D. 1991. *Metallography and Microstructure of Ancient and Historic Metals.* Los Angeles: Getty Conservation Institute.
SHAW, I. 1991. *Egyptian Warfare and Weapons.* Aylesbury: Shire Egyptology.
SWAIN, S. 1995. The use of model objects as Predynastic Egyptian grave goods: An Ancient origin for a Dynastic tradition, in S. Campbell & A. Green (eds.) *The Archaeology of Death in the Ancient Near East.* Oxford: Oxbow Monograph 51: 35-37.
WEINSTEIN, J. 1973. *Foundation Deposits in Ancient Egypt.* Unpublished disertation, University of Pennsylvania.

Table 1. ICP Results

Museum no.	Sb	Pb	Co	Ni	Sn	Fe	Ag
M80-302-31	0.031	0.436	0.017	0.177	8.01	1.37	0.01
M80-203-36	0.092	0.029	0.012	0.015	0.054	0.23	0.032
M80-203-38	0.015	0.048	0.013	ND	0.137	0.436	0.027
M80-203-39	0.028	0.109	0.051	0.045	1.19	1.02	0.028
M80-203-42	0.165	ND	0.012	ND	ND	0.235	0.187
M80-203-43	0.03	ND	0.051	0.007	1.31	0.229	ND
M80-203-46	0.012	ND	0.012	ND	11.05	0.095	0.01
M80-203-48	0.06	0.71	0.027	0.026	8.32	0.04	0.026

Values in percentages. ND (not detected) means less than 0.005%. Analysis for zinc was carried out, but in no case was zinc detected

22

A Study of Ingots and Metallurgical Remains from 'Ein Ziq and Be'er Resisim, Central Negev, Israel

I. SEGAL[1], L. HALICZ[1] & R. COHEN[2]

[1] *Geological Survey of Israel, 30 Malkhe Israel Street, Jerusalem 95501, Israel*
[2] *Israel Antiquities Authority, P.O. Box 586, Jerusalem 91004, Israel*

Copper ingots, slags and lumps of copper found at two Middle Bronze Age archaeological sites in the Central Negev - 'Ein Ziq and Be'er Resisim - were studied chemically, isotopically and metallographically. Investigation of the ingots showed similarity to ores from Timna and Feinan, known to have been exploited for smelting during the Middle Bronze Age I period. Metallographic study of the ingots revealed that secondary casting had taken place on the sites. The presence of manganiferous slags, the high cobalt content of the iron slags and the lead isotope ratios in the slags, all point to the Feinan region as the ore source for the ingots.

Keywords: Middle Bronze Age; Copper Ingots; ICP-AES; ICP-MS; Metallography; SEM-EDS.

Introduction

The existence of local copper production in southern Israel during the Middle Bronze Age I (MB I) was indisputably established by the discovery of mines and a smelting site dating to this period at Timna (Rothenberg & Shaw 1990). Copper metallurgy was also conducted in Transjordan from the Early Bronze Age with continuation on a smaller scale into MB I (Hauptmann 1991, Hauptmann *et al.* 1992). There is ample archaeological evidence for metalworking technology in this region (Merkel & Dever 1989). Bar-shaped ingots, triangular in section and of unalloyed copper, which is distinctive for this period, were found at Lachish (Tufnell 1958), in the Central Negev at Har Yeruham (Kochavi 1967), Be'er Resisim and 'Ein Ziq (Cohen & Dever 1979), and in the Hebron Hills (Dever & Tadmor 1976). Seven bar ingots from Har Yeruham and the Hebron Hills were examined metallurgically by Maddin & Stech Wheeler (1976). The greatest number of these ingots was found at 'Ein Ziq and Be'er Resisim - thirty-five unbroken ingots and fragments. Besides the ingots, some small amorphous pieces of slag and copper lumps were collected from the ground surface at 'Ein Ziq and unidentified copper fragments at Be'er Resisim.

The aim of the present investigation was twofold: i) to attempt a reconstruction of the ancient metallurgical technology for ingot production, and ii) to define the provenance of the ingots. To achieve these goals, the chemical, isotopic and mineralogical composition, as well as the metallographic structure of slags, prills and metal artefacts was determined and analysed.

Brief Geological Information

Because they are both situated at the western margin of the Wadi Arabah rift valley, the type and origin of the copper ore deposits at Timna and Feinan are similar. Both districts therefore have a comparable mineralogical and geochemical structure. According to Hauptmann (1990) and Hauptmann *et al.* (1992), two main types of ores were exploited in ancient Feinan: the first is Massive Brown Sandstones (MBS) with copper mineralization as *malachite* ($Cu_2CO_3(OH)_2$) and the primary copper sulfides *chalcocite* (Cu_2S) and *covellite* (CuS). MBS was used for smelting mainly in the Chalcolithic period. The second is the deeper Cambrian layer known as Dolomite Limestone Shale (DLS), which includes copper silicates (*chrysocolla*, approx. $CuSiO_3.2H_2O$) and *malachite*. Copper ores from the base of DLS were used extensively in later periods, from EB II with continuation, on a smaller scale, into MB I (Hauptmann *et al.* 1992; Hauptmann & Weisgerber 1992; Najjar *et al.* 1995).

At Timna, according to Segev & Sass (1989) and Segev *et al.* (1992), copper ores occur in two stratigraphic layers, the Lower Cretaceous and the Cambrian. The Lower Cretaceous layer comprises the Amir and Avrona Formations. Sandstones in these layers contain *malachite* and *(par)atacamite* ($Cu_2(OH)_3Cl$, but with different structures), which are the result of oxidation, and primary *chalcocite* and *covellite*. These ores were exploited in Timna from the Early Chalcolithic period (Rothenberg 1988; 1990). The deeper Cambrian layer consists of Shehoret and Timna Formations. Sandstones, presenting in Shehoret and partly in Timna Formations, contain *chrysocolla*, *bisbeeite* (approx. $CuSiO_3.H_2O$), *malachite* and minor amounts of *chalcocite*. The Timna Formation is also partly represented by dolomites with *djurleite* (approx. Cu_2S), *paratacamite* and *malachite*. The *djurleite* is surrounded by an oxidized rim composed of dolostones. Manganese and phosphate mineralizations are associated with iron oxide in the sandstones (Segev & Sass 1989; Segev *et al.* 1992).

At Timna there is only a single archaeological site, excavated in 1989-90 by Rothenberg and Shaw, ascribed to MB I (Rothenberg & Shaw 1990). It is located to the northeast of the Timna Valley (Site 149), and its archaeometallurgy is presently under study. Ancient mines were discovered nearby (Site 250), where copper veinlets within the dolomite are widely spread. The main copper minerals in these veinlets are *malachite*, *bisbeeite*, *chrysocolla* and *atacamite*.

Analytical Procedure and Sample Preparation

The chemical composition of the ingots, slags and lumps was determined by Inductively Coupled Plasma Atomic Emission Spectrometry (ICP-AES) using Jobine Yvon JY-38 mono- and JY-48 polychromators. Tin, silver and antimony in the ingots were analyzed using a Perkin Elmer Sciex Elan 6000 Inductively Coupled Plasma Mass Spectrometer (ICP-MS) equipped with flow injection system for sample introduction. Silicon, sulfur and phosphorus were determined using Scanning Electron Microscopy with an Energy Dispersive X-ray analyser (SEM-EDS). The operating conditions for slag analysis are described by Beyth *et al.* (1988) and for metal analysis by Segal *et al.* (1994). The precision, or relative standard deviation (RSD), of the analyses is estimated to be 1% for major elements, 3% for minor elements and about 10% for trace elements. The limits of detection (LOD) for the metallic elements in solution by ICP are between 0.1 and 20 μg/l.

Lead isotope studies were carried out using ICP-MS. The precision of this method was not more than 0.3% (Halicz *et al.* 1996) WHICH RATIOS -204 WORSE?. The mineralogical composition of the slags was examined using a Philips X-ray powder diffraction spectrometer (XRD). The metallographic structure of the ingots and lumps was studied under a Nikon optical metallographic microscope. Detailed micro-petrographic and mineral chemistry studies were performed using a Scanning Electron Microscope (Jeol 840) equipped with a Back Scattered Electron Detector (BEI) and an Energy Dispersive Spectrometer (SEM-EDS). In addition, Vickers standard hardness tests were carried out.

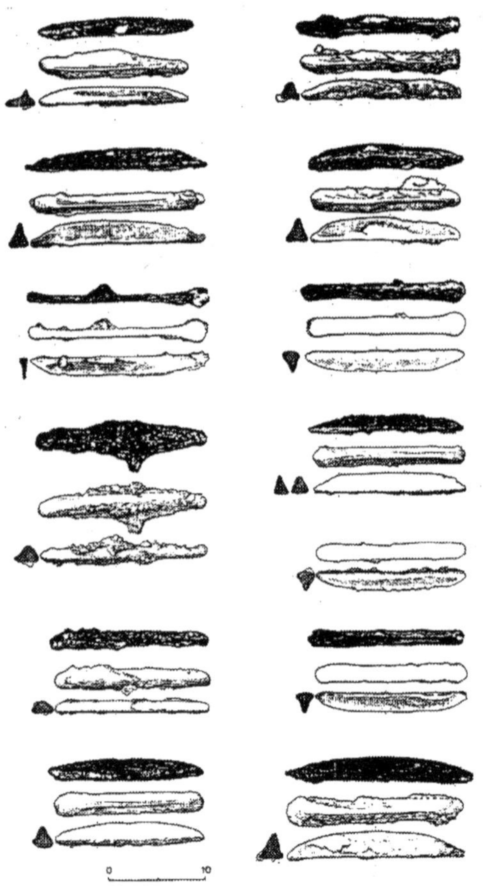

Figure 1. Well preserved copper ingots from 'Ein Ziq and Be're Resisim. Note various form of cross section – traingular, round and thin elongated.

Figure 1 shows a selection of unbroken ingots with various forms of cross sections - triangular, rounded and thin elongated. Almost all of them exhibit shrinkage on the uppermost surface. The slags are amorphous in form, up to 2 cm in size and coloured green. Section drawings of two copper fragments found at Be'er Resisim are shown in Figure 2. They consist of copper with adhering calcitic mould lining. In addition, one of them has a small piece of attached slag.

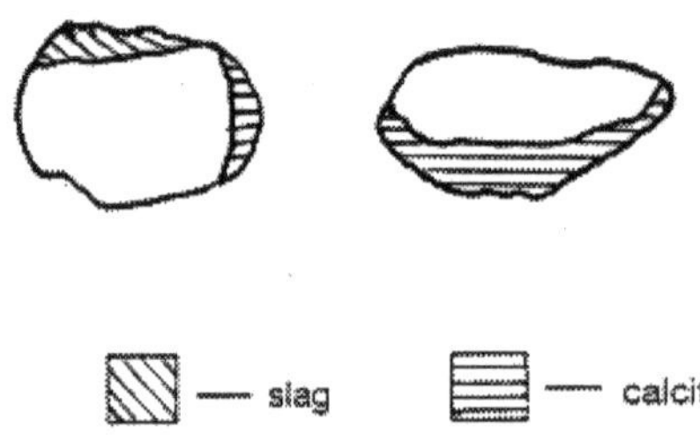

Figure 2. Section drawings of the copper lumps found in Be'er Resisim.

Initially the samples were sectioned, the ingots in transverse section, the slags and lumps in free sections. A slice from all samples was mounted in resin, ground and finely polished. Aqueous ferric chloride and acidified potassium dichromate solutions were used for etching. Additional pieces from each sample were cleaned, drilled or crushed and powdered for the chemical, isotopic and mineralogical examination.

Results and Discussion

Ingots

a) Chemical composition

The chemical composition of the ingots is given in Tables 1 and 2 (at end). Elements which were consistently observed to be below the LOD in all samples are omitted from these tables. The data indicate that for all ingots lead (Pb) and iron (Fe) concentrations are significant, varying from 0.2-3.8% and 0.02-2.84% respectively. Iron is dissolved in the copper and is also present in inclusions as slaggy material (iron oxide or phosphide). Remnants of the latter point to the use of iron or mixed copper/iron ores as a flux. Arsenic, nickel and cobalt occur in all ingots as a trace element. Although in Timna and Feinan ores there are only small amounts of lead and arsenic, these elements are both present in the ingots. As an explanation for this, Merkel (1990) noted that re-melting and refining the black copper leads to the concentration of lead and arsenic in the metal, not in the slag. Ingot 18 (Table 1) contained 0.3% P, as determined by SEM-EDS.

Table 3 presents medians for the ingots from 'Ein Ziq and Be'er Resisim(this study) compared with published medians for Bronze Age copper from Timna (Leese *et al.* 1986) and Feinan (Hauptmann *et al.* 1992). Comparison shows that the MB I ingots from 'Ein Ziq and Be'er Resisim differ from the Timna copper objects but are similar to those from the Feinan region.

In order to ascertain the provenance of the ingots in the present study, we compared their composition with copper ores and slags from the Timna MB I archaeological sites

Element	Timna	Feinan	Central Negev
Ag	0.016	0.001	0.002
As	0.068	0.033	0.054
Sb	0.002	<0.001	0.0007
Co	0.002	0.030	0.042
Ni	0.031	0.025	0.047
Zn	0.008	0.011	0.031
Pb	0.303	0.300	0.620

Table 3. Medians of several elements in Bronze Age copper objects.

149 and 250 (chemical data were provided by B. Rothenberg) and ores from Feinan. Figure 3 shows the cobalt/nickel (Co/Ni) and antimony/arsenic (Sb/As) abundance ratios. The fields for copper ores from Timna (Amir/Avrona Formations) and Feinan (MBS and DLS) are those defined by Hauptmann *et al.* (1992). The ores and slags from Timna sites 149 and 250 are located in the DLS field of Feinan. Taking the archaeological samples as representative of the Timna Formation, it can be concluded that, with respect to these parameters, Timna Formation ores are similar to those from Feinan DLS, rendering it impossible to discriminate between these two ore sources using these ratios. Figure 3 also shows that the ingots in the present study fall into the same region, suggesting that they were manufactured either from Timna Formation or Feinan DLS ores. As shown below, the chemical composition of the slags points to Feinan as the ore source. Contrary to that, the lead isotope ratios of an ingot from 'Ein Ziq, determined by Stos-Gale at the Isotrace Laboratory in Oxford, "was fully consistent with its origin from Timna ores" (Rothenberg *et al.* 1991). It is unclear, however, which copper ores (Amir-Avrona or Timna Formation) were employed by these authors to define the lead isotope ratio field for Timna.

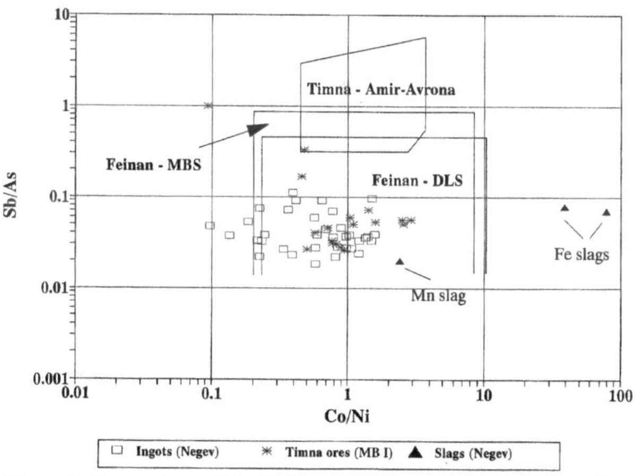

Figure 3. Comparison of the Co/Ni and Sb/As ratios in ingots from the Negev and Cu-ores and slags from Timna sites 149 and 250 (this study) with the ore fields as defined by Hauptmann et al. *(1992).*

b) Metallography

i) Crust

All the ingots (except 17) show a thin upper and, in several cases, a lower crust. Examination of this can be helpful for the provenance study of the ingots (Roman 1990). The upper crust is more porous than the metal and comprises several layers. The external layer consists of iron-free copper oxide (*cuprite,* Cu_2O), which is underlain by a lead layer, and then the copper metal itself. Figure 4 shows penetration of the lead between the copper grains from the middle part of ingot 1 to the surface. The lead layer in the crust occurs only in ingots with triangular or round cross sections, with lead being partly or completely absent inside. This phenomenon may occur during a slow solidification process because of the low melting point of lead, which causes the lead-rich phase to solidify only at the end of the process. In contrast, ingots with a thin elongated section all have the lead dispersed throughout the ingot and no lead layer in the crust. This is the result of rapid solidification due to the thinness of the mould.

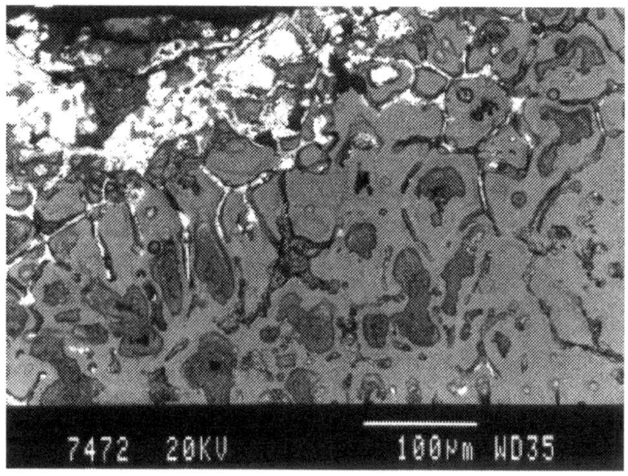

Figure 4. Penetration of lead (white) through the copper grains from the middle part of the ingot to the surface (Ingot 1, for example).

The porous crust is filled with slag material of variable composition, e.g., various silicates in ingots 2 and 3; *barite* ($BaSO_4$) and silver (Ag) inclusions in ingot 9. Relatively large inclusions of *wüstite* (approx. FeO) and iron phosphide, which are slag remnants, are located between the external copper oxide and internal layers in ingots 7, 11 and 18. The presence of manganese in iron-rich slag-like material and arsenic in lead-rich globular inclusions is particularly noteworthy. The thin lower crust, when present, contains *quartz* (SiO_2) inclusions and a small amount of globular lead phases.

It was previously stated that the ores used for smelting in MB I were either from the eastern region of Timna (Site 250) or from Feinan, located in Transjordan to the northeast of Timna. Both ores contained *chrysocolla*, *malachite* and a small amount of *chalcocite*, a product of incompletely oxidized ore. Manganese (Mn) minerals occur in these ores together with *barite*, silver and apatite group minerals (phosphates). The composition of the slag remnants in the crusts of the ingots suggests that they were smelted from this type of ore, followed by casting in a mould.

ii) Metal

Backscattered electron image (BEI) of the samples revealed a granular structure even without etching, because of the presence of various inclusions between the grains. After etching all the samples exhibited an equiaxial grain structure with some random elongation. The grain boundaries are decorated with copper sulfides, copper-iron sulfides and lead inclusions (Fig. 5). Metallic lead also occurs within the copper grains. When lead is located within the copper sulfide inclusions it appears as small white spheroids. Occasionally, lead is associated with arsenic. The presence of rare lead sulfide inclusions indicates that some *galena* (PbS) occurred in the ores. Remnants of copper sulfide show that the oxidized copper ore, from which the ingots were originally produced, did not undergo complete oxidation.

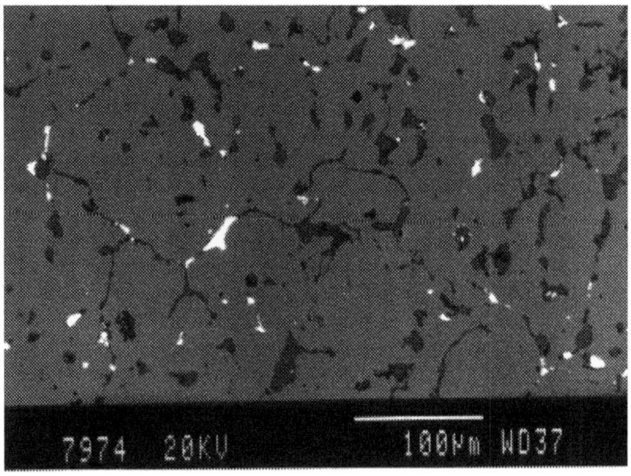

Figure 5. Structure of ingots with thin elongated form of section. Lead is distributed throughout all the ingot.

In the case of three ingots (2, 5 and 6), the ferric chloride etchant was not effective. When etched with potassium dichromate, grains covered by numerous spherical and splatter inclusions were observed. In ingots 2 and 5 these inclusions consist of copper sulfides containing tiny black iron oxide inclusions. The shape and composition of these clusters in ingot 6 is typical of a Cu_2O-Cu eutectic microstructure. Four ingots (7, 11, 18 and 19) contain rhombohedral iron phosphide inclusions and elemental phosphorus in the copper matrix. Microhardness of the ingots varied in the range of 58 - 106 HV.

Slags

a) Chemical composition

Overall chemical and mineralogical composition of the slags (Table 4) shows that the iron content of the 'Ein Ziq slags is high despite the fact that silica (SiO_2) occurs in rather low concentrations. XRD analysis revealed only *magnetite* (Fe_3O_4) in the phase composition. Copper occurs as small metallic and oxidic spheroids sometimes containing copper sulfide inclusions. The cobalt content is surprisingly high (up to 5% in the overall composition, up to 14% in the local *magnetite* analyses, up to 2% in the matrix and up to 0.9% in the copper inclusions). Nickel in these slags is significantly lower. It is known from previous work that cobalt correlates with iron, nickel and manganese in slag (Merkel 1985). In the Timna slags, according to published data, a high cobalt content was not found (Rothenberg 1990). Cobalt in Feinan copper ores is an order of magnitude higher than in Timna ores (Hauptmann *et al.* 1992). Up to 0.1% cobalt has been reported in the copper ores from Wadi Dana and 0.22% from the Khirbet el-Jariye region, which are located in Jordan (Hauptmann *et al.* 1992). Nickel in these ores is very low. Early Bronze Age raw copper produced at Ras en-Naqab (Jordan) was found to contain 0.2% cobalt (Hauptmann 1989). The high cobalt content of the slags reported here therefore suggests Feinan as a probable ore source.

Slag attached to the copper fragment from Be'er Resisim (42010-19-1) contained 28.5% manganese in the form of *tephroite* (Mn_2SiO_4). The chemical and mineralogical composition of this slag is very similar to the Chalcolithic and Bronze Age slags from Feinan (Hauptmann & Weisgerber 1987). The investigation of ores and slags from Feinan has shown that "from the Early Bronze Age until the Late Iron Age predominantly copper-manganese ores from the Dolomite-Limestone-Shale Unit were exploited" (Hauptmann *et al.* 1992). Manganese ore mined in these periods was naturally intergrown with the copper ore and was not added as a flux. At Timna, there is no evidence to date of the use of manganese ores at this time. Slags from the single MB I site 149 excavated by Rothenberg & Shaw (1990) did not contain such high manganese (Rothenberg, pers. comm.). Presence of *tephroite* slag is an additional argument in favour of Feinan. The latter also supports Goren's conclusions from his study of ceramics from the Central Negev (Goren 1996).

b) Mineralogy

All slags from 'Ein Ziq are similar in structure, and all of them are of *magnetite* type. Figure 6 shows tiny round *magnetites* surrounded by a *delafossite* ($CuFeO_2$) rim. The small white spheroids consist of copper and sometimes of copper oxide. Other areas of the same slag show larger, polygonal crystals and dendrites of *magnetite*, and larger copper inclusions. The matrix of all areas is melted *pyroxene* (a family of mixed metal silicate minerals) without crystalline structure. The presence of iron and copper in oxide form suggests that smelting conditions were insufficient for their reduction. The iron was not reduced below the *magnetite* phase, and no *fayalite* (Fe_2SiO_4) was formed. This poor reduction was probably the result of an uncontrolled ore-flux and fuel ratio.

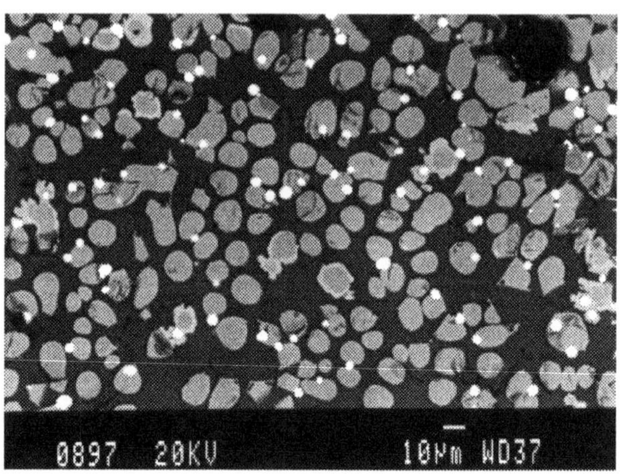

Figure 6. Typical structure of 'Ein Zig' slags. Tiny round magnetites (grey) sometimes is surrounded by delafossite rim (lighter grey).

A small piece of slag attached to the copper lump (see Fig. 2) reveals mainly elongated *tephroite* crystals embedded in a *pyroxene* matrix. Copper inclusions in this type of slag are very rare, which is why the copper content in the bulk composition is rather low.

Copper Lumps

Both copper fragments from Be'er Resisim (Fig. 2) showed a calcite crust (chalk, $CaCO_3$), probably from the mould lining. Pieces of chalk were also attached to some of the ingots. From a palaeontological examination of this chalk, it was concluded it was a type of chalk which is widespread in the Central Negev region, rare in Feinan region and completely absent from Timna.

Chemical composition of the copper lumps (Table 5 – at end) shows that both lumps from Be'er Resisim contain rather high lead and iron. In contrast, the lumps from 'Ein Ziq are made of relatively pure copper containing only copper sulfide and lead inclusions. Grains outlined by copper sulfides at the boundaries can be seen without etching. The lumps from 'Ein Ziq are very similar to the ingots both in their structure and chemical composition, and are waste from the casting process. The copper lumps from Be'er Resisim, on the other hand, are thought to be waste from the smelting.

Description	$^{207}Pb/^{206}Pb$	$^{208}Pb/^{206}Pb$	$^{204}Pb/^{206}Pb$
Be'er Resisim			
3214-387/111	0.8692	2.1177	0.05562
3004-365	0.8712	2.1220	0.05612
F-110-32	0.8703	2.1137	0.05592
3214-385	0.8677	2.1156	0.05562
'Ein Ziq			
85-779 78-443/11	0.8735	2.1221	0.05576
85-812 78-443/10	0.8720	2.1168	0.05556
85-813 78-443/1	0.8693	2.1162	0.05556
85-814 71-338/1	0.8692	2.1200	0.05560
85-815 78-443/2	0.8714	2.1199	0.05551
85-816 78-443/7	0.8702	2.1203	0.05550
85-817 78-443/3	0.8680	2.1102	0.05550
85-818 78-443/9	0.8720	2.1170	0.05571
85-819 78-443/6	0.8735	2.1157	0.05561
85-820 78-443/12	0.8739	2.1194	0.05560
85-822 78-443/4	0.8715	2.1175	0.05551
26-170	0.8701	2.1157	0.05585
1	0.8683	2.1119	0.05552
2	0.8694	2.1238	0.05572
3	0.8728	2.1233	0.05572
5	0.8739	2.1157	0.05571
6	0.8722	2.1189	0.05555
9	0.8687	2.1162	0.05562
11	0.8711	2.1197	0.05592

Table 6. Lead isotope ratios for ingots from the Central Negev.

Lead Isotope Ratio

All the ingots, slags and lumps were analyzed for lead isotope ratios (see Tables 6 and 7). Results for $^{208}Pb/^{206}Pb$ vs $^{207}Pb/^{206}Pb$ are shown in Figure 7a and $^{204}Pb/^{206}Pb$ vs. $^{207}Pb/^{206}Pb$ in Fig. 7b. For comparison with the relevant ores, results from Feinan (Hauptmann *et al.* 1992) and Timna (Gale *et al.* 1990) ores are plotted. In addition, ratios for slags and copper artefacts from Feinan are shown on the same figures. Feinan MBS ores (Hauptmann *et al.* 1992) plot to the lower left of the graphs, whereas values for Feinan DLS ores are significantly higher. Isotopic measurements from the Chalcolithic slags from Feinan cover a wide range showing that both types of ore were used in the Chalcolithic. From the EB II Age onwards only DLS ores were utilized in Feinan. The lead isotope ratios for Timna ores plot in the high right corner and are isotopically indistinguishable from Feinan DLS. As isotope ratios for the most of 'Ein Ziq and Be'er Resisim samples also fall into the high right end of the range, they are in a good correspondence with both the Feinan and Timna ores used in the MB I period.

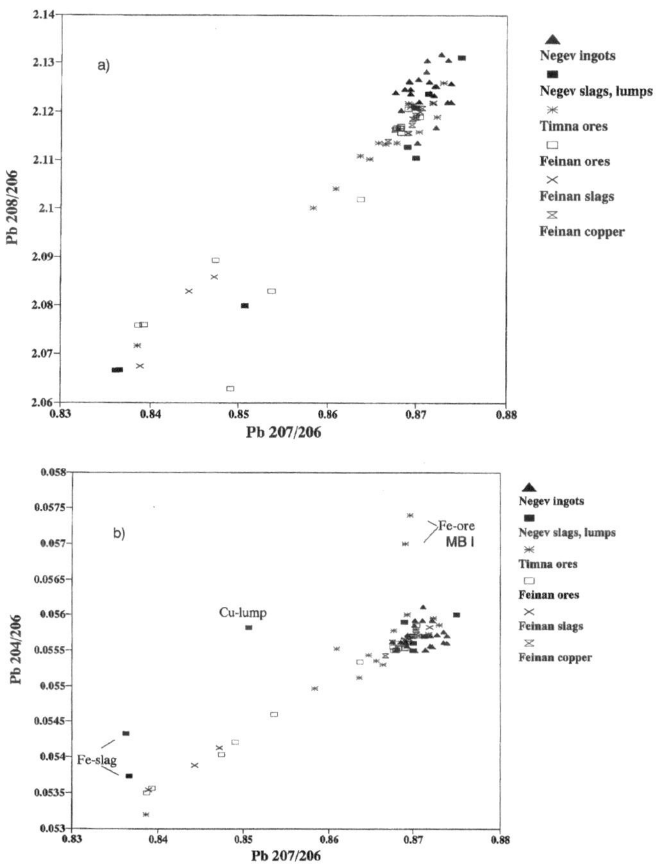

Figures 7a,b. Lead isotope ratios for ingots, slags and copper lumps from the Central Negev in comparison to Timna and Feinan ores, slags and copper objects.

Conclusions

There is no difference in either chemical and isotopic composition or structure between the ingots from 'Ein Ziq and Be'er Resisim. Moreover, they resemble those from Har Yeruham, Hebron Hills and from Har Sayyad, suggesting the same ore source and production technology for all Central Negev sites.

All the ingots are surrounded by a crust, which appears mainly on the upper surface. There are several layers; an external layer of copper oxide; a middle layer of metallic lead extruded to the surface during the solidification process, and an innermost layer of copper with inclusions. The lead layer occurs only in those ingots exhibiting a triangular or rounded section; in those with a thin, elongated, and tapering section, the lead is dispersed throughout the ingot. Chemical analysis of the crust shows variable remnants of manganese, arsenic, apatite, barite, iron silicate, oxide and phosphide. The presence of small amounts of copper sulfide (chalcocite) and copper iron sulfide in the copper itself indicates that the ores used for the smelting were incompletely oxidized. Since the ingots occasionally show remnants of lead sulfide, the ores also contained some galena. The presence of iron in the ingots suggests the use of iron ore as a flux. The latter was either deliberately added to the charge or naturally mixed with copper ores, providing a self-fluxing ore.

Four ingot types were ascertained on the basis of the composition of the inclusions. All the ingots contain copper sulfides and lead. The first type has no additional inclusions. The second (ingots 2, 5, and 10) contain iron oxide inclusions against a copper sulfide background. The third (ingots 7, 11, 18, and 19) contain iron phosphides. The fourth (ingot 6) reveals a Cu_2O-Cu eutectic. Ingots 9 and 12 can be related to the fourth type because of the presence of copper oxide spots.

The metal was secondarily cast in one-sided moulds with a triangular groove, as evidenced by equiaxial grain structure, shrinkage and smoothness on the uppermost surface of the ingots. Some others were cast in thin, elongated, tapering moulds.

The chemical composition of the ingots indicates that the copper ore used in the primary smelting did not originate in Timna's Amir-Avrona Formation. The MB I smelters derived their ore from the Timna Formation which is similar to Feinan DLS. We did not find any differences in the trace element content between these two sources. Lead isotope ratios also were not helpful in this respect. At the same time, medians of several elements in the ingots and slags from 'Ein Ziq and Be'er Resisim are very similar to Feinan slags. High cobalt in the iron slags and the presence of manganiferous slag also point to the Feinan region as ore source for the ingots.

The fact that no specimens of ore or furnaces were found at 'Ein Ziq and Be'er Resisim proves that ore smelting took place in areas close to the ore deposits. The most probable source was the Feinan region. On the other hand, the presence and subsequent examination of slags and copper lumps at both sites gave evidence for metallurgical activity at both sites - probably the re-melting (refining) of imported black copper and ingot casting.

Acknowledgments
The authors would like to thank the Israel Antiquities Authority, especially Dr. R. Cohen for his permission to examine the ingots produced by his excavations. We also wish to express our appreciation to the Geological Survey of Israel for placing at our disposal all the laboratory equipment necessary to carry out this study, and to the staff of the geochemistry division for their support of the work. We are particularly grateful to Drs. A. Segev, M. Bar-Matthews, N. Porat, N. Wolfson, and S. Ilani for helping us to understand the geology of the region under discussion; to Drs. A. Almogi and Ch. Benjamini for palaeontological studies, to Prof. B. Rothenberg for his supervision throughout our study; to Dr. J. Merkel for help and useful discussion, to Dr. I.B. Brenner for valuable comments; to the Graduate School of Applied Science (Hebrew University), particularly to Prof. I. Roman and N. Zacharov for her help in the metallographical study; to Y. Deutsch for his constant assistance; and to M. Dvorachek for help in the SEM investigations

Description	Lab. #	As	Ag	Pb	Ni	Co	Zn	Fe	Cu	Mn	Sn	Sb	Si	S
Be'er Resisim														
3214-387/111	9	0.044	0.0002	2.26	0.065	0.058	0.05	0.48	96.9	0.0003	0.0004	0.0006	ND	0.5
3004-365	10	0.038	0.0023	0.62	0.033	0.006	ND	0.15	98.7	0.0006	0.0001	0.0003	ND	0.5
F-110-32	13	0.071	0.0020	0.34	0.081	0.069	ND	0.33	98.9	0.0011	0.0004	0.0011	ND	1.1
3214-385	14	0.045	0.0019	1.28	0.074	0.051	0.04	0.46	98.0	0.0026	0.0003	0.0005	ND	0.5
'Ein Ziq														
85-779 78-443/11	17	0.110	0.0054	0.95	0.072	0.042	0.08	2.30	96.2	0.0026	0.0007	0.0011	ND	ND
85-812 78-443/10	8	0.021	0.0070	0.54	0.065	0.098	0.15	1.80	95.0	0.0014	0.0008	0.0006	2.4	0.5
85-813 78-443/1	16	0.052	0.0029	0.40	0.046	0.011	0.02	0.29	98.2	0.0004	0.0004	0.0004	ND	0.6
85-814 71-338/1	15	0.053	0.0016	1.70	0.045	0.006	ND	0.02	97.6	0.0001	0.0002	0.0007	0.1	0.2
85-815 78-443-2	1	0.060	0.0050	0.80	0.061	0.074	0.06	1.10	95.4	0.0014	0.0004	0.0007	0.6	2.2
85-816 78-443/7	6	0.060	0.0023	0.20	0.019	0.004	ND	0.30	97.9	0.0002	ND	0.0006	ND	1.2
85-817 78-443/3	4	0.075	0.0030	0.30	0.030	0.010	0.01	0.52	96.5	0.0006	0.0003	0.0006	0.3	1.7
85-818 78-443/9	5	0.052	0.0016	0.70	0.105	0.062	0.05	1.70	94.8	0.0010	0.0004	0.0004	ND	1.3
85-819 78-443/6	3	0.022	0.0007	1.10	0.051	0.033	0.05	0.29	95.9	0.0015	0.0003	0.0004	0.4	1.3
85-820 78-443/12	7	0.084	0.0029	1.20	0.074	0.089	0.07	2.40	95.9	0.0058	0.0007	0.0028	0.5	0.9
85-821 78-443/8	12	0.027	0.0120	0.49	0.068	0.015	0.02	0.45	96.6	0.0012	0.0002	0.0004	0.8	0.5
85-822 78-443/4	2	0.072	0.0040	0.80	0.057	0.033	0.02	1.48	97.5	0.0007	0.0004	0.0009	0.1	0.8
85-823 78-443/5	11	0.056	0.0008	0.62	0.064	0.086	0.14	1.83	96.4	0.0030	0.0008	0.0006	0.4	0.9
61-293	18	0.072	0.0070	0.26	0.054	0.053	0.12	2.50	95.5	0.0007	0.0009	0.0005	0.5	1.7
26-170	19	0.074	0.0007	0.43	0.054	0.058	0.12	2.76	94.8	0.0008	0.0008	0.0005	ND	0.5
84-446	20	0.029	0.0030	0.81	0.048	0.037	0.02	0.73	98.5	0.0019	0.0008	0.0003	ND	0.6

Table 1. Chemical composition of copper ingots from the Central Negev, wt%.

Lab. #	As	Ag	Pb	Ni	Co	Zn	Fe	Cu	Mn	Sn	Sb
1	0.028	0.0020	1.44	0.039	0.014	0.02	0.96	97.3	0.0038	0.0006	0.0004
2	0.022	0.0020	0.62	0.044	0.018	0.02	0.11	98.4	0.0009	0.0004	0.0004
3	0.062	0.0024	0.56	0.026	0.006	0.01	0.22	98.5	0.0003	0.0001	0.0006
4	0.042	0.0020	0.34	0.031	0.001	0.02	0.31	98.5	0.0001	ND	ND
5	0.060	ND	0.57	0.067	0.099	0.10	2.40	96.6	0.0033	ND	ND
6	0.055	0.0050	0.48	0.125	0.098	0.04	2.04	96.5	0.0033	0.0007	0.0008
7	0.091	0.0050	0.32	0.059	0.048	0.04	1.82	97.2	0.0020	0.0024	0.0010
8	0.051	0.0009	0.66	0.057	0.090	0.13	2.84	96.1	0.0020	0.0007	0.0006
9	0.052	ND	3.80	0.071	0.073	0.18	1.98	92.9	0.0170	ND	ND
10	0.086	0.0060	0.34	0.091	0.035	0.03	1.35	97.8	0.0017	0.0008	0.0011
11	0.091	0.0080	0.51	0.045	0.010	0.01	0.45	96.5	0.0004	ND	ND
12	0.054	0.0050	1.35	0.055	0.054	0.03	0.63	96.8	0.0009	0.0002	0.0005
13	0.034	ND	0.52	0.046	0.026	0.02	0.32	97.4	0.0016	ND	ND
14	0.054	ND	2.02	0.085	0.119	0.09	2.76	93.8	0.0015	ND	ND
15	0.018	0.0060	0.51	0.041	0.016	0.03	0.18	98.4	0.0002	0.0005	0.0006

Table 2. Chemical composition of uncleaned ingots from the 'Ein Zig', wt%.

Identification	SiO2	Al2O3	CaO	MgO	Fe2O3	TiO2	P2O5	SO3	Cu	Mo	Zn	Pb	Co	Ni	Ba	Mn	Cr	V	Sr
'Ein Ziq - 1	9.5	2.4	5.1	1.3	62.7	0.11	0.70	1.15	10.6	0.005	1.9	0.007	5.5	0.07	0.012	0.008	0.009	0.005	0.02
'Ein Ziq - 2	12.0	2.5	13.8	1.0	47.3	0.10	0.96	0.45	14.2	0.014	0.1	0.005	3.5	0.09	0.046	0.007	0.002	0.003	0.08
'Ein Ziq - 3	13.4	4.4	12.0	1.6	58.1	0.15	0.58	1.67	6.9	0.008	0.4	0.005	1.1	0.01	0.059	0.012	0.006	0.006	0.05
Mineralogy	Magnetite																		
42010-19-1	30.9	4.3	10.7	1.5	12.1	0.11	3.10	3.6	1.8	0.002	0.06	1.2	0.12	0.05	1.5	28.50	0.007.	0.010	0.11
Mineralogy	Tephroite																		

Notes: slag 42010-19-1 from Be'er Resisim was attached to the copper lump.

Table 4. Chemical and mineralogical composition of slags from the Central Negev, wt%.

Identification	Sn	As	Zn	Pb	Co	Ni	Fe	Mn	Ag	Cu	S
'Ein Ziq lumps											
Ground surface	N.D	0.22	N.D	N.D	0.071	0.027	0.034	0.0003	0.023	97.6	N.D
53-277-1	0.01	0.13	0.06	0.88	0.120	0.040	1.87	0.074	0.004	96.1	0.4
53-277-2	0.01	0.08	N.D	1.4	0.004	0.046	0.022	0.006	0.004	91.2	0.5
Be'er Resisim lumps											
42010-19-1	0.01	0.09	0.20	9.3	0.150	0.097	5.01	0.130	0.07	72.0	0.4
42010-19-2	N.D	0.09	0.16	5.5	0.033	0.053	1.66	0.038	N.D.	83.5	0.5

Table 5. Chemical composition of copper lumps from the Central Negev, wt%.

Description	$^{207}Pb/^{206}Pb$	$^{208}Pb/^{206}Pb$	$^{204}Pb/^{206}Pb$
Iron slag 1 'Ein Ziq	0.8362	2.0624	0.05433
Iron slag 3 'Ein Ziq	0.8366	2.0626	0.05373
Manganese slag 42010-19-1 Be'er Resisim	0.8690	2.1128	0.05590
Copper lump 'Ein Ziq	0.8507	2.0800	0.05582
Copper lump 53-277-1 'Ein Ziq	0.8750	2.1311	0.05600
Copper lump 53-277-2 'Ein Ziq	0.8713	2.1237	0.05570
Copper lump 42010-19-1 Be'er Resisim	0.8700	2.1209	0.05550
Copper lump 42010-19-2 Be'er Resisim	0.8700	2.1105	0.05560

Table 7. Lead isotope ratios for slags and copper lumps from the Central Negev.

References

BEYTH, M., M. SHIRAV, L. HALICZ & R. BOGOSH.1988. *The procedure: geochemical orientation survey*, Geological Survey of Israel Report GSI/15/88.

COHEN, R & W.G. DEVER. 1979. Preliminary report of the second season of the 'Central Negev Highlands Project', *Bulletin of the American Schools of Oriental Research* 236: 41-60.

DEVER, W.G. & M. TADMOR. 1976. A copper hoard of the Middle Bronze Age I, *Israel Exploration Journal* 26: 163-169.

GALE, N.H., H.G. BACHMANN, B. ROTHENBERG, Z.A. STOS-GALE & R. TYLECOTE. 1990. The adventitious production of iron in the smelting of copper, in B. Rothenberg (ed.), *The ancient metallurgy of copper*: 182-191. London: Institue of Archaeometallurgical Studies.

GOREN, Y. 1996. The Southern Levant during the Early Bronze Age IV: the petrographic perspective, *Bulletin of the American Schools of Oriental Research* 303: 33-72.

HALICZ, L., Y. EREL & A. VERON. 1996. Lead isotope ratio measurements by ICP-MS: accuracy, precision, and long-term drift, *Atomic Spectroscopy* 17: 186-189.

HAUPTMANN, A. 1989. The earliest periods of copper metallurgy in Feinan, Jordan, in A. Hauptmann, E. Pernicka & G.A. Wagner (eds.), *Old World Archaeometallurgy*: 119-135. Der Anschnitt, Beiheft 7.

1990. The copper ore deposit of Feinan, Wadi Araba: early mining and metallurgy, in S. Kerner (ed.), *The Near East in antiquity*: 53-62. Amman.

1991. From the use of ore to the production of metal: the discovery of copper metallurgy at Feinan, Wadi Arabah, Jordan, in J.-P. Mohen & C. Eluere (eds.), *Decouverte du Metal*: 397-412. Paris.

HAUPTMANN, A., F. BEGEMANN, E. HEITKEMPER, E. PERNICKA & S. SCHMITT STRECKER. 1992. Early copper produced at Feinan, Wadi Araba, Jordan: the composition of ores and copper, *Archaeomaterials* 6: 1-33.

HAUPTMANN, A. & G. WEISGERBER. 1987. Archaeometallurgical and mining-archaeological investigations in the area of Feinan, Wadi 'Arabah (Jordan), *Annuals of the Department of Antiquities of Jordan, Amman* 31: 419-435.

1992. Periods of ore exploitation and metal production in the area of Feinan, Wadi 'Arabah, Jordan, *Studies in the History and Archaeology of Jordan* 4: 61-66.

KOCHAVI, M. 1967. *The sedentarization wave of the Middle Bronze (Canaanite) Age in the Negev*. Unpublished Ph.D. Dissertation, The Hebrew University, Jerusalem (in Hebrew).

LEESE, M.N., P.T. CRADDOCK, I.C. FREESTONE & B. ROTHENBERG. 1986. The composition of ores and metal objects from Timna, Israel, *Wiener Berichte über Naturwissenschaften in der Kunst* 2: 90-120.

MADDIN, R. & T. STECH WHEELER. 1976. Metallurgical study of seven bar ingots, *Israel Exploration Journal* 26: 170-173.

MERKEL, J.F. 1985. Ore benefication during the Late Bronze/Early Iron Age at Timna, Israel, *MASCA Journal* 3: 164-169.

1990. Experimental reconstruction of Bronze Age copper smelting based on archaeological evidence from Timna, in B. Rothenberg (ed.), *The ancient metallurgy of copper*: 78-122. London: Institue of Archaeometallurgical Studies.

MERKEL, J.F. & W.G. DEVER, 1989. The metalworking technology at the end of the Early Bronze Age in the Southern Levant, *Institute for Archaeo-Metallurgical Studies Newsletter*, 14: 1-4.

NAJJAR, M., A. HAUPTMANN, G. WEISGERBER & H.G. BACHMANN. 1995. The Early Bronze Age at Faynan/Wadi 'Arabah: a period of new technology in copper production, *Studies in the History and Archaeology of Jordan* 5: 519-522.

ROMAN, I. 1990. Copper Ingots, in B. Rothenberg (ed.), *The ancient metallurgy of copper*: 176-181. London: Institue of Archaeometallurgical Studies.

ROTHENBERG, B. 1988. *The Egyptian mining temple at Timna. Researches in the Arabah, 1959-1984, Vol 1.* London: Institue of Archaeometallurgical Studies.

1990. *The ancient metallurgy of copper: archaeology, experiment, theory. Researches in the Arabah, 1959-1984, Vol. 2.* London: Institue of Archaeometallurgical Studies.

ROTHENBERG, B. & C.T. SHAW. 1990. The discovery of a copper mine and a smelter from the end of the Early Bronze Age in the Timna Valley, *Institute for Archaeo-Metallurgical Studies Newsletter* 15-16: 1-8.

ROTHENBERG, B. & Z. STOS-GALE. 1991. Archaeo-metallurgy on the trail of history, *Institute for Archaeo-Metallurgical Studies Newsletter* 19: 1-7.

SEGAL, I., A. KLONER & I.B. BRENNER. 1994. Multi-element analysis of archaeological bronze objects using Inductively Coupled Plasma Atomic Emission Spectrometry: aspects of sample preparation and spectral line selection, *Journal of Analytical Atomic Spectrometry* 9: 737-742.

SEGEV, A., M. BEYTH & M. BAR-MATTHEWS. 1992. *The geology of Timna Valley with emphasis on copper and manganese mineralization - updating and correlation with the eastern margins of the Dead Sea Rift*. Geological Survey of Israel, Report GSI/14/92.

SEGEV, A. & E. SASS. 1989. Copper-enriched syngenetic dolostones as a source for epigenetic copper mineralization in sandstones and shales (Timna, Israel), in R.W. Boyle *et al.* (eds.), *Sediment-hosted stratiform copper deposits*: 547-558. Newfoundland: Geological Association of Canada Special Paper 36.

TUFNELL, O. 1958. *Lachish (Tell ed-Duweir); IV, The Bronze Age*. London.

23

Is Oman the Ancient Magan? Analytical Studies of Copper from Oman

M.K. PRANGE[1], H.-J. GÖTZE[2], A. HAUPTMANN[1] & G. WEISGERBER[3]

[1]Deutsches Bergbau-Museum, Institut für Archäometallurgie,Herner Str. 45, D-44787 Bochum, Germany
[2]Ruhr-Universität Bochum, Lehrstuhl für Analytische Chemie,Universitätsstr. 150, D-44780 Bochum, Germany
[3]Deutsches Bergbau-Museum, Institut für Montanarchäologie, Am Bergbaumuseum 28, D-44791 Bochum, Germany

Contemporary geographical descriptions, the archaeological background of the region, and a considerable number of copper ore deposits support the location of the ancient Sumerian copper source of Magan in the Sultanate of Oman. Extensive copper trade between Mesopotamia and Magan during the third millennium BC is attested to by cuneiform tablets. Hence, chemical and isotopic composition of ores and metal objects from Oman, and from other localities in the Persian/Arabian Gulf were determined to establish a statistical basis for comparison with metal objects from Mesopotamia. In total 330 metal objects from Oman, Umm an-Nar, and from Bahrain, and 100 ores from Oman were analyzed for their chemical composition by ICP-OES. Additionally, lead isotope abundance ratios of 38 samples were determined using Thermal Ionization Mass Spectrometry (TIMS).

The most conspicuous feature of Bronze and Iron Age copper objects from Oman and Bahrain is their high levels of arsenic and nickel (each up to 4%). These result from smelting arsenic- and nickel-rich ores embedded in ophiolitic rocks in Oman. There is no indication of any deliberate alloying. A reasonable correlation of lead isotope abundance ratios between ores and copper was observed. Based upon these data, a comparison with metal objects from Mesopotamia should contribute to a final identification of the Sumerian copper source of Magan.

Keywords: Oman; Magan; Copper; Copper Ore; Arsenic; Nickel; Trace Elements; ICP-OES; Lead Isotopes; TIMS; Ophiolites.

Introduction and archaeological background

During the third millennium, intensive trade connections existed between Mesopotamia, Magan, Dilmun and the cultures of the Indus valley (Fig. 1). Various materials were brought to Mesopotamia such as diorite for sculptures, steatite vessels, sandalwood, furniture, semiprecious stones such as agate, jasper and carnelian, foodstuffs such as onions, dates and goats, and, last but not least, copper metal from Magan. Most interesting and amusing details of this trade are preserved in written documents, namely the cuneiform tablets found at Uruk. For example, there are complaints from King Sargon (*c.* 2300 BC) about the poor quality of copper delivered from Magan. It may well be that this referred to various frauds that happened in those days. Copper was traded in bun shaped ingots to Mesopotamia, and, in fact, in two such ingots from Bahrain and Oman we discovered a core of slag which was surrounded by a thin layer of metal (Fig. 2).

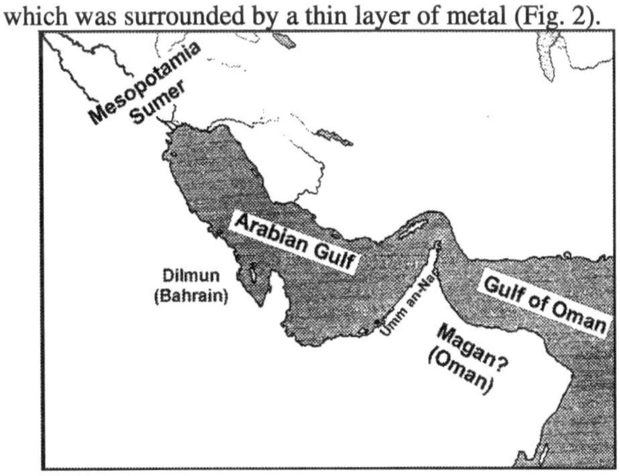

Figure 1. Geographic map of Southeastern Arabia, showing Mesopotamia, Dilman and Magan.

Figure 2. Section through a bun-shaped ingot with a core of slag surrounded by a thin layer of copper. (Al-Aquir, Wadi Bahla, Oman: © Hauptmann 1987). Reprinted from Der Auschn. #39, 1987 and ' Zur frühen Metallurgie des lupters' (1998). Reproduced by permission of Andreas Hauptmann, Deutches Bergban-Museum.

Ancient Dilmun was identified archaeologically by Bibby (1977) as the present day Emirate of Bahrain. Due to its sweet water springs, this small island was an important intermediate and trade station on the sea route to Mesopotamia. Therefore, any metal found there most probably came from the copper lands of Magan. As there is no ore deposit on the island itself, local copper production can be excluded. One of the first attempts to identify Magan in the Sultanate of Oman was made by Peake and Desch in the early 1920s. They analyzed a few Mesopotamian copper artefacts together with several ores and slags from Oman and found unusually high nickel (Ni) contents. For this reason they suggested that the Mesopotamian copper originated from ore deposits in Oman. Combining the contemporary geographical descriptions and the archaeology of the entire region, especially the large number of copper ore deposits and related ancient slag heaps in the Oman mountain range, Weisgerber (1980) concluded that a part of the Sumerian copper source Magan could be located in the modern

Sultanate of Oman. The aim of this study is to test this hypothesis independently by scientific means.

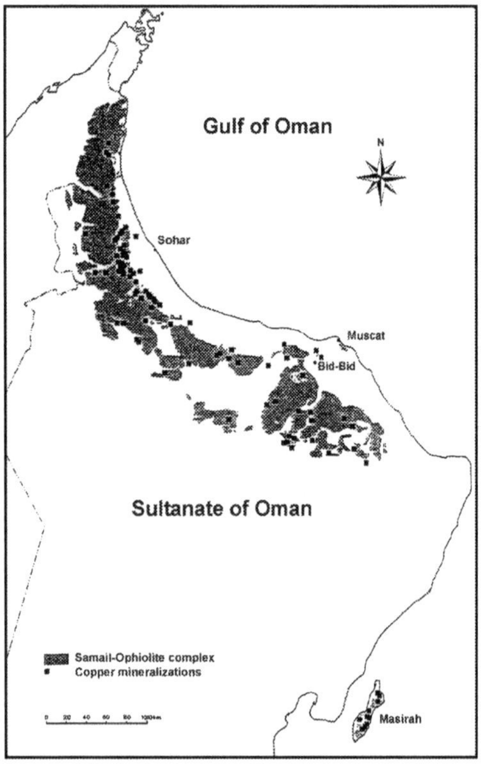

Figure 3. Map of the Sultanate Oman, showing the copper-mineralizations in the Sumail Ophiolite Complex.

In 1974, the Oman Mining Co. (OMCO) started prospection work for copper in the Oman mountain range (Fig. 3). They were successful in discovering nearly 150 copper occurrences, several of them reaching the volume of modern economic standards. Interestingly, the company did the prospection work mostly by following the ancient slag heaps. Altogether, the amount of ancient slag produced in Oman was calculated to be more than 700,000 tons (Hauptmann 1985). During the next two decades, archaeological field work revealed that mining and smelting activities started at least 5000 years ago, during the Bronze Age in the Umm an-Nar period (Weisgerber 1980; 1981). Even though the majority of slags were produced during the early Arabic period (late 1st millennium AD), the finds of metal objects from a number of excavations clearly indicated a surplus production of copper in earlier periods.

Copper production of similar scale seems to be the exception in the Middle East, with only Cyprus being comparable. Information is, however rare about early copper production in Iran, which could also have been a possible copper resource for Mesopotamia. This could be true particularly for the region north of the Strait of Hormuz, which shows comparable geological features (ophiolitic rocks). However, archaeological evidence is lacking.

Geology

The copper ore deposits of Oman are located within the Samail Ophiolite Complex, a special lithological sequence of ultrabasic and basic rocks which forms most of the Oman mountain range. It was formed during the Cretaceous (100 Ma) at a Mid Ocean Ridge, and was obducted later by tectonic movements onto the Arabian Plate (Andritzky and Burgath 1977; Coleman *et al.* 1979). The Samail Ophiolite Complex is probably the best preserved one known so far. It is, however, not the only complex of such rocks in the area. Ophiolitic rocks typically occur all around the northern part of the Arabian Plate, particularly in the orogenic belt of the Zagros and Taurus mountains ("Periarabic Ophiolite Belt", Yilmaz 1993). The ore deposits of the Samail Ophiolite belong to the type of "Massive Sulfide Deposits" including fracture fillings and stockwork mineralizations (i.e., ore bearing veins and impregnations) at the basal part. Their origin is connected to volcanic activities at the Mid Ocean Ridge, which, to a certain extent, is comparable to hydrothermal activities observed recently at the bottom of the Red Sea. These lens shaped "Massive Sulfide Deposits" are usually embedded in basaltic and andesitic volcanic rocks, and they consist mainly of *pyrite* (FeS_2) with minor concentrations of copper (Cu), zinc (Zn) and lead (Pb), and low contents of trace elements like antimony (Sb), arsenic (As), cobalt (Co) and nickel (Ni). The last three elements, however, are enriched in mineralizations in the basal part of the deposits which are embedded in ultrabasic rocks like peridotites. Here, *chalcopyrite* ($CuFeS_2$) is the predominant copper ore with minor amounts of *pentlandite* ($(Fe, Ni)_9S_8$) and nickel arsenides. Cobalt-arsenic ores have also been observed. Similar ore deposits occur at several points in this Periarabic Ophiolite Belt, for example in the region of Ergani Maden in Anatolia. The copper deposits of Cyprus are of the same origin and of the same age. It is especially interesting that at Limassol Forest (Cyprus), Cu-As-Ni-Co-mineralizations occur which are of similar origin to the ores in Oman (Constantinou *et al.* 1972; 1973).

Scientific approach

To confirm the hypothesis that the provenance of the Mesopotamian copper is located in Oman, it is necessary to prove that metal objects from both regions and at localities along the trade route (e.g., Bahrain and Umm an-Nar) are chemically and isotopically identical, and they all must be compatible with ore deposits in Oman. A comparison of ore and metal, however, must consider systematic alteration of element abundances which might occur during smelting. Firstly, of course, the aim of smelting is to separate main components of the ore such as iron (Fe), sulfur (S) and gangue material from the base metal copper. Some elements follow the copper during this process, so that the particular trace element/copper ratio hardly varies. Which elements behave like this depends on the different 'reducibility' for each element. Supposing reaction temperatures of around 1200 °C for Bronze Age smelting processes, and considering the free enthalpy of formation, a list of elements can be prepared which, in smelting oxidic ores, will follow the copper during the process. These elements include cobalt (Co), antimony (Sb), nickel (Ni), arsenic (As), lead (Pb), bismuth (Bi), selenium (Se), tellurium (Te) and silver (Ag), which should therefore be used as 'pathfinder' elements to trace back the provenance of copper (Pernicka 1987).

For this investigation we analyzed 300 copper samples and 100 copper ores from Oman, 30 copper samples from Bahrain and 28 from the island Umm an-Nar. These objects comprise swords, daggers, spear-heads, bun-shaped copper ingots and inclusions of metal in smelting slags. Some of these objects have already been published by Hauptmann (1987). The samples are from different periods, as summarized in Table 1.

			NUMBER OF OBJECTS
Umm An-Nar	BRONZE-AGE	2700 – 2000 BC	40
Wadi Suq		2000 – 1500 BC	55
Late Wadi Suq		1500 – 1200 BC	65
Lizq	IRON-AGE	1200 – 300 BC	125
Samad		300 BC – 900 AC	15

Table 1. Section of cultural epoches in Oman and number of metal objects analysed from different periods.

After partially removing the corrosion crust from the copper objects, 100 mg shavings were drilled with a high-grade steel borer (1.5 mm diameter). The metal turnings were dissolved with brief heating in a mixture of 5 mL concentrated hydrochloric acid and 250 µL nitric acid. Ore samples were taken by selecting 4 - 5 g of high grade (oxidic) copper minerals from 0.5 - 2 kg hand-picked specimens. This material was ball-milled to a grain size below 63 µm. The resulting ore powders were dried for 3 hours at a temperature of 105 °C. A subsample of 100 mg of this powder was decomposed by microwave digestion under elevated pressure in a mixture of 1.5 mL hydrofluoric, 3.5 mL hydrochloric and 3 mL nitric acid. The sample solutions were made up to 50 mL with deionized water.

Analysis was done using an ICP-OES spectrometer (Inductively Coupled Plasma-Optical Emission Spectrometry) at the Institut für Archäometallurgie, Deutsches Bergbau-Museum Bochum. The detection limits range from 3.5 ppb (parts per billion) in solution for silver (Ag) up to 50 ppb for arsenic (As). The technique is able to determine all of the elements of interest for copper provenance studies, as listed above (Prange 1998).

When the ICP-OES was first introduced into chemical laboratories it was argued that analyses were essentially free of interferences, because of the high temperature of the argon plasma (the liquid sample is vaporized, atomized and ionized at temperatures between 6000 and 10000 °C). Although interference effects in ICP-OES are smaller than in any other method of spectrometry, the ICP-OES element determination can be affected by spectral and non-spectral interferences. Spectral interferences result from the inability to resolve a particular emission line because of the presence of lines from other components in the sample. This problem can be solved by using alternative spectral lines or methods of background correction (Boumans 1987). Non-spectral interferences include, for example, sample uptake impairments. These result from different acid concentrations in the analyte, since the viscosity of the liquid depends on the concentration of mineral acids. In order to avoid such errors, the analyte and the standard calibration solutions have to be adjusted to a constant acid concentration, to give the same viscosity (Chudinov *et al.* 1989). Furthermore, highly concentrated sample solutions can also cause interferences, by affecting the flow of the plasma (Prange and Götze 1995).

Interferences on other elements may be expected in the analyzed metal samples from the main component, i.e., copper, and in the ore samples from silicon (Si), iron and copper. In order to prevent this the interfering element must be separated from the solution or the calibration standards adjusted to the correct element concentration in the sample solution (matrix matching). In this study we decided to adopt the latter procedure, in order to avoid additional contamination arising from separation agents.

The elements were determining against matrix matched standards using an ICP-OES spectrometer IRIS/AP (Thermo Jarrel Ash, Franklin MA, USA), with hydrofluoric acid resistant torch and spray chamber. The following measuring conditions were employed - power 1.15 kW, auxiliary flow rate 0.5 mL min^{-1}, nebulizer pressure 32 psi, sample uptake 1.85 mL min^{-1}. These conditions give adequate reproducibility for both the matrix elements and the minor and trace elements. The percentage relative standard deviations are in the range of 0.1 to 2.5 % and are therefore within the normal limits of an ICP-OES determination.

A comparison between metal and ore or between metal and metal is often inconclusive using just the chemical compositions. Therefore, lead isotopic analysis were also obtained. A combination of these two analytical methods is currently the best procedure for provenance studies.

After analysis, the geochemical and the isotopic data are compared by statistical methods. Firstly, the copper objects and the copper ores from the Sultanate of Oman have to be compared in order to assign the copper to the ore. In a second step, copper from Bahrain has to be compared with the Omanian copper to confirm that it does come from Oman. Finally, a comparison between metal objects from Mesopotamia with both is required.

Preliminary Results

At the present stage of our investigation, we report on the abundance of arsenic, nickel and tin, and on some lead isotope ratios in ores and metal objects.

The average nickel and arsenic concentrations of the Oman and Bahrain copper are both about 0.5 %, ranging from 0.01 to 10 % (Fig. 4). Copper with such levels of As and Ni was used from the Early Bronze to the Iron Age. There is one artefact from the Umm an-Nar period (Early Bronze Age, 2700 - 2000 BC) which is of special interest - a dagger with 12 % nickel which attracts attention because of its pale golden colour.

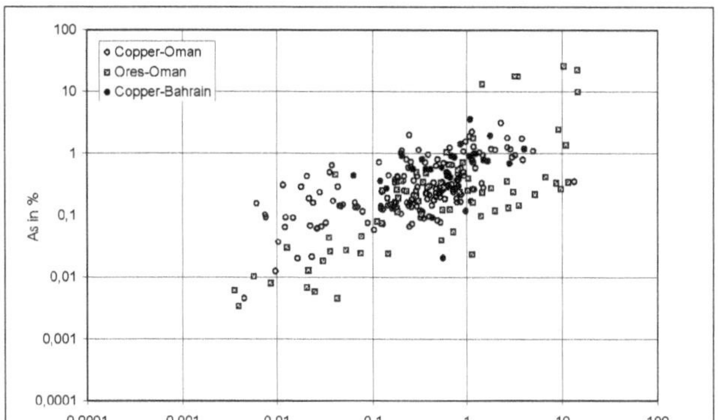

Figure 4. Comparison of arsenic and nickel concentrations in copper and copper ores. All data are recalculated to 100 wt% copper

The average nickel and arsenic concentrations in the ores are in the same range as the metal, but show larger variations. This indicates that the As-Ni copper metal used is a "natural alloy". Even the dagger is probably made from nickel-rich copper ores, such as, for instance, those identified in the Wadi Nujum near the modern village of Bid-Bid (Fig. 3).

In addition, we analyzed some metal inclusions in smelted Bronze Age and Iron Age slags. Even here, the average contents of arsenic and nickel correspond well to the levels of both elements in copper artefacts, ingots and copper ores. This indicates, that most (if not all) of the metal objects analyzed from different localities in Oman must have been produced from local resources.

In the upper right corner of Figure 4, six ore samples from the Wadi Nujum deposit in Oman are shown, with very high nickel and arsenic contents. Apart from these, however, it appears that the arsenic content of the Oman ores which are available today for sampling at the surface (and not in ancient mines!) is somewhat lower on average than in the metal. This is a well known, but poorly understood phenomenon in archaeometallurgy (Merkel 1985; Leese *et al.* 1985/86). A possible but unverified explanation could be that the copper which separates from the slags is enriched in arsenic, in contrast to metal inclusions remaining in the silicate slags, since any concentrations of trace and minor elements will lower the melting point of copper and facilitate this separation.

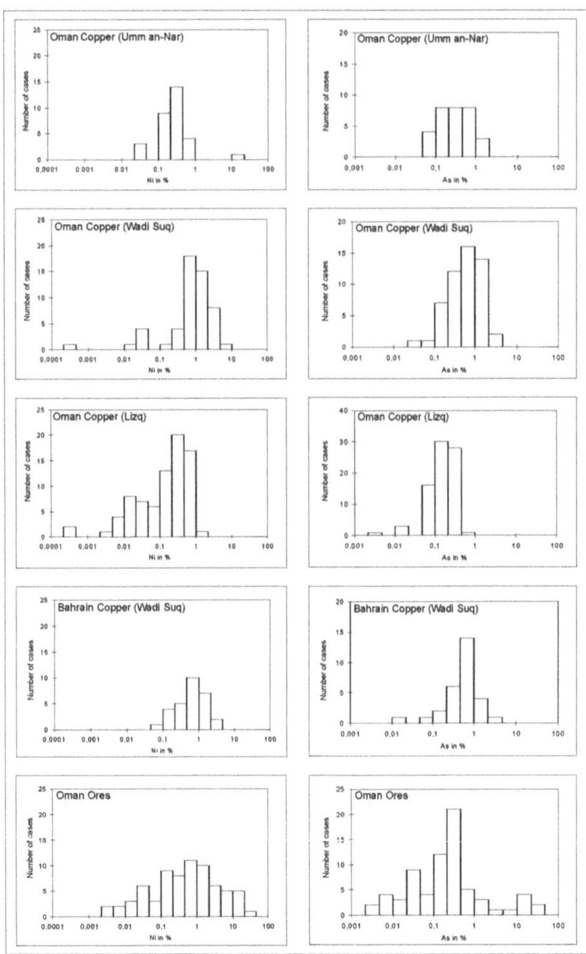

Figure 5. Breakdown of arsenic and nickel concentrations in ores and metal objects for different periods. All data are recalculated to 100 wt% copper. Note the increase of arsenic and nickel in the Wadi Suq period!

Figure 5 shows a breakdown of nickel and arsenic data for metal objects by different periods. Copper produced in the Umm an-Nar period (2700 - 2000 BC) is relatively low in these elements (mean 0.5 %). One possible explanation could be that the objects analyzed were mainly smelted from copper ores in the gossan of the massive sulfide deposits rather than from those in vein mineralizations, because arsenic and nickel concentrations are typically much lower in gossan deposits.

In the Wadi Suq period (2000 - 1500 BC) the average nickel and arsenic concentrations increase considerably. Copper objects now contain up to 4 % of each of these elements. This could indicate that mining and smelting activities moved from the massive sulfide deposits to those ("stockwerk") mineralizations which are located elsewhere in ultrabasic rocks (peridotites) with higher concentrations of minor and trace elements.

The Iron Age samples of the Lizq period (1200 - 300 BC) again show a tendency to lower concentrations of the two elements, caused by a shift of metal production back again to the massive sulfide deposits, where mining and smelting started in the Umm an-Nar period. Another possibility is the recycling of scrap metal, which would fit into the broader archaeological situation for this period.

All the metal objects analyzed from Bahrain are contemporary with the Wadi Suq period in Oman. The nickel and arsenic concentrations in them are in perfect correlation with those from Oman. Hence, at least on the basis of trace element analysis, there is no contradiction that Bahrain copper was produced in Oman. This is confirmed too by the composition of Oman copper ores which are in broad correlation with the metal objects. Additionally, a good correlation between ores and copper was observed for cobalt, antimony, lead, bismuth, silver, selenium and tellurium.

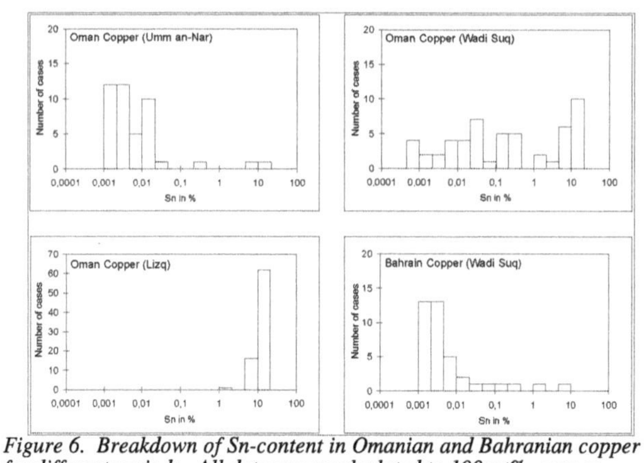

Figure 6. Breakdown of Sn-content in Omanian and Bahranian copper for different periods. All data are recalculated to 100 wt% copper.

Compared to Mesopotamia, tin bronzes appear relatively late in Oman. The first tin bronzes appear in Mesopotamia in the beginning of the third millennium BC (Lutz *et al.* 1995), but until the second half of the second millennium BC tin was not a conspicuous additive in Mesopotamian workshops (Moorey *et al.* 1988). Figure 6 shows a chronological breakdown of the tin content of the Omanian and Bahrainian copper analyzed. In the Umm an-Nar period, we found no tin bronzes. The tin content in the metal is mostly below 100 ppm (0.01%), matching the tin content of the ores. Apparently, alloying practices were not known during this period in southeast Arabia. The Wadi Suq period, in this respect, is one of transition. Tin was added to copper sporadically, but, in most cases, did not exceed 1%. In the Iron Age Lizq period, tin bronzes are widespread over the southeast Arabian peninsula (see also Weeks 1997) and the tin contents reaches 10 % and more. Such high tin bronzes have excellent casting properties and, additionally, are considerably harder than arsenic- and nickel-containing copper. The tin contents of the Bronze Age samples from Bahrain are generally in the range of 50 ppm to 0.5 %. The latter cannot be explained as an intentional alloy; it is probably due to the recycling of scrap metal, as from this period copper objects with up to 6 % were excavated at the Sar burial mound field.

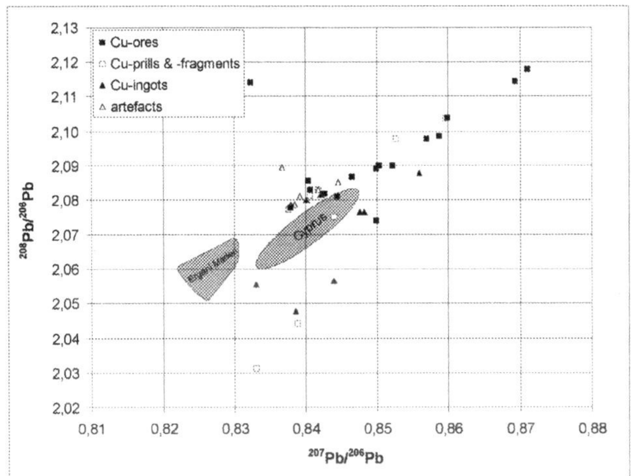

Figure 7. Lead isotope abundance ratios of copper artefacts and ores from Oman. For comparison, the areas of isotopic compositions of Cyprus and Ergani Maden are shown. (Stos-Gale 1997, Wagner et al. 1986). Rreprinted from Der Auschn. #39, 1987 and ' Zur frühen Metallurgie des lupters' (1998). Reproduced by permission of Andreas Hauptmann, Deutches Bergban-Museum.

Lead Isotope Abundance Ratios

In Figure 7, the lead isotope ratios of 22 metal artefacts (copper prills and fragments, bun-shaped ingots and metal objects) and 16 ores are compiled in the isotope diagrams showing $^{208}Pb/^{206}Pb$ vs. $^{207}Pb/^{206}Pb$. The data were analyzed at the Max-Planck-Institut für Chemie at Mainz. For measurements of the isotope abundance ratios a mass-spectrometer (MAT 261, Varian Finnigan) was used (analytical details as described by Pernicka *et al.* 1984). For comparison, the published data from two archaeologically relevant ore deposits are shown, Ergani Maden in Anatolia (Wagner *et al.* 1986) and Cyprus (Stos-Gale *et al.* 1997). The isotope ratios from Oman show a large range of compositions. They extend from ores with $^{208}Pb/^{206}Pb$-ratios up to 2.12 (two plots, upper right) and, hence, comparable to Timna and Feinan (Hauptmann 1998) to those which overlap Cyprus (Stos-Gale *et al.* 1997). Most of the lead isotope data from Oman exhibit a cluster very close and inside the Cyprus-field. From the geological point of view, this is not surprising, since the Samail Ophiolite Complex and the one at Cyprus including the ore deposits in these sequences were formed roughly during the Cretaceous. This cluster comprises both metal artefacts and ores from Oman and thus confirms a copper production from local sources. The lead isotope ratios of ores between $^{208}Pb/^{206}Pb = 2.1$ and 2.2 and $^{207}Pb/^{206}Pb$ near 0.7, however, are not compatible with the formation of the Samail Ophiolite. They seem to be much older and could perhaps be of Cambrian age. Regardless of the exact age, they are proof of a more complicated formation of ore deposits in the Oman mountain range than previously estimated. The composition of ingots and artefacts with $^{208}Pb/^{206}Pb < 2.06$ (i.e., below the Cyprus-field) for the time being are difficult to understand. So far, there are no ores known from Oman from which such metals could have been produced. This may be due to an incomplete sampling of ores from Omani deposits. The reason for this incompatibility, however, could also be caused by importation of copper to Oman from an as yet unknown source, e.g., from the Indus valley.

Conclusions

According to Peake (1928) and Desch (1929), the most conspicuous feature of copper from Oman is its high levels of arsenic and nickel. We confirm this statement on a reliable statistical basis, for metal objects both from Oman and Bahrain, which regularly contain up to 4% each of arsenic and nickel. Although there are certain time dependant differences in the composition of the copper, this is generally true for Bronze and Iron Age objects, and it is also true for metal inclusions in contemporary smelted slags. This composition matches the composition of the Omanian ores, and it appears to be more typical than was realized of the geochemical pattern of ores from Ophiolites.

As-Ni-alloys occur only in the Middle East but not in central Europe. The distribution of such alloys has been mapped originally by Pernicka in 1995, and it was completed by data in this work and by Hauptmann (1998). These new data allow, at least on a preliminary basis, to link the Mesopotamian copper to the Sultanate of Oman. It offers a new important Bronze Age copper source besides the well known circum-Mediterranean deposits (Prange 1998).

We assume that copper objects excavated at Bahrain were smelted from Omani ores, firstly because there are no copper deposits at all on the island, and secondly because the chemical analyses show similar compositions to the Omanian finds.

Based upon the data from Oman and Bahrain, a comparison with analytical results of metal objects from Mesopotamia will possibly contribute to the identification of the Sumerian copper source, and therefore of the famous Magan.

Acknowledgements

We are grateful to Prof. Dr. F. Begemann and Dr. S. Schmitt-Strecker (Max-Planck-Institut für Chemie, Mainz, Germany) for providing us with the lead isotope data. Particular thanks are due to the Ministry of National Heritage & Culture, Department of Antiquities, Muscat, Sultanate of Oman, Director General Dr. Ali Shanfari and to the Ministry of Information, Directorate of Museums and Heritage, Bahrain National Museum, H.H. Shaikha Haya Ali al Khalifa for their co-operation in providing the archaeological sample materials and last not least to W. Steger for his help with the sample preparation and to Dr. Th. Rehren for critically reading the manuscript at various stages.

References

Andritzky, G. & K.P. Burgath. 1977. Results of a survey for copper ore in northern Oman 1976/77. *Interner Bericht E18*. Hannover: Bundesanstalt für Geowissenschaften und Rohstoffe.

Bibby, G. 1977. *Dilmun. Die Entdeckung der vergessenen Hochkultur*. Hamburg: Reinbek.

Boumans, P. W. 1987. Line selection and spectral interferences, in P. W. Boumans (ed.), *Inductively Coupled Plasma Emission Spectrometry, Part I:* 358-456. New York: Wiley-Interscience.

Chudinov, E.G., I.I. Ostroukhova & G.V. Varvanina. 1989. Acid effects in ICP-AES, *Fresenius Zeitschrift fuer analytische Chemie* 335: 25-33.

Coleman, R.G., C.C Huston, I.M. El-Boushi, K.M. Al-Hinai & E.H. Bailey. 1979. The Samail-Ophiolite and associated massive sulfide deposits, Sultanate of Oman, in A M S. Al-Shanti (ed.), *Evolution and Mineralization of the Arabian-Nubian Shield:* 179-192. Institute of Applied Geology, King Abdul Aziz University, Bulletin No. 3. Oxford: Pergamon.

Constantinou, G. & G.J.S. Govett. 1972. Genesis of sulphide deposits ochre and umber of Cyprus, *Transactions of the Institution of Mining and Metallurgy* B81: 34-46.

1973. Geology, geochemistry and genesis of Cyprus sulphide deposits, *Economic Geol*ogy 68: 843-58.

Desch, C.H. 1929. Interim report on Sumerian copper. *British Association for the Advancement of Science, Report of 9th. Meeting*: 437-441. London.

Hauptmann, A. 1985. 5000 Jahre Kupfer im Oman. *Der Anschnitt* Beih. 4. Bochum: Deutsches Bergbau Museum.

1987. Kupfer und Bronzen der südostarabischen Halbinsel. *Der Anschnitt* 39/5-6: 209-218.

1998. Zur frühen Metallurgie des Kupfers in Fenan/Jordanien. *Der Anschnitt.* in press.

Leese, M., P. Craddock, I.C. Freestone & B. Rothenberg. 1985/86. The composition of ores and metal objects from Timna, Israel, *Wiener Berichte über Naturwissenschaft in der Kunst* 2/3: 90-120.

Lutz, J., B. Hellwing, E. Pernicka & H. Hauptmann. 1995. Die Zusammensetzung einiger Metallfunde aus Uruk Warka, *Baghdader Mitteilungen* 26: 1-23.

Merkel, J.F. 1985. Ore benefication during the Late Bronze/Early Iron Age at Timna, Israel, *MASCA Journal* 2(5): 164-169.

Moorey, P.R.S., J.E. Curtis, D.R. Hook & M.J. Hughes. 1988. New analyses of old Babylonian metalwork from Tell Sifr, *Iraq* L: 39-48.

Peake, H. 1928. The copper mountain of Magan, *Antiquity* 2: 452.

Pernicka, E. 1987. Erzlagerstätten in der Ägäis und ihre Ausbeutung im Altertum: Geochemische Untersuchungen zur Herkunftsbestimmung Archäologischer Metallobjekte, *Jahrbuch des Römisch-Germanischen Zentralmuseums Mainz* 34/2: 607-714.

1995. Gewinnung und Verbreitung der Metalle in prähistorischer Zeit, *Jahrbuch des Römisch-Germanischen Zentralmuseums Mainz* 37/1: 21-129.

Pernicka, E., T.C. Seeliger, G.A. Wagner, F. Begemann, S. Schmitt-Strecker, C. Eibner, Ö. Öztunali & I. Baranyi. 1984. Archäometallurgische Untersuchungen in Nordwestanatolien, *Jahrbuch des Römisch-Germanischen Zentralmuseums Mainz* 31: 533-599.

Prange, M.K. & H.-J. Götze. 1995. Analyse früher Kuper- und Eisenschlacken mittels optischer Atomspektrometrie (ICP-OES), *Metalla* 2: 29-41.

Prange, M.K. 1998. *Vergleichende Untersuchungen zur Charakterisierung des omanischen und mesopotamischen Kupfers mittels chemischer Analysenmethoden*. Dissertation Bochum, in prep.

Stos-Gale, Z.A., G. Maliotis, N.H. Gale & N. Annetts. 1997. Lead isotope characteristics of the cyprus copper ore deposits applied to provenance studies of copper oxhide ingots. *Archaeometry* 39/1: 83-123.

Wagner, G.A., E. Pernicka, T.C. Seeliger, I.B. Lorenz, F. Begemann, S. Schmitt-Strecker, C. Eibner & Ö. Öztunali. 1986. Geochemische und isotopische Charakteristika früher Rohstoffquellen für Kuper, Blei, Silber und Gold in der Türkei, *Jahrbuch des Römisch-Germanischen Zentralmuseums Mainz* 33: 723-758.

Weeks, L.R. 1997. Prehistoric metallurgy at Tell Abraq, U.A.E., *Arabian Archchaeology and Epigraphy* 8: 11-85.

Weisgerber, G. 1980. "... und Kupfer in Oman", *Der Anschnitt* 32/2-3: 62-110.

1981. Mehr als Kupfer im Oman, *Der Anschnitt* 33/5-6: 174-263.

Yilmaz, Y. 1993. New Evidence and model on the evolution of the southeast Anatolian orogen, *Geological Society of America Bulletin* 105: 251-271.

The Role of Central Himalayas in Indian Archaeometallurgy

D.P. AGRAWAL

Physical Research Laboratory, Ahemedabad 380009, India

The present paper is focused on the possible role of the Himalayas in the diffusion and spread of ancient copper and iron technologies in India (Indo-Pak subcontinent). It is not meant to be a review of Indian archaeometallurgy; it only brings out the significance of the new data from the central Himalayas and its bearing on the Indian archaeometallurgy. The paper also indicates the probable contacts with south east Asia and the diffusion possibilities of metal technologies from there. The paper also addresses the role of the central Himalayas as possible sources for both copper and iron for north India. The possibility or otherwise of iron technology emerging out of the copper technology is also examined. In summary, the paper presents both some new data and some bold speculations.

1. Preamble

The Indian (read Indo-Pak subcontinent) archaeometallurgical scene appears a bit intriguing, if we see it in the present-day framework of nationalities. The cultural boundaries in the millennia before Christ were different than today. Earlier, on the basis of technology, abundance of metals, mines used and tool typology, I had argued that in Bronze Age India, the 3 main techno-cultural traditions(the Harappa culture, the other Chalcolithic cultures, the Copper Hoard culture) were essentially independent - temporally part-contemporary, but spatially disparate(Agrawal 1982b). It may be noted here, that though both the Harappan and the Chalcolithic cultures used both copper and lithic artefacts, only the non-Harappan cultures distributed over central India, Maharashtra, eastern Rajasthan, etc. are conventionally included under the blanket term, *Chalcolithic cultures.*

The contacts of the Harappans with west Asia are well documented (Piggott 1961). With the recent excavations in north west India (Mehrgarh area) showing a continuous evolution from Neolithic to the Harappan urban phase, however, one need not presume wholesale diffusion of technologies from west Asia, but perhaps some diffusion of ideas. In central India, Maharashtra and east Rajasthan - the distribution zone of the Chalcolithic cultures - the Chalcolithic technology and abundance of metal both appear quite poor in comparison, and, therefore, their technology could be totally indigenous. The Copper Hoard culture, mostly confined to the Ganga basin, as we will discuss below, appears more enigmatic than the other two techno-cultural traditions.

In these dynamics of technological evolution, the Himalayas are ignored as an isolated *cul-de-sac.* Recent evidence, however, seems to indicate that this region may have played a more crucial role in the development of both copper and iron technologies. This paper addresses itself to the possible role of the central Himalayas (the present hilly districts of Garhwal and Kumaun divisions of Uttar Pradesh (UP) in northern India) in the development of both early copper and iron technologies.

I should like to enter a caveat here: I am not discussing on a pan-subcontinental scale. Here I refer specifically to the Copper Hoard culture and zone B of Prakash & Tripathi (1986) of iron technology, as explained below (#7).

But, before I go into such finer details, I should delineate the cultural and archaeometallurgical scenario during the I-III millennia BC in India.

2. Archaeological And Archaeometallurgical Backdrop

Indus Civilization

The Indus civilization or Harappa culture (named after the first discovered site, Harappa) is well known for its uniformity and standardization in weights, measures, ceramics, architecture, town planning and in other arts and crafts (Lal 1997). It spread in the west to Suktagendor (Makran), in the east to Alamgirpur (Uttar Pradesh, UP), to Rupar in the north and to Bhagatrav in the south (Gujarat), covering an area about 2.5 million km^2, bigger than the present-day Pakistan.
Kenoyer (1997) gives the following chronology:

Table I - Provisional prehistoric chronology of Harappa (Kenoyer 1997)

Period 1A and 1B - *c.*3000-2800 Cal.BC
Early Harappan
(Ravi-Hakra Phase)

Period 2 - *c.* 2800-2600 Cal.BC
Early Harappan/Harappan
Transitional (Kot Diji Phase)

Period 3 - Harappan Phase

Period 3A - *c.* 2600-2450 Cal.BC
Period 3B - *c.* 2450-2200 Cal.BC
Period 3C - *c.* 2200-?1900 Cal.BC

Metal Technology of the Harappans

The Harappans were quite rich in copper and also in copper tool types, although they continued to use chert blades. Their products included flat celts, knives, arrowheads, spearheads, chisels, razors, fish-hooks, saws, drills, etc., in addition to vessels and metal figurines (For original excavation reports see Mackay 1938, 1943; Marshall 1931; Vats 1940; Wheeler 1947). It may be noted here, that tubular drills, true saws and needles with eyes at the pointed ends are contributions of the Harappans to the world of instruments (Agrawal 1982b). The Harappans were the only people to use metal vessels in protohistoric India. Metallographic examinations show that they resorted to open and closed casts, slow cooling of casts, annealing and cold work (Agrawal 1971; Agrawal 1982b). Small animal and female figurines were cast by the *cire perdue* (lost wax) process. None of the Harappan moulds is as ornate as the Chinese nor as intricate as the Mesopotamian examples. They used both tin and arsenic alloying. An impurity pattern trace analyses of ores and artefacts indicated that the Harappans probably used the Khetri mines of Rajasthan (Agrawal 1971).

Chalcolithic cultures

The so-called Chalcolithic cultures are characterized by the use of stone and copper artefacts, but metal is rare and the cultures are non-urban (Sankalia 1974). Traditionally, the non-Harappan cultures of the second millennium BC are included in this category. The main cultures (Malva, Banas (Ahar), Kayatha, Jorwe etc.) belonging to this group are distributed from south-eastern Rajasthan to central India and Maharashtra. The general time spread of ^{14}C dates for Chalcolithic cultures seems to be confined to 2000-1500 BC and for the Jorwe culture, 1500-1000 BC (Calibrated).

Chalcolithic Metal Technology

On the whole, about 17% of Chalcolithic artefacts show tin alloying, but the range varies considerably. At Ahar only pure copper was used. A Jorwe axe has only 1.7% tin, a Nevasa chisel has 2.7% tin, at Navdatoli the range of tin alloying is 3.5%; the highest tin (12.8%) recorded was in an axe from Somnath. Unlike the Harappan and Copper Hoard cultures, Chalcolithic groups did not alloy with arsenic. For better fusibility, lead alloying was more common, as 20% of the artefacts analysed contain lead.

Hegde (1965) spectroscopically examined several Chalcolithic artefacts and compared their trace impurities with those of the Khetri (Rajasthan) copper ore. He found a close similarity between the two, thus indicating the possibility of the Khetri mines having been used by the Chalcolithic cultures. The Khetri mines are quite close to Ahar.

The Copper Hoard Culture

Most of the tools of the Copper Hoard culture have been found in hoards or caches mainly from Uttar Pradesh (UP), Bihar and Madhya Pradesh(MP). Since most of the artefacts were chance finds, there is no way to associate a definite archaeological assemblage with them. As a result, finds from such far-flung sites as Shalozan in north-western Pakistan, Bhagrapir in Orissa and Kallur in the south have been included in the Copper Hoard culture zone. A wide variety of tool-types was included in this amorphous collection: rings, a variety of celts, trunnion axes, anthropomorphs, harpoons, swords, double-edged and socketed axes, etc. (Lal 1972; Yule 1985; Chakrabarti & Lahiri 1996).

I have divided the Copper Hoards into two zones. The main Ganga Doab (Uttar Pradesh) types are specialized tools: the anthropomorph, the antennae sword, the harpoon and the hooked sword. The main eastern (Bihar) type is the barcelt. Flat and shouldered celts are common to both zones (Agrawal 1982a). The distribution of the Copper Hoard types is mainly confined to UP, Bihar, and MP. The Copper Hoards appear to be uninfluenced by either the Harappan or the Chalcolithic cultures. This is further supported by the exclusive use of arsenic alloying of the Copper Hoards. Both of these features - exclusive arsenic alloying and characteristic tool types - impart a distinctive individuality to the Copper Hoards.

3. Sources of Copper in India

After this brief survey of the Bronze age cultures of India, we shall look at the mineral distribution (Coggin Brown & Dey 1955). This will help us appreciate the role that I envisage for the central Himalayas in the spread of copper technology.

The main Indian copper minerals are: chalcopyrite, chalcocite, bornite, tetrahedrite, covellite, malachite, azurite. Present mining areas in India for copper are mainly located in Rajasthan, Bihar, Andhra and Kumaun, although there are several other minor deposits. The Singbhum belt of Bihar is 130 km. long and up to 8 km. wide. In 1959, the reserves of copper ore were estimated to be 3.8 million tonnes with an average 2.47% copper content. New explorations have brought to light copper ore resources in the Patharghora, Surdha, Kendadih, Roam-Sidheshar areas. The Andhra Pradesh, Guntur, South Arcot and Hassan districts, too, have yielded copper ore deposits. The Guntur belt is 4.8 km long. Ore deposits of the Jabalpur area (MP) are very poor. The veins of copper minerals are only 15 to 7 cm in width. The minerals are chalcopyrite and tetrahedrite. In Rajasthan, copper ore mines are located in practically all the divisions. The most important of the mines are at Khetri-Singhana in Jhunjhunu district. This belt continues from Singhana to Raghunathpura, about 80 km. In this belt, in the Mardan-Kudan section, there is a reserve of 28 million tonnes of ore with 0.8% copper, and, in the Dariba area, 0.5 million tonnes ore deposit of 2.5% copper. The mineral found in these places is chalcopyrite ($CuSFe_2S_2$) which is usually disseminated in schists and phyllites (For a summary of the mining areas see Agrawal 1971; Chakrabarti & Lahiri 1996: 11-31).

4. Gangetic Copper Hoards and the Sources Of Copper

The Gangetic alluvium has filled the foredeep caused by the Miocene orogeny of the Himalayas. As a result, no rock outcrops are to be seen in this alluvium, hence no ore bodies. The question, then, is from where did the Copper Hoard culture obtain its supplies of copper.

Traditionally, the rich Bihar (Singbhum belt) ores have been identified as sources of copper for the Copper Hoards. It may still be true for the eastern Copper Hoard group. Not many trace element analyses are available to compare the impurity patterns of the Copper Hoards with different mines. No lead isotope analyses have been carried out. However, one trait marks the Copper Hoard artefacts of Central Ganga: about 50% of the artefacts analyzed so far show a significant arsenic content (1-7%), but no tin whatsoever (Agrawal 1982b). Whether arsenic was deliberately added for increasing the hardness of the metal artefacts, or it was an impurity present in the metal used, is difficult to say. As arsenic minerals are rare in India even today, deliberate addition of arsenic seems to be rather a remote possibility. If it was an impurity, the possibility of Kumaun being the source of copper becomes strong as the copper minerals there contain arsenic.

Now we shall look briefly at the copper minerals of Kumaun (Fig.1).

North of Pithoragarh (Fig. 1) near Chandak and east of Kanalichhina, minor veins of pyrite-chalcopyrite are seen within the Gangolihat dolomite. In the Dewalthal area near Almora, the magnesite and dolomite show disseminations of chalcopyrite. All the places with the suffix 'Agar' were ancient mining/metallurgical centres. At Rai Agar, an 18 m long and 0.5 m wide lode of chalcopyrite and pyrite, with an average copper content of 0.04 to 1.5%, has been located. The Bora Agar mineralized zone is about 1.5 km long extending between Tusrani and Dol, and 5.30 m in width, with an average copper content of 1.29% (Krishnamurthy & Krishnaswamy, 1986; Pant, 1988).

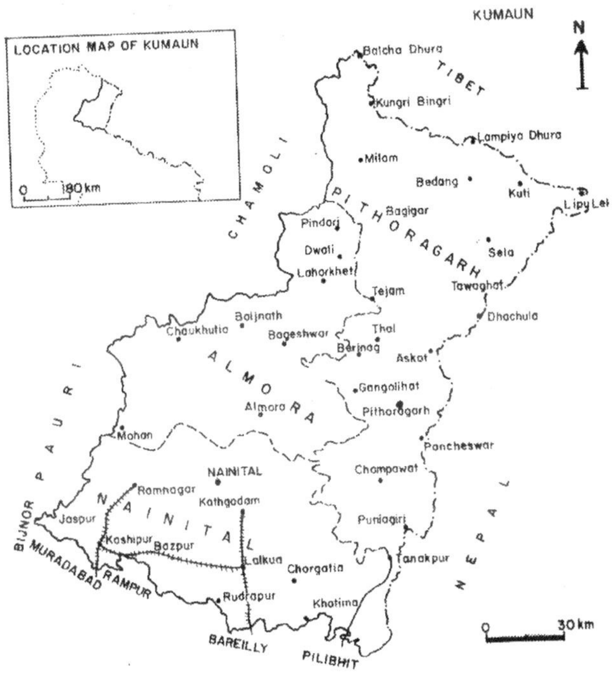

Figure 1. Map of Kumaun.

A polymetallic copper deposit has been located near Askot (Fig.1). This is the only occurrence of significance. The mineralization is mainly represented by sphalerite, galena, chalcopyrite and arsenopyrite, with minor amounts of marcasite, bornite, chalcocite, pyrite, pyrrhotite, cubanite and stinnite (the occurrence of arsenopyrite is to be noted). The thickness of the ore zone is variable, with an average of 4.47 m. The weighted average composition is Cu 2.32%, Pb 2.64% and Zn 3.95% (Krishnamurthy & Krishnaswamy, 1986). A total reserve of 0.77 million tonnes of copper-lead-zinc ore, with an average grade of 9% metal, has been estimated for this polymetallic deposit by the Geological Survey of India.

Kharahi *patti* is located close to the north of Almora town (Fig.1) and extends between Binsar and Bageshwar. The area is known as a traditional centre of metallurgy from times immemorial. Atkinson (1980-81:282) writes that the Gaul mine of Kharahi patti and the Sor Gurang mines produced grey copper in small quantities. Tamtyura, Danochhina, Chaugochhina, Kharak-Tamta, Ghingarkhola, Binsar, Bhatkhola, Simsyari, Bihargaon, Uderkhani, Bilona, Agar, Gair-Sikera, Lob and Borgaon (all in the districts of Almora and Pithoragarh) were some of the places where metallurgy was practised in ancient times, according to the villagers, who now inhabit these places. The coppersmiths are known as Tamtas, and are thought to be responsible for introducing coppersmithy in Kumaun (for the details of the old metal sites see Agrawal & Kharakwal 1998, chapter 4).

The Agri people, who are considered expert miners, have been living in Kharahi patti. It is told that Chanauli, Joshi Palari, Uderkhani, Chaugaonchhina, Bihar-gaon, Naial and Kharak-Tamta (Dist. Almora, Fig.1) are the locations of first settlements of the Tamta people in Kharahi. The following places are believed to be ancient mining sites, where traditional metallurgy is still occasionally practised:

Khan-garh (Khano-ki-garh)	Copper
Agar	Copper
Chaugaonchhina	Copper
Dewaldhar	Copper

5. Copper Hoard Artefacts in Kumaun (see Figure 1 map for the location of main sites)

The possibility of a link between Copper Hoards and Kumauni sources of copper minerals becomes stronger, because at Banakot 8 lugged type anthropomorphs were found in a stone quarry. Bankot village (Long.79¡52'E; Lat.29¡45'N) in Pithoragarh district is situated about 13 km northwest of Pithoragarh town. According to the villagers at Bankot, in 1989 a hoard of 8 anthropomorphic copper objects was unearthed while digging a stone quarry. The Bankot copper anthropomorphs contain 98% copper, 1.22% iron, and arsenic as an impurity. They were probably made in an open, single stone mould. Their obverse sides show nodules and gas cavities, which suggest that they were made in an open mould, without green poling, and left unfinished. The so-called Bankot anthropomorphs could as well be just copper ingots as their weights suggest: their ranges are 3.4 ± 0.1 and 2.4 ± 0.1 Kg.

Another piece of evidence is that in the foothills (at Haldwani) a typical anthropomorph turned up at a scrap dealer's shop. The source of this piece could only be Kumaun (Joshi 1995-96: 23-31). Thus circumstantial evidence connects the Central Ganga group of Copper Hoards with Kumaun.

Linguistic evidence also sheds some new light on the problem.

6. Linguistic Data

Grierson (1967-68), in his linguistic survey of the Himalayas, had identified some residue of Mon Khmer words in the Kumauni dialects. Mon Khmer is part of the Austro-Asiatic language group (Bayard 1996). With the new ^{14}C AMS dates from Thailand, the copper metallurgy there can be pushed back to II-III millennia BC. White (1997) reports, 'The earliest evidence of bronze at Ban Chiang appears by Early Period II/III in both burial and non-burial contexts. The calibrated results...indicate an early II millennium BC date for this part of the sequence.' White further says, 'the 3KUM dating and pollen evidence is broadly consistent with Blust's (1996) linguistic arguments for the expansion of rice-cultivating, Mon

Khmer speaking societies down the Mekong from south-western China beginning by the late 6th millennium BC'. Higham, however, thinks that KUM dates are just burning episodes in the Lake Kumpahawapi pollen cores and have no necessary relationship with agriculture. He says that the earliest date for agriculture in southeast Asia is about 2000-2300 BC from Non Pa Wai, Ban Kao, southern China. Higham (Pers. Comm.) has pointed out that 'Non Nok Tha is a cemetery dated to the period 1500-1000 BC.' This late date, however, is not acceptable to others. The dates quoted by White from 3 KUM come from a sediment core. The evidence is consistent with land clearance in the early 5th millennium BC. Such a date seems to accord well with the linguistic argument of Blust (1996). Our experience from the Rajasthan lakes has also been similar: the advent of *Cerealia* type pollen is accompanied (*c*.7000 BP) by finely comminuted charcoal, which has been interpreted as clearance for agriculture (Singh *et al.* 1990:351-358). Higham seems to conflate the expansion of the Austro-Asiatic language family with the spread of a specific type of pottery, and arrives at the date of 2000-3000 BC as mentioned above. Basically, the difference between the positions of White and Higham lies in the estimation of the gap between the advent of agriculture and metallurgy. White argues for a 6th millennium BC beginning of agriculture accompanied by the expansion of the Mon Khmer family of languages, as argued by Blust (1996), and a substantial time gap between the advent of agriculture and metallurgy. Higham, on the other hand, thinks that this gap could be only a few hundred years.

Unfortunately, in India, the northeast is still archaeologically a *terra incognita.* It will be interesting to define the gap between the advent of agriculture and copper metallurgy through the Himalayan corridor. The movement of Munda speaking people along the Himalayan corridor, however, is generally accepted.

Now we shall see how this evidence is relevant to the Central Himalayas. In Kumaun, the *shilpkars*, earlier called Doms, are supposed to be Kols, speakers of Munda dialects (Sharma 1988). Munda and Mon Khmer dialects are two branches of the Austro-Asiatic family. There are many Austro-Asiatic aboriginal languages still spoken in Thailand (Bayard Pers. Comm.; Bayard 1996). Elsewhere we have discussed (Agrawal & Kharakwal, 1998) that the Kols are said to be the first ethnic group of Kumaun, which had descended from the Munda speaking group and is related to the Austro-Asiatic sub-groups (Majumdar & Pushalkar 1952: 565). They represent the aborigines of Kumaun. Doms (shilpkars) are perhaps the modern representatives of the Kols (Atkinson 1980-81: 370), who today occupy the lowest rung in the hierarchical status in Kumauni society and are now included among the Scheduled Castes. They also claim to have been in Kumaun before either the Khasiya or the immigrants from the plains arrived. The Doms are the traditional smiths in Kumaun.

Higham (1996) discusses the relationship between the Austro-Asiatic and Austronesian languages and whether they belong to a common ancestral Austric phylum, and also the implications for archaeology. In view of the recent work by Reid (Higham 1996) and Blust (Higham 1996), Higham tends to agree with the Austric hypothesis as it would require only one Neolithic expansion and spread of rice cultivation which, according to him, is better supported by archaeological evidence. He describes the SE Asian rivers as being like spokes in a wheel, 'with the hub in the eastern foothills of the Himalayas....In a westward direction, ancestors of the Munda moved, perhaps down the Brahmaputra into eastern India. (Higham 1996:115)' Obviously, the movement did not stop there, but went farther westward to the central Himalayas. This type of south east and westward movements probably explain better the expansion, first of the rice cultivation, and later of metallurgy.

Synthesizing the archaeo-linguistic evidence Bayard (1996) agrees, 'It is very tempting to correlate this with the westward expansion, under the stimulus of agriculture, of Austro-Asiatic (AA) language from Nanhailand through Viet Nam, Laos, Cambodia and Thailand to Burma and ultimately eastern India. Linguists have reconstructed a number of terms related to agriculture in Proto-Munda, and some of these (husked rice, rice husking pestle) appear to go back to Proto-AA as well. It has been generally agreed for some time now that the prehistoric and early historic inhabitants of Thailand and Burma spoke languages in the Mon-Khmer branch of AA (Bayard 1979: 270). Such an expansion would account neatly - in a general sense for the distribution of the AA languages.'

Bayard further says, 'Based on morphological evidence such as that presented by Reid (1988: 31-32, in Bayard 1996), Blust is convinced that Schmidt's Austric hypothesis is correct and postulates an Austric homeland in the eastern Himalayan foothills in the area of the converging valleys of the Salween, Mekong, and Yangzi rivers. From this homeland, one group of rice cultivating people moved down into the middle Yangzi Valley to give rise to the AT superfamily of languages ca. 6500-6000 BC. Somewhat later the now-proto-AA speakers remaining in the homeland began to move down the valleys of the Salween and Mekong (Proto-Mon-Khmer), with pre-Munda speakers moving westward down the Brahmaputra....' The ethnographic evidence, that in Kumaun, (unlike for other cereal crops) rice sowing is accompanied by elaborate rituals, may indicate its remote beginnings.

I admit that considerable miscegenation has been going on in the central Himalayas for several millennia, and the concept of race is now defunct (Kennedy 1995). Yet, in the relative isolation (compared to the Ganga Doab) of the central Himalayas, one can still observe that the Brahmins and Kshatriyas have sharp features and fair complexion. The Bhotiyas and the descendants of the Kiratas bear unmistakable Mongoloid features. The Doms, who practise blacksmithy as a traditional craft, are typically short statured, dark complexioned and answer to Austro-Asiatic features.

We will now examine the iron technology.

7. Iron and the Central Himalayas

It is generally agreed that the beginning of the first millennium BC saw the advent of iron into India, though Chakrabarti (1992:68) has put it much earlier than warranted by the evidence, perhaps to prove an independent origin of iron technology in India. Prakash & Tripathi (1986: 568-569) have identified 5 zones of iron working in ancient India. Zone B is northern India. The Painted Grey Ware (PGW) cultures may be credited with the introduction of iron in the Gangetic basin. The PGW distribution zone extends from Punjab and Himachal Pradesh in the north, to Ujjain in the south, and from the great Indian desert of Bikaner in the west to Kausambi and Vaishali in the east. western boundaries of the PGW have been further expanded to Bahawalpur state in Pakistan.

Although Painted Grey Ware gives the culture its name, PGW is only a small percentage of the total pottery collection. Other wares include black and red ware (without paintings), black-slipped ware, plain grey ware and red ware. The PGW cultural phase, dated roughly 1000-400BC, is characterized by a simple subsistence: use of mud houses; no coinage or writing. Several PGW settlements have demonstrated a trend towards growth in size and technological expertise with the appearance of glass and iron (Prakash & Tripathi 1986).

Here, we will confine ourselves to the Zone B (Painted Grey Ware Zone of north India). Elsewhere, we have shown that the Zone A (north west India) and Zone B (north India) have clear links (Agrawal & Kharakwal, chapter 4, 1998).

Not only archaeological, but also literary evidence indicates that iron was used in early I millennium BC in the Gangetic basin.

Kosambi (1963) says, 'For the iron-ore country, however, we have one ancient record well worth consideration, the Pali *Sutta-nipata*... There is no question but that the work is pre-Asokan and contains the oldest of Buddhist traditions... The word *phala* for plough-share occurs in both the prose and the metrical portions of the discourse. The simile runs: 'like a phala heated for a whole day and plunged suddenly into water...' Iron reduced from ores by any primitive method comes out as a spongy mass which has to be heated and forged repeatedly as well as hardened by sudden chilling before it is of any use for tools.... The date 800-700 BC, therefore, seems to me a reasonable beginning for the iron age in India, meaning thereby specifically the Gangetic basin where iron was most needed and most likely to have been worked'.

Kumaun

Kumaun has an old tradition of iron metallurgy, and high grade ores from Lob and Agar have been reported and analyzed (for details see Agrawal & Kharakwal, Chapter 4, 1998). We have found a number of sites with iron ores, mining shafts, slags, kilns, etc. The British tried to start an iron industry in Kumaun in the 19th Century. There was a strong local iron smithy tradition until recent times. Lohaghat, Pithoragarh District(Fig. 1), was known for a rust-resistant iron made of the local ores.

The iron metallurgical tradition in Kumaun is very old and part of the folklore. As in Bihar and central India, in Kumaun, also, folklore attributes early iron technology to the *Asuras* (Agrawal & Kharakwal 1998). In Kumaun, the folklore tradition is very strong and some of the iron smelting sites are associated with the name *Asura*, e.g., Asurchula (the hearth of the Asuras). It is also interesting to note, that an old fort near Lohaghat in Pithoragarh District is attributed by folklore to Banasur. (Is it a corrupt form of the Assyrian king's name, Asur-bani-pal?) In fact, some historians have referred to the migration of some early Assyrian tribes to this region in the 8th century BC. The famous local goddess Nanda has been traced to the Assyrian goddess Nana(Vaishnav 1977). These issues have been discussed in detail in our book, *Central Himalayas* (Agrawal & Kharakwal 1998), but we have to admit that so far there is no solid archaeological evidence to prove an Assyrian association. On the other hand, very little systematic work has been carried out in the region thus far.

Uleni village, at a height of 1200m AMSL, is located about 8 km north-west of Dwarahat at the base of Dunagiri mountain on the Dwarahat Chaukhutiya road of Ranikhet tehsil in Almora district. A small rivulet flows down to Chaukhutiya (Fig.1) to join the Western Ramganga; on the right bank of this rivulet the village of Uleni and its agricultural terraces are located. On these terraces a large heap of iron slag was found. Though traces of any kind of furnace could not be found on the surface, there must have been some furnaces, as the large amount of slag suggests. It seems that people might have dug out the furnaces to level their fields. The area is rich in ferrugenous material, and occasionally, huge boulders containing iron ore can be seen in the seasonal Uleni channel. Some PGW sites have also been reported and excavated in the Western Ramganga valley (Nautiyal *et al.* 1978-79, and Pers. comm.).

The slag at Uleni was found not only on the agricultural fields, but also in a rock shelter which was located close to the site. A charcoal sample from the site was ^{14}C dated to the early I Millennium BC (PRL-1648 2770+90 BP. Its calibrated date range is 1022 to 826 BC).

The parallels in grave architecture and ceramics between the Swat valley and the Kumauni cists are striking indeed. We have shown that the Swat valley (Tusa 1981; Stacul 1987) has yielded remains of iron and grey ware using cultures with remarkable similarity with the Kumauni cists (Agrawal *et al.* 1995)). PGW sites at Thapli and Purola are known from Garhwal. Recent excavations on the western Ram Ganga terraces by the Garhwal University have yielded evidence of Swat-like graves (Vinod Nautiyal Pers. Comm.) and a proto-PGW type of pottery. The Uleni date of early first millennium BC thus makes it possible that the central Ganga area could have derived its iron from Kumaun (Agrawal & Kharakwal, Chapter 4, 1998).

We shall now look at the sources of iron in Kumaun.

'The iron ores found in Kumaun belong to either of the two varieties, the rhombohedral or the prismatic. The first is a peroxide of iron containing in its best defined type 70% iron and 30% oxygen' (Atkinson 1980-81:287). Scaly iron ore was mined at Ramgarh and Dhaniyakot on the Koshi. Both yielded good quality iron. During the British period, Lusgani, Nathuakhan, Gulla and Satbunga were the principal mines in the Naini Tal district. For the first 18 years of British rule, these mines were leased to the headmen of the Agaris at a nominal tax (Atkinson ibid.) (Agrawal & Kharakwal, 1998, chapter 4).

Many sites are associated with the word *lu* (the Kumauni word for iron), e.g., Lukhani, Lohaghat. Some early metal sites are associated with the name Agar, e.g., Rai-Agar, Banagar. Agar, and the ironsmith communities even today are called *Agarias*. We are emphasizing the folklore tradition, because it points to the antiquity of this craft in Kumaun. Because of the ^{14}C date of the early I millennium BC (PRL-1648) from Uleni, the antiquity of the folk tradition gains support. From this area we have also collected iron slag which has been analyzed and found to be mainly iron silicate, with no traces of copper. We thus find that folklore, historical tradition, the ^{14}C dates and XRD analyses taken together provide the circumstantial evidence that Kumauni iron and iron ores could have been used by the PGW people (Agrawal & Kharakwal 1998).

On the basis of the circumstantial evidence discussed above, we would like to associate the advent of iron technology in Kumaun with the Khasas, a pre-Vedic Indo-Aryan group. Khasas/Khasias still have a strong presence in Kumaun.

8. Ethnographic Evidence: The Khasas

We now have dates also for cist burials from Kumaun (Agrawal *et al.* 1995). The date of cist burials and their typology and ceramics relate them to the Swat valley Gandhar Grave culture (Dani 1967) sites. Stacul (1987) has compared some of the Swat burial complexes with those of Kashmir. From Malari (Garhwal), crouched burial, comparable to Burzahom (Kashmir), has been reported. The main common tool type is the rectangular stone sickle. Linguists (Sankrityayana 1958; Vaishnava 1977) have found affinities between the words Khash, Kasha, Kassite, Kashmir, Kashgar, etc.; the warrior class of Kumaun is still called *Khas* or *Khasia*. They are supposed to be pre-Vedic Indo-Europeans. If these linguistic equations are correct, some of these burial complexes may belong to early Indo-European migrations into the western and central Himalayan regions. It may be mentioned here that Rahul Sankrityayana (1958) and other early writers have traced the Khasa people of Kumaun back to the Kassites of Mesopotamia. The people of Kashmir, Kasgar, etc. are all supposed to be related to the Khasas. These were the tribes who preceded the Vedic Aryans and were looked down upon by them.

The Khasas, who were perhaps of Central Asian origin, came to India through the northwest, like the Sakas and the Kusanas. At the time of their movement, they might have been scattered over almost all the hilly regions of northern India, i.e., Kashgar, Kashmir, Kangra, Kumaun, Kulu, Nepal, even up to Tibet. Though in strength and numbers they might have been stronger than other groups, they were probably only a nomadic-pastoral tribe. In the course of time, the social peculiarities (Joshi 1929) of the Khasas were more or less dissolved in the larger society of the hills. Now, the role of the Caucasoids in spreading iron technology into China is being recognized. Could the same people have been responsible for both PGW and iron in the central Ganga valley? Could they have been related to the Khasas?

An Zhimin (in Pigott 1996:90) states that metal and its technology could have travelled along the Silk Road, and that the technology could have developed in the Qijia culture (2000-1600 BC) of north-western China, a culture which was well ahead of the central plains in its use of copper/bronze. This could also apply to iron technology. With the discovery of the cemeteries, in Xinjiang province, that contained more than 100 mummified individuals who were clearly Caucasian (Mair 1995), datable to 2000-1000 BC, and with the presence of iron in west China in 10th century BC, the possibility of a Eurasian origin can not be ruled out.

Probably, because of the large scale clearance of forests in the Ganga Doab and the needs of the urbanization, the limited iron resources of Kumaun and the difficulty of logistics eventually led to the exploitation of river sands rich in magnetite (Tripathi & Mishra 1977).

9. Iron from Copper Technology ?

In China, some scholars have claimed that iron technology may have developed from copper technology, because pyrites contain both copper and iron minerals. In India, however, there is a distinct break in the two urbanizations, the first based on copper and the second on iron. There is a distinct break in both cultural and technological traditions, as also in the locales of the two urbanizations. The Harappan urbanization took place in the north west of the subcontinent and was based on copper technology, gallery forests, and the pliable alluvium of the Indus (Agrawal 1982a). This copper based urbanization took place in the 3rd millennium BC. But the second urbanization, located in the Ganga valley, took place in the first millennium BC and was based on iron technology. In contrast to the semi-arid Indus valley, the Doab was a dense monsoonal forest and required an abundance of iron to clear forests and to colonize the hard kankary soil of the Doab. Thus, both the temporal and spatial gaps indicate that, in India, the iron technology could not have evolved from the preceding copper technology, but was a new innovation. We have argued that the latter could probably be attributed to the Khasas.

10. Conclusion

Thus the evidence, only suggestive so far, tends to link up the Swat valley (Pirak), central Himalayan iron metallurgy sites, cist burials, Khasa tribes, the Chinese Caucasoid connections, early ^{14}C dates and ancient iron smelting tradition into a consistent pattern which indicates Kumaun to be the probable source of early iron for northern India.

On the other hand, the circumstantial evidence - it is no more than that at this stage of research - does raise strong possibilities of a relationship between the central Ganga

Copper Hoard Group and the Austro-Asiatic language speaking people. The composition of the central Ganga Copper Hoard artefacts, the Munda and Mon Khmer residue in the Kumauni dialects, the finds of Copper Hoards in Kumaun, the Kumauni copper minerals with arsenic, early dates of metallurgy in SE Asia and the new evidence of movements along the Himalayan corridor seem to fit into an internally consistent framework of a relationship between early SE Asian metallurgy, central Himalayas and the Copper Hoards. If nothing else, the hypothesis needs to be verified on an inter-disciplinary basis.

We thus see the possibility of the central Himalayas having played a crucial role in the development and spread of both copper and iron technologies in northern India, and having been a source of supply of copper and iron in the I-III millennia BC. From what I argue above, it seems that the central Himalayas may not have been closed *cul-de-sacs*, but involved in the dynamics of technological diffusion.

For a long time, perhaps due to the British influence, we looked only towards west Asia for prehistoric contacts - real or imagined. Recent years have seen commendable and competent inter-disciplinary research in SE Asia. I think now the excitement lies in the Himalayan corridors and their prehistoric relationship with south China and SE Asia. The central Himalayas, in recent years, have produced some new evidence. The northeast still remains *terra incognita*. It is time that we explore the Himalayan corridors and look east. This will require a paradigm shift in our archaeological thinking.

Acknowledgements
I am grateful to Donn Bayard, Charles Higham, J.S. Kharakwal, V.N. Mishra, Vince Pigott, Vibha Tripathi, Joyce White and Suzanne Young for discussions and help. However, I alone am responsible for the views expressed in the article. I also thank Neha for her help with the input of data.

References

AGRAWAL, D.P. 1971. *The Copper Bronze in India* , New Delhi: Munshiram Manoharlal.
1982a. *The Archaeology of India.* London: Curzon Press.
1982b. The Indian Bronze Age Cultures and their Metal Technology. In (Eds.) F. Wendorf & A.E. Close, *Advances in World Archaeology*, Vol.1. Pp.213-264. New York: Academic Press.
AGRAWAL D.P., J.S. KHARAKWAL, SHEELA KUSUMGAR & M.G., YADAVA. 1995. Cist burials of the Kumaun Himalayas, *Antiquity*, 69(264): 550-554.
AGRAWAL, D.P. & J. S. KHARAKWAL. 1998. *Central Himalayas: an Archaeological, Linguistic and Cultural Synthesis.* Delhi: Aryan Books International.
ATKINSON, E.T. 1980-81. *The Himalayan Gazetteer,* Vols.1,2,3. New Delhi: Cosmo Publication (First published in 1884).
BAYARD, DONN. 1979. Comment on H.L. Shorto, The linguistic protohistory of Mainland South East Asia, In R.B. Smith & W. Watson (Eds.), *Early South East Asia: Essays in Archaeology, History, and Historical Geography*, Pp. 278-80. Oxford: Oxford University Press.
1996, Linguistics, archaeologists and Austronesian origins: comparative and socio-linguistic aspects of the Meacham-Bellwood debate, *Indo Pacific Prehistory Association Bulletin*), 15(2): 71-86.
BLUST, R. 1996. Beyond the Austronesian homeland: The Austric hypothesis and its implications for archaeology. In *Prehistoric Settlement of the Pacific,* (Ed.) W. Goodenough. Transactions of the American Philosophical Society, Philadelphia. 86(5): 117-140.
CHAKRABARTI, D.K. 1992. *The Early Use of Iron In India. Delhi: Oxford.*
CHAKRABARTI, D.K. & NAYANJYOT LAHIRI. 1996. *Copper and Its Alloys in Ancient India.* Delhi: Munshiram Manoharlal.
COGGIN BROWN, J. AND DEY, A.K. 1955. *India's Mineral Wealth.* London.
DANI, A.H. 1967. Timargarh and Gandhar Grave Culture. *Asian Perspectives,* 3:380.
GRIERSON, G.A. 1967-1968. *The Linguistic Survey Of India,* vol. 1(1) and Vol 9(4). Calcutta: Central Publication Branch, Govt. of India.
HEGDE, K.T.M. 1965. *Technical Studies in Chalcolithic Period Metallurgy in India.* Ph.D. Thesis. Baroda: M.S. University.
HIGHAM, C.F.W. 1996. Archaeology and linguistics in SE Asia: implications of the Austric hypothesis. *Indo-Pacific Prehistory Association Bulletin,* 14(1): 71-86.
JOSHI L.D. 1929, *The Khasa Family Law,* Allahabad: Government Press.
JOSHI M.P. 1995-96, The anthropomorphs in the Copper Hoard Culture of the Ganga Valley, *Puratattva,* 26: 23-31.
KENNEDY, KENNETH A.R. 1995. Southeast Asia: biological anthropology and concepts of ancient races. In Erdosy, George (Ed.), *The Indo-Aryans of Ancient Southeast Asia Language, Material Culture and Ethnicity.* Pp.32-66. Berlin: Walter De Gruyter.
KENOYER, JONATHAN M. 1997. Trade and technology of the Indus Valley: new insights from Harappa, Pakistan. *World Archaeology,* 29(2):262-277.
KOSAMBI, D.D., 1963, The beginning of the Iron Age in India, *Journal Of Economic And Social History of the Orient,* 6(3): 309-318.
KRISHNAMURTY, M. & KRISHNASWAMY, V.S.1986. Status of mineral exploration and reserves in Northwest Himalaya, *Proc. International Himalayan Geology Seminar,* Section IV. New Delhi. Calcutta: Geol. Survey of India. pp. 300-352.
LAL, B.B. 1972. The Copper Hoard Culture of the Ganga Valley, *Antiquity* 46: 232-87.
1997. *The Earliest Civilisation of South Asia.* Delhi: Aryan Books International.
MACKAY, E.J.H. 1938. *Further Excavations at Mohenjodaro.* 2 vols. Delhi: Govt. of India.
1943. *Chanhudaro Excavations 1935-36.* New Haven: American Oriental Society.
MAIR, V.H. 1995. Mummies of the Tarim basin. *Archaeology, 48(2):28-35.*
MAJUMDAR, R.C. & A.D. PUSALKAR. 1952. *The Vedic Age.* London: George Allen and Unwin. p. 565.
MARSHALL, J. 1931. *Mohenjo-daro and the Indus Civilisation.* London: Arthur Probsthain.
NAUTIYAL, K.P., B.M. KHANDURI, V. NAUTIYAL & D.L.RAJPUT. 1978-79. Thapli: a Painted Grey Ware site in Garhwal. *Puratattva,* 10:94-95.
PANT, C.C. 1988, Mineral wealth of Kumaun: prospects and problems, In Valdia, K.S. (Ed) *Kumaun: Land and People,* Naini Tal: Gyanodaya. Pp. 213- 222.
PIGGOTT, STUART. 1961. *Prehistoric India.* Middlesex: Penguin.
PIGOTT, V.C. 1996. The study of ancient metallurgical technology: a review, *Asian Perspectives,* 35 (1): 89-97.
PRAKASH, B. & VIBHA TRIPATHI. 1986. Iron technology in ancient India, *Metals and Materials*, September Issue, Pp. 568-579.
REID, L.A. 1988. Benedict's Austro-Tai hypothesis - an evaluation. *Asian Perspectives.* 26(1): 19-34.
SANKALIA, H.D. 1974. *The Prehistory and Protohistory of India and Pakistan.* Poona: Deccan College.
SANKRITYAYANA, RAHUL 1958. *Kumaun* (in Hindi). Varanasi: Jnanmandal.
SHARMA, D.D. 1988. Kumauni language and dialects. In Valdiya K.S.(Ed.), *Kumaun Land And People.* Naini Tal: Gyanodaya. pp. 120-136.
SINGH, G., R.J. WASSON & D.P. AGRAWAL. 1990. Vegetational and seasonal climatic changes since last full glacial in the Thar desert. *Review of Palaeobotany & Palynology,* 64:351-358.
STACUL, G. 1987, *Prehistoric and Protohistoric Swat, Pakistan.* Rome: IsMEO.
TRIPATHI, VIBHA & ARUN K. MISHRA. 1977. Understanding iron technology: an ethnographic model. *Man & Environment,* 22(1): 59-66.
TUSA, S. 1981, Notes on some protohistoric finds in the Swat Valley (Pakistan), *East and West,* 31(1-4): 99-120.
VAISHNAV, Y.D. 1977. *Kumaun Ka Itihas* (in Hindi). Naini Tal: Gyanodaya Prakashan.
VALDIYA, K.S. (Ed.) 1988. *Kumaun Land And People.* Naini Tal: Gyanodaya.
VATS, M.S. 1940. Excavations at Harappa. Delhi: Govt. of India. 2 vols.
WHEELER, R.E.M. 1947. Harappa 1946: The Defences and Cemetery R37. *Ancient India. 3:58-130.*
WHITE, JOYCE C. 1997. A brief note on new dates for the Ban Chiang cultural tradition, *Bulletin of the Indo-Pacific Prehistory Association,* 16: 103-106.
YULE, P. 1985. *Metalwork of the Bronze Age in India.* Munich.

Preliminary Insights into the Provenance of South Indian Copper Alloys and Images Using a Holistic Approach of Comparisons of their Lead Isotopes and Chemical Composition with Slags and Ores

SHARADA SRINIVASAN

Indian Institute of Science, Bangalore 560012, India

An attempt is made to gain greater insights into probable sources of copper alloy images and artefacts from south India by comparing lead isotope ratios and elemental analyses of artefacts with those of ores and slags. About 130 south Indian copper alloys were investigated using lead isotope and ICP-OES analysis. The vast majority of these are historically and artistically important south Indian images from the collections of the Government Museum, Madras; the Victoria and Albert Museum and the British Museum, London. To determine possible ore sources, trace element comparisons, investigated by EPMA-WDS, are discussed alongside those of slags from five sites with evidence of mining and metal extraction in antiquity from southern India. The following classification of images, as ratified by their As, Bi, Ni, Co and Sb element distributions, is: Pre-Pallava (c. 200-600 AD), Pallava (c. 600-850 AD), Vijayalaya Chola (c. 850-1070 AD), Early Chalukya-Chola (c. 1070-1125 AD), Later Chalukya-Chola (c. 1125-1279 AD), Later Pandya (c. 1279-1336 AD), Vijayanagara and Early Nayaka (c. 1336-1565 AD) and Later Nayaka and Maratha (c. 1565-1800 AD) periods. From trace element comparisons it is probable that Kalyadi was a source of smelted bronze for the Vijayalaya Chola group of images and Tintini of copper for Vijayanagara images. Lead isotope studies indicate discrete lead sources for Vijayalaya Chola images (which may have been Ambadongar), Vijayanagara and Early Nayaka images and for most Chalukya-Chola images, while in the Pallava period the use of lead from two sources is indicated. Lead isotope studies indicate that Ambaji was clearly a source of lead, and probably also of metallic zinc, for western India and a minor one for south India by the 12th-13th centuries; that Dariba in northwestern India was a source of lead for south India in the late medieval period; and that for Andhra Satavahana coins (c. 1st-2nd century) both local lead from Andhra and imported Mediterranean lead were used. A source of metallic zinc for south India as early as the 4th century AD is also indicated from a zinc ingot representing some of the earliest known evidence for metallic zinc. Trace element trends of four 'Indo-Roman' copper coins (4th century AD) suggest local manufacture.

Keywords: Provenance; South India; Copper Alloys; Pallava; Chola; Vijayanagara; Early Historic Period; Lead Isotopes; Trace Elements; Slags.

Introduction

Copper alloy images from southern India form one of the most important Indian artistic expressions, as represented by the well known iconography Nataraja, depicting the cosmic dance of Siva, which has been acclaimed by artists like Romain Rolland and Rodin (Coomaraswamy 1924). As with other Indian metal artefacts, there are many problems in making provenance, chronological and stylistic attributions for south Indian copper alloys and images, due to their poor contexts. Many images were recovered from hoards buried to protect them from iconoclasm (at various periods during Islamic raids and also from Portuguese vandalism) (Nagaswamy 1988) or come from temple vaults. While Hindu, Buddhist and Jaina religious images have been made, the majority are Hindu which were seldom inscribed, unlike the others, so making it difficult to give attributions. South Indian image casting also represents one of the longest continuous traditions and is still being practised in the Tanjavur district of Tamil Nadu. This has also compounded problems in the dating of images through imitations of earlier styles and repetitive iconography.

Indeed, some of the above issues are also part of a wider problem in the study and typological classification of all Indian metal artefacts. The fact that a significant proportion are found in hoards rather than in stratigraphical contexts, for example, the copper hoard artefacts of the Indo-Gangetic Valley of about the second millennium BC; the lack of marked burial traditions from at least the early historic period (i.e., Christian era), and the preference for ritual cremations according to Hindu or Brahmanical practice, are probably significant contributing factors. In addition quite a number of artefacts from South India have been uncovered from river beds, as in the case of some 'Indo-Roman' coins discussed in this paper. Hence there is relevance in exploring the usefulness of a scientific approach for gaining greater insights into the archaeology or art history of these artefacts and in building up analytical data on these artefacts against which comparisons can be made.

Methodology

A holistic (integrated) analytical and archaeometallurgical study has explored the implications of analysis and studies in ancient metallurgy for art history, chronology, style and provenancing of south Indian images and related copper alloys (Srinivasan 1996). Images were sampled from the important collections of the Government Museum, Madras, India, the Victoria and Albert Museum, London and the British Museum, London. Compositional and trace element analyses were performed on 106 sets of images (each set comprises one accession number, but can consist of more than one image or include accessories which also may have been analyzed) and ten miscellaneous artefacts (such as coins, charters and bowls) for eighteen elements, using inductively coupled atomic emission spectroscopy (ICP-AES) at Royal Holloway and Bedford New College, UK (TABLE 1a, b (at end)). Lead isotope analysis was done on

60 objects using Thermal Ionization Mass Spectrometry (TIMS) at the Isotrace Laboratories, Oxford (Srinivasan 1996, in press a). Both lead isotope and elemental analysis were performed on 39 objects. Comparisons were also made between traditional image casting practices observed by the author in Tanjavur district in mid 1990, and casting features observed from technical examination of the ancient images. This is reported in Srinivasan (1996).

This is the first comprehensive study using a range of analytical techniques on a significant number of bronze images from south India. The study includes a fairly representative sequence of bronzes of artistic and historical importance from various dynastic periods and their associated styles.

Analyses of less than a dozen early medieval south Indian bronzes (8th-13th centuries AD) have been reported by Werner (1972), and Johnson (1972). Previous provenance studies of religious icons from the Indian subcontinent include about 150 copper alloy images from the Indo-Tibetan region (Reedy 1986). Statistics were employed to discriminate between images from different regions, based on compositional analyses, stylistic criteria, and clay core analyses. For south Indian bronzes, these being solid cast without clay cores, the present study is based solely on metal.

The potential of the lead isotopes in helping to solve archaeological problems is well established and is discussed by Gale (this volume). Lead isotope analysis, which is a more powerful tool for provenancing leaded metal artefacts than elemental analysis, has been used in this study alongside trace element analysis, to determine if analytical signatures of groups of bronzes can be identified as a help in their typological classification.

Lack of information on ore sources, needed before attempting any provenance studies, need not be a deterrent in the use of lead isotopes. These remain a powerful diagnostic device for exploring internal consistencies within groups of artefacts and so may be used to characterize them. For example their use in elucidating the art history of West African figurines (Joel *et al.* 1995) where, although many of the sources could not be identified, discrete lead isotopic groupings were observed between Benin, Udo and Ife castings. Similarly, lead isotopes in conjunction with elemental analysis have been used in the study of Chinese Buddhist religious images now in the Freer Gallery of Art (Sayre *et al.* 1992).

Complete lead isotope data on the analyzed south Indian images as well as data on Indian ore sources are given in Srinivasan (in press a). Lead isotope measurements were made using a fully automated multicollector thermal ionization mass spectrometer (VG 38-54-30) calibrated with NBS SRM 981 standards.

The implications of such studies for stylistic re-analysis of south Indian images are discussed elsewhere (Srinivasan, in press a), while a revised catalogue for the sampled collection is to be published in a forthcoming monograph (Srinivasan, in press b).

This paper discusses some of the inferences that can be made from empirically observed chemical similarities within artefacts in trace elemental and/or lead isotope ratio data about shared sources or histories of processing of metal which may help their typological classification or characterization.

The lead contents of the 60 artefacts selected for lead isotope analysis were semi-quantitatively determined, all but a few had more than 1% lead. It is recognized that even a 1% alloying of lead would swamp the lead isotope ratios from trace amounts of lead in the copper or zinc components of the alloys (Stos Gale pers. comm.). Hence, essentially it is the source of lead in the alloys that is being provenanced.

Trace element patterns on the other hand are more likely to reflect the source of copper, with those like nickel and cobalt being chalcophilic. Indeed, different ore deposits can have characteristic trace element distributions based on ore geochemistry; for instance Mookherjee & Phillips (1979) found that nickel and cobalt values in copper ores from Ingaldhal in south India are linearly correlated. However, the final levels of trace elements in any processed metal would be affected by contributions from fluxes, by the partitioning of trace elements from ore to smelted metal (Merkel 1983) and refining, so that it is difficult to attempt correlations between artefacts and copper ores. This is a topic that requires concerted study. Comparisons with metal remnants in slags should reflect the processed metal more accurately. But in any case consistencies in trace elements could indicate the use of similar sources or at least the use of similar batches of processed metal leading to such uniformity.

To explore these aspects further, a preliminary study was made of ores and slags from old mine workings and slag heaps in south India. These were subjected to trace element analysis using electron probe micro-analysis with wavelength dispersive X-ray spectrometry (EPMA-WDS), at the Wolfson Laboratories of the Institute of Archaeology, University College London. Lead isotope ratios on a few specimens of galena taken from these ore sources were also measured. Firstly, the nature of the base metal smelting that seemed to have taken place was determined from the slags. Secondly, quantitative analyses of the metallic remnants or prills in the slags was performed to see whether characteristic elemental patterns could be observed which then could be compared with finished artefacts. The main reason for attempting chemical comparisons between finished artefacts and metallic prills in slags rather than ores was because a limited number of microanalyses on hand specimens by EPMA showed the ores to be heterogeneous on a fine scale and extensive bulk analyses of a statistically significant number of ore specimens were outside the scope of this study. Furthermore ores undergo great chemical changes during smelting. However, the relationship between the composition of metallic prills and finished artefacts is less tenuous since the prills represent trapped solidified droplets of metal dispersed through slag and so their composition should be representative of the molten ingot produced in the furnace.

Bearing in mind these limitations any foray into exploring sources of metal must be preliminary. The more conclusive body of evidence showing analytical signatures relating to groups of artefacts is discussed here briefly, but given in full in Srinivasan (in press a, b).

Lead isotope ratio and trace element data for a typological classification according to dynastic chronology and implications for provenance

The sampled images and artefacts are categorized under the following groups according to dynasties as listed in Table 1a, these are based on re-assessments made on the published art history literature, plus insights provided by scientific analysis (Srinivasan in press a, b). The lead isotope and trace element trends for different stylistic groups are set out in Srinivasan (in press a). It should be noted that the chronological boundaries of these groups are broadly fixed by what is known of dynastic chronology and are, in that sense, approximate and not absolute. The groups are:

-Andhra & Pre-Pallava (*c.* 200-600 AD)
-Pallava (*c.* 600-875)
-Vijayalaya Chola (*c.* 850-1070)
-Early Chalukya-Chola (*c.* 1070-1125)
-Later Chalukya-Chola (*c.* 1125-1279)
-Later Pandya (*c.* 1279-1336)
-Vijayanagara & Early Nayaka (*c.* 1336-1565)
-Later Nayaka & Maratha (*c.* 1565-1800)

Scatter plots are shown of Pb 208/206 against Pb 207/206 and Pb 206/204 against Pb 207/206 for south Indian images and copper alloys. FIGURE 1 is a scatter plot of Pb 208/206 against Pb 207/206 with symbols for dynastic style/chronology. FIGURE 2 is a scatter plot of Pb 208/206 against Pb 207/206 with symbols showing different metal types. The ellipses marked on both figures indicate groups of artefacts within which there is internal stylistic coherence. Inter-element scatter plots from ICP analysis for cobalt against nickel, bismuth against arsenic and antimony against cobalt are shown in FIGURE 3 a, b and c.

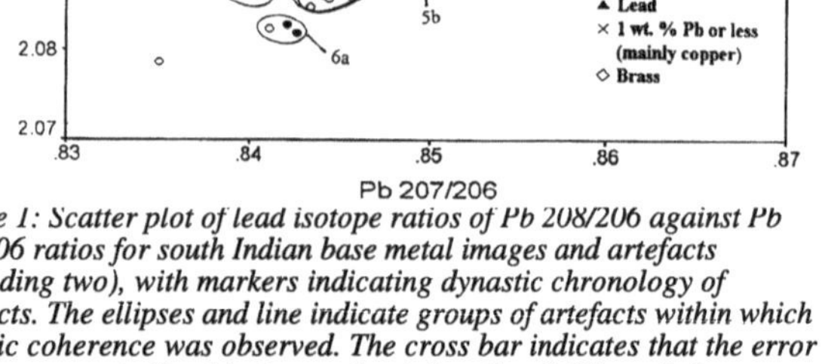

Figure 1: Scatter plot of lead isotope ratios of Pb 208/206 against Pb 207/206 ratios for south Indian base metal images and artefacts (excluding two), with markers indicating dynastic chronology of artefacts. The ellipses and line indicate groups of artefacts within which stylistic coherence was observed. The cross bar indicates that the error in measurement of lead isotope ratios is less than 0.1%, and also indicates the spread in isotopic composition for groups of artefacts.

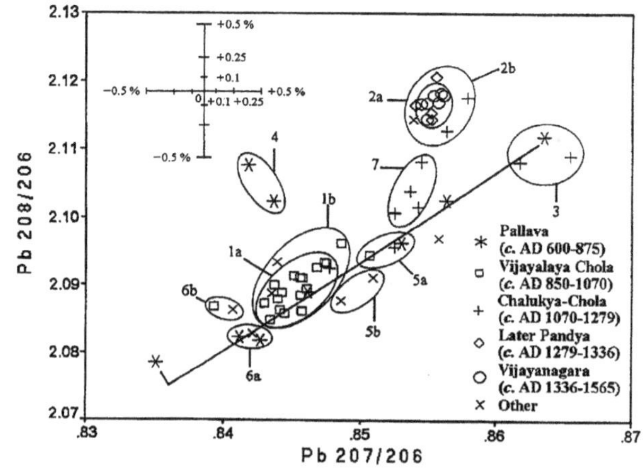

Figure 2: Scatter plot of lead isotope ratios of Pb 208/206 against Pb 207/206 ratios for south Indian base metal images and artefacts (excluding two), with markers indicating metal category of artefacts, whether they are of leaded bronze, lead, copper or zinc or brass. The ellipses indicate groups of artefacts within which stylistic coherence was observed.

FIGURES 1 and 2 show that leaded images of the Vijayalaya Chola (*c.* 850-1070 AD) fall predominantly into Group 1a, and the spread in isotopic composition for this group is narrow enough to suggest that the objects came from a discrete lead source. The Vijayalaya Chola images represent the most artistically significant and prolific group of south Indian bronzes with bronzes attributed to the munificent patrons of the lineage of Vijayalaya Chola such as the remarkable dowager queen Sembiyan Mahadevi, described by Harle (1994) as one of the great patrons of all time, and king Rajaraja Chola (Barrett 1965; Dehejia 1990). Well known images, previously attributed by art historians to this period for which lead isotope and trace element trends fit include a Nataraja, (acc. no. IM-2-1934) (FIGURE 4), from the Victoria and Albert Museum, attributed to Sembiyan Mahadevi's patronage by Dehejia (1990) and Schwindler (1975), and the fine Velankanni Nataraja (acc. no. 234) in the Government Museum, attributed to Rajaraja Chola by Schwindler (1975) and Srinivasan (1963). Indeed Rodin (1921) remarks that this latter Nataraja rivals the Medici Venus in its graceful execution.

Group 2a consists predominantly of Vijayanagara period bronzes (*c.*1336-1565 AD), another powerful dynasty in southern India under which the sprawling capital city of Vijayanagara was built at Hampi in north Karnataka. A published image fitting this group is the Rama (acc. no. IM-71-1927) from the Victoria and Albert Museum attributed to the Vijayanagara period by Michell (1995). This group also seems to represent a discrete lead source from its narrow spread of isotopic ratios, as does group 7. This group consists of images classified as Chalukya-Chola (*c.* 1070-1279 AD) including an image of Kiratarjuna (acc. no. 43-IS-1887) dated by Nagaswamy (n.d.) to the 12th century. The line on FIGURE 2 indicates that images and artefacts (including a lead coin and a copper plate charter) from the Pallava period (*c.* 600-875 AD) have lead isotope ratios plotting roughly along it, suggesting that lead in these artefacts may be the result of the random mixing from two sources leading to a linear combination of their ratios. A published image fitting this trend is the Natesa from Kuram in the Government Museum, Madras (acc. no. 53/38) attributed Nagaswamy (1988)

by
Figure 4: Siva Nataraja (acc. No. IM-2-1934), Tanjavur district, Tamil Nadu, of the period of Chola queen Sembiyan Mahadevi (c.950 AD) which fits lead isotope and trace element trends for the Vijayalaya Chola group (c.850-1070 AD) Reproduced by permission of Victoria & Albert Museum.

Figure 5: Avalokitesvara (acc.no. IM-300-1914) uncovered from the Krishna delta, Andhra, c. 5th C. AD, which fits trace element trends for the Pre-Pallava and Early Andhra group. Reproduced by permission of Victoria & Albert Museum.

to the Pallavas. An important early image, the damaged Buddhist Avalokitesvara (FIGURE 5) (acc. no. IM-300-1914) dredged from the Krishna delta, Andhra, in the Victoria and Albert Museum, fits the trace element trends for the Pre-Pallava and Andhra group (*c.* 200-600 AD). This group can be distinguished from the Pallava group in having antimony contents of not more than 0.1 wt. %. From its style, it had not been certain whether it was South Indian or Sri Lankan, although its early Deccan affinities had been pointed out by Zwalf (1985), while Sivaramamurti (1963: Fig. 2b) had attributed it to the early Andhra Pallavas around the 5th century; and indeed analysis supports such an attribution.

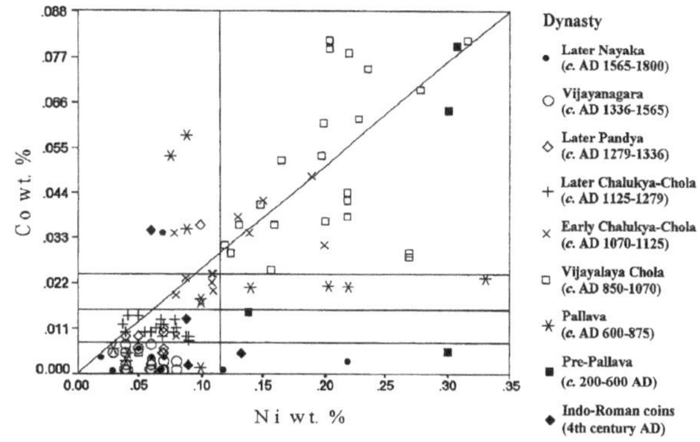

Figure 3a: Scatter plot of Co against Ni contents for south Indian copper alloy images and artefacts analyzed by ICP-OES. The markers relate to dynasty chronology of the artefacts. The diagonal line indicates that a linear trend is noted for bronzes of the Vijayalaya Chola period. The perpendicular lines indicate the boundaries of absolute values for different groups of artefacts. Trends for four 'Indo-Roman' coins analyzed by ICP-MS are also indicated.

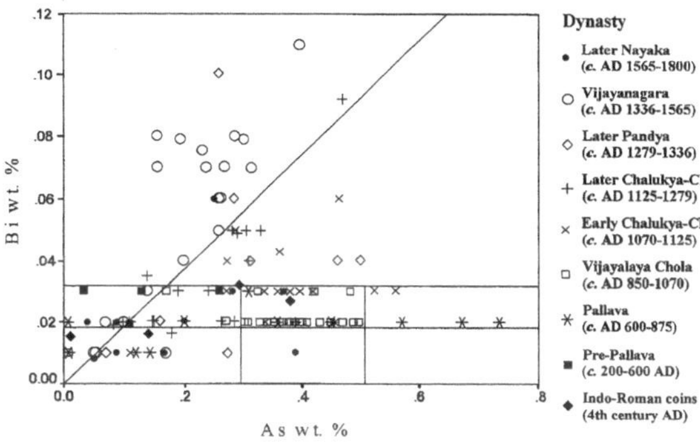

Figure 3b: Scatter plot of Bi against As content for south Indian copper alloy images and artefacts analyzed by ICP-OES. The markers relate to dynasty chronology of the artefacts. The slope is drawn to indicate that the Bi/As ratios for most Vijayanagara and Early Nayaka images is greater than 0.19. The perpendicular lines indicate the boundaries of absolute values for different groups of artefacts. Trends for four 'Indo-Roman' coins analyzed by ICP-MS are also indicated.

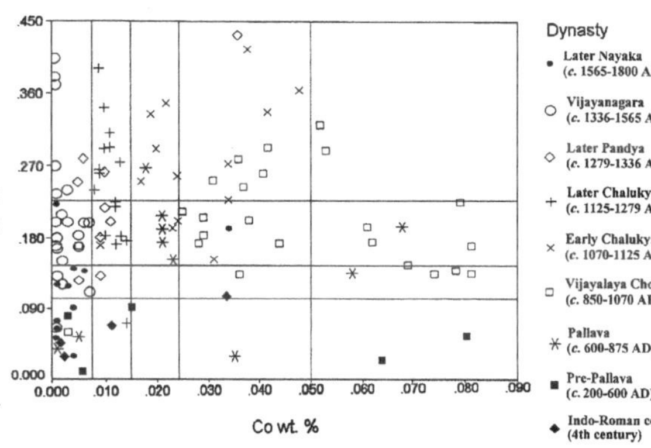

Figure 3c: Scatter plot of Sb against Co content for south Indian copper alloy images and artefacts analyzed by ICP-OES. The perpendicular lines indicate the boundaries of absolute values for different groups of artefacts. Trends for four 'Indo-Roman' coins analyzed by ICP-MS are also indicated.

The discrete trends of the images and alloys for the different stylistic groups as seen in the lead isotope ratio plots, are also generally reflected in the trace element scatter plots (analyzed by ICP-OES, TABLE 1a, b), notably cobalt against nickel, bismuth against arsenic, and antimony against cobalt seen in FIGURE 3a, b and c. Indeed, a linear trend is observed for the Co:Ni ratios of the Vijayalaya Chola group (FIGURE 3a) which may be a characteristic of the copper ore. For the Vijayalaya Chola group, distinct lead and copper sources seem to have been used, which may or may not belong to the same polymetallic deposit. This is also the case for Vijayanagara bronzes. For the Chalukya-Chola bronzes, whilst the lead sources appear to have been similar, the trace element patterns support a separation between Early Chalukya-Chola and Later Chalukya-Chola groups. For most Later Pandya images the source of lead seems to have been the same as the Vijayanagara group. However, for the Early Chalukya-Chola and Later Pandya groups, their trace elements are somewhat intermediary between the trends for the bordering groups: this may be explained if a new copper source was being mixed with earlier metal, which then became more widely exploited in the later periods. However, note that the trace element patterns for different groups need not indicate the use of disparate copper deposits but may indicate the use of metal from different veins/mines within the same mining region.

Examples of images whose compositions suggest re-cycling of metal from their mismatch with earlier groups are discussed in Srinivasan (in press b). However, generally there does not seem to be much evidence for re-cycling, especially between the Vijayalaya Chola and Vijayanagara groups. This may have been because the re-cycling of images was not favoured for religious reasons. This observation is especially true for Hindu images.

That secular scrap was re-mixed is suggested by the trends observed in the lead isotope ratios of the Pallava group (FIGURE 1) where one of the two groups between which mixing occurred (giving rise to the linear trend indicated in the figure) seems to be group 3. This group also contains a Pallava lead coin and two Satavahana lead coins (Srinivasan 1996), suggesting that coins and such utilitarian artefacts were indeed re-cycled. This fits ethnographic observations made in 1990 in Kerala and Tamil Nadu of widespread and organized bartering and re-cycling of utilitarian traditional metalware. By contrast, re-cycling of images was not that common and only to make other images.

In FIGURE 2 the reason for designating groups 1b and 2b as peripheral to the major groups 1a and 2a is to suggest that they consist of artefacts where most of the lead comes from the same sources as the major groups but is mixed with minor amounts of lead from other sources. The same may also be true for other minor peripheral groups including 5 and 6. FIGURE 2 also suggests that some of the smaller clusters comprise artefacts that are also related by metal type. For example group 5b consists of a Deccan zinc ingot, with a Brahmi inscription datable to *c.* 4th century, and a remarkable figurative votive Buddhist brass vessel (acc. no. IM-9-1924) with 14 wt. % Zn from the Andhra region of south India (now in the Victoria and Albert Museum) of about the 5th century. The zinc ingot constitutes some of the earliest evidence for metallic zinc reported anywhere. Group 5a consists of two gilt copper Buddhist images, one from Nagapattinam (acc. no. IPN-2639) in the Tamil region (with trace element trends that match the Later Chalukya-Chola group) and the other dredged off the coast of northern Sri Lanka, namely the famous 1.4 m high gilt Tara (OA-1830-6-12-4) currently in the British Museum (FIGURE 6). Another explanation for the slight shift in lead isotope ratios between 5a and 5b, and their plotting on the periphery of main group 1a (representing a Vijayalaya Chola lead source), could be that it is the result of using different copper, lead or zinc ores from the same mining region. The objects in the above two groups nevertheless share the trace element trends seen in the corresponding chronological groups of south Indian bronzes, increasing the likelihood of their having a south Indian provenance. In fact, out of a dozen images found in Sri Lanka and analyzed by AAS (Von Schroeder 1990), the gilt Tara is the only one which fits with South Indian element analyses, in this case Pallava trends, suggesting, perhaps, re-use of Pallava metal.

Figure 6: Gilt Tara (acc. no. OA-1820-6-12-4) found near Trincomalee off the coast of north Sri Lanka whose analytical signatures suggest a South Indian provenance. (Courtesy of the Trustees of the British Museum, London)

Lead isotope investigations were also undertaken on fewer than ten images found in regions in the southern part of the subcontinent other than the Tamil region (indicated in TABLE 1a) including Karnataka and Andhra Pradesh. Details of these are reported elsewhere (Srinivasan 1996; in press b), but it is worth noting that they either shared trends with, or fell onto, the periphery of lead isotope groups for the corresponding chronological group of Tamil bronzes. This suggests that metals from particular sources were widely distributed over the southern Indian region at particular times.

Ore sources that can be excluded from the analyzed south Indian copper alloys

The identification of exact provenances for leaded artefacts by matching their lead isotope ratios with those of ore sources is complicated by non-availability of all the data for those ore sources. However, it is possible to exclude ore deposits from being the source of an artefact if their lead isotope ratios fall sufficiently outside the field or spread of isotopic ratios for that artefact (Stos-Gale 1993). For the majority of south Indian images and artefacts, precise ore sources could not be pinpointed. There are two exceptions for which ore sources were traced in northwestern India. Other than for some northwestern Indian ores, recent lead isotope data for the Indian subcontinent, as determined from standardized laboratories, are lacking.

Nevertheless, known sources in the old world for which recent and reliable lead isotope data are available, namely Europe, West Asia and parts of Asia such as China, can be eliminated, as the lead isotope ratios of the south Indian images do not fall convincingly within the spread of isotopic compositions for any of the ore deposits from these regions (Stos-Gale pers. comm.) The spread for lead isotope ratios for Indian ore deposits such Zawar and Ambaji in western India is found to be of the order of 1-2% of the mean, as indicated by an inspection of the lead isotope data from the Isotrace data bank. For example, Ambaji has Pb 208/206 ratios ranging from 2.0902-2.14883 for different mines in the same region. This is much higher than that usually observed for ore deposits which is typically not more than about 0.25% of the mean (Stos-Gale 1993). Note that the spread in lead isotope ratios for most of the analyzed south Indian bronzes is around 1% of the mean, so that it is possible that they came from an as yet unidentified mining region in the Indian subcontinent characterized by such a large overall isotopic spread. Another reason for suspecting this might be the case is that there is a chronological shift from groups 1 to 2 with increasing Pb 208/206 and Pb 207/206 ratios.

Lead isotope analyses obtained from galena from two south Indian ore sources were different enough from those of the south Indian images to exclude them. The galenas were from Mamandur in Tamil Nadu (Pb 208/206: 2.39953, Pb 207/206: 1.06335) and the Bandalamottu mine in Andhra Pradesh (mean Pb 208/206: 2.16874, Pb 207/206: 0.94079). Ores from north-western India, some of which are published in Ericson & Shirahata (1985) also do not match. However, there are several other ore deposits in southern India and elsewhere in the Indian subcontinent for which lead isotope data are not available, for example in southern India there are several lead-zinc deposits in Andhra Pradesh including Zangamrajpalle.

Probable sources for Chola and early medieval south Indian images (9th-14th century)

Although an old analysis, lead isotope ratios of galena from an ore deposit in peninsular India, namely Ambadongar, (Venkatasubramanian *et al.* 1982) with Pb 208/206 ratios of 2.07 and Pb 207/206 of 0.84 fall within a spread of ±1% of the mean of group 1a consisting of High Chola or Vijayalaya Chola bronzes and so it is a possible source. TABLE 2 (at end) indicates preliminary results for Ni:Co and As:Bi ratios from analyses of base metal prills in slags from various sites in southern India by EPMA-WDS analysis alongside those of the analyses of trace elements in the south Indian images of the Vijayalaya Chola and the Vijayanagara groups. This indicates that the slags from Kalyadi in Karnataka may have been a source of metal for the Vijayalaya Chola bronzes since the Ni/Co ratios are comparable.

It is notable that the slags from Kalyadi are bronze slags consisting of an average 7% tin (normalized to a 100%, or 5% un-normalized). Based on the presence of metallic iron in the slag, preliminary indications are that the slags are from direct smelting of bronze using copper and tin ores, rather than from casting and alloying copper and tin metals (Srinivasan 1997). In addition, the average tin content of the Vijayalaya Chola bronzes is 6.8% which does not rule out the possibility of Kalyadi being a source of metal for these bronzes. Although the Vijayalaya Chola rulers ruled from their capital of Tanjavur in the Coromandel region of southern India, their influence extended to south Karnataka in the 11th century where Kalyadi is located. Indeed, there is a literary reference of the Imperial Cholas of Karnataka collecting fees on the export of tin (Kuppuram 1986), providing more evidence for the production of bronze from this region at this time.

Figure 7: Brass Buddha (acc. no. IS-44-1966) from Kanchipuram, Tamil Nadu, c. 1280 AD whose lead isotope ratios indicate that the source of lead, and probably zinc, was Ambaji in western India. Reproduced by permission of Victoria & Albert Museum.

A south Indian Buddhist brass image with 24% zinc and 2.8% lead of the 13th century from Kanchipuram (acc. no. IS-44-1966), Tamil Nadu (FIGURE 7), matched lead isotope data for a mine in the west Indian zinc-lead deposit of Ambaji. A spectacular 1 m high Jaina image of a seated Santinatha (acc. no. 930-IS) from western India in the Victoria and Albert Museum, inscribed to 1168 AD, made of brass (25% zinc, 3.6% lead) also has lead isotope ratios matching the same mine in Ambaji. Since lead and zinc are usually associated with each other, a few percent of lead is often found in smelted zinc. This suggests that Ambaji was a minor source of zinc and lead for south India. However, for the south Indian Nagapattinam-style Buddhist brass image, unlike the western Indian Santinatha image, the trace element patterns, such as their nickel and cobalt values (FIGURE 3) and arsenic, bismuth and antimony values, generally fit trends of south Indian bronzes of the Later Pandyan period (*c.* 1280-1336 AD). This difference can be explained since trace elements are more closely related to the provenance of copper, while lead isotope ratios pertain to sources of lead-zinc. There are reasons to believe that the above evidence points more strongly to the use of alloyed smelted zinc metal, rather than to cementation brass. Craddock (1981) suggests that several per cent of lead would be normally retained in smelted zinc. Brass made from cementation of copper with zinc ore need not contain lead. Moreover, for the two artefacts the sources of lead are the same while the sources of copper are different. Given the rather negligible amounts of lead, it seems less likely to have been deliberately alloyed to cementation brass from two different copper sources, so that it is more probable that smelted zinc from the same lead-zinc source retaining a bit of lead was alloyed to copper from different sources. Furthermore, the great size of the western Indian Santinatha image of 1 m, and about 200 Kg in weight and solid cast, also suggests the use of metallic zinc, since cementation brass would usually be produced in small crucibles so that the production of 200 Kg of cementation brass would have required about a dozen crucibles. Hence the casting of metallic zinc and copper seems more likely. It should be noted that Ambaji is less than 200 km south of Zawar, the site of the earliest known convincing evidence for metallic zinc in antiquity (Craddock 1981). Indeed zinc ores could have been brought over from neighbouring regions to be smelted at Zawar (Craddock, pers. comm.).

Probable sources of metal for Vijayanagara and late medieval south Indian images (*c.* 14th-18th centuries)

The trace element patterns and lead isotope ratios patterns are distinctive enough for the Vijayanagara group of images (*c.* 1336-1565 AD) to suggest that the lead and copper came from discrete sources compared to the other images, although these may or may not be from the same source. The ratios of Ni:Co and As:Bi for slags from southern India analyzed by EPMA, compared with those analyzed for the images from the Vijayanagara period, indicate that the slags from the site of Tintini could have been a source of copper for the Vijayanagara images. This is quite plausible in terms of the archaeology of the region, since Tintini in north Karnataka is reasonably close to the Vijayanagara capital of Hampi, and is close to a shrine to a 16th century metal smith and saint known as Moneshwara.

An image of Ganesha (FIGURE 8) (acc. no. 291) from Nellore (with lead isotope ratios of Pb 207/206: 0.95594; Pb 208/206: 2.2046; Pb 206/204: 16.301) of the Later Nayaka and Maratha period (*c.* 16th-18th century) is only one of two south Indian images that could be traced to a metal source. It matches the lead isotope ratios of lead slag from the northwestern Indian deposit of Dariba (with lead isotope ratios of Pb 208/206: 2.21438; Pb 207/206: 0.95271; Pb 206/204: 16.307) (Srinivasan in press a). Indeed the Marathas were a northern dynasty who were prominent rulers from the Tanjavur region about this time and could have introduced new sources of metal.

Figure 8: Ganesha (acc. no. 291) from Nellore, Andhra Pradesh, whose lead isotope ratios indicate that the source of the metal is Dariba in north western India; of a style consistent with the Later Nayaka and Maratha group (c. 1700 AD). (Courtesy of the Government Museum, Madras/Chennai).

Sources for the early historic period (*c.* 2nd-6th centuries AD)

Lead isotope ratios for galena from the Bandalamottu mine, Guntur district, Andhra Pradesh in south India matched ratios of a group of Satavahana coins (1st-2nd centuries AD) reported in Turner (1985) and Srinivasan (in press a). The Satavahanas were a dynasty from Andhra Pradesh with their capital at Amravati in the Guntur district. Under them the remarkable Amravati Buddhist stupa was built, parts of which are now in the British Museum. Another group of Satavahana coins matched Sardinian ore from Fluminese published in Stos-Gale (1993) and Srinivasan (in press a). This suggests that local sources may have been exploited by this period alongside trade in metal with far off regions like the Mediterranean, as discussed by Seeley & Turner (1984).

Four coins (4th century AD) recovered from panning in the Karur river of Tamil Nadu were of uncertain attribution. It was not known whether they were true Late Roman coins or Indo-Roman imitations, made in south India in imitation of Late Roman 'aes' copper coins. Srinivasan (1994) using inductively coupled plasma mass spectrometry (ICP-MS) at the Indira Gandhi Centre for Atomic Research in Kalpakkam found that the trace element plots of these coins, cobalt against nickel, arsenic against bismuth and antimony against cobalt, fitted with those observed for Pre-Pallava south Indian artefacts (ca 200-600 AD). This suggests that these coins may be Indo-Roman imitations made in south India rather than true Late Roman coins. The nickel, cobalt, bismuth and arsenic trends of a Pallava image and two Later Chola images analyzed by ICP-MS (TABLE 1) also fitted those observed for the corresponding groups discussed here from ICP-OES analysis of 116 images and artefacts (Srinivasan 1996).

Summary

From the lead isotope results, the exact provenance of lead in the majority of artefacts cannot be identified, due to insufficient lead isotope data from ore deposits in Asia. However, lead isotope data, together with trace element studies of artefacts, help in the typological classification of south Indian images and copper alloy artefacts within dynastic chronologies. This ultimately suggests the use of different copper and/or lead sources over different periods. During these periods, metals from these sources were apparently not restricted to the Tamil region but were widely distributed/traded over the south Indian region, including the Deccan regions of Karnataka and Andhra. This would makes sense, in the light of the fact that the major base metal deposits in peninsular India are in the Deccan rather than the extreme south. Comparisons between the lead isotope ratios of artefacts and ores are compatible with Ambaji in western India being a source of medieval zinc, and probably zinc metal, for north India and, to a minor extent, south India. Dariba, in northwestern India, was a source of lead for the later medieval period, while a source of Satavahana lead of the early historic period was the Bandalamottu mine in Andhra, together with some imported lead from the Mediterranean. These data also tentatively suggest that Ambadongar in peninsular India was a source of lead for early medieval Chola bronzes. Trace element comparisons of nickel and cobalt from artefacts and metal remnants in slags from old slag heaps tentatively suggest that Kalyadi in south Karnataka was a source for the Vijayalaya Chola bronzes (*c.* 850-1070 AD), the smelted tin-bronze with up to 7% tin. Trace element comparisons of nickel, cobalt, arsenic and bismuth indicate Tintini in north Karnataka may have been a source of copper for Vijayanagara bronzes (*c.* 1336-1565 AD). For some artefacts, for which there was uncertainty as to whether they were south Indian or imported, there is now evidence to suggest they are of south Indian provenance given that their analytical trends (of lead isotope ratios and/or trace element trends) are consistent with south Indian ones. These include Indo-Roman coins (4th century AD) from Karur, a Deccan zinc ingot (4th century AD), and the medieval gilt Buddhist Tara, in the British Museum, found in Batticaloa, Sri Lanka, and the Avalokitesvara image, *c.* 5th century, from Krishna delta, Andhra, in the Victoria and Albert Museum.

Acknowledgements
The author would like to acknowledge the valuable inputs of N.J. Seeley, National Trust, UK, and support of J. F. Merkel, I. Glover and D. Griffiths during the course of PhD. research at the Institute of Archaeology, University College London. Thanks are due to Z.A. Stos-Gale, N. Gale for their interest and support of the lead isotope analyses done at Isotrace laboratories, Oxford Research Laboratory for Art and Archaeology, to N. Walsh and N. Blades for the kind use of ICP-OES facilities at Royal Holloway and Bedford New College, Egham, to K. Reeves for assistance with use of EPMA analysis at Wolfson Technical Laboratories at Institute of Archaeology, London and to C. Mathews and Mrs. Vijayalakshmi, Indira Gandhi Centre for Atomic Research, India for help with ICP-MS analysis of a few coins. Many thanks are due for the access of material to J. Guy and D. Swallow, the Victoria and Albert Museum, R. Knox and P. Craddock, the British Museum, S. Desikan and Devasahayan, Government Museum, Madras and R. Krishnamurthy, former President, Tamil Nadu Numismatic Society. The author is grateful for the financial support for PhD. research from the British Council, New Delhi, the Victoria and Albert Museum, British Museum, Institute of Archaeometallurgical Studies and the Gordon Childe Fund, and currently to the Homi Bhabha Council, and finally to the Indian Council for Cultural Relations and the organisers, S. Young, P. Budd and E. Reber for support to attend the Metals in Antiquity conference at Harvard University.

TABLE 1: PART 1a

Pre-Pallava and Andhra images and artefacts (*c.* 200-600 AD)
IS-8-1989**, IM-9-1924**, Zinc ingot with Brahmi inscription (*c.* 4th C. AD)*, Vishnukundin coins (5th C AD), IM-300-1914, four 'Indo-Roman' coins (4th C. AD)

Middle Pallava (*c.* 600-850 AD)
223**, IM-13-1934**, 53/38*, Lead coin*, 13, OA-1957-2-11-1*, *OA-1967-7-27-1**, Copper Plate 6, Seal 1, Seal 2

Later Pallava period (*c.* 850-875 AD)
752/75*, OA-1969-12-16-1*, 459/61*, IM-127-1927, 14/32, 1527/89
OA-1957-10-12-1(Pedestal inscribed; c. 10th C. AD)*

Early Vijayalaya Chola (*c.* 850-940 AD)
Copper plate 2, 19**, 220**, 240**, IM-137-1927**, IM-158-1929**, IM-149-1927**, 217, IM-75-1935**, IS-2-1951, IM-136-1927, IM-8-1924, 720/73**, *OA-1965-12-14-1**
OA-1905-12-18 (Pedestal inscribed; c. 9th C. AD)*

High Vijayalaya Chola (*c.* 940-1070 AD)
IM-2-1934**, IM-71-1935**, 280** (Pedestal inscribed; *c.* 10th-11th C. AD), 273**, 37**, 234**, 238**, 336**, 450/61*, 233*(Pedestal inscribed; 1511 AD), 286, 337, IM-128-1927, IM-131-1927, 475/36, Velankanni Kali

Early Chalukya-Chola (*c.* 1070-1125 AD)
46**, 43-IS-1887**, IM-70-1935**, 495/65*, IM-135-1957, 215 (2 images), 477/36, 2 (3 images), IM-1065-1873

Later Chalukya-Chola (*c.* 1125-1279 AD)
315/55**, IM-14-1938**, 721/74**, 477/62, 18, 31, 1427/87, IM-15-1939, IM-196-1937, 316/55, 318/55, 634/64, 1593/89, Spouted vessel, IPN-2639**, *OA-1830-6-12-4*, OA-1974-12-9-2**

Later Pandya/Later Pandya-early Vijayanagara transitional (*c.* 1279-1336 AD)
40/36**, 8/27**, *IS-275-1869***, IM-1326-1855**, IM-118-1924*, 260*, 142, 123, 288, IM-731935, IM-138-1927, IM-112-1924, IS-44-1966**, *930-IS* (Image inscribed 1168 AD), OA-1967-10-17-1**

Vijayanagara and Early Nayaka (*c.* 1336-1565 AD)
474/36**, IPN-2657**, IS-204-1959**, IM-6-1924**, IM-72-1935**, IM-71-1927**, IM-1-1934**, 212, IM-2-1948, IM-7-1924, IM-5-1928, IM-361-1924 (Pedestal inscribed; 16th-18th century), IM-356-1924 (Pedestal inscribed; 16th-18th century), 47, 59/39, 224, IM-14-1934, IM-129-1929, 107, Bala-2

Later Nayaka and Maratha (*c.* 1565-1800 AD)
IM-118-1943, IM-120-1924, IM-77-1948, 914-IS, Jertala Devi, Jertala Deva, 730-IS, 188, 338, IM-1062-1873 (Pedestal inscribed; *c.* 19th C. AD), IM-1063-1873 (Pedestal inscribed; *c.* 19th C. AD), 291*

* Indicates those for which only lead isotope analysis was undertaken
(Due to limits on access to equipment both lead isotope and trace element analysis could not be undertaken on all images)
** Indicates those for which both lead isotope and trace element analysis were undertaken
Italics Indicates images found in regions other than the Tamil region of the majority bronzes (from Karnataka, Andhra, Kerala or Sri Lanka) for which lead isotope analysis was done and which have been included under the above categories for convenience since they tend to share lead isotope trends of at least one other object in the corresponding group. These images are discussed elsewhere in Srinivasan (1996; in press b)

Table 1a: Catalogue and accession numbers of objects under their respective stylistic groups.

TABLE 1: PART 1b

	Average Ni/Co	Average Bi/As
Tintini slag: copper prills	70	0.3
Kalyadi slag: bronze prills	3	not applicable (i.e. Bi, As below detection limits of 0.001 wt.%)
Ingladhal slag: copper prills	0.23	0.42
Somalaragada slag: copper prills	2	1.2
Malapadu slag: copper prills	1.5	6

(Analysed for 17 elements by quantitative EPMA-WDS analysis with measurements for each prill averaged over 15 measurements and with measurements made on at least 6 prills in slag specimens from particular sites. Quantitative analysis was done using a JOEL Superprobe JXA 8600 with ZAF corrections with the total composition being within 1-2 wt. % of a 100 wt.%)

	Average Ni/Co	Average Bi/As
Vijayangara images (19) (ca 14th-16th c. AD)	57	0.22
Vijayalaya Chola images (ca 9th-11th c. AD)	4	0.06

(Analysed by ICP-OES for 18 elements including Ni, Co, Bi and As using Philips combined simultaneous sequential ICP-OES spectrometer)

Table 1b: Compositional analysis of 120 South Indian images and copper alloys.

	Average Ni/Co	Average Bi/As
Tintini slag: copper prills	70	0.3
Kalyadi slag: bronze prills	3	not applicable (i.e., Bi, As below detection limits of 0.001 wt.%)
Ingladhal slag: copper prills	0.23	0.42
Somalaragada slag: copper prills	2	1.2
Malapadu slag: copper prills	1.5	6

(Analyzed for 17 elements by quantitative EPMA-WDS analysis with measurements for each prill averaged over 15 measurements and with measurements made on at least 6 prills in slag specimens from particular sites. Quantitative analysis was done using a JOEL Superprobe JXA 8600 with ZAF corrections with the total composition being within 1-2 wt. % of a 100 wt.%)

	Average Ni/Co	Average Bi/As
Vijayangara images (*c.* 14th-16th century AD) (averaged for 19 images)	57	0.22
Vijayalaya Chola images (*c.* 9th-11th century. AD) (averaged for 28 images)	4	0.06

(Analyzed by ICP-OES for 18 elements including Ni, Co, Bi and As using Philips combined simultaneous sequential ICP-OES spectrometer)

Table 2: Preliminary comparisons between trace element ratios of some copper slags from southern India with those from stylistic groups of south Indian images indicating probable sources of copper.

References

BARRETT, D. 1965. *Early Chola Bronzes.* Bombay: Bhulabhai Memorial Institute.

COOMARASWAMY, A. K. 1924. *The Dance of Siva-Fourteen Indian Essays.* New York: Sunwise Turn.

CRADDOCK, P. 1981. The copper alloys of Tibet and their background. In: Oddy, W. A. and Zwalf, Z. (ed.): *Aspects of Tibetan Metallurgy.* London: British Museum Occasional Paper No. 15:1-33

DEHEJIA, V. 1990. *Art of the Imperial Cholas.* New York: Columbia University Press

ERICSON, J. E. & H. SHIRAHATA. 1985. Lead isotope analysis of ancient copper and base metal ore deposits in western India. In: P.A. England and L. Van Zelst (ed.): *Application of Science in Examination of Works of Art. Proceedings of the seminar: September 7-9, 1983.* Boston: Museum of Fine Arts, 207-12.

HARLE, J.C. 1994. *The Art and Architecture of the Indian Subcontinent.* New Haven: Yale University Press.

JOEL, E.C., E.V.SAYRE, R.D.VOCKE & F.WILLET. 1995. Stable lead isotope characterisation of various copper alloys used in West Africa: an interim report. *Journal of Historical Metallurgy,* 29: 25-34.

JOHNSON, B. 1972. *Krishnarajamannar Bronzes - A Technical Report.* Los Angeles: Los Angeles County Museum of Art Monograph Series, 1: 45-53.

KUPPURAM, G. 1986. *Imperial Cholas of Karnataka.* Delhi: Sundeep Prakashan.

MERKEL, J. F. 1983. Reconstruction of Bronze Age Copper Smelting. Unpublished PhD. thesis. University of London.

MICHELL, G. 1995. *Architecture and Art of Southern India.* Cambridge: Cambridge University Press.

MOOKHERJEE, A. & R.PHILLIPS. 1979. Distribution of copper, cobalt and nickel in ores and host rocks, Ingaldhal, Karnataka, India. *Mineralium Deposita,* 14: 33-55.

NAGASWAMY, R. 1998. South Indian bronzes. In: Khandalavala, K. (ed.): *The Great Tradition -Indian Bronze Masterpieces.* New Delhi: Festival of India, 142-79.

NAGASWAMY, R. n.d. Kongu bronzes in the Victoria and Albert Museum. In: Khandalavala, K. and Chandra, M. (ed.): *Lalit Kala,* 13: 41-2.

REEDY, C. 1986. Technical Analysis of Medieval Himalayan Copper Alloy Statues for Provenance Determination. Unpublished PhD. thesis, University of California, Los Angeles.

RODIN, A. 1921. Sculpture Caivites de l'Inde: La Danse de Civa. *Ars Asiatica,* 7-13.

SAYRE, E. V., P.JETT. & E.C.JOEL. 1992. A technical examination of the Chinese Buddhist bronzes in the Freer Gallery of Art, Part B: Stable lead isotope analysis. In: P.B.Vandiver and J.R. Druzik. (ed.): *Material Issues in Art and Archaeology III.* Materials Research Society Symposium Proceedings. Pittsburgh: Materials Research Society, 267: 205-22.

SCHWINDLER, G. J. 1975. The Dating of South Indian Metal Sculptures. Unpublished PhD. thesis, University of California, Los Angeles.

SEELEY, N. J. & P.J.TURNER. 1984. Metallurgical investigations of three early coinages: implications for metal trading and dynastic chronology. In: B. Allchin (ed.): *South Asian Archaeology 1981. Proceedings of the Sixth International Conference of the Association of South Asian Archaeologists in Western Europe.* Cambridge: Cambridge University Press, 331-3.

SIVARAMAMURTI, C. 1963. *South Indian Bronzes.* Bombay: Lalit Kala Academy.

SRINIVASAN, P.R. 1963. *Bronzes of South India.* Madras: Government Museum.

SRINIVASAN, S. 1994. Appendix-III. Metallurgical investigations on Late Roman or Indo-Roman coins from Karur. In: Krishnamurthy, R. *Late Roman Copper Coins from South India. Karur and Madurai.* Madras: Garnet Publications, 121-132.

1996. The Enigma of the Dancing 'Pancha-Loha' (Five-Metalled) Icons-Archaeometallurgical and Art Historical Investigations on South Indian

Bronzes. Unpublished PhD. thesis. Institute of Archaeology, University College London.
1997. Composition of bronze slags from Kalyadi in south India and the implications for the problem of tin in south Indian antiquity. In: A.Sinclair,
1998. E. Slater and J. Gowlett (ed.): *Archaeological Sciences 1995. Proceedings of a conference on the application of scientific techniques to the study of archaeology, Liverpool, July 1995.* Oxbow Monograph Series 64: 136-142.
in press a. Lead isotope and trace element analysis in the stylistic study of South Indian metal icons. *Archaeometry.*
in press b. *Archaeometallurgical and Art Historical Investigations on South Indian Bronzes.* ed. Indicopeustoi Series, Lyon. Brussels: Brepols.
STOS-GALE, Z. A. 1993. Lead isotope provenance studies-do they work ? *Archaeologie Polona,* 31: 149-80.
TURNER, P. J. 1985. Roman and Indigenous Silver Coins in Central and Southern India in the First Three Centuries AD. Unpublished PhD. thesis. Institute of Archaeology, University College London.
VENKATASUBRAMANIAN, V.S., S.JAYARAM & V. SUBRAMANIAN, 1982. Lead ages of galena from peninsular India. *Journal of the Geological Society of India,* 23: 219-25.
VON SCHROEDER, U. 1990. *Buddhist Sculptures of Sri Lanka.* Hong Kong: Visual Dharma.
WERNER, O. 1972. *Spektralanalytische und Metallurgische Untersuchungen an Indischen Bronzen.* Leiden: E. J. Brill.
ZWALF, W. (ed.) 1985. *Buddhism-Art and Faith.* London: British Museum Publications.

26

Ice Archives of Atmospheric Pollution from Mining and Smelting Activities during Antiquity

CHRISTOPHE P. FERRARI[1,2], SUNGMIN HONG[3], CLAUDE F. BOUTRON[1,4]

[1] *Laboratoire de Glaciologie et Géophysique de l'Environnement du C.N.R.S., 54, rue Moliére, Domaine Universitaire, B.P. 96, 38402 Saint Martin d'Héres cedex, France*
[2] *Institut des Sciences et Techniques de Grenoble, Université Joseph Fourier, 28, Avenue Benoît Frachon, B.P. 53, 38041 Grenoble cedex*
[3] *Polar Research Centre, Korea Ocean and Development Institute, Ansan, P.O. Box 29, Seoul 425-600, Korea*
[4] *U.F.R. de Mécanique, Université Joseph Fourier de Grenoble (Institut Universitaire de France), Domaine Universitaire, B.P. 68, 38041 Grenoble, France*

More than two decades ago, it was hypothesized that valuable information on the past history of lead and copper mining during antiquity could be found in the well preserved dated ice layers which have been accumulating in Greenland and other cold ice caps over the Holocene (past 10 000 years). This has been verified recently by analysing lead (including lead isotopes) and copper in various sections of the 3028.8 m European GRIP ice core which was drilled in 1990-92 at Summit in Central Greenland. These sections were dated from seven millennia ago onwards, and covered especially the periods of the Iron Age, the Roman Republic and Empire, and medieval times. They were analyzed using sophisticated ultraclean and ultrasensitive analytical techniques which allow reliable measurements to be made at the extremely low concentration levels (pg/g, i.e., 10^{-6} ppm) observed in polar ice.

Lead and copper were of totally natural origin seven millennia ago. From around 3000 to 1500 years ago, enhanced values were observed in part of the samples, with mean values about 4 and 2 times higher than natural values for lead and copper respectively. These high levels result from early anthropogenic lead and copper emissions to the atmosphere. These emissions are attributed to the crude, highly polluting, mining and smelting technologies used for silver, lead and copper production during Roman and medieval times. Lead and copper were emitted in the atmosphere as small size aerosols, which were transported to the Greenland ice cap by air masses and then incorporated into the snow by in-cloud and below-cloud processes. After the fall of the Roman Empire, lead and copper values decline before increasing again during medieval times.

These data show that Greek and Roman civilizations polluted the atmosphere two millennia ago, long before the Industrial Revolution. They open up the possibility of future quantitative studies of the history of metal production during Antiquity.

Keywords: Ice Archives; Greek and Roman Time; Lead and Copper Pollution; Mining and Smelting Activities; Greenland.

Introduction

More than two decades ago, Clair Patterson, a famous geochemist at the California Institute of Technology (Epstein, 1981; Patterson, 1971, 1972, 1980, 1994a, b) was the first to speculate that data of interest for archaeology could be obtained from the investigation of metals such as lead (Pb), Copper (Cu), Silver (Ag) and Tin (Sn) in the successive well preserved dated ice layers which have been accumulating in Greenland, Antarctica and possibly other cold ice caps during the past millennia (Patterson *et al.*, 1970). These ice layers had already proven to be fantastic frozen archives of our past atmospheric environment, allowing for instance reconstruction of the climatic history of our planet during the successive glacial-interglacial cycles for the last hundreds and thousands of years, and past natural or anthropogenic changes in the composition of our atmosphere (Boutron, 1995).

Amongst the most interesting elements to be considered are heavy metals, especially the toxic metals lead, cadmium, and mercury. They have indeed been dispersed in large quantities into the atmosphere and into various other compartments of our environment by human activities for many centuries (Nriagu & Pacyna, 1988; Nriagu, 1996) and have become a major ecological and health problem (see for instance Patterson, 1980; Needleman, 1984; Friberg *et al.*, 1985/86; Davis and Svendsgaard, 1987; Fitzgerald & Clarkson, 1991).

Chemical species emitted in our latitudes can be transported to Greenland in a few days. When species like heavy metals, arrive above Greenland for example, they can form condensation nuclei in clouds. They may also be directly incorporated into the snow flakes during their fall or deposited by sedimentation. At Summit, Central Greenland, the snow accumulation rate is about 40 cm/year near the surface.

This field of research has, however, proved to be very demanding and various challenges had to be faced by the leading research group. Logistical and engineering skills are formidable for obtaining deep ice cores in such remote and inhospitable environments. Moreover, the remoteness of Greenland makes the precipitation exceedingly pure: in fact, polar snow and ice are, by far, the purest natural water found on our planet. Many elements and compounds are therefore present in these very valuable samples at extremely low concentration levels. Ultraclean and highly sensitive analytical procedures are then required to get fully reliable data, as was beautifully illustrated more than two decades ago by Clair Patterson and his co-workers at the California Institute of Technology in their pioneering work on lead in Greenland snow and ice (Murozumi *et al.*, 1969).

Clair Patterson's idea was that heavy metal emissions to the atmosphere from mining and smelting activities during Antiquity (Tylecote, 1992) were probably intense enough to have resulted in a significant increase of metal concentrations in the ice above natural concentration levels (Patterson *et al.*, 1970). This would open up the possibility of quantitative studies of the history of metal production during Antiquity.

In this paper, we will present the method used for working on ice core sections from Central Greenland (decontamination procedures, clean room work, dating of the samples and analysis) and the results obtained for lead and copper from ice dated from the Greek, Roman and medieval times.

Material and Methods

Field Sampling

The G.R.I.P. (Greenland Ice Core Project) deep ice core was drilled during three summer field seasons from 1990 to 1992 in central Greenland at a site named Summit. Summit is located at the top of the ice cap, at an altitude of 3238m (Dansgaard *et al.*, 1993). The ice thickness is about 3000m, which means that the present day altitude of the bedrock is about 200m. The mean annual temperature is -32°C, which allows ice to be obtained which has not suffered from any summer melting. The 3028.8m long ice core (diameter: ~10cm was obtained using a large electromechanical drill. The core was divided into sections 55cm long which were packed into tight polyethylene bags. The core sections were transported frozen to the laboratory for analysis.

Fourteen sections were selected from the period between 2360 and 1775 years ago (depths of 511.0 to 349.3 meters), which are the ~600 years corresponding to the flourishing of the Roman Republic and Empire. Two sections were selected from ice deposited 3000 to 2500 years ago (depth of 619.3 to 569.3 m, respectively), a period that includes the rise of Greek civilization. Five sections were selected from ice deposited from 1520 to 470 years ago (depth of 349.3 to 129.3m, respectively), during the medieval and Renaissance periods. Finally, one section was chosen in ice near of the period pre-dating lead smelting (7760 years ago, depth of 1286.5m) and two sections for the pre-copper smelting period (7760 and 7260 years ago) to asses natural metal concentrations in the ice.

Dating of the samples

Dating is a crucial matter, since samples lose most of their scientific value if good quality dating is not available. It is a complicated issue, which has been discussed in some detail by various authors (Lorius, 1980; Hammer, 1989). Here we briefly present the different methods than can be relied upon.

For the most recent time periods (decades to thousands of years), the dating can be obtained by counting the successive annual layers through the measurement of isotopic abundance or chemical species for which pronounced seasonal variations do exist (Hammer, 1989). This is the case for the $^{18}O/^{16}O$, and D/H isotopic ratios, which are temperature dependent and exhibit clear annual cycles. This is also the case for various species such as hydrogen peroxide which peaks in summer, sodium, nitrates, calcium, dust, acidity. These methods have been used to date a large number of ice cores for periods as long as the Holocene.

Another approach is the identification of fallout from well known events such as volcanic eruptions. Amongst the well known volcanic events are the eruptions of Vesuvius in 79 BC, Tambora in 1815 and Krakatoa in 1877. They resulted in acidity, sulfate and heavy metals peaks which can be used as time markers (Zielinski *et al.*, 1994; Hong *et al.*, 1996a). For longer time periods (more than 100 000 years), the ice dating becomes rather difficult and need to combine glaciological models of the ice sheet with other methods (Reeh, 1989). The accuracy in dating ice archives is better than for other environmental archives such as peat bogs, or lake sediments. This accuracy is about 10 years for ice from the Roman time.

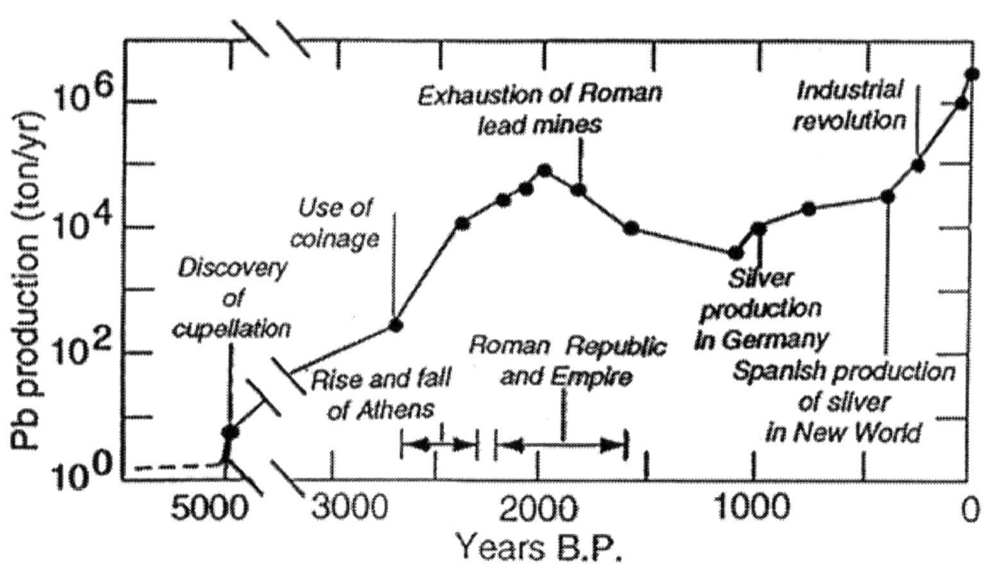

Figure 1. Changes in world-wide lead production over the past 5500 years. From Settle & Patterson (1980).

Decontamination of the ice core sections

A key point to obtain reliable data is in the cleanliness of the available samples. A major problem is that deep ice cores are always more or less contaminated on their outside. This contamination originates from various contributions (wall retaining fluid, body of the drill, the cable, and handling, packing and storage). In order to obtain the real concentration of heavy metals in the ice, each section must be decontaminated using ultraclean procedures such as those described in Candelone *et al.* (1994). Each core section is decontaminated by mechanically chiselling successive veneers of ice from the outside towards the centre inside a laminar flow clean bench at -15°C to isolate the innermost part of the core (Candelone *et al.*, 1994). In order to check the efficiency of the decontamination, changes in the concentrations of each metal from the outside to the inside each section must be systematically investigated. Good plateau of concentration were observed for lead and copper in the central part of the cores in all cases, which indicates that external contamination has not reached the centre of the core: the plateau concentrations then represent the original concentrations in the ice.

Heavy Metal analysis

The analysis of heavy metals in ice samples from polar ice caps is a real challenge for analysts. Actually only few laboratories have successfully faced it. The main reason for that is that heavy metal concentrations in polar ice are extremely low (about pg/g). During the last few years, a large number of incorrect data, erroneous by orders of magnitude, have been published. In order to remove this contamination, the decontamination procedures, melting operations of the ice, pre-concentration steps and analysis all have to be performed in a very clean environment. This clean environment can only be created in clean rooms similar to those used in microelectronics.

Inside the clean laboratory (Boutron, 1990), each veneer layer and the inner core are melted separately in ultraclean polyethylene bottles. Heavy metals are then determined in the melted samples using ultrasensitive analytical techniques. For lead, both Graphite Furnace Atomic Absorption Spectrometry (GF-AAS) (Hong *et al.*, 1994 ; Candelone *et al.*, 1995) and Thermal Ionization Mass Spectrometry (TIMS) (Chisholm *et al.*, 1995; Rosman *et al.*, 1993, 1997) are used. A distinct advantage of TIMS is that it allows the simultaneous determination of both lead abundance (using the isotope dilution procedure with a ^{205}Pb spike) and lead isotopes (^{204}Pb, ^{206}Pb, ^{207}Pb and ^{208}Pb). For copper, the GF-AAS technique was used. For each step of the analytical procedure, blank determinations must be performed. For example, the blank corresponding to the chiselling operation has been evaluated (Candelone *et al.*, 1994) by preparing an artificial core made by freezing ultrapure water. The amount of lead added to the sample during the chiselling operations was found to be typically about 0.15 pg/g (Candelone *et al.*, 1994). The precision of the concentration values so obtained is typically about 10-20%, which is fairly good, considering the exceptionally low concentrations measured.

Results and Discussion

Results concerning lead, lead isotopes and copper measured in Greenland ice dated from the Roman time are presented in the following paragraphs.

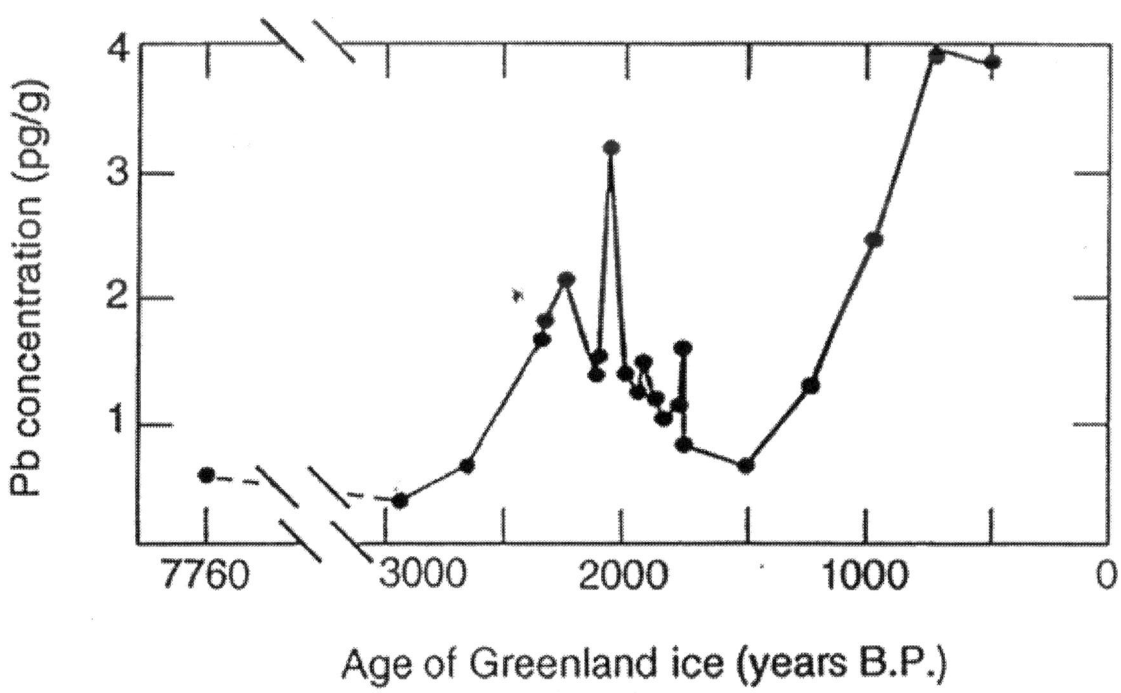

Figure 2. Changes in lead concentration in central Greenland ice from 7760 to 470 years ago. Each data point has been obtained from the analysis of a core length corresponding to exactly two years of ice accumulation (except one, which was only one year because of poor core quality) to eliminate the influence of possible short-term (seasonal) changes in lead deposition. From Hong et al. (1994).

Lead

The history of lead production (Figure1), documented by Settle & Patterson (1980) and Nriagu (1983b), began about six millennia ago, during the Copper age, when lead was first produced as a by-product of silver. But the real production of this metal began only one millennium after, with the discovery of new techniques has permit man for smelting lead-silver alloys from galena and cupeling silver from the alloys (Nriagu, 1983c). Lead production then increased continuously during the Copper, Bronze and Iron ages especially after the introduction of silver coinage. Lead production reached a maximum of about 8×10^4 tonnes/year (Figure 1) during the Roman Republic and Empire (Settle & Patterson, 1980). Then it decreased rapidly during the decline of the Empire. It remained very low during the early medieval time before increasing again one millennium ago. It then increased continuously until the present (Figure 1). During the Greco-Roman period, it has been clearly documented that several hundred thousand of people, especially slaves, died from lead poisoning (saturnism) during mining and smelting operations (Nriagu, 1983a, 1983b, 1983c). These mining and smelting activities have generated important local and regional pollution (see Renberg *et al.*, 1994), but was this pollution sufficient to reach remote areas of our planet such as Greenland? Figure 2 clearly demonstrates that this was indeed the case.

The Figure 2 shows changes in lead concentrations in Greenland ice during the past seven millennia (Hong *et al.*, 1994). Lead concentrations were very low (about 0.5 pg/g) seven millennia ago before significant lead production started. These values, ~0.5 pg/g, correspond to natural lead emitted from natural sources such as rock and soil dust (Nriagu, 1989). No particular change has been noted until ca. 3000 BP. During this period therefore, anthropogenic sources were not significant enough to generate a detectable signature in ice archives. From 2500 to 1700 years ago, lead concentrations increase markedly up to about 2 pg/g (four times the natural values (Hong *et al.*, 1994)). This 800 years period correspond to the flourishing of Greco-Roman civilization. During the period of the decline of the Roman Empire, lead concentration also declined to values very close to the natural values observed seven millennia ago (Figure 2). After this period, lead levels started to increased again in medieval times, up to about 4 pg/g at the time of the Renaissance (Hong *et al.*, 1994). They then continued to increase into the 20th century, with concentration values of ~10pg/g at the time of the Industrial Revolution and ~50pg/g in the 1900s (Candelone *et al.*, 1995).

Lead isotopes

Lead occurs naturally as four stable isotopes: ^{204}Pb, ^{206}Pb, ^{207}Pb, ^{208}Pb. In the Earth, the abundance of ^{204}Pb is fixed, but the other three increase with time because they are products of the radioactive decay of ^{238}U (half life 4.5 10^9 year), ^{235}U (half life 7.1 10^8 year) and ^{232}Th (half life 1.4 10^{10} year) respectively. Dust forming aerosols originate from the erosion of surface rocks the isotopic composition of which reflects parent U and Th abundances, the age and geological history of the rock. Lead ores are usually free of the U and Th parent nuclides and therefore have distinctive isotopic compositions relating to the time of their formation. While terrestrial rock-dust generally displays a ^{206}Pb/^{207}Pb isotope abundance ratio greater than ~1.20 (Chow & Patterson, 1962; Hamelin *et al.*, 1990) lead ores have ratios which are frequently lower. For instance Spanish lead from the Rio Tinto mines is typically ~1.16 (Stos-Gale *et al.*, 1995).

The lead isotope data recently obtained for Greenland ice (Rosman *et al.*, 1997) show that there is a depression in the ^{206}Pb/^{207}Pb ratio between 600 BC and 300 AD, giving unequivocal evidence of early, large scale, atmospheric pollution by this toxic metal. The ratio changes from ~1.201 in ~8000 year old ice to ~1.183 during Roman times (Rosman *et al.*, 1997). When combining these ice data with the isotopic fingerprint of the different lead mines which were worked during Roman times (Rosman *et al.*, 1997), it appears clearly that the mining districts in South West and South East Spain were the dominant sources of this lead (Cortizas *et al.*, 1997). This gives for the first

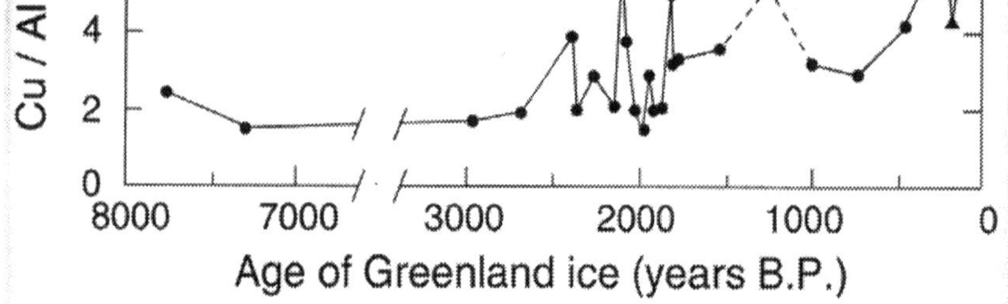

Figure 3. Changes in the copper/aluminium ratio in central Greenland ice from Summit, central Greenland, dated from 7760 to 470 years ago (black circles). Also shown (triangles) are representative mean values for ice and snow dated from the Industrial Revolution to present, collected at the same location. From Hong et al. (1996b,c).

time quantitative evidence of the Carthaginian and Roman civilizations (Rosman *et al.*, 1997). On the whole lead with a Rio Tinto type fingerprint is found to represent 2/3 of the lead found in Greenland ice between ~150 BC and 50 AD.

Copper

Figure 3 shows changes in the occurrence of copper in Greenland ice during the past seven millennia (Hong *et al.*, 1996b, 1996c). Instead of showing changes in copper concentrations, we have preferred to use changes in copper/aluminium concentration ratio. (aluminum, which is one of the major components in earth's crust, is used to normalize copper concentrations in order to compensate for year to year changes in parameters such as atmospheric transport or precipitation rate, Wedepohl, 1995).

Seven millennia ago, copper/aluminium ratio in the ice was $\sim 2\times10^{-4}$ (Figure 3). This value characterises the natural occurrence of copper in Greenland ice for the Holocene interglacial conditions before man started to mine and smelt copper. At that time, copper originated entirely from natural sources such as rock and soil dust and volcanoes (Nriagu, 1989).

2500 years ago, the ratio was still unchanged, showing, that copper emissions to the atmosphere linked with copper mining and smelting were still too small to significantly enhance copper concentrations in the ice above natural values. No data were obtained for the period 7000 to 3000 years ago as it corresponds to the 'brittle zone depth interval' where the quality of the core was very poor and does not allow reliable measurements of heavy metals to be performed in the ice.

From ~2500 to 1800 years ago, something new happens: the lowest values remain close to the natural values, but higher values, up to $\sim 6\times10^{-4}$, are observed in part of the core sections. The mean value is $\sim 4\times10^{-4}$ i.e., 2 times the natural value. Then, from ~1800 to ~200 years ago, all values are ~2 times higher than natural values. Finally, a spectacular increase up to $\sim 16\times10^{-4}$ (i.e., about 8 times the natural values) is observed during the past two centuries.

These values, high above natural, from ~2500 years ago onwards cannot be explained by changes in contributions from natural sources such as rock and soil dust, seaspray, volcanoes and biomass burning. This can be verified by estimating these different natural contributions for each core section from the measured concentrations in the ice of tracers of the different sources as aluminum for rock and soil dust, sodium for marine source for example. These high values can only be explained as resulting from early anthropogenic copper emissions to the atmosphere of the Northern hemisphere. From ~1800 years ago onwards, these emissions became intense enough to have enhanced copper concentrations in Central Greenland ice above natural values. The data confirm, for the first time, for copper the hypothesis of Clair Patterson that ancient anthropogenic copper emissions to the atmosphere were probably intense enough to have left detectable signals in Greenland ice long before the Industrial Revolution.

Are the observed changes in the occurrence of copper in Greenland ice in agreement with what is known from archaeological and historical sources about past variations in anthropogenic copper emissions to the atmosphere? In view of the lack of published data on the history of world production during the past seven millennia, we decided to tentatively reconstruct it from an extensive review of available literature, as described in detail in Hong *et al.*, (1996b, 1996c). Figure 4 shows the reconstruction so obtained. Three maxima of world copper production are observed. The first (~15000 tonnes per year) took place ~2000 years ago (peak of copper production in the Roman Empire at the zenith of the Roman power). The second (~15000 tonnes per year) is observed ~900 years ago (peak of Copper production in China at the zenith of the Northern Sung dynasty). The third (~9 million tonnes per year) is the present day peak. To evaluate the corresponding copper emissions to the atmosphere, we estimate that the emission factors were very high (~15%) during the Antiquity (Hong *et al.*, 1996b; Tylecote, 1976). The Copper sulfide ores required numerous roasting, smelting, oxidizing and refining steps. It is only after the Industrial Revolution that improved furnaces and metallurgical techniques were developed, resulting in a pronounced decrease in Copper emission factors, which were reduced to ~1% at the beginning of the 20th century (present day values are ~ 0.25 %) (Nriagu, 1979).

Are the Greenland data in agreement with these estimates of Copper emissions to the atmosphere? To answer that question, it is important to keep in mind that natural emissions of Copper to the atmosphere are ~2800 tonnes per year for the interglacial conditions of the past ~10000 years (Nriagu, 1989). Three millennia ago, anthropogenic copper emissions of Copper were estimated to be ~20 tonnes per year which correspond to only nearly ~1% of the natural emissions. This explains why Copper concentrations in Greenland ice were still not enhanced above natural values. Two millennia ago, anthropogenic Copper emissions were ~2300 tonnes per year (Hong *et al.*, 1996b), that is a value which is comparable to the natural emissions. Total (natural plus anthropogenic) emissions were then ~2 times the natural emissions. This is in excellent agreement with the Greenland ice data which show a mean Copper/Al ratio for that period ~2 times above natural values (Figure 3). Present day anthropogenic Copper emissions to the atmosphere are ~ 23000 tonnes per year, which is ~8 times the natural emissions. This is again in excellent agreement with the ice data, which show present day copper/aluminium ratios of $\sim 16\times10^{-4}$, Figure 3, i.e., ~8 times the natural values.

It appears therefore that the Greenland ice data faithfully reflect changes in the past anthropogenic Copper emissions to the atmosphere and past changes in Copper production, as hypothesized by Clair Patterson.

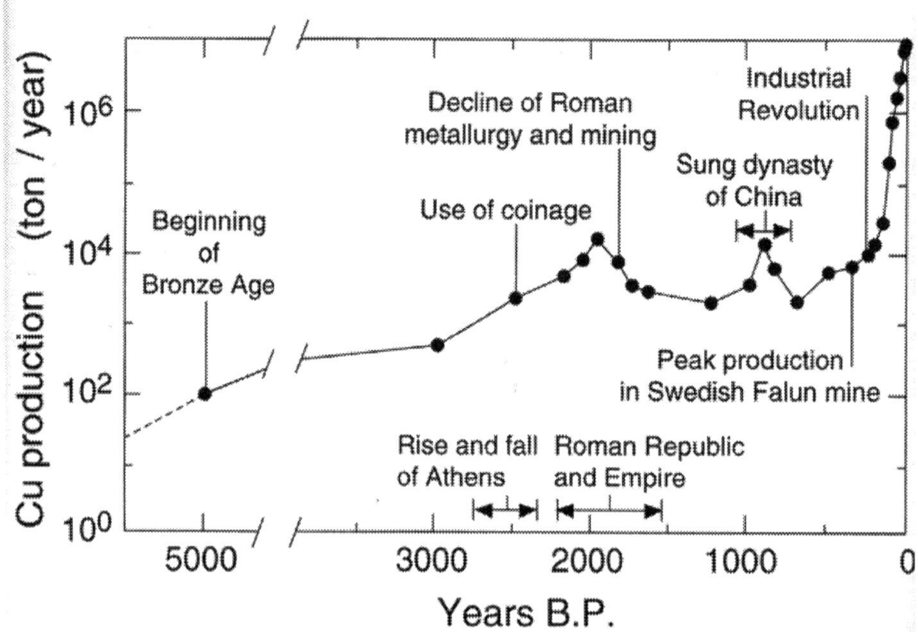

Figure 4. Variations in copper production rate during the past 5000 years as reconstructed from an extensive review of the available literature. From Hong et al. (1996b,c) and Candelone et al. (1995).

Conclusion

The Greenland data confirm that very interesting information on the past history of early metals production can be found in Greenland ice cores as predicted more than two decades ago by Clair Patterson (Patterson *et al.*, 1970). This multidisciplinary approach will need to be continued in the future. It will be especially interesting to explore these ice archives for various other metals such as arsenic (As) and tin (Sn) to obtain more precise estimates of when the process of smelting copper alloyed with arsenic to make hard weapons and tools began, as well as the techniques for smelting copper with tin, which marked the onset of the Bronze Age. Another interesting metal is zinc (Zn) which was used to make brass, the important binary copper-zinc alloy which was probably produced as early as ~ 4000 years ago (Darling, 1990; Hong *et al.*, 1997).

It will also be important to analyse ice cores obtained from other locations to read the exciting archives of mining and smelting in the different parts of the world (Nriagu, 1996). Of particular interest will be ice cores from high altitude cold ice caps in South America and China to reconstruct the history of the past metallurgy of ancient South American and Chinese civilizations.

Acknowledgements
This work is a contribution of GRIP organized by the European Science Foundation. We thank the national funding agencies and organizations in Belgium, Denmark, France, Germany, Iceland, Italy, Switzerland and the United Kingdom as well as the XII Directorate of the Commission of European Communities for financial support. Supported by the French Ministry of the Environment (grants 92295 and 94074), the Institut National des Sciences de l'Univers, the University of Grenoble and the Institut Universitaire de France. This paper is dedicated to the memory of Clair Patterson. His exceptional broadness of mind will remain a continual stimulus for us.

References

Boutron, C.F. 1990. A clean laboratory for ultralow concentration heavy metal analysis. *Fresenius Z. Anal. Chem* 337: 482-491.

1995. Historical reconstruction of the earth's atmospheric environment from Greenland and Antarctic snow and ice cores, *Environ. Rev.* 3: 1-28.

Candelone, J.P., Hong, S. & Boutron, C.F. 1994. An improved method for decontaminating polar snow and ice cores for heavy metals analysis. *Anal. Chim. Acta* 299: 9-16.

Candelone, J.P., Hong, S., Pellone C. & Boutron, C.F. 1995 Post-industrial revolution changes in large-scale atmospheric pollution of the Northern Hemisphere by heavy metals as documented in central Greenland snow and ice. *Journal of Geophysical Research* 100: 16605-16616.

Chisholm W., Rosman K. J. R., Boutron C .F., Candelone J. P., & Hong S. 1995. Determination of lead isotopic ratios in Greenland and Antarctic snow and ice at picogramm per gram concentration. *Analyt. Chim. Acta* 311: 141-151.

Chow T. J. & Patterson C.C. 1962. The occurrence and significance of lead isotopes in pelagic sediments., *Geochim. Cosmochim. Acta* 26: 263-308.

Cortizas, X., Pontevedra Pomba, J.C., Novo Munoz & Garcia- Rodeja E. 1997. Four thousand years of atmospheric Pb, CD, and Zn deposition recorded by the ombrotrophic peat bog of Penido Vello, northwestern Spain. Water Air and Soil Pollution 100: 387-403.

Dansgaard, W., Johnsen, S.J., Clausen, H.B., Dahl-Jensen, D., Gundestrup, N.S., Hammer, C.U., Hvidberg, C.S., Steffensen, J.P., Sveinbjörnsdottir, A.E., Jouzel, J. & Bond, G. 1993. Evidence for general instability of past climate from a 250 kyr ice core record. *Nature* 364: 218-220.

Darling S.A. 1990. Non ferrous metals. In *An Encyclopaedia of the history of technology*, Edited by I McNeil, Routledge, London, pp 47-115.Davis J. M. & Svendsgaard D. J. 1987. Lead and child development. *Nature* 329: 297-300.

Epstein S. 1981. Introduction of Clair Patterson for the V.M. Goldsmith medal 1980. *Geochim. Cosmochim. Acta* 45: 1383-1387.

Fitzgerald W. F. & Clarkson T. W. 1991. Mercury and monomethyl mercury present and future concerns. *Environ. Health Persp.* 96: 159-166.

Friberg L., Elinder C. G., Kjellström T., & Norberg G. F., 1985/86. eds.,. Cadmium and Health: A Toxicological and Epidemiological Appraisal, Vols. 1 and 2. CRC Press, Boca Raton, Fla.

Hamelin B., Grousset F., Sholkovitz E. R. 1990. Pb isotopes in surficial pelagic sediments from the North Atlantic., *Geochim. Cosmochim. Acta* 54: 37-47.

Hammer, C.U. 1989. Dating by physical and chemical seasonal variations and reference horizons. In *The Environmental Record in Glaciers and Ice Sheets*, Dahlem Konferenzen. *Edited by* H. Oeschger and C.C. Langway, Jr. Wiley, Chichester, U.K., pp. 99-121.

Hong, S., Candelone, J.P., Patterson, C.C. & Boutron, C.F. 1994. Greenland ice evidence of hemispheric pollution for lead two millennia ago by Greek and Roman civilizations. *Science* 265: 1841-1843.

Hong S., Candelone J. P., Soutif M., & Boutron C. F. 1996b. A reconstruction of changes in copper production and copper emissions to the atmosphere during the past 7000 years, *The Sci. Tot. Environ.* 188: 183-183.

HONG, S., CANDELONE, J.P. & BOUTRON, C.F. 1996a. Deposition of atmospheric heavy metals to the Greenland ice sheet from the 1783-1784 volcanic eruption of Laki, Iceland. *Earth Planet. Sci. Lett.* 144: 605-610.

1996c. History of ancient copper smelting pollution during Roman and Medieval times recorded in Greenland ice, *Science* 272: 246-249.

1997. Changes in zinc and cadmium concentrations in Greenland ice during the past 7760 years. *Atmos. Environ.* 31: 2235-2242.

LORIUS, C. 1980. Polar ice cores and climate. In *Climate and Geosciences. Edited by* A. Berger, Kluwer, Amsterdam. pp. 77-103.

MUROZUMI, M., CHOW, T.J., & PATTERSON, C.C. 1969. Chemical concentrations of pollutant lead aerosols, terrestrial dusts and sea salts in Greenland and Antarctic snow strata. *Geochim. Cosmochim. Acta* 33: 1247-1294.

NEEDELMAN H. L., 1984. The hazard to health of lead exposure at low dose. In *Changing Metal Cycles and Human Health* (edited by J. O. Nriagu), Springer-Verlag, pp 311-322.

NRIAGU, J.O. 1979.Copper in the atmosphere and precipitation. In: J.O. Nriagu (Ed.), Copper in the environment, Part I, Wiley, New York, pp. 43-75.

1983a. Saturnine gout among roman aristocrats. Did lead poisoning contribute to the fall of the empire? *N. Engl. J. Med.* 308: 660-663.

1983b. Occupational exposure to lead in ancient times. *The Sci. Total Environ.* 31: 105-116.

1983c. Lead and Lead poisoning in Antiquity. (Wiley, New York)

1989. A global assessment of natural sources of atmospheric trace metals, *Nature* 338: 47-49.

1996. A history of global metal pollution, *Science* 272: 223-224.

NRIAGU, J.O. & PACYNA J. M. 1988. Quantitative assessment of worldwide contamination of air, water and soils by trace metals, *Nature* 333: 134-139.

PATTERSON C. C. 1980. 'Lead in the Human Environment', Report prepared by the Committee on Lead in the Human Environment (National Academy of Sciences, Washington, DC), pp 265-349.

1971. Native copper, silver and gold accesible to early metallurgists. *American Antiquity* 36: 286-321.

1972. Silver stocks and losses in ancient and medieval times. *The Economic History Review* 25: 205-235.

1994a. Historical changes in integrety and worth of scientific knowledge. *Geochim. Cosmochim. Acta* 58: 3141-3143.

1994b. Delineation of separate brain regions used for scientific versus engineering modes of thinking. *Geochim. Cosmochim. Acta* 58: 3321-3327.

PATTERSON C. C., CHOW T.J. & MUROZUMI M. 1970. The possibility of measuring variations in the intensity of world-wide lead smelting during medieval and ancient times using lead aerosol deposits in polar snow strata. In *Scientific Methods in Medieval Archeology* Edited by R. Berger. University of California Press, Berkeley, pp 339-350.

REEH N., 1989. Dating by ice flow modeling: a useful tool or an exercise in applied mathematics? In *The Environmental Record in Glaciers and Ice Sheets*. Dahlem Konferenzen. *Edited by* H. Oeschger and C.C. Langway, Jr. Wiley, Chichester, U.K. pp 141-159.

RENBERG, I., PERSSON, M.W., & EMTERYD, O. 1994. Pre-industrial atmospheric lead contamination detected in Swedish lake sediments. *Nature* 368: 323-326.

ROSMAN, K.J.R., CHISHOLM, W., BOUTRON, C.F., CANDELONE, J.P. & GÖRLACH, U. 1993. Isotopic evidence for the sources of lead in Greenland snows since the late 1960s. *Nature* 362: 333-335.

ROSMAN, K.J.R., CHISHOLM, W., HONG S., CANDELONE, J.P., & BOUTRON, C.F. 1997. Lead from Carthaginian and Roman Spanish mines isotopically identified in Greenland ice dated from 600 BC to 300 AD. *Environ. Sci. Technol.* 31: 3413-3416.

SETTLE, D. M., & PATTERSON, C.C. 1980. Lead in albacore : guide to lead pollution in Americans. *Science* 207: 1167-1176.

STOS-GALES Z., GALE N. H., HOUGHTON J. & SPEAKMAN R. 1995. Lead isotope data from the isotrace laboratory, oxford: archaeometry data base1, ores from the western mediterranean. *Archeometry* 37: 407-415.

TYLECOTE, R.F., 1976, A History of Metallurgy, Mid-County, London.

TYLECOTE R. F. 1992. A history of metallurgy, 2nd edition, The Institute of Materials, London

WEDEPOHL, K. H. 1995. The composition of the earth crust. Geochim. Cosmochim. Acta 59: 1217-1232.

ZIELINSKI, G.A., MAYEWSKI, P.A., MEEKER, L.D., WHITLOW, S., TWICKLER, M.S., MORRISON, M., MEESE, D.A., GOW, A.J., & ALLEY, R.B. 1994. Record of volcanism since 7000 B.C. from the GISP2 Greenland ice core and implications for the volcano-climate system. *Science* 264: 948-952.

27

Archaeometallurgy: Helping Archaeology Bridge the Gap Between Science and Anthropology

ROBERT M. EHRENREICH

National Research Council, National Academy of Sciences, 2101 Constitution Ave., NW, Washington DC 20418, USA

Few fields straddle the sciences and humanities so intrinsically as does archaeology. As opposed to being detrimental, the combination of these symbiotic approaches imbues archaeology with a vitality and fortitude that is enjoyed by few other fields. I hope to show in this article how the combination of the physical and anthropological sciences within archaeological interpretation can increase our understanding of the human role in the development of early technologies and how the application of archaeometallurgical studies can advance our understanding of ethnicity, industrial organization, and technical accomplishment within a cultural context for early societies.

Specifically, I will discuss in this article how scientific studies of ancient ironwork can further our understanding of coexisting groups when correlated with general archaeological knowledge. The advent and diffusion of quenching during the Pre-Roman and Roman Iron Ages in southern Britain will be used as an indicator of the level and type of interactions that occurred between the independent and Roman-controlled ironworking communities of this region. The advent of quenching in Britain will be discussed within the context of a database of 511 metallurgical analyses of Iron Age artefacts from 22 sites in central, southern Britain and 429 analyses of Roman samples from 12 sites in central, southern Britain.

The Pre-Roman Iron Industries and the Advent of Quenching

The iron industries of Middle and Late Iron Age Britain were predominantly divided into two segments that had very different industrial organizations and levels of craft specialization (Ehrenreich 1991b). Since historical records are rare for this period, the level of craft specialization must be inferred from the archaeological record. Jeanne Arnold (1987) states that the level of craft specialization can be identified based on the evidence of five criteria: (1) the manufacture of products beyond self-sufficiency, (2) the existence of dedicated workshop areas, (3) the export of products, (4) the standardization of product and production methods, and (5) the control of resources, including technology. Based on four of these criteria, iron-smelting in Iron Age Britain can be seen to have been a specialized industry.

First, the limited historical evidence suggests that iron was being both produced beyond self-sufficiency and also exported. Strabo (1923) stated that iron was exported from Britain to the continent during the Iron Age, and Caesar (1980) mentions that iron was available near the southeastern coast of the island.

Second, the archaeological evidence indicates that a sophisticated stock structure was in existence. A mature, specialized craft-system requires the existence of a standard by which a commodity can be consistently and efficiently exchanged. For Middle and Late Iron Age Britain, at least two standardized forms of stock--known as currency bars--were traded and hoarded (Allen 1967). The first currency bar type is spit-shaped, low in phosphorus, and thought to have been produced in the Forest of Dean area, based on its distribution (Allen 1967; Cleere 1981). The second currency bar is sword-shaped, high in phosphorus, and believed to have been produced along the Jurassic Ridge in the Northants region of Britain, since the ore lodes of this area have a high phosphorus concentration. The distribution of these currency bars tends to concentrate in this region (Allen 1967), and the Hunsbury hillfort--situated in this region--contained an extensive metalworking assemblage and a potential currency-bar precursor (Ehrenreich 1991b). Since iron's mechanical properties are affected by phosphorus content (Davies & Oelmann 1983; Kaloyeros & Ehrenreich 1991), it is important to note that the iron manufacturers may have deliberately selected these distinctive shapes so that the smithing community could distinguish between their metals (Ehrenreich 1991b).

The final evidence of the high level of craft specialization of the iron smelting community is the apparent control of raw resources and basic technologies by distinct communities. Although iron ore is common throughout Britain (Tylecote 1986), large-scale iron production during the Middle and Late Iron Ages seems to have been predominantly restricted to the island's three richest ore sources: the Weald of southeastern Britain, the Jurassic Ridge of middle Britain, and the Forest of Dean of southeastern Wales. Most Middle and Late Iron Age smelting remains are located within these three areas, with 12 situated in the Weald and seven found in the region of the Forest of Dean (Ehrenreich 1991b). Only two sites were discovered along the Jurassic Ridge, but the existence of the sword-shaped currency bars and the fact that 60 percent of the Middle Iron Age artefacts examined were made of metal smelted from these ores indicate that Northants was a significant iron-manufacturing area (Ehrenreich 1985). Very few sites were found to contain evidence of iron smelting in the rest of ore-rich central, southern Britain, however, and these only seem to have resulted from limited attempts to satisfy the occasional, individual requirements of particular sites. Thus, this region does not appear to have been producing iron in quantity for trade as stock, but obtained its metal from the three large iron production areas.

The iron manufacturing community was undoubtedly controlled by specialists, because the technology is extremely time-consuming and demanding, and requires specialized and detailed knowledge of the ores and techniques to be used (Ehrenreich 1991b). Thus, the restriction of iron manufacturing to the three regions may not have been due to a lack of access to natural resources, but rather to a lack of knowledge of the smelting technique.

It is also important to note that there is trace-element evidence that some iron was being produced throughout the Iron Age in the Devon-Cornwall region of Britain (Ehrenreich 1985). Although the sourcing of iron artefacts by trace element analysis has generally proven unreliable, the combination of trace element analysis with archaeological evidence can occasionally reveal the general regions from which an artefact may have originated, as was done for the high-phosphorus iron currency bars discussed above. Among the Iron Age and Roman artefacts sampled, there is a small number of artefacts with cobalt concentrations in excess of 0.1 percent (Ehrenreich 1985). The number of these artefacts is highest during the Early Iron Age and decreases thereafter. The sources of high-cobalt iron ore are extremely rare in Britain. The only currently known source of ore that could produce this iron is the Great Perran lode near Newquay in Cornwall (Groves 1952). The Devon-Cornwall region of Britain has an extensive metalworking history that spans the Bronze and Iron Ages. First, Devon-Cornwall was one of the few regions where localized tin smelting probably occurred during the Late Bronze Age (Pearce 1981). Second, southwest Britain possessed most of the early British iron smelting furnaces (Cleere 1981). Third, the Devon-Cornwall region contained three of the oldest iron manufacturing sites discovered, including the Early Iron Age promontory fort of Trevelgue, which was situated near Newquay and did contain sufficient quantities of slag to suggest large scale smelting (Cleere 1981; Salter & Ehrenreich 1984). This evidence may suggest that iron was first smelted and smithed in the Devon-Cornwall region of Britain, with the technology possibly having arrived in this metallurgically sophisticated region via the traditional trade routes of the Atlantic seaways from northwest France (Rowlands 1980). If the Devon-Cornwall region were one of the initial points of entry for ironworking in Britain and the location of some of the earliest large-scale iron production, then it would be reasonable for high-cobalt iron to have been best represented during the Early Iron Age--when iron production was still limited elsewhere--and then decrease in proportion thereafter as new production centers were established.

Blacksmithing appears to have been a more generalized industry during the Iron Age, with few indications of craft specialization. First, there is little evidence of the manufacture of product surpluses for trade (Ehrenreich 1991b). Few caches or hoards of iron artefacts, other than currency bars, have been found dating from the Iron Age. The lack of evidence of the hoarding of finished iron artefacts versus the wide-spread hoarding of currency bars suggests that finished ironwork was not being generally produced in surplus for export, but rather that iron was being traded as raw stock.

Second, few dedicated smithing work-areas have been found on Iron Age sites. Millennia of plowing in this region have erased most floor levels and left only those artefacts discarded in pits to be recovered. Thus, possible workshop locations must be inferred from the clustering of metal-working debris within certain regions of sites. The smithing remains of most sites appear random, however. For instance, in the Danebury Hillfort, the ironworking by-products discovered show no obvious concentration within any specific region of the hillfort, suggesting that ironworking was being generally performed throughout the site (Sellwood 1984).

Third, resources or technologies do not appear to have been controlled during the Iron Age. Hoards of currency bars have been discovered on the two major site types of this period (i.e., hillforts and settlement sites), suggesting that access to raw stock was not restricted to particular site types or that inhabitants of certain site forms were unable to obtain iron through trade networks. Similarly, no technological imbalances are apparent for Wessex during this period. Metallurgical analyses revealed that the techniques used for the production of elite artefacts were the same as those used for tools (Ehrenreich 1985). In fact, the only artefacts found to have been generally produced in a more sophisticated manner were chisels (Ehrenreich 1985). Two quenched chisels were discovered within the Danebury hillfort assemblage, and one quenched wedge was discovered among the Worthy Down settlement-site assemblage (Ehrenreich 1985).

Although the existence of only three quenched artefacts from a total of 511 might suggest that these artefacts were accidentally produced, when placed within the context of the larger sample, it becomes clear that this technique was actually understood and deliberately applied during this period by at least a small portion of the blacksmithing community (Ehrenreich 1985; Ehrenreich 1991a). First, not all of the tools sampled should have been quenched. Of the 511 artefacts sampled, the properties of only 116 tools--including the three quenched artefacts--would have benefited by the application of this technique. Second, quenching is only valuable for artefacts with medium or high carbon contents. Thus, the three artefacts must also be placed within the context of the further subset of tools that have high-carbon contents. Whereas only 40 percent of the 116 tools had high carbon concentrations, ten of the fifteen Iron Age chisels sampled (i.e., 67 percent) possessed high carbon concentrations, showing that steel was being deliberately selected during the Iron Age for these artefacts either through the use of carborization or bloom selection. Third, the blades of both the wedge and one of the quenched chisels are composed of a high-carbon strip embedded in a low-carbon body, which is the finest configuration for a cutting edge (Ehrenreich 1985). Thus, when these artefacts are taken within the context of the subset of tools that would have benefited from quenching, the further subset of tools that would have had a suitable carbon content for quenching, and the even further subset of tools that were properly designed to take full advantage of quenching, it is safe to conclude that these three artefacts were not accidents, but were made by a small segment of the domestic blacksmithing community that practised this technique and perhaps kept it as proprietary. The fact that the more advanced tools were also found on both hillforts and settlement sites indicates that neither superior artefacts nor advanced smiths were restricted to one site form.

The evidence presented would suggest that the Iron Age ironworking community was probably not a specialized industry, but had a more subtle and complex organization, indicative of what Carole Crumley terms a heterarchy. Crumley (1987) defines heterarchy as a structure in which "each element is either unranked relative to other elements or possesses the potential for being ranked in a number of different ways". The lack of craft specialists and advanced technologies does not mean that the ironworking community was either stagnant or simplistic, however (Ehrenreich 1991b). The discovery of quenched artefacts on both a hillfort and a settlement site shows that ironworking was actually in a highly dynamic state. Smiths on many sites must have been attempting to improve their ironwork through experimentation, and some innovations were being discovered. The limited number of advanced artefacts would suggest that these new techniques were proprietary and not for general dissemination, however.

The Roman Iron Industries and the Diffusion of Quenching

Iron manufacturing was completely restructured with the Roman incursion (Cleere 1984). In 43 AD, the Romans began annexing large portions of the three primary ore sources of Britain previously mentioned--the Weald, the Forest of Dean, and the Jurassic Ridge--and licensing the remainder to private entrepreneurs. Iron production was dramatically increased by the third century AD. The iron, copper, and tin sources of the Devon-Cornwall region of Britain were neglected by the Romans, however. The archaeological evidence suggests that these ore sources were not annexed and exploited by the Romans until the disappearance of Spanish tin supplies in the 3rd century AD (Pearce 1981).

Although iron was still being traded in bar form, there is clear evidence that blacksmithing was becoming increasingly specialized during the Roman period (Potter 1983). The material needs of the Roman legions and towns and the growing demand for luxury goods with the spread of Roman habits resulted in a marked increase in specialized craftsmen, including blacksmiths. Evidence exists of the presence of such specialists as armors, locksmiths, and cutlers, some of whom even stamped their names or trademarks upon their products. Most of these blacksmiths probably served only limited regional markets, however, and almost all sites in southern Britain still contained some evidence of small-scale smithing activities.

The restructuring of the iron smelting and smithing industries appears to have had only limited effect on British blacksmithing technology, however. Metallurgical analyses revealed that the techniques used for the production of elite artefacts were still the same as those used for tools, and that the deliberate use of steel was rare (Ehrenreich 1991a). The ratio of quenched artefacts to the total number requiring advanced heat-treatments did increase, however, from 2.6 percent in the Iron Age to 11.9 percent during the Roman Period, with a total of 12 of the 101 tools sampled having been quenched. Of greater interest in the context of the quenched artefacts, however, is that one third of these tools had cobalt contents in excess of 0.1 percent (Ehrenreich 1991a).

As stated previously, the proportion of high-cobalt iron used in Britain decreased throughout the Iron Age and Roman Period, as the other iron sources were developed. Only two percent of the entire Roman-Period assemblage had high cobalt concentrations, suggesting that the increased output of iron from the Roman-regulated regions continued to reduce the demand for this stock (Ehrenreich 1991a). The existence of some high-cobalt iron artefacts within the Romano-British assemblage indicates, however, that this iron was still being smelted and smithed during the Roman Period and that this ironwork was still being traded into southern Britain, albeit rarely and via unknown mechanisms. Although little is currently known about the production and trade of this metal, its trickle into central, southern Britain would have continued during the Roman period only if some property justified it. The large number of quenched, high-cobalt iron artefacts may suggest that this technique was more generally known in the Devon-Cornwall region. Perhaps the iron tools then remained in demand in southern Britain because of their higher hardness and durability.

The evidence presented would suggest that the ironworking industry of southern Britain during the Roman Period contained two segments. The first segment consisted of the large-scale, Roman-controlled iron smelting industry and the increasingly specialized iron-smithing industry that together mass-produced iron stock and implements for both regional consumption and continental export (Cleere 1984). The second industrial segment existed both within, and at, the periphery of the Roman-controlled ironworking community. This industry consisted of small-scale iron smelters and smiths who were more closely tied to the previous indigenous ironworking traditions and produced stock and implements for more-localized consumption. The restricted knowledge of quenching in Roman Britain suggests that the interaction between these two communities was limited. If the large-scale industry had obtained the technical expertise to successfully quench artefacts, the quantity of quenched tools would have quickly spread and many more quenched artefacts should have been found in the archaeological record. The limited quantity of quenched artefacts and the existence of small numbers of similar artefacts during the Iron Age suggest that this technique was possibly a closely guarded secret within the remnants of the previous indigenous ironworking tradition and was not generally disseminated.

These results are preliminary, however, and much more research is required before any definitive conclusions can be reached. For example, Iron Age and Roman Period iron artefacts from the Devon-Cornwall region of Britain need to be analyzed to determine whether this was actually the source of the high-cobalt iron artefacts and could have been the source of the quenched ironwork. The organization of other industries in ancient Britain should also be examined to determine whether this model of Romanized industries versus indigenous technological communities can be further generalized for this period.

The results of this preliminary study show, however, that archaeometallurgical research can transcend the simple construction of technological progressions, increase our understanding of early sociotechnical systems within the context of their associated social and symbolic aspects, and aid the construction of archaeological societal models. First, archaeometallurgical studies can reveal the coexistence of sociotechnical systems within seemingly unified communities. Second, the regulation or diffusion of more advanced technologies can indicate the level of interaction and extent of communication between coexisting sociotechnical systems. The archaeometallurgical research then needs to be correlated with further archaeological models to lead to increased understanding of the technological or social conditions that would permit an earlier industrial community to survive and coexist with a larger, restructured, formalized industry as well as the over-all societies.

Discussion

I hoped, in this article, to present the problem of sciences versus humanities as a positive dialogue between two symbiotic approaches. This task would have been far easier, of course, if central, southern Britain were to have the same type of ironworking oral traditions as Ireland or areas of Africa. I find it difficult to understand anyone arguing that the technical and methodological underpinnings of archaeology are not science, however. Archaeologists are adept at analyzing site lay-outs and formulating excavation strategies. We scrutinize topographies, environmental conditions, and prehistoric and historic trends; employ advanced remote sensing techniques; and identify the tools, site crews, and schedules that are required to excavate sites and ensure that the maximum amount of information is obtained. Archaeologists are also experts at determining the basic structure and subsistence of an excavated site by the analysis of its construction, site plan, and artefact assemblages. We take all the finds recovered, and translate them, using a range of scientific and archaeometric techniques, into a variety of data sets that can be statistically analyzed and build hypotheses about how the inhabitants of a site survived: for example, what the inhabitants ate, what they made, when they made it, and how they obtained their tools and weapons. Archaeologists then test these hypotheses by comparing them with the results of other sites.

The dichotomy between the sciences and humanities in archaeology is thus not an issue in terms of the fundamental archaeological techniques and methodologies, but in terms of our ability to interpret the archaeological record once the artefacts have been catalogued and the original excavation objectives have been met. Every object can stimulate a wealth of interpretation about the original population. To what extent can we vouchsafe these interpretations? After all, did we get into this field just to find out how people smelted metal? The production of materials alone has much larger implications. Was metalworking constant or did it change with seasons and time? Were new materials and techniques introduced that permitted the population to increase, or were they a response to some external stress and adopted because of necessity? Who obtained the metal and manufactured the tools? Were there rituals associated with both the material and its production? What do these data tell us about the inhabitants' subsistence strategies, societal organization, and personal beliefs? Is there anything we can learn from these data that could help us better understand our own societies? These questions form a logical progression that moves archaeology from its scientific to its humanistic underpinnings.

There are two dangers to the perpetuation of this argument within archaeology. The first is the potential elimination of specialist fields. Archaeology has become increasingly specialized over the past twenty years, as more and more techniques were introduced that could provide new insights into the archaeological record. Unfortunately, this trend now directly conflicts with the job market. Many people are painfully aware that the over-all decrease in the number of archaeological positions at universities has meant that more institutions are now searching for generalists who can teach a broad range of subjects. Thus, specialization can now be seen as detrimental to a career prospect. If this trend continues, much of the specialization that has become such an integral part of the new archaeology--and the new insights that it provides--will be lost.

The second danger to archaeology is that this argument will so divide the field, that both components will end up engaging in bad research and open the field to less qualified individuals and charlatans. In an attempt to make the discipline more quantitative, archaeologists and pseudo-archaeologists could begin accumulating inordinate quantities of irrelevant data and submitting them to advanced statistical analyses without properly formulating the research objectives or really understanding the biases and limitations of the analytical techniques and statistical packages. Conversely, in an attempt to make the discipline more humanistic, excavation data could be increasingly marginalized in favour of interpretations relying on personal experience and pseudo-science. At best, such results would be irrelevant and ignored. At worst, they could be spurious, inflammatory, and destructive to both archaeology and possibly even society as a whole.

The key is for archaeologists to learn to bridge these distinctions through understanding and collaborations, and to produce a new systemic approach that provides the right admixture of humanities and scientific ingredients necessary to extract an optimal amount of credible information and support new levels of shared interpretation. The distinction between the sciences and the humanities within archaeology would then become less defined as more archaeologists realize the strengths of the two approaches, borrow components from each other's work, and collaborate to identify and accumulate the data required to support higher levels of interpretation.

The history and development of archaeometallurgy places it in an excellent position to serve as a model for this new systemic approach. Archaeometallurgical studies have long been a battle-ground for these two, supposedly conflicting, views of archaeological interpretation. Initially, the field tended to be populated by scientifically oriented archaeologists, while the humanistically oriented archaeologists were inclined to deride the research as just being histories of technology and ineffectual for the development of mainstream archaeological models. After all, Colin Renfrew (1982) once said of the field, "there seems to be some uncertainty whether we are talking about the history of materials or whether we are talking abut the history of the human race...there is a difference, and the latter is more interesting". Unfortunately, many archaeologists had a point. Archaeometallurgical research used to emphasize the discovery of ancient smelting techniques, the development of histories of technologies, and the technical appreciation of specific artefact types. Unfortunately, the results of most of these studies were of little use to archaeologists attempting to model sociotechnical systems.

Archaeometallurgy has made great strides over the past 15 to 20 years, however, as more and more humanistically oriented archaeologists have entered the field. Current studies generally do combine the physical and anthropological scientific approaches, and archaeometallurgical papers and sections have become increasingly common in national and international archaeological journals and conferences. Archaeometallurgists must not become complacent, however. It is vital that we continue to demonstrate the relevance of our research to the mainstream archaeological community and show the positive aspects of combining the scientific and humanistic aspects of archaeology.

References

ALLEN, D. 1967. Iron currency bars in Britain. *Proceedings of the Prehistoric Society* 33:307-335.

ARNOLD, J.E. 1987. *Craft Specialization in the Prehistoric Channel Islands*, California. Berkeley, California: University of California Press.

CAESAR, J. 1980. *The Battle for Gaul*, A. Wiseman & P. Wiseman (trans). Boston: David R. Godine.

CLEERE, H.F. 1981. The Iron Industry of Roman Britain. Ph.D. dissertation. London: London University.

1984. Ironmaking in the economy of the ancient world: The potential of archaeometallurgy. Pp. 1-6 in *The Crafts of the Blacksmith*, B.G. Scott & H.F. Cleere (eds). Belfast: Comite pour la Siderugia Ancienne.

CRUMLEY, C. 1987 A Dialectical critique of Hierarchy. In *Power Relations and State Formation*, T.C. Patterson & C.W. Gailey (eds.), pp. 155-169. American Anthropological Association, Washington, DC.

DAVIES, D.J., & L.A. OELMANN. 1983. *The Structure, Properties and Heat Treatment of Metals*. London: Pitman.

EHRENREICH, R.M. 1985. Trade, Technology, and the Ironworking Community of Southern Britain in the Iron Age. *BAR British Series* 144. Oxford: British Archaeological Reports.

1991a. Archaeometallurgy and archaeology: Widening the scope. Pp. 55-62 in *Recent Trends in Archaeometallurgical Research*, vol. 8, part 1, P. Glumac (ed). Philadelphia: MASCA.

1991b. Metalworking in Iron Age Britain: Hierarchy or heterarchy? Pp. 69-80 in *Metals in Society: Theory Beyond Analysis*, vol. 8, part 2, R.M. Ehrenreich (ed). Philadelphia: MASCA.

GROVES, A.W. 1952. *Wartime Investigations into the Haematite and Manganese Ore Resources of Great Britain and Northern Ireland*. Permanent Records of Research and Development Monograph 20-703. London: Ministry of Supply.

KALOYEROS, A. & EHRENREICH, R.M. 1991 The distribution of phosphorus in Romano-British ironwork. Pp. 725-730 in *Materials Issues in Art and Archeology II*, vol. 185, P.B. Vandiver, J. Druzik, & G.S. Wheeler (eds). Pittsburgh: Materials Research Society.

PEARCE, S.M. 1981. *The Archaeology of South West Britain*. London: Collins.

POTTER, T.W. 1983. *Roman Britain*. Cambridge: Harvard University Press.

RENFREW, C. 1982. Quote from transcript of round table discussion. Page 42 in Future Directions in Archaeometry: A Round Table, J.S. Olin (ed). Washington, D.C.: Smithsonian Institution.

ROWLANDS, M.J. 1980. Kinship, alliance, and exchange in the European Bronze Age. Pp. 15-56 in *Settlement and Society in the British Later Bronze Age*, J. Barrett & R. Bradley (eds). British Series 83. Oxford: BAR.

SALTER, C., & R.M. EHRENREICH. 1984. Iron Age iron metallurgy in central southern Britain. Pp. 146-161 in *Aspects of the Iron Age in Central Southern Britain*, B. Cunliffe & D. Miles (eds). Committee for Archaeology Monograph No 2. Oxford: University of Oxford.

SELLWOOD, L. 1984. Manufacturing activities. Pp. 436-439 in *Danebury: An Iron Age Hillfort in Hampshire*, vol. I, B.W. Cunliffe (ed). CBA Research Report 52. London: Council for British Archaeology.

STRABO. 1923. *Geography, vol. 2*, H.L. Jones (trans). London: W. Heinemann.

TYLECOTE, R.F. 1986. *The Prehistory of Metallurgy in the British Isles*. London: The Institute of Metals.

A graphical method to determine furnace efficiency and lining contribution to Romano-British bloomery iron making slags (Bristol Channel orefield, UK)

G.R. THOMAS & T.P. YOUNG

Department Of Earth Sciences, University Of Wales Cardiff, PO Box 914, Cardiff, CF1 3YE, UK

A graphical technique is presented for determining the ore:furnace lining mixture in the melt and the yield of the reaction during smelting, using the concentrations of the "immobile" elements, particularly the rare earth elements (REE). Data are presented from slags, furnace linings and an ore fragment recovered from the Roman iron working town of Ariconium, *situated to the north-east of the Bristol Channel iron ore field. The ores from this region are characterized by their high purity and thus low trace element concentration. The mass balance approach is used to gain an assessment of the efficiency of the smelting operations at this site and to demonstrate that the chemical composition of the bloomery iron slags found in this region is dominated by a component contribution from the furnace lining which was assimilated into the melt during the smelting operation.*

Keywords: Slag; REE; Furnace Efficiency; Mass Balance.

Introduction

In recent years techniques for the chemical analysis of iron smelting materials have improved, being able to analyze rapidly and accurately for a large number of elements, yet authors still place great reliance upon single elements. Aluminium, being relatively immobile, is frequently used along with SiO_2 in calculating the ore:slag ratios used in ancient smelting operations (Tylecote *et al.* 1971), whereby the recorded levels within a slag can be directly related to those within an ore source.

The iron ores found within the Bristol Channel orefield are of high grade and as such contain only minor quantities of Al; even minor levels of contamination from an Al-rich source can result in an overestimation of the amount of ore required to produce a slag of known composition. Indeed Fulford & Allen (1993), working on slags recovered from an iron working site at Woolaston, to the east of the orefield, noted that the majority of the non-ferrous material present in the slags would have been derived from the consumption of furnace lining. Thus to determine the efficiency of smelting operations within the Bristol Channel orefield and to determine the effect of the assimilation of furnace lining upon the final chemistry of the slag a multi-elemental approach was required. The advantage of this over more traditional methods being that it would largely avoid the minor uncertainties associated with not having true ore:slag:lining combinations.

In this study 21 slag samples consisting of both tap and furnace slags (total weight 5 kg) along with 3 lining samples (0.4 kg) and a single ore fragment (40 g) were obtained from 220 kg of material excavated by the Hereford and Worcester County Council Archaeology Service during their 1993 excavation at the major Roman iron working site of *Ariconium* (Figure 1) These were subjected to geochemical analysis using X-ray fluorescence of fused beads (major elements) and powdered pellets (minor and trace elements) along with Inductively Coupled Plasma - Mass Spectrometry of solutions (minor and trace elements).

Geological Background

The Bristol Channel iron ore field is closely associated with the post-Variscan unconformity with most of the ores lying below this level within cavernous porosity (caves, joints and faults), normally within the Carboniferous Limestone. The best known areas of this widespread mineralization are in the Vale of Glamorgan and the Forest Of Dean, the latter having provided a major component of the British iron industry in Roman and Medieval times, and there is widespread evidence for small-scale early iron mining across the orefield. At some localities, particularly within the Bristol and Mendip areas, the mineralization is dominantly hosted above the unconformity within the Dolomitic Conglomerate, of Triassic age. The published literature on most of these occurrences is now rather old (Cantrill *et al.* 1919) and geochemical data of the ores are sparse, the most recent account of a deposit within the area being that of Llanharry by Gayer and Criddle (1970). Shepherd and Goldring (1993) gave a brief discussion of the Bristol Channel orefield during a description of the similar Cumbrian orefield. It is to this latter orefield that the material within the Bristol Channel region bears its closest comparison within the UK. The ores from both these regions are of a higher purity then material found within the rest of the UK, being dominated by the anhydrous iron oxide hematite and the hydrated oxide goethite. These ores typically contain in the region of 74% Fe compared to 33% from the Cretaceous claystone ironstones from the Weald of Sussex and 25% from the ooidal ironstones that occur within the Jurassic ridge down the centre of England (Young, 1993). The high grade of the ores found within the field area indicates a minimal detrital content and as a consequence these ores posses a relatively low total REE (ΣREE) value (<40 ppm) when compared to the detrital rich claystones (typical claystone from the Weald, ΣREE = 150 ppm).

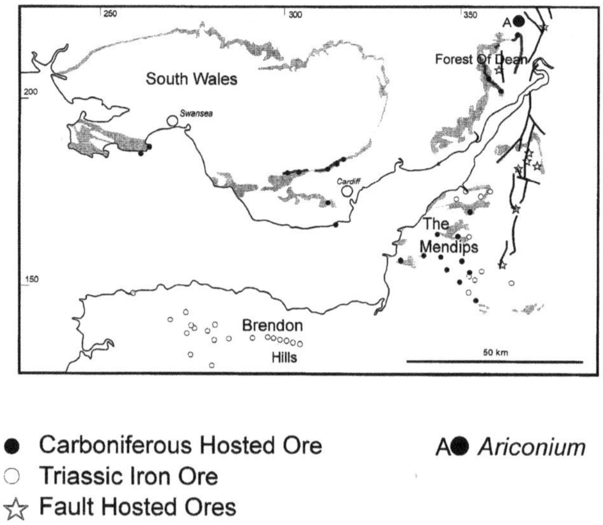

Figure 1. Distribution of iron ore in the Bristol Channel area, showing the outcrop of Carboniferous Limestone (pale grey) and the fault system bounding the western margin of the Worcester margin (shown in black).

The use of REE data in provenancing the ore fragment from *Ariconium*.

The stalactitic morphology of the *Ariconium* ore fragment can be used as a basis to provenance it to the Bristol Channel orefield. Trace element geochemistry has been investigated as a tool for provenancing within an orefield. Due to the total absence of published comparable data a database was constructed from 71 ore samples obtained from deposits throughout the orefield. The immobile REE along with U appear to provide the best means by which to compare material within this area. An origin for the *Ariconium* ore fragment from within the fault hosted ores of the Worcester Graben can be dismissed as the ores within this region have a lower uranium content (m= 0.85 ppm, s= 0.47 ppm, n= 4) compared to the value of 2.4 ppm for the *Ariconium* material. Ores from Glamorgan, South Wales are variable in their uranium content, though most outcrops yield material with a higher uranium content than the *Ariconium* specimen, except Llanharry (m= 2.73 ppm, s= 0.47 ppm, n=4).

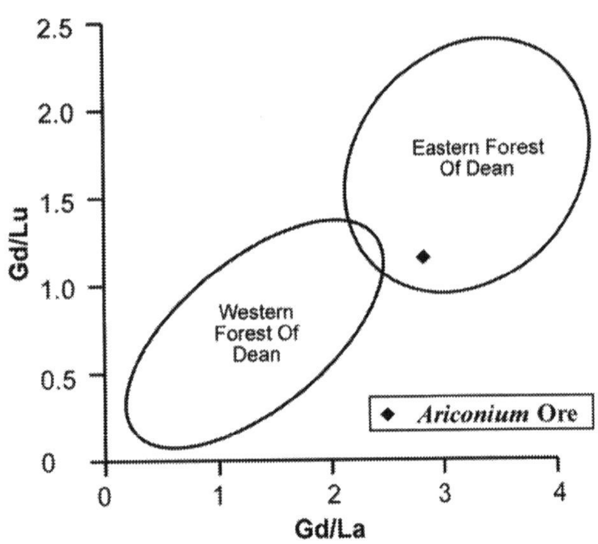

Figure 2. Diagram of Gd_N/Lu_N vs. Gd_N/La_N (normalized against upper crust, Taylor & McLennan, 1981), showing analyses of the Ariconium ore fragment together with the fields of analysis from the western and eastern sides of the Forest of Dean.

An origin within the Mendip Hills is unlikely as the ores from this area are characterized by a Ba content of at least an order of magnitude greater then the 40.6 ppm recorded in the *Ariconium* specimen. The trace element geochemistry of the *Ariconium* ore fragment is broadly similar to analyses of material from both the eastern and western outcrops of the Forest Of Dean. Analyses from these two areas can however be discriminated on the basis of the relative abundance of the various rare earth elements. This can be illustrated by plotting a graph (Figure 2) of Gd/La against Gd/Lu (data normalized against an upper crustal standard, Taylor & McLennan, 1981) This graph represent the degree of Light REE (LREE) depletion against the degree of Heavy REE (HREE) depletion. The *Ariconium* ore fragment plots within the field of material derived from the eastern part of the Forest Of Dean.

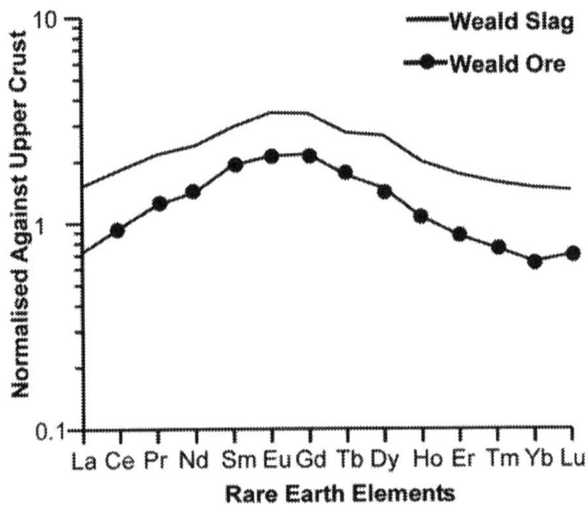

Figure 3. Plot of REE data (normalized against upper crust, Taylor & McLennan, 1981) for slag and iron ore samples from the Weald of Sussex.

Comparison of REE profiles of slags and their parent ores from regions of high and low grade iron ores

If the REE content of a slag is derived solely from that of the parent ore then it will produce parallel profiles on a standard REE diagram. Any deviation from parallel will indicate the addition of other materials during smelting, e.g., furnace lining or charcoal.

A slag (ΣREE = 288 ppm) produced from a low grade iron ore (ΣREE = 160 ppm) originating from the Weald of Sussex produces a similar REE profile to that of the parent (Figure 3). This indicates that in these circumstances the slag REE chemistry is less influenced by the furnace lining or that the lining is of a similar REE chemistry to that of the ore and therefore does not produce a significant deviation in the REE profile of the slag when compared to that of the ore.

Data obtained from the average slag composition (ΣREE = 59 ppm), a sample of furnace lining (ΣREE = 144 ppm) and the ore fragment (ΣREE = 18 ppm) from *Ariconium* shows that the slag is significantly enriched in the LREE with respect to the ore and that its REE profile is closer in shape to that of the furnace lining (Figure 4). This suggests that the slag represents a mixture of material derived from

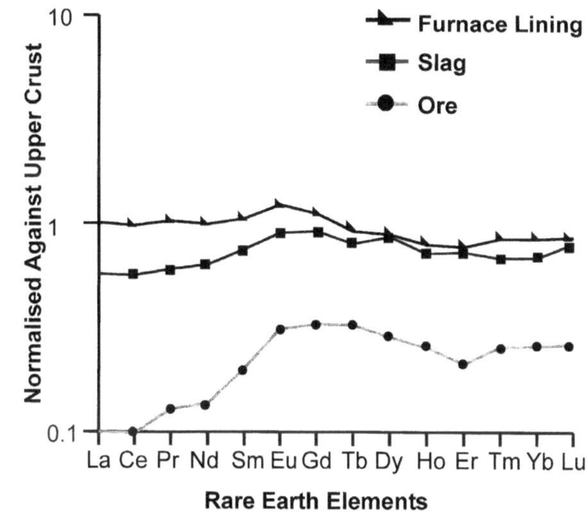

Figure 4. Plot of REE data (normalized against upper crust, Taylor & McLennan, 1981) for furnace lining, slag and iron ore samples recovered from the Roman iron smelting town of Ariconium.

the ore (low ΣREE, high LREE depletion) and from the lining (high ΣREE, no LREE depletion).

Contribution of furnace lining to the REE profile of *Ariconium* slag

In order to examine the contribution of the furnace lining to the melt a multi-element approach was chosen in an attempt to avoid problems associated with variation of slag, lining and ore compositions. The elements utilized are principally the REE, these having the advantage of being relatively immobile, so their concentrations are less affected by weathering of samples and also the fact that they are only present in charcoal in extremely low concentrations, so the influence of charcoal on the calculation can be ignored as a first approximation.

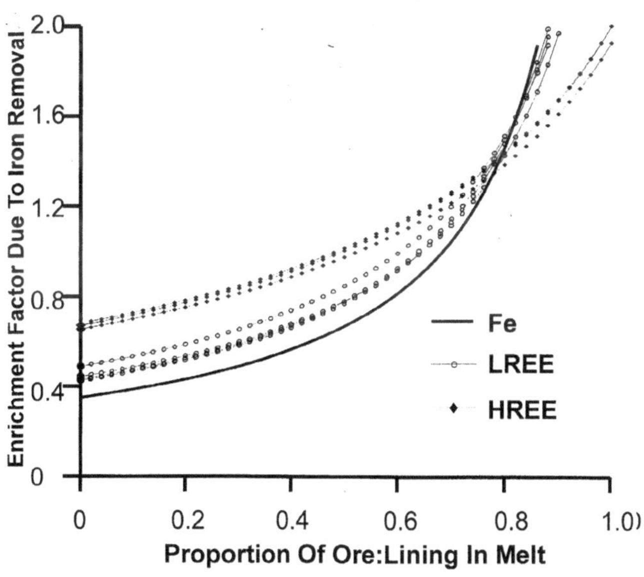

Figure 5. Graphical representation of the enrichment factor due to iron removal vs. the proportion of ore:lining in the melt for Ariconium ore, slag and furnace lining. The intersection represents the mixture of ore and lining required to generate the observed slag composition.

A graphical technique was then used to solve simultaneously the mass balance equations for the individual elements. The degree of enrichment produced by iron removal required for each element in the melt to reach the concentration observed in the slag, was calculated for the melt at 2% intervals of ore:lining mixture ranges from 0% (pure molten lining) to 100% (pure molten ore). The results of this were then plotted (Figure 5) such that each line represents a single element. A perfect result would be represented by all elements showing a common intersection, at which all can be produced at their appropriate concentrations in the slag from a single melt composition by the same degree of enrichment. In this example the elemental intersection corresponds to an ore:lining mix value (X) of 0.78 and an enrichment factor (E) of 1.4.

Therefore:

Proportion of Melt Removed as Bloom

$$(1-1/E) \times 100 = 29\% \quad (M)$$

1 kg of Reduced Anhydrous Ore Requires

$$(1-X) / X = 0.28 \text{ kg lining } (L)$$

And will produce

$$(1+L) \times M / 100 = 0.36 \text{ kg FeO } (I) = 0.28 \text{ kg iron}$$

And

$$1 + L - I = 0.91 \text{ kg slag}$$

Thus to produce a typical bloom of 2 kg, as found in the Forest of Dean, 7.02 kg of ore will be required, 1.81 kg of furnace lining will be consumed during the process and 6.43 kg of slag will result.

This gives an overall efficiency (iron in bloom/iron in ore) of 49%; larger then the figure of 30 - 40% generally quoted for furnaces of this period. Even though the data relating to the effect of charcoal upon this relationship have yet to be finalized it can be said that the dilution effect produced by adding large quantities of charcoal with a ΣREE lower then that of the ore will produce an efficiency value slightly in excess of that quoted above.

The total contribution of the furnace lining towards the final ΣREE content of the slag can be calculated thus:

ΣREE_{ore} (ΣO) = 20.9

ΣREE_{lining} (ΣL) = 144.4

ΣREE_{mix} (ΣM) $= \Sigma O \times X + (1-X) \times \Sigma L$ = 48.1

Recalculated for removal of iron during the process:-

$$= \Sigma M \times E = 67 \text{ ppm}$$

Of the 67 ppm ΣREE in the slag generated from the model 44 ppm ($\Sigma L \times (1-X) \times E$) is derived from the furnace lining whilst just 23 ppm ($\Sigma O \times X \times E$) originates from the ore used. Total contribution from the charcoal to this total is estimated not to exceed more than 2 ppm.

Conclusions

The analysis of REE concentrations can be a useful adjunct to more traditional approaches (such as physical form and major element chemistry) when attempting to provenance ore material recovered from archaeological sites.

In the case of the high grade (low ΣREE) ores found within the Bristol Channel orefield the assimilation of minor quantities of furnace lining during the smelting process can produce a slag with a lining-dominated chemistry (as suggested by Fulford & Allen, 1993). This is in contrast to the slags produced from the smelting of low purity (high ΣREE) iron ore whereby the REE profile produced is much closer to that of the parent ore, under which conditions detection of furnace lining contribution may be more difficult. Therefore it is essential to investigate the chemistry of the clay used to construct the furnaces in which the slags were produced.

References

CANTRILL, T.C., R.L. SHERLOCK & H. DEWEY. 1919. Iron ores (contd.) Sundry unbedded ores of Durham, East Cumberland, North Wales, Derbyshire, The Isle of Man, Bristol District and Sommerset, Devon and Cornwall. *Special Reports on the Mineral Resources of Great Britain.* vol IX. London. HMSO.

FULFORD, M.G. & J.R.L. ALLEN. 1993. Iron making at the Chesters villa, Woolaston, Gloucestershire: Survey and excavation 1987-1991. *Britannia* 23: 159-215.

GAYER, R.A. & A.J. CRIDDLE. 1970. Mineralogy and genesis of the Llanharry iron ore deposits, Glamorgan. *Proceedings of the Ninth Commonwealth Mining and Metallurgical conference, Mining and Petroleum geology section.* London. Institution of Mining & Metallurgy. 605-626.

SHEPHERD, T.J. & D.C. GOLDRING. 1993. Cumbrian hematite deposits, North-west England, in R.A.D. Patrick & D.A. Polya (eds.) *Mineralization in the British Isles.* London. Chapman & Hall.

TAYLOR, S.R. & S.M. MCLENNAN. 1981. The composition and evolution of the continental crust: rare earth element evidence from sedimentary rocks. *Philisophical Transactions of the Royal Society* A301: 381-399.

TYLECOTE, R.F., J.N.AUSTEN & A.E. WRAITH. 1971. The mechanism of the bloomery process in shaft furnaces. *Journal of the Iron & Steel Institute* 209: 342-363.

YOUNG, T.P. 1993. Sedimentary iron ores, in R.A.D. Patrick & D.A. Polya (eds.) *Mineralization in Britain.* London. Chapman & Hall.

29

Roman Iron Production in the East Midlands, England

IRENE SCHRÜFER-KOLB

School of Archaeological Studies, University of Leicester, University Road, Leicester LE1 7RH, UK

This paper aims to provide an overview of the Roman iron industry of the East Midlands, England, from the archaeologist's point of view. A complex network of mining, smelting and smithing sites is investigated along the Jurassic Ridge from South Humberside to Oxfordshire. Information is given about the archaeology of iron production sites, their distribution and level of industrialization. Light is shed on the individual technologies employed and the organization of the industry as a whole. Presentation of interim results of scientific slag analysis complements the assessment of the archaeological evidence.

In the East Midlands, iron production has an Iron Age tradition and is mostly settlement-related. The large number of comparatively small production sites shows a great individuality in applied technology, but the aggregate production of these sites would suggest that this was indeed a highly important element of the regional economy throughout the Roman period (AD 43 - c.409).

Two case studies from different parts of the Jurassic Ridge illustrate this very well. Iron smelting at Whitwell and Piddington, situated in the west and south of the Jurassic Ridge respectively, shows different levels of industrialization. Whereas Piddington can be interpreted as an industrial villa, iron production at Whitwell appears to have been operated on a smaller scale. First analytical results of slag from these two sites offer some insights into the properties and composition of the samples as well as technological aspects of the production.

Keywords: Archaeometallurgy; Roman Period; Iron Production; East Midlands; Slag Analysis.

Introduction

Iron making in the East Midlands, England, is roughly defined by the course of the Jurassic Ridge, a crescent from South Humberside to Oxfordshire (Figure 1). Evidence for extensive iron production in the region in antiquity has been recognized from the last century onwards (Artis 1828: plates 1, 25, 38). Considerable amounts of slag, furnace remains and smithing tools retrieved from archaeological excavations and field walking over the last few decades suggest that the ancient iron industry was indeed a highly important element of the regional economy. Current excavations continue to add new data, indicating substantial iron production in the area (Beamish 1997, with contributions by Schrüfer-Kolb).

In the East Midlands iron production continued from the Iron Age up to the late 1970s. Pre-Roman industrial activity was mainly confined to local centres like Hunsbury, Northants. (Jackson 1993/4). The arrival of the Roman army in the East Midlands soon after the Conquest in AD 43 no doubt had an impact on the development of the local iron production. For a full account on the present knowledge of the Roman iron industry in the East Midlands see Condron 1997 and Schrüfer-Kolb 1997. Iron-working in an early Roman military context is attested for a works depot associated with the fort at Longthorpe (Dannell & Wild 1987) and for Margidunum (Oswald 1927; Todd 1969). A short-term garrison probably ended soon after the Boudiccan revolt of AD 60 had been successfully subdued (Dannell & Wild 1987: 61).

Slow, but peaceful, development of the region from the 2nd century onwards up to the end of the Roman period at the beginning of the 5th century AD meant a continuously increasing demand in iron implements to be satisfied by the local industry. A growing network of iron production sites that focused on the regional infrastructure of roads and waterways developed through time up to late antiquity. This indicates a clearly market-orientated industry for the Roman period. By analogy with the Nene Valley potteries and perhaps coastal salt production, surplus trade extended beyond the nearest market centres (Jones & Mattingly 1990: 209-10; Condron 1996: 172-3, Condron 1997: 11-12) and may have placed the Roman Period iron industry in the East Midlands alongside other recognized Romano-British iron production regions such as the Weald and the Forest of Dean.

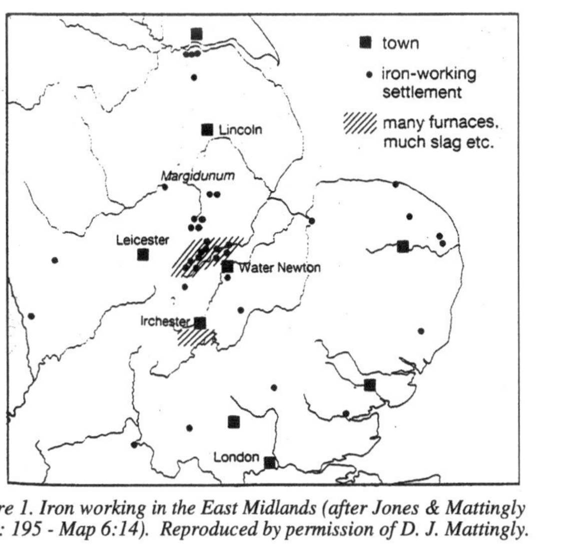

Figure 1. Iron working in the East Midlands (after Jones & Mattingly 1990: 195 - Map 6:14). Reproduced by permission of D. J. Mattingly.

Recent excavations have augmented the evidence for Saxon iron smelting (D. Jackson and G. Foard, pers. comm. 4/98 (7th cent. AD Laxton/Northants.); HMSNews 35, Spring 1997, 2-3 (Southwick/Northants.); Jackson 1993/4; Tylecote 1970), while the Domesday Book mentions a number of sites in the area. Iron making regained importance in medieval times, and iron and steel production are a vital part of the East Midlands industrial heritage.

Although much evidence now has been lost due to modern quarrying, most ancient sites must have been closely related to open-cast mining. Occasionally, a considerable number of pits has been preserved in woodlands such as Old Sulehay Forest near Yarwell, Northants. However, due to the lack of precise dating evidence, these can rarely be safely assigned to a specific period, especially if iron production is attested for more than one period in the immediate vicinity. The bell pits preserved in several places in the East Midlands are very likely to originate from the medieval period.

The geology of the East Midlands is characterized by sedimentary deposits of Liassic and Jurassic date, overlain by boulder clay from the ice ages. Iron ore deposits belong to the carbonate type, iron is present essentially as ferrous carbonate (Tylecote 1986: 124-25). Outcrops of siderite ore from the lower-lying sedimentary strata exposed at the slopes of river valleys could have been easily exploited by open-cast mining. Surface outcrops have often weathered to limonites. The clay soils and woodland covering the higher parts of the region supplied fuel and clay for furnace construction. East of the Jurassic Ridge the Fenlands have been formed by maritime silts, sands and alluvial deposits. Iron production in this area may have used local bog ore rather than rock ore brought to the area from the West. Future analysis of the chemical composition of slags from this area may verify this hypothesis (Condron 1997: 2 and Figure 1).

The present research project is based on interdisciplinary approaches to archaeometallurgical questions and its key aim is the combination of both archaeological and scientific analytical methods for the assessment of the Roman iron industry of the East Midlands. As slags are process indicators their analysis should shed more light on individual production technologies already deduced from the archaeological record. The data can be used to reconstruct individual melting conditions and slag formation processes and to get an insight into the quality, type and thus - to some degree - the origin of the ore used. Where possible, comparative analyses of slags and related ore samples will be carried out to obtain a relation between the ore used and the final waste product. Such information, and the corresponding archaeological evidence of structures, may verify the possible use of advanced production methods. Comparison of the results with samples from outside the East Midlands is vital as very few examinations have been carried out for the East Midlands themselves.

The following sections briefly discuss the archaeology of Roman iron production in the East Midlands, giving information on the distribution of sites, levels of industrialization and administrative organization as well as the individual technologies employed. Two case studies and the presentation of preliminary results of scientific slag analysis will complement the archaeology.

The archaeological evidence

Our present database of c.150 iron production sites is expected to increase as research is ongoing (Condron 1997, 13-16; Schrüfer-Kolb 1997: Figures 1a & b, 4). As yet, particular clusters of industrial activity can be distinguished, notably in north-eastern Northamptonshire and the upper Witham valley (Figure 1). Further data collection needs to be undertaken for the fringes of the Jurassic Ridge in particular. Evidence from South Humberside and the Lincolnshire Wolds (Dudley 1949: 130, 142-3, 193-6; Whitwell 1970: 113) contradicts the view that Romano-British iron smelting did not extend northwards much beyond Goadby Marwood (Condron 1997: 2)

Apart from early Roman military production, iron smelting and smithing in the East Midlands has always been settlement-related. Most iron smelting was undertaken on rural or suburban sites on varying economic levels from small-scale domestic production up to large manufacturing units. The overall pattern of industrial site location, taken together with the lack of written evidence from epigraphic and literary sources for any imperial involvement, suggests a model of *collegia* and *societates* responsible for the organization of the local iron production (Schrüfer-Kolb 1997).

The archaeological evidence points towards a great variety in applied technology for Roman iron production sites in the region. Both tall-shaft and low-shaft (?domed) furnaces for slag-tapping were used during Roman times (Jackson *et al.* 1978; Jackson & Tylecote 1988). The earliest closely datable slag tapping shaft furnaces in the East Midlands date from the early 2nd century AD (Aitken & Weaver 1962: 10, 16, 19). The prehistoric domed furnace remained in use, probably throughout the Roman period. There are also a few examples of the 'advanced bowl furnace': a small number of late Iron Age/early Roman shaft furnaces with no provision for slag-tapping which allowed the molten slag to cool slowly on the bottom of the furnace (Jackson *et al.* 1978: 151-53, 155, 157, 163f; Jackson & Tylecote 1988: 285). Finds of slags resembling those from slag pit furnaces definitely deserve further investigation as, at present, they cannot be clearly attributed to a specific furnace type (Schrüfer-Kolb in Beamish 1997).

The local carbonate ore required roasting prior to smelting, and two different roasters are known for the area: the circular roaster and the channel hearth. Some smelting sites also show possible evidence for ore washing. Apparently, refining of the bloom was also carried out on the smelting site. Smithing hearths can be found both in the context of rural settlements and villa estates and are most likely to indicate iron working for domestic use. Occasionally, on rural sites smelting is associated with smithing but professional blacksmiths' workshops are restricted to small-towns (Condron 1997: 10).

The scientific evidence

At present, examination of the East Midlands iron production material involves; a) a visual inspection by naked eye and microscope to determine sample appearance, type and internal structure; b) the measurement of basic physical parameters like sample weight and specific gravity; c) geochemical analyses using X-ray fluorescence analysis (XRF) to determine the chemical bulk composition. Basic metallographic examination of polished sections imbedded in resin are carried out under natural light by reflected light microscopy. Results are documented by micrographs.

In the following section recent work on slag samples from two Romano-British iron production sites in the East Midlands, Whitwell and Piddington, will be presented. As these two sites are situated in different parts of the region the results may allow a comparison between specific parts of the Jurassic Ridge.

East Midlands case studies

Whitwell

The site is situated on the western fringes of the Jurassic Ridge (Figure 2). It lies on a junction of Northampton Sand Ironstone with Upper Lias Clay, just on the northern shores of Rutland Water in the county of Rutland, close to Empingham (NGR SK9259 0815). The site that produced the slag sample (A22.1996 Whitwell Rescue Centre; PA 95/0209/9) has been subject to development in March 1996, work has been overseen by a half-day archaeological watching brief on 5/3/96 (Pollard 1997 and pers. comm.). The north baulk of the east-west foundation trench yielded Feature 1 (F1), a shallow clay-lined U-profile cut into the natural Marlstone (2.56m narrowing to 1.97m wide and 35cm deep as excavated) and containing unburnt limestone fragments. F1 comprised of Contexts 1 and 2 (C1 and C2). These contexts had been partially disturbed by ploughing and recent bulldozing respectively. C2 showed a central pit C1. In section, this pit was 51cm wide (narrowing down to 21cm) and 35cm deep as excavated. C1 consisted of a dense, 17cm thick layer of slag in its upper part, in brown, loose, silty clay with limestone fragments c.20cm in size. C2 contained a number of limestone fragments flanking C1. No other diagnostic archaeological feature or dating evidence has been recorded. F1 generally showed very little burning. This contradicts the interpretation of the feature as a furnace (R. Pollard, pers. comm.).

The collected slag sample was handed over to the Archaeology Materials Research Laboratory (G. Morgan) for analysis. The entire sample weighs 1009.5g and comprises of 24 pieces of tap slag and 7 minor pieces. Individual sample weights of the 10 examined pieces (down to 2cm^3) varied from 377.6g down to 20.3g. One of the biggest samples contained a small piece of iron ore (c.2cm^3) imbedded in the slag. This was initially thought to be hematite, but subsequently identified as roasted siderite/limonite. This sample also shows two distinct layers, indicating two separate phases of consecutive tapping. Externally, all pieces feature the characteristics of tap slag: a dark-grey to blackish colour, a smooth surface with a flowing structure, a rather rugged bottom showing impressions and some macro-inclusions of the soil onto which the slag had been tapped. In section, the samples appear rather dense and show comparatively few vesicles. Density measurements yielded a specific gravity of c.4g cm^{-3} on average. As expected, checking the magnetic properties of the samples with a hand magnet proved them virtually nonmagnetic. Occasional minimal reaction of some portions of the sample surface to the magnet can be due to subsequent reoxidization of the slag surface during cooling. This was confirmed by the presence of magnetite close to the surface in the polished section.

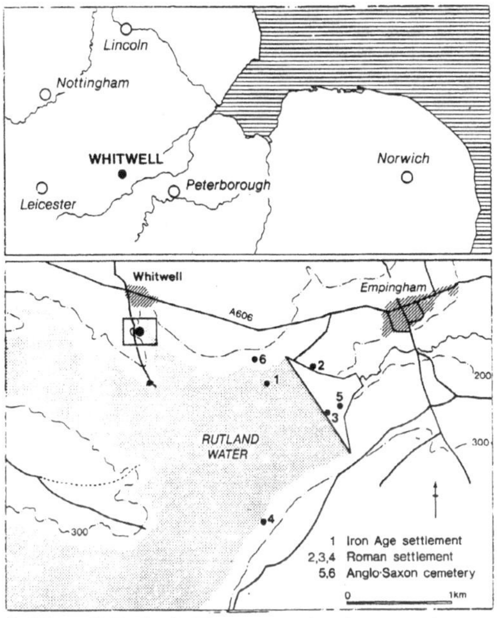

Figure 2. Whitwell (Rutland). The site in its local setting (after Todd 1981: 2 - fig. 1). Reproduced by permission of M. Todd.

Microscopic examination of two polished sections from Sample 1 and 2 respectively was undertaken under reflected natural light up to a magnification of 400 times. Samples were imbedded in resin and ground and polished with a water lubricated grinding wheel, a hand grinder for wet grinding and a polishing wheel lubricated with metallographic polishing alumina. FeO is present in the slag mainly in the form of dendritic structures. From a central stem, a system of two branches can be observed branching off perpendicularly from one another. Also, some rounded grains of wüstite can be seen. Both dendrites and rounded grains are surrounded by light grey fayalite and a fine, grainy, mixture of eutectic glass. The small size of the dendrites indicates rapid cooling. Sample 1 proved highly dendritic with a number of wüstite grains, imbedded into fayalite and a glassy matrix. Occasionally, areas of densely spread tiny rounded wüstite grains can be found in the matrix. Some inclusions of metallic iron were also detected. Clearly defined areas showing large grains of fayalite - occasionally incorporating some linear matrix inclusions - surrounded by dendrites intersect areas of primarily dendrites crossing each other. Sample 1 showed bigger dendrites than Sample 2. A long narrow line of wüstite runs across a large proportion of Sample 2. In general, Sample 1 showed a more varied structure than Sample 2. Although this description of the internal structure is far from conclusive and the mineralogy deserves specialist examination, it hopefully gives a basic idea of the internal composition of the examined slag samples.

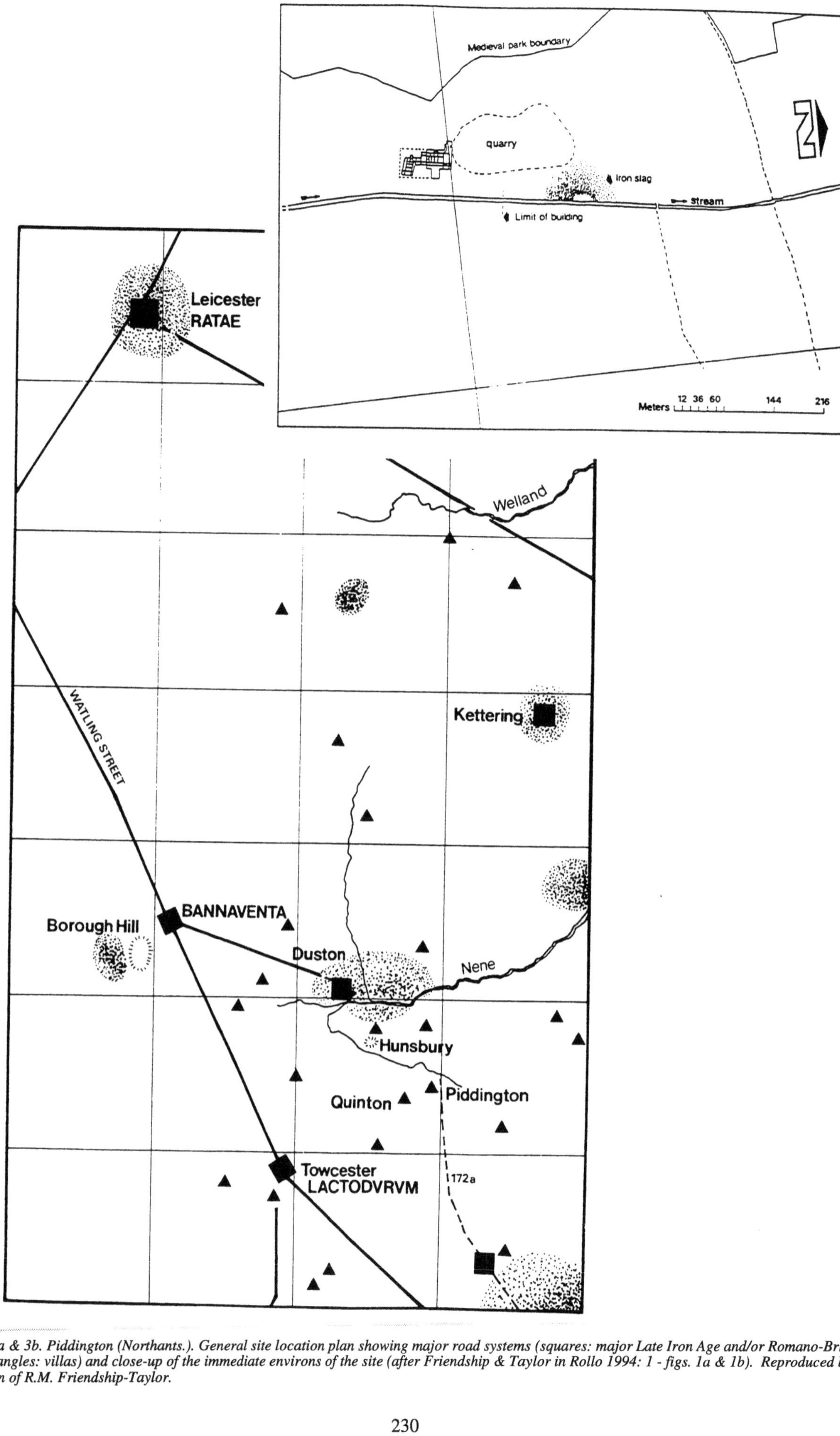

Figures 3a & 3b. Piddington (Northants.). General site location plan showing major road systems (squares: major Late Iron Age and/or Romano-British towns, triangles: villas) and close-up of the immediate environs of the site (after Friendship & Taylor in Rollo 1994: 1 - figs. 1a & 1b). Reproduced by permission of R.M. Friendship-Taylor.

Although no definite furnace structure has been associated with the material from Whitwell external and internal examination of the slag samples allowed them to be classified as tap slag that had cooled rapidly outside the furnace. Although Anglo-Saxon activity is known for Empingham, the slag from the Whitwell Rescue Centre watching brief is much more likely to have been related to iron smelting evidence from the Romano-British villa site at Whitwell close-by (G. Morgan in Todd 1981: 15-20). Excavation area 5 lies c.25m to the west (R. Pollard, pers. comm.; Todd 1981: Figure 2). Two furnaces have been excavated on the site, centrally located in Room 4 of the villa building, on Area 1a. Furnace 1 can be safely interpreted as a slag-tapping shaft furnace. Furnace 2 was less well preserved and could have functioned either as a low-shaft furnace, or a refining hearth for consolidation of the bloom. It is rather unlikely to have been a roaster, taking into account the archaeological evidence for ore roasting from other sites in the East Midlands. If this feature is interpreted as a furnace the term low-shaft furnace may be more suitable than bowl-furnace as this feature clearly shows a level bottom suitable for slag-tapping. However, no tap slag has been found associated with this structure.

Although the two furnaces from the villa site are certainly of Roman date their location within a building suggests they either ante- or post-date its occupation. A late, 4th century date is favoured by the excavator (G. Morgan in Todd 1981: 19). However, there appears to be evidence for iron smelting also for the mid-1st century AD (Todd 1981: 43). Some of the elongated features in Area 1b and 6 may also have been connected with metalworking.

Piddington

Piddington, is situated 6 miles south-east of Northampton in the southern part of the Jurassic Ridge (NGR SP7965 5400). Ongoing, long-term, excavation of the site brought to light the remains of a complex Iron Age settlement that had developed into a Romano-British villa. The circular Iron Age timber buildings were occupied from the second half of the 1st century AD. The first rectangular house - similar to that from Whitwell - appeared at the end of the 1st century AD and, step-by-step, was enlarged by annexes until the end of the 2nd century to form a winged courtyard villa. The late 2nd century villa was destroyed by a fire and subsequently remodelled. It reached its peak in the second half of the 3rdcentury AD and was continuously inhabited until the last quarter of that century. After being deserted for a few decades the site was transformed into six separate modest buildings, in use until the end of the 4th century AD (Friendship-Taylor in Rollo 1994: 1-7).

Although showing indications of several kinds of metallurgical activity, including small-scale early Roman copper smelting and perhaps even mid/late-4th century lead or pewter working, the main archaeological evidence is for iron production. In addition to slag being commonly found during excavation of the villa site itself, a concentration of iron slag c.100m in diameter has been recognized in the fields immediately to the north of the villa site, close to a stream (Figure 3). No detailed examination of the scatter has been carried out as yet, although a small amount of field-walking and some inconclusive geophysical prospection have been undertaken. Slag samples collected from the scatter included tap slag. The evidence for substantial iron production on the site integrates well with other industrial activities known to be associated with the villa: 2nd century stamped tiles appear to have been produced on-site, and sherds of sizeable mortaria found during excavation indicate large-scale food processing.

A number of different slag and ore samples from Piddington have been brought to the laboratory for examination, in order to shed more light on the iron production processes that have been carried out on the site and to understand better the role of the villa in the context of regional iron production. As work on the material is still ongoing and current excavation may yield further structural information no final picture can be drawn as yet.

Sample I/1 Pid. H/F 93 is an unstratified find from Piddington Grid System JJ/005, just to the south of the big well north of the newly excavated bath-house (grid square U), situated towards the north-east of the north wing. The sample comprises of 39 pieces of tap slag and 4 minor pieces. Most pieces show a smooth or ropy surface - some have big gas holes on the surface - and/or a rugged bottom. Some pieces have a corrugated appearance and therefore may have been raked out of the furnace while still soft. One of these pieces contains a fragment of charcoal or wood, one has some fuel ash attached. The entire sample weight is 352.815g. Density measurements yielded a specific gravity of c.3g cm^{-3} on average. The samples proved virtually non-magnetic. Microscopic examination of a polished section taken from the largest piece of Sample I/1 was undertaken up to a magnification of 400 times. Again, dendrites of wüstite can be seen surrounded by fayalite and a glassy matrix. Occasionally, areas of densely spread tiny rounded fayalite grains can be found in the matrix, but no wüstite grains. Just a few iron droplets were detected. Examination of the bottom area of the slag sample proved it to be dendritic throughout, with small dendrites of a rather rounded shape; wüstite is also present in the form of rounded grains. Dendrites could be observed as perpendicular systems of up to three branches branching off from a central stem and one another.

Some large fayalite grains can be seen, and there are several areas showing ladders of elongated narrow fayalite grains. A small number of iron inclusions are concentrated in one corner of the bottom. Some fuel ash pieces are pressed into the bottom side. A considerable number of rounded vesicles can be seen near the top surface of the sample. Some fuel ash inclusions - larger ones at the bottom and smaller ones at the top, can be found in the sample.

Sample I/2 Pid. H/F 93 is a flat piece of tap slag weighing 105.69g. It originates from MM/001 L2 (U), near the southern wall of the bath-house where a late tile kiln has been superimposed on the structure. The top surface of the sample again shows the characteristic smooth appearance with a number of gas holes concentrated diagonally on one side. Some reddish and white inclusions can be seen even at a macroscopic scale. The bottom side is rugged as usual, the sides of the sample show needle-like structures rising from the bottom and some iron inclusions. Oblong vesicles concentrate near the top surface and become gradually smaller towards the bottom. The sample is non-magnetic save a small area on the top surface that showed a very slight reaction to the hand magnet. This may be due to re-oxidation during cooling. The specific gravity of the sample is c.3g cm^{-3}. Examination at a magnification of 10 times proved the reddish inclusions to be iron and the white ones fuel ash. In one surface gas hole many needles can be seen, signs of the crystallization process on a macroscopic scale. On the bottom side, some iron inclusions can also be observed. In section, the cut side again shows needles on a macroscopic scale, and some iron inclusions.

Sample I/3 Pid. H/F 93 comes from LL/1 F701/711 L2 (L), near the south wall of the bath-house. It consists of two pieces of tap slag; total sample weight 66.33g. Density measurements yielded a specific gravity of c.2.4 g cm^{-3} on average. The smooth top surface shows a few white inclusions and several gas holes. Occasionally reddish corrosion products concentrate in them. The rugged bottom contains some charcoal inclusions and a piece of fuel ash. In section the cut sides show gas holes near the top, the number of which gradually decreases towards the centre. The smaller piece of the sample has an irregular drop-like shape and may have dropped down from the furnace mouth or tapping pit. Examination of the two pieces at a magnification of 10 times showed for the larger piece charcoal and fuel ash inclusions at the top. The sides again show rounded gas holes from the centre towards the top and some charcoal and iron inclusions as well as corrosion products inside them. The bottom in turn contains several charcoal impressions and inclusions (one vitrified), one iron droplet and one piece of fuel ash. The smaller piece shows two iron inclusions and crystalline needles in one area of the top surface and on the sides.

While these are only basic external descriptions, microscopic examination of polished sections taken from the above samples and documentation by micrographs are currently being undertaken. It is also planned to examine the composition of the iron inclusions in the slag in more detail.

XRF analyses of the slags

Preliminary XRF analyses have been carried out to give an insight into the chemical bulk composition of the material. Since these are our first geochemical check of the East Midlands material the analyses certainly need refinement and hopefully will improve as experience grows. Results are shown in Table 1 (at end).

The total iron content was first calculated as Fe_2O_3 and then recalculated for FeO. Totals shown are for total Fe as FeO. Also LOI - together with weight loss and gain in relation to $Fe/FeO/Fe_2O_3$ - has not been considered at present and will further change the individual values. However, some basic deductions can already be made. The iron content is comparatively high, even in comparison with the few other samples analyzed for the East Midlands that are published in the literature (Table 2 (at end)). Comparison of the values for the remaining major and minor elements with the published samples show only a small difference. The CaO values are normal, the low MnO values clearly indicate that no bog iron ore has been smelted. Bog ores from the British Isles normally contain relatively high concentrations of manganese, sometimes exceeding 10 wt. %.

This goes almost entirely into the slag during the smelting process. Thus, as expected, along the Jurassic Ridge the local rock ore deposits were exploited, and mining within a Romano-British environment was probably controlled by local entrepreneurs or companies, if not by state officials.

The P_2O_5 values indicate the use of phosphorus-enriched iron ore. Implements made of phosphoric iron show a considerable hardness and durability, similar to carbonized iron. Though a bit brittle in the cold state, phosphoric iron can be successfully forged to form tools of high quality. Trace element analysis has not yet been carried out on East Midlands material so that only the V value can be compared with the analysis from Wakerley; it matches that value quite well. The high values point towards sedimentary rocks. The W values are relatively low and indicate the hard nature of the material which has absorbed a little of the tungsten coating of the geological stone mills during XRF sample preparation. The chrome values cannot be interpreted at present and may be a characteristic of the East Midlands.

Interpretation and conclusions

Although no furnace structure is known for Piddington at present, examination of the above slag samples proved they can be clearly assigned to primary smelting perhaps similar to that carried out at Whitwell. Whilst external examination of the samples from both sites yielded similar results as expected, the ones from Whitwell show a higher specific gravity and prove more structured under the microscope. Chemical composition varies in detail, too: the total iron content is higher for the Whitwell samples, the silica content lower than in the Piddington samples. Most interestingly, the calcium content is much lower in the Whitwell samples, as is the potassium content. This could be matched by the observation of several charcoal and fuel ash inclusions in the Piddington samples. As for trace elements, sulfur is lower for Whitwell whereas vanadium is higher. Further analyses need to be undertaken to enable us to provide secure interpretations of the data. One of the next steps will be analysis of East Midlands ore from both the Whitwell area and Piddington that will hopefully enable us to see an ore-slag relationship.

These small steps towards an interdisciplinary approach to archaeometallurgical research hopefully can be intensified over time and attract specialist attention. Given the current state of research on the Roman iron production in the East Midlands, the analytical work presented here is meagre and can only be provisional. A comprehensive collection and appraisal of the archaeological data indicating extensive iron production all over the East Midlands in the Roman period is now being undertaken. Much more scientific analytical work needs to be carried out to complement this in-depth examination of the archaeological record. Hopefully, renewed excavations of a range of iron production sites will bring fresh material to light. Recovered with appropriate method and the necessary awareness, this offers a great potential to add further to our knowledge of ancient metallurgy.

This research project is funded by a British Academy Postgraduate Studentship.
Conference attendance was facilitated by an R F Tylecote Travel Grant of the Historical Metallurgy Society.

Acknowledgements
In addition to David Mattingly (University of Leicester, Archaeology), Graham Morgan (Analytical Conservation) and Frances Condron (Urban Hinterlands Project) I am much indebted to Nick Marsh (Geochemistry) and Colin Cunningham (Thin Section and Microprobe Preparation Unit - Geology) and all members of their team for their generous technical assistance as well as invaluable help and advice. I would also like to thank Richard Pollard (Leicestershire County Council) and Roy Friendship-Taylor (Upper Nene Archaeological Society) for the provision of slag samples and archaeological background information from Whitwell and Piddington respectively.

XRF Results for LAR1 Fe smelting slags

Pellet	Sample Site	Sample Code	SiO_2	TiO_2	Al_2O_3	Fe_2O_3*	FeO*	MgO	CaO	Na_2O	K_2O	MnO	P_2O_5	Total
L31951	Whitwell	S1/95 - 0209 - 9 (1)	20.10	0.49	3.37	87.36	78.61	0.36	0.55	0.00	0.52	0.27	0.63	104.90
L31952	Whitwell	S1/95 - 0209 - 9 (2)	20.67	0.68	4.10	79.76	71.77	0.36	0.68	0.00	0.50	0.32	0.60	99.68
L31953	Piddington	H/P 93 I/1	23.99	0.58	4.29	70.04	63.02	0.43	2.71	0.08	0.91	0.31	0.58	96.90
L31954	Piddington	H/F 93 I/3	20.41	0.42	3.97	84.24	75.80	0.46	2.22	0.00	0.70	0.29	0.88	105.15
L31955	Piddington	H/F 93 I/5	21.95	0.57	3.93	74.22	66.78	0.46	2.95	0.00	0.92	0.31	0.66	98.53
L31956	Piddington	H/F 93 I/2	28.32	0.54	4.90	69.54	62.57	0.35	1.69	0.07	1.22	0.31	0.41	100.38

*Total Fe has been reported as Fe_2O_3 and then converted to FeO. Totals on the basis of FeO.
Much of the Fe will be present as Fe metal and FeO

As	Ba	Cr	Cu	Ni	Pb	S	Sr	V	W	Y	Zn	Zr
8	255	400	39	10	0	264	158	752	45	102	45	168
7	217	535	30	24	3	287	380	1026	47	130	66	232
7	207	408	31	26	8	414	125	723	42	107	60	205
3	236	367	50	27	21	613	98	675	52	141	29	147
4	216	459	35	19	18	365	139	810	56	111	63	211
10	266	418	30	17	0	322	170	709	81	99	57	201

Table 1.

Sample Site	Category	SiO_2	TiO_2	Al_2O_3	Fe total	MgO	CaO	K_2O	MnO	P_2O_5	SO_3	V_2O_5	LOI	CO_2+H_2O	Total
Clipsham	tap slag	25.28	n/d	5.42	62.01	0.86	1.3	n/a	0.05	0.16	n/d	n/a	n/a	n/d	95.08
Great Casterton	slag	26.2	0.45	9.53	49.23	1.08	7	n/a	0.7	3.32	n/a	n/a	n/a	n/a	97.51
Great Casterton	slag	18.2	n/a	3.7	72.1	0.1	1.8	1.1	1.4	0.1	0.02	n/a	n/a	1.4(C)	99.92
Great Casterton	tap slag	20.9	n/a	1.9	67.4	0.9	2.4	n/a	n/a	n/a	n/a	n/a	n/a	n/a	93.5
Longthorpe	slag	23.93	n/a	5.38	66.39	n/a	2.33	n/a	n/a	0.95	n/a	n/a	n/a	n/a	98.98
Pickworth	tap slag	28.76	n/d	5.32	61.62	0.18	2.5	n/a	n/a	0.19	n/d	n/d	n/a	n/d	99.53
Sacrewell	slag	32.8	0.57	14.82	48.44	tr	1.85	n/a	0.33	1.11	0.05	n/a	n/a	n/a	99.97
Wakerley	slag	15.7	0.34	4.6	76.8	0.2	1.1	n/a	0.3	n/a	n/a	0.12	0.41	n/a	99.57
Wakerley	smelting slag	7.5	n/a	1.3	60.5	n/a	2.4	n/a	1.3	0.37	0.08	n/a	n/a	n/a	73.45

Table 2.

References

AITKEN, J. & WEAVER, G. H. 1962. Some archaeomagnetic measurements in Britain, *Archaeometry* 5: 4ff

ARTIS, E. T. 1828. *The Durobrivae of Antoninus.* London.

BEAMISH, M. G. with contributions by I. E. SCHRÜFER-KOLB. 1997. *Wing to Whatborough Trunk Main - Interim Report & Post-excavational Assessment and Project Design.* University of Leicester Archaeological Services Report 97/137.

CONDRON, F. M. 1996. *Roman Small Towns in the East Midlands: a regional study of settlement development and interaction.* Unpublished PhD thesis. University of Leicester.

1997. Iron production in Leicestershire, Rutland and Northamptonshire in antiquity, *Transactions of the Leicestershire Archaeological and Historical Society* 71, 1-18

CORDER, P. 1951. *The Roman town and villa at Great Casterton, Rutland.* 1st report.

1954. *The Roman town and villa at Great Casterton, Rutland.* 2nd report.

1961. *The Roman town and villa at Great Casterton, Rutland.* 3rd report.

DANNELL, G. B. & WILD, J. P. 1987. *Longthorpe II - The military works-depot: an episode in landscape archaeology.* Britannia Monographs Series No. 8. London.

DUDLEY, H. E. 1949. *Early days in North-West Lincolnshire: a regional archaeology.* Scunthorpe.

FRIENDSHIP-TAYLOR, R. M. & D. E. 1994. *Iron Age and Roman Piddington. An interim report on the excavation of a late Iron Age settlement & Romano-British Villa in Northamptonshire.* Upper Nene Archaeological Society Fasc. 1.

JACKSON, D. A. 1993/4. Iron Age and Anglo-Saxon settlement and activity around the Hunsbury hillfort, Northampton, *Northamptonshire Archaeology* 25: 35-46.

JACKSON, D. A. & AMBROSE, T. M. 1978. Excavations at Wakerley, Northants. 1972-75, *Britannia* 9: 115-242

JACKSON, D. A. & TYLECOTE, R. F. 1988. Two new Romano-British iron-working sites in Northamptonshire - A new type of furnace? *Britannia* 19: 275-298.

JONES, G. D. B. & MATTINGLY, D. J. 1990. *An Atlas of Roman Britain.* Oxford.

OSWALD, F. 1927. Margidunum, *Transactions of the Thoroton Society of Nottinghamshire* 31: 55-84.

POLLARD, R. J. 1997. Whitwell. Rutland History in 1995-96. *Rutland Record* 17: 315.

ROLLO, B. A. 1994. *Iron Age and Roman Piddington: the mortaria 1979-1993* (With an interim report on the site by R. M. & D. E. Friendship-Taylor). Upper Nene Archaeological Society Fasc. 2.

SCHRÜFER-KOLB, I. E. 1997. Metallurgy of Roman Britain - the East Midlands perspective. In: A. Vatandoust *et al.* (eds.) *Proceedings of the International Symposium on Archaeometallurgy in Central and Western Asia, 18-25 April 1997, Tehran, Iran* (in press).

TODD, M. 1969. The Roman settlement at Margidunum: excavations 1966-8, *Transactions of the Thoroton Society of Nottinghamshire* 73: 7-104.

1981. *The Iron-Age and Roman settlement at Whitwell.* Leicestershire Museums, Art Galleries and Record Service Archaeological Report No. 1.

TYLECOTE, R. F. 1970. *Recent work on early ironworking sites in the Stamford area, Bulletin of the Historical Metallurgy Group* 4: 24-27.

1986. *The Prehistory of Metallurgy in the British Isles.* London.

WHITWELL, J. B. 1970. *Roman Lincolnshire.* Lincoln.

30

A study of Roman mining and metallurgy and their environmental consequences at Plasenzuela, Extremadura, Spain

ROBERT G. SCHMIDT[1], CATHY M. AGER[2] & JUAN GIL MONTES[3]

[1] *Scientist Emeritus, U.S. Geological Survey, Reston, VA, and Research Associate, Anthropology Department, National Museum of Natural History, Smithsonian Institution, Washington, DC, USA*
[2] *Geobotanist, U.S. Geological Survey, Denver, CO, USA*
[3] *Geologist, Instituto Bachillerato el Brocense, Cáceres, Spain*

Significant amounts of lead, zinc, arsenic and copper continue to enter several sectors of the environment from Roman industrial wastes in the Plasenzuela mining district, Cáceres Province, Spain. The deposits were exploited by Roman miners for roughly 100 years starting about 30-20 BC, and by modern miners from the 1850's to 1908. The heavy metals persist in several environments besides those related to modern mining: in mine waste-rock dumps of probable Roman origin, in dumps of smelter slags well-dated as Roman, in associated soils and plants, and in river alluvium. The Plasenzuela Project, a geologic, geochemical, geobotanical, and archaeometallurgical study, conducted research here because Roman mine waste and smelter slag heaps provide a unique opportunity to study the retention and dispersion of metals from industrial sources over a time span not available in the Western Hemisphere.

Additional areas of elevated soil metal that seem unrelated to natural mineralization or directly to any ancient or modern mining industry activity are present north and north-east of the Roman smelters. Our interpretation of the latter areas as the result of fallout from the plume of the nearby smelters is still tentative. There was no smelting in the district in modern times. The possibility of residues of atmospheric lead here is interesting in light of recent investigations of lead in Greenland ice cores and Swiss peat bogs, because anomalous lead in both is attributed to Greek and Roman smelting.

The slags here are generally similar to the Roman slags of the Iberian Pyrite Belt as well as to some Bronze Age slags, but details of surface textures, probably reflecting viscosities, are unlike those from the Pyrite Belt, where somewhat different types of ores were processed using more iron-rich fluxes.

Introduction

The Plasenzuela silver-lead district in Cáceres Province, Extremadura, in west-central Spain (Figure 1) was exploited by Roman-era miners for perhaps 100 years starting about 30-20 BC, and by modern miners from about 1850 to 1908 (Domergue, 1987). Roman mineral industry wastes provide us with a unique opportunity to study the retention and dispersion of metals from industrial sources over a time span not available in the Western Hemisphere. This report summarizes preliminary results of our studies, with emphasis on the Roman mining and smelting industry developed in this small district, much less complex than that of the extensive Iberian Pyrite Belt (Rio Tinto, etc.).

No pre-Roman mining activity has been documented in the Plasenzuela district, but excavations at the hill fort, Villasviejas, indicate that a considerable population lived here in the five centuries preceding the Christian era (Hernandez, 1972). Domergue (1987) reported extensive Roman mining by surface pits and trenches, now mostly obscured, as well as by shafts that reportedly reached depths of at least 137 m.

An old report has suggested Arabic activity in some of the mines (Roswag, 1853), but no substantiating evidence is cited and no Arabic artefacts are known. There is no evidence of smelting later than the Roman era, and any significant mining, without concomitant smelting, seems improbable before modern transportation. The slag dump was in place when modern exploitation began (Roswag, 1853) and local residents confirm that no smelting in the district accompanied the 1850-1908 mining. Modern mining was entirely underground, and the last mine, la Serafina, ceased operation in 1908. For the most part however, the prehistory and history of the Plasenzuela district are very sketchy.

Field work in 1994, 1995, and 1996 resulted in a regional soil geochemistry data base, detailed surveys of an extensive Roman smelter slag dump, systematic sampling of several mine waste piles believed to be Roman, and collection of river alluvium over a distance of 30 km.

Sampling and analyses

Soil samples were collected from the surface, as an A horizon is poorly developed or absent at most sites. Samples for analysis of soils, river alluvium, and bedrocks were composites of 6-15 sub-samples from an area up to 10 m across, except in the close-spaced transects on waste-rock and slag dumps, where a few sub-samples or only one sample were taken at a site. Sampling of soils and river alluvium were initially undertaken on the assumption that the areas affected by Roman industry were much smaller than they proved to be, and the area covered by the program had to be enlarged in each successive field season. Soils, bedrocks, and alluvium were analyzed by XRD or ICP-AES. All slag analyses were made on a composite of 6-10 approximately equal fragments broken from separate pieces of slag at the site. Slags were analyzed by ICP-AES.

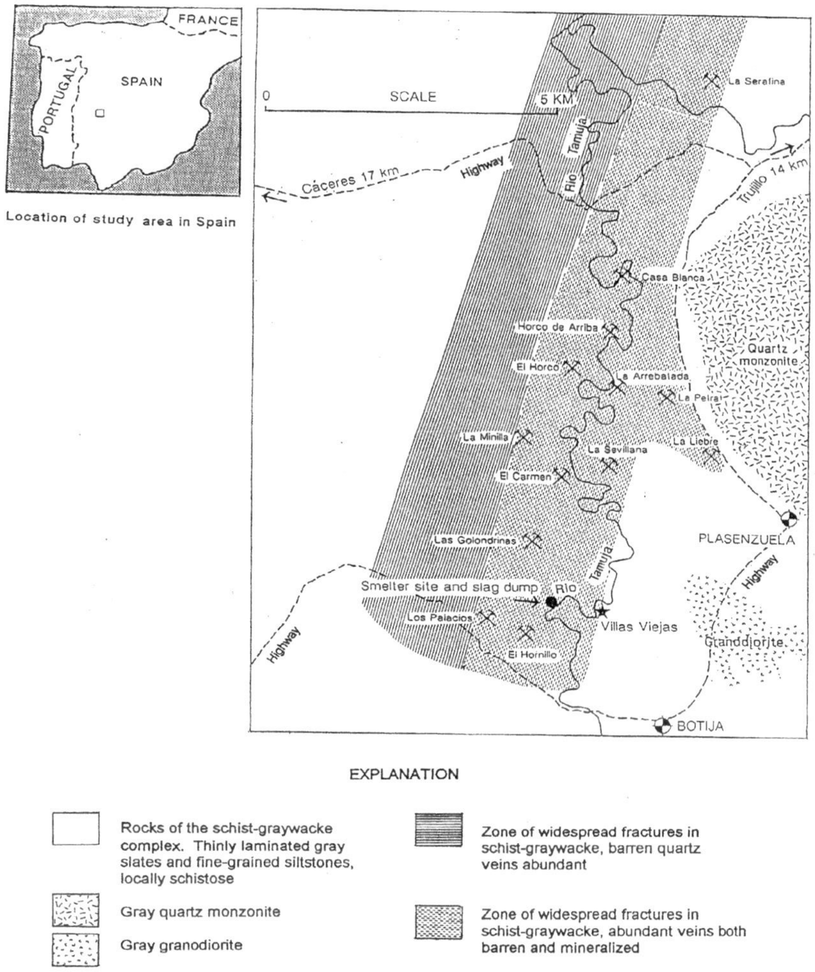

Figure 1. Location of the Plasenzuela study area in Extremadura, Spain, and map of the major ancient and modern mines in a generalized geological setting. The deepest known Roman mine workings are in Mina la Serafina at the opposite end of the district from the only significant smelting site. Large waste-rock dumps of undetermined age at the ranch, las Golondrinas, are suspected to be of Roman origin.

A concentration of 60 ppm was selected as a threshold value for anomalous soil lead where there is no indication of mineralization or industrial contamination at the sample site. The sites with at least this much lead are noticeably grouped together in the southern part and to the east of the mining district, and 60 ppm is 2X the 90 percentile maximum lead content in a group of 112 samples collected by the Instituto Geológico y Minero de España from an area of similar bedrock 5 km to the west.

Geographic Setting And Geology

The Plasenzuela mining district is situated in the broad intermontane basin of the Rio Tajo, between the Sierra de Gredos on the north and the Sierra de Guadalupe on the south. The basin is largely underlain by slate and schist, less resistant to erosion than the mainly granitoid rocks of the mountains. Most of the broad basin is largely cleared ranch and farm land devoted to grain cultivation or pasture. River valleys are partly wooded pasture, and the uplands and mountains are primarily oak forests, wooded pastures, and cork oak groves. The Plasenzuela mining district is small and mostly confined to the valley of the north-flowing Rio Tamuja, a tributary of the Tajo.

Prevailing winds at the present time are from the west and we assume that they were generally the same in Roman times; however, there may have been more rainfall then than now (Linés Escardó, 1970). Late summer rain storms tend to come from the south-west.

The geology of the mining district has been summarized by the Instituto Geológico y Minero de España, (1978), and by Domergue, (1987). Rowan and others, (1986), and Lago Rodriguez and others, (1989) have presented more detailed descriptions. All known mineralization is in veins within the "schist-graywacke complex", a generally monotonous sequence of slates and metasiltstones of Cambrian-Precambrian age (Gumiel *et al.*, 1995), folded along axes that trend north-eastward. A 3 km-broad zone of abundant veining trends in a north-east direction for 12 km. Mineralized veins seem to be confined to the eastern 2 km of this zone (P. Gumiel, personal communication). An elliptical quartz monzonite pluton, 5 x 7 km in surface dimension, lies east of the mineralized trend, and is generally thought to have furnished the heat energy to drive circulation of the mineralizing solutions (Lago Rodriguez *et al.*, 1989).

The mineralized zone is made up of veins parallel to the general N. 15 E. trend as well as cross-cutting veins at N. 65 W. The veins are generally small, but the sulfide/gangue ratio is high. The common sulfides are sphalerite, arsenopyrite, galena, and pyrite, probably in that order of abundance. Minor copper minerals are present. Native silver is reported to have been an important part of the exploited ore, probably in the upper oxidized parts of the veins, but in two mines it was present in small crystals accompanying sulfide minerals (Instituto Geológico y Minero de España, 1978). Quartz and carbonates make up most of the gangue. The depth of oxidation on the veins would have controlled the proportion of oxidized ore available and consequently the smelting methods needed. Oxidation of veins in other mines in the region extends to about 60 m (P. Gumiel, pers. comm.).

The Roman Mineral Industry And Its Waste Products

The metal-bearing mine waste-rock dumps

Gently rounded, low, earthy mounds are common in many areas in the mining district. Their weathered and vegetated appearance and their isolation from modern mine openings, suggest they may be Roman mine waste, either from an obscured mine nearby or the waste of a concentrating operation separated from the mines themselves. Higher than normal soil metal confirms their relationship to the mining industry. The dump material is now generally small rock chips in an earthy matrix, part of which is probably derived from weathering of finely broken rock. Ore minerals surviving within the dumps are now likely to be entirely oxidized.

The waste rock dating problem

Definitive dates have not yet been obtained from any of the potentially Roman waste rock piles. Metallic objects are fairly common in the plow zone, but none of certain Roman age were found in mine waste dumps. Roman ceramic material is widespread in the district, and present on the surfaces of some waste-rock dumps, but the chance of lateral transport on the surface is too high for it to be trustworthy. To date the best indicator of possible Roman age of the waste-rock is scraps of lead metal encountered in small excavations at depths of 10-20 cm in a low dump near Mina la Serafina. The fragments are corroded deeply like the lead in Roman associations elsewhere in the province, and much more corroded than lead from recent centuries. Systematic excavations to try to confirm finally the ages of waste-rock dumps by stratigraphy, interbedded artefacts, or with wood or charcoal datable by radiocarbon methods is a future goal.

Metals in the waste-rock dumps

Four sites were selected for study as potentially Roman. Surfaces of five dumps at these sites were searched for materials of Roman origin, and soil samples were taken along transects beginning above or at the top of each dump and extending across it and down the slope below. Soils in the la Serafina dump and other possibly Roman mine waste dumps contain significant amounts of heavy metals: up to 2.0 % Pb, 0.42 % Zn, 0.024 % Cu, and 0.54 % As. Soils downslope from the dumps are enriched in the metals from erosion of the dump soils and probably from solution and redeposition of the elements as well.

Roman smelting and the metal-bearing slag dumps

Information gathered to date about the smelting process is limited to examination and chemical analyses of the slags, observation of pieces of mostly granite furnace wall, and a study of the anomalous soil lead provisionally interpreted to be fallout from Roman smelter plumes.

The Roman smelters were concentrated in one area at the southern (upstream) end of the area of major mining (Figure 1). The scattered slag forms a discontinuous mantle with lenses up to perhaps 3 m thick, extending along the Rio Tamuja for 700 m, and as much as 300 m up the gentle slope from the river. Nineteeth century estimates put the total quantity of slag here at 30,000-40,000 tons (Roswag, 1853). Some of this may have been removed subsequently (Domergue, 1987), but there is no visible indication of this, and many thousands of tons remain.

Very small amounts of slag were found at three other sites in the district. These are considered insignificant. There was inadequate archaeological evidence to date them and we regard their ages as undetermined.

Figure 2. Surface of Roman slag dump, Plasenzuela, with weathered and slightly rounded roof tile fragments. The tile shape is unique to Roman and possibly early Visigothic construction. Similar fragments found below the slag dump are unrounded and preserve the fine original surface textures and sharp edges, thus establishing the undisturbed nature and Roman age of the dumps. The dump was ascribed to Roman industry before the start of significant modern mining (Roswag 1853).

Evidence for Roman origin of the major smelters and slags

Rounded and weathered Roman roof tile fragments are common on the surface of the slag dumps but they may have been eroded from an extensive settlement located upslope from the smelters (Figure 2). In excavating small pits to obtain samples for analysis however, roof tile fragments were obtained which retain their sharp edges and uneroded surfaces, thus establishing a Roman age for the slags. The adjacent settlement is characterized by Roman coins, common bits of lead scrap and small ingots of lead, ceramic materials, building foundations, and scattered transported building stones. Roswag (1853), who had access to the ancient mines before modern mine development began, regarded the slag as certainly Roman.

SAMPLE NAME OR NUMBER	Fe	Fe2O3	Si	SiO2	Ca	Pb	Zn	Cu	As	Ag ppm	S
	Analyses, in percent (except Ag, in ppm)										
	See footnote										
Composite samples from Plasenzuela, collected and analysed on our project ICP-AES analyses, P.H. Briggs, U.S. Geological Survey, Analyst											
2763-W	22	31.5	18	37.7 *	4.5	6.4	7.4	0.12	0.035	71	
2764-E	22	31.5	19	41.7 *	4.5	6.1	4.4	0.12	0.030	45	
2764-W	34	48.6	20	42.1 *	2.4	1.7	0.0	0.03	0.026	66	
2763-E	20	28.6	20	42.9 *	5.3	4.9	5.6	0.11	0.022	35	
2763	20	28.6			5.0	6.2	7.0	0.15	0.035	70	
2764	19	27.2			4.7	5.2	3.7	0.12	0.036	31	
2765-S	20	28.6			4.7	5.9	7.1	0.15	0.033	41	
2765-G	16	22.9			4.8	7.2	3.4	0.10	0.019	52	
Pyrite Belt samples of Roman slags, analyses from Rothenberg, (1989):											
Cu-1	43.9	62.8		25.1		0.11		1.00	tr	11.7	1.4
Cu-2	48.0	68.6		22.0		0.11		0.67	tr	2.8	1
Ag-1	45.7	65.4		26.2		1.40		0.30	0.430	63.2	1.28
Ag-2	46.3	66.2		22.0		0.64		0.15	0.600	189.2	2.91
Pyrite Belt samples of Roman slags, analyses from Salkield, (1982):											
Bella Vista	32.5	46.5		32.9		1.05		0.14	0.480	99	
Brown-H5	52.5	75.1		15.6		1.51		0.05	0.430	31.4	0.82
Tab-1, Cu-1	48.0	68.6		22.0		0.11		0.67	tr	2.8	1.0
Tab-1, Cu-2	48.2	68.9		21.7		0.11		0.73	tr	6.5	1.1
Tab-1, Cu-3	42.4	60.6		25.5		0.32		0.71	0.300	15.5	0.8
Phoenician	25.1	35.9		45.0		0.84		0.10	0.060	575	

Fe2O3 for each sample has been calculated from Fe in original analysis.
* These SiO2 values are calculated from the Si analysis given.

Table 1. Compositions of the slags from Plasenzuela and other archaeometallurgical sites in Spain.

Figure 3. Small mound of slag that flowed from the furnace in ropy streamlets that adhered to form a large mass, Rio Tinto, Iberian Pyrite Belt. This type of slag has not been recognized at Plasenzuela. The pocket knife is 7.5cm long.

The lead-zinc smelter slag

Although overall appearance of the slag is similar to the products of Roman smelting in the Iberian Pyrite Belt (described by Salkield, 1982, and Rothenberg and others, 1989), the ropy flow textures common on Roman slags from there seem generally lacking here (Figures 3 and 4). Plasenzuela slags also differ from Pyrite Belt slags in composition and heavy metal content (Table 1).

Most of the slag is dense and dark gray to black and stony; sparse fragments are glassy. That with stony appearance is also largely glass, but contains many included microscopic sheaves (crystal bundles) of the iron silicate fayalite. The slag consists of platy pieces 1-3 cm thick (Figure 5), the largest pieces up to 25 cm long, and thicker irregular chunks as much as 10 cm thick. It is all tapped slag permitted to flow from the furnace before hardening, and the thin platy pieces suggest that it was relatively fluid. Neither adhered accumulations of successive taps forming larger masses, nor untapped slag masses from furnace bottoms were observed. Upper surfaces are smooth to slightly wrinkled and many are pitted by bubbles that escaped during cooling. Open pockets as much as 10 cm across formed in some of the pieces, which were probably large bubbles formed under the surface crust, or possibly hollows from which the liquid drained after the surfaces cooled (Figure 6). The upper surfaces of some pieces have a finely wrinkled flow texture. There are a few highly vesicular pieces and the undersides of some of the plates are rough or vesicular, but most of the slag is dense and solid. Only a very thin skin of the upper surface of the plates appears weathered or leached, but weathering may have penetrated highly vesicular pieces and bottom surfaces more thoroughly. Sparse, localized, very glassy slag shows no sign of weathering.

The lower surfaces of many platy fragments are generally smooth, gently irregular, and totally lacking adhering sand or soil particles, nor are there bits of earlier-cooled slag attached. These are best interpreted to have cooled on top of ponded lead-silver metal, the major product of the smelting process. We also considered the possibility that some had cooled on pooled matte or speiss, sulfide and arsenide smelter products sometimes made as an intermediate product, as at Rio Tinto. No trace of matte or speiss has been reported from Plasenzuela, or observed during the current study.

Heavy metals in the slags and associated soils

Most slag samples contain 5-7 % lead, 3-7 % zinc, and 0.02-0.04 % arsenic (Table 1). Ores here are generally arsenic-rich, but much arsenic would have been vaporized in the smelting process, and not be retained in the slag. Table 1 compares slag compositions at Plasenzuela with those of the Pyrite Belt at Rio Tinto.

Soils interstitial to the slag fragments contain 1.5-2.3 % lead, 0.3-0.6 % zinc, and up to 0.031 % arsenic. These metals may be contained in fine smelter waste in the soil, or perhaps be leached from the slag fragments. The ratios of lead to zinc and lead to arsenic are much higher in interstitial soil than in the slag, and the Pb/As ratio is many times greater than in soils at a distance from the mining industry.

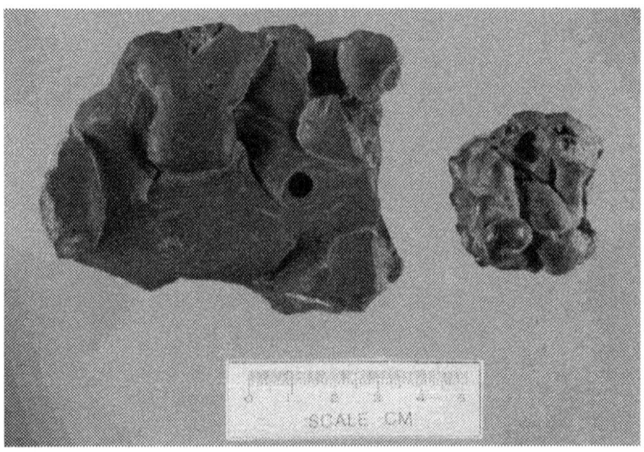

Figure 4. Detail of upper surface of ropy-surfaced slag from Rio Tinto, Iberian Pyrite Belt.

Figure 5. Typical flat-surfaced slag fragment, main slag dump, Plasenzuela. The slag remained sufficiently fluid after tapping to allow some gas bubbles to escape.

Located close to the Rio Tamuja, slag and the soil components of the dumps continue to be swept from the piles in major floods; however many thousands of tons remain. Slag pebbles are present in river gravels as far as 18 km downstream.

Additives to the ores in smelting

Depending on the composition of the ore material in the furnace feed, other constituents (fluxes) may have been added to obtain a slag of proper composition to incorporate the unwanted elements and to flow freely when the furnace was tapped. Most slags contain 4.5-5.3 % calcium, more than 5 times the calcium content of most wall rocks of the ore veins. If the ores used were calcite-rich (and hence mostly unoxidized), they may themselves have contained enough calcite ($CaCO_3$) to account for the calcium in the slag. The most likely additive to have supplied calcite would have been limestone, and several pieces of a limestone that is not present in the mining district were seen in the slag dumps but we do not know whether it was used in the smelting process. Plasenzuela slags contain 16-34 % iron, generally much less than the Roman slags of Rio Tinto (Table 1), but possibly more than would have been supplied by the natural ores used here, in which case an iron oxide flux material such as hematite would have been added to the furnaces.

Figure 6. Edge view of Roman slag fragment with large cavities from the main slag dump, Plasenzuela. These were large bubbles formed within the slag or possibly openings left when the surfaces cooled and stiffened, permitting some of the interior fluid to drain.

The furnace locations

The areas at the tops of the slag-strewn slopes where the furnaces should have been were examined, but without excavation only a few pieces of granite furnace wall with slag fused to them were found.

The location of the major smelters at the southernmost end of the district may indicate the relative importance of the southern mines in Roman times. The deep mines, la

Serafina and la Sevillana, 10 and 2.2 km to the north, may have been developed late in the Roman exploitation, or have been of lesser importance, or the major smelter site may have been chosen for entirely different reasons. The site is immediately downstream from a low rocky rapids that might have been a source of water for grinding mills, if the Romans were using water mills so early, or it may have fed a ditch that supplied washeries close to the smelters. Remains of either would have been exposed to periodic torrential flooding, and probably erased.

The age of part of the large mine complex at the Finca las Golondrinas, a kilometre from the smelters, is undetermined. Though modern mines were operated at this site, a large area of waste dumps has the weathered, earthy, rounded appearance that suggests great age. Roswag's description of his pre-1853 examination of extensive underground workings at las Golondrinas shows that major mines existed somewhere on the property before modern mining (Roswag, 1853), yet the exact location of those mines remains uncertain. We speculate that the mine associated with these dumps was one of the most important Roman ore sources in the district, and the real reason for the siting of the smelters close by.

It is also probable that the source of hematite, if it was indeed used as a flux ingredient, and charcoal for fuel lay in the mountains to the south or south-west, making this a logical smelter site. Roswag (1853) has stated that charcoal within the slags was made from both Quercus ilex (encina or holm oak) and Q. suber (cork oak). As the latter now grows in the area only at higher elevations, one can speculate that a slightly moister climate (as posited by Linés Escardó, 1970) permitted both species to grow nearby in Roman time, or perhaps the local Q. ilex trees were exhausted by the mining industry and Q. suber charcoal had to be brought from the mountains.

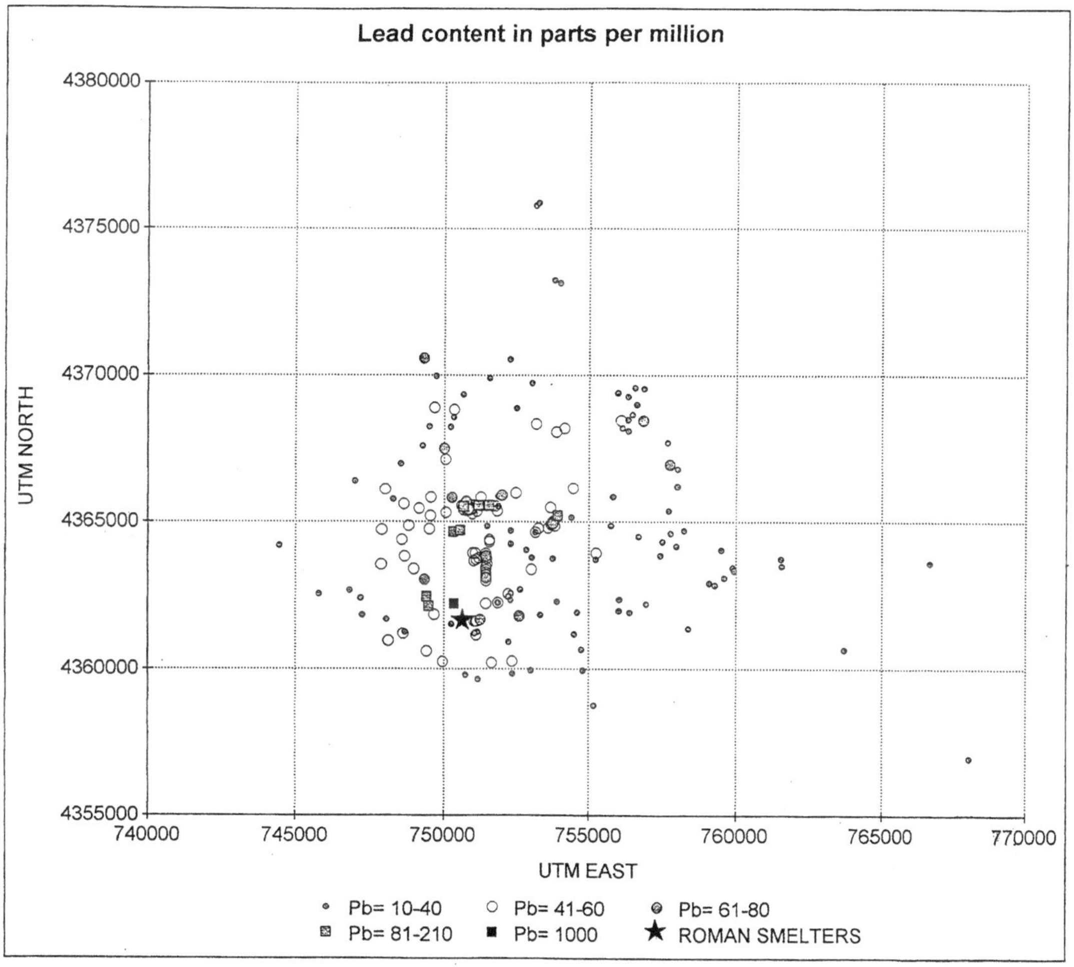

Figure 7. Map of the lead content of soils at sites selected to be free from direct contamination by mining operations or natural mineralization. Soils with lead contents greater than 60ppm are anomalous and may have been contaminated by fumes and airborne particles from the smelters. Several Roman and modern mines are present north of UTM 4366000, but seem to have had little influence on lead soil content. Each grid square is 5 km on a side.

The use of cupellation at Plasenzuela

Roswag (1853) reported that pieces of litharge (lead oxide) and crucible fragments, indicators of the use of the cupellation process, had been found across the river from the slag dumps and near the hill fort, Villasviejas. Cupellation was a metallurgical method well known to the Romans in which a lead and precious metal alloy was exposed to an oxidizing blast in the furnace, oxidizing the lead, zinc, etc., to form litharge, and leaving the silver in metallic form. It probably would have been performed in a crucible. Neither litharge or crucibles have been found during this investigation.

Termination of Roman mining and smelting

It is not known why Roman mining stopped at about 80 AD. Unfortunately, it does not seem clear from Roswag (1853) whether he found ore remaining in the Roman workings of a grade and in sufficient quantity to have supported further exploitation by them. Some of his tables of analyses imply that about 12 veins were sampled systematically by up to 10 samples from each, and that several samples contained more than 2000 ppm Ag, one exceeding 5200 ppm. However, it is unclear how careful Roswag was to obtain truly representative samples of the vein, nor is the minimum grade that the ancient miners considered workable known.

On the other hand, other factors, such as consumption of all nearby oak for mine supports, ore roasting and charcoal may have led to the closure. Allen (1970) mentioned the need to import lead to the Rio Tinto smelters to enable extraction of the silver from those ores. If the industries of the two mining districts were closely linked, a change in the economy or the type of ore processed at Rio Tinto might have led to the cessation of mining at Plasenzuela.

Study of anomalous soil heavy-metals possibly related to Roman smelter gases, and timing of erosion

Soil analyses at 92 non-mine-related sites selected to be well removed from mining or transportation activity, and where there is no indication of natural mineralization, delineate an area of anomalous lead and zinc content extending north and north-eastward from the Roman smelting site (Figure 7). Of this same data set, 55 of the composite soil samples were each paired with a composite bedrock sample collected from the same 10 m diameter area. Analyses of these sample pairs (Figure 8) show that part of the soil lead is independent of the bedrock lead, which has little variation, indicating a probable outside source for the anomalous soil lead. Sites near several important Roman mines in the northern part of the district far from known smelters have only normal amounts of soil metals.

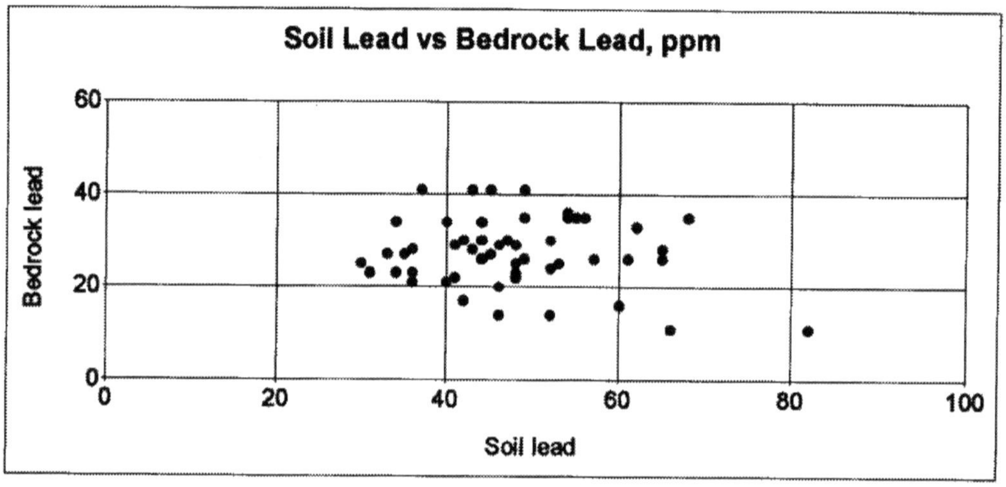

Figure 8. Lead content of soil compared with lead present in the underlying bedrock at 55 sites selected to be free from direct contamination by mining operations. The lack of correlation of soil metal with bedrock metal is a significant indicator of an outside source for the element. Each sample in each pair was a composite made of several sub-samples, both taken from an area about 10m in diameter. Collection of the rock samples was discontinued before their importance in studying the suspected smelter plume fallout was realized. Hence, the pairs are restricted to the central part of the distrct.

We think that the higher-than-normal metal concentrations are best explained as airborne deposits from the smelting furnaces, preserved in the soil blanket. If this interpretation is correct, preservation of patches of the contaminated surface soil tells us that only limited soil erosion has taken place here since the end of Roman mining. Furthermore, it suggests that significant soil contamination from smelter fumes may persist in soils for many centuries. The considerable contribution of Roman smelter gases to world-wide atmospheric lead has been documented by studies of Greenland ice cores (Hong *et al.*, 1994), and peat from a Swiss bog (Shotyk *et al.*, 1997).

Heavy metal contamination of Rio Tamuja sediments

We collected and analyzed Rio Tamuja alluvium from 37 sites (Figure 9), beginning 2.5 km upstream from the first known mines, and ending 11 km below the last known mine, in order to assess the downstream contamination, as demonstrated in other mining districts (Baumann, 1984, Johns, 1995). Most of the increase in metal content upstream from point B (Figure 9) is probably related to Roman mining and smelting, although part has certainly come from small modern mines and prospects in the tributary watersheds as well. Extensive dumps at the Finca las Golondrinas supply almost all of the metals entering at point B. If the greater part of the dumps there prove to be Roman, their contributions plus the metals leaching from the Roman slag dumps may make Roman industry responsible for a significant part of the heavy-metal load down to 8.7 km, (C, Figure 9), where the first major influx from a modern mine enters the river. The greatest influx of heavy-metals is below that point and coincides with water courses draining areas of abundant modern mine waste.

Notwithstanding the length of time the Roman slag dumps have been subjected to erosion, they still shed a significant amount of material into the river during heavy rains. The abundance and sizes of slag fragments in the river gravels diminish downstream, but we found some small pieces as far as 18 km below the dumps.

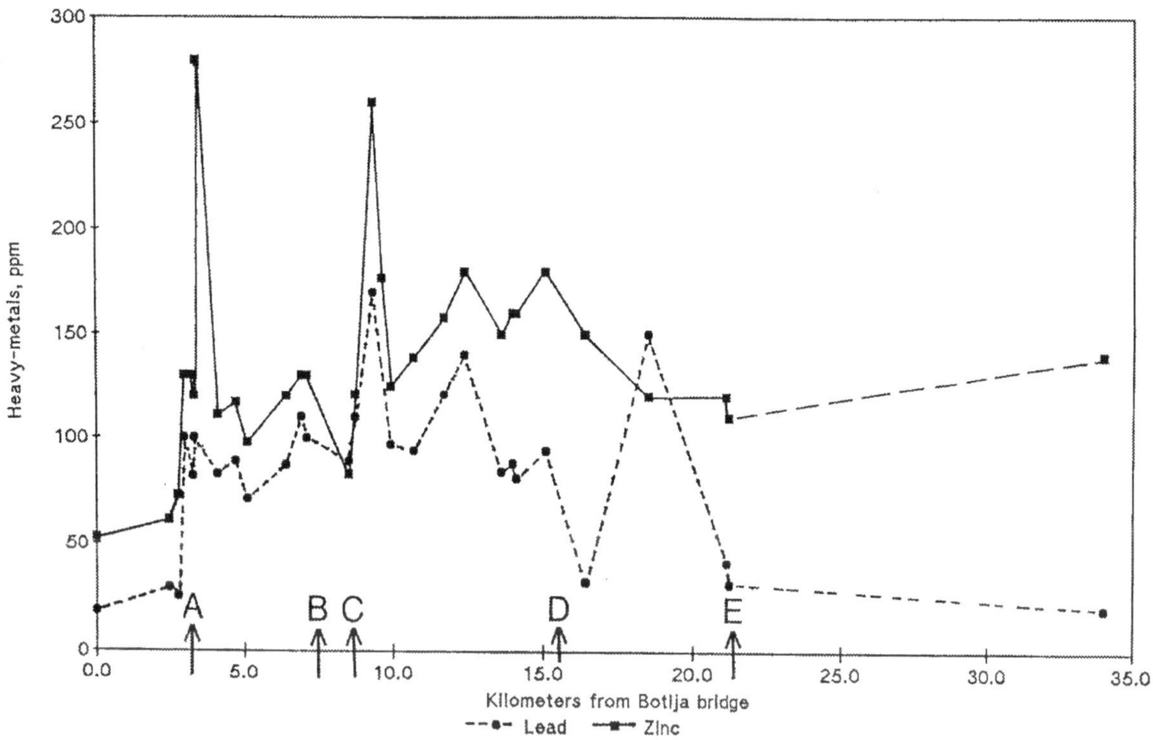

Figure 9. Toxic elements in alluvium of the Rio Tamuja as it crosses the Plasenzuela silver-lead mining district from south to north. Samples were taken from the highway bridge near Botija, upstream from the first mines, to the bridge at Sta. Marta, about 10 km below the district. Major landmarks and points of influx of mining and metallurgical waste are shown as follows: A. Group of Roman smelters along river back. B. Las Golondrinas mine dump complex, which may be predominantly Roman or entirely modern. C. La Sevillana and el Carmen mining complex, mostly modern wastes, but with a small Roman component. D. Entry of Arroyo de la Colmena draining several modern mines in its tributary headwaters. E. Trujillo-Caceres road bridge. The final zinc analysis upstream from Sta. Marta bridge seems anomalous, but has been confirmed with a separate sample.

Vegetation studies

Vegetation is another medium affected by both ancient and modern metallurgical wastes. Samples of Rumex acetosella (sheep sorrel), a forb tolerant of metal-rich soils, were collected and analyzed to assess its metal uptake, and maxima of 445 ppm lead and 1404 ppm zinc, dry weight, were found. The same species on uncontaminated soil contained 0.7 ppm lead and 31 ppm zinc. The plant was observed growing freely on very acidic soil that contains as much as 14,000 ppm Pb, and 800 ppm Zn, and normal-appearing Quercus ilex (encina or holm oak) were noted on soil with 13,000 ppm Pb, and 480 ppm Zn. Beyond this, the effects of high metal content on trees or small plants, such as the degree to which species may have adapted to this additional stress in an already harsh environment where times of ample rainfall are punctuated by long periods of drought, have not been examined.

Other problems noted in our studies

The processing methods employed by the Romans must have been separately designed for the near-surface oxidized ores and again for the sulfide ores obtained from deeper in the mines. It has not been possible to determine the depth of oxidation in these ore bodies, but in similarly situated mines in the region an average depth is perhaps 60 meters. Certainly the first ores mined were oxidized, but some Roman mines extended to twice the probable oxidation depth. Future excavation in the slag dumps may recover unused bits of ore, although the question will remain of how much small pieces of sulfide may oxidize when left close to the surface for two millennia. A lump of galena about 5 cm in diameter was found near a building foundation close to the smelters, but there is no way to know how or when it got there.

Ores may have been roasted to drive off much of the sulfur and simplify the smelting process, but no evidence of roasting has been observed in the district.

Attempts are being made to learn the nature, amounts and sources of the materials added to the furnaces as fluxes to form the slag. A few pieces of limestone found in the slags may have been part of the material added, but the real purpose is unsure and the source unknown since none is present in outcrop for several kilometres from the smelters. The increased level of iron in the slags may have resulted from the addition of an iron oxide flux material such as hematite, and a source for this material is also unknown, although the Montanchez Mountains to the south-west are a potential origin.

Conclusions

Studies of the Roman metallurgical industry at Plasenzuela, Spain, have shown that a variety of forms of metal contamination may persist for millennia. Significant amounts of lead, arsenic, zinc, and copper have been measured, not only in the mine waste and smelter slag dumps, but in soil trains leading downward from them, in vegetation growing on them, and in river alluvium for many kilometres downstream. It is probable that some Roman mines still drain acidic metal-bearing solutions into the Rio Tamuja. Anomalous quantities of lead in soils north and north-east of the smelters are interpreted as a relict of a smelter fallout plume.

The authors would welcome comments, suggestions, or relevant observations, especially from others working with similar problems in ancient metallurgy. Please contact: R. G. Schmidt, U.S. Geological Survey, MS 954, Reston, VA 20192, USA; e-mail: rschmidt@usgs.gov; fax: (703) 648-6383, or C. M. Ager, U.S. Geological Survey, DFC Box 25046, MS 973, Denver, CO 80225, USA; e-mail: cager@usgs.gov; fax: (303) 236-3200).

Acknowledgements

Research grants from the National Geographic Society Committee for Research and Exploration and from the U.S. Department of the Interior, Human Resources Initiative, made the project possible by supporting our three data collecting trips, and paying part of the analytical costs. The U.S. Geological Survey continued Ager's salary while she was in the field. Carmen Anton-Pacheco and Pablo Gumiel, of the Instituto Tecnológico GeoMinero de España, Madrid, helped us in the field and introduced us to local scientists who gave us much assistance and made the work much more enjoyable. Colleagues at the University of Extremadura, Enrique Cerrillo Martin, Jose Maria Fernandez Corrales, Dionisia Gomez-Amelia, and Susanne Schnabel were most generous with their advice and time, and to them we are exceedingly grateful.

References

ALLEN, J.C., 1970. Considerations on the antiquity of mining in the Iberian peninsula. Royal Anthropological Institute, Occasional Paper 27.

BAUMANN, ALBRECHT, 1984. Extreme heavy metal concentrations in sediments of the Oker - a river draining an old mining and smelting area in the Hartz Mountain, Germany. in J.O. Nriagu (ed.), *Environmental impacts of smelters*: 579-591. New York: John Wiley.

DOMERGUE, CLAUDE, 1987. Término municipal: Plasenzuela, in Catalogue des mines et des fonderies antiques de la Peninsule Iberique. Publications de la Casa de Velazquez, serie archeologie, Fac. VIII, T. 1: 53-57.

GUMIEL, P., R. CAMPOS, D.J. SANDERSON, & S. ROBERTS, 1995. Geometría y fractalidad de los sistemas filonianos de la mina de La Parilla (Cáceres): Conectividad y percolación. *Boletín Geológico y Minero* 106: 316-337.

HERNANDEZ, F., 1972. Excavaciónes en el castro de las Villasviejas del Tamuja (Botija, Cáceres). *Zephyrus* XXI-XXII: 321-330.

HONG, SINGMIN, J-P. CANDELONE, C.C. PATTERSON, & C.F. BOUTRON, 1994. Greenland ice evidence of hemispheric lead pollution two millennia ago by Greek and Roman civilizations. *Science* 265: 1841-1843.

INSTITUTO GEOLÓGICO Y MINERO DE ESPAÑA, 1978. Mapa Geológico de España, hoja no. 705, Trujillo (Cáceres). Madrid: Servicio de Publicaciones.

JOHNS, C., 1995. Contamination of riparian wetlands from past copper mining and smelting in the headwaters region of the Clark Fork River, Montana, USA. *Journal Geochemical Exploration* 52: 193-203.

LAGO RODRIGUEZ, A., R. CASTROVIEJO BOLIBAR, & T. NODAL RAMOS, 1989. Las mineralizaciones argentíferas de Plasenzuela, Cáceres, España. *Boletín Geológico y Minero* 100: 1059-1074.

LINÉS ESCARDÓ, A., 1970. The climate of the Iberian Peninsula, in Climates of Northern and Western Europe, in C. C. Wallén, (ed.), *World Survey of Climatology* 5: 195-226. Amsterdam: Elsevier.

ROSWAG, CLEMENTE, 1853. Apuntes sobre las minas de Palacios y Golondrinas, y refutacion del informe del señor ingeniero Letellier. Cáceres.

ROTHENBERG, BENO, FÉLIX GARCÍA PALOMERO, HANS-GERT BACHMAN, & J.W. GOETHE, 1989. The Rio Tinto enigma, In Minería y Metalurgia en las Antiguas Civilizaciónes Mediterraneas y Europeas: Coloquio International Asociado: 57-70. Madrid, 24-28 Octobre, 1985, II.

ROWAN, R.C., CARMEN ANTON-PACHECO, D.W. BRICKEY, M.J. KINGSTON, ALBA PAYÁS, NORMA VERGO, & J.K. CROWLEY, 1986. Digital classification of contact metamorphic rocks in Extremadura, Spain using Landsat thematic mapper data. *Geophysics* 52: 885-897.

SALKIELD, L.U., 1982. The Roman and pre-Roman slags at Rio Tinto, Spain, in T.A. Wertime & S.F. Wertime, (eds.), *The evolution of the first fire-using industries*. 137-147. Washington, D.C., Smithsonian Institution Press.

SHOTYK, W., D. WEISS, A.K. CHEBURKIN, M. GLOOR, R. FREI, & D.J. KRAMERS, 1997. A complete post-glacial record of the changing atmospheric fluxes and isotopic composition of lead recorded in a peat bog profile, Jura Mountains, Switzerland, [abs.], Metals in Antiquity International Symposium, Harvard University/Bradford University, Cambridge MA, 10-13 September, 1997.

31

Early historical iron production in the Netherlands: estimations of the output

INEKE JOOSTEN & HENK KARS

State Service for Archaeological Investigations in the Netherlands (ROB), PO Box 1600, NL-3800 BP, Amersfoort, the Netherlands

This study discusses two early historical, large scale, iron production sites, Heeten in the Vecht area (first half of the fourth century AD) and the Orderbos in the Veluwe area (9th century AD), in terms of iron output and impact of the production process on the environment. Iron at the Heeten site was produced in slag-pit furnaces from bog-iron ore and at the Orderbos site rattlestones were used in slag-tapping furnaces. Using a recently developed model (Joosten et al. *1997a) the ore:slag:iron ratios of the Heeten and Orderbos sites are calculated to be 1.4:1:0.2 and 1.8:1:0.4, respectively. The higher yield of the process at the Orderbos site is probably due to the use of a different type of furnace. The surplus production at the native settlement of Heeten could have been motivated by the historically known integration of the Salian Franks into the Roman Empire during the first half of the fourth century AD (Groenewoudt & Van Nie 1995). The iron from the Orderbos site must have been exported since traces of the re-heating of blooms in the Veluwe area are absent. The burning of charcoal for iron production at the Heeten site must have caused the deforestation of a large area around the settlement. The end of the production could have been the result of fuel shortage. The smelters at the Orderbos site probably made use of coppiced oak-wood to meet the demand for charcoal.*

Keywords: The Netherlands; Early Historical Iron Production; Supra-Regional Production; Environmental Impact; Archaeometry.

Introduction

In the Netherlands, within a relatively small area, three major early historical iron production areas have been recognized (Figure 1). In the Vecht area several settlements dating from the 1st to the 4th centuries AD with numerous traces of iron production are found. In the Early Middle Ages, from the 7th to the 9th centuries AD, iron was produced in the central part of the Netherlands, the Veluwe area. In the Montferland area iron production took place from the 9th to the 11th centuries AD (Figure 1).

The production areas differ not only in age but also in type of furnace and use of ore types. The iron in the Vecht area was produced from bog-iron ore in slag-pit furnaces; for the production in the Veluwe and Montferland areas rattlestones were used in slag-tapping furnaces. Bog-iron ore is locally developed in stream valleys while a certain type of iron oxide concretions, mineralogically similar to bog-iron ore, the so-called rattlestones, were quarried in the Veluwe and the Montferland areas. A recently developed model (Joosten *et al.* 1997a) allows quantification of the smelting activities in terms of the ore:slag:iron ratio. Characterization of the chemical composition of ores and slags makes it possible to study a genetic relationship between ores and slags, to calculate the ore:slag:iron ratios, and to establish to what extent the furnace lining and fuel ash contributed to the chemistry of the slag.

The model is used in this study to discuss two early historical, large scale, iron production sites, Heeten in the Vecht area and the Orderbos in the Veluwe area, in terms of iron output and impact of the production process on the environment.

The model

In the early historical iron production process iron oxide in the ore is reduced to metallic iron in the solid phase while the gangue minerals, mainly quartz (SiO_2), are separated from the iron and form the slag. Non-metallic elements from the ore are mainly concentrated in the slag phase, whereas metallic elements are collected in the iron phase. The chemical composition of the slag is mainly determined by the original chemical composition of the ore. However, the addition of fluxes and the contamination with fuel ash and furnace lining will also influence the composition of the slag.

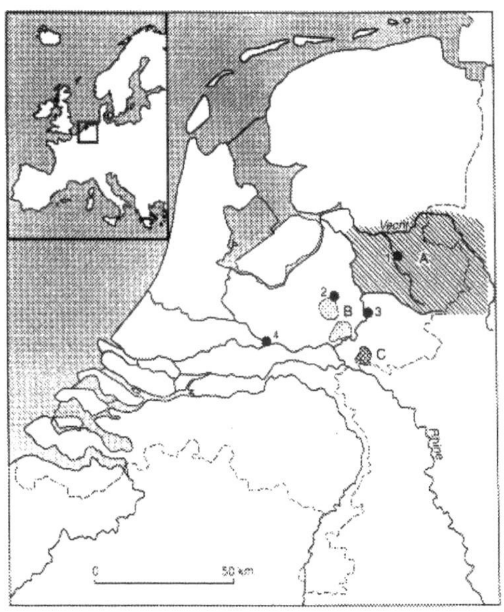

Figure 1. The Netherlands and locations of the major sites discussed in the text. A. Vecht area: B. Veluwe area: C. Montferland area: 1. Heeten: 2. Orderbos: 4. Dorestad.

To calculate the amount of ore needed to produce the slag, the slag:ore ratio has to be computed. Therefore, the chemical components that originate exclusively from the ore must be determined (Espelund 1995; Eschenlohr & Serneels 1991; Serning 1979; Tylecote *et al.* 1971). This means that a linear relationship between the components in the ore and slag through the origin is expected when plotted in a variation diagram. The ratio is calculated by dividing the sum of the chemical components that show such a relationship (X) of the slag by that of the ore (Joosten *et al.* 1997a).

$$\frac{(X)_{slag}}{(X)_{ore}} = \text{slag:ore ratio} \qquad (1)$$

The yield of the production process is defined by Tylecote *et al.* (1971) as the amount of produced iron divided by the amount of iron in the ore (Equation 2).

$$\frac{(FeO_{ore} \times ratio_{slag:ore}) - FeO_{slag}}{(FeO_{ore} \times ratio_{slag:ore})} = \text{yield} \qquad (2)$$

The chemical components in the ore and slag that do not show the linear relationship (Y) must originate either from the charcoal ash, the furnace lining, a flux, or from a combination of those. The contribution of these materials to the chemistry of the slag is reflected by its enrichment in Y with respect to the ore. The surplus of Y in the slag is calculated by subtracting the amount of Y the ore contributed to the slag from that in the slag (Joosten in press).

$$Y_{slag} - ratio_{slag:ore} \times Y_{ore} = \text{surplus } Y_{slag} \qquad (3)$$

The origin of every component of which Y consists can be investigated by plotting the components of ore, slag, furnace lining, charcoal ash and possible fluxes in a suitable variation diagram. The chemistry of the slags is determined by both the efficiency of the production process and the mixing of the ore and the other material(s). If it can be established that certain components of Y originate unambiguously from the charcoal ash, the charcoal to ore ratio can be calculated by dividing the surplus of these components by their amounts in the ash. It must be noted that the charcoal to ore ratio calculated by the model probably reflects only the amount of charcoal used during smelting. It does not account for the fuel used to pre-heat the furnace and to burn down the last charge. Moreover, it is possible that the slag would not have absorbed all of the ash available. Therefore, the calculated charcoal to ore ratio is assumed to describe the *minimum* charcoal consumption during smelting.

The model has been tested by experimental studies in which iron was smelted in reconstructions of slag-pit- and slag-tapping furnaces. The theoretical ore:slag:iron ratios predicted by the model are in good agreement with the measured ratios and yields of these studies (Joosten *et al.* 1997b). The calculated charcoal:ore ratio also corresponds with the ratio as charged. It must be noted however, that the range of ratios is large due to the variation in the chemical composition of the charcoal that was used (Joosten in press).

Analytical method

Samples from two sites were analyzed, Heeten in the Vecht area and the Orderbos in the Veluwe area. Thin sections were made of some of the slag samples and studied by transmitted and reflected light microscopy. Samples of rattlestones, bog-iron ore, slags and furnace lining were crushed in a tungsten carbide swing mill. The volatile fraction (LOI) was determined by means of weighing before and after drying the samples in a furnace at 1000°C for twelve hours. Major elements, Al, Ca, Fe, Mg, Mn, Na, P and Si were analyzed by X-ray fluorescence (XRF). International standards with high iron contents (Fer-3 and Fer-4) were analyzed together with the samples to check the accuracy of the analyses. The accuracy was better than 10% for all oxides. FeO was determined by titration of the samples with $K_2Cr_2O_7$ in the presence of an indicator (sodium diphenylamin sulfonate). The mineralogy of the ores is studied by means of X-ray diffraction (XRD).

The Heeten site: a native settlement in the Vecht area

From 12 BC to the fifth century AD the southern part of the Netherlands belonged to the Roman Empire (Van Es 1981). The Roman border was formed by the river Rhine (Figure 1). The earliest traces of iron production in the Netherlands date from this period. Along the river Vecht, in the area of *Germania Libera* which was inhabited by Frankish tribes, several native settlements with small scale iron production are found (Van Nie 1995). Small scale is defined as production that provided the needs for iron of only small regions, e.g., a few neighbouring settlements (Van Nie 1997). Recently however, the site of Heeten was excavated and showed traces of iron production on a large scale (Figure 1) (Verlinde & Erdrich in press). The expression 'large scale' refers to a supra-regional production in which a large amount of iron was produced (Groenewoudt & Van Nie 1995).

The Heeten site is situated on the very edge of a steep and visually prominent cover-sand ridge, by far the highest in the immediate surroundings. The ridge is surrounded by damp soils (Spek 1995). The settlement traces are covered with a so-called plaggen soil (Pape 1970). The site is dated between the late second to the fourth or fifth centuries AD, but the majority of the finds date to the first half of the fourth century AD. The settlement traces belonging to this period are found inside a large rectangular area of 110 by 75m enclosed by a palisade trench. The enclosed area contains the remains of several houses, some twenty sunken huts and a dozen granaries, some of them with exceptional heavy foundations, indicating that they had a large storage capacity.

The structure of the Heeten settlement is, apart from the enclosure, similar to other contemporaneous rural settlements in the Netherlands (Groenewoudt & Van Nie 1995). However, a huge amount of iron production slag indicates that its economy was mainly based on the production of iron. Several hundred large pieces of slagblocks and *in situ* remains of at least 17 slag-pit furnaces were found within a large area surrounding the enclosure. In the pit of one of the furnaces both production slag and bog-iron ore have been found. The amount of slag has been calculated to be at least 45 to 50 tons (Groenewoudt & Van Nie 1995). The figures are based upon a core sampling programme in the surroundings of

Bog-iron ores							Slag and furnace lining								
Sample	He110	Ra110	mrOoiA	mrOoiB	Barv8	mrH	HeIA	HeIB	He2A	He3A	He4A	HeG	HeH	HeJ	He54
wt.%															
SiO_2	28.7	20.0	14.1	17.1	16.1	10.9	22.8	26.1	17.9	17.6	17.2	15.0	51.2	18.1	78.8
TiO_2	0.1	0.1	0.1	0.1	0.1	0.03	0.2	0.2	0.1	0.1	0.1	0.1	0.6	0.2	0.5
Al_2O_3	2.2	1.8	0.9	1.6	1.4	1.4	2.1	2.1	1.1	1.2	1.1	2.0	9.9	3.0	8.5
FeO	10.3	<0.01	0.1	<0.01	0.2	0.2	58.4	52.8	62.8	62.1	64.0	56.1	21.2	60.7	0.4
Fe_2O_3	46.9	72.5	78.8	71.9	78.0	78.5	2.8	2.3	2.8	4.2	2.0	13.1	3.9	4.2	4.6
FeO_t	52.5	65.2	70.9	64.7	70.3	70.8	60.9	54.8	65.3	65.8	65.8	67.8	24.8	64.6	4.6
MnO	4.8	2.7	0.8	5.8	0.6	6.1	1.4	1.9	1.3	1.4	1.4	0.7	0.3	0.6	0.2
MgO	0.2	<0.02	0.2	0.1	<0.02	<0.02	0.2	0.2	0.2	0.2	0.2	0.3	0.7	0.4	0.6
CaO	1.3	<0.2	0.8	1.0	<0.2	1.1	2.2	3.3	2.3	2.3	2.3	3.0	1.5	3.3	0.4
Na_2O	<0.1	0.2	<0.1	0.2	<0.1	<0.1	<0.1	<0.1	<0.1	<0.1	<0.1	<0.1	0.2	<0.1	1.2
K_2O	0.6	0.4	0.2	0.2	0.3	0.3	0.4	0.6	0.3	0.3	0.3	<0.1	1.9	0.6	1.9
P_2O_5	1.5	1.7	3.6	3.1	3.1	1.0	2.3	2.7	2.9	3.0	3.0	2.3	1.1	2.1	0.5
Total	97.9	99.4	99.6	100.6	99.6	99.5	99.2	98.1	98.7	99.2	98.6	99.0	95.6	100.0	97.6
LOI	8.7	16.7	16.9	13.6	16.4	18.3	0.8	0.6	0.3	0.5	0.5	2.2	1.9	0.3	1.5

Table 1: Chemical composition of ores, slags and furnace lining from Heeten (Vecht area). The total given is from the XRF, i.e., sum of the dry oxides, with all the iron as Fe_2O_3. Legend: Bog-iron ores: He110: burnt ore Heeten; Ra110: unburnt ore Heeten; mrOoiA-B: Ooyerhoek; Barv8: Barvoorde, mrH: Hengelo, Slags and furnace lining: HeIA-B, 2A and 3A: loose slag from furnace pit2; He4A: piece of the slagblock from furnace pit2; HeG, HeH and HeI: pieces of slagblock18; He54: furnace lining; LOI: volatiles.

the excavation. The great majority of the slag consists of production slag. Only a few re-heating and smithing slags were found. The ore and the building material for the furnaces (clay) were probably dug in the marshy areas in the immediate vicinity of the settlement. At least two charcoal burning pits, so-called *Grubenmeilers*, with charcoal made from Ash (*Fraxinus*) and Alder (*Alnus*), were unearthed, and 9 other charcoal burning pits have been identified. The wood is assumed to have been cut in the surroundings of the settlement.

The following material was collected from the Heeten site: bog-iron ore (e.g., from furnace pit 1), loose slag and a slagblock from furnace pit 2, a complete slagblock (18) and a piece of furnace lining. To gain a greater insight into the chemistry of bog-iron ore, samples were obtained from Barvoorde, Hengelo (Prov. of Overijssel) and Ooyerhoek (Prov. of Gelderland).

Results

The bog-iron ore from Heeten consists of loosely packed brown grains. One of the ore-samples from Heeten, He110, has a deep red colour and is magnetic. The ore from Ooyerhoek, Barvoorde and Hengelo contain reddish brown to shiny black grains. The main constituent of the bog-iron ores is the mineral goethite, an iron hydroxide with the formula FeOOH. The slags from Heeten have a homogeneous texture of large fayalite crystals with a few specks of metallic iron and fine-grained dendritic and globular wüstite in a glass matrix. Pieces of furnace lining are incorporated in slag samples HeG, H and J. The furnace lining consists of clay tempered with organic material, sand and small pebbles (Groenewoudt & Van Nie 1995). The results of the chemical analysis of ores and slags are listed in Table 1. The bog-iron ores contain up to 79 wt.% Fe_2O_3. The MnO content varies from 0.6 to 6.1 wt.%. The CaO content is mainly below 1.3 wt.%. SiO_2

and Al_2O_3 are correlated in the ores and slags and range from 11 to 26 and 0.9 to 3 wt.%, respectively. The slags contain up to 68 wt.% FeO (all iron calculated as FeO), up to 1.9 wt.% MnO and up to 3.3 wt.% CaO.

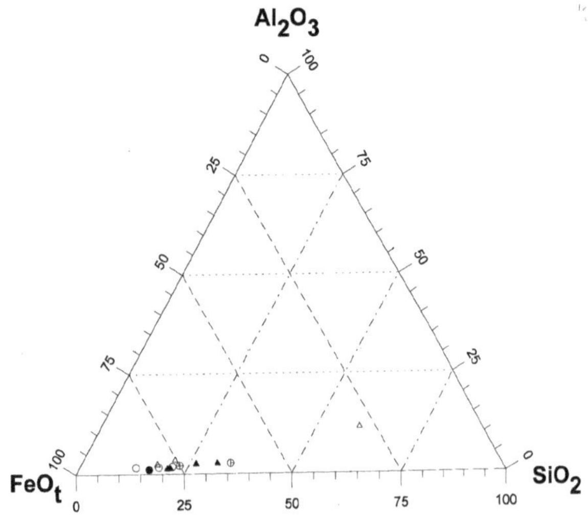

Figure 2: Vecht area: Heeten. Triangular variation diagram of FeO_t, Al_2O_3 and SiO_2. Cross in circle: bog-iron ore Heeten; filled circle: high-grade bog-iron ore Ooyerhoek; open circle: bog-iron ore Ooyerhoek, Hengelo and Barvoorde; filled triangle: slag from furnace pit 2; open triangle: slag block 18.

Yield of the production process

The percentage of iron in the slags is still quite high, as could be expected from the mineralogical composition of the main component (fayalite contains 70 wt.% FeO). The iron content of the bog-iron ore found at Heeten is, however, lower than that of the slags (Table 1). Because of the low iron content of the ore the analyzed pieces cannot be characteristic for the ore used for the production. The ore must have been dressed carefully by hand-picking and washing, leaving less iron-rich pieces aside. Some of the iron in ore He110 was reduced to FeO (Table 1). Since the ore was found in a furnace pit together with slag it probably was charged in a furnace. The reduction process, however, must have been stopped before all the ore had reacted.

The oxides FeO_t, Al_2O_3 and SiO_2 have been normalized and plotted in a triangular diagram. Both Al_2O_3 and SiO_2 enter the slag phase, whereas FeO_t enters the slag as well as the metallic iron. Therefore, related slags and ore should plot on a line originating from the FeO_t-corner, with the ores plotting nearest to the corner. The bog-iron ores and the slags plot approximately on the same line (Figure 2). The high-grade ore from the Ooyerhoek and the ore from Barvoorde (Table 1: mrOoiA and Barv8) seem to be the most promising candidates for a representative ore analysis because of their high iron content and ratio of gangue elements comparable to that of the slags. The contribution of the furnace lining to the chemical composition of the samples from slagblock18 is clearly shown in Figures 2 and 3.

The slags are enriched in CaO compared with the ore. Also the elements K, Mg and P do not show a relationship between the ores and the slags passing through the origin. Therefore only SiO_2 and Al_2O_3 can be used to calculate the slag:ore ratio (Figure 3). The ratio of SiO_2 and Al_2O_3 of the high-grade ore from the Ooyerhoek seems to be the most comparable to that of the slags. The exact composition of the ore that was used is unfortunately not known and the ratio will only be an estimate. However, the ratio of SiO_2 and Al_2O_3 is reasonably constant for all the examined bog-iron ores, therefore the calculated slag:ore ratio can be used. Slags that are contaminated with furnace lining are not included in the calculation of the ore:slag:iron ratio since the clay consists of about 90% of SiO_2 and Al_2O_3. If the volatile fraction of the ores and slags (mainly water) is also taken into account then the slag:ore ratio describes the amount of ore used to produce one unit of slag. The most conservative ore:slag:iron ratio of the Heeten site, calculated with the most iron-rich slag, He4A, is 1.4:1:0.2 (by weight). The difference in weight between ore, and slag and iron is caused both by the volatile content of the ore and the difference in weight between metallic iron and Fe_2O_3. The yield of the process is calculated to be 24%. Using equations 1 and 2 it is calculated that the amount of slag present in Heeten represents at least 9 to 10 tons of metallic iron.

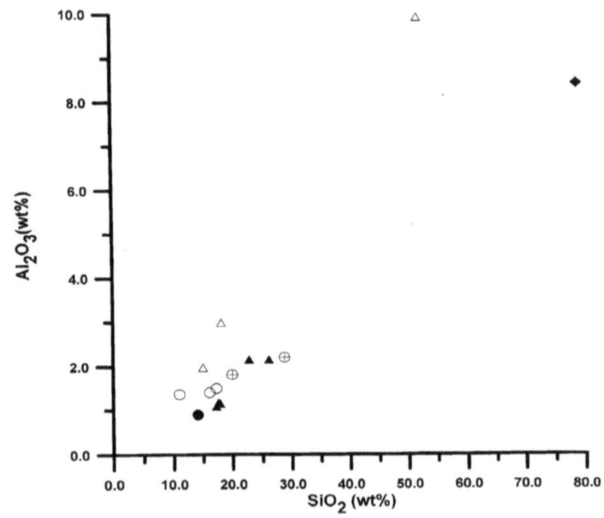

Figure 3: Vecht area: Heeten. Chemical relationship between ores and slags: Al_2O_3 vs. SiO_2. Cross in circle: bog-iron ore Heeten; filled circle: high-grade bog-iron ore Ooyerhoek; open circle: bog-iron ore Ooyerhoek, Hengelo and Barvoorde; filled triangle: slag from furnace pit 2; open triangle: slag block 18. Filled diamond: furnace lining.

Charcoal consumption

The oxides SiO_2, Al_2O_3 and CaO listed in Table 1 have been normalized and plotted in a triangular variation diagram. Related ores and slags should plot in the same area (all five oxides enter the slag phase) unless a surplus amount of the oxides was added to the slag by another source such as furnace lining or charcoal ash. The slags are displaced from the ores towards the CaO corner (Figure 4). The CaO presumably originates from the charcoal, which can contain 0.26 wt.% of CaO (Table 3). However, the CaO content of bog-iron ore varies considerably (Table 1). The variation introduces an unacceptable large error in the calculation of the CaO-surplus of the slag (equation 3). Charcoal contains also a considerable amount of K_2O. The ratio between K_2O and CaO in the slags and charcoal could indicate the amount of charcoal that was absorbed. The K_2O-surplus in the slags, however, is within the analytical error, except that in those that are contaminated with furnace lining. Therefore, it must be concluded that in the case of the Heeten site the calculation of the charcoal:ore ratio by the chemistry of ore and slag is not possible.

The Orderbos site: an Early Medieval slagheap in the Veluwe area

On two of the ridges that consist of sediment pressed up by an advancing glacier, so-called ice-pushed ridges, in the central part of the Netherlands the occurrence of early iron production is evidenced by enormous slagheaps and several rows of open-cast mining pits. The earliest evidence for iron production on the Veluwe is found at Hog Buurlo where, in the 7th century AD, blooms have been worked on a considerable scale (Heidinga 1987). The slagheaps are dated to the 9th century AD. The iron ore, so-called rattlestones, is present in certain strata in the ice-pushed ridges. These rattlestones are ellipsoidal envelopes of iron hydroxide containing a loose-fitting core of clay or silt, that makes a rattling noise when shaken. The visible remains of mining activities consist of rows of pits with a total length of over 82 kilometres from which the rattlestones have been extracted. The pits are up to four metres deep. So far, the presence of over seventy slagheaps could be established in the Veluwe area (Van Nie 1995). It is historically known that many heaps once present were removed for metalling roads and used as ore in the 19th century blast furnaces, for example in nearby Deventer (Van Nie 1997). The tapslag demonstrates that furnaces with a slag-tapping device were used. Charcoal must have been burnt on an enormous scale but only one *Grubenmeiler* has been discovered so far.

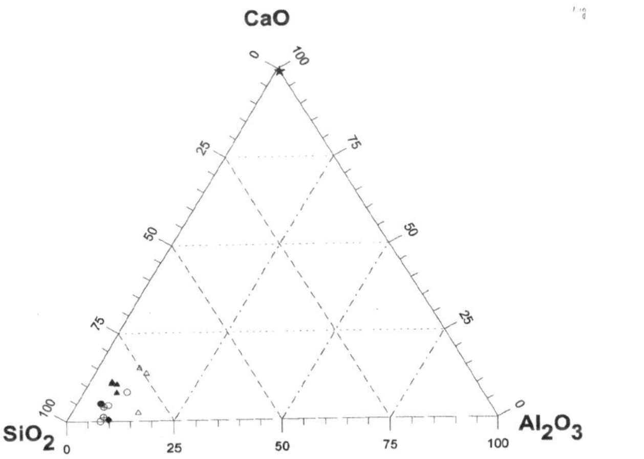

Figure 4: Vecht area: Heeten. Triangular variation diagram of SiO_2, Al_2O_3 and CaO. Cross in circle: bog-iron ore Heeten; filled circle: high-grade bog-iron ore Ooyerhoek; open circle: bog-iron ore Ooyerhoek, Hengelo and Barvoorde; filled triangle: slag from furnace pit 2; open triangle: slag block 18. Filled star: charcoal.

The largest slagheap of the Veluwe area (Orderbos; Figure 1) consists of at least four densely packed smaller heaps and the minimum amount of slag present is estimated to be about 650 tons (Van Nie 1990). The heap is mainly composed of tapslag, but also charcoal, furnace remains and pieces of bloom are found. The charcoal is exclusively derived from oak, *Quercus* sp. (Musch 1991). Also the hearths of three furnaces, partly projected on top of each other, were excavated.

The following material was collected from the Orderbos site: rattlestone ore, slag (tapslag and furnace bottom, i.e., slag that has accumulated at the bottom of the smelting furnace) and furnace lining. Samples of rattlestone ore are also derived from open-cast mining pits (Asselse Veld) situated approximately 3500m from the smelting site.

Results

The size of the ore from the Veluwe, the rattlestones, usually varies between 5 and 20 cm. Rattlestones consist mainly of the mineral lepidocrocite, an iron hydroxide with the formula -FeOOH. The rattlestone from the slagheap (Table 2: Or6klI) has a reddish purple colour and is magnetic. The structure and mineralogy of the tap slags of the Orderbos is quite similar to the slags from Heeten apart from the more acicular and smaller fayalite crystals. The tap slags also show ribbon-like zones of pure wüstite that are assumed to represent the former surface of the slag. Fayalite crystals in the furnace bottom are more equidimensional in shape. The furnace bottom contains sand or silt. The bottom of tap slags OR5 and OR6 also contains sandy material. The furnace lining consists of subangular quartz grains embedded in a clay matrix.

The results of the chemical analysis of ores and slags from the Orderbos are listed in Table 2. The rattlestones contain up to 88 wt.% Fe_2O_3. The samples from the Asselse Veld contain less MnO than the rattlestone from the slag heap in the Orderbos, respectively 1.7-1.9 and 2.6 wt.%. From the ore to the slags SiO_2 and Al_2O_3 are clearly correlated and range from 8 to 25 wt.% and 2 to 4.5 wt.%, respectively. The furnace bottom and two slags (OR5 and 6) are enriched with SiO_2 when compared to the ore. CaO is present in trace amounts in the ore, mainly below 0.2 wt.%, whereas most tap slags contain up to 1.0 wt.% CaO.

Yield of the production process

The iron content of the rattlestones is only somewhat higher than that of the tap slag from the Orderbos (Figure 5). Traces of selection and ore dressing, consisting of a thin layer of crushed rattlestones mostly with adhering gravel and coarse sand and cores of rattlestones, have been found adjacent to one of the open-cast mining pits

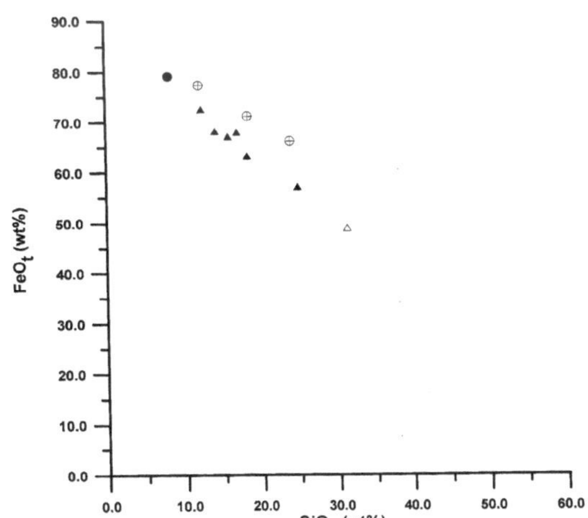

Figure 5: Veluwe area: Orderbos. Chemical relationship between ores and slags: FeO vs. SiO_2. Filled circle: rattlestone found in the slagheap of the Orderbos; cross in circle: rattlestones Asselse Veld; filled triangle: tap slag; open triangle: furnace bottom

	Rattlestones				Slags and furnace lining							
Sample	OR6kl	kl3B	kl2C	kl4B	OR5	OR6	OR8	OR21	OR24	OR25	OR10	OR1222
wt.%												
SiO_2	8.1	12.0	23.9	18.3	24.8	16.8	18.2	12.3	15.7	14.0	31.2	78.4
TiO_2	0.4	0.1	0.1	0.1	0.2	0.2	0.2	0.1	0.2	0.1	0.3	0.6
Al_2O_3	2.0	2.5	2.5	2.4	4.5	3.4	4.6	3.8	3.8	3.7	5.9	8.0
FeO	<0.01	<0.01	<0.01	<0.01	51.0	56.6	53.8	64.3	62.7	60.6	39.0	1.7
Fe_2O_3	87.9	85.9	73.4	79.1	6.5	12.3	10.2	9.0	4.6	8.1	10.7	5.3
FeO_t	79.1	77.3	66.0	71.1	56.8	67.7	62.9	72.3	66.8	67.9	48.6	6.4
MnO	2.6	1.8	1.9	1.7	4.4	4.2	5.5	4.8	4.9	5.2	5.6	0.1
MgO	0.3	0. 2	0.2	0.2	0.4	0.3	0.4	0.3	0.4	0.4	0.5	0.6
CaO	<0.2	<0.2	<0.2	<0.2	0.9	0.4	0.8	0.2	0.7	0.8	1.0	0.2
Na_2O	<0.1	<0.1	<0.1	<0.1	<0.1	<0.1	<0.1	<0.1	<0.1	<0.1	<0.1	0.4
K_2O	0.3	0.4	0.4	0.4	0.9	0.3	0.4	0.7	1.0	1.0	1.0	1.7
P_2O_5	0.4	0.6	0.6	0.6	0.8	0.8	0.8	0.7	0.8	0.8	0.7	0.1
Total	102.0	103.5	102.9	102.7	100.1	101.6	101	103.1	101.7	101.4	100.2	97.3
LOI	11.1	13.0	11.4	12.2	0.5	0.3	0.7	0.7	0.6	0.5	0.3	0.9

Table 2: Chemical composition of ores, slags and furnace lining from the Orderbos and ores from the Asselse Veld (Veluwe area). The totals given are from the XRF, i.e., sum of the dry oxides, with all the iron as Fe_2O_3. Legend: Rattlestones: OR6kl: magnetic piece of burnt rattlestone from the Orderbos; kl3B-4B: Asselse Veld; Slags and furnace lining: OR5-6, OR8, OR21, OR24-25: tap slag; OR10: furnace bottom; OR12: furnace lining; LOI: volatiles.

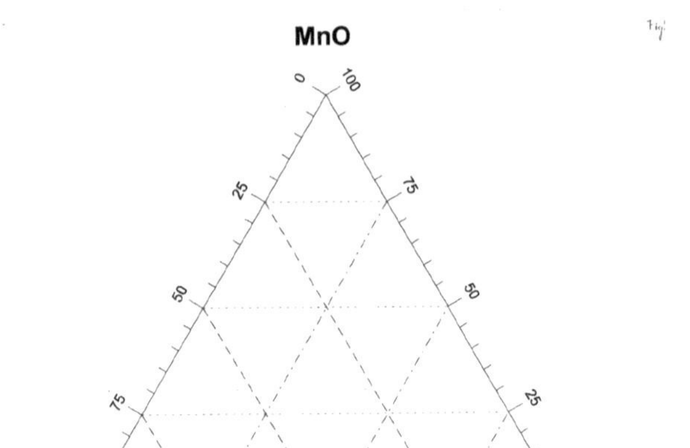

Figure 6: Veluwe area: Orderbos. Triangular variation diagram of FeO_t, SiO_2 and MnO. Filled circle: rattlestone found in the slagheap of the Orderbos; cross in circle: rattlestones Asselse Veld; filled triangle: tap slag; open triangle: furnace bottom.

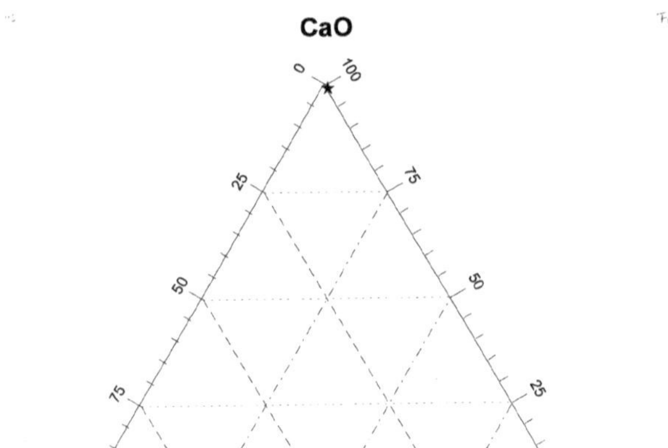

Figure 7: Veluwe area: Orderbos. Triangular variation diagram of SiO_2, Al_2O_3, and CaO. Filled circle: rattlestone found in the slagheap of the Orderbos; cross in circle: rattlestones Asselse Veld; filled triangle: tap slag; open triangle: furnace bottom. Filled diamond: furnace lining, filled star: charcoal.

(Van Nie 1990). The chemistry of rattlestone Or6kl (Table 2), found in the slagheap, seems to be the most representative for the ore parental to the slags (Figure 5). The three samples from the Asselse Veld contain on average too little MnO to be related with the Orderbos slags.

The slag:ore ratio can be calculated with SiO_2, Al_2O_3 and MnO (Figure 5 and 6). The most conservative ore:slag:iron ratio, calculated with rattlestone Or6kl and the most iron-rich slag, OR21, is 1.8:1:0.4. The yield of the process is calculated to be 44%. The slagheap in the Orderbos is calculated to represent at least 260 tons of metallic iron.

Sample	Alder	Oak
wt.%		
SiO_2	N.A.	N.A.
Al_2O_3	0.0021	0.0015
FeO	0.0025	0.001
MnO	0.0061	0.0018
MgO	0.065	0.108
CaO	0.258	0.192
Na_2O	0.0026	0.012
K_2O	0.178	0.214
TiO_2	<0.0005	<0.0005
P_2O_5	0.0765	0.076

Table 3: Chemical composition of charcoal. The charcoal samples are analyzed with AAS by I. Devine. Legend: Alder: charcoal of alder; Oak: charcoal of oak; N.A.: not analyzed.

Charcoal consumption

The slags are enriched in CaO compared to the ore (Figure 7). The CaO is not derived from the furnace lining since this plots near the SiO_2-corner. Some slags are also enriched in MgO (OR10) and K_2O (OR24, 25 and 10) compared to the ore. Probably, Ca, Mg and K originates from the charcoal that was used for the production (Table 3; oak). The most conservative charcoal to ore ratio, calculated with OR21 and charcoal of oak, is 1.3:1. The figure depends largely on the amount of CaO in the charcoal. However, the variation in the chemical composition of the charcoal of oak is not known. Large variations have been recorded in the CaO content of the charcoal of birch and beech (Joosten in press). The chemistry of charcoal depends not only on the type of tree but also on the part of the tree it is made from and on the type of soil the tree has grown on (Fengel & Wegener 1989). Therefore, the chemical composition of charcoal made from different types and parts of trees is currently being studied, taking the kind of soil the trees were grown on into account.

Thermodynamics of the production process

The calculation of the charcoal to ore ratio is difficult due to variations in the chemical composition of ore and charcoal, therefore the problem is also approached thermodynamically.

Stoichiometrically the reduction of 1 mole Fe_2O_3 to 2 moles of metallic iron consumes 3 moles C, and the reduction of 1 mole Fe_2O_3 to 2 moles of FeO consumes another mole C. Using the ore:slag:iron ratio determined for the Orderbos site, 1.8:1:0.4, it can be calculated that the reduction of 1 kilogram of ore to 0.38 kg FeO (fayalitic slag) and 0.22 kg metallic iron requires about 100 grams of carbon (charcoal). Theoretically a fuel:ore ratio of 1:10 should be sufficient to reduce the iron oxide. However, experiments with reconstruction of slag-pit and slag-tapping furnaces favour a fuel to ore ratio (by weight) of at least 1:1 (Joosten & Van Nie 1996; Dörfler 1995; Crew 1991). This means that also the composition of the atmosphere in and the heat balance of the furnace have to be considered carefully.

The complete reduction of the ore is affected by the residence time in the reducing zone of the furnace and by the partial pressure of CO (pCO). The residence time is, among others, determined by the size of the ore and the amount and size of the charcoal. A low fuel to ore ratio will most likely shorten the residence time of the ore and hence lower the yield of the process. The pCO in the furnace that is required to reduce the iron oxide to metallic iron is dependent on the temperature and the pressure in the furnace. From the equilibrium diagram of Schürmann (1958) it can be derived that the pCO has to be at least about 58% for a total pressure of 0.4 atmosphere. During the production process the temperature in the furnace has to be sufficient high to liquefy the slag. Also heat losses through the furnace wall, escaping gas and the removal of a slag with a temperature of 1200 C from the furnace have to be taken into account. The temperature in the furnace is established by the strongly exothermic oxidation reactions of C to CO and CO_2 by gaseous oxygen. However, the oxidation of C to CO, the reducing agent of the iron oxide, liberates only 30% of the total heat value of the carbon (Eketorp 1972). Most probably the furnace also has to be heated by the oxidation of CO to CO_2 by oxygen. This implies that a fuel to ore ratio of 1:10 is certainly not enough to both heat the furnace sufficiently and reduce the iron oxide in the ore.

Taken the above stated arguments into account then the fuel to ore ratio during smelting is assumed to be 1:1 (by weight). This figure has also been used by Voss (1995) and Bielenin (1995) to calculate the charcoal consumption of slag-pit furnaces in, respectively, Denmark and Poland.

Using a charcoal to ore ratio of 1:1 it is estimated that the amount of slag established in the Heeten site represents a consumption of at least 63 to 70 tons of charcoal. The production of the slagheap of the Orderbos site in the Veluwe area would have consumed an estimated amount of at least 1170 tons of charcoal.

Discussion

The yield of the iron production at Heeten is lower than that at the Orderbos site. A comparison of Heeten with other iron production areas using the same technology, like the Holy Cross mountains in Poland (Bielenin 1995) and Snorup in Denmark (Voss 1995), shows that the yield of the Heeten site corresponds with the yield of the other sites, i.e., 30%.The yield of the Orderbos is in good agreement with that of other production areas using slag-tapping furnaces, like Sunnanäng in the Dalarna region in Sweden (Serning 1979).The difference in yield between the two sites is thought to be mainly the effect of the use of a different type of furnace.

At both sites only evidence of the primary production of iron is found and traces for the re-heating of blooms are absent. Some four kilometres from Heeten a contemporaneous site, Wesepe, with a large amount of re-heating slag was found. Probably, at least part of the blooms that were produced at Heeten were worked into bar iron at Wesepe (Groenewoudt & Van Nie 1995). The spatial separation of primary production and re-heating perhaps could have enforced a better control of the production. However, in the different stages of production, the primary production consumes the most fuel, therefore, also the availability of wood could have been of importance. Heeten was probably located in one of the few forested areas in the surroundings and, therefore, the obvious place to smelt iron. Re-heating and forging the bloom requires a relatively small amount of fuel and could have occurred at places with smaller stretches of wood. This way the available economic resources were used as economically as possible (Groenewoudt & Van Nie 1995).

In the Veluwe area nearly no traces of re-heating activities have been found. Only one settlement (the Braamberg near Hog Buurlo, dating from the early 7th century AD), revealed re-heating slag. The settlement is interpreted as a special purpose site where blooms were worked (Van Nie 1997). However, the site was only a short period in use, maybe only one generation, after which it was deserted as a result of sand-drifts (Heidinga 1987).

The timespan of production in Heeten has been estimated at 30 to 35 years at most (Groenewoudt & Van Nie 1995). This gives at least 0.3 tons of raw iron a year, an amount which exceeds by far the annual need for iron in the rural settlements in the region, i.e., about 2 kg of iron tools (van Nie forthcoming). Other settlements, in the same region and dating to the same period, seem to have been self-supporting in their need for iron (Joosten & Van Nie 1993). This means that there must have been another reason for the surplus production. Groenewoudt & Van Nie (1995) have postulated that the large scale production seems to be motivated by an increasing demand for military and other equipment needed by the Salian Franks for their historically known integration into the Roman Empire during the late third and first half of the fourth centuries AD.

The timespan of the production at the Orderbos site is assumed to be 150 years (Van Nie 1997). This gives at least a production of 1.7 tons of iron a year. The absence of re-heating slag in the Veluwe area indicates that the iron must have been exported. The important *port of trade* Dorestad (Kars 1984; Van Es & Verwers 1980) revealed some tap slag and possibly re-heating slag. The chemical composition of the tap slag found at Dorestad is similar to that of the tap slag from the Veluwe. Possibly, raw iron produced in the Veluwe area, was worked at Dorestad. The significance of the iron trade is further stressed by the high amount of imported ceramics found in the village of Kootwijk (the Veluwe area, 8th-10th centuries AD; Heidinga 1987).

The environmental impact of the burning of charcoal for the production of iron is for both sites quite different. The valley surrounding Heeten was most probably too small to have supplied enough wood, so a much larger area must have been used. The iron production must have caused the deforestation of a large area around Heeten by the middle of the fourth century AD. This suggests that the end of the production in Heeten could have been a result of fuel shortage.

The smelters in the Orderbos most probably used coppice oak-wood to cope with the demand for charcoal. The amount of charcoal that could have been produced per hectare of coppice forest can be estimated to be 10 ton +/- 50% per hectare (Voss 1995). This estimation is partly based upon accounts written around 1800 for the blast furnaces in Port-Brillet, France; they got 12 tons of charcoal from each hectare of coppiced oak-wood. However, after a few years of chopping wood for the charcoal production, it theoretically gets exhausted. This means that there probably must have been a shift in production from one location to another. The fact that the slagheap in the Orderbos consists of four smaller ones might indicate that from time to time the smelters revisited locations used earlier. Assuming a time span for the Orderbos slagheap of 150 years, the heap would have been revisited every 20 years (Van Nie 1997). This would give the coppice wood enough time to regenerate. The end of the production in the Veluwe area could be caused by a diminished concern of the area as a result of feudal changes and the beginning of iron production along branches of the river Rhine (Van Nie 1997).

Acknowledgements

Grateful thanks are expressed to: Ian Devine for the AAS analyses of charcoal, which were carried out as part of the Plas Tan y Bwlch experimental iron project (Wales). I should like to acknowledge Matthijs van Nie for his perceptive comments on this paper. The project was granted by the Dutch Organization for Scientific Research (NWO).

References

BIELENIN, K. 1995. Frühgeschichtliches Eisenhüttenwesen im Heilig-Kreuz-Gebirge (Góry Swietokrzyskie) und seine Beziehungen zu Nowoklinowo in der Transkarpatischen Ukraine. *_tudijné Zvesti Archeologického Ústavu sav* 31: 89-98.

CREW, P. 1991. The Experimental Production of Prehistoric Bar Iron, *Historical Metallurgy* 25.1: 21-36.

DÖRFLER, W. 1995. Versuch einer Modellierung des Energieflusses und des Rohstoffverbrauchs während der römisch-kaiserzeitlichen Eisenverhüttung in Joldelund, Ldkr. Nordfriesland, *Probleme der Küstenforschung im südlichen Nordseegebiet* 23: 175-185. Oldenburg.

EKETORP, S. 1972. Fundamental basis for iron- and steelmaking processes. *Journal of the Iron and Steel Institute*, January 1972: 1-9.

VAN ES, W.A. 1981. *De Romeinen in Nederland*. 2nd rev. ed., Haarlem: Fibula-Van Dishoek.

VAN ES, W.A. & W.J.H. VERWERS. 1980. Excavations at Dorestad I. The Harbour: Hoogstraat I, *Nederlandse Oudheden* 9. Amersfoort: ROB.

ESCHENLOHR, L. & V. SERNEELS. 1991. *Les bas fourneaux mérovingiens de Boécourt, les boulies (JU/Suisse)*. Cahier d'árchéologie jurassienne 3. Porrentruy: Office du patrimoine historique et Société jurassienne d'Emulation.

ESPELUND, A. 1995. *Iron Production in Norway during Two Millennia. From the ancient bloomery to the early use of electric power*. Trondheim: Arketype.

FENGEL, D. & G. WEGENER. 1989. *Wood. Chemistry, Ultrastructure, Reactions*. Berlin-New York: Walter de Gruyter.

GROENEWOUDT, B.J. & M. VAN NIE. 1995. Assessing scale and organisation of Germanic Iron Production at Heeten, the Netherlands, *Journal of European Archaeology* 3: 187-215.

HEIDINGA, H.A. 1987. *Medieval Settlement and Economy North of the Lower Rhine, Archaeology and history of Kootwijk and the Veluwe (the Netherlands)*. Assen.

JOOSTEN, I. In press. Early iron production: calculation of the charcoal:ore ratio, in P. Crew and S. Crew (eds). *Early Ironworking in Europe. Archaeology and Experiment. Volume 2: Experimental and Technical studies*. Plas Tan y Bwlch.

JOOSTEN, I. & M. VAN NIE. 1993. Inheems-romeinse ijzerproduktie. Een proefonderzoek te Ooster-Dalfsen. *Overijsselse Historische Bijdragen* 108: 145-147.

1996. Experiments with Different Types of Dutch Iron Smelting Furnaces, in G. Magnusson (ed.), *The importance of Ironmaking. Technical Innovation and Social Change, papers presented at the Norberg Conference on May 8-13, 1995* Volume 2: 87-94. Stockholm: Jernkontoret.

JOOSTEN, I., J. B. H. JANSEN & H. KARS. 1997a. The use of a model to describe bloomery production in the Netherlands, in L.C. Nørbach (ed.), *Early Iron Production- Archaeology, Technology and Experiments. Technical Reports*. Nr. 3. 1997: 37-46. Lejre: Historical-Archaeological Experimental Centre.

JOOSTEN, I., M. VAN NIE & P. RIJK. 1997b. Experiment with a slag-tapping furnace at the Historisk Arkæologisk Forsøgscenter, Lejre, in L.C. Nørbach (ed.), *Early Iron Production- Archaeology, Technology and Experiments. Technical Reports*. Nr. 3. 1997: 81-92. Lejre: Historical-Archaeological Experimental Centre.

KARS, H. 1984. Early-Medieval Dorestad, an Archaeo-Petrological Study. Heerhugowaard: Casparie.

VAN NIE, M. 1990. *IJzerkuilen en slakkenhopen op de Veluwe, aanzet tot nader onderzoek [Ore extraction and Iron production on the Veluwe. Introduction to further research], unpublished MA thesis, University of Amsterdam.*

1995. Three Iron-producing Areas in the Netherlands: Contrasts and Similarities, in G. Magnusson (ed.), *The Importance of Ironmaking. Technical Innovation and Social Change, papers presented at the Norberg Conference on May 8-13, 1995* Volume 1: 100-106.

1997. Early Medieval Iron Production and its Organisation in the Veluwe Area, the Netherlands. in G. De Boe & F. Verhaeghe (eds), *Material Culture in Medieval Europe. Papers of the 'Medieval Europe Brugge 1997'* Conference. Volume 7: 33-41. Zellik: I.A.P. Rapporten 7.

Forthcoming. Early historical iron production in the Netherlands. An integrated archaeological/technical study. The archaeologically based results (provisional title).

MUSCH, J. 1991. Houtskool van de Veluwe: een onderzoek naar de economische exploitatie van de natuurlijke rijkdommen van de 'woestijn' van Nederland in de Middeleeuwen [Charcoal from the Veluwe: an investigation of the economical exploitation of the natural resources of the 'desert' of the Netherlands in the Middle Ages], *unpublished B.A. thesis, University of Amsterdam.*

PAPE, J.C. 1970. Plaggen soils in the Netherlands, *Geoderma* 4: 229-255.

SERNING, I. 1979. Prehistoric Iron Production, in H. Clarke (ed.), *Iron and Man in Prehistoric Sweden*: 50-89. Stockholm: Jernkontoret.

SCHÜRMANN, E. 1958. Die reduktion des Eisens im Rennfeuer. *Stahl und Eisen* 78, no. 19: 1297-1308.

SOUCHOPOVA, V & K. STRANSKY. 1989. Experimental iron smelting in the Blansko region, Czechoslovakia. *Archeomaterials* 3: 149-161.

SPEK, T. 1995. Het natuurlandschap van Salland. Ontstaan en opbouw van relief, bodem en vegetatie van Westelijk Overijssel, in: *Geschiedenis van Salland*. Kampen.

TYLECOTE, R.F., J.N. AUSTEN, & A.E. WRAITH. 1971. The Mechanisms of the Bloomery Process in Shaft Furnaces. *Journal of the Iron and Steel Institute* 209: 342-363.

VERLINDE, A.D. & M. ERDRICH. in press. Eine germanische Siedlung der späten Kaiserzeit mit umwehrter Anlage und umfangreicher Eisenindustrie in Heeten, Provinz Overijssel, Niederlande. *Germania*.

VOSS, O. 1995. Snorup- An Iron producing Settlement in West Jutland, 1st- 7th Century AD, in G. Magnusson (ed.), *The Importance of Ironmaking. Technical Innovation and Social Change, papers presented at the Norberg Conference on May 8-13, 1995* Volume 1: 132-139.

32

"The same ... ... but different": A juxtaposition of Roman and Medieval brass making in Central Europe

TH. REHREN

Deutsches Bergbau-Museum Bochum, Herner Str. 45, D-44787 Bochum, Germany

European brass making relied for more than 1500 years on the cementation of zinc oxide with copper metal, conducted in crucibles as reaction vessels. Both archaeological and documentary evidence now allows us to distinguish two variants of this one process, namely a Roman method, conducted as a solid state reaction in neatly closed vessels, and a Medieval method, employing open vessels at higher temperatures, resulting in liquid brass of lower quality. The apparent differences and the time gap between the two variants raise doubts as to whether the Medieval process is a direct offspring of the Roman one. Alternatively, an independent, possibly eastern, origin of the later technique is to be considered. Furthermore, a paradigm shift in the understanding of the nature of matter becomes apparent. Ironically, the more 'modern' view of alloy handling is associated with the poorer process, both economically and ecologically, and probably yielded a lower quality of metal.

Keywords: Brass Making; Roman; Medieval; Zinc Oxide; Crucibles; Cementation.

Introduction

Brass is the most recent of the ancient metals and alloys. Its regular production started only in the first century BC, probably in the eastern part of the Roman empire. The early use of brass was limited to coinage and military implements, but soon brass merged into the metal stock in circulation (Caley 1964, Hook & Craddock 1996). Beginning in the 2nd century AD, the purity and quality of brass decreased, resulting in an ill-defined quaternary alloy of copper, tin, zinc and lead ('leaded gunmetal') for many everyday implements and castings. It appears that the production of fresh brass largely ceased even before the collapse of the Roman empire (Caley 1964, Dungworth 1997a, b:907). The Middle Ages saw a revival of brass making, and soon it became the dominant copper-based alloy of the Old World, except for more specialized objects such as bells, which remained high-tin bronze. The outstanding technological question is whether medieval brass making is or is not a direct continuation of the Roman tradition.

Figure1. Sketch map of central Europe, showing the location of Roman and medieval brass factories mentioned in the text. Reprinted from Haedeke, 1973. Reproduced by permission.

The same ...

It is generally accepted that during both periods brass was made by the cementation process, i.e., by the reaction of zinc ore, charcoal and copper metal in a crucible. Whilst the metallurgical outlines of this process are well studied, archaeological evidence for it was absent until less than twenty years ago. The investigation of recent finds from Roman (Picon *et al.* 1995, Rehren in press) and medieval settlements (Rehren *et al.* 1993, and unpublished results) now allows for the first time further differentiation between two variants of this process. The location of the sites discussed here are shown on the map in Figure 1.

Brass cementation

In brass cementation, zinc oxide is reduced to zinc metal by charcoal at a temperature of about 900 to 1,000 °C. At this temperature, zinc metal forms a vapour phase, which is absorbed easily by copper metal. The upper limit of zinc uptake in cementation brass is about 30 wt% Zn, with the precise limit being controlled by a range of factors, including temperature, pressure, atmosphere composition (Haedecke 1973), and other alloying elements present in the metal phase. It appears that the maximum zinc uptake occurs at temperatures around 900 to 950 °C, while higher temperatures result in lower upper limits of zinc concentrations in the alloy, with as little as 20 wt% at 1,200 °C (Fig. 2; Haedecke 1973:232). At lower temperatures the diffusion of zinc into the solid metal is slow and thus high zinc uptake is restricted to a thin surface layer. The lower uptake at high temperatures is due to the higher partial pressure of zinc vapour, which promotes the volatilization of the zinc from the solid phase. In consequence, brass cementation is ideally done at temperatures at or below 1,000 °C, with the recipient copper metal in a suitable form, e.g., thin sheet or foil with a large surface relative to the volume and with as little impurities like lead, tin or arsenic as possible. A full discussion of brass making kinetics is not attempted in this paper, although details of the process are still a matter of debate (e.g., Grothe 1973).

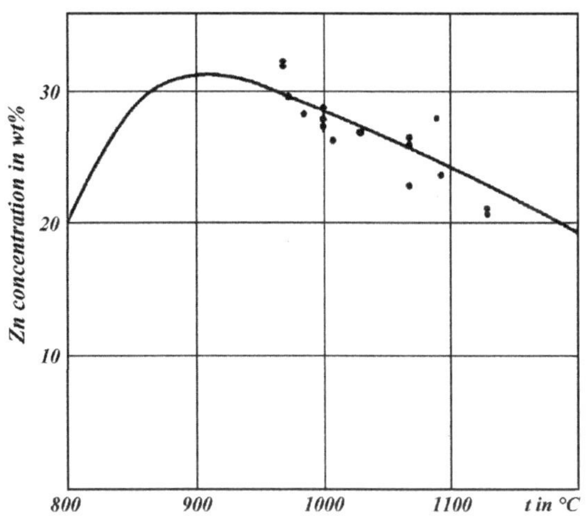

Figure 2. Experimental results (dots) and calculated equilibrium concentrations (vurved line) of zinc in brass. From Haedecked (1973: 232).

Brass cementation generally requires closed crucibles in order to keep the highly reactive and elusive zinc vapour in close contact with the copper metal, but away from any oxidizing agents. These constraints were only completely overcome long after the introduction of metallic zinc into brass making (Day 1990), when the crucible step was transferred to the distillation of zinc in retorts, i.e., crucibles providing a temperature gradient suitable to condense the metal vapour within the closed system.

Zinc oxide used in cementation can be either secondary zinc ores ('calamine'), i.e., zinc carbonates etc., transformed to zinc oxide by roasting (in the mid-13th century, Albertus Magnus reports that at Cologne and Paris they "... *convert copper into brass by means of the powder of a stone called calamina.*" Wyckoff 1967:224), or artificial zinc oxide obtained from a smelting process (again Albertus Magnus: "*Tutty, which is frequently used in the transmutation of metals, is an artificial, not a natural compound. It is made from the smoke that rises upwards and solidifies by adhering to hard bodies, where copper is being purified from the stones and tin which are in it. And a better kind is made by resubliming this.*" Wyckoff 1967:250). Literary evidence furthermore indicates the use (and therefore production) of large quantities of zinc oxide (*'tutia'*, *'spodos'*, etc.) as a pharmaceutical in the Roman empire, and Dioscorides explicitly mentions it condensing on iron rods put over smelting furnaces (Gunther 1934:234). Similar evidence exists for 'the East', i.e., Persia to India, from the early Middle Ages onwards (Allan 1979). Archaeological evidence for the production of such material is scarce (but cf. Hezarkhani-Zolgharian *et al.* 1994). Only recently, however, Eckstein *et al.* (1996) suggested a large-scale zinc oxide production in north west India from mixed, sulfidic lead-zinc ores, leaving behind a dense, lead-rich silicate slag.

Both types of zinc compounds, the natural calamine and the artificial *tutia*, are known to occur in different qualities, usually indicated by specifying their colour or origin. Impurities may contain earthy, non-reactive contaminants as well as metalliferous compounds, mainly lead oxides and sulfides. In particular lead enters the metal phase during the process (Maréchal 1938), probably more so if the alloy in forming is liquid rather than solid. Systematic studies of the behaviour and significance of these impurities, however, are lacking.

Brass working

Having made the brass, there are little characteristics in the further working of the alloy that distinguish it from the general contemporary copper-based metallurgy. Metal was melted in pear-shaped crucibles and cast, often with an addition of lead to increase fluidity and to lower the melting temperature. Sheet metal was being made, and both hammered and cut into shape. A range of surface decoration techniques were known, including the application of foil and leaves of gold and silver (Anhaeuser 1997), tinning, and artificial patination (La Niece & Craddock 1993). Non of these techniques seem specifically to be related to a certain alloy, although alloy preferences are known for particular types of objects.

Certainly, trade and recycling of copper and copper alloys were widely practised throughout the periods under consideration here. The extent to which recycling was done separately for different alloys, and whether there were any means to determine alloy composition except by colour, is still a matter of debate, though often based upon little factual evidence. Similarly, evidence for Roman or medieval brass trade is limited to a surprisingly small number of ingot finds (Bachmann & Jockenhövel 1974, Bollingberg 1995). Hence, there is again no sufficient base for an in-depth discussion of details of these aspects of archaeometallurgy, let alone for the discussion of differences between or characteristics of individual periods.

... but different

Within the general framework of brass cementation briefly given above, there are some parameters which vary to some extent. Reaction temperature, crucible design and raw materials employed are the most important ones, leading to different variants (and thus products?) of the same process. In order to reveal these differences, it is necessary to first characterize sufficiently the various processes. For the time being, this can be done only for two of the parameters, namely temperature and crucible design, while the exact nature and condition of the raw materials employed still remain enigmatic.

Roman brass making

"The metal is also got from a coppery stone called by a Greek name *cadmea*, a kind in high repute coming from overseas and also formerly found in Campania ... and it is also reported to have been recently found in the province of Germany. In Cyprus, where copper was first discovered, it is also obtained from another stone also, called *chalcitis*, copper ore; this was however afterwards of exceptionally low value when a better copper was found in other countries, and especially *aurichalcum*, which long maintained an outstanding quality and popularity, but which for a long time now has not been found, the ground being exhausted. The next in quality ... ... The highest reputation has now gone to the Marius copper, also called Cordova copper; next to the Livia variety this kind most readily absorbs *cadmea* and reproduces the excellence of *aurichalcum* in making sesterces and double-as pieces, the single as having to be content with its proper Cyprus copper. That is the extent of the high quality contained in natural copper (alloy)." (Pliny, translated by Rackham 1952:127-9)

This is the most detailed classical account of brass making, and certainly not a very precise one. It does show, however, that brass is considered a high quality variant of copper whose supply has been mainly depended on the availability of suitable ore. Only later, and after the exhaustion of that particular ore, the absorption of *cadmea* by copper was utilized to make brass. This report matches nicely the observation that brass was a rare, highly valued natural alloy during most of the first millennium BC, and only during the first century BC was made on a regular scale by a controlled process (Caley 1964, Pernicka 1990). Attention should be paid to the mention that certain variants of copper are more suitable for brass making than others.

Archaeological evidence for Roman brass production was first published by Bayley (1984, 1990), briefly describing closed, friable crucible fragments employed in a cementation process. More detailed studies of crucible remains, including chemical analyses of the ceramic and tentative process reconstructions, are given by Picon *et al.* (1995) and Rehren (in press). Common features of these crucibles are that they were mass-produced, though individually formed, sealed vessels, made from clay of no particular refractoriness, into closely similar size and shape at each findspot, but with huge variations among different sites. As far as can currently be said they were used at temperatures at or below 1,000°C, and their fabric contains only a few percent of zinc. The luted crucibles (Fig. 3) imply that they were charged and sealed while cold, while the conspicuous absence of any metal prills or slag remains within the vessels strongly indicate that cementation was done as a solid-vapour reaction, with no liquid metal/alloy or slag formation.

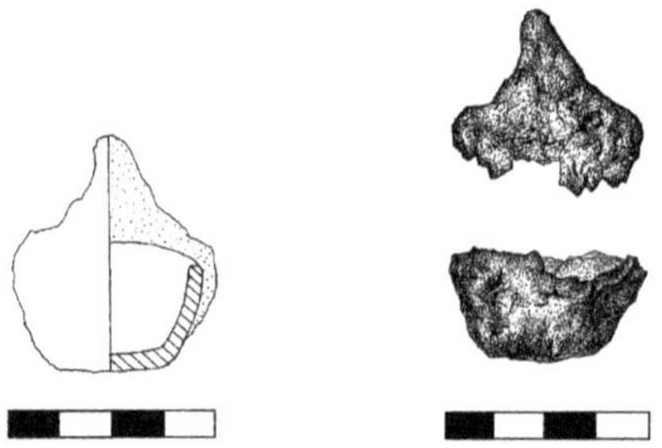

Figure 3. Roman brass making crucibles from Xanten, about 1st c. AD (n.b.: the lids and cups are made from the same clay). Drawing K. Engel.

Medieval brass making

"And when the crucibles are red-hot take some calamine, that has been ground up very fine with charcoal, and put it into each of the crucibles until they are about one-sixth full, then fill them up completely with the above-mentioned copper, and cover them with charcoal. ... Now, when the copper is completely melted, take a slender, long, bent iron rod with a wooden handle and stir carefully so that the calamine is alloyed with the copper. Then with long tongs raise each crucible slightly and move them a little from their position so that they may not stick to the hearth. Put calamine in them all again as before and fill them with copper and cover them with charcoal. When it is once more completely melted, stir again very carefully and remove one crucible with the tongs and pour out everything into furrows cut in the ground. Then put the crucible back in its place. Immediately take calamine as before and put it in, and on top as much of the copper that you have just cast as it can hold. When this is melted as before, stir it and add calamine again and fill it again with the copper you have just cast and allow it to melt. Do the same with each crucible. When it is all thoroughly melted and has been stirred for a very long time pour it out ... and keep it until you need it." (Theophilus, translated by Hawthorne & Smith 1963:143-4)

This medieval description of how to make brass comes much closer to the heart of an archaeometallurgist than the Roman equivalent. Together with the description of how to make the crucibles, the necessary furnaces and even how to refine the copper when making high-grade brass, all given in adjacent chapters of the treatise, there remain only a few uncertainties about the process.

Archaeological evidence for such brass making, however, was widely absent until quite recently. The first published investigation of medieval brass making crucibles is from Rehren *et al.* (1993). A voluminous waste deposit of crucible fragments (Fig. 4) from Dortmund, dating to the late first millennium AD, confirmed Theophilus' description in almost every detail, even adding further information. In the meantime, two more broadly contemporary brass making sites were identified in Westphalia, namely in Schwerte and Soest. In all these cases, cylindrical, template-formed crucibles were made from a highly refractory clay. Whilst the vessels from Dortmund and Schwerte are flat bottomed and of a height similar to their width, the Soest crucibles are plano-convex bottomed and appear, on average, much taller than wide. Most significantly, they are all open vessels, with the possible exception of Soest where matching clay disks were found, tentatively identified as loose crucible lids. The crucible fabric is now very rich in zinc oxide, and regularly contains inclusions of brass both within the fabric, and trapped in an internal slag layer.

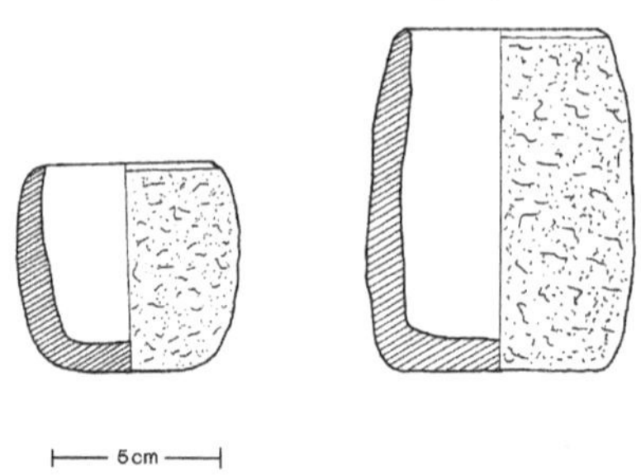

Figure 4. Medieval brass making crucibles from Dortmund, about 10th c. AD. Drawing N. Zieling. Reprinted from "Schlachen und Tiegel aus dem Adlertaun in Dortaund" in Moutanarchaologie in Europa 1993. Reproduced by permission of Jan Thorbecke Verlag Sigmaringen.

From the combination of literary and archaeological evidence, it is highly probable that the crucibles were charged only when hot, heated to a temperature well above the melting point of the resulting alloy, and operated in an 'open mode' except for a top cover of charcoal or possibly sometimes a protective lid. From this, a liquid-vapour reaction is inferred, based upon a kind of counter-current with rising zinc vapour and downward dripping molten metal.

Noteworthy, though still unexplained is the intimate association of the crucible fragments with dense, glassy lead silicate slag lumps in the archaeological record at Dortmund and Soest. However, there are only traces of similar slag within the crucibles, making it highly unlikely that this slag formed inside the vessels during the process.

Juxtaposition

Crucible fragments are regularly found in many urban archaeological sites. A common and probably diagnostic feature of the brass cementation vessels studied here is that they consist of small vessels of identical size and shape, found in large quantities within closely defined areas. This is interpreted as evidence for the use of these vessels in a standardized mass production process both in the Roman examples (Lyon and Xanten) and in the medieval ones (Dortmund, Schwerte and Soest). Significantly, the fabric of these crucibles always contains elevated levels of zinc oxide, absent in 'ordinary' melting vessels. Contemporary melting crucibles, in contrast, are typically individually formed vessels, found in much smaller quantities and/or spread over wider areas, representing occasional casting-on-demand in less specialized workshops.

The basic differences between the two brass cementation variants discussed here are firstly the lower temperature of the Roman process and secondly the use of open crucibles in the Middle Ages. These two parameters have wide-ranging implications for the actual conduct of the process. The use of a closed reaction vessel minimizes the loss of volatile zinc vapour, while a reaction below the melting point of brass ensures maximum exposure of potentially reactive metal surface and hence a maximum uptake of zinc. Thus, the zinc vapour has every chance of being absorbed by the metal, slowly but steadily. Since all the materials are contained in tightly closed vessels, loss of zinc vapour is restricted to the inevitable pressure release. Without evidence for appropriate openings in the crucibles, one can only speculate if this was achieved through the porosity of the extremely thin-walled ceramic (Rehren in press).

The medieval variant, in contrast, was conducted at a higher temperature and in open vessels. It relies on a kind of counter-current of zinc vapour rising from the bottom of the crucibles, and liquid copper alloy dripping downwards. To protect the zinc vapour from re-oxidation, and to hinder its escape, the vessels are covered either with a top layer of charcoal, as described by Theophilus, or with a loose lid as indicated by the Soest finds. As soon as all the metal has settled in a pool at the bottom of the vessel, and particularly after stirring, the incorporation of zinc into the alloy ceases even if the generation of zinc vapour continues. Eventually, the metal pool is poured out, the vessel recharged with fresh calamine/charcoal and the metal fed again into the crucible, until a satisfactory alloy quality is reached.

On a first glance, the medieval variant appears more sophisticated (and hence technologically more impressive) than the Roman one, using better quality ceramics, higher temperatures and experienced personnel. It employed crucibles of much higher refractoriness (Table 1 (at end)), able to withstand temperatures well above 1,000 °C while being attacked by zinc oxide, which is a strong flux, and craftsmen handling them with tongs for pouring and recharging. The Romans, in contrast, used mere clay for their crucibles, charged and luted them while cold. The firing of these vessels probably did not even demand any particular furnace constructions, but would have been possible in heating devices like those used in underfloor heating or thermal baths. The only disadvantage of the lower temperature lies in a relatively slow diffusion of zinc into the copper alloy. On the other hand, the higher temperature of the medieval process inevitably led to a lower upper limit of zinc content in the alloy (Haedecke 1973).

When comparing the economic aspect of the medieval variant with that of the Roman technique, the apparently 'high-tech' process falls badly behind. As a consequence of its higher temperature, iterative and open mode of operation, the medieval process required a much greater input of qualified labour to manipulate the red-hot crucibles, of thermal energy to reach the higher temperatures in the furnace and to maintain them over an extended period of time, and of costly raw materials for high quality crucible production and the zinc-rich charge feed than the Roman one.

Know-how transfer and the Nature of Things

Medieval technology and philosophy are deeply rooted in late Roman traditions (Gies & Gies 1994). The church, and in particular its monasteries, served as the major time tunnel, bridging the 'Dark Ages', thus providing the wealth of classical knowledge, often *via* Islamic intermediaries, to the Middle Ages. Theophilus (Hawthorne & Smith 1963) and Albertus Magnus (Wyckoff 1967) are amongst the most important authors in this process with regard to technical know-how and the medieval view of the Nature of Things. Does this hold true for the tradition of brass making technology as well?

Probably not. A fundamental problem in the tradition of classical technology lies in the lack of comprehensive technical literature. Apparently, the crafts were generally not considered worthy of being discussed in literature; with exceptions like Pliny only confirming this rule. Thus, the transmission of traditional technology was restricted to direct teaching and individual apprenticeship. In regard to brass making, we are faced with a particular problem, namely the strong possibility that this particular knowledge was already mostly lost before the collapse of the Roman empire in the west (Caley 1964, Craddock 1978, Dungworth 1997a, b). Therefore, the usual ways of know-how transfer probably did not work in this particular case.

Furthermore, the fundamental differences in crucible design and mode of operation make it unlikely that the

	SiO_2	Al_2O_3	FeO	MgO	CaO	Na_2O	K_2O	ZnO	Cu
Dortmund (2)	50.3	22.9	2.8	0.5	2.8	0.9	2.9	11.5	0.6
Soest (4)	62.5	16.1	2.5	0.4	0.9	0.4	1.5	8.2	1.5
Schwerte (3)	63.6	16.2	1.5	0.2	1.2	0.6	1.4	12.3	1.0
Xanten (4)	*75.7*	*11.0*	*3.8*	*1.3*	*1.3*	*1.7*	*2.8*	*1.6*	*0.1*

Table 1: ICP-OES data of the fabric of brass cementation crucibles from the Middle Ages (top three rows) and Roman (bottom row). Note the significantly higher alumina and lower alkali concentrations in the medieval samples, resulting in a higher refractoriness of the ceramic. The zinc oxide content of the Roman crucibles is much lower than in the medieval ones, indicating a better control of the process. Data are averages of multiple analysis (number of individuals from each site are given in brackets) in wt%, analysis W. Steger.

medieval variant is just a derivative of the Roman brass making technology. An independent origin is entirely possible. In view of a broad flow of know-how and materials into Europe from 'the East' during the Viking Period and the Middle Ages, a link to traditional Indian brass making appears not at all unlikely. Documentary indications for this are to be found in the mid-13th century work *Mineralia* by Albertus Magnus (Wyckoff 1967:250), when he refers to *tutia* from India, and half a century later, when Marco Polo reports the production of zinc oxide in Persia (Latham 1972).

Very recently, Eremin *et al.* (1998) presented data on late first millennium AD brass objects from Scotland, indicating that those finds which were identified as local on stylistic grounds were of a low-zinc brass, often with significant tin contents, whilst imported Scandinavian objects were of brass with a much higher in zinc and without indications of the recycling of tin bronze. In view of the known Viking contacts to Central Asia, this would also seem to support the proposed 'Eastern' origin of medieval brass making technology.

Another possible hint to an 'Eastern' origin of the medieval brass cementation technique lies in the frequent association of the Westphalian crucibles with large amounts of dense lumps of lead-silicate slags (Rehren *et al.* 1993 for Dortmund, unpublished data for Soest). These slags have their closest parallels in finds from north-west India (Table 2 (at end)). These are related to a tremendous underground mining of sulfidic zinc ore, and are tentatively interpreted as the remains of a smelting operation leading to the production of zinc oxide (Eckstein *et al.* 1996). Without more detailed information about the kind of raw materials used in brass making, however, one can only speculate whether or not these European slags are really zinc oxide slags.

Another fundamental difference between the Roman and the medieval variant lies in the approach to metals in general. The Romans accepted different qualities of copper as a natural phenomenon, while the medieval craftsmen actively controlled the quality of copper or scrap to adjust it to their needs. Pliny mentions that the good quality copper/*aureichalcos* of former times is now exhausted, and that certain kinds of copper are more suitable for brass making than others. Theophilus, in contrast, tells us in his next chapter how to refine copper for making better quality brass, suitable for gilding work. This reveals a shift from a passive, receptive understanding of nature to an actively manipulating one. Ironically, the inferior, wasteful

medieval technology comes together with a more modern, 'superior' philosophy, which in the long run eventually enabled significant progress beyond the Roman technology. In brass making, however, this was not achieved before the Modern Period. Even in 1546 Agricola (Fraustadt 1958:231-2) describes brass making in open crucibles which are only half-heartedly covered by lids, more than half a millennium later than the Westphalian finds, and cementation served as the major method of brass making well into the Industrial Period (Day 1990).

Acknowledgements
The title of this paper was inspired by Graham Philip's 1995 paper on MBA metalwork from Palestine and Egypt, and I am grateful that he allowed me to make use of this phrase in a similar manner.

	SiO_2	Al_2O_3	FeO	MgO	CaO	K_2O	PbO	ZnO	CuO
Dortmund (9)	32.8	3.8	4.1	0.6	3.3	2.2	44.9	3.2	1.8
Zawar (47)	43.0	6.2	12.1	4.6	13.1	1.9	6.9	11.0	0.1

Table 2: ICP-OES data of dense slag lumps from medieval Dortmund (top row) and Zawar in NW India (bottom row). Data from Rehren et al. (1993) and Eckstein et al. (1996), resp. Both sites are in intimate context with brass making (top) and zinc ore mining and smelting (bottom). Data are averages of multiple analysis (number of individuals from each site are given in brackets) in wt%.

References

ALLAN, J. 1979. *Persian Metal Technology 700-1300 AD*. London: Ithaca Press.

ANHAEUSER, K. 1997. *An Investigation of Amalgam Gilding and Silvering on Metalwork*. Unpublished PhD thesis. Oxford.

BACHMANN, H.G. & A. JOCKENHÖVEL. 1974. Zu den Stabbarren aus dem Rhein bei Mainz, *Archäologisches Korrespondenzblatt* 4: 139-144.

BAYLEY, J. 1984. Roman brass-making in Britain, *Historical Metallurgy* 18: 42-43.

1990. The production of brass in antiquity with particular reference to Roman Britain, in P.T. Craddock (ed.), *2000 Years of Zinc and Brass*: 7-27. British Museum Occasional Paper No. 50. London: British Museum.

BOLLINGBERG, H. 1995. Copper-alloyed artefacts from Roman imports in Scandinavia: elemental analysis, in P. Vandiver, J. Druzik, J. Galvan Madrid, I. Freestone & G. Wheeler. (eds.), *Materials Issues in Art and Archaeology IV*: 621-629. Pittsburgh (Pa): Materials Research Society.

CALEY, E.R. 1964. *Orichalcum and related ancient alloys*. American Numismatic Society Notes and Monographs No. 151. New York: American Numismatic Society.

CRADDOCK, P.T. 1978. The composition of copper alloys used by the Greek, Etruscan and Roman civilisations. 3. The origins and early use of brass.

Journal of Archaeological Science 5: 1-16.
DAY, J. 1990. Brass and zinc in Europe from the Middle Ages until the 19th century, in P.T. Craddock (ed.), *2000 Years of Zinc and Brass*: 123-148. British Museum Occasional Paper No. 50. London: British Museum.
DUNGWORTH, D. 1997a. Caley's 'zinc decline' reconsidered, *Numismatic Chronicle* 156: 228-234.
1997b. Roman copper alloys: Analysis of artefacts from Northern Britain, *Journal of Archaeological Science* 24: 901-910.
ECKSTEIN, K., P. CRADDOCK & I. FREESTONE. 1996. Silber, Blei oder Zink? Frühe Bergbau- und Verhüttungsaktivitäten in Zawar, Dariba und Agucha (Nordwestindien), *Metalla (Bochum)* 3: 11-26.
EREMIN, K., J. GRAHAM-CAMPBELL & P. WILTHEW. 1998. Analysis of copper-alloy artefacts from pagan Norse graves in Scotland. Paper presented at *Archaeometry '98*, Budapest.
FRAUSTADT, G. 1958. *Agricola De Natura Fossilium Libri X*. Berlin: Deutscher Verlag der Wissenschaften.
GIES, F. & J. GIES. 1994. *Cathedral, Forge, and Waterwheel. Technology and Innovation in the Middle Ages*. New York: Harper Collins.
GROTHE, H. 1973. Entgegnung zu Haedecke, 1973, *Erzmetall* 26: 555.
GUNTHER, R.T. 1934. *The Greek Herbal of Dioscorides*. Oxford: Oxford University Press.
HAEDECKE, K. 1973. Gleichgewichtsverhältnisse bei der Messingherstellung nach dem Galmei-Verfahren, *Erzmetall* 26: 229-233.
HAWTHORNE, J.G. & C.S. SMITH. 1963. *Theophilus On Divers Arts*. New York: Dover Publications.
HEZARKHANI-ZOLGHARNIAN, Z., I. KEESMANN, A. KRONZ & P. MAIER. 1994. Zink in alten Schlacken und technischen Produkten. In: *Berichte der Deutschen Mineralogischen Gesellschaft, Beihefte zum European Journal of Mineralogy* 6, Nr. 1, 107.
HOOK, D. & P.T. CRADDOCK. 1996. The scientific analysis of the copper-alloy lamps: aspects of classical alloying practices, in D. Bailey, *Catalogue of the Classical Lamps in the British Museum*: 144-163. London: British Museum Publications.
LA NIECE, S. & P. CRADDOCK. (eds.). 1993. *Metal Plating and Patination*. Oxford: Butterworth-Heinemann.
LATHAM, R. 1972. *The Travels of Marco Polo*. Harmondsworth: Penguin.
MARÉCHAL, J. 1938. Manufacture of brass before the discovery of the Dony process of zinc extraction, *Cuivre et Laiton* 11: 539-544.
PERNICKA, E. 1990. Gewinnung und Verbreitung der Metalle in prähistorischer Zeit, *Jahrbuch Römisch-Germanischen Zentralmuseums Mainz* 37/1: 21-129.
PICON, M., M. LE NEZET-CELESTIN & A. DESBAT. 1995. Un type particulier de grands récipients en terre réfractaire utilisés pour la fabrication du laiton par cémentation, in L. Rivet (ed.), *Actes du Congrès de Rouen*: 207-215. Société Francaise d'Étude de la Céramique Antique en Gaule.
RACKHAM, H. 1952. *Pliny Natural History in ten volumes, Volume IX, libri XXXIII-XXXV*. London: Heinemann.
REHREN, TH. In press. Small size, large scale: Roman brass production in Germania Inferior, *Journal of Archaeological Science* 25.
REHREN, TH., E. LIETZ, A. HAUPTMANN & K.H. DEUTMANN. 1993. Schlacken und Tiegel aus dem Adlerturm in Dortmund: Zeugen einer mittelalterlichen Messingproduktion, in H. Steuer & U. Zimmermann (eds.), *Montanarchäologie in Europa*: 303-314. Sigmaringen: Thorbecke.
WYCKOFF, D. 1967. *Albertus Magnus Book of minerals*. Oxford: Clarendon Press.

The investigation and archaeological applications of anthropogenic heavy metal isotope fractionation

RONA A. R. MCGILL[1], PAUL BUDD[1], BRETT SCAIFE[1], PAUL LYTHGOE[2], A. MARK POLLARD[1], RANDOLPH HAGGERTY[1], & SUZANNE M. M. YOUNG[3]

[1]*Department of Archaeological Sciences, University of Bradford, Bradford, BD7 1DP, UK*
[2]*Department of Earth Sciences, The University of Manchester, Oxford Road, Manchester, M13 9PL, UK*
[3]*Archaeometry Laboratories, Peabody Museum, Harvard University, 11 Divinity Avenue, Cambridge, Massachusetts 02138, USA*

The paper reviews a number of recent studies of established and novel isotope systems and their application to archaeometallurgical studies. Lead, tin and zinc isotope fractionation in natural and anthropogenic processes are considered. There is no evidence for process induced fractionation of lead isotopes on a scale comparable with their natural variation. Similarly, no measurable fractionation has been noted in the case of tin, either in nature or as a result of metallurgical processes; probably because of the element's relatively low vapour pressure and tendency to form complex oxides even at high temperature. New experimental evidence, however, suggests that systematic zinc isotope fractionation occurs as a result of zinc evaporation from molten alloys. The significance and potential applications of the observed effect are discussed in the context of ancient processes and their reconstruction.

Keywords: Lead; Tin; Zinc; Isotope Fractionation; Archaeometallurgy; ICP-MS.

Heavy isotope studies in archaeometallurgy

Both compositional and lead isotope analyses have been widely deployed in archaeometallurgy with the aim of provenancing metal artefacts (see Budd *et al.* 1995a, Budd and Thomas 1997, Pollard and Heron 1996, 302-340 for some recent reviews). Trace element data have also been used to shed light on production technology, where certain chemical changes are thought to be characteristic of particular smelting processes for instance (Budd *et al.* 1992). More recently, there has been a developing interest in the possible use of additional heavy metal isotope systems, less thoroughly investigated than lead, which may also display a degree of isotopic heterogeneity: either in nature, or as a result of production or processing. Copper, tin and zinc have been considered (Budd *et al.* 1995b, 1997, Begemann *et al.* in press, Gale 1997b, Maréchal *et al.* 1997, Yi *et al.* in press).

Lead

Lead isotope analysis has been widely used in science-based archaeology in studies of the provenance of metal artefacts (primarily lead, silver and copper). The nature of this paper precludes a detailed discussion of lead isotope methodology for which the reader is referred to a recent review by Pollard and Heron (1996, 302-40). Data interpretation has been the subject of much recent discussion in the literature (Budd *et al.* 1995c, 1995d and references therein) which will not be reproduced here, but the situation maybe briefly summarized. The analyses of copper ores known, or thought, to have been utilized in antiquity has been used to build up a database for comparison with data from metal artefacts of unknown origin. In examining the ore data, attempts have been made to identify "fields" with a relatively restricted range of lead isotope ratios considered to be characteristic of particular mining regions. More recently it has been suggested that characteristic and separable fields maybe identified even for individual mine sites within a single mining region. Difficulties which remain concern the type of material which should be used to construct such fields (copper ores, fluxes, slags etc.); the degree of overlap between them; the degree of mixing of metals derived from different sources; and over the size, and more importantly archaeological significance, of the geological entities represented by such fields.

The possibility that some mines or ore bodies might have very small natural lead isotope variations and that they might be characterized and separated from similar entities by very high precision measurements raised the possibility that provenancing attributions might be effected by small, but measurable, lead isotope fractionation in the production of archaeological artefacts (Budd *et al.* 1995a). Recent experimental work (Scaife, 1997), shows that such lead isotope fractionation is not likely to be measurable with current instrument precision, at least for a number of key metallurgical processes including smelting (see also Macfarlane, this volume). However, the work raised the possibility of metal isotope fractionation in processes involving significant evaporation from the liquid state. In the case of lead smelting, losses appear to be mainly a result of the sublimation of galena. Evaporative losses from molten metal are minimal as a result of the low vapour pressures of metallic lead and litharge. The principal loss of lead in melting appears to be as a result of slagging due to the high reactivity of lead oxide at elevated temperature. This combination of circumstances is fortuitous for archaeological provenancing work, but does deny us the opportunity to use lead isotope measurements with a view to reconstructing process details. Fortunately however, the same is not true of other isotopic systems.

Tin

The tin isotope system appears to have a great potential to evaluate the manufacturing history of copper-alloys. Estimating the extent to which ancient copper-based artefacts were recycled in prehistory is highly topical in archaeometallurgy and has a significant bearing on the interpretation of lead isotope and trace element data. Attempts have been made to detect recycling in a systematic manner using a combination of compositional and lead isotope data (Pernicka *et al.*, 1990). However, the model considered mixing from only two sources and, although it applied to the particular site for which it was developed, the method has not detected recycling in a wider context (Budd *et al.*, 1995a).

Kinetic theory relating to isotope separation has been applied to the tin isotope system (Budd *et al.* 1995b) to give an estimate of the fractionation that might be expected with tin loss by evaporation. Although the degree of fractionation with vapour loss can be modelled in this way, it can only be quantified by reference to a coefficient, the 'fractionation factor (α)' (Budd *et al.* in press). The *maximum* value of α is given by the square root of the ratio of masses of the isotopes of interest, but the coefficient can only be determined empirically for any particular process. Reasonable estimates of α suggested the possibility that tin isotope measurements might be used to quantify the degree of re-melting and, by implication, recycling to which ancient artefacts were subjected. Tin has the greatest number and mass range of stable isotopes of any element making it, potentially, particularly susceptible to fractionation despite its relatively high mass. The theoretical consideration of Budd *et al.* (1995b) suggested that operations involving significant loss of volatile tin compounds from copper alloys being manipulated at high temperature might produce alterations in tin isotope ratios exceeding natural variations.

A series of experimental bronze samples were prepared to investigate fractionation effects induced by metal processing. These, and a number of bronze artefacts thought likely to be subject to recycling, were analyzed by Multi-Collector ICP-MS at the University of Michigan (Yi *et al.*, in press). None of the samples measured were significantly fractionated. The absence of systematic fractionation with prolonged heating, indicated that α was considerably lower than initial estimates (Budd *et al.*, 1995b). It may be that the relatively low vapour pressure of tin at smelting/melting temperatures and the rate of metal vapour condensation at atmospheric pressure combine to dramatically reduce the predicted effect. It also appears likely that tin, even at high temperature, tends to form more complex oxides than the tin(II) oxide on which the predictive model was based. The absence of isotope fractionation even in the rather extreme metallurgical processes investigated suggests that significant tin isotope variation in nature is unlikely (Yi *et al.*, in press). Subsequent work by Gale (1997a) and Begemann (in press) on a number of artefacts and experimental samples are in agreement with these results.

Zinc

Although less relevant to tin than originally suspected, liquid-vapour fractionation is highly significant in the case of volatile metals. Of these, zinc is of obvious archaeometallurgical interest as a significant impurity element in some early copper-based artefacts and as a common alloying element from later pre-history to the present day. Systematic zinc isotope fractionation in smelting, melting and re-casting operations for instance, raises the possibility of using these measurements to reconstruct recycling and metal-circulation patterns. It might also be possible to use zinc isotope measurements to distinguish between brass production processes.

The fractionation of zinc as a result of its evaporation from liquid metal at elevated temperature arises from the slight difference in vapour pressure between its isotopes. This ensures that, in equilibrium, the vapour will be marginally enriched in the lighter isotopes with respect to the liquid. In various industrial and metallurgical processes involving the manipulation of zinc at high temperature, vapour may be lost from the liquid to condense elsewhere. If vapour is progressively lost in this way there will be a systematic change in the isotopic composition of the residual liquid and correspondingly, of the evolving vapour.

The effect may be modelled using kinetic theory, but as with tin, the magnitude is given by the fractionation factor (α). This is related to temperature, time, isotopic mass difference and vapour pressure. The model assumes that equilibrium is established instantaneously; complete mixing occurs within the liquid; and that separation between the vapour and liquid is entirely efficient. These conditions are not achieved in practise leading to a reduction in the efficiency of fractionation. Liquid brass (Cu-Zn alloy) is also clearly different from the pure, single element, systems with which the theory deals. The value of α is therefore process dependant, and the magnitude of expected fractionation must thus be derived experimentally.

A preliminary investigation of zinc isotope fractionation in metallurgical processes

Detailed explanation of the theoretical model for zinc fractionation during evaporation is reported elsewhere (Budd *et al.* in press). A brief summary is given here to illustrate the progress and potential of the work. The experiment involved zinc evaporation from brass in the manner expected during the direct process of brass making and subsequent handling of the liquid alloy at high temperature. The direct process, by which zinc metal is added to molten copper at high temperature, results in considerable zinc evaporation with consequent scope for fractionation. A small quantity of a homogenous pure brass standard (29.99% Zn) (*BNF C30.05*) was heated to 1100°C in a muffle furnace, under excess charcoal for variable periods between 5 to 30 minutes (Table 1). Zinc evaporation from the liquid alloy was monitored by weight loss on cooling. This was found to give a useful approximation for the loss of zinc as vapour as the vapour pressure of copper is insignificant at this temperature. Eight samples with zinc weight losses of between ~20 and ~80% were selected for zinc isotope analysis by ICP-MS.

Samples were pre-treated to separate zinc from copper to eliminate potential isobaric interferences from ^{63}Cu and ^{65}Cu, using methods described in Budd *et al.* (in press). Zinc isotope ratio measurements were made using a *VG Elemental PQ2+* ICP-MS at the Department of Earth Sciences, University of Manchester. All stable isotopes of zinc were measured, but results are given as the $^{68}Zn/^{64}Zn$ ratio maximizing mass difference and minimizing analytical error. ^{65}Cu was monitored in case of isobaric interference. The PQ2+ ICP-MS was operated in rapid peak jumping mode. Sample uptake time was 115 seconds (acquisition time 60 seconds) with a 120 seconds wash of 2% HNO_3 (*Primar*) between samples. Six replicate measurements were made for each sample. Two standard

Sample	Heating Time (mins)	Zn loss Wt. %	Measured $^{68}Zn/^{64}Zn$	$\delta^{68}Zn$ ‰ u^{-1}	Error (2 SE)	α
Standard	0	0.00	0.4114	0	0.57	
1100-19	5	19.38	0.4117	0.18	0.61	1.0034
1100-20	10	20.03	0.4117	0.19	0.58	1.0033
1100-33.6	10	33.60	0.4116	0.15	0.34	1.0012
1100-38.1	12	38.15	0.4120	0.35	0.86	1.0030
1100-51	15	50.95	0.4131	1.03	0.44	1.0058
1100-61.6	18	61.60	0.4137	1.39	0.92	1.0059
1100-69.7	23	69.77	0.4151	2.18	0.60	1.0075
1100-76.5	30	76.52	0.4153	2.36	0.40	1.0066

Table 1. ICP-MS data for the brass samples following open crucible re-melting under charcoal. Fractionation is given as $\delta^{68}Zn$ relative to the standard brass (BNF C30.05).

solutions were also run, the laboratory ~30%-zinc-brass standard (*BNF C30.05*) and a zinc solution prepared from 99.999% pure zinc. The latter was found to be isotopically lighter than the other values, suggesting the possibility that zinc purification processes may also lead to zinc isotope fractionation. Similar observations of fractionation in metal purified by vacuum evaporation have been detailed in the work of Russell *et al.*, (1978) for calcium.

The results of the zinc isotope analyses are presented in Budd *et al.* (in press) from which Table 1 and Figure 1 are reproduced. Measurements of $^{68}Zn/^{64}Zn$, for samples undergoing progressive Zn loss by evaporation, give $\delta^{68}Zn$ values between 0.18 and 2.36‰ u^{-1} with respect to the ~30% Zn-brass standard. The relationship between $\delta^{68}Zn$ and zinc weight loss (Figure 1) appears systematic, but is clearly close to the resolution of the quadrupole instrument; operated, in this preliminary work, without mass bias correction. Nevertheless, the two samples with the greatest Zn loss have significantly different Zn isotope ratios to the brass standard.

Values of α for each of the experimental samples are given in Table 1. All of the calculated values of α are estimates applicable to the brass melting process. However, the accuracy of those relating to relatively small zinc losses are the least reliable because expected variations are likely to be lower than the measurement errors. This is reflected in the data. The calculated values of α for those samples within measurement error of the standard are considerably more variable than those relating to zinc losses over ~50%. The mean of calculated values of α for the four samples with zinc losses over 50% is 1.0064 ($\sigma = 0.0008$) and this is considered the best estimate for the fractionation factor relating to brass melting at 1100°C.

These results have been used (Budd *et al.*, in press) to calculate $\delta^{68}Zn$ for the residual liquid using this estimate of the fractionation factor to give an approximation of the isotopic fractionation for any proportion of zinc loss (Figure 1). This curve intersects the 2 SE error on the measurement of the brass standard at ~30% Zn loss, providing a rough estimate of the evaporation required to produce a measurable shift in the isotope ratio of the residual liquid using the quadrupole ICP-MS method. Similarly, this data may be used to estimate the isotopic composition of the vapour (Figure 1). This suggests that the initial vapour emitted will be fractionated by approximately -1.5‰ u^{-1} ($\delta^{68}Zn$) with respect to the liquid, approximately three times the 2 SE error on the measurement of the brass standard.

Conclusions

The large mass of lead and consequently small mass difference between its isotopes, and the relatively low vapour pressure of the element at the temperatures of ancient metallurgical processes, combine to ensure that process induced lead isotope fractionation is insignificant compared with natural (radiogenic) variation. Nevertheless, recent advances in instrumentation, particularly with the development of ICP-source multi-collector mass spectrometry, are leading to greatly increased measurement precision. Ultimately, if increased precision were to result in the definition of very tightly constrained fields, for individual mines for instance, it might be necessary to reconsider the limitations imposed by possible fractionation in processing.

Although tin appeared to offer scope for fractionation in metal melting and casting operations, because of the large mass range, and the significant relative mass difference of its isotopes, this has not proved to be the case in practise. As with lead, the relatively low vapour pressure of the metal at temperatures typical of ancient metal working appears to preclude significant fractionation. Furthermore, it seems likely that tin forms relatively complex oxide molecules during evaporation, even at relatively high temperatures. This also lowers the magnitude of any predicted fractionation.

The results of the work on zinc however are the first demonstration of unintentional anthropogenic fractionation of a metal in a simple metallurgical process. This observation raises many possibilities, but its significance in terms of the reconstruction of production processes depends on the extent to which those processes involve zinc loss by evaporation. At present measurements have not been made at the precision necessary to distinguish between the direct and cementation brass making processes. For the direct process alone there is wide variation in estimates of zinc loss depending on the method employed; estimated as 5% in modern brass making (Alexander & Street 1976, 61), but as high as 25% in 17th century Chinese texts (Pollard & Heron 1996, 204).

To date, with the limited measurement precision offered by quadrupole ICP-MS, significant fractionation has been noted only under rather extreme conditions of vapour loss. The data have, however, allowed a rough estimate of the fractionation factor to be made. A Ga:Ge double spiking methodology is currently under development for higher precision ICP-MS zinc isotope measurements, but it is possible that the technique may ultimately require magnetic sector, multi-collector, ICP-MS analysis for accurate measurement. High precision measurements of this sort offer the potential to effectively discriminate between the two brass making processes and perhaps also to comment on the recycling of Zn-bearing alloys.

Acknowlegements
The authors would like to record thanks to Robert Cliff, Brendan Derham, Sarah James, Nick Walsh, Professor Alex Halliday, Wen Yi and colleagues for their help throughout the tin isotope project. Thanks also to Dr Richard Thomas for his help and advice on all aspects of the lead and novel isotope work and to Dr Dave Polya for his help in relation to the zinc isotope measurements. Much of the work reported here was supported by the NERC under research grant GR3/10057. One of the authors (PB) is supported by the NERC under the Advanced Fellowship programme.

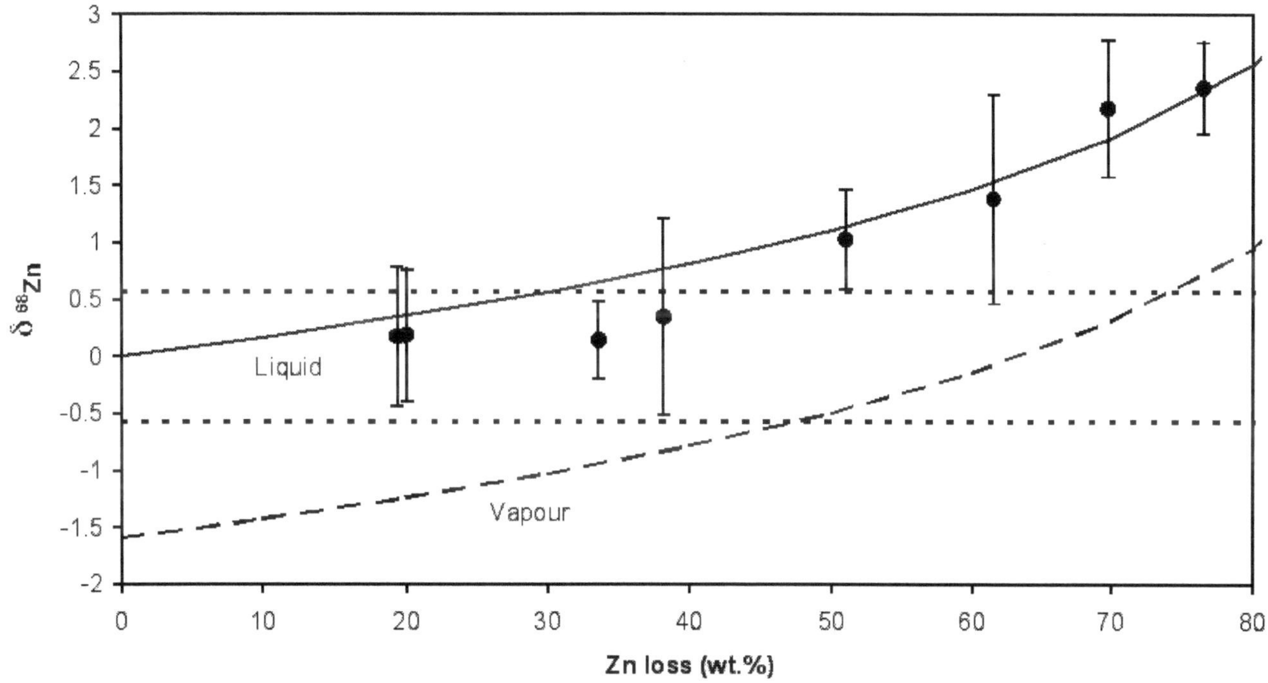

Figure 1. ICP-MS data showing zinc isotope fractionation in the residue of the Cu-Zn alloy after varying zinc losses by evaporation at 1100°C. Bars represent two standard errors. The short-dashed line represents the limits of error (2 SE) for the standard (BNF C30.05). The solid line represents the theoretical curve for fractionation of the remaining liquid with $\alpha = 1.0064$, as estimated from the experimental data. The long-dashed line shows fractionation within the corresponding vapour.

References

Alexander, W. & Street, A. 1976. *Metals in the Service of Man.* 6th Edn. Harmondsworth: Penguin.

Begemann, F., Kallas, K., Schmitt-Strecker, S., & Pernicka, E. In Press. Tracing ancient tin via isotope analyses. In A. Hauptmann & E. Pernicka (eds.) *Proc. Int. Symp. Beginnings of Metallurgy, Bochum, April 1995.*

Budd, P., D. Gale, A.M. Pollard, R.G. Thomas & P.A. Williams. 1992. The early development of metallurgy in the British Isles. *Antiquity* 66: 677.

Budd, P., Haggerty, R., McGill, R.A.R., & Pollard, A.M. 1997 The reconstruction of early brass production processes by Zn isotope ratio measurements Abs. *213th ACS National Meeting, San Francisco, April 13-17, 1997.*

Budd, P., Haggerty, R., Pollard, A.M., Scaife, B., & Thomas, R.G. 1995b. New heavy isotope studies in archaeology. *Israel Journal of Chemistry* 35(2), 125-130.

1995d. Lead isotopes and oxhide ingots: a final comment. *Journal of Mediterranean Archaeology* 8(1): 70-75.

Budd, P., McGill, R.A.R., Lythgoe, P & Pollard, A.M. In Press. Zinc isotope fractionation in liquid brass (Cu-Zn) alloy: potential environmental and archaeological applications. *Journal of the Geological Society.*

Budd, P., Pollard, A.M., Scaife, B., & Thomas, R.G. 1995a. The possible fractionation of lead isotopes in ancient metallurgical processes. *Archaeometry* 37(1): 143-150.

1995c. Oxhide ingots, recycling and the Mediterranean metals trade. *Journal of Mediterranean Archaeology* 8(1): 1-32.

Budd, P. & R.G. Thomas. 1997. Archaeometallurgy and the pursuit of provenance: thousands of answers in search of a question. Abstract 82. *Australian Archaeometry: Retrospectives for the Millennium,* Sydney 10-13 February 1997.

Gale, N.H. 1997a. The isotopic composition of tin in some ancient metals and the recycling problem in metal provenancing. *Archaeometry* 39: 71-82.

1997b. The measurement of copper isotopes in ancient artefacts. Poster presentation at 'Metals in Antiquity' conference, Harvard University. September 1997.

Maréchal, C.N., Albarède, F, & Emmanuel, N. 1997. Natural variations of copper and zinc isotopic compositions. Abs. *7th Annual V.M. Goldschmidt Conference, Tuscon, June 2-6, 1997.*

Pernicka, E., Begemann, F., Schmitt-Strecker, S. & Grimanis, A.P. 1990. On the composition and provenance of metal artefacts from Poliochni on Lemnos. *Oxford Journal of Archaeology* 9(3): 263-298.

Pollard, A.M. & Heron, C. 1996. *Archaeological Chemistry.* Cambridge: Royal Society of Chemistry.

Russell, W.A., Papanastassiou, D.A., & Tombrello, T.A. 1978. Ca isotope fractionation on the Earth and other solar system materials. *GCA* 42: 1075-1090.

Scaife, B 1997. Lead Isotope Analyses in Archaeology. Unpublished. PhD Thesis University of Bradford.

Yi, W., Budd, P., McGill, R.A.R., Young, S.M.M., Halliday, A.N., Haggerty, R., Scaife, B., & Pollard, A.M. In Press. Tin isotope studies of experimental and prehistoric bronzes. In A. Hauptmann & E. Pernicka (eds.) *Proc. Int. Symp. Beginnings of Metallurgy, Bochum, April 1995.*

34

The gold from Ghana and the Muslim expansion. A scientific enquiry into the Middle Ages using ICP-MS combined with an UV laser

A. GONDONNEAU & M. F. GUERRA

Centre de Recherches Ernest-Babelon, C.N.R.S., 3D, rue de la Férollerie - 45071 Orléans, France

The combination of proton activation analysis with LA-ICP-MS provides quantitative determinations of major, minor and trace elements. ICP-MS in laser ablation configuration achieves detection limits below the ppm level and determines a wide range of elements so permitting the fingerprinting of gold ores.

The development of this technique for the quantitative analysis of gold is presented here as well as its application to the characterization of monetary-gold sources exploited for coinage in African mints during the Muslim conquest.

Preliminary results show that the first dinars were struck with a Byzantine-type gold obtained by re-melting earlier coinage. The Aghlabids minted coins using gold from two different African sources. The coins struck in Spain and in Africa by the Almoravids also used African-sourced gold.

Keywords: Gold; ICP-MS; Laser Ablation; Nuclear Activation; Characterization; Provenance; Ghana; Sudan; Muslim.

Introduction

Establishing the manufacturing technology of artefacts and the provenance of the ores used in their manufacture are common problems posed in the study of archaeological metals. The rarity or abundance of ore deposits, the high temperatures required to purify metals and their subsequent re-melting and re-use makes the identification of the origin of many metals difficult (Guerra 1995).

From an analytical point of view it is hard to identify the origin of archaeological ores even for native metals. Although for silver and copper, lead isotope ratios are routinely determined, for pure and native gold where the Pb content is below 100 ppm, this technique requires a very large sample. Hence trace element fingerprinting of gold ores may be the only possible solution.

The analysis of gold using classical analytical methodology is often not sensitive enough to determine many of the elements at the concentration levels in which they occur (Guerra,1998). Consequently, ICP-MS with its multi-element capability and sub ng. g^{-1} detection limits offers an extremely viable alternative. Problems associated with dissolution are largely overcome by the use of laser ablation and the dry plasma required by this technique overcomes many of the problems associated with molecular ions, which complicate analyses in solution mode.

A Fisons PlasmaQuad PQXS was combined with a 2 mJ pulsed VG Nd:YAG laser. The laser wavelength was frequency quadrupled to work the UV wavelength of 266 nm with a maximum shot frequency of 10 Hz (Gondonneau *et al.* 1996). Samples must fit inside the 5cm diameter and 2.5 cm high ablation chamber. The main parameters of the laser and of the ICP-MS were determined. Using gold coins certified by nuclear activation analysis as standards and with internal standard corrections a quantitative analytical protocol was produced.

The association of LA-ICP-MS with 12 MeV proton activation analysis gives the concentrations of major, minor and a large number of trace elements within good detection limits. These two techniques are here applied to the study of Muslim West African gold.

Development of the LA-ICP-MS technique

LA-ICP-MS has been used for the analysis of industrial and geological gold samples (Kogan *et al.* 1994; Watling *et al.* 1994). However, unlike gold standard producers who make pure gold and silver samples, archaeometrists face the problem of determining objects having variable matrix compositions (Guerra *et al.* in press). Indeed, this is the main problem with the analysis of native gold (Watling *et al.* 1994)

For the quantitative evaluation of archaeological gold artefacts optimization of the ICP-MS operating conditions was needed in order to give the maximum analyte sensitivity for minimum interferences from polyatomic and refractory oxide ions. The LA optimized parameters under investigation were; the mode of laser operation, the number of shots per site and the number of replicate analyses. The system was operated in Q-switched mode as it leads to a greater proportion of finely divided material, to more reproducible transport and to less significant memory effects (Kogan *et al.* 1994). The advantage of using a UV instead of an IR laser for sample introduction into the ICP was shown by Geertsen *et al.* (1994).

In order to use inter-element associations to produce a fingerprint of each ore a protocol was adopted that allowed representative sampling together with multiple determinations of high sensitivity and good reproducibility.

Data collection parameters

Optimization of the data collection parameters with the ICP-MS and laser was made by using the S.I.M. mode spectra. Parameters were established for a given (maximum) laser energy and were chosen to minimise the

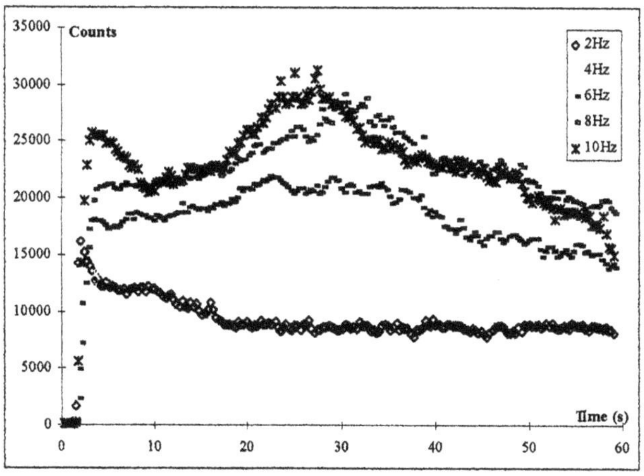

Figure 1: S.I.M. mode signal for copper at different shot frequencies for 10 ms dwell-time for coin Fran.

ablation of the archaeological objects, namely the size of the crater, whilst retaining good reproducibility and low detection limits.

Almost all elements give the same type of signal namely an initial high intensity signal lasting between five and seven seconds followed by a less intense but stable signal. Hence a ten-second pre-ablation phase (to avoid the first high intensity signal and to clean the surface of the sample) was followed by 60 seconds of ion collection. For statistical reasons a 10ms dwell-time was chosen. Under these conditions different shot frequencies from one to ten Hz, for copper, were studied as shown in FIGURE 1 and the optimum value was found to be 6 Hz.
FIGURE 2 shows the S.I.M. mode signal for platinum obtained with the three collimators of 0.5, 1 and 2 mm diameter. It is clear from the figure that using the largest collimator gives the best signal. Under those conditions the crater size is of about 40µm wide and 130µm deep. The crater depth was estimated by piling 20µm thick, certified 80Au/20Cu foils (made by Goodfellow SARL) upon each other.

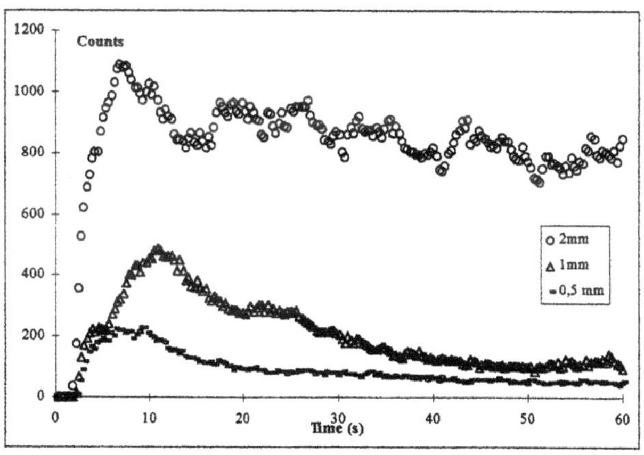

Figure 2: S.I.M. mode signal for platinum obtained with the three collimators of 0.5, 1 & 2mm diameter for coin M2.

Using these parameters the reproducibility of the analyses were tested by performing 30 replicate ablations and measuring about 20 elements on a gold coin. Except at high concentrations of osmium, ruthenium and iridium, (for Os-Ir-Ru precipitates in gold see Meeks & Tite 1980) elements display good reproducibility, as shown in FIGURE 3.

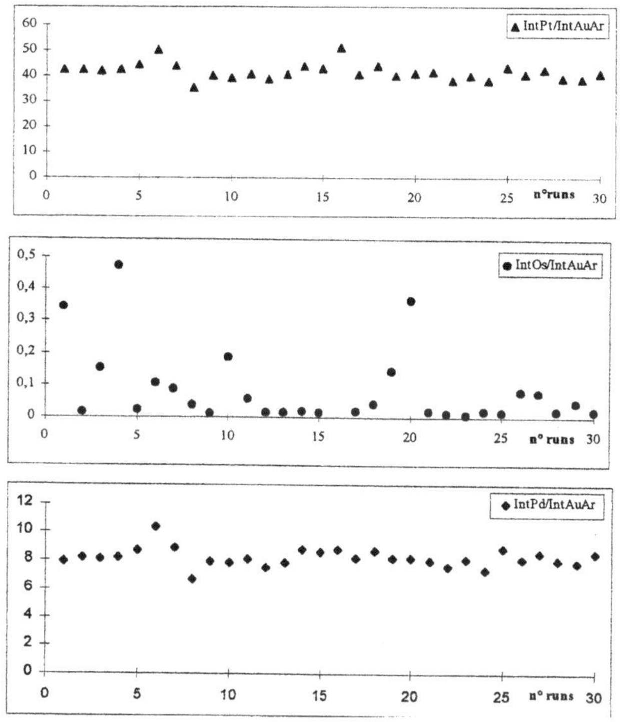

Figure 3: Reproducibility tests for platinum, osmium and palladium normalized to gold concentrations as $AuAr^+$. The test was performed on coin Fran.

In order to avoid heterogeneity problems three analyses were made for each element. Optimum data collection parameters and instrument operating conditions are given in TABLE 1 (see end).

Quantitative calculations

Ancient gold coins, which were certified by nuclear activation analysis, were chosen as the gold standards. These were chosen in order to be as close to the investigated objects as possible. As the sensitivity from one analyte to another may vary by as much as 50% both between samples and within a single sample (Jarvis *et al.* 1995), the use of this last correction improves the reproducibility of the measurements and corrects for instrumental drift.

An internal standard must be present at high, constant and known concentrations, be homogeneously distributed and its behaviour during ablation and transport to the plasma must be representative of the other elements of interest. In this study only gold fulfils these criteria. As gold is a mono-isotopic element and in order to be detected in pulse counting mode, one of the argon species generated in the plasma, namely $AuAr^+$ was used.

Gold was measured by 12 MeV proton activation analysis using the CERI/CNRS laboratory cyclotron facilities (Barrandon 1986). Gold produces by (p,n) reactions two isomers of mercury-197, both emitting gamma rays of energy lower than 300 keV. The interposition of a lead sheet between the sample and the detector strongly absorbs low energy gamma rays and so gamma-ray spectrometry can be performed on the gold object after irradiation. Quantitative calculations were made by using a basic equation containing the proton range in the sample and solved using the Ricci & Hann (1965; 1967) mean _

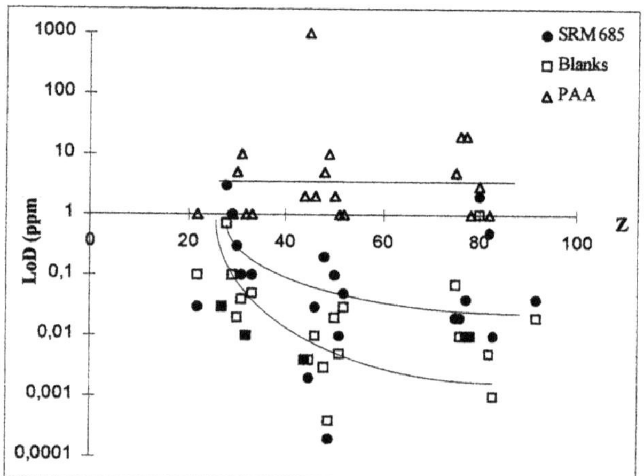

Figure 4: Detection limits for 12 MeV proton activation analysis (PAA) and for LA-ICP-MS on replicate blanks (Blanks) and of the NBS gold standard (SRM685).

method; concentrations are normalized to 100%. The detection limits are shown in FIGURE 4.

Experiments to measure the other two major elements in gold, namely silver and copper, by LA-ICP-MS were made. The machine was operated with the same laser energy but without the second vacuum pump, which achieves a pressure of 10^{-6} mb (the first pump only reaches 10^{-2} mb between the two Ni cones). Both analogue detection mode to measure the direct peaks of the elements and pulse counting mode to measure the $AgAr^+$ and $CuAr^+$ argon species generated in the plasma were used. It was noticed that for both modes when the composition of the standard was different from that of the sample, then the measured concentrations started to diverge from the true values although there were some exceptions to this. To deal with the difficulties posed by major elements other experiments, using different operating parameters optimized for copper, silver and gold, are being performed. In the meanwhile these elements are determined by activation analysis.

Isotopes		
Ti48	Ru102	Os189
Cr52	Rh103	Os192
Co59	Pd105	Ir191
Ni60	Pd106	Ir193
Cu65	Pd108	Pt195
Zn66	Cd111	Hg202
Ga69	In115	Tl203
Ga71	Sn118	Tl205
Ge70	Sn120	Pb208
Ge73	Sb121	Bi209
As75	Te125	Th232
Se82	Te126	U238
Ru99	Sm147	
Ru101	Re185	

Table 2: Isotopes used for quantitative evaluation of the measured elements in gold.

Trace elements concentrations were determined by LA-ICP-MS using gold coins certified by proton activation analysis as standards. However, the limited number of certified elements in these standards requires the use of interpolation calibration based on the assumption of a 'linear' response in certain mass regions. The most abundant isotope that was free from isobaric overlap was chosen for each element as is shown in TABLE 2. Interference was measured for isotopes having one low abundance isotope free from interference like platinum and isotopes, which are mono-isotopic, like rhodium.

A program developed in Visual Basic makes automatic quantitative evaluations as well as provides data storage and treatment. This program is based on a formula, which considers for a certain element x a relative to internal standard E_i sensibility coefficient K_x in standard and in sample:

$$K=I*[E_i]standard/[C_x]standard.$$

where I represents the measured intensities I= $[I_x]$sample/$[I_{Ei}]$sample and C_x the concentration of the element x.

It is assumed that the coefficient K for each element is the same for both the sample and the standard. Thus the concentration of the element in the sample can be written as:

$$[C_x]sample= \{[I_x]sample/[I_{Ei}]sample\}*\{[E_i]sample/[E_i]standard]\}*\{[I_{Ei}]standard/[I_x]standard\}*[C_x]standard$$

As a first approximation detection limits were estimated by considering the concentration equivalent to three times the standard deviation of 10 replicate blank analyses (Jarvis *et al.* 1995; Williams & Jarvis 1993). The background produced by laser firing on the sample was then considered; here the detection limits were defined as the concentration equivalent to three times the standard deviation of three replicate analysis of the high-purity gold standard NBS RMS685 (certified composition is given in TABLE 3) and assuming a perfectly homogeneous distribution of the analytes. The latter conditions were the standard ones normally used for the analysis of the samples.

SRM 685	
Element	*ppm*
Cu	0,10
In	0,01
Fe	0,20
Ag	0,10
Sn	<0,07
Zn	<0,04
Ni	0,05

Table 3: Certified composition of gold NBS SRM685 standard used for the determination of the detection limits.

FIGURE 4 shows detection limits for the most important elements (elements present in the NBS SRM685 standard influence the detection limits) as evaluated by both techniques. This figure shows that the blanks and the gold standard techniques differ by a factor of 10 and there is a factor of 10 to 100 difference between the LA-ICP-MS and proton activation analysis results.

How representative is the analysis?

Carr & Horlick (1982) state that: "In general laser vaporization is a very complex process and difficult to successfully model at a fundamental level. The condition of the crater, depth, size and shape as well as the amount of sample removed depends on the properties of the sample, the condition of the surface and the properties of the laser all of which undergo complex time dependent changes during the laser pulse".

Four certified gold coins of different compositions were routinely used for the quantitative experiments. Their composition is variable for several elements like Pt: 100, 300, 560 and 3000 ppm. The compositions of the certified coins used as standards are given in TABLE 4. At least two different standards with trace elements present at different concentrations were analyzed with each group of three samples to ensure good results.

Réf	*M2*	*Ang*	*Fran*	*Rus*
Au %	98,4	92,9	90,1	91,2
Ag %	1,5	3,0	6,3	0,23
Cu %	0,04	3,9	2,9	8,5
ppm				
Pb	33	146	196	79
Pt	98	562	2924	307
Pd	7	191	149	16
Fe	550	662	1998	103
Sn	3	56	272	1
Sb	0,4	59	112	28
As	0,1	26	82	63
Zn	9	74	146	5
Te	0,5	4	15	1
Ti	2	1,5	2	0 ,3

Table 4: Composition determined by proton activation analysis of the gold certified coins used as standards.

It was discovered that it was vital that the composition of the standards needed to be similar to the unknown artefact and that the concentration of an individual element had to be greater in the standard than in the unknown. These phenomena are currently being explored in order to find an explanation for them.

To check the technique, in some gold coins the concentrations obtained by proton activation and LA-ICP-MS techniques were compared. It was noticed that almost all the elements behaved in the same way as, for example, platinum and palladium, which are shown in FIGURE 5. Elements such as zinc (low melting point) may have very good sensitivity but poor reproducibility even if the measured concentrations have the same order of magnitude. Lead is difficult to measure by LA-ICP-MS as FIGURE 5 shows. It is noteworthy that lead also shows relative errors of more than 70% in silicates (Williams & Jarvis, 1993). As lead is an important element in the study of the manufacturing technologies of gold its concentration was obtained by proton activation analysis.

Application to the economic history of the western Muslim Empire

Middle Ages literary sources state that the gold fields of ancient Sudan (now Mali, Mauritania, Niger etc) were the main factor in trans-Saharan commercial relations during the western Muslim conquest (8th-12th century). To cite Fomin (1990) "The route to the wealth of the fabulous Ghana ran across Tahert, Wargla, Tadmekka and Sijilmasa". FIGURE 6 shows the main trade routes across the Sahara. However, no proof of these assumptions has ever been found and no theories have ever been advanced as to the quantity of gold that was exploited or on possible changes in gold supplies with time.

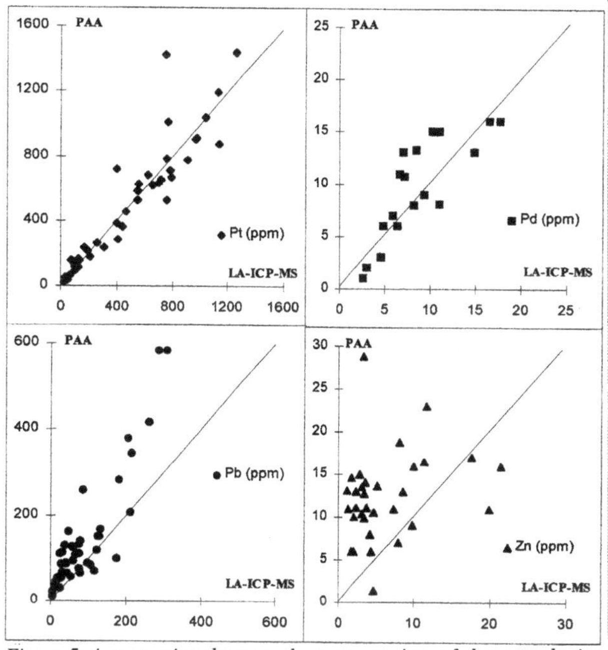

Figure 5: A comparison between the concentrations of elements obtained by proton activation analysis and LA-ICP-MS on a set of gold coins.

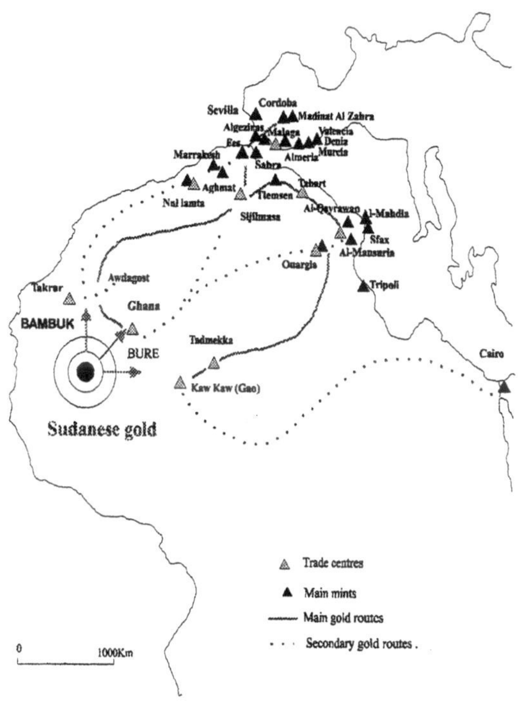

Figure 6: Trade routes of Sahara in the 8th-10th centuries (after Fomin, 1990).

The dinar, minted in the Muslim Western Caliphates was the object most sensitive to political and economical changes. Golden dinars have been analyzed by Ehrenkreutz, (1959; 1963), Oddy, (1980) and Messier, (1973). They have described a limited number of examples of devaluation of the coinage at different times and under different dynasties. Gold alone was analyzed for most of the coins but for a few copper and silver were also measured.

Combining proton activation analysis with LA-ICP-MS allows both the study of the manufacturing technologies of the coins and the provenancing of their gold ores.

Proton activation analysis measures the concentrations of major elements as well as trace elements like lead. This technique can help to determine the evolution of the debasement of the dinar as a result of the deliberate addition of silver or through the re-melting of impure native gold. Since galena is a major, silver-bearing, lead mineral and native gold generally has less than 100 ppm lead, the addition of silver to gold can be identified by an associated rise in lead. Trace elements determined by LA-ICP-MS relate the coins to their gold ores and so help to demonstrate changes in gold supply. TABLE 5 (see end) gives the concentrations of representative elements for coins from different dynasties.

Manufacturing technology

Study of the changes in fineness of the coins and knowing their date of issue shows that, apart for Arab/Latin solidi, all the coins were struck to a standard high fineness. This confirms previous workers (Oddy 1980; Messier 1973). TABLE 6 shows the mean and standard deviation of the gold content for each dynasty.

Dynasty	*mean Au value %*	–	*Number of coins*
Umayyad	98,9	0,3	11
Aghlabid	97,8	1,1	26
Fatimid	96,0	3,4	34

Table 6: Mean and standard deviation of the gold content for each dynasty.

If the concentrations of silver as a function of those for lead for the coins are considered, then it is seen that most of the coins were minted with native gold. However, some coins struck in Ifriqyia (no other mint identification) show high concentrations of lead as well as a high fineness. The correlation between silver and lead suggests that the debasing of the gold could have been achieved by adding silver with a high lead content of about 4 - 5%.

There are too few analyses of coins showing high lead concentrations to reach any firm conclusion. Hence more gold, plus some silver coins from these periods, will be analyzed further to explore this phenomenon.

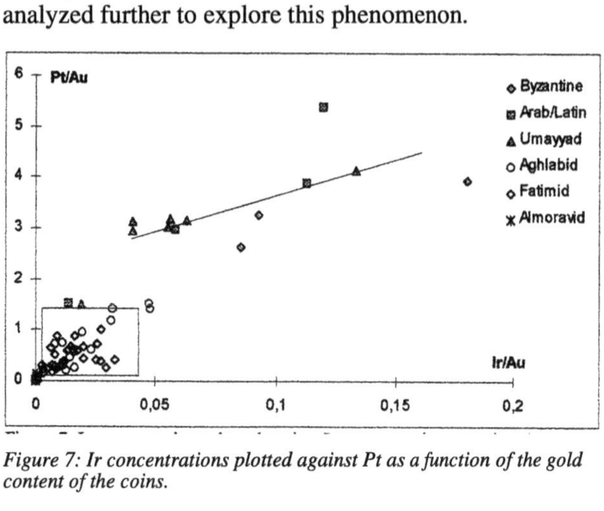

Figure 7: Ir concentrations plotted against Pt as a function of the gold content of the coins.

Provenance of the gold ores

In order to examine if there was a change of gold used in African mints immediately after the Muslims arrival, Byzantine and Arab/Latin solidi minted in Carthage as well as Ifriqiyan dinars minted under and after the Umayyads were analyzed. FIGURE 7 shows the iridium concentration plotted against the platinum concentration as a function of gold for these coins. Two main groups can be distinguished. Byzantine, Arab/Latin solidi and Umayyad dinars all fall in the same group, showing a correlation between iridium and platinum. These results agree with the chronological sequence determined by Bates (1992, 1995) for coins struck in Ifriqiya (based on the assumption that the same people worked in the Carthage mint under the Byzantines and the Muslims) from early solidi having Byzantine images until 715 when the eastern type Muslim dinar began to be coined.

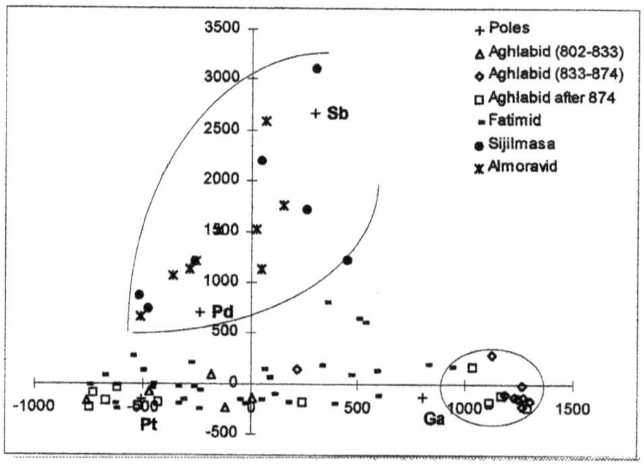

Figure 8: Principal component factor analysis for the dinars. All the elements were normalized to the gold content. Sijilmasa coins were struck under the Almoravids and the Almohads.

Principal component factor analysis was performed on all the analyzed dinars for the following elements: Pt, Pd, Ga and Sb all plotted against Au content. FIGURE 8 shows that several groups are present. This suggests that gold supplies (gold ores) changed under the Aghlabids on three different occasions.

The coins will be discussed in their chronological sequence. In general, the first gold is characterized by a high platinum content like the Ummayads' gold but with lower absolute values than the latter. The second gold, very different from the rest, has high gallium concentrations and very low platinum group element concentrations and corresponds to a stable historical period. The third gold type, plots between the first and the second and seems to correspond to a difficult time and re-melting and re-use of existing metals is an acceptable hypothesis to explain its spread of compositions.

Under the Fatimids it is possible that there were several ore supplies and/or re-melting of existing dinars and/or other gold objects took place. FIGURE 8 shows that Fatimid gold coin compositions approach those with high gallium and platinum concentrations. It can also be seen that a few coins were struck with a gold that approaches the composition of the Almoravid coins, which have higher concentrations of palladium and antimony. Nevertheless it is important to remember that after 1050/51 AD the Fatimids only held the eastern gold routes of Africa as the Spanish Umayyads controlled the western routes. Hence the question, did the Fatimids gold supplies come from the Red Sea mines during the Hilalian invasion when there was unrest in Ifriqiya?

The dinars, struck under the Almoravids and Almohads in the North African Sijilmasa mint (in the south of Morocco on the most important gold route of the Sahara) show a composition which seems to be different from those coined by the earlier African dynasties. Spufford, (1994) states that: "At the end of the 10th century the key route ran due north across the western Sahara from the gold fields of Bambuk and Bure to Sijilmasa, and then through the Atlas Mountains to Fez and finally to the caliphate of Cordoba in southern Spain". The few analyses of dinars struck in different Spanish mints by the Almoravids show a composition that is similar to that of the Sijilmasa dinars. Additional Spanish Caliphate coins will be analyzed to check this in order to further clarify the relationship between Africa and Al-Andalous. Perhaps the same gold ore was used to manufacture all of these coins and it came from Sudan?

Conclusion

LA-ICP-MS is a fairly new technique that facilitates rapid multi-element quantitative analysis of solid gold samples. When combined with activation analysis, major, minor and a large number of trace elements can be measured, so allowing the fingerprinting of gold metal. The potential for fingerprinting metals places this technique among the most powerful ones for solving the provenance of archaeological gold.

As only a small amount of ablated sample is required the non-destructive label given to this technique is apt. Hence this opens the possibility of studying assemblages of artefacts rather than individual archaeological objects.

The limiting step of dissolution for very insoluble materials is solved by the use of laser ablation. This sample introduction technique reduces problems associated with liquid solutions like potential contamination and low detection limits (Tykot & Young 1996).

A quantitative analytical method for the LA-ICP-MS analysis of pure gold matrices has been developed. The ICP-MS and LA parameters were optimized in order to give the best sensitivity and reproducibility for the lowest interference and crater size. Quantitative evaluation was done using an internal standard and determining the constant ratio between the number of counts and the concentration of the element in each sample and using gold coins as standards.

The techniques have been applied to the study of Muslim North African gold. For the first time the number of elements that have been measured has allowed the problems of devaluation, manufacturing technology and provenance for the dinar, which was one of the most important mediums of commercial exchange in the Mediterranean basin during the Middle Ages, to be tackled.

The first Muslim dinars were struck with the same gold as Byzantine solidi. However under the Aghlabids three different 'types' of gold appear. Under the Fatimids there are certainly two different golds, one similar to Aghlabids gold and another closer to Almoravids and Almohad one(s?). Coins struck under the latter dynasties are made with a different gold. The analysis of a few Spanish Almoravid coins seems to show that they were made with the trans-Saharan Sudanese gold.

A similar study to this one on Spanish coins will check the importance of Sudanese gold as well as one on Oriental coins in order to determine the importance of Red Sea and Middle East gold sources during the Middle Ages.

References

BARRANDON, J.-N. 1986. Metals analysis by charged particles: evaluation and prospects. *Nuclear Instruments and Methods* B 14: 133-141.

BATES, M. L. 1992. The coinage of Spain under the Umayyad Caliphs of the East. 711-750, *Actas III Jarique de numismatica hispano-arabe*, Madrid, Museo Arqueologico Nacional, Museo Casa de la Moneda: 271-298.

1995. Roman and Early Muslim Coinage in North Africa. *Antiquity to Islam: Papers of the Conference of Bristol,* October 1994, M. Horton and T Wiedeman (ed.): Centre for Mediterranean Studies, University of Bristol, Occasional Paper 13.

CARR, J. W. & G. HORLICK. 1982. Laser vaporization of solid metals samples into an inductively coupled plasma. *Spectrochimica Acta* 37B (1): 1-15.

EHRENKREUTZ, A. S. 1959. Studies in the monetary history of the Near East in the Middle Age. *Journal of the Economic and Social History of the Orient* II: 128-160.

1963. Studies in the monetary history of the Near East in the Middle Ages. II- The standard of fineness of western and eastern dinars before the Crusades. *Journal of the Economic and Social History of the Orient* 6: 243-267.

FOMIN, A. 1990. Silver of the Maghrib and gold from Ghana at the end of the VIII-IXth centuries A.D. *Proceedings of the Sigtuna Symposium on Viking Age coinage* K.Jonsson & B.Malmer (ed.) 6: 69-75.

GEERTSEN, C., A. BRIAND, F. CHARTIER, J.-L. LACOUR, P. MAUCHIEN, S. SJÖSTRÖM, & J.-M. MERMET. 1994. Comparison between infrared and ultraviolet laser ablation at atmospheric pressure - implications for solid sampling inductively coupled plasma spectrometry. *Journal of Analytical Atomic Spectrometr* 9: 17-22.

GONDONNEAU, A., M. F. GUERRA, & J.-N. BARRANDON. 1996. Sur les traces de l'or monnayé: recherche de provenances par LA-ICP-MS. *Revue d'Archéométrie* 20: 23-32.

GUERRA, M. F. 1995. Elemental analysis of coins and glasses. *Applied Radiation and Isotopes* 46: 583-8.

1998. The analysis of archaeological metals. The place of XRF and PIXE in the determination of technology and provenance. *X-ray Spectrometry* 27:73-80.

GUERRA, M. F., C. SARTHRE, A. GONDONNEAU & J.-N. BARRANDON. In press. Precious metals and provenance enquiries using LA-ICP-MS. *Journal of Archaeological Science*

JARVIS, K. E., J. G. WILLIAMS, S. J. PARRY & E. BERTALAN 1995. Quantitative determination of the platinum-group elements and gold using NiS fire assay with laser ablation-inductively coupled plasma-mass spectrometry (LA-ICP-MS). *Chemical Geology* 124: 37-46.

KOGAN, V. V., M. W. HINDS & G. I. RAMENDIK. 1994. The direct determination of trace elements in gold and silver materials by laser ablation inductively coupled plasma mass spectrometry without matrix matched standards. *Spectrochimica Acta* 49B: 333-343.

MEEKS, N. D. & M. S. TITE. 1980. The analysis of platinum-group element inclusions in gold. *Journal of Archaeological Science* 7: 267-275.

MESSIER, R. A. 1973. The Almoravids. West African gold and the gold currency of the Mediterranean basin. *Journal of the Economic and Social History of the Orient* 16: 31-41.

ODDY, W. A. 1980. The gold contents of Fatimid coins reconsidered. *Metallurgy in Numismatics I,* Royal Numismatic Society Special Publication 13: 99-118.

RICCI, E. & R. L. HANN 1965. Theory and Experiment in Rapid, Sensitive Helium-3 Activation Analysis. *Analytical Chemistry* 37: 742-748.

1967. Sensitivities for Activation Analysis of 15 Light Elements with 18-MeV Helium-3 Particles. *Analytical Chemistry* 39: 794-797.

SPUFFORD, P. 1994. *Money and its use in medieval Europe.* Cambridge University Press, 5th edition, Cambridge.

TYKOT, R. H. & S. M. M. YOUNG. 1996. Archaeological applications of inductively coupled plasma-mass spectrometry. *Archaeological Chemistry: organic, inorganic and biochemical analysis*, M. V. Orna (ed.): American Chemical Society: 116-130.

WATLING, R. J., H. K HERBERT., D. DELEV, & I. D. ABELL 1994. Gold fingerprinting by laser ablation-inductively coupled plasma-mass spectrometry.

Spectrochimica Acta 49B: 205-219.
WILLIAMS, J. G. & K. E. JARVIS. 1993. Preliminary assessment of laser ablation inductively coupled plasma mass spectrometry for quantitative multi-element determination in silicates. *Journal of Analytical Atomic Spectrometry* 8: 25-34.

ICP-MS instrument	
Instrument	Fison Elemental PlasmaQuad PQXS
Plasma	argon
Forward radio frequency power	1.350 kW
Reflected radio frequency power	< 2 W
Coolant gas flow	14.0 $l.min^{-1}$
Carrier gas flow	1.1 $l.min^{-1}$
Auxiliary gas flow	1.4 $l.min^{-1}$
Sampling cone	Ni 0.9 mm orifice
Skimmer cone	Ni 0.7 mm orifice
Data collection parameters	
Acquisition mode	pick jumping, 3 points per peak
Dwell time per channel	10.24 ms
Ion lens tuning	$5x10^5$ for ^{115}In in NIST612
Acquisition time	60 s after 10 s pre-ablation
Laser operating parameters	
Laser type	VG Nd:YAG (continuum)
Maximum output	2 mJ
Fundamental wavelength	266 nm
Laser operating mode	switched-Q
Number of shots per site	3
Shot repetition rate	6 Hz

Site pattern for each analysis 3 by 3 raster, i.e. 9 sites per analysis.

Table 1: LA-ICP-MS operating conditions and data collection parameters.

References	*Date AD*	*Mints*	*Au %*	*Ag %*	*Cu %*	*Pb ppm*	*Pt ppm*	*Pd ppm*	*Ir ppm*	*Ga ppm*	*Sb ppm*
Byzantine											
1968.798	659-661	Carthage?	98,3	1,4	0,3	47	320	16	9	2	4
Sch2831	668-674	Carthage?	95,5	3,7	0,8	nd	252	12	8	3	18
Af725	675-678	Carthage?	96,7	2,6	0,7	95	380	21	18	1	2
Arab/Latin											
L109	704-706	Carthage?	88,6	10,4	3,0	125	344	24	10	2	11
L110	706-707	Carthage?	90,5	7,6	1,8	121	269	18	5	1	6
L111bis	710	Carthage?	84,9	12,0	3,2	84	457	31	10	2	8
L114	717-718	Carthage?	60,6	37,4	2,0	nd	92	8	1	8	9
Umayyad											
1978.51	720-721	Ifriqiya	98,8	1,0	0,04	nd	310	18	4	2	1
L408bis	720-721	Ifriqiya	98,7	1,1	0,1	18	315	18	6	2	1
1966.341	721-722	Ifriqiya	98,8	1,1	0,04	nd	312	33	6	0,4	0,3
L425	722	Ifriqiya	99,6	0,4	0,01	nd	149	12	2	0,5	1
1978.52	722-723	Ifriqiya	99,0	0,8	0,2	nd	301	19	5	1	1
L464	732-733	Ifriqiya	98,8	1,0	0,1	15	291	17	4	3	2
L465	735-736	Ifriqiya	98,4	1,2	0,2	21	406	18	13	nd	nd
Aghlabid											
L823	807-808	Ifriqiya	93,8	5,7	0,4	nd	30	2	1	12	3
L830	814-815	Ifriqiya	98,2	1,6	0,05	372	25	2	2	17	1
1972.1319	815-816	Ifriqiya	97,7	2,1	0,1	235	139	9	5	3	1
L202bis	817-818	Ifriqiya	96,2	3,2	0,5	nd	147	7	5	74	1

References	*Date AD*	*Mints*	*Au %*	*Ag %*	*Cu %*	*Pb ppm*	*Pt ppm*	*Pd ppm*	*Ir ppm*	*Ga ppm*	*Sb ppm*
L836	828-829	Ifriqiya	97,2	2,5	0,2	61	116	9	3	22	3
1974.31	833-834	Ifriqiya	97,9	1,8	0,2	51	61	4	2	57	1
1974.32	837-838	Ifriqiya	97,4	1,8	0,2	638	20	2	1	19	3
L845	845-846	Ifriqiya	99,1	0,8	0,1	133	0,1	0,2	0,01	25	0,5
1972.1321	850-851	Ifriqiya	98,2	1,3	0,3	nd	0,2	0,4	0,003	22	1
L846	854-855	Ifriqiya	98,4	1,3	0,3	nd	1	1	0,04	26	0,5
1978.53	856-857	Ifriqiya	98,2	1,2	0,5	37	2	1	0,01	78	2
1974.35	863-864	Ifriqiya	99,1	0,6	0,2	nd	1	0,4	0,02	14	2
1974.36	864-865	Ifriqiya	98,7	0,8	0,3	nd	0,3	0,2	nd	27	1
1970.269.2	867-868	Ifriqiya	97,9	1,5	0,5	751	0,4	0,4	0,01	23	0,1
L857	875-876	Ifriqiya	98,5	0,7	0,7	nd	1	0,5	0,02	21	0,5
1967.210	882-883	Ifriqiya	98,6	0,8	0,5	nd	2	0,3	0,04	126	1
L865	884-885	Ifriqiya	98,5	0,7	0,6	3	1	0,3	0,02	12	1
L867	889-890	Ifriqiya	98,4	1,0	0,3	nd	4	0,4	0,03	45	1
L871	891-892	Ifriqiya	97,8	1,2	0,3	33	46	3	1	48	1
1972.1326	893-894	Ifriqiya	98,4	1,3	0,3	48	71	1	1	2	1
1976.189	895-896	Ifriqiya	98,4	1,4	0,1	nd	95	9	2	7	0,2
L873	897-898	Ifriqiya	97,1	1,4	1,4	nd	3	1	0,1	51	1
1972.1328	901-902	Ifriqiya	97,3	2,2	0,4	95	73	3	1	12	1
L883	903-904	Ifriqiya	98,4	1,2	0,3	126	55	5	2	2	1
1967.570	906-907	Ifriqiya	97,1	2,3	0,6	97	139	8	3	32	2
L887	909-910	Ifriqiya	97,5	2,0	0,5	62	60	6	2	5	2
Fatimid											
1977.51	910-911	Al-Qayrawan	99,0	0,8	0,1	nd	67	14	2	2	2
1972.1331	917-918	Al-Qayrawan	98,6	1,2	0,2	23	37	8	1	0,2	0,04
1972.1332	921-922	Al-Mahdia	97,7	1,9	0,3	267	86	4	2	9	0,2
1975.12	930-931	Ifriqiya	93,9	4,3	1,7	99	38	11	2	31	2
1972.1333	931-932	Al-Mahdia	93,9	4,6	1,5	73	25	9	3	9	1
L69	934-935	Ifriqiya	93,4	5,0	1,6	nd	3	8	0,03	42	2
1970.269.5	937-938	Al-Mahdia	97,9	1,5	0,6	nd	3	22	0,01	63	1
1984.3	938-939	Ifriqiya	97,9	1,5	0,4	5	64	6	1	39	0,2
1970.269.6	941-942	Al-Mahdia	97,6	1,9	0,3	nd	57	12	1	23	0,2
1970.269.7	943-944	Al-Mahdia	98,9	0,9	0,1	nd	44	4	2	13	0,05
1975.13	944-945	Al-Qayrawan	98,3	1,4	0,1	29	86	34	1	14	0,01
F10204(1904)	944-945	Ifriqiya	98,6	1,2	0,2	nd	37	8	3	8	0,1
1967.239	945-946	Al-Mahdia	98,6	1,1	0,2	nd	57	4	2	5	0,1
L87	946-947	Al-Qayrawan	98,7	1,1	0,1	60	58	6	2	19	0,4
L89	948-949	Al-Mahdia	99,1	0,7	0,1	42	30	10	1	71	0,4
1972.1334	950-951	Al-Mansuria	98,6	1,2	0,1	nd	28	7	1	10	0,1
L91	951-952	Al-Mahdia	98,9	0,9	0,05	59	28	4	1	21	0,1
L92	952-953	Al-Mahdia	98,3	1,2	0,1	nd	98	4	3	37	0,3
FSN	953-954	Ifriqiya	98,8	0,9	0,2	nd	40	6	3	24	0,1
L120	954-955	Al-Mahdia	97,4	2,1	0,4	137	71	13	3	21	1
L107bis	955-956	Al-Mansuria	97,8	1,9	0,2	nd	24	6	1	5	0,1
L96bis	956-957	Al-Mansuria	97,7	1,7	0,4	90	51	7	1	42	1
L96ter	962-963	Ifriqiya	97,6	1,7	0,4	88	60	4	1	54	1
L108	962-963	Al-Mansuria	97,7	2,0	0,1	194	2	1	0,01	28	nd
1967.572	968-969	Al-Mansuria	98,3	1,4	0,2	nd	22	13	0,5	20	0,3
L96(5)	974-975	Ifriqiya	97,8	1,6	0,4	85	38	4	1	52	1
L124	975-978	Ifriqiya	96,3	3,0	0,5	nd	8	8	0,01	21	5
1974.1162	983-984	Al-Mansuria	97,7	1,8	0,4	2	19	18	1	45	0,2
L152	993-994	Al-Mansuria	97,7	1,7	0,2	nd	31	35	0,2	62	0,4
1970.17	1029-1030	Al-Mahdia	97,2	2,2	0,2	nd	2	14	0,01	12	2
L382	1044-1045	Al-Mansuria	84,2	13,4	2,2	235	12	11	0,2	38	1

References	*Date AD*	*Mints*	*Au %*	*Ag %*	*Cu %*	*Pb ppm*	*Pt ppm*	*Pd ppm*	*Ir ppm*	*Ga ppm*	*Sb ppm*
L383	1045-1046	Al-Mansuria	85,1	12,3	2,4	321	14	10	0,3	35	8
L389	1062-1063	Al-Mahdia	97,9	1,9	0,1	39	23	13	0,3	2	0,2
FSSN	1094	Ifriqiya	93,9	5,3	0,7	nd	62	12	1	5	1
Midrarid											
L929	947-948	Sijilmasa	94,9	4,5	0,4	37	1	2	0,01	57	2
L930	952	Sijilmasa	96,4	3,1	0,4	nd	3	3	0,01	7	4
Almoravid											
L510	1075-1076	Sijilmasa	95,8	3,8	0,2	64	2	2	0,01	1	4
L511ter	1080-1081	Sijilmasa	94,8	4,8	0,3	nd	1	2	0,01	2	6
L512	1083-1084	Sijilmasa	95,2	4,4	0,2	84	1	2	0,004	2	1
L540	1093	Cordoba	87,2	10,9	1,9	70	12	15	0,04	2	3
L543	1100-1101	Almeria	90,2	8,9	0,8	103	5	6	0,03	5	3
L535	1103-1004	Seville	85,9	12,6	1,5	nd	10	10	0,03	1	1
L536	1103-1104	Valencia	90,1	9,1	0,6	111	7	7	0,1	2	3
L538	1103-1104	Denia	84,8	13,6	1,4	124	6	7	0,03	2	2
L542	1106-1107	Malaga	84,8	14,4	0,7	120	3	4	0,01	2	3
L599	1111-1112	Murcia	75,4	22,1	2,3	177	4	5	0,03	2	9
L585	1113-1114	Algeciras	80,4	18,0	1,5	72	5	5	0,02	2	3
Almohad											
1967.142.5	1150-1151	Sijilmasa	96,9	3,0	0,1	7	1	8	0,01	1	0,5
L507	1150-1151	Sijilmasa	90,6	7,5	1,9	nd	13	14	0,1	1	1
1967.142.1	1202-1217	Sijilmasa	96,9	2,9	0,1	76	51	100	1	1	2

Table 5: Composition of all the analyzed coins determined by PAA and LA-ICP-MS. Only elements present in the diagrams are given. Elements below detection limits are indicated by nd (detection limit for Pb by proton activation analysis is 1 ppm).

35

'Brave at heart': clanship and the work of the Highland Smith

JOHN A ATKINSON & EFFIE PHOTOS-JONES

Department of Archaeology, University of Glasgow, Glasgow G12 8QQ, UK

This paper explores the state of wrought iron making in Scotland during the Medieval to post-Medieval periods by concentrating on aspects of technology as well as organization of the industry. Archaeological evidence comes from recent excavations of iron working sites in the southern Highlands, undertaken as part of the Scottish Bloomeries Project.

Clan society in the Scottish Highlands prior to the sixteenth century was essentially atextual in nature; a culture that revolved around agrarian production, in particular cattle rearing, and tribal warfare. Social development within the Highlands centred on three key groups of individuals within the clan structure: the fine *(elite), the* doineuaisle *(lesser gentry, including the tacksmen and leaseholders) and the* bunnachan *(warrior class). The interrelationship among these groups and their requirements for weapons and tools essentially placed the smith in an important position in Highland society.*

It is the intention of this paper to develop the role of Highland smiths and place them in their cultural milieu, whilst at the same time presenting the archaeology and archaeometallurgy of iron production in this sphere. The juxtaposition of lowland society with its anglicized language and role to play in European politics with Highland society, a peripheral Celtic culture with its own language (Gaelic) led to a very different social development during the Medieval period.

The very nature of Highland society and its culture means that a unique opportunity arises to study the evidence of the iron metals industry in a 'pre-documentary' context, while at the same time drawing from evidence provided by folk traditions and Lowland sources.

Keywords: Scotland; Highlands; Iron; Smiths; Archaeology; Metallurgy.

I. Introduction

Recently, a three year project has been initiated to assess the level of iron technology associated with the numerous bloomery mounds – small slag heaps in the vicinity of furnaces/working floors – scattered throughout the Scottish Highlands (Atkinson & Photos-Jones *et. al.* in press). Most are undated but a small number have been dated to the Medieval or post-Medieval periods.

Highland culture during the high Medieval period (c 11th to 16th centuries) was essentially atextual in nature. Most of the contemporary documentary sources are derived from the Lowland Royal burghs of Edinburgh, Perth and Aberdeen. There are considerable pictorial evidence and numerous folk traditions concerning the making of iron weaponry, which fill the existing documentary void. This paper will discuss the archaeological and archaeometallurgical evidence for iron production, as derived from the excavation of two bloomery mounds in the southern Highlands, namely Tamheich Burn, and Allt na Ceardaich, in the Cowal Peninsula, Argyll (Figure 1). At this stage, no direct archaeological or laboratory evidence exists which can securely relate the recovery of artefacts (such as swords or other forms of weaponry) with these particular sites.

Furthermore, there exist no metallographic analyses of such artefacts in museum collections which, combined with newly derived analytical data from bloomery sites, can provide the important missing link. Therefore, the purpose of this paper is to integrate new field and laboratory evidence with the existing "documentary" accounts of clanship and Highland smiths during the Medieval period. So that this can occur, four main issues will be addressed:

i. The social context of Medieval society and warfare in Scotland

ii. Clanship and the role of the smith as 'armourer' to the clans

iii. The archaeological evidence of iron working installations

iv. The metallurgical evidence of iron production as derived from the examination of the industrial waste

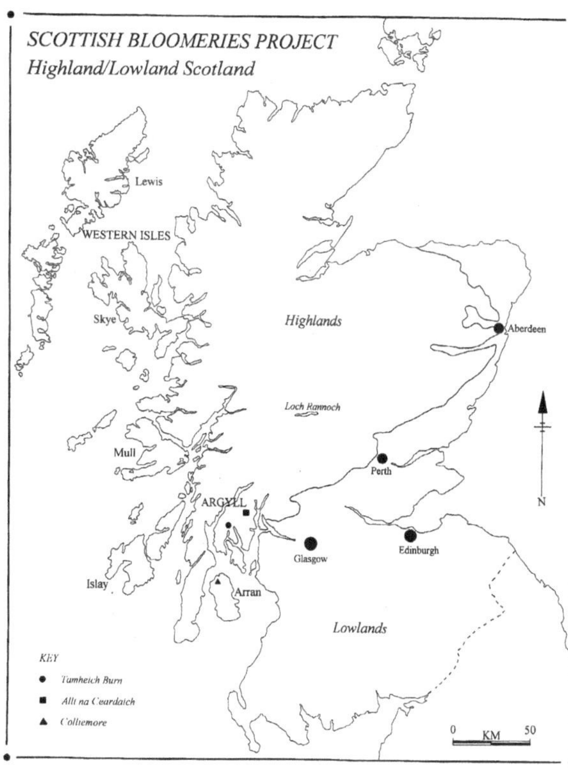

Figure 1. Location of Highlands and Lowlands of Scotland.

The fundamental aim of this paper is to view the archaeology of a non-documentary culture via its physical remains and relate this to the folk culture of the Scottish Highlands during the Medieval period. To achieve this, it is essential that the archaeological and archaeometallurgical evidence of iron working be integrated with an understanding of the clan-based system and the work of rural smiths, whose craft and position in society have been retained within folk traditions and pictorial representations of that culture.

II. The social context of Medieval society and warfare in Scotland

Highland iron working, and indeed, culture as a whole, need to be juxtaposed against their Lowland counterpart to allow a clearer understanding of the social position and role that smiths held within Gaeldom during the Medieval period. Therefore, it is necessary to view the work of the Lowland urban smiths in parallel with the Highland rural smiths if any degree of understanding is to be achieved. In contrast to the wealth of information derived from recent excavations at Medieval Lowland centres, such as the royal burghs of Edinburgh (Driscoll & Yeoman 1997), Perth (Holdsworth *et al.* 1987; Bowler *et al.* 1995; Cox 1996) and Aberdeen (Murray 1989), relatively little archaeological evidence has been forthcoming about rural life in the Scottish Highlands during the same period. A clear contrast exists between the two cultural spheres, the anglicized Lowlands and the Gaelic Highlands. There are two main reasons for this: a) Highland society was basically atextual. The early 'history' of the Highlands is based on records written by Lowlanders or oral traditions, which related the exploits of the clans. Formal documentary records only begin to appear in any numbers by the 16th century, and even then, they are limited in extent. In contrast to this, Lowland Scottish history is based directly on existing Crown, local government and mercantile sources. b) Archaeological excavations within the Highlands have been unable, as yet, to target Medieval settlement sites, but have been more successful in identifying and excavating iron working sites. With few exceptions, the Medieval settlement pattern in the Highlands is still to be properly defined and dealt with, whilst excavations in the Lowlands have been successful in characterizing medieval lifestyles, in both rural and urban centres.

Despite this, the work of the smith within the warrior culture of the Scottish Highlands can be inferred from the depiction of chain mail, swords and daggers in numerous items of architectural sculpture, particularly grave markers (Steer & Bannerman 1977). This class of architectural sculpture predominated in the Western Highlands in particular, where a series of schools developed, including the Iona school of sculpture (RCAHMS 1982). Effigies of armoured warriors like the one from Abbey Museum, Iona (RCAHMS 1982, 234) which commemorates Gilbride, a chief of the MacKinnons, have been dated to the second half of the 14th century (Figure 2). He wears a quilted surcoat or aketon, a pointed bascinet and an aventail or coif (chain mail). On his feet, he wears sabbatons, and over his shoulder is a shield bearing a "coat of arms". Suspended from his waist belt is a claymore (*claidheamh mor*). For most of the Medieval period and probably until the 16th century, the Highlander's weapon was the claymore – the long, cross-hilted, two-handed sword. William Wallace's – the Brave Heart – claymore certainly ranks among the most famous. Claymores like those in the collection at the Kelvingrove Museum and Art Gallery, Glasgow, are over 1.5m long and weigh a considerable amount (5-7kg) which require a strong man, indeed, to wield it. Although the purpose of the claymore and the two-handed swords was primarily to cut, it is likely that the force of the blow, when directed properly must have stunned the opponent and/or broken his bones, effectively incapacitating him.

The short dagger or dirk and the leather covered wooden shield (*targe*) made up the rest of the Highlander's military attire.

Figure 2. Late 14th century effigy of Gilbride Mackinnon. Reproduced with permission of the Royal Commission on the Ancient and Historical Monuments of Scotland.

In most European countries, the high Medieval period (11th to 16th centuries) saw a change over from chain mail to armoured plate, which in turn led to an adaptation of weaponry. Rather than a flat sectioned blade with a blunt tip aimed primarily at cutting, swords were made with a lozenge section and a sharp point, so that they could pierce gaps in the armour plate by a thrusting action of the user.

By contrast, Highland weaponry was slow to respond to this form of fighting. Claymores were still produced as late as the 17th century. Two-handed swords retained their status within Highland culture long after they had gone out of use elsewhere. Only with the change in Highland fighting practices in the 17th and 18th centuries, did claymores finally become obsolete. The adoption of the "Highland charge" – where clansmen with minimal armour charged their foes at speed – led to the adoption of the basket hilt sword, imported from the Lowlands or elsewhere in Europe. The evidence of medieval forms of Highland weapons, such as the two-handed claymore, the spear or intricate chain mail leads to two fundamental questions:

i. Were weapons imported from outside the highlands, or were they produced at rural bloomery sites?
ii. Can proof of place of production be inferred from the surviving material culture?

Prior to addressing these questions, it is necessary to understand the social context of the clan system and the pivotal relationship held by smiths within that culture.

III. Clanship and the role of the smith as 'armourer' to the clans

A '*Chlann* or, in English, the clan, meaning "literally the children, as a political, social and cultural entity was the collective product of Feudalism, kinship and local association" (MacInnes 1996, 1). Clanship, the social system in which the Highlanders operated, differed considerably from the anglicized Lowland social system of the period, a system which was controlled and managed by central government. The differences between Highland culture and Lowland culture were clearly noticeable from the late 14th century onwards, in the attitude of early writers, such as John of Fordun. Highland society was viewed as barbaric, "a savage and untamed nation, rude and independent, given to raping" (Hume-Brown 1893, 11).

The juxtaposition of different language, different dress and different culture was, for the Lowlanders, an indication of an uncivilized race. In reality, this was not the case.

Although Highland society differed from that of the anglicized Lowlands, it operated on an equally structured level. Each Highland clan had a chief and a leading elite group known as the *fine*, below this, was a group of lesser gentry (*doine-uaisle*) composed of tacksmen and leaseholders who managed and supervised the ordinary clansmen; the clansmen farmed the townships (*baile*) and made up the warrior class (*buannachan*). In certain areas, particularly along the western seaboard and Hebrides, after the collapse of the Lordships of the Isles in 1493 substantial numbers of *buanachan* were known to have operated as mercenary forces for the Irish, especially during the 16th century. "One English estimate of the 1590s suggests that there was a military class of well over 6000 men in the Western Isles, quite distinct from those clansmen who were usually employed in labouring the ground" (Devine 1994, 7).

Within the clan structure, a number of other individuals and groups held important positions; for example, the bards and the craftsmen served the clan elite (*fine*).

Although Highland culture was atextual, oral traditions were predominant with the clan bard or poet. A tight structure was retained through eulogies and elegies depicting the exploits of the clan and chiefs in the past and present. The position of craftsmen, such as smiths, wrights and millers, within clanship stood somewhat to the side of the other groups already discussed. Although they lived and operated within the clan system, they did not hold land within the township, but were generally given individual small holdings, known as crofts or pendicles, which were attached to individual *bailes* (Grant 1961, 241). A managerial system operated in which clansmen gave 'presents' in kind to the *fine* "for the clan elite's provision of specialist services, notably by smiths, farriers and millers whose charges were strictly controlled" (MacInnes 1996, 20-21). In return, clansmen were obliged to use those specialist services. Smiths, in particular, held positions of esteem within Highland society, "the armourer and smith were adherents of the chief and the office was generally hereditary" (Grant 1961, 240).

In a culture which was focused through land, cattle and warfare, smiths held a fundamental role as toolmakers and armourers to the clans. Interestingly, this does not seem to have been as simple a relationship as one might expect. Although the skills of the smiths appear to have been held within specific family groups, not all clans had their own dedicated smiths. Clans like the MacEacherns who "were famous sword-makers to the Lord of the Isles" (Grant & Cheape 1987, 82) appear to have operated as smiths and armourers for a number of other clans (Grant 1961, 240). The MacNabs of Dalmally were another such family. The MacRurys were hereditary smiths and armourers to the MacDonalds of Sleat (Grant & Cheape 1987, 197). The fact that these names, in association with the craft practised by these clans, survive so well must be seen as proof that weapons were made in the Highlands. In other words, although the control of production of iron was managed by the *fine* (clan elite), the social position of the smiths was such that they had hereditary control of their craft.

The contrast between Highland culture and Lowland culture is most readily seen in medieval Perth, which sat at the cross-roads between the two cultures. Recent work within the burgh has indicated that a thriving metals industry, which produced knives, was active from the 13th century onwards (Cox 1996). It has recently been argued that this industry held an independent position and was not reliant on rural craftsmen to produce the raw metal prior to re-working into artefacts within the burgh (Photos-Jones & Atkinson in press). The concomitant position regarding to the rural smiths within the Highlands must therefore also be one of independent craftsmen, who controlled their own manufacturing and trading base. Differences may be apparent in respect to the markets targeted; however, in general, the wide distribution of bloomery sites, the limited scale of those sites and the readily available raw materials utilized (charcoal and bog ore) imply that the manufacturing of weapons and tools was undertaken by rural smiths in the Highlands. In other words, smelting, smithing and artefact production were undertaken on a specific needs basis, rather than to compete in commercial markets – a feature which was redolent of urban centres.

IV. The archaeological and archaeometallurgical evidence for iron working

Since the end of the last century, numerous bloomery mounds have been known to exist throughout the Highlands, primarily through the antiquarian interest stimulated by MacAdam and Dixon in the 1880s (MacAdam 1887; Dixon 1886). In the intervening period, the only other concentrated programme of work was related to the Loch Rannoch area of Perthshire, which was finally published in 1970 (Aitken 1970). The limited nature of the previous programmes of work have, in many senses, left a poorly understood and defined archaeological resource in Scotland. Although many archaeologists and historians had commented on the need for further work in this sphere, interest had primarily been focused through the later periods, particularly the early seventeenth and eighteenth century charcoal blast furnace sites in the Highlands (e.g., see Lewis 1984; Lindsay 1977). The problematic nature of the position in Scotland was summarized by the economic historian John Butt, who stated in 1966 that "today prospects of discovering more about the origins of the iron industry in Scotland seem to depend less on formal documentary research and more upon the use of the techniques of the classical archaeologist, especially excavation" (Butt 1966, 194). Ironically, it has taken nearly thirty years for an archaeological response to be formulated.

The work of Scottish Bloomeries Project over the last two and a half years has sought to redress the balance somewhat, by utilizing the skills of the classical archaeologist with those of the metallurgist, the geologist and the geophysicist, in an attempt to unravel the early iron industry of the Highlands. This work has included assessment of the known sites in the Highlands, geophysical and geological prospection for new sites, and excavation of a number of bloomery locations. Excavation was targeted on a temporally and thematically wide range of sites, from shaft furnace bloomeries such as Tamheich Burn, Argyll to large scale processing and production sites such as the early seventeenth century ironworks around Loch Maree, Wester Ross (Atkinson & Photos-Jones 1995; 1997; Photos-Jones *et al.*, in press). However, for the purpose of this paper, two sites in Argyll (Allt na Ceardaich and Tamheich Burn) will be discussed in more detail.

a. The Tamheich Burn furnace

The site of Tamheich Burn was located in the neuk of a small meandering burn (Tamheich Burn) within an open area of rough pasture on the edge of a modern forestry plantation. The immediate vicinity of the furnace and slag heap had been overgrown by scrub woodland, in particular, Scots oak. The scale of production at the site is reflected in the small slag heap adjacent to the furnace. The heap was approximately 3 m in diameter and stood to 0.5 m in height. In reality iron production, from this site and its adjacent sister site further along the same burn, can only be viewed as small scale, the seasonal activities of one or two smiths.

Figure 3. Photograph of Tamheich Burn shaft furnace

The Taimheich Burn furnace was a small, almost square, shaft furnace, some 320 mm by 300 mm in size and standing to 300 mm high (Figure 3). The shaft itself was composed of fired clay, which was surrounded by a small mound of large overlapping flat slabs which formed a surrounding structure. The substantial amounts of slabs within the vicinity and within the bowl itself indicated that the furnace had probably stood substantially higher prior to abandonment. Air supply was provided via a tuyere hole.

The study of a polished thin section of the Tamheich Burn furnace was particularly illuminating (Photos-Jones & Hall 1998). The wall fabric consisted of two layers – an internal vitrified layer (dark grey and vescicular, Figure 4a), and an external unvitrified one made up primarily of red clay (Figure 4b). The red clay consisted of fine (less than 100 microns) micaceous material (muscovite and chlorite) with a scattering of angular quartz grains (up to 1 mm) and small bright inclusions (no more than 50 microns) which are probably rutile. There is a fragment of metallurgical slag within the layer of red clay (Figure 4a, left hand side), suggesting the intentional addition of such fragments for the purpose of enhancing the refractoriness of the local clay. In the internal vitrified layer, mica has melted while the quartz grains were rounded and/or melting at the edges. Mica melts at 500-700^{0}C; the exact temperature is determined by localized conditions of composition, air pressure and water pressure.

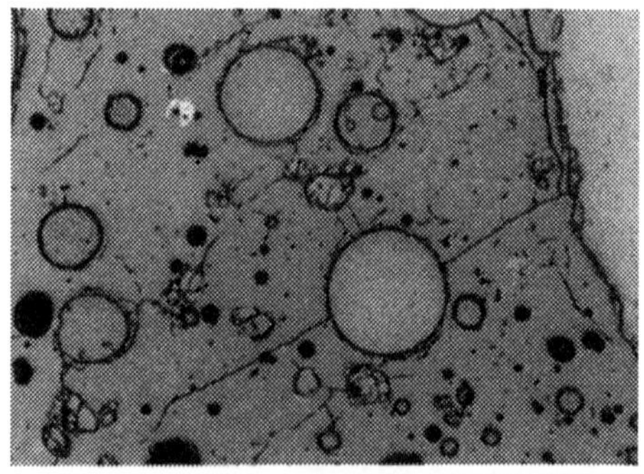

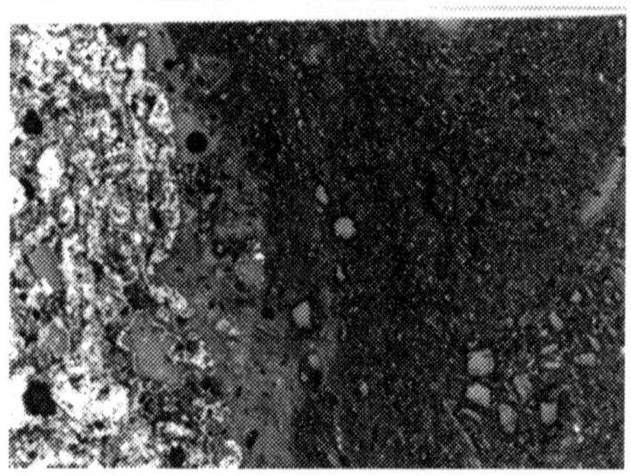

Figure 4a. Reflected light photograph of the inner vitrified section of the Tamheich Burn furnace lining; grey glass matrix with rounded quartz grains and rounded gas holes (field: 2.6 mm across).
Figure 4b. Reflected light photograph of the outer (red clay) section of the Tamheich Burn furnace lining; angular grey inclusions are quartz; the matrix is fine micaceous clay; bright amorphous areas metallurgical slag, globular wustite (field 2.6 mm across).

b. The Allt na Ceardaich furnace

The site of Allt na Ceardaich was located within a mature forestry plantation, some 50 m from the edge of Loch Eck. The topographic position of the site was very different from that witnessed at Tamheich Burn. Here the site was located on a plateau in the side of a steep valley, adjacent to a fast flowing burn (Allt na Ceardaich). The plateau was surrounded by a natural terrace to the north and east, thereby sheltering the working area on three sides. As in the case of Tamheich Burn, a sister site was also present at Allt na Ceardaich. It was positioned some 40 m further up the valley side.

Production at Allt na Ceardaich was of a much larger scale than at Tamheich Burn. the volume of the slag heaps immediately to the east of the furnace, although considerably eroded by tree roots, indicate substantial deposition of waste materials. It is most likely that this represents the results of cumulative activity over a more prolonged period of time.

Figure 5. Photograph of Allt na Ceardaich bloomery hearth.

The Allt na Ceardaich furnace sat within a semi-circular arc of slag on its southern side, and a similar arc of drystone walling formed its northern extent. Excavation of the site and, particularly the hearth, indicated a more sophisticated set up with the hearth located within a stone and timber structure, and a charcoal storage area to the south and a smithing hearth to the west (Figure 5). The hearth itself had a similar corbelled structure surrounding a heavily vitrified northern wall with lining rising to a height of 350 mm. Drystone base slabs, about 360 mm thick outlined the bowl of the hearth, the floor of which was slightly inclined towards the south. No evidence for a clay lined wall rising over these slabs was apparent. the internal size of this hearth was 415 mm by 375 mm. On the northern vitrified walling the hole for the tuyere was readily visible. Excavation of the interior of the hearth allowed a radiocarbon date for the final firing of this site of AD 1450-1660 (GU-4415).

The majority of the industrial waste from Tamheich Burn *(BCP 16; BCP22)* and Allt na Ceardaich *(BCP10; BCP13; BCP14; XSB1; XSB2; XSB3; AC21; AC26; AC33; AC35; AC36)* consisted of small fragments of bluish black slag of the tap slag morphology. They were manganese-rich in composition with a very high percentage of iron oxides (see Table 1 (at end)). Analyses were undertaken with the Scanning Electron Microscope attached to an energy dispersive analyser (SEM-EDAX). Area analyses (at low magnification x25-x100) are used as representative of the "bulk" chemical composition while spot analyses of wustite (FeO), fayalite (2FeO.SiO2), hercynite (Al2O3.FeO), interstitial glass and other possible inclusions are also represented in Table 1, which also includes a description of the sample context. Each entry in the Table represents the mean of two spot analyses per mineralogical phase and three area analyses at three different sections of the sample. Barium is present as a minor constituent of the Allt na Ceardaich slags, concentrating primarily in the glassy phase. The four distinct mineralogical phases can be seen in Figure 6a.

The presence of high amounts of both manganese and iron would have made the slag fluid at relatively lower temperatures (c.1150°C), resulting in good separation between metal and slag. Indeed, metal inclusions within the slag were very sparse. At Tamheich Burn the ore was manganese rich bog ore. At both Tamheich Burn and Allt na Ceardaich the sources of bog ore remain elusive. The inability to locate the source of ore for any of the twenty three sites surveyed in the course of the Scottish Bloomeries Project was the cause of much "anxiety", which resulted in the compilation of a directory of possible sources of iron ore in Scotland (Hall & Photos-Jones *in press*).

The working floor at Allt na Ceardaich, located to the north west of the furnace, consisted of a compacted floor with loose stores set in a rough circle (Figure 7). A section of the flooring material was sampled and analyzed (Figure 6b). It consisted of fragments of metallurgical debris (wustite and fayalite) and charcoal cemented together by banded iron oxyhydroxide. Amorphous and particulate matter was embedded within the gaps. The nature of this distinct flooring material is still under investigation.

Similar fragments were retrieved from the slag heap to the south of the 'charcoal storage' area. It is almost certain that they must have originated from smithing or billet making.

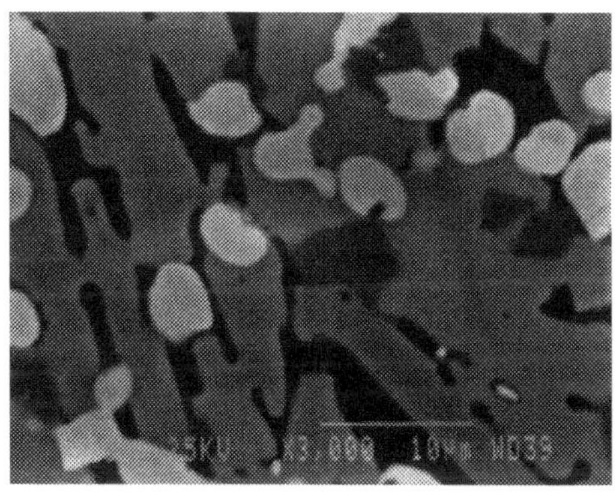

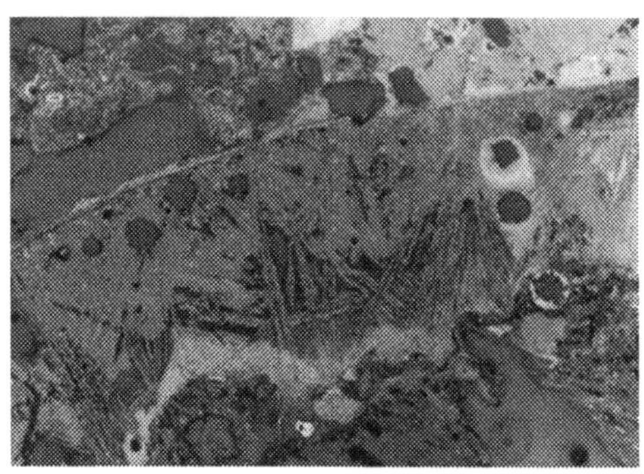

Figure 6a. SEM-EDAX backscattered image of Allt na Caerdaich tap slag (bar = 10 microns, x3000); bright globular dendtrites of wustite, light grey needled offayalite, dark grey angular grains of hercynite and interstitial glass (black).
Figure 6b. Optical Microscope (reflected light) photograph of Allt na Caerdaich working floor "slag". Composite material consisting of fayalite needles, round metallic iron, quartz grains in an advanced state of weathering (field: 2.6 mm across).

V. Interpretation

Interpretation of Tamheich Burn and Allt na Ceardaich and its integration with the supporting cultural evidence is the crucial step in the understanding of the two sites.

Tamheich Burn has the classic form of a low shaft furnace, with the exception of the surrounding corbelled stonework which would appear to have a functional use within its architecture. Although the western face of the furnace had been badly damaged, the other three sides retained the outer drystone structure. The western face must have included the tapping arch, since it lies opposite the tuyere-bearing eastern side, and therefore was probably damaged during its last firing. Allt na Ceardaich also exhibited a similar corbelled structure, though its function within this hearth was clearly different. The Allt na Ceardaich hearth was clearly not a shaft furnace and exhibited features which suggest it may have operated in a similar fashion to the Catalan Hearth described by Percy (in Rostocker & Bronson, 1990). The Catalan Hearth appears to have had a forerunner in the north of England in the late 1600's as can be deduced from a letter presented in the *Philosophical Transactions* of 1694 by Mr John Sturdie of Lancashire who gave a detailed description of the type of bloomery hearth in operation at Milthrope Forge (Fell 1908, 204). The eyewitness accounts of the bloomery hearth given there are very similar to the archaeological remains at Allt na Ceardaich. Given Scotland's many technological connections with the north of England, it is not improbable that such hearths may have found a venue in the Highlands. The overall architectural layout of Allt na Ceardaich finds parallels in the site of Minepit Wood in Sussex (Money 1971).

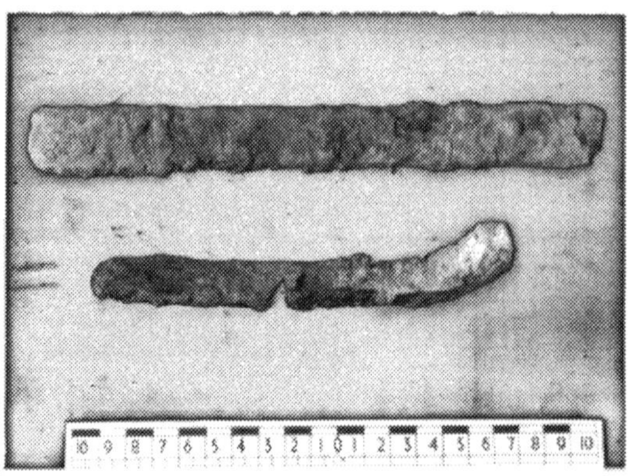

Figure 8. the Arran (bottom) and Woodend, Dumfries (top) "currency bars". The latter is of Romano-British context.

The Allt na Ceardaich drystone structure may well be related to the need to insulate the hearth structure, whereas the Tamheich Burn shaft could well have operated without any further insulating outer structure. Each furnace represents a different architectural tradition – one eminently small scale and piece-meal (Tamheich Burn) and the other more substantial and probably used over a long period (Allt na Ceardaich). Interpretation of these structures implies that the small scale shaft furnace may have been used by a smith for production of a small amount of raw material, whereas the larger hearth at Allt na Ceardaich could well be the site of a much larger scale of production and manufacture, possibly over a number of seasons. When the historical evidence is combined with the archaeological results, it implies that a site like Allt na Ceardaich may well have been the focus for production of goods (weaponry or tools) on a grander scale, and by conjecture may have been the focus for a clan-based smith. This may also find support in the toponomy of the site which translates from the Gaelic as 'the burn of the smith'.

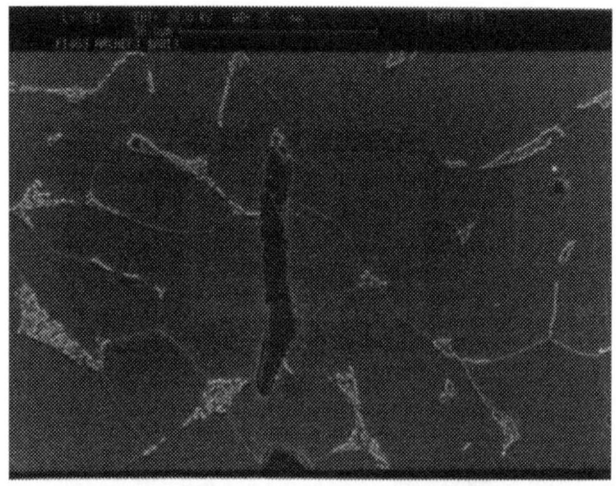

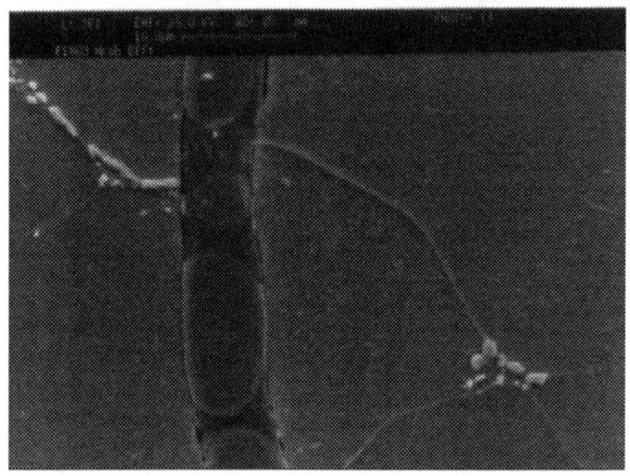

Figure 9a. SEM-EDAX backscattered image of the Arran "currency bar". Fine lamellar pearlite growing amidst ferrite grains and elongated slag inclusion (bar = 50 microns).
Figure 9b. SEM-EDAX backscattered image of the elongated slag inclusion in Arran bar showing phosphorous and manganese rich fayalitic phase and globular wustite (bar = 10 microns).

During the excavations, no finished objects were recovered, and consequently no relationship between the bloom and metal can be derived at this stage. Indeed, this was the case in most of the bloomery sites visited in the course of the project. However, there is one exception to this general rule. At the site of Colliemore, on the island of Arran, Argyll (Figure 1) a bar of iron (Figure 8, bottom), perhaps a 'currency bar' has been recovered amidst the finds of a bloomery mound. Slag was also recovered from the site (see Table 1 *BAR4, BAR8, BAR9*). Although both slags and artefact (see Table 1 *BAR13*) were not stratified, there is a tradition of iron working around the hill side which includes a charcoal burning platform and a bloomery slag heap. One knoll to the south-east of this heap of slag was known as 'Hammer Head' and a second knoll as 'Smith's Hill' (MacAdam 1887, 98). The metallographic investigation of two sections of this 'currency bar' is shown in Figures 9a and 9b. The Arran bar consists of a ferritic matrix with fine pearlite growing at the grain boundaries. The fine submicron specs within the ferritic matrix must correspond to undissolved carbides. Hardness testing on the ferrite grains ranged between Hv (100)=136-142 suggesting phosphric iron. The analysis of its slag inclusions were undertaken, in order to establish the elements which are expected to be carried from the slag to the artefact in the context of Highland bloomeries. The high level of the phosphorus in the fayalite within the slag inclusions was unexpected, but the manganese was within expected levels. Thus, while Highland bloomery slags are characterized by their manganese, phosphorus and alumina content, slag inclusions within Highland bloomery artefacts may be 'fingerprinted' albeit in a general way by their manganese and phosphorus contents.

VI. The work of the Highland smith in Gaeldom – some early conclusions

The ubiquitous nature of bloomery mounds in the Scottish Highlands testify to the continuous need for iron making and extraction. Whether for weaponry or agricultural tools, a small scale bloomery furnace could be easily set up, given the ready availability of raw materials (both ore and charcoal). The historical evidence for production of weaponry and tools in the Medieval period appears to support the supposition that Highland smiths were the source of production in Gaeldom, from initially winning the ore and forming the bloom to creating the finished article This, however, has not been reflected so far in the sites under consideration within this project.

The absence of artefacts, allowing the assessment of the level of carbonization or extent of presence of phosphoric iron, is one of the reasons for the present inability to establish the type of iron technology involved with these sites. This is not the case in the Lowlands, where there are analyses of phosphorus-rich knives (Cox 1996) and metallurgical waste (Photos-Jones & Atkinson *in press*). In the Lowlands, there is a strong tradition of sword blades being imported from the continent and the urban centres providing expert smiths for the basket hilt swords (Grant & Cheape 1987, 197). Indeed there are swords which are signed by two different individuals.

The contrast between the sites such as Tamheich Burn and Allt na Ceardaich in Argyll with those smithies excavated within Lowland burghs introduces an interesting dichotomy. Recent excavations at Edinburgh Castle (Driscoll & Yeoman 1997) and Meal Vennel in Perth (Cox 1996) indicate fairly sophisticated smiths' workshops operating from at least the 13th century. When this is contrasted with the scale and relatively unsophisticated nature of rural furnace sites, it would appear to support Yeoman's contention that "the winning and smelting of metals took place in the countryside, the unsmithed products being supplied to blacksmiths in the burghs" (1995, 76). If Yeoman is correct, this would suggest that the ethnographic and folk history relating to smiths within clanship is inaccurate, as their role would simply have been to produce billets of iron which would be traded in Lowland burghs for finished weapons. Although trading for weapons may well have been achievable in Highland areas fringing on the Lowlands, it was clearly not such a simple matter to deliver quantities of goods to the far north of the country. In this scenario the *fine* and the *buannachan* would have been severely limited as a fighting force in times of warfare. In this case, Yeoman's assertion appears improbable and a generalization at best.

How has the position of the smith within the clan system been established as a result of this multidisciplinary work associated with the project?

a) Bloomery mounds show a high uniformity in slag chemical composition and mineralogy and this would probably extend to iron as well. The type of phosphorus or carbon-rich iron produced at those sites impinges on the recovery of metal artefacts or fragments of bloom. Although the former is less likely, the latter may yet be forthcoming.

b) Contrary to the Tamheich Burn smith who may have smelted his iron 'here one day, somewhere else the next', in a sense following his ore and using local clays for his furnace, the Allt na Ceardaich smith built quasi-permanent installations with the intention of producing iron to meet specific local demands. The nature of these installations and the surviving place names (e.g, both the Gaelic *Ceardaich* and *Gobha* are translated as 'smith' and commonly appear in place names in the Highlands) most likely echo the establishments that the MacEacherns or other known families of hereditary craftsmen might have set up.

c) Although working/smithing floors were not evident at Tamheich Burn, they were indeed present at Allt na Ceardaich. Could Allt na Ceardaich smiths be producing claymores or other forms of weaponry or armour for the clan chief and his men? Working floors for the making/ shaping of billets can be set up relatively easily, the furnace floor can potentially act as one. The special care taken for the preparation of the working floor at Allt na Ceardaich suggests that smithing was taking place on site. Therefore, at present the evidence suggests that smithing as testified by both dedicated areas and slag typology was an integral part of the operation.

In conclusion, the producers of iron within the Scottish Highlands during the Medieval periods, although small scale in nature, clearly held a pivotal role within the cultural milieu of clanship. The need for a readily available source of weaponry and tools for the *fine* and *buannachan* essentially made the rural smith an independent, but indispensable, member of the community. This, of course, has been reflected in the folk traditions and toponomy of the Scottish Highlands, however, it has not, as yet, been observed within the archaeological record. Although this paper goes some way to altering that perception, further work is still clearly necessary in establishing the 'history' of early production in the Scottish Highlands.

Acknowledgements

Thanks are due to a number of individuals and organizations for funding and services which have made this paper possible: the fieldwork and analyses discussed in this paper were funded by Historic Scotland, Glasgow University Archaeology and Geology and Applied Geology Departments and Strathclyde University Metallurgy Department. We are also grateful to Historic Scotland for providing a grant to allow the paper to be delivered at the Metals in Antiquity conference at Harvard. Mrs Fiona Gorman of Arran supplied samples of the Arran material for analysis and gave freely her time. Figure 2 is reproduced by kind permission of the Royal Commission on the Ancient and Historical Monuments of Scotland and Figure 1 was produced by JillSievewright of GUARD. We are also grateful to Dr Alex Morrison for fruitful discussions on the topic of Highland history and Dr Iain Banks for permissions to reproduce the Woodend currency bar.

Table 1													
		Na20	MgO	Al2O3	SiO2	SO3	P2O5	K2O	CaO	TiO2	MnO	FeO	OTHER
BCP10	AREA	1.84	0.27	7.82	19.79	0.41	0.63	1.49	2.11	0.39	8.75	56.54	
	WUSTITE	0.43	0	0.42	0.99	0	0.23	0.07	0.06	0.38	6.56	90.67	
	FAYALITE	2.23	0.6	9.06	29.03	0.19	0.27	2.14	2.96	0.26	10.91	42.04	
	INTERST	0.54	0.18	10.49	31.54	0	0	2.15	3.41	0.32	10.85	40.07	
	HERCYN	0		28.41	3.13	0	0	0.2	0.24	1.15	6.4	60.49	
BCP13	AREA	1.69	0.63	5.14	21.83	0.41	0.81	0.68	1.21	0.41	7.12	60.11	
	WUSTITE	1.12	0.63	0.86	1.06	0.62	0.48	0	0	0.91	3.15	90.89	
	FAYALITE	0.71	0.79	0	24.38	0	0.87	0.06	0.32	0	10.15	62.61	
	INTERST	2.31	0	16.85	35.33	0.81	0.91	4.88	4.4	1.14	2.35	27.04	
BCP14	AREA	1.01	0	7.18	20.12	0.65	0.63	1.17	0.73	n	4.72	63.79	
	WUSTITE	1.84	0.41	1.06	2.01	0	0.59	0.06	0.05	0.63	2.65	90.8	
	FAYALITE	1.3	0.59	1.76	21.04	0.06	0.47	0.25	0.3	0.2	6.16	67.85	
	INTERST	3.88	0	15.42	37.77	1.72	1	5.99	4.29	1.47	2.09	26.4	
BCP16	AREA												
	WUSTITE	0	0	0.64	1.01	0	0.24	0	0.12	0.36	7.27	90.37	
	FAYALITE	0	0	0	26.91	0.3	1.1	0	1.23	0.11	17.71	52.65	
	INTERST	2.85	0	17.08	34.71	0.61	0.86	6.28	8.16	1.89	5.48	22.07	
XSB1	AREA	0	0.36	9.22	32.14	0.39	0.53	2.43	2.12	0.9	15.57	36.33	
	WUSTITE	0	0.57	0.33	27.26	0	0	0	1.28	0.07	23.63	47.77	
	FAYALITE	0	0	21.03	51.29	0	0	15.51	0.86	0.16	0.79	3.27	
	INTERST	0	0	24.33	21.87	0.4	0	4.6	2.68	5.71	6.05	33.04	
		Na20	MgO	Al2O3	SiO2	SO3	P2O5	K2O	CaO	TiO2	MnO	FeO	OTHER
													Ba
AC21	AREA	O.65	0.06	4.93	17.66	0.56	0.19	0.52	0.35	0.29	3.75	70.82	0.21
AC33	AREA	0.39	0.28	8.1	20.06	0.41	0.33	1.43	0.75	0.21	11.54	54.96	0.42
	WUSTITE	0.02	0.06	0.16	0.35	0.03	0.07	0.01	0	0.36	9.05	89.87	0
	INTER	0.11	0.72	46.65	0.64	0.03	0	0	0.02	0.71	6.99	44.11	0
	HERCYN	3.86	0	19.26	38.62	1.04	1.97	11.26	3.46	0.07	3.18	15.04	2.28
AC26	AREA	0.93	0.32	7.21	24.42	0.23	0.35	1.53	1.29	0.23	8.25	54.56	0.68
	FAYA	0.05	0.43	0.22	27.37	0	0.13	0.08	0.62	0	11.12	59.81	0.09
	INTERS	0.14	0	18.61	34.98	0.95	1.46	6.49	3.69	0.18	2.97	23.23	2.81
	WUSTITE	4.63	0	0.36	1.55	0.06	0	0.06	0.04	0.57	3.78	93.53	0
BAR4	AREA	0	0	3.97	19.52	0.2	0	0.64	1.65	0.12	5.14	66.77	Fe/Ni/Cr
	WUSTITE	0	0.18	0.9	1.37	0.32	0.34	0.03	0.3	0.4	2.03	93.86	
	FAYALITE	0	0.79	0.13	24.08	0.2	0.54	0	0.7	0.08	7.17	66.33	
	INTERST	2.69	0	14.27	35.02	1.77	1.42	6.15	8.51	0	1.85	27.85	
BAR8	WUSTITE	0	0	0.22	1.13	0.16	0	0.08	0	0.34	2.41	95.24	Zr inclusions
	FAYALITE	1.8	1.02	0	27.53	0	0	0	0.81	0.03	7.67	61.15	
	INTERST	0	0	13.64	38.49	1.33	0.31	4.21	7.23	0.43	2.1	31.89	
BAR9	WUSTITE	0	0.27	5.15	9.58	0.29	0	1.11	0	1.34	0.23	81.76	Zr inclusions
	HERCYN	0	0	22.68	41.12	0	0	6.96	0.07	0.04	0	29.11	
BAR13	WUSTITE	0.69	0	0.45	1.05	0.04	0.32	0	0.06	0.07	3.27	94.48	0.06
	FAYALITE	0.34	0.84	1.89	20.01	0.09	9.87	0.73	3.21	0.04	8.93	54.62	0.02

Table 1. Sample specification and SEM-EDAX analyses of Tamheich Burn, Allt na Caerdaich and Isle of Arran metallurgical waste; concentration in normalised weight percent.

References

AITKEN, W. G. 1970. Excavations of bloomeries in Rannoch, Perthshire and elsewhere. *Proceedings of the Society of Antiquities of Scotland* 101: 188-203

ATKINSON, J. A. & PHOTOS-JONES, E. 1995. *Medieval metallurgical waste and the preservation of bloomery sites: a feasibility study*, GUARD/SASAA interim report, Glasgow.

1997 *Scottish Bloomeries Project: phase 2, the excavations*, GUARD/SASAA interim report, Glasgow

BOWLER, D., COX, A. & SMITH, C. 1995. Four excavations in Perth 1979-84. *Proceedings of the Society of Antiquaries of Scotland* 125: 917-1000

BUTT, J. 1966. The Scottish iron industry before the hot-blast. *Journal of the West of Scotland Iron and Steel Industry* 73(6): 193-220.

COX, A. 1996. Backland activities in Medieval Perth: excavations at Meal Vennel and Scott Street. *Proceedings of the Society of Antiquities of Scotland* 126: 733-822.

DEVINE, T. M. 1994. *Clanship to the Crofters' War: The Social Transformation of the Scottish Highlands*. Manchester University Press.

DIXON, J. H. 1886. *Gairloch in north-west Ross-shire*. Edinburgh.

DRISCOLL, S. T. & YEOMAN, P. A. 1997. *Excavations within Edinburgh Castle in 1988-1991*, Society of Antiquaries of Scotland Monograph Series no 12, Edinburgh.

FELL, A. 1908. *The Early iron Industry of Furness and District*, Ulverston.

GRANT, I. F. 1961. *Highland Folk Ways*, London.

GRANT, I. F. & CHEAPE, H. 1987. *Periods in Highland history*, London.

HALL, A. J. & PHOTOS-JONES, E. in press The bloomery mounds of the Scottish Highlands: part II, a review of iron mineralisation in relation to the investigation of bloomery iron making in the Scottish Highlands. *The Journal of the Historical Metallurgy Society*.

HOLDSWORTH, P., BLANCHARD, L.M., FORD, B., HODGSON, G. W. I., MACASKILL, N. L., ROBINSON, D., ROSS, L., SMITH, C. & SPEARMAN, R. M. 1987. *Excavations in the Medieval Burgh of Perth, 1979-1981*. Society of Antiquaries of Scotland Mongraph Series no 5, Edinburgh.

LEWIS, J. 1984. The charcoal-fired blast furnaces of Scotland. *Proceedings of the Society of Antiquities of Scotland* 114: 433-79.

LINDSAY, J. M. 1977. The iron industry in the Highlands. *Scottish Historical Review* 56: 49-63.

MACADAM, W. I. 1887. Notes on the ancient iron industry of Scotland. *Proceedings of the Society of Antiquities of Scotland* 21: 89-131.

MACINNES, A. I. 1996. *Clanship, Commerce and the House of Stuart, 1603-1788,* East Lothian.

MURRAY, J. C., BROOKS, C., DENT, J., DICKSON, J. H., FRASER, M., GREIG, J. C., HODGSON, G. W. I., JONES, A., MURRAY, H. K. & STONES, J. A. 1982. *Excavations in the Medieval Burgh of Aberdeen, 1973-1981*, Society of Antiquaries of Scotland Mongraph Series no 2, Edinburgh.

PHOTOS-JONES, E., ATKINSON, J. A., HALL, A. J. & BANKS, I. in press The bloomeries of the Scottish Highlands: Part 1, the archaeological background. *The Journal of the Historical Metallurgy Society*.

PHOTOS-JONES, E. & ATKINSON, J. A. in press Iron working activities in Medieval Perth: a case of Town and Country? *Proceedings of the Society of Antiquaries of Scotland* 128.

PHOTOS-JONES, E. & HALL, A. J. 1998. *Scottish Bloomeries Project: phase 2, the analyses,* GUARD/SASAA interim report, Glasgow.

ROSTOCKER, W. & BRONSON, B. 1990. *Pre-Industrial Iron*, Pa.

RCAHMS. 1982. *Inventory of Argyll: volume 4, Iona*, Edinburgh.

STEER, K. A. & BANNERMAN, J. W. M. 1977. *Late Medieval Monumental Sculpture in the West Highlands,* Royal Commission of Ancient and Historic Monuments, Edinburgh.

YEOMAN, P. 1995. *Medieval Scotland: an archaeological perspective,* London.

36

"The manner how it grew was like unto the haire of a man's head": the early 1600's discovery and exploitation of native silver at Hilderston in Scotland.

E. PHOTOS-JONES[1,2], A.J. HALL[2,3], T. POLLARD[4], T.K. MEIKLE[5] & A. NEWLANDS[3]

[1] *Scottish Analytical Services for Art and Archaeology, Glasgow, UK*
[2] *Department of Archaeology, University of Glasgow, Glasgow, UK*
[3] *Department of Geology and Applied Geology, University of Glasgow, Glasgow, UK*
[4] *Glasgow University Archaeological Research Division, Dept of Archaeology, University of Glasgow, Glasgow, UK*
[5] *13 Tassie Place, East Kilbride, UK*

Deposits of native silver around the world are rare and most were probably exploited and exhausted by prehistoric times. However, one such unusual deposit was discovered and worked in the early 1600s at Hilderston in the Bathgate Hills in the Midland Valley of Scotland. At Hilderston as well as nearby Linlithgow there is historical and archaeological evidence for "melting, fining and stamping mills". Such evidence seems contrary to the conventionally accepted view that silver was recovered from the parent rock by simply crushing, grinding and washing. This paper presents an assessment of the Hilderston site where, in addition to the working of the native silver, there is evidence for continuous exploitation and development of coal seams, lead mining, limestone quarrying and lime-making. Attributing extant remains and industrial waste to these diverse activities is a major problem in reconstructing the Hilderston industrial landscape at that time. Establishing the methods used to process the silver is also problematic in the absence of visible remains and full scale excavation.

Keywords: Native silver; mining; processing; Hilderston; Scotland.

I. Introduction

In the autumn of 1606, Sandy Maund , a Scottish collier, while prospecting for coal by the "*Silver bourne, under the hill called Kerne-Popple*" in the Bathgate Hills, now Hilderston (NGR: NS 991717) (Figure 1), made a chance discovery of native silver. The event was vividly described thirteen years later by Stephen Atkinson, a metals assayer who became actively involved at the mine. *"And this Scotsman, by meanes of digging the ground, hitt upon the heavy peece of redd-mettle; no man thereabouts ever saw the like. It was raced with small stringes, like unto haiers, or thredds."* (Atkinson 1619: 47). Atkinson proceeded to provide a description of the ore and, being an assayer, also assessed its value: "*the manner how it grew was like unto the haire of a man's head, and the grasse in the feilde. And the vaine thereof, out of which I had it, was once two inches thicke, by measure and rule: the mettle there was both malliable and toughe. It was course silver, worth 4s vid the ounce weight; not fine silver, as is made by the art of man."* (Atkinson 1619: 49). From that momentous day in 1606, there followed four years of intensive exploration of this ore deposit with a substantial capital outlay provided by the Crown. However, the entire venture collapsed in early 1610.

Metals exploration, particularly of that scale, should always be seen in the political and historical context of its time, of the people behind those ventures and of the events that justified their actions. In Scotland the Union of the Crowns in 1603 had just led to James VI of Scotland becoming King James I of England and VI of Scotland. In the popular mind King James's reputation has rested on greed, the Scottish King who turned his back on his native land once he tasted the finery and intellectually stimulating high life of the London court. Yet "few sovereigns have equalled his literary output and no other prince ever carried to the English throne as extensive a knowledge derived from reading and study" (Donaldson 1994: 256). His interests were particularly theological in nature, the translation of the King James Bible being one of them. On less academic grounds and more relevant to the present case, he also gave his wholehearted support to "projects", through the granting of patents and monopolies to individuals, usually over-confident adventurers who were willing to undertake risky enterprises or concentrate on patents for new processes. The move was perhaps a reaction to "the conservatism of the burghs, jealous as they were of the exclusive rights of the inhabitants" (Donaldson 1994: 245). The "conservatism of the burghs" must really be interpreted as the conservatism of the guilds which certainly by that time must have dominated most craftsmen's activities. The guild of the Perth Hammermen, to mention one (Hunt 1889), included blacksmiths, plumbers, armourers, gunsmiths, brass and pewter workers, silver and gold smiths (Simpson & Stevenson 1982: 5). The extent to which their members included workers associated with raw metal production or indeed the assaying/refining stage, is a point worth pursuing further in the future.

This paper focuses on one particular venture or "project", that of the extraction and processing of silver at Hilderston in the Bathgate Hills. It presents an overview of the documentary evidence and a topographic survey undertaken by the authors. The purpose of the research was to update the 1772 plan which includes details of the further mining activities of the late 1700's. The paper also presents the scant mineralogical and industrial waste-based evidence from the Hilderston site relating to the extraction of the native silver. These materials resulted from metallurgical operations carried out locally prior to the setting up of the Linlithgow stamping, fining and melting mills, which were presumably larger and more elaborate installations for the processing of the Hilderston ore. The activities at

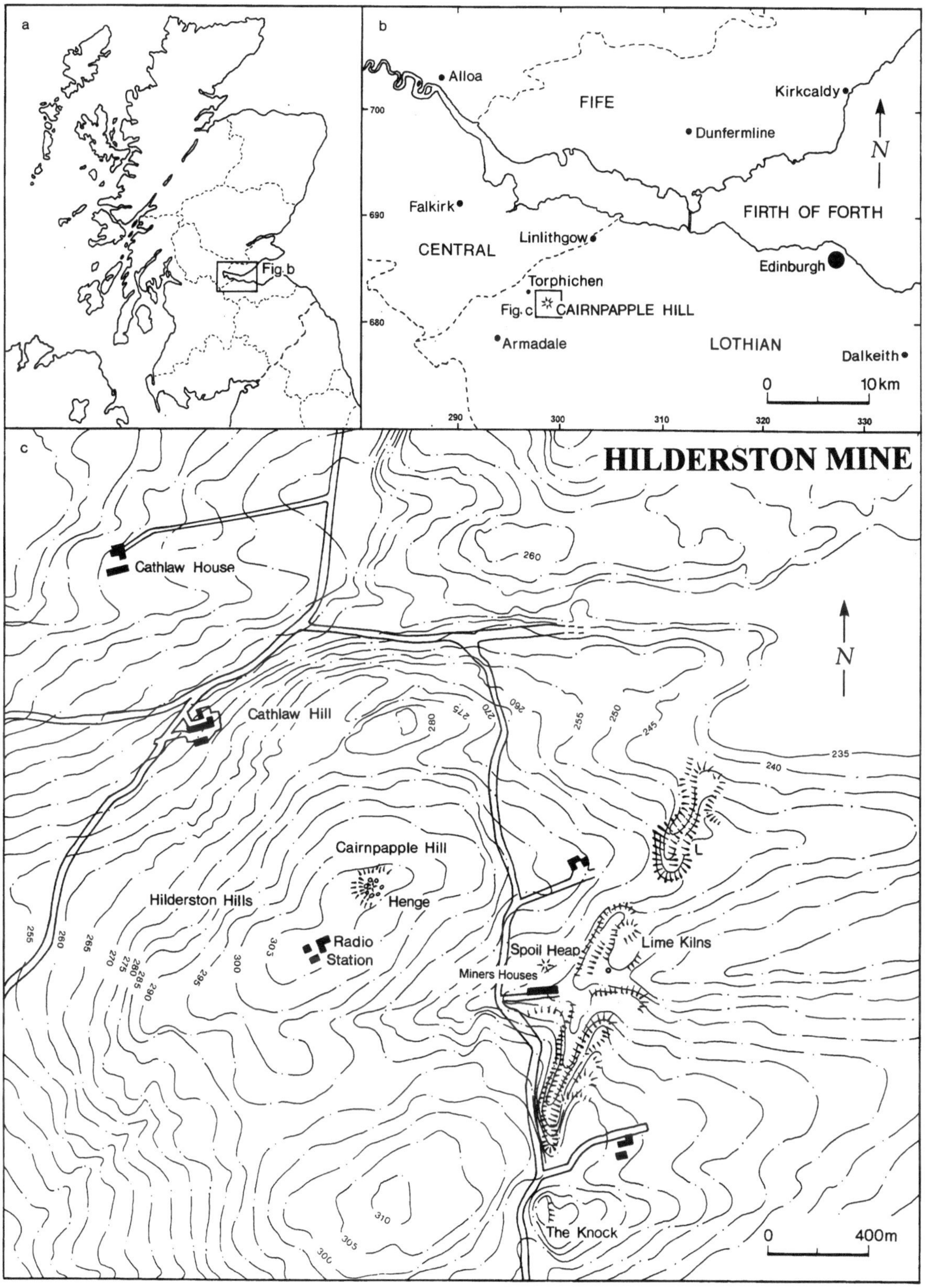

Figure 1. Locality map for Hilderston mine, close to Cairnpapple Hill in the Midland Valley of Scotland.

Linlithgow, about 7 km north of Hilderston, (Figure 1), although well documented, are difficult to confirm archaeologically given that a major motorway has obliterated the probable site. Hence the need to concentrate at the Hilderston site. The present assessment has succeeded in highlighting two points: a) the problem posed in reconstructing the Hilderston industrial landscape at that time in view of the continuous exploitation and development of coal seams, lead mining, limestone quarrying and lime-making; and b) the problems associated with the identification of the methods used to process the native silver in view of the complexity of the ore and the lack of sufficient archaeological evidence.

II. Hilderston: the historical evidence

It was Sir Bevis Bulmer, of Leadhills, well known metallurgist and assayer of his time (Tyson 1996) who first analyzed Sandy Maund's original sample of the silver ore.

He was considerably surprised by the result of his first assay "*because it proved, riche .. (he) went to it againe, and againe, and still it proved rich, and wondrous rich.*" (Atkinson 1619: 48). Sir Bevis's results were sufficient incentive for Sir Thomas Hamilton, the landowner but also the King's Advocate, a notable lawyer and 'a very clever man', to proceed on January of 1607 to acquire a lease from the King for all the minerals in the district (Cadell 1925). He promptly commenced mining at Hilderston with Bulmer in charge assisted by Atkinson as his refiner. At Martinmas (14th November) 1607, Hamilton informed the King's Privy Council that within the previous 12 months he had mined 30 tons of ore. The King's tenth part reserved by Act of Parliament amounted to 20 stones of "clean silver"; the cost of the year's working amounted to £700 sterling. Hamilton had calculated that the profit was likely to be about £500 sterling per month clear of all charges (Cochran-Patrick 1878: 117).

As the King's Advisers became aware of the apparent richness of the mine, Sir Bevis and five other Commissioners were ordered to Hilderston to evaluate the ore. To be assured of proper assaying the King issued a *'warrand for trying of the quantitie and qualitie of the ure wone be his Majestie his Commissioners'* and the *'Lordis of Secreit Counsell'* authorized ore to be barrelled but *'three to four stane weight or theirby'* of representative ore to be sampled and smelted. The ingot produced was to be divided in the presence of commissioners into two - half for Sir Bevis Bulmer, master and surveyor and half for the *'Lordis of Secreit Counsell'*. (Cochran-Patrick 1878: 120). Enthusiasm was running high: "*And for the riches it far exceeds annye that euer was in Garmanie.....*" (Cochran-Patrick 1878: 128). In the light of analyses of the ore at the Tower of London this enthusiasm was well justified. 100 lbs of ore produced 22 oz silver with lead being used to extract the silver. (Cochran-Patrick 1878: 130). A further 10 tons of ore were barrelled and sent by sea for trials at the Tower of London and elsewhere, followed later by a further 20 tons for smelting in iron furnaces in Sussex.

In April 1608 the workings were officially repossessed and the mineral rights were reverted to the King with a compensation to Sir Thomas of £60,000 Scots (£5000 Sterling) and the appointment of Sir Bevis Bulmer as head (Cochran-Patrick 1878: 126). Meanwhile, in order to process the large quantities of the rich ore expected, a mineral processing plant was designed and constructed at Linlithgow. The exact localities of the mills have been lost and with extensive road and building development in the area, there is little chance now of locating the sites. A complete series of accounts spanning the twelve month period between May 1608 - May 1609 illuminate all aspects of the activities of the *'Melting Fining and Stamping Mills'* at Linlithgow following their transcription from original records (Meikle 1996). "Slag", "cinder" and other types of industrial waste are scattered throughout the area but the analyses of a fair number of samples thereof collected in the course of two surveys remained inconclusive (Photos-Jones & Hall 1997).

By the end of October 1608, the results of the first trials were reported. In the opinion of Sir Richard Martyn , Master of the London Mint, "*... there is in ye Scottish ure, a substance of a matter which some call a marquisit, and other some an arsenick, and others a sulfurous matter wch houldeth the silver, y^t it cannot easily be gathered out of the same together, and also there is a stone called a white sparr, wth holdeth some smale quantinty of silver, wch poisoned matters if they could be destroyed, and the silver gott cleane out of ye same, it were a good and profitable service for his Matie wch is to be don by men if art and knowledge experienced in ye practice of ye same*". (Cochran-Patrick 1878: 135). The tone is dry and factual but the warning is clear that mineral separation is required if the silver is to be recovered and this should really be undertaken by men of art and knowledge experienced in the practice of the same.

By December of the same year it became clear that things were not going according to plan. Smelting silver in iron ore furnaces in Scotland was not followed by successful results in London and Sussex where it was also more expensive to undertake (Atkinson, 1619).

The main problem was evidently the variation in ore grade. "*Until the same redd-mettle came unto 12 faddomes* (18m) *deepe, it remained still good; from thence unto 30 fathome* (55m) *deepe it proved nought: the property thereof was quite changed miraculously in goodness, it was worth little or nothing; and more, uppon an instant, after the Bronswicks entered, it was quite altered in quality, but not in colour, fashion, and heaviness.*" Atkinson (1619:150).

By May 1609 smelting operations were probably moved to Linlithgow and were probably in full swing when, by August 1609, an estimated 120 tons of undressed ore containing 31 lbs of silver per ton had been processed, producing 3720 lbs of silver worth £9920 Scots (Meikle 1996). Silver was coined but the weight of coins produced is unknown. Total running costs were £45,838 Scots for the mine and mills to 7 May 1609. The operation would not have been profitable until the summer of 1610 if the grade had been maintained (Meikle 1996). However, it seems that this proved not to be the case and the accounts for the whole period of 1 May 1608 to December 1610 demonstrate how unprofitable the nationalized silver mine was (Meikle1996). It was inevitable that operations would soon cease and so they did in December 1610.

In March 1613 the Crown gave up all efforts and the mine was let to Sir William Alexander of Menstrie, Thomas Foulis, goldsmith in Edinburgh, and a Portuguese, Paulo Pinto. (Cochran-Patrick 1878: 159). The conditions were the payment of 1/10 of the ore to the Crown and the right to buy back the mine for £100,000 Scots, on six months notice. The ore could be refined by the said individuals with 1/3 of the silver given to the Crown. All the silver was to be brought to the Edinburgh Mint (Cunyie House) to be melted, fined and coined (Cochran-Patrick 1878: 159). Needless to say , this option was never exercised by the Crown.

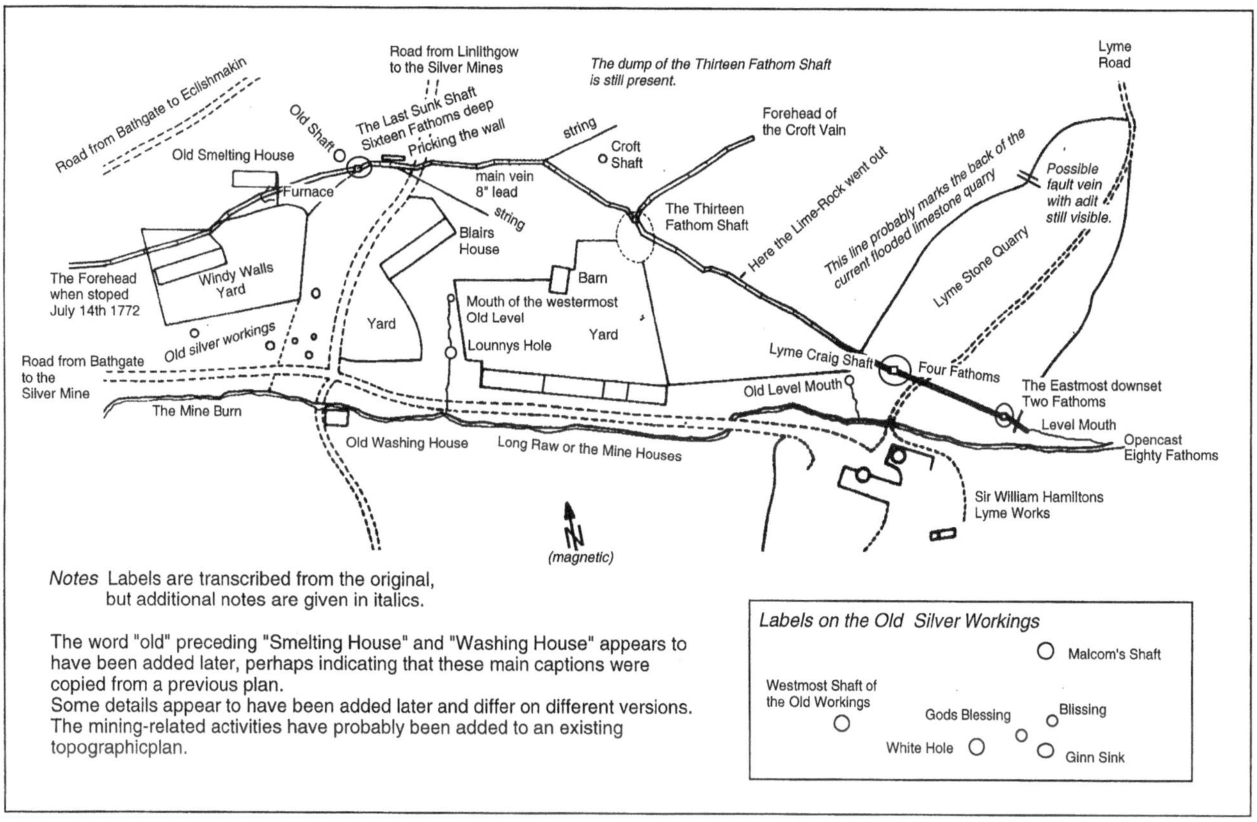

Figure 2. Details from the plan of the "Works of Silver Mines of Hilderston and Tartraven" surveyed by Joseph Udny 14 July 1772.

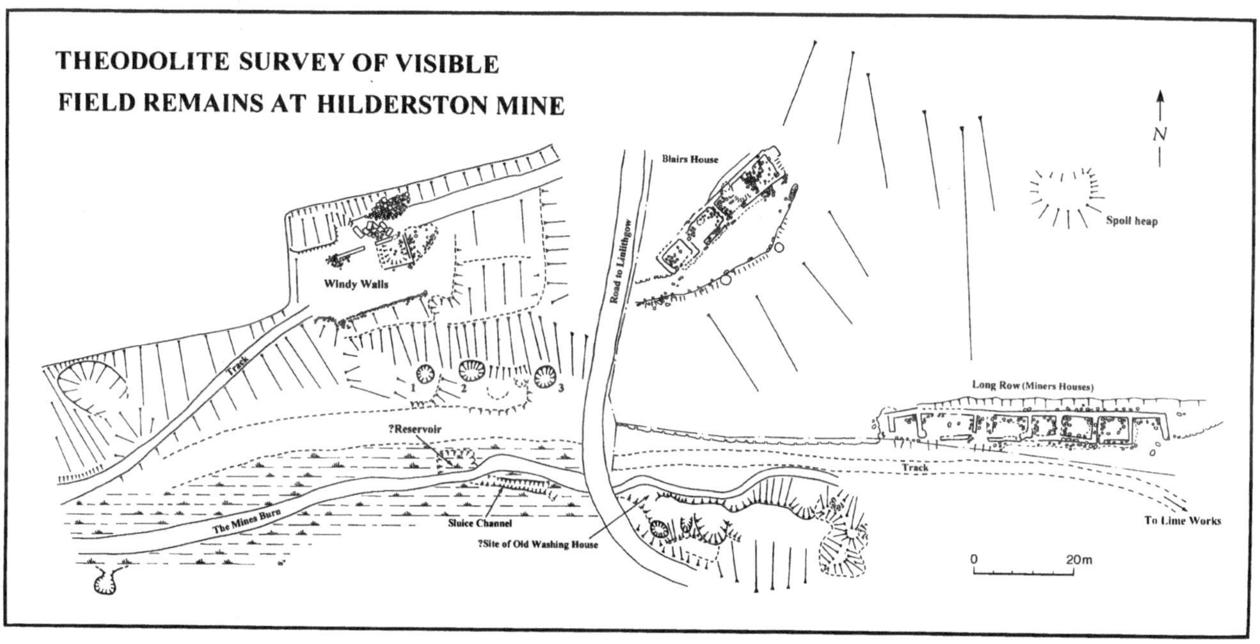

Figure 3. Theodolite survey of visible field remains at Hilderston mine, May 1996.

III. Udny's map of 1772 and the 1996 archaeological survey of the area

The assessment of an archaeological site normally involves the matching of the field remains, be it features or materials, with known documentary sources referring to both structures and processes. One of the most useful documents about the overground and underground remains at Hilderston is undoubtedly the plan of the site, titled *"plan of the Works of the silver Mines of Hilderstoun and Tartraven"* prepared by the surveyor Joseph Udny, dated 14th July 1772 (Figure 2) and held in the archives of the Hopetoun Estates. This date evidently marks the end of late 18th Century mining activities started in 1766.

Udny's plan shows not only surface features but also the underground levels and shafts. On initial inspection the resemblance between this plan and the area as it appears today is a close one. The modern N-S road follows the same course as that labelled *"the road from Linlithgow to the silver mine"* but the other old roads are now only tracks or minor topographic features. It was this network of roads which provided the infrastructure for both the silver mine and the associated workings, representing routes along which the products of extraction were transported out of the area. The most important of these routes would inevitably have been the road from Hilderston to Linlithgow.

a. The mining shafts and pits

Udny's plan clearly shows six open shafts distributed around the southern and eastern perimeter of the yard which surrounds *Windy Walls*. These are*: Westernmost shaft of the old workings, White Hole, Gods blessing, Blissing, Ginn sink, and Malcolm's shaft.* Atkinson (1619) reports that when Henry Starchy or Starkie, arrived from Germany in October 1608 three additional shafts were sunk and the shaft names at that time were: *Blissing Shaft* (also known as *Grace and Benison*), *Germans east shaft, Harliers shaft, Serve them all (set down by David the German), Long Shaft*, and *Blak* (sic) *shaft.* From the foregoing it can be concluded that at least *God's Blessing, Westernmost shaft of the old workings (Germans West shaft?) and Blissing* relate to the earliest use of the mine.

The topographic survey revealed positive evidence for only three pits (Figure 3), all of which were represented by shallow circular depressions or craters. The most easterly (3) of these is probably *Blissing* while the central pit (2) is *God's Blessing* and the most westerly (1) *White Hole*. There was no trace of *Ginn sink* or *Western*. The slightest suggestion of two pits could be seen between *God's Blessing* and *White Hole*. These do not appear on Udny's plan and may therefore be earlier and filled by the time Udny carried out his survey. However, it is equally possible that they are later since there was activity on the site up to the beginning of this century. No sign of Malcolm's shaft was found during the survey but examination of photographs taken does suggest the presence of a slight depression in the side of the terrace in roughly the same place shown on Udny's plan.

Several other pits are shown on Udny's plan but no sign of these can be seen today. Udny's plan also shows underground tunnels. An opening, presumably towards the main shaft, is currently visible in the sandstone face above the limestone in the back of the quarry. The tunnel appears to be well preserved apart from some collapse of the shale roof and indeed relatively modern timber pit props were visible some short distance from the mouth of the shaft.

b. The ore dressing areas : the 'Old Washing House'

This building has been drawn by Udny as an open rectangle situated immediately to the south of the burn (shown as the *Mines Burn*) where it crosses the north-south road (suggesting that the burn today is culverted beneath the road where there was originally a ford). Today, no trace of this building is visible, although an oval/sub-rectangular depression can be seen immediately to the south of the area shown as the building on the plan. However, this may represent a quarry scoop related to the construction of the road which curves around to the west immediately behind the pit. Further, more irregular, scoops are located to the west of this feature and may also relate to the construction of the road, or indeed work related to the construction of the mine building. The use of the term *"old washing house"* on Udny's plan may suggest that it had fallen into decay by 1772, its ruinous state and lack of roof perhaps being the reason behind the use of an open rather than and inked-in rectangle to portray it.

Although no sign of the washing house was detected, closer inspection of the burn's banks did reveal archaeological remains. The southern edge of the burn on the other side (west) of the road showed distinct signs of modification. Here a well preserved trough with a V-shaped profile ran parallel with the burn for some 15m. The presence of large slabby stones in places suggests that this feature may have been stone-lined (Figure 3). This trough feature appears to be related to a rectangular depression cut into banks of the burn on its northern side. The eastern terminus of the trough appears to join with this depression running away from it westwards along the side of the burn. This feature may have served as a header-tank or reservoir supplying water to the trough, perhaps including a sluice gate at its junction with the trough. It is possible that the trough at one time continued further to the east, perhaps running through or past the washing house which sat beside the burn on the other side of the road. However, the construction of the modern tarmac road and the culverting of the burn beneath it appear to have truncated the trough. It is also possible that the site of the washing house has been partially buried beneath the road, which today is undoubtedly wider than it was at the time of Udny's survey.

This series of burn-related features may be envisaged as serving an important function during the life of the mine. Water channelled through the trough would have been used to wash ore removed from the nearby pits prior to being broken up with hammers. Both of these processes are illustrated in Agricola (Figure 4) with water being diverted from a main channel over a trough containing the ore. In the background workers use flat headed hammers to crush the ore on circular stone platforms prior to washing . Similar platforms may have been needed assuming the ore was found embedded in gangue. However, none were presently visible.

c. The smelting area: the "Smelting House"

No sign of the smelting house shown on Udny's plan survives on the ground today. Udny's plan shows the furnace sitting within an enclosed area, but this does not appear to represent a building as such, more closely resembling the yards seen elsewhere on the plan. At the time of the present survey, the site of the smelting house lay within a gently sloping tilled field recently seeded with grass. In the absence of permission from the land owner, there was no opportunity to carry out a detailed surface evaluation of the area through field-walking. It is hoped that the site of the furnace will be located during a later phase of work when a magnetic susceptibility survey will be carried out in this area. Nevertheless, samples of "industrial waste" were collected from the area for further analysis (see section on materials).

d. Other structures

A number of other structures are shown on Udny's plan where they are annotated as *"Windy Walls"*, *"Blairs house"* and *"long row"* or the *"miner's houses"*. At least partial evidence for all of these features was identified during the survey, with *"Blairs house"* and *"long row"* being the best preserved of these stone built features. It is likely that these buildings relate to the accommodation of miners and their overseers, perhaps during the later years of the mine's use. However, the open fronted character of some of the units within the *"long row"* may indicate that they represented workshops, but only excavation will clarify this matter. Other features marked on Udny's map, such as the *"barn"*, have left no physical traces within the landscape, a factor which may suggest that they were constructed from timber rather than stone. A fuller discussion on these structures will be included within forthcoming papers on the site .

IV. The Materials Evidence: the geology of the Hilderston area and the mineralogy of the Hilderston ore

a. The geological setting of the silver mine site at Hilderston in the Bathgate Hills

Considering the small aerial extent (<0.5 km^2) of the study area surrounding the old silver mines, the geological setting has had a surprisingly strong impact on the history of the site. The geological parameters which have influenced the area can be classified as a) the general topography, soil, drainage and land use of the area as influenced by the underlying geology, b) the nature and distribution of the silver mineralization and its control on the location and extent of the silver mining activity, c) lead mineralization, and d) other mining/quarrying of resources provided by the local geology.

a1. The general topography, soil, drainage and land use of the area as influenced by the underlying geology

A diagrammatic geological map is provided in Figure 5. The dominant rock types in this area are Lower Carboniferous (about 340 Ma) shallow fresh water shales, sandstones and volcanics deposited on the southern flank of a major volcanic complex centred on Linlithgow, about 7 km to the north (Cadell 1925). The limestone represents a major marine excursion.

Figure 4. Breaking up the ore on stone platforms in the vicinity of water channels, sluice gate and trough, after Agricola (Hoover 1950: 272). A: area paved with stones, B: broken ore, C: area covered with broken ore, D: iron tool, E: its handle, F: broom, G: short strake, H: wooden hoe. Reproduced by permission of Mining Journal Ltd.

This mixture of lithologies underlying relatively gentle slopes might be expected to produce fertile soils but the area is of low agricultural quality as glaciation has stripped soil and degraded rock, leaving a thin smearing of a clayey soil. As a result the area became one of variable land use with mixed agriculture (mainly cattle and sheep) as well as mining and quarrying; the latter possibly practised interchangeably or in parallel.

a2. The nature and distribution of the silver mineralization and its control on the location and extent of the silver mining activity

There is relatively little known for certain about the details of the nature and distribution of the silver mineralization. The silver workings, marked on old maps (Figure 2) are close to a prominent geological feature - the N-S dike-like structure of quartz dolerite related to the Stephanian (295 Ma) sill/dike intrusive complex of the Midland Valley of Scotland. The N-S dike forms a high topographic feature to the immediate north and especially south of the silver mine site. The lack of its presence at the workings in the valley can be explained in two ways:

i) An E-W fault at this position provided a zone of weakness due to fragmentation, dislocation and some alteration of the dolerite dike. This was exploited by a stream and the stream valley was enhanced by glaciation.

ii) The dolerite dike was intensely altered (carbonated and hydrated to "white whin") by the hydrothermal mineralization event and independently of, or combined with faulting, this produced a weak zone exploited by the stream. There are many examples of fragments of such altered dolerite near the old pits and on the dump to the east of the road. This alteration was noted by Gallagher (1964). Iron- and magnesium-silicate rich rock such as dolerite is well known to be prone to hydrothermal alteration.

If it is assumed that the alteration of the dolerite is related to the silver mineralization event, then the alteration (a wall-rock reaction) would have produced a marked change in the chemical composition of upward-migrating hydrothermal solutions, perhaps driven by the heat of the dike (Russell 1985). This could well have led to the precipitation of silver as well as the associated nickel arsenide although the details of the chemistry of the process are not known. The silver mineralization is likely to be confined to the vicinity of the focal point provided by the intersection of the N-S dike and the E-W fault. The silver-mineralization is therefore likely to form an irregular pipe-like zone extending to an unknown depth.

a3. Lead veining

Minor lead-zinc mineralization is quite common in the Bathgate Hills in particular in association with the marine limestone. There are also small baryte veins. The 18th century and both earlier and later mine levels, mainly in the area east of the tarmac road and extending to the old limestone, quarry are likely to be exploration and/or workings of galena veins. The baryte might also have been worked and/or assessed since baryte is mentioned on a plan dated 1871/73 held in the Geological Survey office, Edinburgh and large fragments of baryte are common in the stream and adjacent ground on either side of the N-S road. However, this baryte could be remains of crushed ore dating from the 1600's.

The presence of such mineralization, and mining activities related to it, near the silver deposit contributes to the difficulty in understanding the nature of the silver mineralization, the history of its working and the characterization of any relicts of its processing and smelting. The relationship of the lead/baryte mineralization to the silver mineralization is not obvious because of lack of exposure and appropriate samples. However, limited petrographic evidence suggests that the silver mineralization preceded the baryte/galena phase of mineralization .

a4. Other mining/quarrying of resources provided by the local geology

Although at the outset of the investigation, it was evident that limestone had been quarried in the area, the extent and diversity of possible mining/quarrying near the silver mine site was not realized. Such activity impacts on the interpretation of the nature of the silver mineralization and its 17th century exploitation. The full history of rock working has not been established, but the following are the main observations.

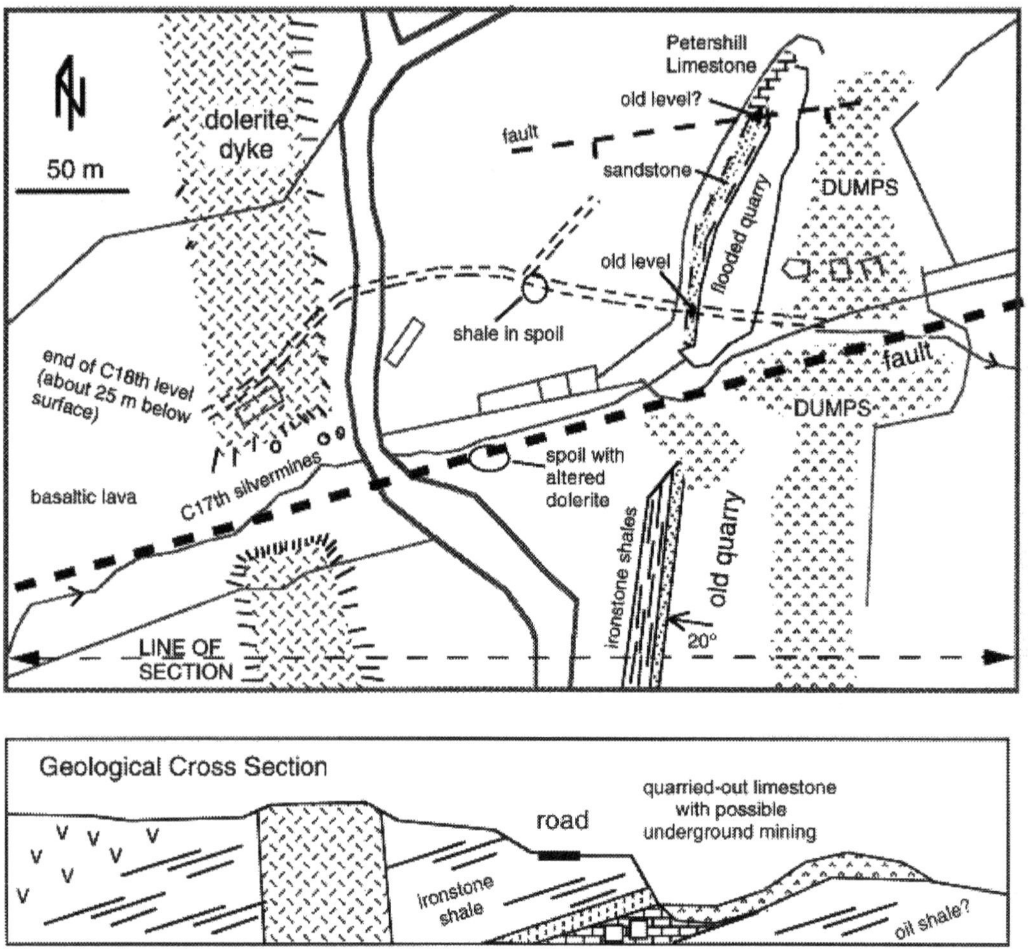

Figure 5. Geological sketch map and cross section showing the main features of topography, mining and quarrying at the Hilderston silver mine site.

i. Limestone

The major limestone worked in the Bathgate Hills is a thick Carboniferous marine limestone, the Petershill Limestone and this lies in the east of the area of interest (Figure 5). Historic maps of the area held in the archive of Hopetoun House Estate provide evidence that this limestone was worked after and during metalliferous mining on the site. Slaggy material, originally thought to be metallurgical slag, has proven on inspection to be associated with the operation of lime kilns (Photos-Jones & Hall 1997). This calc-silicate waste material is spread throughout the site and usually has a distinctive white to grey colour. Metallurgical slag can have the same colour but slag from the silver processing would only be confirmed if analysis determined the presence of a diagnostic chemical signature of Ag, Pb, Ni, As and/or Ba.

ii. Coal

There are old coal workings several hundred metres to the west on the western flanks of Cairnpapple Hill (Figure 1). Fine coal was found to be impregnating the lower crags immediately to the south of the old silver mine site. It is possible therefore that coal may have been washed on this site, perhaps using the mineral washing plant. Coal is also quite common in the soils in the vicinity of the old silver mines. Also, some of the old workings to the east of the road, e.g., *Croft Shaft* (Figure 2) are still unexplained and might represent coal workings. A coal seam, only a few centimetres thick, outcrops above the limestone on the western wall of the old quarry (Figure 5).

iii. Ironstone

The shale occupying the area to the west of the Petershill Limestone (Figure 5) contains ironstone in the form of layers and nodules of siderite, $FeCO_3$. Such material provided iron ore for the industrial development of Scotland in the 18th Century. It is possible that some mining was attempted, at least to assess the potential of this ironstone.

b. The mineralogy of the Hilderston ore

At Hilderston, as well as at Alva (Hall *et al.* 1982; Moreton 1996) in the Ochils, about 25 km to the NNW near Alloa (Figure 1) silver is found in its native state in an unusual and quite rare type of mineralization characterized by the presence of arsenic, with cobalt (particularly at Alva) and nickel (particularly at Hilderston). Silver is more commonly produced as a by-product of lead (galena, PbS) ore. Silver-rich galena usually contains some Sb, As or Bi because monovalent Ag cannot replace divalent Pb in the galena structure and silver-rich minerals may be present as micro-inclusions such as the ruby silver pyrargyrite, Ag_3SbS_3 or as more complex sulfides. Native silver can form as a result of near surface-oxidation of such silver-rich galena veins but there is no evidence for this secondary origin at Hilderston.

Native silver is now exceedingly rare at Hilderston but a sample of filiform silver with the following minerals was found and described by Meikle (1994: 86): nickeline (or niccolite) NiAs, annabergite $Ni_3(AsO_4)_2.8H_2O$, erythrite $Co_3(AsO_4)_2.8H_2O$, sphalerite ZnS, galena PbS, acanthite Ag_2S, pyrite FeS_2, dolomite $CaMg(CO_3)_2$, native Hg and amalgam Ag,Hg; also calcite $CaCO_3$, quartz SiO_2, chalcopyrite $CuFeS_2$, bravoite $(Fe,Ni)S_2$; unidentified Ni_2As_2S, and hydrocarbon (albertite). The limited paragenetic evidence indicates that there was an Ag-Ni-As phase of mineralization before a phase of Pb-Zn-S-baryte-hydrocarbon mineralization. As noted above, this latter phase is of common occurrence along the strike of the Petershill limestone in the Bathgate Hills.

The site of old silver workings is the focus of a marked geochemical soil anomaly (Chung 1991) with elevated levels of Ni, As, Pb, Ba, Zn, Co, Cu and Mn. The nickel and arsenic attain levels of 600 and 480 ppm respectively (Chung 1991). It is inevitable that the anomaly has received contributions from the natural outcrop of the mineralization, mining waste and smelting.

V. The Materials Evidence: The metallurgical waste from the furnace area, Hilderston

The materials evidence relating to both ore and metallurgical waste from mining and smelting respectively has been scarce at Hilderston and when available, at best inconclusive. Evidence for smelting must have centred around the site of the *Furnace* which lies currently at the edge of a ploughed field. The field has been ploughed repeatedly leaving few traces but fragments of heated remains of "furnace walls" are suggestive of attempts to smelt . Despite the paucity of the artefactual evidence three fragments are worth mentioning here.

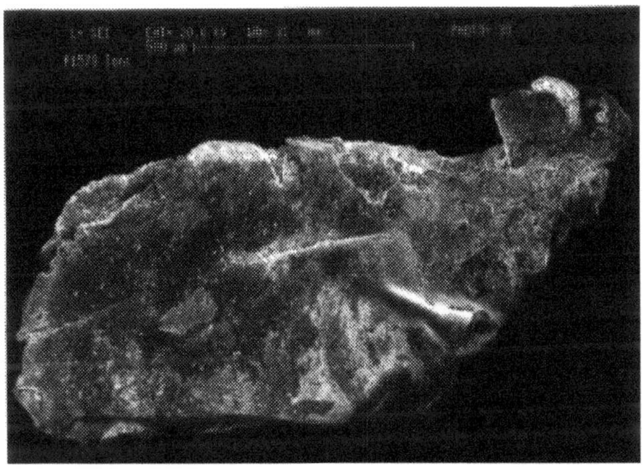

Figure 6. Scanning electron micrograph of a 1mm platelet of crumpled silver found near the site of the 'furnace' (Figure 2). EDX analysis confirmed silver with lead-rich areas. Scale bar is 500µm.

A 1 mm silver platelet with small fragments of translucent phase is most certainly the relict of silver processing. The thickness of the platelet does not exceed 20µm (Figure 6). The SEM X-ray analysis confirms silver and demonstrates the lead-rich nature of the translucent grains, presumably lead oxide or lead carbonate. Although the shape of the platelet suggests that it may have been one of these cuttings perhaps generated by the same type of shears illustrated in Agricola , the presence of lead "trapped" within the platelet implies cupellation with the peculiar shape best seen in polished section (Figure 7) resulting from liquid immiscibility. This is the only evidence so far for this type of activity at Hilderston.

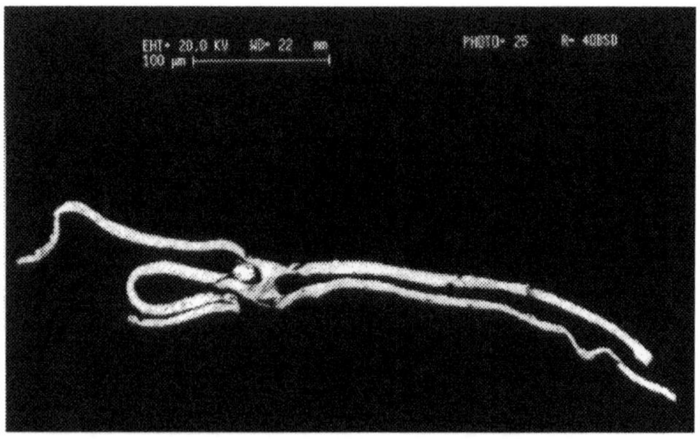

Figure 7. Polished section cross section of the silver platelet viewed by SEM in backscattered mode. There are two layers joined by a narrow bridge. The round grain (translucent in polished section) between the two layers is very rich in lead and is probably PbO or PbCO3. Further polishing showed the round, lead-rich, area and addition slithers to be closely bonded to the silver. The platelet is believed to represent a relict of the cupellation process, the textural feature resulting from liquid immiscibility between silver and PbO. Scale bar is 100µm.

The second sample consisted of a fragment of "slag" from the same area (Figure 8). The XRD diffraction pattern consisted of gehelenite/ akermanite with major calcite. Similar types of "slag" were recovered from the area thought to be that of the processing plant at Linlithgow. However, their less conventional mineralogy and the absence of significant Pb, Ba, Ni and other indicator elements within the matrix remains to be explained.

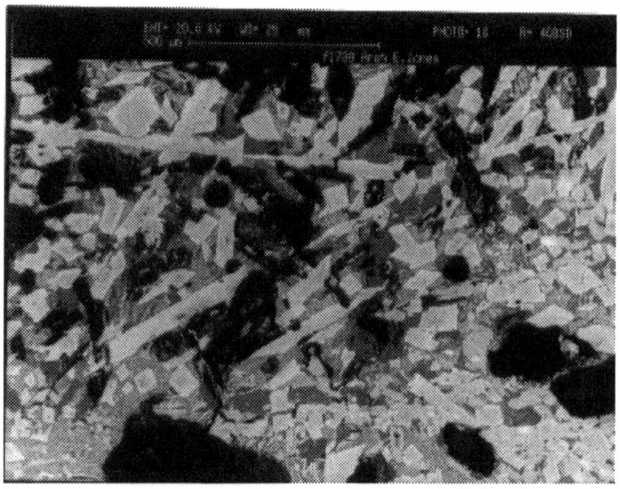

Figure 8. Polished section of fragment of slag found near the 'furnace' (Figure 2) viewed by SEM in backscattered mode. The crystalline gehlenite/akermanite-rich nature (light areas) was confirmed by XRD. The lack of Pb, Ba, Ni or other indicator elements means that the slag is not confirmed as relating to the smelting of silver ore. Scale bar is 500µm.

A third sample was identified as "ore". It consisted largely of quartz with a large number of metallic inclusions such as iron sulfide, copper-iron sulfide, lead-copper-iron sulfide, and baryte, as well as traces of silver. Nickel- and chromium-rich iron inclusions were also revealed and traces of chromium probably related to alteration of the dolerite. "Slag" and "ore" are yet to be identified as such on the basis of further analyses.

VI. Discussion

The assessment of the Hilderston mining and metals working site was aimed at bringing together documentary evidence and the geological background of the area with the survey of the extant features and the analyses of relevant metallurgical waste. By its very nature, an assessment is limited in the conclusions that it can reach, the only way to resolve crucial questions being by excavation and post-excavation desk and laboratory work. However, excavation of multi-phase and multi-process heavily worked industrial sites is not always easy to justify to funding organizations. Therefore a great number of such sites are lost to developers with a corresponding loss of the valuable information they carry. The present work suggests that, although excavation and full scale understanding of such complex sites may not be achievable in practice, an assessment of the extant remains can act as a powerful incentive to collate all evidence, past and present. Since relatively little is known about the mining and working of native silver, at least in relation to its argentiferous sulfidic and other ores, the need to assess the evidence at Hilderston became paramount.

Hilderston native silver mineralization is complicated due to its association with other metallic components such as nickel and arsenic. Although arsenic was well known, nickel had not been discovered and would therefore have eluded Sir Bevis Bulmer and his colleagues at the Mint in the Tower of London. Fortunately, an impression of the native silver can be obtained from the statements of Atkinson: *"It was raced with many small stringes, like unto haiers or threads. It had descended from a vaine thereof where it had engendered with sparr stone, which in foraine province is called... cacilla. And he sought further.. and found a peece of brownish sparr-stone, and it was mossie. He broke it with his mattocke and it was white and glittered within like unto small white copper-keese which is to be found in may common free-stones"* (Atkinson 1619: 47)

Since no good samples of silver "ore" are presently available, there is some uncertainty about the exact nature of the gangue minerals. Silver at Hilderston was hosted in '*redd mettle*'. Meikle (1994) assumed that the term refereed to nickeline which is red. However, *redd* is Scottish mining term for waste, e.g., redd bing - a debris heap. 'Metal' is used for hard rock and 'metals' is general name for strata in which minerals, e.g., coal, occur. Hence 'redd mettle' could mean "gangue" without any implications of red colour. '*Cacilla*' is equated with white spar stone which is presumably calcite, but may also have been or included dolomite and baryte. '*Brown spar or mossie*' could well be ferroan dolomite which is particularly prone to oxidation on weathering to produce a brown-stained carbonate or soft brownish limonite. '*Marqusit*' (Cochran-Patrick 1878: 135) is most likely the niccolite, but its exact identity could not have been known in the early Seventeenth Century.

Contemporary understanding of metal ores can be appreciated from the writings of Agricola and in the case of silver it is evident that a number of silver "ores" could be distinguished mainly on the basis of colour in Agricola (1556: 108). *Rudis silver, ruby silver, gray silver, black silver, liver silver and silver glance* are listed. Rudis silver is presumably the native silver; ruby silver is proustite, Ag_3AsS_3; gray silver is cerargyrite, AgCl; black silver is stephanite, Ag_5SbS_4; liver silver is bromyrite, AgBr; and silver glance is argentite, Ag_2S. Agricola argues that it requires an experienced miner to tell the difference between galena and native silver:

"Lead coloured rudis silver...is occassionally so like galena (plumbago) in colour although it is darker, that one who is not experienced in minerals is unable to distinguish between the two at sight, but in substance they differ greatly from one another...whereas this silver is malleable under the hammer, may be dented by the teeth, and cut with a knife.. galena when struck with a mallet or bitten or hacked with a knife splits and breaks to pieces.". (Agricola 1556: 109).

Figure 9. Native silver processing showing lumps of metal being hammered into sheets and a large pair of shears embedded in a stump "for the cutting of the metal". After Agricola (Hoover 1950: 209). A: masses of metal, B: hammer, C: chisel, D: tree stumps, E: iron tool similar to a pair of shears. Reproduced by permission of Mining Journal Ltd.

Then follows a passage which refers to the cutting of the native silver with long shears (Figure 9) . *"The metal which is dug out in a pure or crude state to which class belongs native silver, silver glance and gray silver is placed on a stone by the mine foreman and flattened out by pounding with heavy square hammers. The masses when they have been thus flattened out like plates are placed either on a stump of a tree, and cut into pieces by pounding an iron tool similar to a pair of shears. One blade of these shears is three feet long, and is firmly fixed in stump and the other blade which cuts the metal is six feet long. These pieces of metal are afterwards heated in iron basins and smelted in the cupellation furnace by the smelters."* (Agricola 1556: 269). This practice was perhaps devised as a means of controlling the quantity of material which was being charged in the "iron furnace" where silver was added to the lead. However, the size of the shears seems unnecessarily large. The process assumes the presence of non-rock or earthy material associated with the ore which must have hardly been the case with the Hilderston ore.

The evidence for lead (lead carbonate or lead oxide) inclusions within the platelet does suggest cuppelation. Indeed recently Smith and Meikle (*in press*) have shown that this must have been the practice at the "melting" and "fining" mills at Hilderston and Linlithgow. It does however, seem a scanty bit of evidence compared to the possibility for evidence of bone ash crucibles, litharge and similar equally and more substantial pieces of evidence.

While there is ample documentary evidence for native silver mining and processing at Hilderston, there is to-date little in terms of archaeological evidence. The present investigation has highlighted the problems associated with extraction of native metal in the context of a complex mineralization. The major problem in reconstructing the Hilderston industrial landscape and establishing the method of silver processing is the 18th and 19th Century exploitation of other minerals near the site, in particular limestone.

Acknowledgements
The authors wish to thank the following institutions for the funding of this project.: the Russell Trust, the Department of Archaeology, University of Glasgow and the Metallurgy and Engineering Materials Group, Department of Mechanical Engineering, University of Strathclyde. Particular thanks are extended to the following individuals for their comments, and assistance in the field and the laboratory: Mrs Crichton, archivist, Hopetoun House Preservation Trust; Mr A. Aitken , Ballencrief Farm, Linlithgow; Mr J. Graham, Tartraven Farm, Bathgate; Mrs Potter, Knock Farm, Bathgate; John Gilleece and Peter Ainsworth, Polished Section Laboratory and SEM Laboratory , Department of Geology and Applied Geology, Glasgow University. Mining Journal Ltd, are thanked for granting permission to use extracts from the 1912 version of Agricola's De Re Metallica.

References

AGRICOLA, G. 1556. *De Re Metallica,* Transl. by H.C. & L.H. Hoover, London: The Mining Magazine, 1912.
ATKINSON, S. 1619. *The Discoverie and Historie of the Gold Mynes in Scotland,* Edinburgh: J. Ballantyne and Co. 1825.
CADELL, H.M. 1925. *The Rocks of West Lothian,* Edinburgh Oliver & Boyd.
CHUNG, P.Y. M. 1991. *Silver mineralisation in Scotland: Genesis and Exploration,* Unpubl. M. Phil. thesis. Univ. of Strathclyde.
COCHRAN-PATRICK, R.W. 1878. *Early Records relating to Mining in Scotland,* Edinburgh: David Douglas.
DONALDSON, G. 1994. *Scotland: James V - James VII. The Edinburgh History of Scotland vol 3,* Edinburgh: Mercat Press.
GALLAGHER, M.J. 1964. Rock alteration in some mineralised basic dykes in Britain, *Trans. Instn. Min. Metall.* 73: 825-840.
HALL I. H. S ., J. J. GALLAGHER, B. R. H. SKILTON & C. E. JOHNSON. 1982. Investigation of polymetallic mineralisation in Lower Devonian volcanics near Alva, Central Scotland, *Mineral Reconnaissance Program. Rep. Inst. Geol. Sci.* No 53. HMSO.
HUNT, C.A. 1889. *The Perth Hammermen Book, 1518 to 1568,* Perth.
MEIKLE, T.K. 1994. Native silver from Hilderston Mine, West Lothian, Scotland, *J. Russell Soc.* 5(2): 83-90.
1996. The Melting, Fyning and Stamping Mylnis at Linlithgow, *British Mining* 57: 106-124.
MORETON, S. 1996. The Alva Silver Mine: silver Glen, Alva, Scotland, *The Mineralogical record,* 27: 405-414.
PHOTOS-JONES, E & A.J. HALL. 1997. *The characterisation of surface "industrial waste" from the Linlithgow and Hilderstone area,* SASAA Internal Report, September 1997.
RUSSELL, M.J. 1985. The evolution of the Scottish mineral sub-province, *Scott. J. Geology* 21: 513-545.
SIMPSON, A.T. & STEVENSON, S. 1982. *Historic Perth: the Archaeological Implications of Development,* Scottish Burgh Survey, Univ. of Glasgow.
SMITH, R. & MEIKLE, T.K. (*in press*) The smelting of silver ore from Hilderston Mine at Linlithgow, Scotland. *British Mining.*
TYSON L.O. 1996. Sir Bevis Bulmer: an Elizabethan adventurer, *British Mining* 57: 47-69.

LA-ICP-MS evidence for the distribution of lead and strontium in Romano-British, medieval and modern human teeth: implications for life history and exposure reconstruction

JANET MONTGOMERY[1], PAUL BUDD[1], ALAN COX[2], PETRA KRAUSE[2] & RICHARD G. THOMAS[3]

[1]*Department of Archaeological Sciences, University of Bradford, Bradford, BD7 1DP, UK*
[2]*Centre for Analytical Sciences, University of Sheffield, Sheffield, S3 7HF, UK*
[3]*Department of Civic Engineering and Environmental, University of Western Sydney, Nepean, PO Box 10, Kingswood, 2747 NSW, Australia*

Laser-ablation inductively coupled plasma mass spectrometry (LA-ICP-MS) is a relatively new analytical technique offering the possibility of in situ elemental analysis in a variety of matrices with minimal sample preparation or destruction. This paper describes an application of the technique to Romano-British, medieval and modern skeletal material, principally tooth enamel, from the UK. Lead and strontium are the elements of interest because of their systematic isotopic variation in nature and the potential that they offer in reconstructing both human migration and palaeo-pollution. The distribution of the two elements is of considerable interest in assessing the integrity of the biogenic signal accumulated during the life time of the individual as distinct from diagenetic, post-depositional, contamination.

New semi-quantitative Time Resolved Analysis (TRA) data are presented which highlight significant differences in lead and strontium distributions between dental tissues. A characteristic lead distribution with enhanced surface enamel peaks, low and consistent core enamel levels and higher dentine levels is interpreted as fundamentally physiological in nature with core-enamel tissues preserving biogenic lead in archaeological samples. Strontium is shown to be largely homogenous throughout modern teeth, but enhanced especially in the dentine of some archaeological teeth. Strontium accumulation in the archaeological material is considered to be diagenetic and dentine is found to be unreliable as a reservoir of biogenic metals. Implications for isotopic studies of prehistoric migration and historic pollution exposure are briefly discussed.

Keywords: Lead; Strontium; Isotopes; Laser Ablation; ICP-MS; Human Teeth; Archaeology.

Introduction

There have been a number of studies of the concentration and isotope ratios of lead and strontium in both archaeological and modern human tissues in order to reconstruct individuals' dietary and pollution exposure histories. Lead isotope analysis of archaeological tooth enamel may be used to complement and corroborate recent strontium isotope studies (Price *et al.* 1994; Sealy *et al.* 1995; Grupe *et al.* 1997) and further enhance traditional archaeological life-history reconstructions and inferences on migration drawn from studies of grave goods, burial practices and osteology.

Isotopic environmental, forensic and clinical studies of modern populations have been used to pinpoint sources of lead exposure (Yaffe *et al.* 1983; Rabinowitz 1987; Delves & Campbell 1993; Gulson & Wilson 1994; Gulson *et al.* 1994). The approach has recently been extended to identify first generation immigrants by virtue of isotope ratios characteristic of tetra-ethyl lead fuel additives used in their country of origin and remnant in both blood (Gulson *et al.* 1995) and teeth (Gulson *et al.* 1997). Combined isotopic studies of elephant ivory and rhinoceros horn have also proved successful in provenancing material to individual National Parks in southern Africa, in a bid to control poaching (van der Merwe *et al.* 1990, Hall-Martin *et al.* 1993). A slightly different environmental approach has involved lead isotope analysis of walrus and beluga cementum to identify changes in historical sources of pollution in marine mammals in the Canadian Arctic (Outridge *et al.* 1997).

Environmental and clinical studies on modern material have met with widespread acceptance, but the successful application of similar techniques to archaeological skeletal remains suffers from the not inconsiderable difficulty of confidently identifying, isolating and measuring metals of *biogenic*, as opposed to *diagenetic*, origin. We aim to evaluate the potential of combined Sr- and Pb-isotope and concentration measurements to identify individuals with non-local origins among sedentary populations in the archaeological record. The work may also yield insights into the place of origin of such individuals, diet, sources of prehistoric lead exposure and diagenetic change in the burial environment. However, before such work can be confidently performed and accepted there is a need for more information on the origin of the metal cations measured in skeletal material that has been buried post-mortem.

In order to distinguish between biogenic and diagenetic metals, it is necessary to study both the distribution and the incorporation of these elements into human skeletal remains. Laser-ablation inductively coupled plasma mass spectrometry (LA-ICP-MS) has the ability to provide high-resolution spatial information to assess the first and give unique insights into the second of these objectives. LA-ICP-MS is a relatively new analytical technique offering the possibility of *in situ* elemental analysis in a variety of matrices with minimal sample preparation or destruction. It is, therefore, an ideal method for the analysis of irreplaceable archaeological resources. Pioneering *in situ* LA-ICP-MS studies have recently been published investigating micro-spatial lead distribution in both human dental tissues (Cox *et al.* 1996, Budd *et al.* 1998) and walrus cementum (Evans *et al.* 1994, 1995). This paper describes the application of the technique to archaeological and modern skeletal material, principally tooth enamel and focuses on lead and strontium as the elements of interest.

Lead and strontium incorporation in skeletal tissues

Lead isotopes vary in different lithologies and rock units as a function of parent U- and Th-content and age (Faure 1986, 309). Weathering introduces this lead into soil and groundwater from where it finds its way into the food chain and living organisms. It has been suggested that lead substitutes for calcium in the apatite structure of the carbonated hydroxyapatite of skeletal tissues (McConnell 1973, 23), becoming incorporated into the mineral phase. Lead does not fractionate in biological systems so that its isotopic composition remains characteristic of the geological environment from which food sources are drawn. Strontium can be considered analogous. Complications are obviously introduced in later prehistory with metallurgy and exposure to its products as well as the importation of exotic foodstuffs.

In contrast to bone and dentine, mature tooth enamel is acellular, not remodelled *in vivo* and is highly resistant to diagenesis (Boyde *et al.* 1988, 1488), thus preserving an archive of lead ingestion during mineralization of the tooth crown in early life. Deciduous tooth enamel is representative of *in-utero* exposure with various permanent teeth representing exposure at ages between birth and as late as about 12 years of age (Hillson 1986, 189). As tooth enamel appears to offer the greatest resistance to post-mortem change the current study focuses upon this tissue with other skeletal elements analyzed for comparison.

Despite the static nature of the enamel tissue some studies have produced data taken as evidence of *in-vivo* incorporation of lead at the enamel surface (Brudevold & Steadman 1956; Brudevold & Söremark 1967). Given the absence of cellular tissue and a blood supply, this suggests the additional possibility of diagenetic alteration of the enamel surface during burial. Such surface enhancement might be associated with structural differences within the tissue. It should be noted however that strontium results from the same work exhibited no such surface enhancement.

Methods and samples

Archaeological teeth (canines and premolars) were obtained from two archaeological sites of different periods. The site of Blackfriars, Gloucester was excavated during 1991 by the Gloucester Archaeology Unit. It formed part of the burial ground of the Dominican Friary of Blackfriars, now in the centre of the city, which was founded in AD 1239 and dissolved in AD 1538. Only a small portion (140) of the estimated 2000 burials have been excavated. Results presented here are from a simple juvenile soil burial (19/91-8). Teeth were also obtained from two individuals excavated by the Avon Archaeological Unit in 1997 from Mangotsfield near Bristol. The burials were removed from a single large Bathstone (limestone) sarcophagus filled with soil and excavated within a late-Roman period site. Upon excavation the stone coffin was found to contain the remains of two complete skeletons, one male, one female and a clavicle and vertebra from a third individual (G. Barber, pers.comm.). Data presented here was obtained from the middle-aged male (MN97-SK1b). Modern samples of shed deciduous and extracted permanent teeth were also analyzed for comparison. The profiles presented here represent a small sub-sample of the teeth used in this, ongoing, study.

Sample preparation

By far the major problem in trace lead analysis is the prevention of anthropogenic contamination from tools, reagents and airborne particulate. Initial sample preparation and instrument cleaning was carried out in a positive pressure glove box with a filtered air supply. An acid-cleaned, flexible diamond-impregnated stainless steel saw was used for initial bisection of the teeth, one half being reserved for lead and strontium ID-TIMS analysis and oxygen isotope analysis, the results of which are reported elsewhere (Barreiro *et al.* 1997; Budd *et al.* 1998, in press).

Samples for LA-ICP-MS were cleaned ultrasonically in high-purity water (*Millipore Alpha Q)* to remove surface dirt and then embedded in low-lead epoxy resin blocks (*Buehler Epo-thin*), from which thin sections of variable thickness could be cut as dictated by individual tooth morphology. These were usually about 0.5-1mm as thinner slices are easily cracked and broken by the laser beam. Sections were cut using a *Malvern Microslice 2* precision annular microtome saw fitted with an aluminium oxide blade lubricated with a *Millipore Alpha Q* water supply. No further preparation of the cut sections was performed in order to minimize opportunities for laboratory contamination. Lead contamination of sections cut in this manner was investigated by LA-ICP-MS analysis of mount blanks and found to be insignificant.

Sample analysis

LA-ICP-MS was used to investigate the distribution of lead and strontium in human teeth *in situ*. The analysis was performed at the Centre for Analytical Sciences, University of Sheffield, UK, using *CETAC LSX-100* laser ablation system coupled to *Hewlett Packard HP4500* ICP-MS. The system uses a Nd:YAG frequency quadrupled laser operating in the UV range at 266nm. Time Resolved Analysis transects were ablated across longitudinal sections of the tooth crown by firing the laser at 10Hz at a constantly moving sample. This gives an interconnected series of ~20μm diameter craters to collect a continuous trace element profile.

Results

Elemental profiles are presented for three modern teeth: two deciduous teeth, one shed *c.*1970 and one shed in 1996 together with a permanent, unerupted premolar. The data obtained is, as yet, only semi-quantitative as suitable matrix matched standards with which to calibrate the instrument are still under development. Results are, therefore, presented as Pb/Ca and Sr/Ca ratios calculated by simultaneous collection of the minor ^{46}Ca isotope. The length of the laser tract varied between teeth depending on tooth size and location of the tract but was generally 3-6 mm in length. Enamel thickness also varied between the teeth but was always <2mm. Laser tracts were ablated from left to right and repeated right to left to check that the profiles obtained were not the result of ablation characteristics rather than sample heterogeneity. In every case a reproducible, mirror image, tract was obtained.

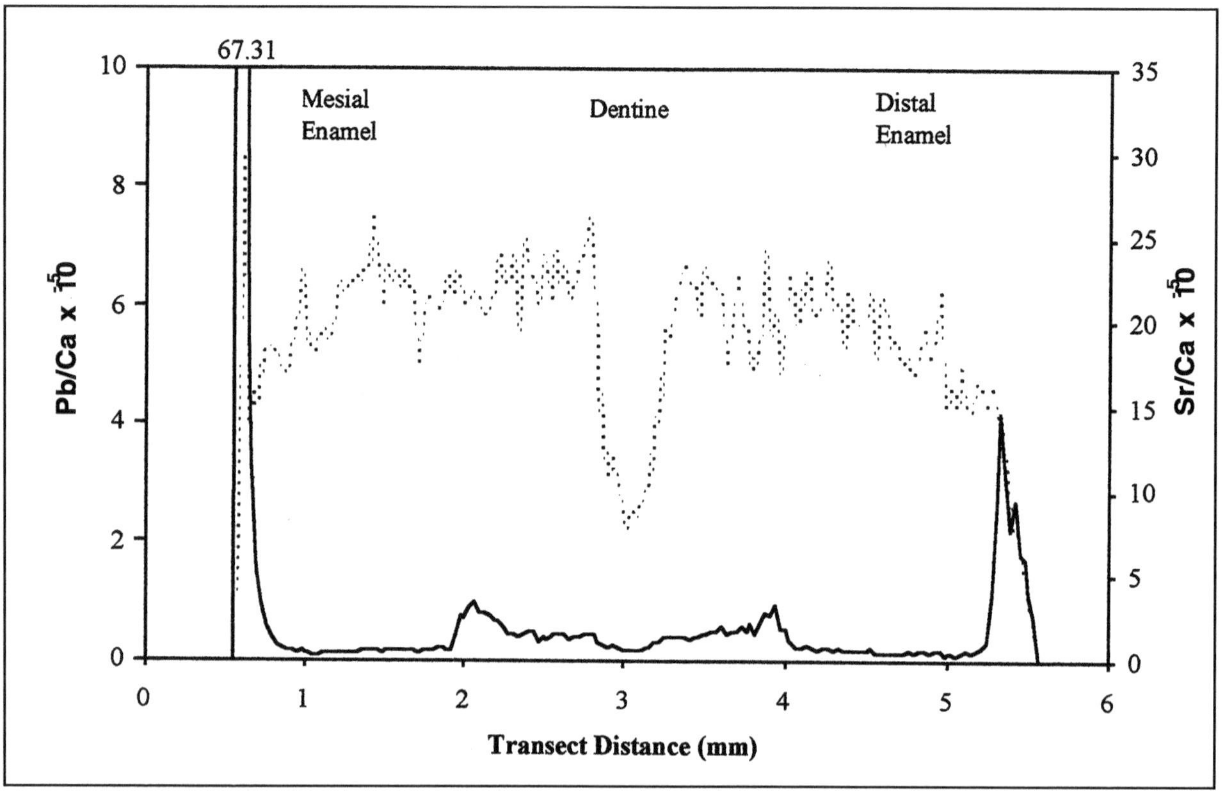

Figure 1.Modern permanent unerupted premolar. The dentine is quite clearly delineated by symmetrical shoulders in the Pb/Ca profile. Significant lead peaks in the surface enamel are visible on both sides of the tooth and very low and relatively consistent levels of lead are characteristic of the core enamel. The Sr/Ca profile is very different from Pb/Ca showing relatively consistent levels throughout the entire profile apart from a drop in the mid-dentine due to elevated Ca.

Modern permanent unerupted premolar

The graph (Figure 1) clearly shows the dentine delineated by symmetrical shoulders in the Pb/Ca profile and the enamel dentine junction (EDJ). Significant lead peaks in the surface enamel are visible on both sides of the tooth and very low and relatively consistent levels of lead are characteristic of the core enamel. The Sr/Ca profile is very different from Pb/Ca showing relatively consistent levels throughout the entire profile. One feature of the Sr/Ca profile, the apparent drop in Sr at the tooth centre, is actually an artefact created by elevated calcium. This was probably caused by differing ablation characteristics resulting from drop in tissue density in the region immediately adjacent to the pulp cavity.

Modern deciduous canine shed 1971

This tooth (Figure 2), together with the great majority we have examined to date displays a strikingly similar Pb/Ca profile to the modern permanent tooth shown in Figure 1 with large peaks at the enamel surface, low core enamel levels and prominent shoulders at the EDJ. Again, the Sr/Ca profile shows a very different distribution to that of Pb/Ca and is relatively consistent throughout the tooth section.

Modern deciduous incisor shed 1996

Once again the Pb/Ca profile (Figure 3) shows similar surface enamel peaks to the other modern teeth, but in this case the enamel layer was very thin, falling entirely within the area of the surface peak on the mesial side. The most striking feature of this profile however is the extremely low and consistent lead throughout the dentine. This could be due to much lower early childhood lead exposure than the subject in Figure 2, but may also be related to the positioning of the laser tract, which, in this case, passed through the dentine very close to the EDJ in the labio-lingual plane. In general we have found that profiles passing closer to the pulp cavity display higher Pb levels in the core dentine.

Late mediaeval period permanent canine c. 1450AD

The Pb/Ca profile obtained from this tooth (Figure 4) shares many similarities with the previous modern examples, although the different scale should be noted. Surface enamel peaks are obvious and the dentine is clearly delineated by the small rises in the profile. The large Pb/Ca peak in the middle of the dentine, where the laser tract passes close to the pulp cavity, could well be of diagenetic origin. The most interesting point however, is the presence of very low consistent levels in the core enamel which do not appear to show any incorporation of lead either from the enamel surface or the dentine.

The Sr/Ca profile is again quite different from that of Pb/Ca but does appear to show two distinct levels in what could correspond to enamel and dentine tissues. Whereas the enamel Sr/Ca ratios are comparable with modern values, they show a considerable increase in the dentine. This increase could be diagenetic in origin or arise from differing ablation characteristics of the two tissues. The potential for both of these is greater in archaeological teeth as the dentine is more susceptible to post-mortem degradation than the enamel.

Late Roman period permanent canine c. 300AD Once again the Pb/Ca profile (Figure 5) displays surface peaks and low core enamel ratios. Again the EDJ is associated with a clear rise in the Pb/Ca ratio with rather higher lead levels in the dentine. Sr/Ca ratios are an order of magnitude higher than found in modern specimens and again show a significant rise between the enamel and dentine.

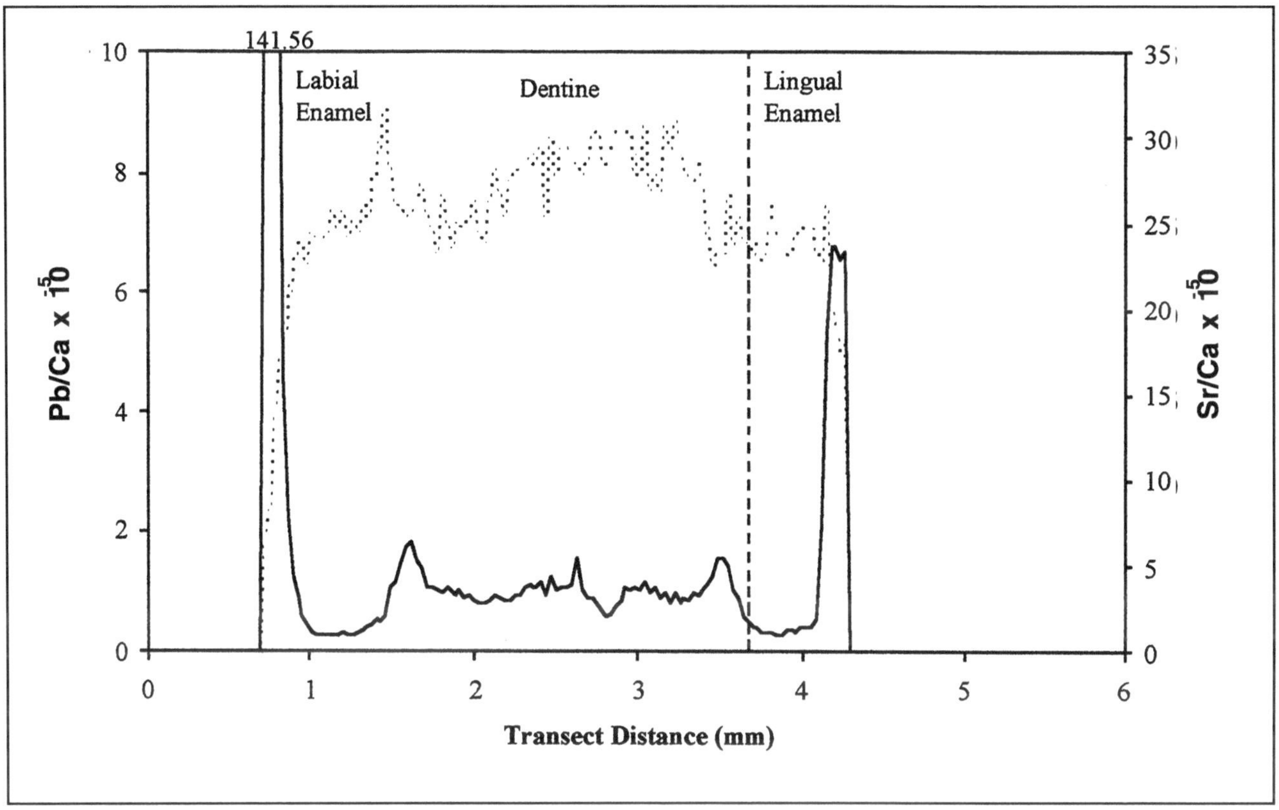

Figure 2. Modern deciduous canine shed 1971. The tooth shows large peaks at the enamel surface, low core enamel levels and prominent shoulders at the EDJ. The Sr/Ca profile shows a very different distribution to that of Pb/Ca being considerably more consistent throughout the tooth section.

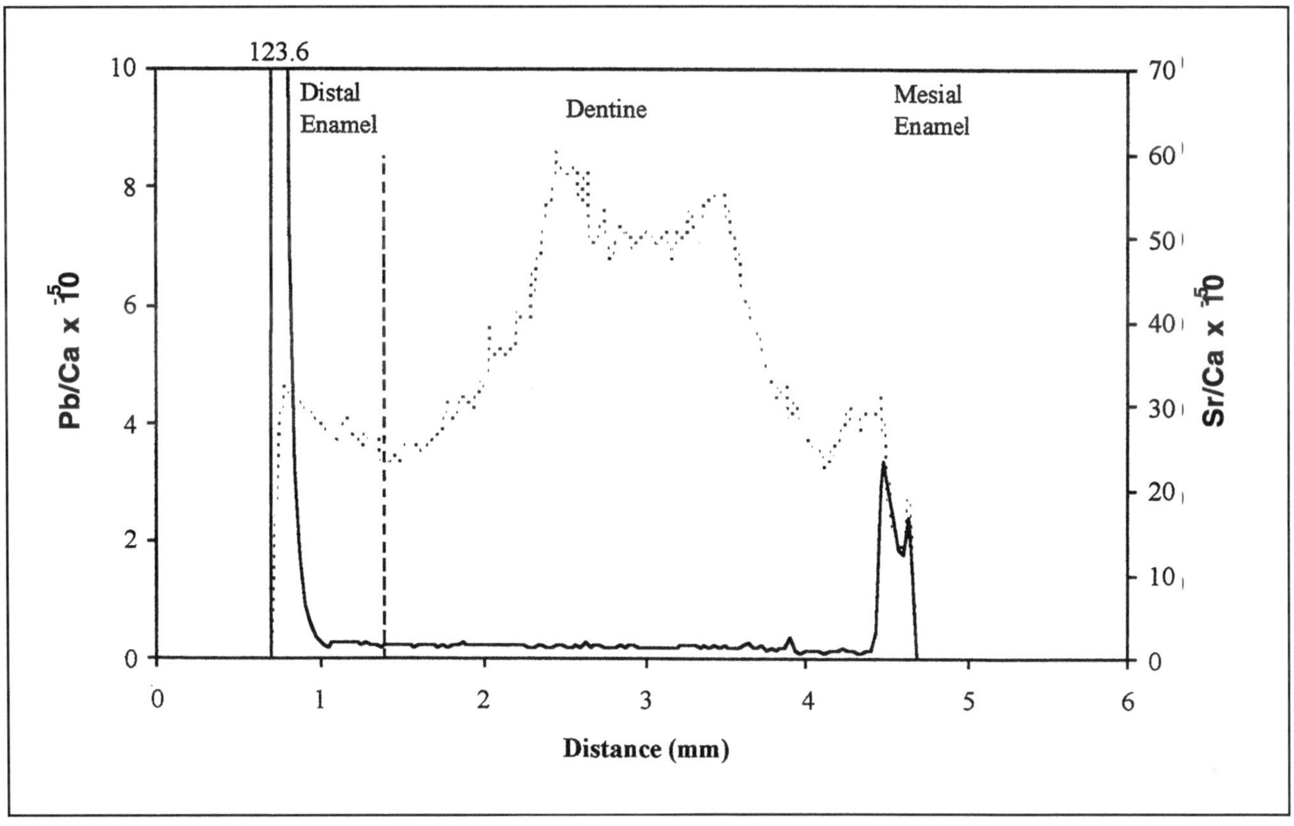

Figure 3. Modern deciduous incisor shed 1996. The Pb/Ca profile shows surface enamel peaks, but in this case the enamel layer was very thin and falls entirely within the area of the surface peak on the mesial side. Note the extremely low and consistent lead throughout the dentine.

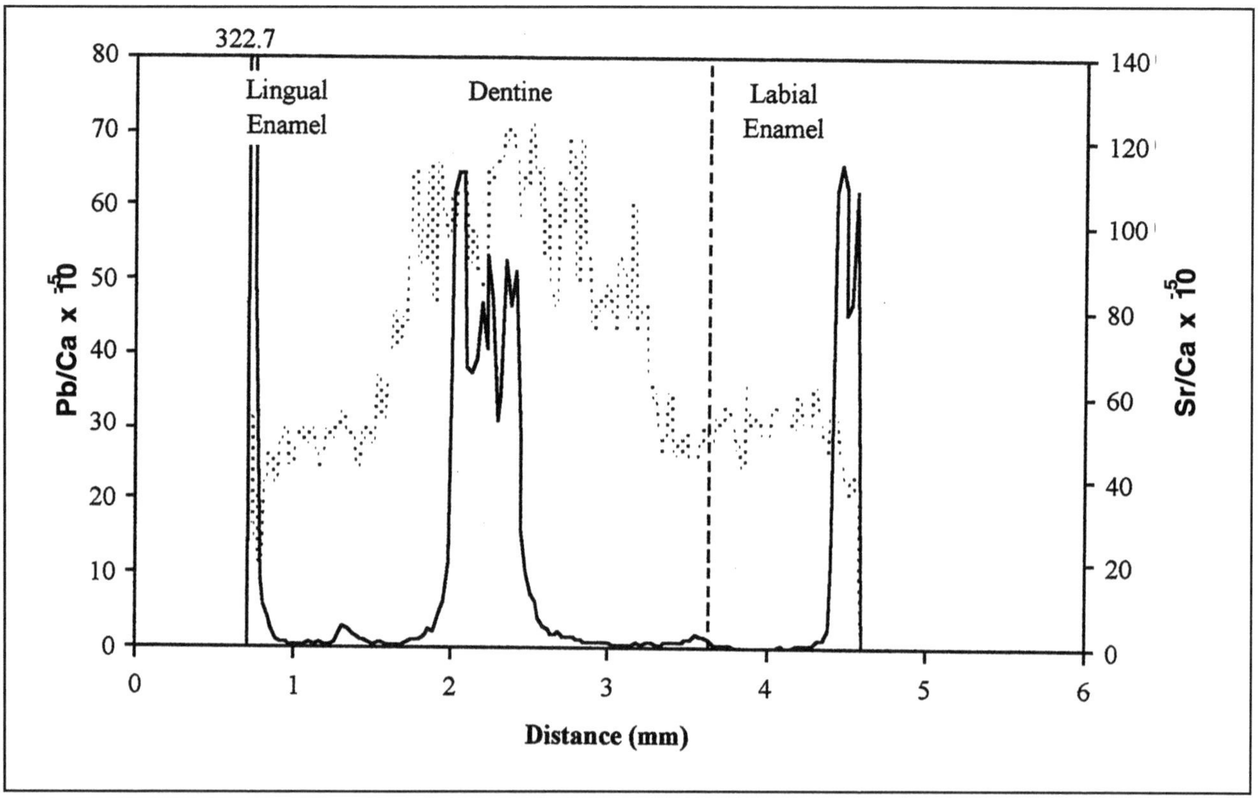

Figure 4. Late mediaeval period permanent canine c. 1450AD. Surface enamel peaks are obvious and the dentine is clearly delineated by the small rises in the profile. There is a large Pb/Ca peak in the middle of the dentine, where the laser tract passes close to the pulp cavity. Note the very low and consistent levels in the core enamel. The Sr/Ca profile appears to show two distinct levels in what could correspond to enamel and dentine tissues.

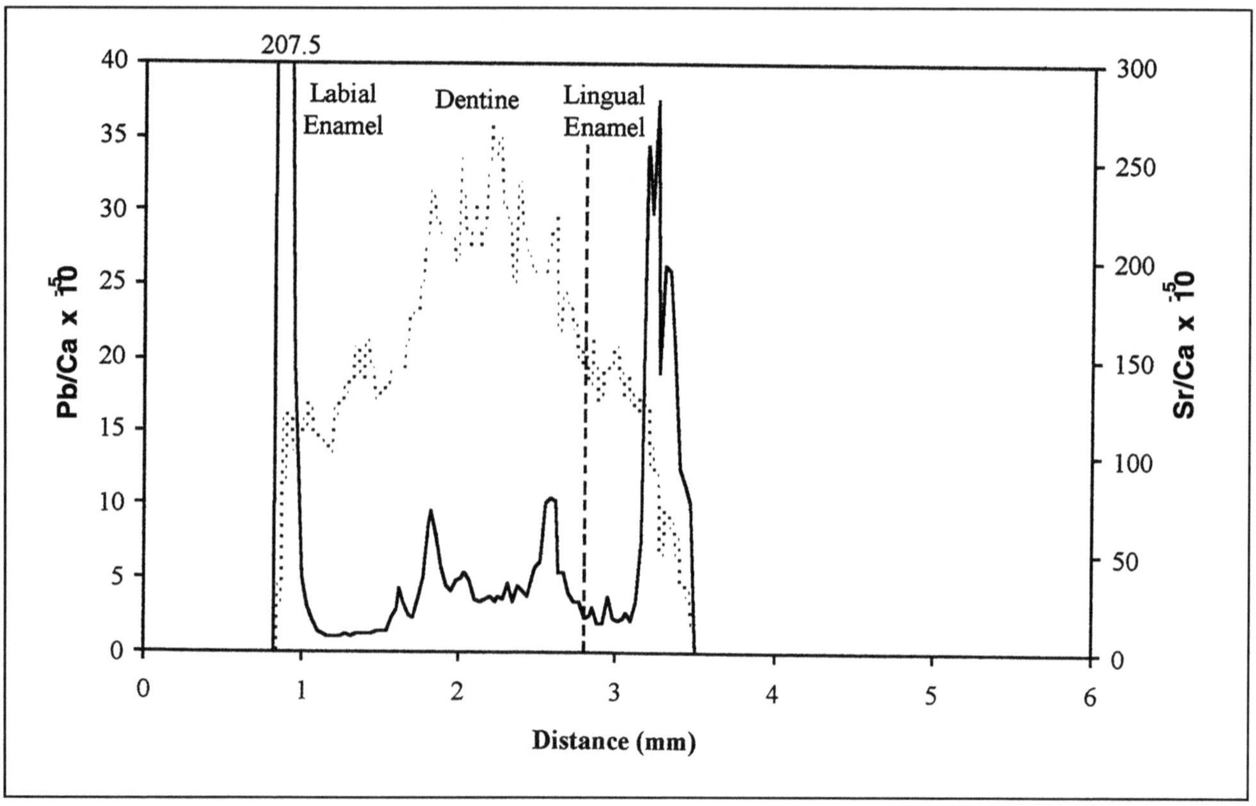

Figure 5. Late Roman period permanent canine c. 300AD. The Pb/Ca profile displays surface peaks and low core ratios. The EDJ is associated with a clear rise in the Pb/Ca ratio with rather higher lead levels in the dentine. Sr/Ca ratios are an order of magnitude higher than found in modern specimens and again show a significant rise between the enamel and dentine.

Summary

To date, all of the tooth sections analyzed, with the possible exception of the modern deciduous incisor (Figure 3), which has exceptionally low dentine levels, have displayed the following characteristics:-

a) Sharply defined lead-rich surface enamel peaks;
b) A significant and abrupt increase in lead concentration at the enamel dentine junction (EDJ) with elevated lead in the dentine compared with that in core enamel;
c) Consistent and low lead levels in core enamel tissue;
d) Strontium distributions which are considerably different from those of lead with relatively consistent levels throughout all tissues in modern samples whereas some archaeological samples show significantly increased levels in dentine compared with enamel.

It has also been noted that all of the Pb/Ca profiles show a considerably higher surface enamel peak on one side of the tooth than the other. The observation can be rejected as an artefact of the analysis as it is found to be reproducible in transects run in the opposite direction. Most of the sections presented were cut in the labial-lingual plane and the differences may simply be due to differential, orientation dependent, wear of the tooth surfaces. There is some support for this suggestion from the 1996 deciduous incisor (Figure 3). This was sectioned in the mesial-distal plane, but the subject had a pronounced mesial diastema leading to enhanced wear on the mesial surface of the tooth. This corresponds to the lower lead peak.

Discussion

It is clear from the LA-ICP-MS results that lead distributions in both modern and archaeological teeth are highly heterogeneous and contrast with those of strontium. Surface enamel tissues are enriched in lead compared with other dental tissues in both archaeological and modern teeth whereas core enamel has a much lower and relatively constant lead level. The data corroborate and clarify previous results by other researchers (Brudevold & Steadman 1956; Shapiro *et al.* 1972; Waldron 1982; Whittaker & Stack 1984; Molleson 1987). The core enamel profiles of the archaeological teeth examined here, when compared with those of the modern teeth, demonstrate little evidence of diagenetic alteration either from the surface or EDJ. This suggests that unaltered, biogenic, lead is present in this tissue (Budd *et al.* 1998, in press).

Previous researchers suggested that surface enamel lead accumulated *in vivo* in the oral environment independently of metabolic processes (Brudevold & Steadman 1956). It seems possible that enamel immediately adjacent to the tooth surface may indeed be more chemically active than core tissue. This might be linked with its higher organic content (D. Whittaker pers. comm.). There is some evidence that as the enamel wears *in vivo* the lead enriched tissue is reduced, but the fact that the phenomenon is still seen even where wear is considerable suggests that *in vivo* lead incorporation in surface enamel tissue is likely. However, the presence of the surface peak even in the unerupted modern tooth examined suggests that the incorporation of lead at the enamel surface is not exclusively associated with the oral environment and might be partly related to the tissue formation process. Whatever the explanation, there is no evidence that core enamel tissues incorporate lead after their formation and these would appear capable of retaining biogenic material ingested in early life.

Although the chemical incorporation of lead and strontium in skeletal material is generally regarded as analogous and resulting from the direct substitution for calcium in the mineral lattice, their profiles in all of the teeth examined are quite different. Lower strontium levels in modern teeth and an absence of elevated strontium in their dentine suggests that much of the strontium in the dentine of the archaeological teeth examined here is diagenetic and would not be representative of biogenic incorporation. Strontium profiles in modern teeth closely match those for calcium whereas lead appears to be more dependent on tooth structure than calcium distribution.

Conclusions

For archaeological teeth, LA-ICP-MS can provide relative concentration data and a quick and simple surveying technique to assess lead and strontium diagenesis in different tooth tissues prior to other forms of analysis. It seems clear that for strontium and probably also for lead, dentine is an unreliable reservoir of the biogenic signal and core enamel should be the preferred tissue for analysis. Our suggestion that the structure of core enamel allows the preservation of an unaltered biogenic lead signature is still being tested. However, if substantiated it offers archaeology the exciting prospect of overcoming the long-standing uncertainty about post-mortem contamination in lead analysis of archaeological skeletal material and will thus allow future studies of both prehistoric migration and historical exposure histories to be undertaken with a much greater degree of confidence.

Acknowledgements

The authors would like to thank their numerous colleagues engaged in related areas of the project for their help and advice on various aspects of the analytical programme, especially Barbara Barreiro and Carolyn Chenery, David Whittaker and Suzanne Young. We would also like to record our thanks to Dave Lucy and Jason Maher for their help and advice on sample preparation techniques and Charlotte Roberts for her advice and support. We would also like to thank all of those who made samples and supporting information available, especially Phil Greatorex, Andy Young, Geraldine Barber, Julian Richards, Olivia Lawrenson and Constance Montgomery. One of the authors (JM) would like to thank the NERC for provision of a Research Studentship. One of the authors (PB) is supported by the NERC under the Advanced Fellowship scheme.

References

BARREIRO, B., P. BUDD, C. CHENERY & J. MONTGOMERY. 1997. Combined Pb-, Sr- and O-isotope compositions of human dental tissues in life history reconstruction. Abstract. Paper presented to the Applied Isotope Geochemistry Conference, Lake Louise, Canada, September 1997.

BOYDE, A., M. FORTELIUS, K.S. LESTER & L.B. MARTIN. 1988. Basis of the structure and development of mammalian enamel as seen by scanning electron microscopy, *Scanning Microscopy* 2: 1479-1490.

BRUDEVOLD, F. & R. SÖREMARK. 1967. Chemistry of the mineral phase of enamel, in A.E.W. Miles (ed.), *Structural and chemical organization of teeth*: 247-277. London: Academic Press.

BRUDEVOLD, F. & L.T. STEADMAN. 1956. The distribution of lead in human enamel, *Journal of Dental Research* 35: 430-437.

BUDD, P., J. CHRISTENSEN, R. HAGGERTY, A.N. HALLIDAY, J. MONTGOMERY & S.M.M. YOUNG. 1997. The measurement of biogenic lead within archaeological mammalian dental enamel for life history reconstruction and pollution exposure monitoring. Abstract. Paper presented to the 213th ACS National Meeting, San Francisco April 13-17 1997.

in press. The measurement of biogenic lead within archaeological mammalian dental enamel for life history reconstruction and pollution exposure monitoring, *Journal of Archaeological Science.*

BUDD, P., J. MONTGOMERY, A. COX, P. KRAUSE, B. BARREIRO & R.G. THOMAS. 1998. The distribution of lead within ancient and modern human teeth: implications for long-term and historical exposure monitoring, *The Science of the Total Environment* 220, 121-136.

COX, A., F. KEENAN, M. COOKE & J. APPLETON. 1996. Trace element profiling of dental tissues using laser ablation-inductively coupled plasma-mass spectrometry, *Fresenius Journal of Analytical Chemistry* 354: 254-258.

DELVES, H.T. & M.J. CAMPBELL. 1993. Identification and apportionment of sources of lead in human tissues, *Environmental Geochemistry and Health* 15: 75-84.

EVANS, R.D., P.M. OUTRIDGE & P. RICHNER. 1994. Applications of laser ablation inductively coupled mass spectrometry to the determination of environmental contaminants in calcified biological structures, *Journal of Analytical Atomic Spectrometry* 9: 985-989.

EVANS, R.D., P. RICHNER & P.M. OUTRIDGE. 1995. Micro-spatial variations of heavy metals in the teeth of walrus as determined by laser ablation ICP-MS: the potential for reconstructing a history of metal exposure, *Archives of Environmental Contamination and Toxicology* 28: 55-60.

FAURE, G. 1986. *Principles of isotope geology*. New York: John Wiley & Sons Inc.

GRUPE, G., T.D. PRICE, P. SCHRÖTER, F. SÖLLNER, C.M. JOHNSON & B.L. BEARD. 1997. Mobility of Bell Beaker people revealed by strontium isotope ratios of tooth and bone: a study of southern Bavarian skeletal remains, *Applied Geochemistry* 12: 517-525.

GULSON, B.L., D. HOWARTH, K.J. MIZON, M.J. KORSCH & J.J. DAVIS. 1994. The source of lead in humans from Broken Hill mining community, *Environmental Geochemistry and Health* 16: 19-25.

GULSON, B.L., C.W. JAMESON & B.R. GILLINGS. 1997. Stable lead isotopes in teeth as indicators of past domicile: a potential new tool in forensic science?, *Journal of Forensic Sciences* 42: 787-791.

GULSON, B.L., K.R. MAHAFFEY, K.J. MIZON, M.J. KORSCH, M.A. Cameron & G. Vimpani. 1995. Contribution of tissue lead to blood lead in adult female subjects based on stable lead isotope methods, *Journal of Laboratory and Clinical Medicine* 125: 703-712.

GULSON, B.L. & D. WILSON. 1994. History of lead exposure in children revealed from isotopic analyses of teeth, *Archives of Environmental Health* 49: 279-283.

HALL-MARTIN, A.J., N.J. VAN DER MERWE, J.A. LEE-THORP, R.A. ARMSTRONG, C.H. MEHL, S. STRUBEN & R. TYKOT. 1993. Determination of species and geographic origin of rhinoceros horn by isotopic analysis and its possible application to trade control. Paper presented to the International Rhino Conference: Rhinoceros Biology and Conservation, San Diego, California, USA, 1991.

HILLSON, S. 1986. *Teeth.* Cambridge: Cambridge University Press.

MCCONNELL, D. 1973. *Apatite, its crystal chemistry, mineralogy, utilization and geologic and biologic occurrences.* Wien: Springer-Verlag.

MOLLESON, T. 1987. Trace elements in human teeth, in G. Grupe & B. Herrmann (eds), *Trace elements in environmental history. Proceedings of the Symposium held from June 24th to 26th 1987, at Göttingen*: 67-90. Berlin: Springer-Verlag.

OUTRIDGE, P.M., R.D. EVANS, R. WAGEMANN & R.E.A. STEWART. 1997. Historical trends of heavy metals and stable lead isotopes in beluga (*Delphinapterus leucas*) and walrus (*Odobenus rosmarus rosmarus*) in the Canadian Arctic, *The Science of the Total Environment* 203: 209-219.

PRICE, T.D., G. GRUPE & P. SCHRÖTER. 1994. Reconstruction of migration patterns in the Bell Beaker period by stable strontium isotope analysis, *Applied Geochemistry* 9: 413-417.

RABINOWITZ, M.B. 1987. Stable isotope mass spectrometry in childhood lead poisoning, *Biological Trace Element Research* 12: 223-229.

SEALY, J.C., R. ARMSTRONG & C. SCHRIRE. 1995. Beyond lifetime averages: tracing life histories through isotopic analysis of different calcified tissues from archaeological skeletons, *Antiquity* 69: 290-300.

SHAPIRO, I.M., H.L. NEEDLEMAN & O.C. TUNCAY. 1972. The lead content of human deciduous and permanent teeth, *Environmental Research* 5: 467-470.

VAN DER MERWE, N.J., J.A. LEE-THORP, J.F. THACKERAY, A.J. HALL-MARTIN, F.J. KRUGER, H. COETZEE, R.H.V. BELL & M. LINDEQUE. 1990. Source-area determination of elephant ivory by isotopic analysis, *Nature* 346: 744-746.

WALDRON, H.A. 1982. Lead in bones: a cautionary tale, *Ecology of Disease* 1: 191-196.

WHITTAKER, D.K. & M.V. STACK. 1984. The lead, cadmium and zinc content of some Romano-British teeth, *Archaeometry* 26: 37-42.

YAFFE, Y., C.P. FLESSEL, J.J. WESOLOWSKI, A.D. ROSARIO, G.N. GUIRGUIS, V. MATIAS, T.E. DEGARMO, G.C. COLEMAN, J.W. GRAMLICH & W.R. KELLY. 1983. Identification of lead sources in California children using the stable isotope ratio technique, *Archives of Environmental Health* 38: 237-245.

38

Authenticity of a Korean Iron Warrior on Horseback

R.P. BEUKENS[1], L.A. PAVLISH[2], G.C. WILSON[1,3], & R.M. FARQUHAR[4]

[1] *IsoTrace Laboratory, University of Toronto, Toronto, Ontario M5S 1A7, Canada*
[2] *Archaeometry Laboratory, Department of Physics, University of Toronto, Toronto, Ontario M5S 1A7, Canada*
[3] *Turnstone Geological Services Ltd., P.O. Box 130, Station "B", Toronto, Ontario MST 2T3, Canada*
[4] *Department of Physics, University of Toronto, Toronto, Ontario M5S 1A7, Canada*

The textures and major element variations within a millimetre scale metal shard of a Korean metal figurine are described in some detail, using instrumental neutron activation analysis (INAA), reflected light microscopy, secondary electron imaging, both energy dispersive spectrometry (EDS) and wavelength dispersive (WDS) electron microprobe analysis, and AMS. Results show there is a two stage oxidation rind on the surface of the 0.6 cm^2 shard. The fresh metal is identified as a cast iron, probably a chilled, grey cast iron, the best microprobe data suggesting that the bulk composition includes ≈1.26% silicon and <3.3% carbon. The texture and the chemical contrast between the pale and dark components of the interior are consistent with their identification as pearlite, a eutectic intergrowth of pale cementite (enriched in C, P and Cr) and darker ferrite (higher totals, higher Fe, Si and perhaps Mn). The microtextures display both granular and lamellar variants of pearlite. No graphite was positively identified, most carbon apparently residing in cementite, perhaps with additional traces on grain boundaries. Trace (Mn,Fe)S in the metal seems too rare to be a deliberate addition. The rim shows progressive oxidation of the interior to a hydrated Fe oxide, goethite, which displays characteristic chemical variability. The outer rim contains traces of a K-aluminosilicate, and of fine grained (<100 μm) copper granules. Some of the Cu is now in oxide form, implying that Cu was present when the piece was made in a silicate (clay?)-lined mould. Traces of copper probably resided in the iron ore, were incorporated into Fe alloys on smelting, and were then gradually released as the host phase(s) weathered. A case of this type has been reported of Cu enrichment within rust layers on the surface of a Japanese sword, dated at 471 AD (Nippon Steel 1985). Based on the available evidence, and provided that coal was used for smelting in that era (as suggested by the AMS dating of the carbon), the Korean iron figurine is most probably genuine.

Keywords: Authenticity; Korea; Japan; China; Iron Figurine; Horse; Coal; Neutron Activation Analysis; A M S Radiocarbon Dating; Metallography.

Introduction

This object, without provenance, is a cast iron representation of a "Warrior on Horseback" (Fig. 1), of a form identical to 5th and 6th century AD ceramic vessels from Korea, and measuring approximately 10 cm in width, 23 cm in height and 28 cm in length.. The ceramic figurines function both as container and dispenser of liquids, whereas this cast iron object does not have these functions, since the horse's core, once filled with casting sand, is open and does not allow the storage of liquids. This lack of functionality initially cast some doubt on the authenticity of this object, but the appearance of both clay and iron figurines has been documented in the Korean archaeological record. Clay and iron horse figurines, wearing saddles and bits but without obvious functionality, are presumed to be votive objects. Archaeological evidence from tombs in Japan has been found which is consistent with an elite class of horseriding warriors having entered Japan from Korea in the 5th Century (Douglas 1978: 364-366).

A first attempt to authenticate this object was made by the Oxford Thermoluminescence Laboratory, using the sand from the casting plug in the belly of the horse. They were unsuccessful because of the presence of iron oxide in the sand.

Methods

Herein we present the results of Instrumental Neutron Activation Analysis (INAA), Scanning Electron Microscope (SEM) analysis using energy dispersive spectrometry (EDS) and wave length dispersive (WDS) electron microprobe analysis, reflected light microscopy and Accelerator Mass Spectrometry (AMS) radiocarbon analysis.

Fig 1 here. Caption: Korean cast iron representation of a 'warrior on horseback'.

Sampling

The samples analyzed were taken from the fragment at the base of the saddle blanket (Fig. 1). This fragment was broken off by the owner of the object.

Discussion and Interpretation

Metallography

Figure 2 shows a photomicrograph of a polished section of the sample in reflected plane polarized light. It shows a corrosion layer consistent with a slow, natural process rather than an electrolytically accelerated corrosion. The microprobe analyses (below) support this interpretation. The structure of pale and dark laths in the uncorroded core is identified as *pearlite,* a microstructure consisting of alternating lamellae of *ferrite* (α-Fe, iron) and *cementite* (Fe_3C, alias the mineral *cohenite*). This eutectic intergrowth, which also occurs in a granular form, may be found in alloys with ≈0.02 - 4.3% carbon (C), and forms by exsolution from austenite, a solution of carbon in γ-Fe, on cooling below 727°C (Gulyaev 1978:160-173, 242). From this observation it follows that the pale, more reflective form is a variant of *cementite* with higher C, phosphorus (P) and chromium (Cr), whilst the darker interstices are *ferrite,* with higher analytical totals, higher Fe, silicon (Si) and perhaps manganese (Mn). Both Si and Mn can de-oxidize steel, the Si being fully soluble in ferrite, and Mn dissolving in both ferrite and cementite (Gulyaev 1978:178). Many cast irons contain forms of *graphite* (carbon) which were not detected in this study. The sample is a *chilled cast iron,* a *grey cast iron* in which rapid cooling has forced most of the C into Fe alloys (Shrager 1969:63). The presence of thin films of graphite on fractures and grain boundaries remains a possibility.

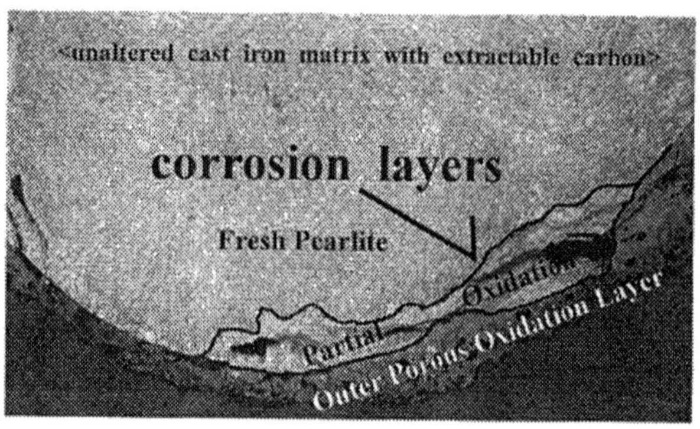

Fig. 2 here. Caption: Photomicrograph of sample in plane polarized reflected light (160X, field if view 0.7 mm)

Electron Microprobe

a) The inner bulk of the sample

The components of the microstructure are discussed in terms of their chemistry, as determined by the electron microprobe. The fresh metal contains both lamellar and granular *pearlite* domains, which together comprise >99% of the metal. Within the pearlite there is a clear chemical distinction between pale reflective *cementite* and adjacent dull *ferrite.* Eleven points were analyzed, and all yielded EDS spectra with dominant iron (Fe) and minor silicon (Si) peaks, with or without other detectable elements. The range of compositions as determined by EDS for the cementite and ferrite phases are shown in Table 1.

In all 11 samples, nickel (Ni) and cobalt (Co) are <0.1%. Sulfur (S) was detected on only one pale domain, at 0.11 wt.%. Five other points were also analyzed, representing the oxide rim (three points) and a tiny inclusion and microfracture in the interior (one each). Si is concentrated in the darker, carbon-poor phase, hence Si content correlates with higher Fe and higher analytical totals. Cr is concentrated in the brighter cementite, hence there is a negative correlation with Si. Ni, Co and S are ≤0.1 wt.% in both phases. The carbon content of the sample was estimated at ≈4% by comparison of the ^{13}C current (in the high-energy faraday cup of the IsoTrace accelerator mass spectrometer) with a standard steel. The analytical totals of the 11 analyses average 96.7 ± 1.8% (one standard deviation error) suggesting a 'missing' component averaging ≈3.3% which is probably largely carbon, a result statistically indistinguishable from the AMS estimate. One analysis of thin dark fractures in dull metal yielded a total of ≈87%, including 1.4% Si. The low total may indicate a concentration of carbon along fractures, but it is also possible that the result may be due to scattering of X-rays from rough microtopography along the fractures.

b) Inclusions in the metal

A medium-gray 10x8 μm nodule in the metal core yielded on analysis a very low total and low minor element abundances (except Si) with no trace of additional elements in the EDS spectrum. The total is much below the ≈78% Fe that might be anticipated for *wüstite* (FeO), or the ≈63% Fe in *goethite* (α-FeOOH). This tiny inclusion may be Fe oxide with a local concentration of C, or a carbon (*graphite*) inclusion in which most of the Fe signal is derived from X-rays emanating from the subjacent metal host. Two adjacent rounded grains with ragged margins, 10x10 and 8x5 μm in size, are largely MnS (the mineral *alabandite*) with S, Mn and lower Fe in the EDS spectrum.

c) The oxide rim and incorporated inclusions

The fresh inner metal yields EDS spectra which always display Fe and minor Si, and perhaps traces of P, Mn or S. In comparison, the rim oxide may display all these elements plus more volatile elements such as halogens and alkalis (chlorine (Cl), potassium (K) and calcium (Ca)). The outer rim oxide has a lower analytical total, consistent with a higher degree of oxidation and hydration than the inner oxide rim, in which clear vestiges of the host microstructure are preserved.

Element	*Cementite*	*Ferrite*
Total	93.6-96.5	96.5-99.4
Fe	90.8-94.5	94.1-97.1
Si	0.1-1.1	1.3-2.1
Cr	0.14-0.26	≤0.09
Mn	≤0.1, except one 0.15	0.10-0.17
P	0.15-0.50 (3), n.d. (1), 4.25 (1)	≤0.19

Table 1.

Two kinds of inclusion occur in the outer oxide rim. A rhombic 120x60 μm grain was seen by EDS to be a pure K aluminosilicate (K-feldspar or *mica*, or a clay mineral, such as *orthoclase*, *muscovite* or *illite*). This may be a remnant of the medium in which the metal was cast. There are also metal inclusions in the rim - a reddish copper metal, in which EDS reveals major Cu and minor Fe peaks. The copper occurs as single rounded to tabular grains as well as granular aggregates up to 70x50 μm in area. An EDS spectrum of oxide near a copper grain revealed high S and all the other minor elements seen in the rim oxides, as well as trace magnesium (Mg) plus appreciable Cu and zinc (Zn). If the Cu is a component of the oxide then the copper metal is clearly not an artefact of the polishing process. A rounded 7x5 μm nodule and a diffuse 15x10 μm patch of Cu oxide (Cu with minor Fe in EDS spectrum), found in the goethite outer rim also point towards long residence of the copper in the sample.

INAA

The University of Toronto SLOWPOKE reactor was used for a bulk INAA analysis of the object (Table 2). The "less than" figures represent concentrations below the detection limits. Phosphorus and silicon could not be determined because of interferences. These results imply a relatively pure cast iron, most likely refined pig iron. The only notable constituents are manganese (Mn) and copper (Cu) implying the use of a copper-containing *magnetite* (Fe_3O_4) ore and not a *hematite* (Fe_2O_3) ore.

Element	Concentration
Fe	84 ± 8 %
Mn	3100 ± 100 ppm
Cu	2200 ± 100 ppm
As	430 ± 10 ppm
Al	120 ± 10 ppm
V	84 ± 2 ppm
Sb	65 ± 4 ppm
Co	33 ± 5 ppm
Na	31 ± 4 ppm
W	30 ± 4 ppm
Mg	< 2500 ppm
Ca	< 550 ppm
Ti	< 320 ppm
Cl	< 170 ppm
Br	< 41 ppm
I	< 17 ppm

Table 2.

AMS Radiocarbon Analysis

The IsoTrace AMS Radiocarbon Facility at the University of Toronto was used for radiocarbon analysis of the carbon dissolved in the steel. Radiocarbon analysis of steel is the only way an unequivocal date can be obtained for a steel or cast iron object, in the absence of a datable casting plug. It dates the age of the charcoal used in the smelting and refining process of steel, and usually dates the actual casting of the objects. However, complications arise if, instead of charcoal, coal was used in the smelting process, producing unacceptably old apparent radiocarbon ages. Coal came into regular use in steel industries around the world in the late 19th or early 20th centuries AD, but there is evidence of earlier use in China (see below).

Two methods can be used for radiocarbon analysis of steels:
i) the carbon can be extracted as CO_2 by the oxidation of a molten aliquot of the steel (Cresswell 1991:78-85; 1992:898-905),
ii) if the carbon content is high enough, the carbon can be dated directly by placing a small plug of the iron directly in the sputter ion source of the AMS equipment. Since the bulk sputter rate of iron is much higher than that of carbon, after a few minutes mainly carbon is sputtered by the primary cesium ion beam.

Because of the relatively high carbon content of this object the second method was chosen for analysis. Three separate measurements were performed, yielding an *average* ^{14}C concentration of 1.69 ± 0.14 pMC (percent Modern Carbon). This concentration implies an apparent radiocarbon age of 32,780 ± 670 BP, demonstrating clearly that coal was used in the production of this object. This apparent age cannot be attributed to 'old' carbon from limestone as the INAA data shows negligible quantities of Ca (Shrager 1969:54).

Although in a Western context, the use of coal would generally assign this object to the 19th or 20th century, several authors (e.g., Shiba 1979, Yoon 1984:14) have mentioned the use of coal as a substitute for charcoal in ancient Chinese steel making. All evidence within a Korean context, however, for the use of coal appears to be either anecdotal or indirect. Nevertheless, the use of coal in steel making appears to have been known in China with respect to the Chinese crucible process from at least the 4th Century AD:
"The early use of coal in China for smelting is extremely important and goes back to the 4th Century AD at least. Reference to the method in Chinese texts is relatively infrequent probably because it was confined to certain provinces (Read 1934; 1939). Read noted that a small amount of black earth (hei thu) was habitually added to the ore by the smelters to promote full fusion of the metal and its perfect running into moulds (Read 1937:31)..." (Needham 1958:14). Needham further notes that *vivianite* ($Fe_3(PO_4)_2.8H_2O$), present in some coals, if present in sufficient quantity could drop the melting point of the iron from 1130°C to 950°C (Needham 1958:14).

There are two Western literary sources which attest to the widespread use of coal in China. They are the travels of Marco Polo (1254-1323) and Ibn Battuta (1304-1377). Their travels to China are separated by half a century, but both allude to its use.

Marco Polo noted coal burning:
"... It is a fact that all over the country of Cathy there is a kind of black stone existing in beds in mountains which they dig out and burn like firewood... It is true that they

have plenty of wood also, but they do not burn it, because these stones burn well and cost less..." (Yule 1871:395).

Although critics say he never visited China, Ibn Battuta certainly knew about the use of coal in general, but his technological knowledge of the role of coal in porcelain manufacture is muddled: *"The Chinese pottery (porcelain) is manufactured only in the towns of Zaytun and Sinkalan. It is made of soil of some mountains in that district which takes fire like charcoal, as we shall relate subsequently. They mix this with some stones which they have, burn the whole for three days, then pour water over it. This gives a kind of clay which they cause to ferment"* (Gibb 1929:282-283). Ibn Battuta obviously confused coal and porcelain clay. possibly owing to a custom followed in China of powdering the coal and mixing it with clay to form a 'patent fuel' (Gibb 1929:369): *"All the inhabitants of China and Cathay use in place of charcoal a kind of lumpy earth found in their country. It resembles our 'fuller's earth', and its colour too is the colour of 'fuller's earth'. ... They break it up into pieces about the size of pieces of charcoal and set it on fire and it burns like charcoal, only giving out more heat than a charcoal fire. When it is reduced to cinders, they knead it with water, dry it, and use it again in cooking, and so on over and over again until it is entirely consumed. It is from this clay that they make the Chinese porcelain ware after adding to it some other stones as we have related..."* (Gibb 1929:284).

Conclusions

The sample is identified as a pearlitic, hypereutectoid Fe-C alloy, a chilled grey cast iron rather than a steel. The complex nature of the oxidation rind appears consistent with a long, but undeterminable period of weathering, and thus argues for the statuette's authenticity, provided that coal was in use during the 5th and 6th centuries in China and Korea. There is sufficient evidence from Chinese and European sources to show that coal was clearly used in some technological processes by this time. The work presented here suggests that the same may be true of Korea.

References

CRESSWELL, R.C. 1991. The radiocarbon dating of iron artefacts using accelerator mass spectrometry. *Journal of Historical Metallurgy* 25: 78-85.
1992. Radiocarbon dating of iron artifacts. *Radiocarbon* 34: 898-905.
DOUGLAS, J.H. 1978. The horsemen of Yamato. *Science News* 113: 364-366.
FLOHR, M.J.K. 1992. Geochemistry and origin of the Bald Knob manganese deposit, North Carolina. *Economic Geology* 87: 2023-2040.
GIBB, H.A.R. (trans.). 1929. *Ibn Battuta, Travels in Asia and Africa 1325-1354.* London: Routledge.
GULYAEV, A. 1978. *Physical Metallurgy, Volume 1.* Moscow: Mir Publishers. (English translation 1980).
NEEDHAM, J. 1958. *The development of iron and steel making technology in China.* London: Newcomen Society for the Study of the History of Engineering and Technology, the Science Museum.
NIPPON STEEL. 1985. Iron analysis unveils Man's ancient past - 1. The Inariyama sword: flakes of rust tell the story. *Nippon Steel News* 175: 1 (January).
READ, T.T. 1934. The early casting of iron, a stage in Iron Age civilization. *Geographical Review* [In Needham 1958].
1937. Ancient Chinese casting. *Transactions of the American Foundryman Association.* p. 30 [In: Needham 1958].
1939. The earliest industrial uses of coal. *Transactions of the Newcomen Society* 20: 119. [In Needham 1958].
SHIBA, R. 1979. Iron in the history of Japan. *Weekly Asahi* (March 28, 1979) (in Japanese).
SHRAGER, A.M. 1969. *Elementary Metallurgy and Metallography.* New York: Dover (3rd edition).
YOON DONG-SUK. 1984. *Metallurgical Study of Early Iron Age Artifacts found in Korea.* Pohang, Korea: Pohang Iron and Steel Co.
YULE. H. (trans.) 1871. *Marco Polo, The book of Ser Marco Polo, the Venetian, concerning the kingdoms and marvels of the East.* London: John Murray.

39

Sicán alloying, working and use of precious metals: an interdisciplinary perspective

IZUMI SHIMADA[1], ADON GORDUS[2], JO ANN GRIFFIN[3], & JOHN F. MERKEL[4]

[1]*Department of Anthropology, Southern Illinois University, Carbondale, IL 62901, USA*
[2]*Department of Chemistry, University of Michigan, Ann Arbor, MI 48109, USA*
[3]*6206 Vanderbilt Ave., Dallas, TX 75214, USA*
[4]*Institute of Archaeometallurgy, University of London, 31-34, Gordon Square, London, WC1H 0PY, UK*

A range of analytical and interpretive perspectives have been brought to bear on the large variety and quantity of gold-silver-copper alloy objects recovered from recent excavations of Middle Sicán (ca. AD 1000) elite tombs at the site of Sicán, on the north coast of Peru. Metallography, EPMA, NAA, study of tool marks, and replicative experimental goldworking together show important covariation among artifact design, manufacturing technique and finish, as well as chemical composition, coloration, and mechanical and working properties of alloys. In addition, detailed contextual analysis of base and precious metals (including scraps) and their placement inside the tombs reveals insights into the social significance of metal in this culture. Our study also questions imposed analytical and conceptual separation of base and precious metals in understanding their cultural significance.

Introduction

Archaeometallurgy strives toward an integrated understanding of both the intrinsic and extrinsic qualities of ancient metallurgy. The extrinsic or cultural dimension encompasses the organization and technology of metal production as well as its distribution, economic and social role and symbolic meaning. Yet, it has been difficult to achieve comprehensive coverage and understanding of the cultural dimension. Unintentionally or not, the technological study of metal objects tends to be undertaken largely divorced from concern with their cultural dimensions. Such studies may take on a life of their own, pursuing ever narrower foci as investigation proceeds. Thus, analysis becomes more and more distant from behaviorally and culturally meaningful contexts and issues. Such analysis is often focused on a specific metal without considering either its relationship with other metals that were concurrently produced and/or used in relative proximity or other crafts that utilized metal components (Shimada in press a). Metal objects, like any other category of artifact, must be examined from a multitude of perspectives and in their proper cultural contexts.

The 1100-year old Sicán metallurgy of the northern coast of Peru has been a major research focus of the Sicán Archaeological Project (SAP). This paper is largely methodological and describes SAP's "holistic" approach, underlying philosophy, and some of the findings to date on the cultural dimension of Sicán metallurgy. It illustrates how a number of parallel lines of inquiry dovetail to yield important data and insights. Results presented here pertain primarily to the alloying, working, and use of precious metals. The paper begins with a summary characterizing the archaeological setting of the research.

The archaeological setting of the Sicán culture

The Sicán culture emerged in the Lambayeque region on the northern coast of Peru *ca.* AD 750-800 following the demise of the northern Mochica polity (e.g., Shimada 1990, 1994a, in press b). It remained viable until *ca.* AD 1375-1400 when it was conquered by the expanding Chimú Kingdom with its capital at Chan Chan to the south. The Sicán cultural chronology consists of Early, Middle, and Late periods. During its classic Middle period, dated AD 900-1100, the Sicán polity attained state-level socio-political

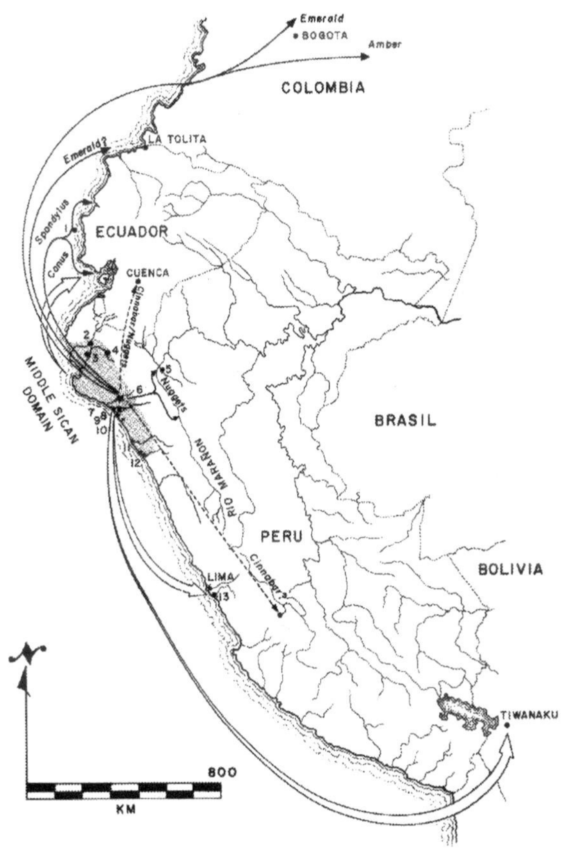

Figure 1. Map of the middle La Leche Valley with the location of the site of Sicán with over a dozen monumental mounds (e.g., La Merced and Loro). Drawn by I. Shimada..

complexity and dominated some 400 km of northern coast (Shimada 1990, 1995, in press b). Its capital was the ceremonial city of Sicán in the mid-La Leche valley, a T-shaped settlement (1.6 km east-west x 1 km north-south) dominated by over a dozen monumental, adobe truncated pyramids and platforms (Figure 1).

The Middle Sicán culture had a thriving, multi-faceted economy. In addition to subsistence-oriented fishing and large-scale irrigation agriculture, its economy featured arsenical copper and gold-alloy production on a scale heretofore unseen in Peru. The Middle Sicán became the harbinger of the "bronze age" in northern Peru (Cleland and Shimada 1992; Shimada and Merkel 1991). Without doubt, the large scale production and diverse and pervasive use of varied alloys are among the most distinguishing qualities of this culture. Even 'commoner', burials are invariably found with a few arsenical copper objects such as *tumi* knives (knives with crescent-shaped blades and rectangular, vertical handles), cast 'digging stick tips', and needles. Those of elite individuals are endowed with impressive quantities and varieties of arsenical copper and precious metal objects which are primarily ornamental and ritual in character. In addition, elite tombs contain ceramic vessels completely wrapped with *tumbaga* (copper-gold-silver alloys, typically of 10 or lower karats) sheet metal and painted cotton cloth with *tumbaga* sheet metal backing (Shimada in press a).

Local arsenical copper products were apparently traded for exotic, luxury and ritual items from northern neighbors, particularly the contemporaneous Manteño culture on the south-central coast of Ecuador, some 500 km to the north (Shimada 1985, 1987, 1991, 1995, in press b). For example, emeralds and amber are believed to have been imported from the northern coast of Ecuador or even farther north from the Muzo region of Colombia, and Colombian Amazonia, respectively (Shimada 1995, in press b, c; Shimada *et al.* 1997). The same maritime trade route brought large numbers of *Spondylus princeps* (colorful "thorny oyster") and *Conus fergusoni* (large spiral shell), both valued as ritual items throughout prehispanic Peru (Cordy-Collins 1990; Shimada 1990, 1991, 1995). The Sicán polity was well situated to have controlled the inland and southward distribution of these high-status and ritual goods. Overall, the Middle Sicán was arguably the most powerful, influential, and materially rich culture in Peru for its time and it left a lasting technological legacy.

Approach and underlying philosophy

Sustained, interdisciplinary investigation into prehistoric mining and metallurgy in the Batán Grande region has been an important component of the SAP since its onset in 1978. Prior to this project, it was already widely known that late prehispanic tombs in the region, particularly deep shaft tombs at the site of Sicán, often contained numerous metal objects, both precious and base (e.g., Pedersen 1976; Reichlen 1941, 1962; Tello 1937a,b,c; Válcarcel 1937). Not surprisingly, SAP's 1978 and 1979 surveys revealed the ubiquitous presence of metal objects and metallurgical byproducts (e.g., slag and slagged sherds) in the La Leche and adjacent valleys (Shimada 1981). Often, grave looters left behind cast and sheet metal arsenical copper objects as not worth gathering (see Pedersen 1976). The name of the principal modern settlement in the region, Batán Grande, is derived from the numerous *batanes* (large, ~1m diameter, stone anvils) which are found around the town and which we now know were used prehistorically, together with heavy rocking stones called *chungos*, to grind ore and slag.

The abundance of metal objects and production debris was especially significant given their general rarity at archaeological sites in the New World, and hinted at the presence of an important *regional* prehispanic metallurgical tradition. By the end of 1983, it was becoming evident from the excavation of various contexts (habitational, ceremonial, and funerary) in and outside of the capital site of Sicán that metal artifacts were pervasive and served diverse functions in the Middle Sicán culture (see below). Excavations at Cerro Huaringa (also called Cerro de los Cementerios), Cerro Sajino, both sites with many *batán-chungo* sets, and at Huaca del Pueblo Batán Grande, revealed prehispanic metallurgical workshops. These were engaged in intensive smelting of arsenical copper during the span of Middle Sicán to Sicán-Inka, AD 900 to 1532. Each workshop was equipped with a row of small 'bowl-type' smelting furnaces, the only such structures excavated to date in the New World. Survey work showed that the site of Cerro Huaringa was connected by a prehispanic road to a nearby copper mine on the eastern flank of Cerro Blanco (Shimada 1981, 1994b; Shimada *et al.* 1982, 1983). The workshops and associated smelting remains and technology, as well as the associated mines are described in detail elsewhere (Epstein and Shimada 1984; Merkel *et al.* 1994; Shimada 1985, 1994b; Shimada *et al.* 1982, 1983; Shimada and Merkel 1991).

The abundance of diverse, well-preserved metallurgical remains and indications of varied and widespread use of metals added up to a rare research opportunity. Recognizing this unique research potential, the SAP initiated what would become a sustained effort to achieve a "holistic" understanding of Sicán metallurgy. The term "holistic" is used in various senses. In one sense, it implies reconstruction of the technology and organization of metal production and definition of the meaning and role of metal products within the culture. In this respect the investigation encompasses the entire process of production, from the perception and procurement of raw materials to manufacture of artifacts and the mundane and/or ritual use of metal products (Shimada 1985, 1994b). The holistic approach is also taken to imply an integrated understanding of all the metals which were produced and used; rather than assuming *a priori* the validity of the analytical distinction between base and precious metals. The project seeks to define whether such a distinction was meaningful in Middle Sicán culture (Shimada in press a).

These efforts to attain a holistic understanding of Middle Sicán metallurgy had important methodological implications. Firstly, fieldwork had a clear regional focus with the mid-La Leche valley as the core study area. The work was also of long duration. Originally, it was anticipated that the project would require 15 years to complete, but today, after 20 years, we still have many years of research to complete. Middle Sicán sites of varied environmental setting, complexity, size and/or inferred function (e.g., ceremonial, craft production, and residential - both commoners and elite) have been excavated to varying extents. The work has yielded diverse metallurgical samples from a wide range of social and functional contexts. Discarded metal artifacts were scarce in habitational settings and much of the information pertaining to the use and role of metal was derived from systematic sampling of tombs of varied social status.

Included in this regional sampling was a concerted search for mines, smelting, and metalworking sites. To date, five Middle to Late Sicán arsenical copper smelting sites and two inferred precious metalworking sites have been excavated. Nine prehispanic mines (essentially yielding secondary copper minerals and arsenopyrite and its weathered products) have also been surveyed.

The second methodological implication was the pursuit of a number of parallel investigative avenues, including step-wise archaeometric analysis, replicative field and laboratory experiments, and stylistic, iconographic and contextual analysis of metal artifacts. Various techniques have been employed for the analysis of both surface-collected and excavated metallurgical remains. These include AAS, EPMA, NAA, PIXE, SEM, XPS (X-ray photoelectron spectroscopy), XRF, and XRD in addition to metallographic techniques.

Archaeometric data were incorporated in the design of a series of replicative smelting and sheet metal making experiments. Analyses of the resulting products then assisted in further refining metallurgical models. In regard to the arsenical copper smelting technology, experiments by J. Merkel and I. Shimada, assisted by S. Epstein were conducted at Cerro Huaringa. These used a full-scale replica furnace and tuyéres that were made using local materials, as identified by petrographic analysis of ancient Sicán originals. The final phase of our experiments was conducted using a well-preserved 600-year old prehispanic furnace (see Merkel and Shimada 1988 and Shimada and Merkel 1991 for details of experimental results). Copper oxide ore from the Cerro Blanco mine was used for all experimental runs.

Gold-alloy sheet metal making was investigated by J. A. Griffin, an expert goldsmith and metal conservator, and I. Shimada. Literature on prehispanic Andean gold working commonly emphasizes masterful sheet metal making (e.g., Lechtmen 1984a,b, 1988). However, there is a relatively poor understanding of the specific steps taken to make uniformly very thin (as thin as 0.025 mm) sheets using hammers made of hard, dense fine-grained stone such as magnetite and hematite. How did prehispanic metalworkers prepare an ingot or blank for large gold alloy objects such as masks and beakers, some of which weigh well over 0.5 kg? How was such a large object evenly annealed? If gold was effectively dissolved for the inferred electrochemical plating of copper objects (Lechtman *et al.* 1982), what sort of durable container was used? These and many other technical and practical questions had not been addressed in previous studies of prehispanic metalworking.

There are no modern goldsmiths who continue to work sheet metal in the prehispanic manner, and no ethnoarchaeological information is available which might have allowed us behavioral insights relevant to these questions. In this case, the integration of replicative experiments, the study of ancient tools and tool marks, and archaeometry held out the greatest potential for understanding the prehispanic metal working. Sheet metal working experiments were therefore undertaken using modern materials. Gold alloys, the compositions of which had been determined by NAA of Middle Sicán objects (see below), were produced. Metal sheets were then prepared using replica stone tools (basalt anvil and magnetite hammers). Subsequently, tool marks on the reproduction sheets were compared with those found on Middle Sicán objects. Concurrently, metalworking tools, predominantly in museum collections in Lima and Quito, were examined.

Finally, the holistic study required the close collaboration of a team of specialists. Included in the SAP team were archaeologists, an art historian, a chemist, a geologist, a goldsmith, a historian, metal conservators, metallurgists, mineralogists, and a local mining engineer. Many participated actively in the research design formulation, sampling of ore and other metallurgical remains, and excavation. Although it may seem an obvious point, it cannot be overstated that any productive collaboration requires good communication, patience, compromise, and diplomacy.

Precious metal samples and their proveniences

Recent SAP excavation of two large Middle Sicán elite tombs in 1991-2 and 1995 at Huaca Loro in Sicán (Figure 2) yielded, for the first time, a substantial number of well preserved and documented Sicán precious metal objects. The overwhelming majority of samples for our investigation into Sicán precious metal alloying derive from these tombs. The most important, in terms of the diversity and quantity of samples offered, was the East Tomb excavated in 1991-2. The excavation revealed an intact 3 x 3m burial chamber containing five individuals (One adult male, two young adult females, and two juveniles of indeterminate sex) and ~1.2 tonnes of diverse grave goods. The chamber was located at the bottom of an 11m vertical shaft (Shimada 1992, 1995; Shimada and Merkel 1993; Shimada and Montenegro 1993). Goods were placed carefully around the centrally placed body of a robust male individual 40-50 years of age (Figure 3). Objects include 489 cast arsenical copper implements distributed in 15 bundles (~200kg in total weight) and roughly 500kg of *tumbaga* scraps in 4 large piles. The tomb also contained an estimated 1500 bundles of *naipes,* inferred primitive currency and status markers cut into standardized shape and sizes from arsenical copper sheets (weighing ~25 kg). There were some 80 significantly high (12 to 18) karat gold objects, and well over 30 large *tumbaga* objects. The last two categories consisted primarily of personal adornments and ritual paraphernalia Adornments included, ear spools (large circular ornaments whose cylindrical shafts were inserted through perforations in enlarged earlobes), nose clips, a face mask, head bands, crowns and associated ornaments, gloves, and forehead ornaments. The assemblage also included a large sleeveless tunic, long since perished, that was originally decorated with some 2000 small gold foil squares. Ritual equipment included staffs, ceremonial dart throwers, and rattles.

The West Tomb of Huaca Loro is symmetrically situated with respect to the East Tomb across the north-south longitudinal axis of the Hu*aca*. The tomb contained far fewer metal objects than its eastern neighbor, but provided

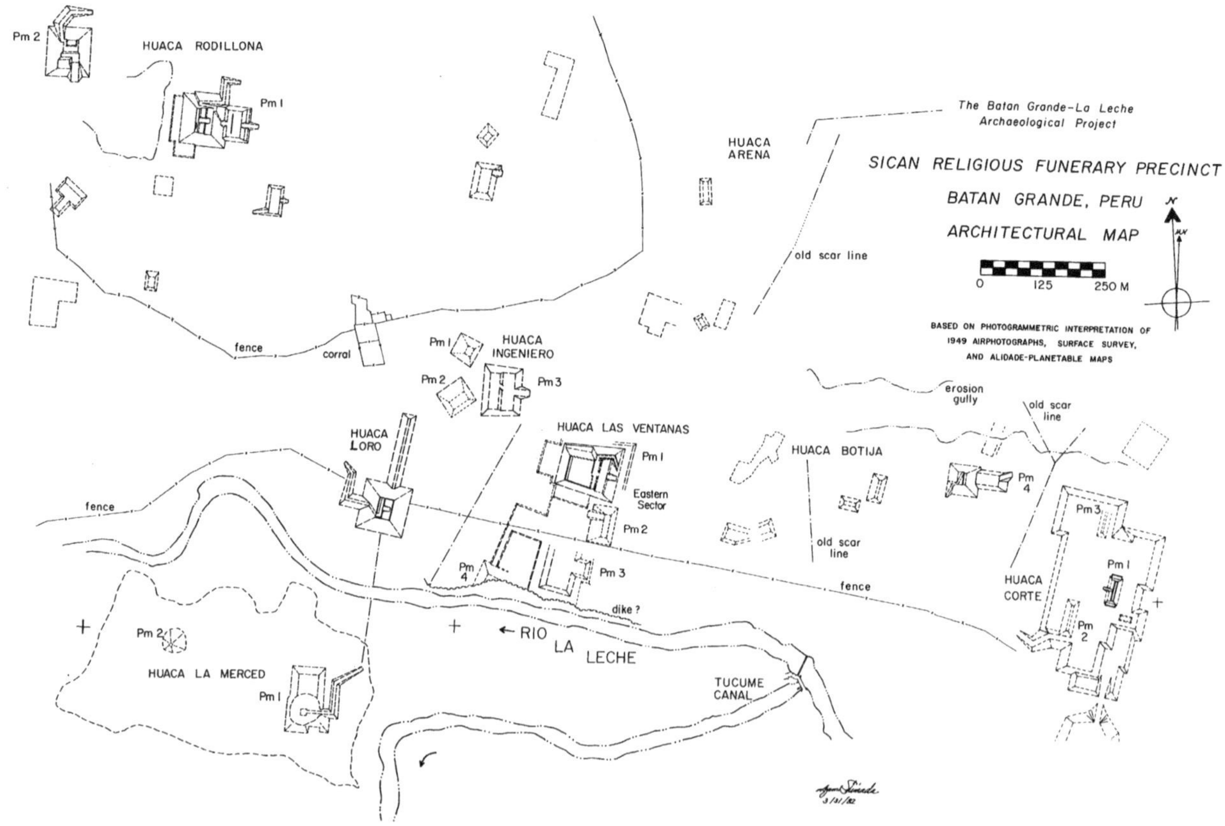

Figure 2. Reconstructed plan of the Huaca Loro platform mounds with the locations of the East and West Tombs. Drawn by I. Shimada.

important comparative data and perspective. The tomb was excavated to test the hypothesis that the East and West tombs were part of a planned cemetery of a Middle Sicán elite lineage placed under and around the Huaca with its the north-south axis serving as the organizational basis. The placement of the East and West tombs below the north base of the Huaca Loro mound, along with the relatively close biological distances (as measured by 23 inherited dental traits) between the individuals interred in these tombs effectively confirmed the hypothesis (Shimada et al. 1998).

Excavation revealed that the West Tomb had a two-tier, nested construction. The antechamber was located 12m below surface and measured some 10m north-south and by 6m east-west. It featured 10 wall niches and 12 rectangular sub-floor pits. Situated at the center of the tomb was the central chamber, a 3 x 3m shaft that descended another 3m to a depth of 15m below the surface. Six pits formed two symmetrically opposing groups on the south and north sides of the central chamber. Each pit contained one or two females with textiles, ceramics, and other goods.

The central chamber appears to have been reserved for placement of the principal individual and his grave goods. This individual, a relatively robust male of ~30-35 years who had suffered a serious stab wound on his hip, was placed at the center of the mat-lined floor in a cross-legged, seated position. He wore full regalia, including a large *tumbaga* mask, *tumbaga* and gold headdress, *tumbaga* gloves, and pectorals of mineral beads. He was surrounded by a diverse range of grave goods including eight camelid heads, and the articulated feet of at least 25 camelids, both large and small forms, earthen casts of nine rolls of strip cloth long since perished, and four *tumbaga* sheet-covered ceramic bottles and jars, among other items.

As if to flank the principal individual, an adult female was placed in each of the symmetrically opposing niches on the north and south walls of the Central Chamber. Placed beneath the female on the north side was a rectangular chest that contained some two dozen personal ornaments (e.g., four cylindrical crowns and five *tumi*-shaped crown ornaments), and ritual paraphernalia (a rattle and two ceremonial darts).

Overall, in spite of its impressive size (some seven times the volume of the East Tomb), complex internal structure and many bodies (24 as opposed to five in the East Tomb), the quantity and quality of metal goods (both precious metal and arsenical copper) in the West Tomb pale in comparison with those of the East Tomb. The West Tomb contained only 2.5 kg of arsenical copper *naipes,* 2.5 kg of *tumbaga* ornaments and ritual objects, and about 0.1kg of high karat gold objects (mostly small bangles).

Analytical data and their significance

The abundance and diversity of precious metal objects from the tomb excavations presented a dilemma. The material represented an unprecedented research opportunity to elucidate technological, social, symbolic, and other aspects of Sicán goldworking, but there were major restrictions as to the manner and amount of sampling that would be permitted by the National Institute of Culture of Peru.

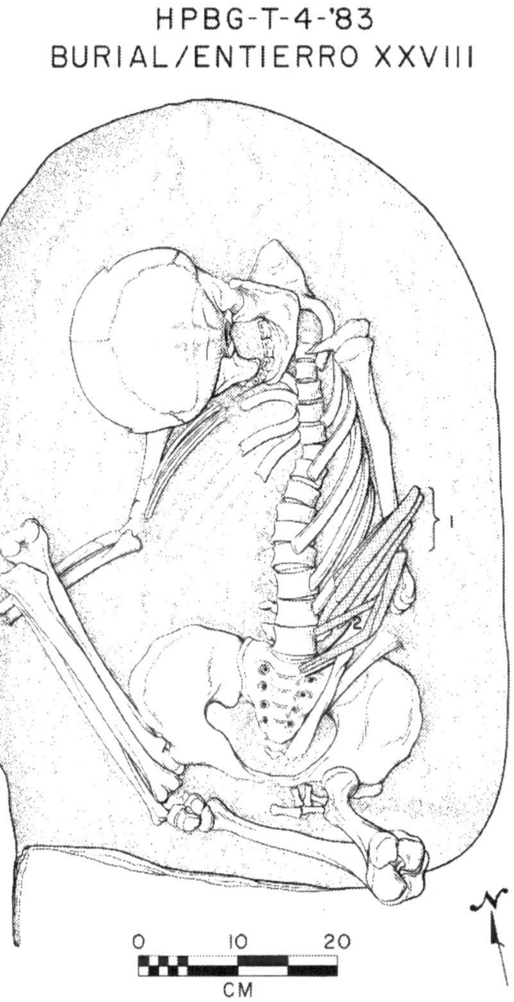

Figure 3. Third and fourth levels of grave goods surrounding the principal individual of the East Tomb. TS refers to tumbaga scrap, SP, Spondylus shells, and C, Conus shells. Drawn by César Samillán and I. Shimada.

The solution was, to the fullest extent possible, to conduct visual examination of excavated objects in Peru and devise a sampling scheme that would be acceptable to the National Institute of Culture and appropriate to the project research interests. The strategy involved different combinations of the authors working together in Peru. The aim was to carry out a systematic analysis of alloy composition and relate this to the mechanical properties and workability of alloys, as well as the inferred functions and symbolism of manufactured objects. In order to obtain accurate data with minimal damage to the artifacts, a methodology previously developed by Gordus was adopted which entailed the removal of rubbing samples for subsequent NAA (Gordus and Shimada 1995). Over 1,000 ~10 microgram rubbings were collected on quartz glass tubes (mostly by A. Gordus) from high-karat gold alloy objects from both the East and West Tombs. The overwhelming majority of *tumbaga* objects were too corroded to be sampled. Instead, tiny broken fragments were collected for EPMA, metallographic, and other analyses conducted at the Institute of Archaeometallurgy, University College London by J. F. Merkel and his students (Boissonnas 1997; Brazil 1995; McLoughlin 1996; Merkel *et al.* 1995).

To compensate for any surface depletion of copper and/or silver, three rubbings were collected from each sampling position, reaching an estimated depth of ~30 microns. Further, EPMA and NAA were carried out on tiny broken fragments that were allowed to be exported from Peru. These provided comparative compositional data. Rubbing samples for NAA were first taken from these fragments, which were later sectioned to expose the alloy core for EPMA (Gordus *et al.* 1996). Although minor surface copper depletion was detected in the metal fragments, the comparison established the general reliability of the NAA data for the study of alloy composition.

The analyses have shown that:-

(1) Essentially all of the over 1,000 samples taken from some 350 artifacts and fragments are ternary alloys of gold, silver, and copper.

(2) The Au:Ag:Cu ratios of the artifacts vary widely, ranging from a rough equivalent of 18-karat gold to barely 2 karats, and include single and multi-phase alloys. Most of the observed alloys are multi-phase. On the ternary phase diagram for Au-Ag-Cu, there is a large immiscibility field between the Au-Ag and Au-Cu phases. For example, 22-karat alloys are single phase, but 14-karat and 9-karat alloys are two phase. The two-phase structure is a dominant factor for control of mechanical and working properties of gold alloys.

(3) The sheet metal objects had been extensively cold-worked.

(4) Each type of metal object from the East Tomb, such as bangles, 'feathers' (believed have been parts of an elaborate headdress), and ear spools had its own relatively consistent alloy content, with the ear spools being very high in gold (~18 karats: 59-69 wt.% Au, 30-38wt.% Ag and 2-6 wt.% Cu) compared with the other.

(5) Surface depletion of copper or copper and silver was detected in nearly all of the samples analyzed (Gordus *et al.* 1996; McLoughlin 1996). However, the extent of depletion varies considerably and correlates with the percentage of copper in the alloy composition. EPMA of a cross-section of a high karat bangle (~45wt.% Au, 37wt.% Ag, 18wt.% Cu) and its wire attachment showed that only the copper had been notably depleted (15 wt.% Cu in the interior but only 4 wt.% Cu at the edge) and only in a surface layer a few microns thick (Gordus *et al.* 1996). The data suggest that depletion resulted largely from the repeated heating, annealing, and weak-acid surface cleaning or quenching involved in sheet metal making and/or other preparatory metalworking tasks. However, in the case of silver- and copper-rich ternary alloys (e.g., 20 wt.% Au, 30 wt.% Ag, 50 wt.% Cu), intentional depletion gilding is likely given that surface depletion is typically on the order of 10-20 microns (McLoughlin 1996; Merkel *et al.* 1995).

This study is not the first to document variability among the gold-silver-copper alloys used to make prehispanic objects from the north coast of Peru (Kroeber 1944; Lechtman 1973, 1984b, 1988; Patterson 1971). The variability is not surprising given that copper, gold and silver effectively melt together to form a complete solid solution series throughout the entire range of possible alloy composition. This study is distinguished, however, for its large scale, systematic analysis of samples from excavated, primary contexts and in documenting not just a wide range of variability in alloy composition, but also correlations between composition and artifact type.

Correlation between alloy composition and artifact type was anticipated from the design, function(s), and techniques employed in manufacturing the objects (Gordus and Shimada 1995; Shimada and Griffin 1994). Functions were inferred from the placement of the object in the tomb (e.g., on the body of the principal personage of the East Tomb) and aided by artistic depictions (e.g., murals, painted cloth, and on ceramic and metal figurines; see Carcedo 1989; Carcedo and Shimada 1985). Manufacturing techniques were determined from direct evidence such as tool marks and repairs (see Shimada and Griffin 1994), radiographic examination, as well as metallographic and other physical analyses.

A good illustration of the co-variation between composition and function is seen in the over a dozen *tumi*-shaped headdresses. Each of these measures ~28-29 x 16-17 cm and weighs 90-100g. Their average alloy composition is ~50 wt.% Au, 38wt.% Ag and 11wt.% Cu. The tang, or quill, of each headdress had to be narrow, but of sufficiently thickness (~0.6-1.5 cm wide and 1.0-1.24 mm thick at the tip) to stay upright when inserted into a turban or crown socket. The sheets were hammered on a flared gradient; i.e., in such a way to gradually taper in thickness from the stem to the upper tip. The design dictated that the finished piece be thin and flexible; in fact, at the top, the sheet is only ~0.15-0.18 mm thick. The goldsmith knew exactly how much to planish the sheet to give it just the right amount of springiness (tensile properties) so that it would wave in the air with each movement of the head, but not so much that the piece would crack.

Results from microhardness tests on samples from other similar *tumi*-headdresses support these observations (McLoughlin 1996). The tip of the tang (110 - measured in Vickers Microhardness values for the silver-rich phase and 243 for the copper rich phase) is the hardest, followed by the crescent-shaped blade (121 for the silver-rich phase and 181 for the copper rich phase) and finally the lateral ornaments (*ca.* 120 for the 2-phase mixture). Although present in a low concentration of 1 to 2%, arsenical copper probably contributed to the reported hardness. Overall, the study indicates that, while using the same composition throughout, specific parts of the headdresses were worked to produce specific, required mechanical properties.

The above comments apply to five sets of gold 'feathers'. A requisite balance between thickness and tensile properties was also achieved with a flared gradient in hammering and relatively high copper content. All but one of the five sets analyzed have an alloy composition of ~43 wt.% gold, 40 wt.% silver and 16 wt.% copper.

The systematic study of gold alloy objects from the East and West Tombs shows that the gold content of an alloy only generally correlates with the complexity and degree of technical skill involved in artifact manufacture. Iconographic detail correlates in a more consistent manner.

Figure 4. Large gold mask that covered the face of the principal individual of the Huaca Loro East Tomb shown after lengthy conservation work. The mask was found badly deformed. Photo by Y. Yoshii.

The first conclusion is best illustrated by comparing the masks that covered the faces of the principal individuals of the Huaca Loro East and West Tombs and their *tumi*-shaped crown ornaments.

The East Tomb mask is a *tour de force* of Middle Sicán gold working (Figure 4; see Shimada and Griffin 1994). It was fashioned out of a single gold alloy (52 wt.% Au, 31 wt.% Ag, 17 wt.% Cu) sheet measuring 46 x 29 cm and ~0.6 mm in thickness. It has no visible stress cracks even on and around the realistically shaped nose. Obscuring the view of the nose was a set of four high-karat gold bangles hanging from two wires which project through holes at the base of the nose. In addition, a U-shaped ornament with small circular bangles surrounds the base of the nose. The eyes were made to resemble human eyes. On the silver alloy "whites" of the eyes were large, highly polished, semi-spherical amber beads representing the irises and pierced, spherical emerald beads for the pupils. The stylized earlobes are furnished with a pair of masterfully crafted high karat (59 wt.% Au, 35 wt.% Ag, 6 wt.% Cu) ear spools.

In contrast, the West Tomb mask (Figure 5) was made by mechanically joining four separate *tumbaga* panels (their fragile condition precluded taking rubbing samples for NAA). Its eyes resembled those of the East Tomb mask in construction, size, and shape, but they lacked piercing gold wires and emeralds. It had no ear spools, though corroded staples indicate that originally there had been perishable decorations on the earlobes. The nose ornaments were limited to four bangles that appear to have been made out of a silver-rich gold alloy with turquoise inlay. In spite of general similarities in size and shape, the above two masks differ markedly in alloy composition, technical difficulty in the preparation of raw materials (e.g., gold sheets), assembling components, and even iconographic details.

The West Tomb mask pales in comparison with its East Tomb counterpart in regard to labor, materials, and skills required in manufacture.

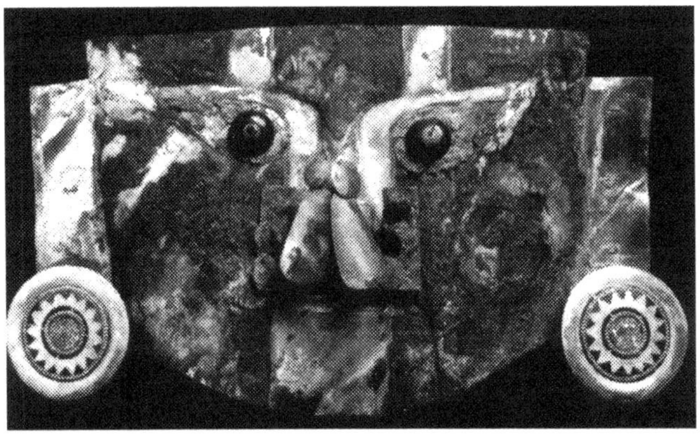

Figure 5. Large tumbaga mask as found covering the face of the principal individual of the Huaca Loro West Tomb. In spite of its general similarity to the mask shown in Figure 4, there are various important differences as discussed in the text. Photo by I. Shimada.

At the same time, the West Tomb mask is better appreciated in comparison with the one that covered the face of a seated adult male in a small pit (~1.5 x 0.8 m and 3 m deep) between Huaca La Botija and Huaca El Corte at Sicán. The latter was a simple, totally mineralized 'copper' mask (Alva 1986). It appears to have been made only with copper sheets. The metallic composition of this mask has not been determined and it may well have been a copper-silver alloy.

These observations regarding the masks hold for many ornaments and ritual items contained in similar rectangular chests found in both the East and West Tombs. It is noteworthy that, though metal objects in the West Tomb largely duplicate in artifact type those found in the East Tomb, they are made predominantly of relatively low-gold *tumbaga*, are considerably fewer in number of examples per type (e.g., one rattle as opposed to 4), and in many ways inferior in technical quality and iconographic detail.

At the same time, *tumi*-shaped, *tumbaga* crown ornaments in the West Tomb chest hardly differ from those of gold in the East Tomb chest in regard to size, shape, iconography and manufacturing techniques. Both sets are cut out of sheetmetal, probably using a template. While the original thickness of the *tumbaga* sheetmetal cannot be precisely determined due to corrosion, it does not appear to differ much from that of gold.

The preceding discussion shows that the quality of metal (i.e., proportions of precious metals) in the excavated Sicán objects *in general* appears to vary positively with the complexity, difficulty, and innovative character of manufacturing techniques employed (e.g., fashioned out of a single sheet versus multiple sheets mechanically joined together). It also appears to be related to the presence and quality of shell or mineral inserts, as well as to iconographic details.

However, these conclusions must be regarded as preliminary as the size of the relevant sample is still small and restricted to the site of Sicán. It must be emphasized that we are speaking only of a general correlation. High karat gold does not guarantee high craftsmanship. Conversely, *tumbaga* objects are not always inferior to gold objects in technical quality and/or iconographic contents.

A systematic comparison of grave goods found with the some 50 Middle Sicán burials excavated thus far in and outside of the site of Sicán shows that the different alloys have clearly differentiated distributions. The burials can be divided into four groups:-

(a) extended burials in simple, shallow pits without metal objects; may have one or two ceramic vessels;

(b) extended or flexed burials in simple shallow pits accompanied by arsenical copper objects; usually have a number of ceramic and other classes of artifacts;

(c) seated burials in deep conical pits or shaft tombs associated with arsenical copper and *tumbaga* objects, in addition to a variety of other classes of grave goods, and;

(d) seated individuals in deep shaft tombs accompanied by arsenical copper, *tumbaga,* and high-karat gold objects, in addition to a variety of other classes of grave goods (Shimada 1995).

This grouping *generally* correlates with the size and complexity of tombs. However, the gigantic West Tomb is an anomaly. What correlates well with them is *the iconographic details and range of alloys present.* The most explicit and detailed iconographic representations are found on precious metal objects. Overall, internal analysis of Sicán metal objects indicates that prestigious high-karat gold objects were emulated using less noble metals and that the Sicán ternary alloys of different compositions had specific functions and values.

The wide range of ternary alloys and different degrees and kinds of surface depletion (both intentional and coincidental) which have been documented among metal objects from the East and West Tombs indicate that Sicán metalworkers had the means to produce a correspondingly diverse range of hues and tints, from reddish and pale gold to whitish silver. But, what role did the metal color play in Sicán alloying? Was the mechanical property of a given alloy secondary to the color or vice versa? Although gold and silver are commonly linked to the sun and the moon (e.g., Emmerich 1977; Kauffman 1989; Lechtman 1984a,b), what hues and tints represented them? Was color variation meant to mimic similar variation in the sun and the moon? In fact, the sun's color may be described as white, yellow, orange and/or red by different cultures with correspondingly different symbolic meanings. The complexity of symbolism of the two metals, one white and one yellow, among the Tukanoan tribes of Colombian Amazonia recorded by Reichel-Dolmatoff (1981) effectively illustrates the difficulty of deciphering color symbolism. In Tukanoan conception, these colors related more to the cosmic creative forces and the male-female and sun-moon dichotomies than to gold and silver.

Among high-karat gold alloys, specific hues and tints of the surface appear to have been secondary to obtaining desired mechanical properties. As seen earlier, the extent of surface depletion is relatively limited and there is good correlation between alloy composition, on the one hand, and artifact types and their mechanical properties, on the other. In addition, in terms of the logical sequence of goldworking, surface treatment would follow alloying and sheet metal making. However, the question of the role of color in *tumbaga* is much more difficult to answer (see Boissonnas 1997; Brazil 1995; Gordus and Shimada 1995; McLoughlin 1996; Merkel *et al.* 1995; Perkins 1997; Shimada and Griffin 1994). The considerable variation in alloy composition of *tumbaga* (1-10 karats) and its intentional surface depletion of copper or copper and silver resulted in essentially a continuum of color between what are conventionally thought to be gold and silver. Further, consideration needs to be given to the potential symbolic role of high-arsenic arsenical copper that has a silvery appearance. Shimada (in press a) concludes that "It would be difficult to determine what constituted prestigious gold color to prehispanic Andean people. We are still a long way from understanding emic perceptions (i.e., perceptions of the aboriginal makers and users) of the relative importance of the multiple factors that contribute in producing given colors in precious metals."

Conclusion

Exploration of the extrinsic dimensions of ancient metallurgy is rewarding, but cannot be easily undertaken. Based on our experiences with Sicán metallurgy, we recommend that such pursuit be firmly based on sustained regional study that samples a wide range of contexts. Those contexts without evidence of metal use are just as informative as those with it in elucidating the role and value of metals. At the same time, a multitude of investigative avenues, such as replicative experiments, archaeometry, ethnoarchaeology, stylistic and iconographic studies - should be brought to bear. It is these intersecting lines of evidence and perspectives that hold promise in unraveling the diverse information embodied in ancient metal objects.

Acknowledgements

We are grateful to the Shibusawa Ethnological Foundation of Tokyo for research grants issued to Izumi Shimada to support field and laboratory work that provided the data for this paper. The Museo de la Nación in Lima and the Bruning National Archaeological Museum in Lambayeque kindly provided us with space for our documentation, analysis, and conservation of excavated artifacts. Lastly, we are indebted to Paul Budd and Melody Shimada for their editorial assistance.

References

ALVA, WALTER. 1986. Una Tumba con Máscara Funeraria de la Costa Norte del Perú. *Beiträge zur Allgemeinen und Vergleichenden archäologie* 7 (1985): 411-421. KAVA, Deutschen Archäologischen Instituts, Bonn.

BOISSONNAS, JEAN-VALENTIN. 1997. *The Use of Low-Pressure Hydrogen Plasma Reduction on Depletion Gilded Copper Alloys.* Unpublished B.Sc. thesis in Archaeological Conservation, Institute of Archaeology, University College London, London.

BRAZIL, RACHEL B. 1995. *The Application of X-Ray Photoelectron Spectroscopy to the Study of Depletion Gilded Alloys from Batán Grande, Peru.* Unpublished M.Sc. thesis in Archaeometallurgy, Institute of Archaeology, University College London, London.

CARCEDO, PALOMA. 1989. Anda Ceremonial Lambayecana: Iconografía y Simbología. In *Lambayeque,* edited by José Antonio de Lavalle, pp. 249-270. Colección Arte y Tesoros del Perú. Banco de Credito del Perú, Lima.

CARCEDO, PALOMA & IZUMI SHIMADA. 1985. Behind the Golden Mask: Sicán Gold Artifacts from Batán Grande, Peru. In *Art of Precolumbian Gold: Jan Mitchell Collection,* edited by Julie Jones, pp. 60-75. Weidenfeld and Nicolson, London.

CLELAND, KATE M., & IZUMI SHIMADA. 1992. Sicán Bottles: Marking Time in the Peruvian Bronze Age. *Andean Past* 3: 193-235. Ithaca, New York.

CORDY-COLLINS, ALANA. 1990. Fonga Sigde, Shell Purveyor to the Chimu Kings. In *The Northern Dynasties: Kingship and Statecraft in Chimor,* redited by Michael E. Moseley and A. Cordy-Collins, pp. 393-417. Dumbarton Oaks, Washington, D.C.

EMMERICH, ANDRÉ. 1977. *Sweat of the Sun and Tears of the Moon: Gold and Silver in Pre-Columbian Art.* Hacker Art Books, New York.

EPSTEIN, STEPHEN M., & I. SHIMADA. 1984. Metalurgia de Sicán: Una Reconstrucción de la Producción de la Aleación de Cobre en El Cerro de los Cementerios, Peru. *Beiträge zur Allgemeinen und Vergleichenden archäologie* 5 (1983): 379-430. KAVA, Deutschen Archäologischen Instituts, Bonn.

GORDUS, ADON A., HENDERSON, CARL E., & IZUMI SHIMADA. 1996. Electron Microprobe and Neutron Activation Analysis of Gold Artifacts from a 1000 AD Peruvian Gravestite. In *Archaeological Chemistry: Organic, Inorganic, and Biochemical Analysis,* edited by Mary V. Orna, pp. 83-93. American Chemical Society Symposium Series 625. Washington, D.C.

GORDUS, ADON, & IZUMI SHIMADA. 1995. Neutron Activation Analysis of Microgram Samples from 364 Gold Objects from a Sicán Burial Site in Peru. In *Materials Issues in Art and Archaeology IV,* compiled by P. B. Vandiver, J. R. Druzik, and J. L. Galvan , pp. 127-142. Materials Research Society Proceedings, Vol. 352. Pittsburgh.

KAUFFMANN, FEDERICO. 1989. Oro de Lambayeque. In *Lambayeque,* edited by José Antonio de Lavalle, pp. 163-222. Colección Arte y Tesoros del Perú. Banco de Credito del Perú, Lima.

KROEBER, ALFRED. 1944. *Peruvian Archaeology in 1942.* Viking Fund Publications in Anthropology 4. Chicago.

LECHTMAN, HEATHER N. 1973. The Gilding of Metals in Precolumbian Peru. In *The Application of Science In Examination of Works of Art,* edited by W. J. Young, pp. 38-52. Museum of Fine Art, Boston.

1984a. Andean Value Systems and the Development of Prehistoric Metallurgy. *Technology and Culture* 25(1): 1-36.

1984b. Pre-Columbian Surface Metallurgy. *Scientific American* 250(6):56-63.

1988. Traditions and Styles in Central Andean Metalworking. In *The Beginning of the Use of Metals and Alloys,* edited by Robert Maddin, pp. 344-378. MIT Press, Cambridge, MA.

LECHTMAN, HEATHER N., ERLIJ, ANTONIETA, & EDWARD J. BARRY, JR. 1982. New Perspectives on Moche Metallurgy: Techniques of Gilding Copper at Loma Negra, Northern Peru. *American Antiquity* 47:3-30.

MCLOUGHLIN, SEAN D. 1996. *A Metallurgical Characterisation of Precious Metal Alloys from Batán Grande, Peru.* Unpublished M.Sc. thesis in Archaeometallurgy, Institute of Archaeology, University College London, London.

MERKEL, J. F., A. I. SERUYA, DAFYDD GRIFFITHS, & IZUMI SHIMADA. 1995. Metallographic and Microanalysis of Precious Metal Objects from the Sicán Period at Batán Grande, Peru. In *Materials Issues in Art and Archaeology IV,* compiled by P. B. Vandiver, J. R. Druzik, and J. L. Galvan , pp. 105-126. Materials Research Society Proceedings, Vol. 352. Pittsburgh.

MERKEL, JOHN F., & IZUMI SHIMADA. 1988. Arsenical Copper Smelting at Batán Grande, Peru. *Newsletter of the Institute for Archaeo-Metallurgical Studies,* No. 12: 4-7. London.

MERKEL, JOHN F., IZUMI SHIMADA, C.P. SWANN, & R. DOONAN. 1994. Investigation of Prehistoric Copper Production at Batán Grande, Peru: Interpretation of the Analytical Data for Ore Samples. In *Archaeometry of Pre-Columbian Sites and Artifacts,* edited by D. A. Scott and P. Meyers, pp. 199-227. The Getty Conservation Institute, Marina del Rey, *CA.*

PATTERSON, CLAIR C. 1971. Native Copper, Silver, and Gold Accessible to Early Metallurgists. *American Antiquity* 36: 286-321.

PEDERSEN, ASBJORN. 1976. El Ajuar Funerario de la Tumba de la Huaca Menor de Batán Grande, Lambayeque, Perú. *Actas del 41 Congreso Internacional de Americanistas* 2: 60-73. Mexico City.

PERKINS, RACHAEL N. 1997. *The Technology, Deterioration and Benzotriazole Treatment of Electrochemically Gilded Copper Alloy Artifacts from Peru.* Unpublished B.Sc. thesis in Archaeological Conservation, Institute of Archaeology, University College London, London.

REICHEL-DOLMATOFF, GERARDO. 1981. Things of Beauty Replete with Meaning. In *Sweat of the Sun, Tears of the Moon* (exhibition catalog), pp.17-33. Natural History Museum of Los Angeles County, Los Angeles.

REICHLEN, HENRY. 1941. Etude Technologique de Quelques Objets d'or de Lambayeque, Peróu. *Journal de la Société des Américanistes* 33: 119-154.

1962. Un Bijou d'or de Lambayeque, Pérou. *Objets et Mondes* 2(2): 77-84.

SHIMADA, IZUMI 1981 The Batán Grande-La Leche Archaeological Project: The First Two Seasons. *Journal of Field Archaeology* 8: 405-446.

1985 Perception, Procurement and Management of Resources: Archaeological Perspective. In *Andean Ecology and Civilization,* edited by Shozo Masuda, I. Shimada and Craig Morris, pp. 357-399. University of Tokyo Press, Tokyo.

1987. Horizontal and Vertical Dimensions of Prehistoric States in North Peru. In *The Origin and Development of the Andean State,* edited by J. Haas, T. and S. Pozorski, pp. 130-144. Cambridge University Press, Cambridge.

1990. Cultural Continuities and Discontinuities on the Northern North Coast, Middle-Late Horizons. In *The Northern Dynasties: Kingship and Statecraft In Chimor,* edited by Michael E. Moseley and A. Cordy-Collins, pp. 297-392. Dumbarton Oaks, Washington, D.C.

1991. *Pachacamac Archaeology: Retrospect and Prospect.* University Museum Press, University of Pennsylvania.

1992. La Sépulture Inviolée d'un Riche Seigneur Sicán. In *Les Dossiers d'Archeologie,* Special Issue, " Le Pérou de l'Origine aux Incas", edited by Daniel Lévine (Hors-serie No. 2), pp. 46-57. Paris.

1994a. *Pampa Grande and the Mochica Culture.* University of Texas Press, Austin.

1994b. Prehispanic Metallurgy and Mining in the Andes: Recent Advances and Future Tasks. In *In Quest of Mineral Wealth: Aboriginal and Colonial Mining and Metallurgy in Spanish America,* edited by Alan K. Craig and Robert West, pp. 37-73. *Geoscience and Man* 53. Louisiana State University, Baton Louge.

1995. *Cultura Sicán: Dios, Riqueza y Poder en la Costa Norte del Perú.* Banco Continental, Lima.

In press a. Sicán Metallurgy and Its Cross-Craft Relationships. *Boletín del Museo de Oro,* Bogotá.

In press b. Late Prehispanic Coastal States. In *Pre-Inka States and the Inka World,* edited by Laura Laurencich Minelli. University of Oklahoma Press, Norman.

In press c. Exotic Goods and Archaeological Methodology: How Viable Are Our Data on Interregional Interaction? In *Commerce and Exchange in the Andes: Highland-Lowland Interaction from an Archaeological and Ethnohistorical Perspective,* edited by J. F. Cárdenas-Arroyo and Tamara L. Bray. Universidad de los Andes, Quito.

SHIMADA, I., ANDERSON, K. B., HAAS, H., & J. H. LANGENHEIM. 1997. Amber from 1000-year Old Prehispanic Tombs in Northern Peru. In *Materials Issues in Art and Archaeology V,* compiled by P. B. Vandiver, J. R. Druzik, and J. F. Merkel, pp. 3-18. Materials Research Society Proceedings, Pittsburgh.

SHIMADA, IZUMI, STEPHEN M. EPSTEIN, & ALAN K. CRAIG. 1982. Batán Grande: A Prehistoric Metallurgical Center in Peru. *Science* 216: 952-959.

1983. The metallurgical process in ancient north Peru. *Archaeology* 36(5): 38-45.

SHIMADA, IZUMI, & JO ANN GRIFFIN. 1994. Precious metal objects of the Middle Sicán. *Scientific American* 270(4): 60-67.

SHIMADA, IZUMI, & JOHN F. MERKEL. 1991. Copper Alloy Metallurgy in Ancient Peru. *Scientific American* 265: 80-86.

1993. A Sicán Tomb in Peru. *Minerva* 4(1): 18-25.

SHIMADA, IZUMI, & JORGE MONTENEGRO. 1993. El Poder y la Naturaleza de la Elite Sicán: Una Mirada a la Tumba de Huaca Loro, Batán Grande. *Boletín de Lima* XV(90): 67-96.

TELLO, JULIO C. 1937a. Los trabajos arqueológicos en el Departamento de Lambayeque. *El Comercio* 29, 30 y 31 de Enero, Lima.

1937b. La búsqueda de tesoros ocultos en las huacas de Lambayeque. *El Comercio* 11 de Marzo, Lima.

1937c. El oro de Batán Grande. *El Comercio* 18 de Abril. Lima.

VALCÁRCEL, LUIS E. 1937. Un valioso hallazgo arqueológico en el Perú. *Revista del Museo Nacional* VI: 164-168. Lima.

Editors note: One table and 3 figures had to be removed from the paper as they were never provided. Please contact the authors if a full set of figures and tables is required.

40

The lead isotope method for tracing the sources of metal in archaeological artefacts: strengths, weaknesses and applications in the Western Hemisphere

ANDREW MACFARLANE

Department of Geology, Florida International University, Miami, FL 33199, USA

Pronounced isotopic variations of ore lead with increasing distance from the geological trench in continental arc-type mining regions provide a potentially powerful tool for determining the sources of metals in archaeological artefacts. Similar zoning may persist in other areas of archaeological interest, which are not presently related to trench margins and could also be exploited. Although the lead isotope ratios of metal artefacts reflect those of the source ores, care must be taken to ensure that isotope data on ore deposits directly represent the material smelted by ancient metallurgists and that the data are not biased by heterogeneity among ore minerals. Measurable modification of ore lead isotope ratios by smelting processes is very unlikely unless evaporation exceeds 40% of the original lead.

Introduction

Lead isotope ratios of metal artefacts are increasingly used to constrain the sources of ores used in their manufacture (e.g., Gale & Stos-Gale 1989; Gale & Stos-Gale 1992; Hosler & Macfarlane 1996). The popularity of this method stems from 1) the presence of lead at measurable levels in all ore and metal artefact samples, including those in which lead was not deliberately added; 2) the much greater availability of published lead isotope measurements on ore deposits than other potential isotopic tracers; 3) the wide range of lead isotope ratios of geological materials, including ores; 4) the correspondence of lead isotope ratios of artefacts to the ores they were derived from. These attributes make the lead isotopic system nearly ideal for tracing metal sources in artefacts, as it has long been used by geologists to study the sources of lead in the ore deposits themselves. Effective use of the lead isotope method requires an understanding of its strengths and limitations, particularly items 3 and 4 above. Although many archaeological studies to date have made little reference to the geological context of the ores, even a rudimentary understanding of the geological context can greatly enhance the effectiveness of lead isotope studies.

Lead isotope ratios of rocks and ores vary widely, depending on the U/Pb and Th/Pb values of the rocks and on their age. The U/Pb and Th/Pb values of rocks depend on their chemistry and thus their geological history, and rocks with higher U/Pb will have higher 206Pb/204Pb and 207Pb/204Pb values than rocks with lower U/Pb of the same age (Faure 1986). The average ages of rocks also vary greatly at the continental scale. Bennet & DePaolo (1987) have published a map of the average age of continental crust in North America, based on variations in Nd isotope ratios. Pb isotopes vary along broadly similar boundaries (Zartman & Doe 1981). Regional- and continental-scale isotopic patterns in rocks and ore deposits appear to be common, and often permit definition of isotope provinces distinguished by similar isotopic compositions or trends.

Pb isotope studies of archaeological artefacts must always contend with a sampling bias toward modern ore deposits. Ore lead isotope data are only available for mines and prospects that are known at the present time. Some of these deposits may have been mined in antiquity, and others not and some ore sources used in antiquity may be unknown today. Lead isotope analysis of an artefact may correspond to a mineral deposit that is unknown today, or for which lead isotope data are not yet available. Provided that an adequately large database on the lead isotope ratios of ores from a region exist, the recognition of lead isotopic provinces within a study area can provide a partial solution to the sampling bias problem by indicating the probable lead isotope characteristics of unrecognized deposits in a given isotope province. There may be isotopically anomalous deposits in any isotopic province, because some provinces contain ore deposits of widely differing ages, formed under drastically different geological conditions, but isotopic provinces can provide a useful frame of reference for planning provenance studies and interpreting the results. This paper addresses the implications of lead isotope province studies generally, as well as certain problems concerning the relationship of the lead isotope ratios of ore samples to those of artefacts produced from them.

Discussion

Lead isotopic provinces in South America

Large-scale ore lead isotope patterns in the Central Andes were studied by Macfarlane *et al.* (1990) and Petersen & Macfarlane (1993). FIGURE 1 shows the lead isotopic compositions of ores from the Central Andes and the corresponding geographical provinces as defined by Petersen & Macfarlane (1993). Andean ores can be divided into three major provinces, regardless of the type of deposit, or the metals contained in that deposit (Au, Ag, Cu, Sn, Fe, and others). Province I forms a belt parallel to the continental margin nearest the Peru-Chile trench, and is dominated by young volcanic rocks and related igneous intrusions. Ores from province I define small lead isotope fields, which were divided into subprovinces Ia, Ib and Ic with slightly different isotopic characteristics. Province I ores become generally more radiogenic from south to north. Subprovince Ia, corresponding to central and northern Chile, is the most isotopically homogeneous part of province I. Magmatism related to subduction is an efficient way of mixing lead from many sources and producing fairly homogeneous lead isotope signatures over

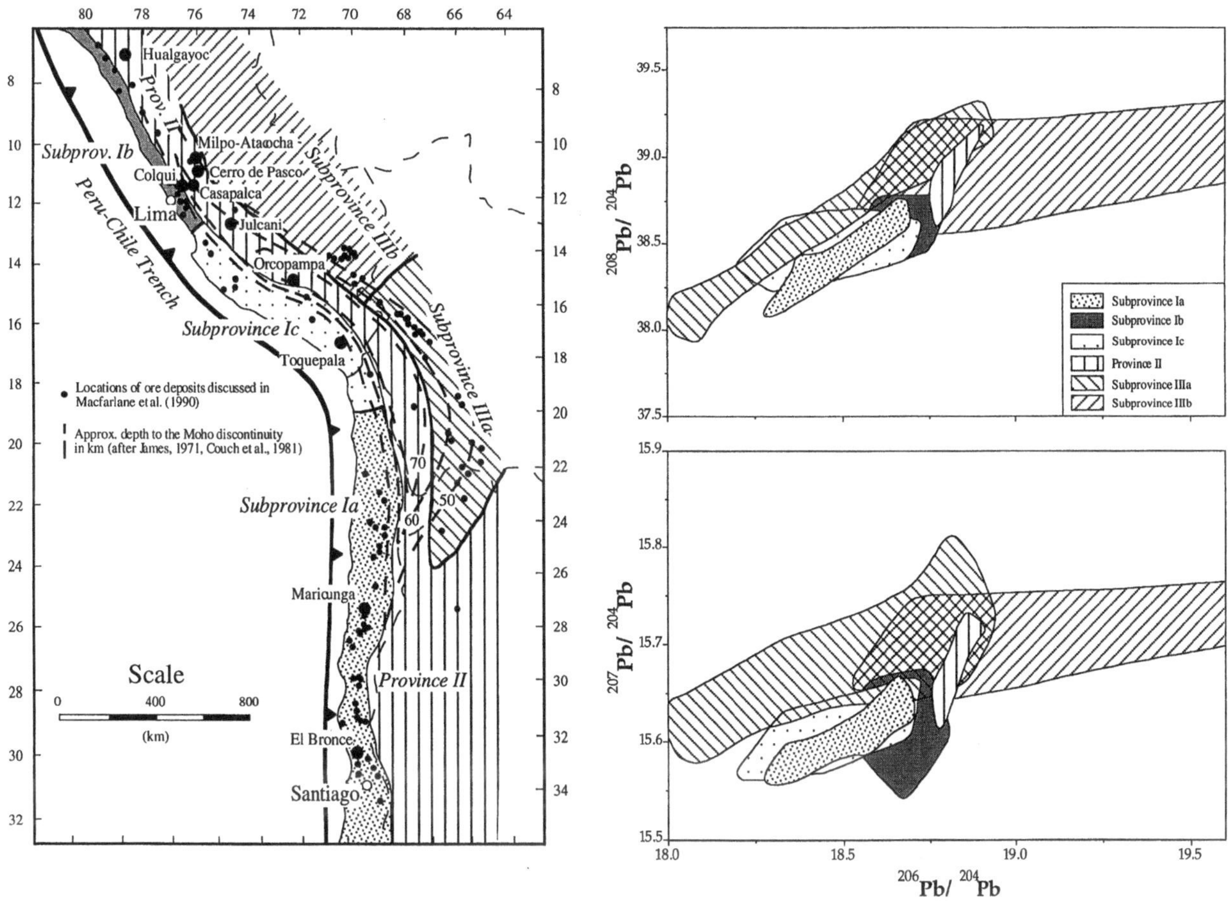

Figure 1: A) Geographical distribution of lead isotopic provinces in the Central Andes. Dots indicate the locations of ore deposits with lead isotope analyses; large dots are deposits in figure 3.
B) Lead isotopic compositions of Central Andean ores; fields are shaded to match the corresponding Pb isotope provinces in A. Figure modified from Macfarlane and Petersen (1990).

a large area (Zartman & Doe 1981). Macfarlane *et al.* (1990) interpreted the lead isotopic compositions of province I ores to reflect a well-mixed source of metals below the crust. Province III is the furthest inland of the provinces, encompassing the eastern altiplano and subandean ranges. Ores from province III have substantially higher 207Pb/204Pb and 208Pb/204Pb for given values of 206Pb/204Pb than province I ores, and were interpreted to indicate extraction of most of the ore lead from the crust.

Ores from province II define steep arrays between province I-type compositions and province III-type compositions, and seem to reflect mixing between deeper sources and shallow sources in this region, which corresponds roughly to the area presently underlain by the thickest continental crust. FIGURE 1 shows only province II data from northern Peru (Macfarlane *et al.* 1990); subsequent data on other province II deposits (FIGURE 2a) show that the steep mixing trends are a general feature of province II ores. In contrast, where multiple ore lead isotope analyses are available for individual deposits in province I, the data form small clusters or clouds rather than steep arrays (FIGURE 2b).

The large-scale isotopic zoning of the Central Andes, in addition to having important implications for the processes of ore formation in one of the world's largest ore provinces, provides a powerful means for discriminating between sources of ores used to make metal artefacts. Ores from provinces I and III are isotopically distinct in nearly all cases. Province II ores can overlap isotopically with both of the other two provinces, but are distinguished by forming steep isotope arrays (FIGURE 2). Data for artefacts made from ores from a province II deposit would form the same steep arrays, and would therefore be distinguishable from province I or III sources even if the same isotope ratios of some artefacts overlapped with compositions of the other provinces. Compositional differences among isotope arrays from different province II deposits suggest that artefacts derived from different deposits would often be distinguished if enough analyses were available. Identification of ores and artefacts that have been moved from one province to another may also be possible.

The most effective use of the lead isotope method would compare ores and artefacts from areas with distinct lead isotope compositions. Under favourable circumstances, the lead isotope method could distinguish among ore sources in close proximity to one another, for example, between province II ores, or province Ia and IIIa ores. Conversely however, the lead isotope method may be incapable of distinguishing among sources within the same province even where separated by hundreds of kilometres, such as in some areas of province I.

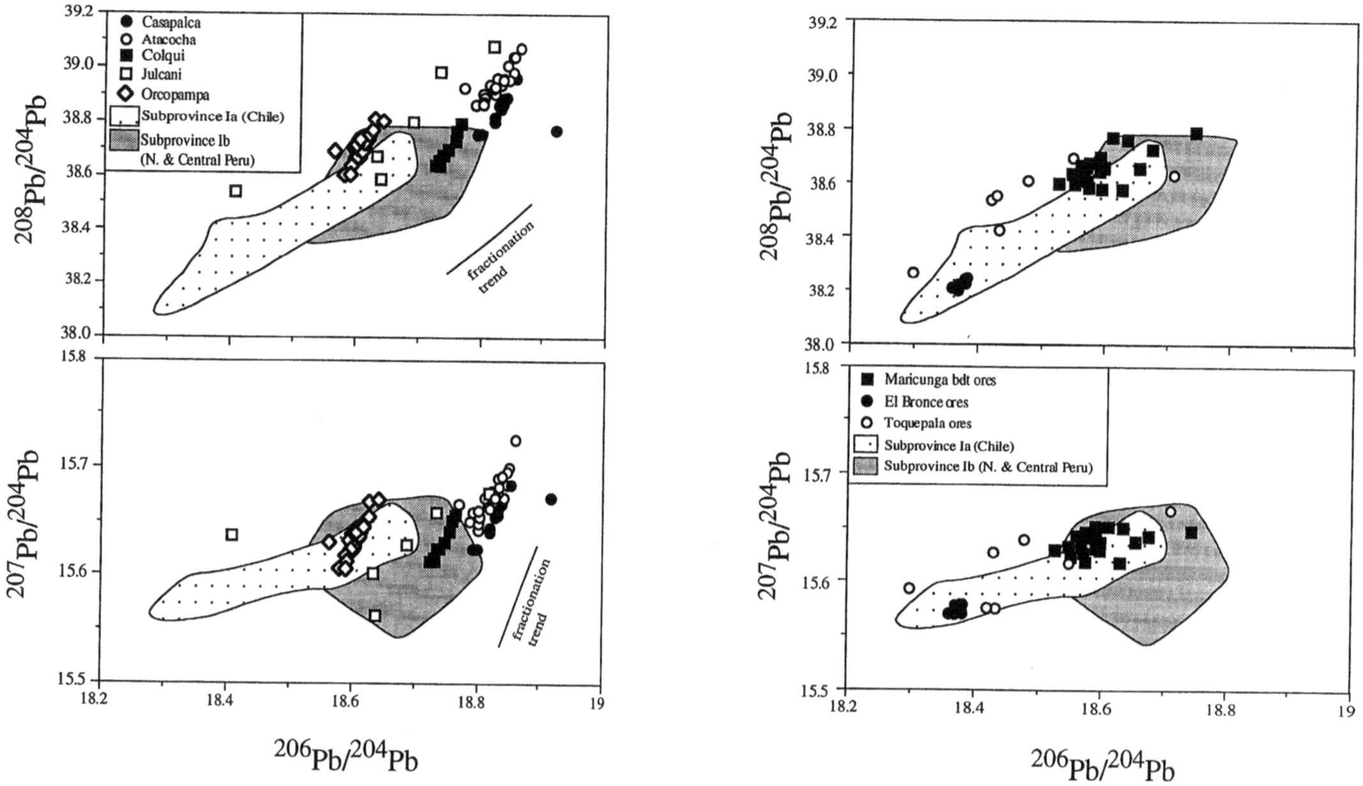

Figure 2: Lead isotope ratios of Central Andean province II ores; fields of province Ia (Chile) and Ib (northern and central Peru) ores are shown for comparison (from Macfarlane et al, 1990). A) Multiple analyses of ores from deposits in province II form steep arrays between province I-type lead at their lower ends to province III-type, upper crustal lead at their upper ends. B) Multiple analyses of ores from individual deposits in province I do not form steep arrays, but clouds within the total range of province I ores

Lead isotope variations in southern Mexico

Cumming *et al.* (1979) published the first comprehensive survey of lead isotope ratios in Mexican ore deposits; they described several isotopically distinct lead isotope provinces and a general increase in radiogenic lead from west to east across Mexico. Recently, Hosler & Macfarlane (1996) examined the lead isotope ratios of copper deposits throughout southern Mexico, with the goal of identifying the source or sources of ores used to make late post classical bronze artefacts in the region. Several workers (Campa & Coney 1983) have proposed that southern Mexico consists of a number of geologically distinct terranes that have been assembled over the past 100 million years (FIGURE 3). These terranes are thought to have had greatly different origins and their rocks are thought to vary greatly in average age. For example, rocks in province II (the Guerrero terrane) are thought to be primarily Mesozoic marine and continental sediments and volcanic rocks, rocks in province VII (the Oaxaca terrane) are thought to be primarily about 1.1 billion years old, and rocks in province VI (the Mixteca terrane) are thought to be primarily about 350 million year old metamorphic rocks with younger overlying sediments (Ruiz *et al.* 1988; Yanez *et al.* 1991; Ceteno-Garcia *et al.* 1993). Rocks in such geologically distinct terranes should have different lead isotopic characteristics. It was hoped that ores from different terranes would also be distinct from one another, so that lead isotopes could be used to uniquely identify the source of ores used to make an artefact with a specific geological terrane or mine.

Analysis of hundreds of samples from copper mines and prospects revealed that there was no simple correlation between ore lead isotope ratios and geological terrane. For example, ores from Ayutla in the geologically young province II and from El Taviche/Los Ocotes in the relatively ancient province VII are isotopically indistinguishable (Hosler & Macfarlane 1996). Instead, a simpler variation was found from southwest to northeast, with distance from the Middle America Trench (MAT). Combining the data of Hosler & Macfarlane (1996) and Cumming *et al.* (1979) demonstrates a clear pattern of increasing 206Pb/204Pb (averages of all data available from each deposit) with distance from the MAT (FIGURE 3). Like the Central Andean example, this isotopic zoning applies to all metal deposits regardless of deposit type (e.g., skarn, porphyry or vein) or contained metal (Cu, Sn, Pb, Zn, Ag or polymetallic ores).

Geologically, this result suggests that either the terrane hypothesis for southern Mexico is in serious error and should be re-examined or discarded, or that the sources of metals in these deposits are below the crustal terranes exposed at the surface, as suggested for province I of the Central Andes. Both suggestions may be correct, and the question requires further study. Two exceptions to the pattern were found, the large La Verde porphyry copper prospect and the smaller La Esmeralda copper- bearing

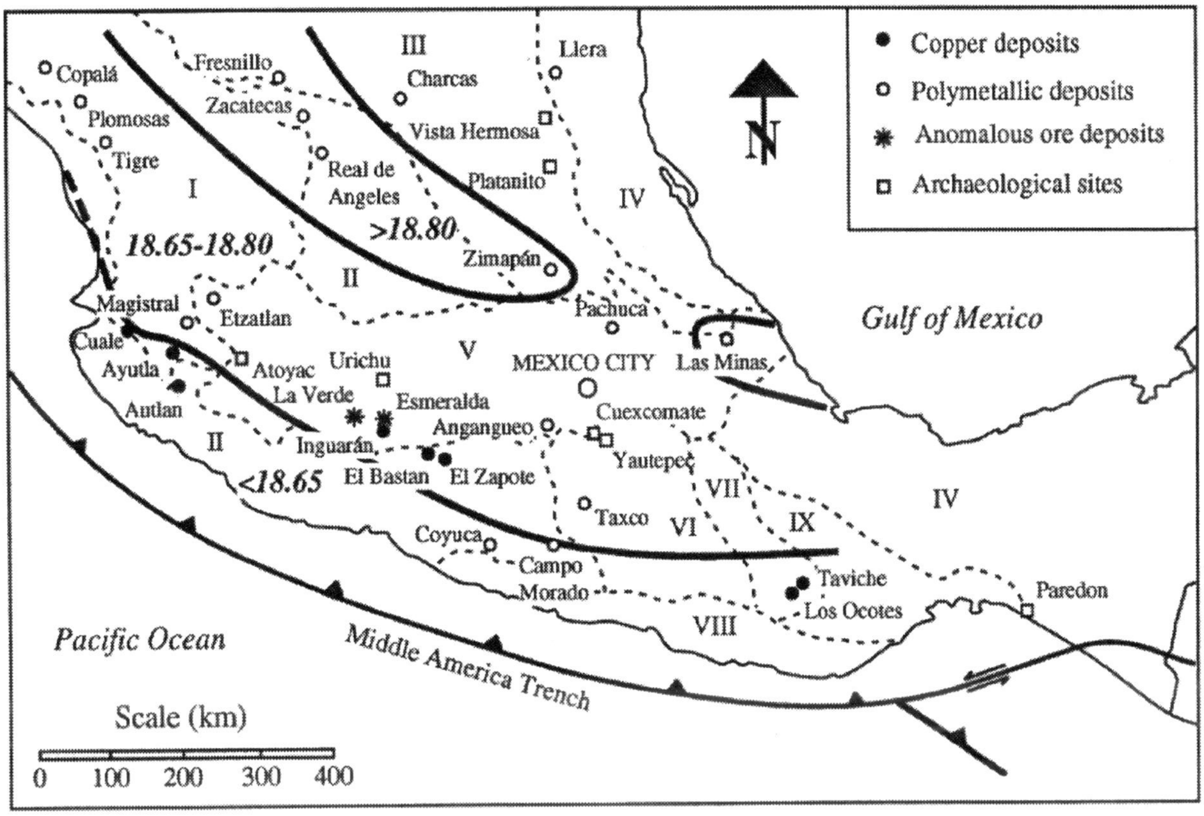

Figure 3: Geographical variability of 206Pb/204Pb in southern Mexico. Ores nearest the Middle America Trench have consistently lower ratios than those further inland, and there is a belt of higher values (> 18.80) in the interior. Modified from Hosler and Macfarlane (1996). Province boundaries (Roman numerals I-IX) from Campa & Coney (1983); geological descriptions of rocks in these provinces are found in that publication.

breccia pipe, which showed large lead isotope variations. The cause of these variations is not well understood, but must be related to local processes of extraction of metals from the host rocks.

Archaeologically, these results (together with data on 208Pb/204Pb not shown in FIGURE 3) allowed Hosler & Macfarlane (1996) to show that artefacts from throughout southern Mexico were made from ores from a small area of Jalisco and Michoac« n. The main value of lead isotope studies is to rule out potential ore sources that have lead isotope ratios that differ from those of the examined artefacts. Matching lead isotope ratios do not prove a unique source of ore, because lead isotopes can show little variation over large geographical areas; in southern Mexico, these large areas form belts parallel to the MAT (FIGURE 3). Additional data on the chemistry of ores and artefacts and on the physical and stylistic characteristics of the artefacts are needed to discriminate among potential sources within given isotopic provinces.

Can lead isotope ratios be altered by smelting?

Budd *et al.* (1995) have proposed that the smelting process may shift the isotope ratios of the resulting metal from those of the source materials, and that such changes render the lead isotopic method fundamentally unreliable as a tracer of the sources of artefacts. These authors suggest that kinetic fractionation of isotopes of different masses during the evaporation of lead during smelting will measurably alter the isotope ratios of the remaining metal; because lead is made up of four isotopes--204Pb, 206Pb, 207Pb and 208Pb--the lighter isotopes will evaporate at a very slightly faster rate than the heavier isotopes. Continued extraction of vapour will lead to a continuous shift in the composition of the remaining liquid.

The proposed process is a simple Rayleigh distillation, which may be modelled using the equation:

$$R_l = R_{l_i} f^{\frac{1}{\alpha}-1} \qquad eq.\ 1$$

where:

R_l is the lead isotope ratio of the liquid

R_{li} is the starting lead isotope ratio of the liquid

f is the fraction of lead remaining, and

α is the fractionation factor.

This equation incorporates the assumption that all evaporated lead is removed from the liquid; if some vapour condenses and mixes back with the liquid, the rate of isotopic change will decrease. The main difficulty in calculating the isotopic evolution of evaporating lead is that the kinetic fractionation factor (α) is unknown. Because the isotopes involved are heavy (204 to 208, or 220 to 224 if evaporation takes the form of PbO), and the mass differences are relatively small, the magnitude of fractionation is expected to be very small, relative to the precision with which the isotope ratios can be measured. Because of this, no reliable data are published on the fractionation of lead isotopes under these conditions.

Lacking direct data on evaporative fractionation of lead isotopes, the expected magnitude of alpha may still be reasonably constrained. Isotopic ratios of some metals in meteorites are strongly anomalous, and these isotopic variations have been attributed to evaporation of metal as the meteorite is heated passing through the atmosphere. Studies of isotopic fractionation of iron and oxygen under these conditions have been published by Wang *et al.* (1994), who found that the fractionations were well described by the square root of the mass ratio between isotopes. Measured fractionations were sometimes smaller than this value, but never greater (TABLE 1). This value therefore probably represents the best estimate of the maximum possible fractionation of lead isotopes during evaporation (R. Clayton pers. comm.). Relevant values are shown in TABLE 1 for both Pb and PbO vapour.

Another estimate of isotopic fractionation of lead during evaporation comes from the analytical procedure. Lead isotope ratios fractionate during evaporation of lead from the filament in the source of the mass spectrometer; this fractionation is well known and affects all isotopic measurements made by thermal ionization methods. Instrumental fractionation during analysis of standards of known composition is routinely determined to correct measured ratios of unknown samples. Fractionation in the mass spectrometer takes place under a high vacuum (approximately 10^{-7} torr), so the isotopes are likely to fractionate more strongly than under one atmosphere pressure; there are fewer molecules of gas around the filament to cause the slightly more abundant lighter isotopes of evaporating lead to recoil into the substrate. However, evaporation during modern analyses takes place from a substrate of silica and phosphoric acid, precisely because this method reduces the magnitude of fractionation. Instrumental fractionation can therefore be considered an estimate, although not a firm estimate, of the minimum value of evaporative lead isotope fractionation. Measured values of this fractionation factor vary somewhat from laboratory to laboratory and from instrument to instrument, but a typical value would be about 0.1% per mass unit.

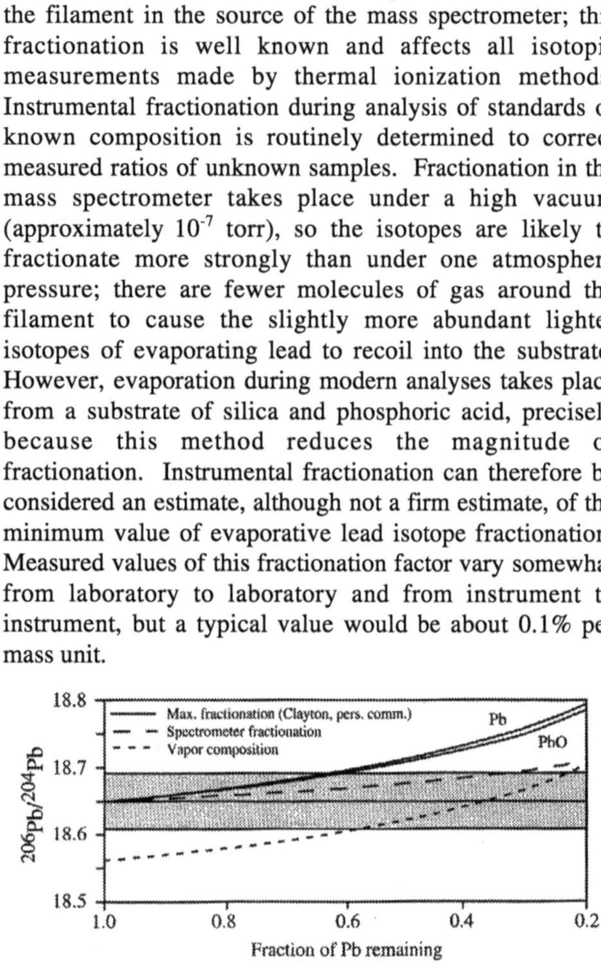

Figure 4: Isotope ratios of lead undergoing evaporation plotted against the fraction of lead remaining, calculated using eq. 1 and the data in table 1. Lead isotope ratios of lead vapour evaporating from lead metal are calculated using eq. 2. The starting material is lead with a typical 206Pb/204Pb value of 18.65. Isotope ratios of lead and lead vapour outside the grey envelope are measurably different from the starting composition.

Trajectories for the isotopic evolution of a sample of evaporating lead are shown in FIGURE 4, calculated using the values in TABLE 1. The curves are shown against a box representing the reproducibility of the 206Pb/204Pb value. The reproducibility of lead isotope measurements is usually given as ± 0.05% per mass unit at the 95% confidence level, or ± 0.1% for 206Pb/204Pb (DM = 2). Any two measurements must differ by at least twice this much to be measurably different. Calculations for 207Pb/204Pb, 208Pb/204Pb, 207Pb/206Pb or 208Pb/206Pb would yield identical results.

Ratio	Measured α	$\sqrt{\frac{M_2}{M_1}}$
$^{56}Fe/^{54}Fe$	1.0183	1.0184
$^{57}Fe/^{54}Fe$	1.0274	1.0274
$^{58}Fe/^{54}Fe$	1.0345	1.0364
$^{18}O/^{16}O$	1.0217	1.0308
$^{206}Pb/^{204}Pb$		1.0049
$^{206}PbO/^{204}PbO$		1.0045

Measured values are from Wang et al. (1994), and are constant from 1450 to 1750°C. Measurements of instrumental fractionation values are typically near 0.1% per mass unit, corresponding to α = 1.002 for $^{206}Pb/^{204}Pb$, 1.003 for $^{207}Pb/^{204}Pb$ and 1.004 for $^{208}Pb/^{204}Pb$.

Table 1: Fractionation of isotopes during evaporation.

FIGURE 4 shows the isotopic composition of remaining lead liquid plotted against the fraction of lead remaining. Using the estimated maximum possible fractionation factors from meteoritic studies the composition of lead would become isotopically distinct from the composition of the starting material after at least 40% of the lead had evaporated. Fractionation due to evaporation of PbO would be slightly slower because 206PbO/204PbO has the same average mass difference as 206Pb/204Pb, but the species involved are about 16 mass units heavier. The fractionation factor derived from instrumental fractionation measurements produces a more gradual shift: at least 70% of the starting material must evaporate before a measurable shift is seen. Production of a small but measurable shift in the isotopic composition of lead during smelting therefore requires evaporation of a very large fraction of the starting material, perhaps as much as 70%. No solid evidence of such high evaporative losses of metals (as opposed to sulfur, carbonate or other ore and gangue constituents) has been presented to date. While mixing of lead from different ore and gangue minerals (including those added as fluxes) may cause the lead isotopic composition of smelted metal to differ from the composition of individual ore minerals, evaporation of lead during smelting appears unlikely to do so.

Similar calculations can be made of the lead isotope composition of the vapour evolved from evaporating lead, using the equation:

$$R_V = (R_{l_i} f^{\frac{1}{\alpha}-1})/\alpha \qquad eq.\ 2$$

where R_V is the isotopic composition of the vapour emitted at a given value of f. Isotope ratios of vapour given off at low degrees of evaporation are measurably shifted from the lead liquid (FIGURE 4). Isotopic studies of lead in bog and lake deposits, carried out to constrain the sources of metal contamination in historical and prehistoric times, may therefore reflect values slightly shifted from the geological sources of the ore lead in the source region of the airborne pollution. The fractionation may be too small to be of interest, however.

What do ore lead isotope analyses represent?

Although evaporative fractionation of lead during smelting is very unlikely to produce measurable changes in the isotopic compositions of artefacts, other factors can cause the lead isotopic compositions of metal artefacts to differ from published analyses of the source ore deposits. Geological studies of the sources of lead in a given ore deposit often use analyses of lead in individual minerals. Nearly all samples of base or precious metal ores are heterogeneous mixtures of minerals with widely varying lead contents. Many silicate, carbonate and oxide minerals in ores contain less than five ppm lead, copper, iron and zinc sulfide minerals typically contain between 10 and 200 ppm lead, and galena (PbS) is 80 wt. % lead. U/Pb and Th/Pb ratios are similarly variable. Ore deposits also vary widely in age from young deposits of one million years to deposits two billion or more years old. Any group of minerals with variable U/Pb values will develop measurably heterogeneous lead isotope ratios if enough time has passed since formation of the deposit. Minerals in ore deposits less than about 20 million years old will not show such heterogeneity due to *in-situ* decay of U and Th after formation, but minerals in older deposits may. A further complication arises from the fact that the ore minerals may not have homogeneous lead isotope ratios at the time of formation. Although this is probably not common, some striking instances of initial isotopic heterogeneity have been found, such as in Mississippi Valley-type deposits in the US (Crocetti *et al.* 1988).

Mineral	wt. % mineral	Pb conc.	Pb content (mg)	Fraction of Pb contained	$^{238}U/^{204}Pb$	$^{206}Pb/^{204}Pb$ (20 Ma)	$^{206}Pb/^{204}Pb$ (500 Ma)
Pyrite	29	50ppm	0.725	0.00172	4.0	17.232	17.543
Chalcopyrite	25	250ppm	3.12	0.00738	2.0	17.226	17.381
Arsenopyrite	20	25ppm	0.250	$5.92*10^{-5}$	7.0	17.242	17.785
Galena	25	80%	400	0.947	0.02	17.220	17.222
Quartz	1	50ppb	0.000625	$1.48*10^{-6}$	15	17.267	18.430
Average						17.220	17.224

Pb contents assume a total sample weight of 50gm.

Table 2: Hypothetical Pb isotope ratios in 20 and 500 Ma ore samples.

TABLE 2 shows uranogenic Pb isotopic compositions of minerals in a hypothetical sample of copper-arsenic ore. Representative lead contents of these minerals vary widely, as do the U/Pb values. Most sulfide ore minerals will have significantly lower U/Pb than an average crustal rock (approximately nine ppm), but the value will not be zero. Galena, which makes up only 1% of the ore by volume and therefore might not be noticed during sample preparation, contains most of the lead in the sample, overwhelming contributions from copper and arsenic minerals. The isotope ratio of the galena therefore dominates that of the mixed sample.

The temporal evolution of lead isotopes in the hypothetical sample in TABLE 2 may be described assuming that all of the ore minerals had the same 206Pb/204Pb value at the time of formation (17.22). After 20 million years, the lead isotope ratios of all minerals are still within analytical error of each other (±0.017), except the quartz which contributes a vanishingly small proportion of lead and would not be analyzed separately in any case. Analysis of any other individual mineral would not yield a result measurably different from the bulk sample. After 500 million years, all of the ore minerals in the sample have developed measurably different lead isotope ratios. As before, the galena dominates the Pb budget of the sample; the bulk composition of the sample is therefore still essentially the same as the galena (17.224), but is different from the composition of any other mineral, including the copper and arsenic ore minerals. If no galena were present in the 500 million year old sample, the lead budget would be dominated by trace lead in the chalcopyrite; in this case, the average composition of the bulk sample would be 17.435, which does not agree within error with the composition of any individual mineral in the sample. Such problems are most likely to be significant in areas where the ore deposits are geologically ancient, such as in shield areas of Canada, northern Europe or Asia. Similar calculations could be carried out for thorogenic lead (208Pb/204Pb) values of ore minerals, but very little is known about their thorium contents.

The dominance of galena over the lead isotope ratio of the bulk sample in this example also highlights the effect of mixing ores of different lead contents. If ancient metallurgists were deliberately mixing ores of different provenance, one of which was high in lead and the other low in lead, the presence of the low-lead component may be undetectable in the lead isotope ratio of the mixture. For example, if metallurgists were mixing copper carbonate or sulfide ore with a low lead content with arsenopyrite-rich ore containing abundant lead (perhaps as traces of galena), the lead isotope ratios of the resulting bronze would indicate only the source of the arsenic.

A third complication arises from the fact that published geological analyses of ores almost always represent unweathered ore minerals, because these are most closely related to the ore-forming process. However, archaeological artefacts may have been made from ores that were exposed and weathered at the surface (e.g., copper carbonates like malachite or azurite). Because Pb is easily carried away by groundwater during weathering, such weathered ores are often low in lead. If lead in the surrounding rocks contrasts isotopically with that in the lead ore, lead weathered from those rocks can mix with ore lead in the weathering zone, shifting the isotope ratios of the resulting surficial ores away from the composition of the deeper, unweathered ores, so geological analyses of unweathered ore may not reflect the composition of oxidized ores used by early cultures. Geological ore lead isotope studies deliberately avoid analysis of any weathered ores for this reason.

The foregoing examples illustrate the importance of understanding the mineralogy of samples analyzed for lead isotope ratios, whether they represent the material that was smelted by early metallurgists, and especially whether or not galena is present. For example, it would be a mistake for an archaeologist trying to determine the sources of copper in a suite of artefacts to use only lead isotope measurements of copper minerals in potential polymetallic source deposits. Chalcopyrite or malachite may not dominate the lead isotope composition of the bulk sample, producing a misleading result. Potential ore sources that are low in lead and/or geologically old must be studied with particular care. None of this means that lead isotopes cannot be effectively applied, but they can produce erroneous interpretations if complicating factors are not considered.

Conclusions

The lead isotope method is used increasingly to determine which mines or mining regions were used by ancient cultures. Effective use of this method requires a clear understanding of its strengths and weaknesses. Some ore provinces, including copper, gold and polymetallic mining regions in the Central Andes and Southern Mexico, display marked geographic zoning of lead isotopes related to distance from nearby subduction zones. Carefully-planned studies that take account of such natural variations will most effectively reveal the sources of archaeological ores and artefacts. Lead isotope variations in the Central Andes and southern Mexico can help to discriminate among potential ore sources at varying distances from the geological trench, such as those between coastal and inland or highland areas. Combined with other archaeological data, lead isotopes can help to distinguish the source of metals used to make artefacts. Potential sources at comparable distances from the trench (along strike of each other) may show little or no measurable lead isotopic variation.

Care must also be taken in evaluating lead isotope data from ore deposits, first to ensure that the ore material analyzed represents the ore that was used by the ancient metallurgists; geological lead isotopic studies normally focus on primary sulfide mineralization rather than oxidized or supergene minerals that may have been mined by early miners. Analyses of individual minerals also may not represent the bulk composition of ores smelted by early miners, and may not reveal isotopic heterogeneities among minerals in a given ore, particularly if the ore is geologically old (greater than 20 million years). Information on the ages and mineralogies of deposits that may have been used by early smelters will help determine what analyses will be most appropriate to characterize their lead isotope ratios.

Another potential problem, the possibility that lead isotope ratios may be altered by evaporation of lead during the smelting process, is impossible to evaluate rigorously without data on the evaporative fractionation of lead isotopes under appropriate conditions. Experimental studies of Fe and O evaporation under analogous conditions suggest that the square root of the mass ratio of the isotopes undergoing evaporation provides a good estimate of the maximum possible fractionation factor. This value corresponds to 1.0049 for 206Pb/204Pb and 1.0045 for 206PbO/204PbO; fractionations in the source of mass spectrometers during isotopic measurements are typically somewhat less than half this large. Rayleigh distillation calculations using the above fractionation factors indicate that the isotopic composition of lead will not be measurably shifted until 40-70% of the original lead in the ore has evaporated. Such extensive evaporation has not yet been documented.

References

BENNET, V.C. & D.J.DEPAOLO. 1987. Proterozoic crustal history of the western United States as determined by neodymium isotopic mapping: *Geological Society of America Bulletin* 99:674-685.

BUDD, P., A.M. POLLARD, B. SCAIFE & R.G.THOMAS. 1995. The possible fractionation of lead isotopes in ancient metallurgical processes: *Archaeometry* 37: 143-150.

CAMPA, M.F. & P.J.CONEY. 1983. Tectono-stratigraphic terranes and mineral resource distributions in Mexico: *Canadian Journal of Earth Science* 20: 1040-1051.

CETENO-GARCIA, E., J.RUIZ, P.J. CONEY, P.J. PATCHETT & F. ORTEGA-GUTIERREZ. 1993. Guerrero terrane of Mexico: its role in the Southern Cordillera from new geochemical data: *Geology* 21: 419-422.

COUCH, R., R. WHITSETT, B.HUEHN & L.BRICENO-GUARUPE. 1981. Structures of the continental margin of Peru and Chile; In: L.D. Kulm, E.J.Dasch, and D.M.Hussong (ed.): Nazca Plate: Crustal Formation and Andean Convergence. Geological Society of America, 703-726.

CROCETTI, C.A., H.D.HOLLAND & L.W. MCKENNA. 1988. Isotopic composition of lead in galenas from the Viburnum Trend, Missouri. *Economic Geology* 83: 355-376.

CUMMING, G.L., S.E.KESLER & D.KRSTIC. 1979. Isotopic composition of lead in Mexican mineral deposits: *Economic Geology* 74: 1395-1407.

DAMON, P.E., M.SHAFIQUILLA & K.F.CLARK. 1983. Geochronology of the porphyry copper deposits and related mineralization of Mexico: *Canadian Journal of Earth Science* 20: 1052-1071.

FAURE, G. 1986. *Principles of Isotope Geology,* 2nd ed., New York, John Wylie & Sons.

GALE, N.H. & Z.A.STOS-GALE. 1989. Bronze Age archaeometallurgy of the Mediterranean: the impact of lead isotope studies: in, Archaeological Chemistry IV R.O.Allen, (ed.): Advances in Chemistry series 220: 159-198.

1992. Lead isotope studies in the Aegean (the British Academy project): in *New Developments in Archaeological Science* A.M.Pollard, (ed.), Proceedings of the British Academy, Oxford University Press, 77: 63-108.

HOSLER, D. & A.W. MACFARLANE. 1996. Lead isotope studies of the sources of ores used in the manufacture of Precolumbian bronze artefacts, SW Mexico: *Science* 273: 1819-1824.

JAMES D.E. 1971, Andean crustal and upper mantle structure. *Journal of Geophysical Research* 76:3246-3271.

MACFARLANE, A.W., P. MARCET, A.P. LEHURAY & U. PETERSEN. 1990. Lead isotope provinces of the Central Andes inferred from ores and crustal rocks: *Economic Geology* 85: 1857-1880.

PETERSEN, U. & A.W. MACFARLANE. 1993. Extended abstract. Lead isotopic provinces in Peru, Bolivia and northern Chile: 2nd International Symposium on Andean Geodynamics, Oxford, UK 477-480.

RUIZ, J., P.J. PATCHETT & F. ORTEGA-GUTIERREZ. 1988. Proterozoic and Phanerozoic basement terranes of Mexico from Nd isotopic studies: *Geological Society of America Bulletin* 100: 274-281.

WANG, J., A.M. DAVIS, R.N. CLAYTON & T.K. MAYEDA. 1994. Kinetic isotopic fractionation during the evaporation of iron oxide from the liquid state: *Lunar and Planetary Science* XXV: 1459-1460.

YANEZ, P., J.RUIZ, P.J. PATCHETT, F. ORTEGA-GUTIERREZ & G.E.GEHRELS. 1991. Isotopic studies of the Acatlan complex, southern Mexico; implications for Paleozoic North America tectonics: *Geological Society of America Bulletin* 103: 817-828.

ZARTMAN, R.E. & B.R.DOE. 1981. Plumbotectonics-the model: *Tectonophysics* 75: 135-162.

41

Chronological Markers? Chemical Analysis of Copper-based Trade Metal Artefacts From Petun Sites in Southern Ontario, Canada

C. WALKER,[1] R.G.V. HANCOCK,[2] S. AUFREITER,[2] M.A. LATTA[3] & C. GARRAD[4]

[1]*Geography Department, York University, Toronto M3J 1P3, Canada*
[2]*SLOWPOKE Reactor Facility and Department of Chemical Engineering and Applied Chemistry, University of Toronto, Toronto M5S 3E5, Canada*
[3]*Department of Anthropology, University of Toronto, Toronto M5S 1A1, Canada*
[4]*Petun Research Institute, 103 Anndale Drive, North York, Ontario M2N 2X3, Canada*

This paper builds on the chemical seriation model of Hancock et al. *(1994) for addressing chronological problems in northeastern North American archaeology. 1176 copper-based artefacts from the Iroquoian Petun People, from AD 1580-1650, were first classified into groups based on their metal composition using a scratch-test and the results analyzed. Chemical analysis by INAA of 130 samples, chosen by a stratified random sampling method, identified seven native and 19 European copper artefacts. Cluster analysis, multiple discriminant analysis and univariate F-tests were applied to the copper data. A dendrogram, based on the copper metal groups identified, was constructed to hypothesize migration and trade routes, and compared with our understanding of Petun chronology from historical and other archaeological sources. Subsequently, INAA analysis of a further 89 copper samples provided evidence that the results of the original cluster and discriminant analysis and the metal groupings were stable over repeated samples.*

Keywords: Copper; 17th Century; Iroquoian; Petun; Chemical Analysis; INAA; Multivariate Analysis.

Petun Chronology

The Petun were an Iroquoian-speaking agricultural people, closely related to the Huron, who flourished from AD 1580-1650, in the Collingwood area of Ontario, Canada. They contributed furs, agricultural products and, reportedly, tobacco to the regional trade network with the French, acting as middlemen between the Ottawa, the Huron and the Neutral. Their chronology, based on historic sources and our knowledge of their ceramics, pipes, lithics and trade goods, is not yet fully understood. It is commonly presented in terms of the three Trade Glass Bead Periods (GBP), which are based on the appearance of European-made trade beads on Iroquoian sites (Fitzgerald *et al.* 1995).

In 1616, Champlain visited them, mentioning ten villages and naming the people who lived there the "Petun," the Tobacco People. When the Jesuits, in 1639, established missions to the Petun, they assigned the apostles names to nine villages, indicating which ones were occupied at the time (Garrad 1989:16). They said that the Petun belonged to two *"Nations,"* the southern "Wolf" and the northern "Deer" (Jesuit Records 33:143, in Garrad and Heidenreich 1978:394) (Figure 1). Archaeology indicates that the Petun villages formed at least two geographical clusters, giving support to the theory of the two nations.

The Huron and the Petun came from the Pickering cultural phase of the Terminal Woodland period. Migrants from the Humber Valley most likely settled the earliest site, Sidey-Mackay. They would become the Wolf Nation. By 1610, they had moved to Melville and Glebe sites, along with people from Latimer. These villages later moved on to Hamilton-Lougheed (*Ehwae* in the *Jesuit Records*), Connor-Rolling and Graham Ferguson sites. In 1642, a thousand people abandoned *Ehwae,* after a raid. Their place of refuge, *Etharita* (Kelly-Campbell), was destroyed by the Five Nation Iroquois in December 1649.

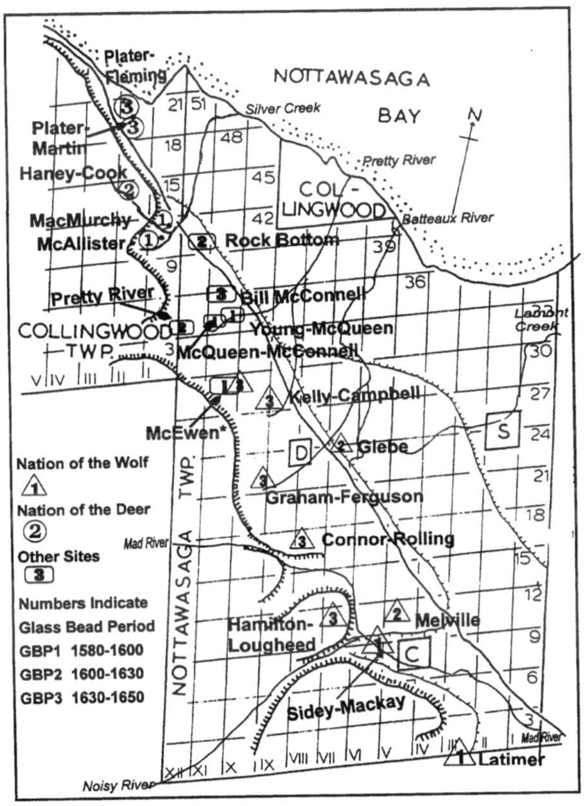

Figure 1: Petun Sites, 1580-1650, by Petun Nation and by Trade Glass Bead Period (GBP)

Ceramic analysis indicates that the early Deer Nation sites, MacMurchy and McAllister, were contemporary with Sidey-Mackay. These villages relocated to Haney-Cook Upper (the first *Ekarenniondi)* and Haney-Cook Lower, during the "Champlain era," probably after 1615. Later, they moved on to Plater-Martin, which was the final *Ekarenniondi,* and to Plater-Fleming.

Sites in the Pretty River drainage have not yet been assigned to either nation. The McQueen-McConnell and Young-McQueen sites are roughly contemporary with GBP I Deer and Wolf sites. Pretty River and Rock Bottom are contemporary with the two Haney-Cook sites, and Bill McConnell with the latest sites. Figure 2 shows the chronology of these Petun sites, as suggested by historical reports and the analysis of ceramics and European trade goods.

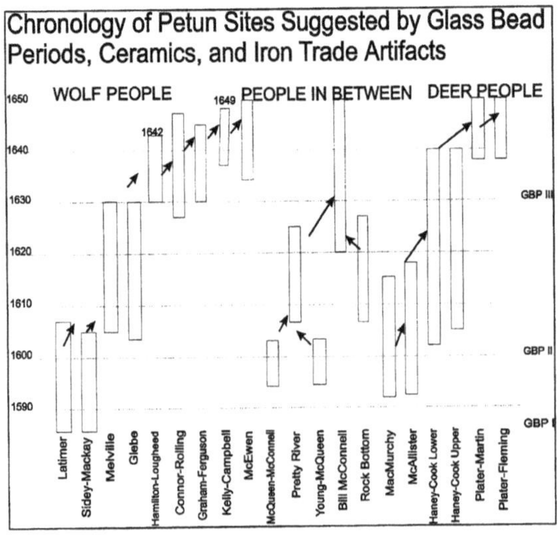

Figure 2. Chronology of Petun Sites. Arrows indicate hypothesized village movements.

The Plater-Martin and Plater-Fleming villages sheltered refugees from the south. 1649 had reduced the Petun to five villages and 3,000 people. In 1650, they abandoned their homeland. Some were adopted into other Iroquoian tribes and others joined Huron and Neutral refugees. They continually moved west and eventually become the present-day Wyandot of Kansas and Oklahoma.

European Trade Goods as Chronological Indicators

The relative frequencies of European trade goods may be useful in establishing chronologies for North American sites. The time of their first manufacture, when known, provides a possible starting point. Events, such as the British capture of Quebec in 1629, may also be related to local assemblages. However, deposition factors, provenance and sampling problems, mean they are unlikely to be aligned exactly with European sequences.

The first European visitors may have usually given North American peoples gifts of "trash" but a copper kettle was a valuable item. The Basques, along with the Normans, dominated fishing and whaling in the Gulf of St. Lawrence from 1540-1600, trading the red copper kettles, high-quality glass beads, long-bladed knives, and the heavy iron axes that predominate in the early assemblages (Bradley and Childs 1991:15, Fitzgerald *et al.* 1993:55).

Fig 3 Basque kettles, picture to come from Nova Scotia Museum (not included: permission not granted).

These kettles, (Figure 3) found in Micmac burials and on a few early Iroquoian sites, are of thick copper, with flat bottoms and flaring sides (Fitzgerald *et al.* 1993). Neat concentric circles of peening rise in parallel rows from the bottom centre to the rim. The rims jut outward, perpendicular to the wall, forming a shelf 17-22 mm wide, and then folding tightly under. Iron bands were riveted beneath the rim and a bale attached. None of the identifiable rims from Petun sites are of this type.

After 1600 the Basques were gradually forced out of the North American trade. There was a marked change, a tendency to standardization in European trade goods, as the French, Dutch and English took over the fur trade. By the 1630s, complete copper kettles are no longer found on North American sites and copper scrap and finished pieces come to comprise only a small part of trade metal assemblages (Fitzgerald *et al.* 1993).

However, the Spanish, French, English, Dutch and Portuguese were also active from Florida to Labrador in the 16th-century. They tended to trade the smaller, thinner, machine hammered brass kettles of northern Europe. Whilst this chronology of "Basque copper early, northern European brass later" has been, and still is, generally accepted, an increasing number of studies, including this one, show that the sequence is likely to be more complicated than that.

The 16th Century European Copper/Alloy Metal Trade

Looking for the sources of glass beads in Europe has met with some success. Similar efforts for trade metals have just begun. There were four main copper mining centres in continental Europe: the Swiss Tyrol, Neusohl in Hungary, Mansfeld-Eisleben in Germany, and the Balkans (Westermann 1981:71ff). The main distribution centre for Tyrolean and Neusohl copper was Venice, bound for France, the Iberian Peninsula and Central Europe. Most Mansfeld-Eisleben copper went overland via Frankfurt to Aachen/Stolberg. The central Balkan mines remained an important supplier to Venice even after they fell under Turkish control in 1453. In 1580 Swedish copper re-entered the continental market, dominating it after 1600.

Shipping and warfare boomed in the 16th century. Copper and its alloys were widely used for armaments, navigation instruments, vats, coverings, fittings, screws, and wire, as well as for housewares and decorative objects. The Spanish Crown minted copper coins in 1599, 1617 and 1621. Demand was strong, although uneven. Copper prices collapsed in 1510-15, but doubled between 1530 and 1560 and then again between 1560-1620. The Portuguese started the first boom in the industry in 1504 when they brought Sudanese gold to Antwerp in exchange for brass and copper wares to trade with western Africa and India (Westermann 1981:71ff). Portugal also needed silver for its Asian spice trade. As copper was a byproduct of the silver-rich ores of Neusohl and Mansfeld, its production increased along with silver. Rising demand attracted investment from the great trading houses and encouraged innovation.

However, African markets were soon saturated and years of war eventually decreased the demand for housewares. After 1545, Mexican silver permanently depressed European silver markets. Copper became the main foundation for the fading prosperity of the traditional European mining areas. By 1600, Frankfurt, Hamburg and Amsterdam had replaced Antwerp, becoming the main European metal trading centres. Aachen and Stolberg were the main brass production centres in the 16th and 17th centuries. But, in the 1580s, the fur trade with Native America offered new opportunities. Copper and brass kettles could be traded for beaver pelts.

Tracing metal objects back to their ore source is difficult, if not impossible. The changes in the chemical composition of metals, from ore body, to smelter, to foundry and battery, to finished product greatly complicate this research. Tyrolean and Swedish coppers are thought to have contained no silver, whereas the Neusohl and Mansfeld ores were reported to contain a great deal (Westermann 1981:71). Swedish ore was low in nickel (Pollard and Heron 1996:218ff). Hungarian copper was high in nickel, but also in antimony and arsenic.

Linking Trade Metals to Social Change in the Americas

The uses Native Americans made of kettles varied and changed over time. George Hamell (1983) describes the mythic values of shell, crystal and native copper in all Woodland cultures. Whilst copper seems to share the powers of shell and crystal for a long life, well being and success, little is known about its ritual expression. At earliest contact, European copper, and perhaps brass, may have been accepted for sacred use. The kettles were seemingly broken up and small pieces, often made into ornaments, moved far into the interior along established trade networks. As they became more numerous, kettles became prestige goods, for feasting, gifts, dream fulfilment and for crime compensation, as well as repositories and offerings for the dead (Moreau 1998, Ramsden 1981). Later, there is ample ethnographic evidence that kettles were used for cooking, especially for feasts, although pottery continued to be made.

The evidence from the Petun, who lived far from the Atlantic, is consistent with this. Early sites have few cuprous artefacts and none of the lugs, doubled-over pieces of metal riveted to each side of the brass kettle with a hole drilled to hold the handle. Lugs are increasing found on later sites, indicating that whole kettles may have been reaching them.

Scratch Testing Cuprous Artefacts

Analysis began with scratch-testing all the copper-based artefacts in the collection of the Petun Research Institute, directed by Charles Garrad. Objects yielding a "yellow" scratch were recorded as brass, whilst those which were "reddish-orange" were regarded as copper (Fitzgerald and Ramsden 1988). The artefacts were classified into lugs and "European objects," ornaments, tools, and "blanks and scrap." A chi-squared test, comparing the frequencies of distribution of these artefact categories over time as defined by Glass Bead Period (Table 1), found them to be significantly different, with the greatest changes coming after GBP I (X^2 (6) = 43.687. $p < .0005$). The directions of change also differ. The percentage of artefacts that are ornaments decreases over time, and each of the other categories of artefacts increases as a proportion of total artefacts over time.

Perio	Lugs, European Objects			Tools: points, awls, blades			Ornaments: beads, hoops			Blanks and Scraps			Total
GBP	n	%Type	%GBP	n	%Type	%GBP	n	%Type	%GBP	n	%Type	%GBP	
I	6	6%	9%	1	2%	1%	35	16%	46%	35	4%	45%	77
II	20	18%	7%	11	19%	4%	50	23%	19%	188	24%	70%	269
III	82	76%	10%	46	79%	6%	132	61%	16%	570	72%	69%	830
Total	108	100%	9.0%	58	100%	5%	217	100%	19%	793	100%	67%	1176

Table 1. Copper-based artefacts by Glass Bead Period and Category.

The percentages of copper and brass artefacts, from those sites subsequently selected for INAA, were calculated in order to test the 'copper early/brass late' hypothesis (Table 2). The results for each period were compared to those obtained by Fitzgerald and Ramsden (1988:156, 158ff) on a collection of 1205 Micmac, Neutral and Huron artefacts from 1550 - 1650. The observed numbers of copper and brass artefacts (Table 3) differed significantly from those that would have been expected on the basis of the proportions obtained by Fitzgerald and Ramsden (1988:156, 158ff) (X^2 (3) = 203.6, $p<.0005$).

Glass Bead Period	Fitzgerald & Ramsden (1988) suggested proportions of Brass	Observed Proportions of Brass [a]	Site	No. Brass artifacts (%) [b]	Artifact Sample Size
I	13%	55%(77)	Sidey-Mackay (SM)	3(50%)	6
			McAllister (McA)	6(35%)	17
			MacMurchy (McM)	14(42%)	33
			McQueen-McConnell (McQ)	19(90%)	21
II	62%	79%(277)	Latimer (LS)	3(30%)	10
			Glebe (GL)	35(61%)	57
			Melville (MV)	27(77%)	35
			Haney-Cook (HC)	83(83%)	100
			Pretty River	71(95%)	75
III	83	94%(835)	Connor-Rolling (CR)	46(81%)	57
			Plater-Fleming (PF)	50(89%)	56
			Graham-Ferguson(GF)	26(90%)	29
			Hamilton-Lougheed (HL)	220(91%)	241
			Kelly-Campbell (KC)	44(94%)	47
			Plater-Martin (PM)	399(99%)	405
a. Total sample of artifacts for each bead period are reported in brackets. b. Percentages of the total artifacts at each site are reported in brackets.					

Table 2. Observed Proportions of brass artifacts by Glass Bead Period, compared to Fitzgerald and Ramsden (1988) suggested proportions.

Glass Bead Period	Artifact Type[a]	
	Copper	Brass
I	35(67)	42(10)
II	58(87)	219(190)
III	50(142)	785(693)
Predicted on the basis of Fitzgerald and Ramsden (1988) in brackets		

Table 3. Comparison of copper and brass artifact distributions in samples from each Bead Period. . . .

We hypothesized a linear relationship between the proportion of brass artefacts in a site's sample and the final date of the site. To test this hypothesis, we first transformed the proportion of brass artefacts by taking the arcsin of their square roots. This is the traditional transformation applied to percentages to fit a linear model to percentage data. There was evidence of a significant linear relationship that explained 75% of the variation among the transformed observed proportions (Figure 4).

All sites except for Pretty River, which appears to contain a high percentage of brass for its date (it lies just outside the 95% confidence interval (c.i.) for the regression line), are adequately described. There is seemingly a long time interval between early Deer sites, McAllister and MacMurchy, and later GBP II Deer sites at Haney-Cook. Both Pretty River and the Haney-Cook sites may have been occupied longer than was previously thought (Garrad 1989). Connor-Rolling and Latimer sites may also have been earlier. The width of the confidence interval, however, is too large to distinguish between the relationship found here and that predicted by Fitzgerald and Ramsden (1988).

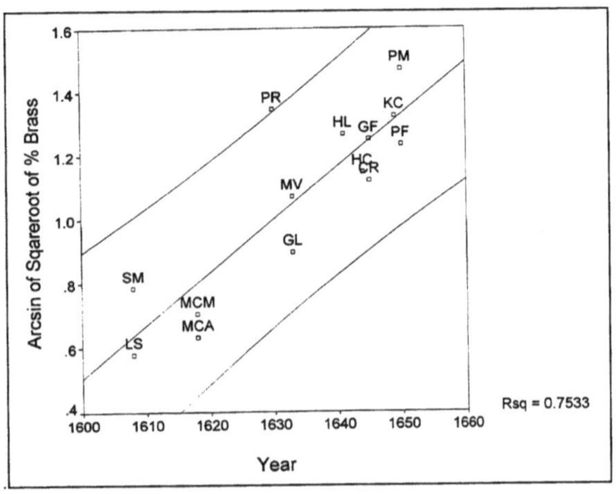

Figure 4. Scatterplot of arcsin of square root of % brass artifacts against estimated final Petun site dates, 95% c.i.

Conclusions from scratch testing must however be carefully qualified. Figure 4 assumes that each metal artefact represents a separate kettle or trading transaction, and yet some, or all, of the artefacts from several sites might have come from a single broken-up kettle. If chemical testing indicates that some artefacts may be from just one kettle or kettle batch, the percentage of brass for that site should decrease. Also, the early site, Sidey-Mackay, on other evidence firmly placed in GBP I, has 50% of brass artefacts, but this figure is based on a sample size of only six. Such problems decrease as sample sizes increase. However, early sites have fewer trade artefacts and they vary greatly in the degree to which they have been excavated. More serious is the inability of scratch testing to distinguish between native and European coppers. European coppers indicate a post-contact date for a site, and are evidence for interactions between Europeans and aboriginal North Americans (Hancock *et al.* 1994:263, their Table 4).

The Chemical Seriation Method

The number of artefacts to be sampled from each site was determined based on allocating the total sample among the sites as if the sites represented separate strata of the population of artefacts. This method does not assume the hypothesis of the glass bead periods, making it possible to test the usefulness of trace element analysis as a chronological tool. Because the scratch data is count data, the required number of samples was calculated using the binomial standard error method (Cherry 1978). Based on a proportion of 12% copper artefacts, 169 samples would be representative of 1065 artefacts, at a 95% confidence interval ($N = p\%.q\%/(S.E.)^2$). The S.E. (95% c.i.) is 50% (5%) or 2.5. Hence $N = 12 \times 88/2.5^2 = 1056/6.25 = 169$.)
The late site Plater-Martin was not included in this initial testing, reducing the sample size to 130 and yet maintaining a reasonable number of artefacts from each site. Eight artefacts were sampled twice to act as controls. The sampling factor for each site was based on its percentage of total artefacts and samples were chosen using random numbers. 17% of the samples were eventually shown to be copper, a satisfactory approximation of the assemblage at a 95% c.i. To give a more complete representation of the early Petun sites, unpublished INAA results from R.G.V. Hancock, for the three samples from Sidey-Mackay, were added.

INAA was done at the University of Toronto's SLOWPOKE Reactor Facility. Samples of 10 - 30 mg were cleaned with nitric acid. Elements tested and procedures were similar to Hancock *et al.* (1991). Copper, tin and zinc were reported as a percentage, by weight while the remaining elements are in parts per million (ppm). Where readings were below detection limits, the detection limit was substituted and marked <, indicating a value of less than or equal to that limit at a 68% c.i. For the statistical tests, zero was substituted, considering a value below the detection limit to be indistinguishable from zero. Hodson (1969:98) and Baxter (1989:49) use a similar methodology for dealing with values below detection limits. The brass samples, analyzed separately, are not reported here.

Distinguishing Native and European Coppers

The sorting of the 26 copper samples into native and European, and their division into subgroups, was first done by the inspection method pioneered by Hancock *et al.* (1991). Study of native and European copper has suggested that diagnostic elements include gold (Au), silver (Ag), antimony (Sb) and arsenic (As) and, sometimes, cobalt (Co), indium (In), nickel (Ni), tin (Sn) and zinc (Zn). Recurring elements and their respective ranges form groups or "chemical fingerprints" based on an internal group consistency of ± 10 - 20% (68% c.i.) for readily measurable elements (Hancock *et al.* 1991:82). They further conclude that native coppers fall into a minimum of three categories: "clean," iron-rich, and arsenic-rich (over 5000 ppm). All of the Petun native coppers tested here are "clean." Confirmation of these groupings was sought by statistical methods.

All Petun sites considered here had brass artefacts, including the earliest, Sidey-Mackay. All of the early sites, with the exception of Sidey-Mackay, contained European coppers, again indicating contact. None of the native coppers came from post 1630 sites, suggesting that it was no longer available. Nineteen samples were European coppers, seven each from GBP I-II, and five from GBP III. The inspection method produced two broad groups: i) low levels of antimony and arsenic but high gold, silver, cobalt and indium levels; and ii) high antimony and arsenic levels with lower gold and silver levels, with many samples lacking nickel, indium and cobalt. For the former group, sites were equally divided between GBP II and I. The latter group spanned all periods. Although the sample size is small, the "native coppers early, European coppers later," hypotheses is suggested. However, the "dirty European copper early, clean late" theory is less clearly demonstrated.

Case	Sampl No.	Mangan ppm	Indium ppm	Arsenic ppm	Copper %	Silver ppm	Gold ppm	Cobalt ppm	Nickel ppm	Antimony ppm	Tin %	Zinc %	Site	GBP N	Type
Cluster 1: "Native Coppers"															
Group I															
5	17	<11.3	<1.0	<1.4	99.9	113	0.02	3.2	<20	<0.1	0.05	0.003	GL	2 W	strip
7	*200	14.2	<0.7	<5.2	99.9	112	<0.10	2.2	<44	<1.0	0.05	<0.002	SM	1 W	bead
2	2	<15.8	<1.2	<1.9	99.9	143	<0.03	2.7	<58	<0.3	0.09	0.009	McM	1 D	scrap
1	1	<0.7	<0.8	<7.0	99.9	153	<0.15	8.3	<104	<1.5	2.16	0.034	McM	1 D	scrap
6	*100	8.7	<0.6	<4.4	99.9	108	<0.09	3	<23	7	0.02	<0.001	SM	1 W	bead
Group II															
4	11	<9.5	<1.1	<3.9	99.9	289	<0.03	2.7	<63	<0.3	0.12	0.027	McQ	1,3 ?	bead
8	*300	54.6	<0.6	<5.5	99.6	511	<0.10	4.5	<72	<1.1	0.1	0.0125	SM	1 W	bead
European Coppers															
Cluster 2: Low arsenic/antimony coppers															
Group IIIa															
22	45	<13.9	21.8	119	99.6	963	35	18.2	<141	552	<0.04	0.041	HC	2 D	scrap
24	19	<10.4	19.3	163	100	1104	36.7	14.2	<181	672	<0.04	0.029	GL	2 W	hoop
25	12	<10.7	20.1	191	100	827	29.5	12.4	<160	470	<0.05	0.017	McA	1 D	bead
Group IIIb															
19	21	<16.3	77.7	68.9	99.6	936	37.1	31.8	<88.0	179	<0.02	<0.003	GL	2 W	rim
20	15	<10.4	36.1	93.3	100	1121	41.2	57.3	<149	150	<0.04	0.011	McA	1 D	hoop
21	4	<15.4	22.3	109	99.6	798	40.2	13.1	<107	210	<0.06	0.031	McM	1 D	strip
Group IV Outlier															
23	6	<11.9	51	125	100	1062	37.9	40.8	241	262	<0.03	0.008	McM	1 D	hoop
Cluster 3: High antimony/arsenic coppers															
Group V															
9	5	<0.9	14	457	99.6	688	24.9	9.8	<227	2165	<0.05	<0.008	McM	1 D	hoop
13	7	<11.9	<1.1	529	101	651	11.9	<7.2	<342	2287	<0.08	<0.023	McM	1 D	detritus
12	56	<13.8	4.5	795	99.5	872	18.1	<4.7	<213	1649	<0.07	0.181	LS	2 W	trap
17	128	<9.9	<0.9	886	99.7	313	28.9	<5.3	<230	2693	<0.06	<0.010	HL	3 W	strip
Group VIa															
16	119	<15.3	<1.0	1020	100	286	22.1	<4.2	563	2239	<0.06	<0.006	HL	3 W	trap
18	129	<11.7	<1.1	847	100	351	22.9	<4.6	554	2062	<0.07	<0.018	HL	3 W	trap
14	41	<13.6	<1.0	997	99.6	353	12.1	<4.8	825	2914	<0.05	<0.011	HC	2 D	strip
15	91	<12.5	<0.9	1007	99.7	360	16.6	<2.5	480	2800	<0.02	0.186	PF	3 D	hoop
Group VIb															
10	13	<7.5	11.4	421	99.9	767	28.5	8	486	1502	<0.04	<0.007	McA	1 D	strip
26	22	<10.1	12.6	273	101	731	28.3	7.6	300	848	<0.02	<0.003	GL	2 W	trap
11	32	<7.4	4.3	871	99.4	716	18.2	<3.8	712	1715	<0.05	<0.009	PR	2 ?	detritus
Cluster 4, Extreme Outlier															
27	109	<5.8	<1.8	3526	99.9	931	16.3	<3.9	<176	1235	<0.05	<0.007	HL	3 W	rattle

* samples provided by R. G. V. Hancock

Table 4. Elemental values from INAA of 26 copper artifacts grouped according to the cluster analysis.

Cluster Analysis of Copper Artefacts

The SPSS 8.0 for Windows package was used. As each clustering algorithm predetermines the types of clusters that will be formed, the choice of algorithm is crucial. A hierarchical, rather than a partitioning clustering, method was used. Hierarchical agglomerative procedures begin by considering each artefact as a separate cluster, successively amalgamating them at decreasing levels of similarity until a single cluster remains. Average between-group linkage considers the average similarity or dissimilarity between clusters.

The phi-squared measure of distance was used, considering the data to be counting data. The spectrometer counts gamma emissions that accumulate into peaks. The areas of the peaks are then converted into elemental concentrations by means of a computer program. The resulting data can be considered a rescaled version of the original counts and, thus, count data. However, we bypassed the default clustering method, forcing a matrix of phi-square distances to be used in the hierarchical clustering.

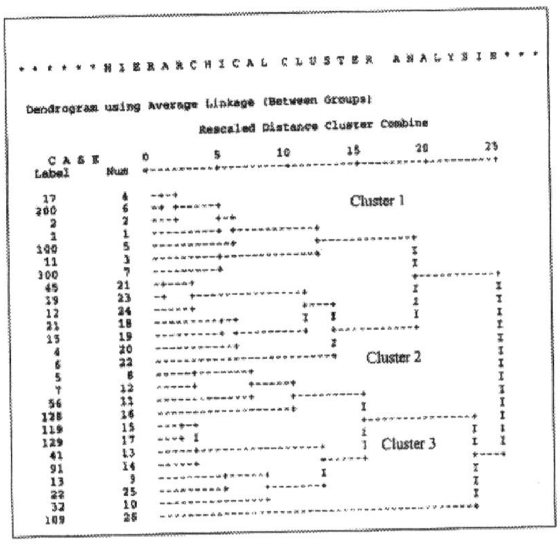

Figure 5. Hierarchical cluster analysis of copper artifacts from Petun sites.

The dendrogram produced by the cluster analysis (Figure 5) indicates that, at the stage where three groups or clusters can be distinguished, these clusters are similar to those identified by the inspection method of Hancock *et al.* (1991). Cluster 1 includes all those previously identified as native coppers (Samples 17 - 300). Cluster 2 is the previously identified low antimony/arsenic group of European coppers (Samples 45 - 6), except for Sample 22. Cluster 3, with high values for antimony/arsenic, contains all other European coppers. Cluster 2 joins with Cluster 1 earlier than with Cluster 3 because it appears that their chemistries correlate more strongly. However, the final clustering is very late suggesting that the artefacts in Cluster 3 are very different from those in the other clusters. Generally, there is considerable variability, with very few samples combining at low distance levels.

Reference to Figure 5 and Table 4 further reveals these relationships. In Cluster 1, Samples 17, 200 and 2 cluster at the lowest level, joining Sample 1 at level 5, perhaps indicating migration from Sidey-Mackay to MacMurchy and then to Glebe. Samples 300 and 11, from Sidey-Mackay and McQueen, also join at level 5. However, these two groups join above level 10, indicating only a very distant relationship.

In Cluster 2, two late-joining subgroups link early Deer sites to GBP II Wolf site, Glebe, and Deer site, Haney-Cook. Group IIIa is closely related, with Samples 45 and 19 clustering at the first level and Sample 12 (McAllister) joining at level 3, perhaps indicating migration.

In Cluster 3, Group V is a distantly related group spanning earliest to latest sites. Group VIa forms the most closely related of all the subgroups identified in the dataset.

To determine which elements had made the most significant contributions in distinguishing between the three main clusters, multiple discriminant analysis was performed. This produced two linear discriminant functions, each of which was shown to be responsible for independent identifiable decisions in separating the three clusters. Discriminant function 1 separates native coppers from European coppers. Discriminant function 2 separates Clusters 2 and 3, the European coppers.

Univariate F-tests, for each of the chosen elements individually, indicated that significant contributions to distinguishing between the clusters were made by: indium ($p < 0.0005$); arsenic ($p = 0.0016$); silver ($p < 0.0005$); gold ($p < 0.0005$); cobalt ($p < 0.0005$); nickel ($p = 0.0045$); and antimony ($p < 0.0005$). Since there was no reason to assign any particular artefact to any one of these three clusters without reference to the proportions of the elements within the artefact, an equal prior probability model was chosen. This model implies that, without reference to the data set, every artefact is just as likely to be in one cluster as another.

An examination of the standardized canonical discriminant function coefficients showed that antimony (1.70), silver (1.47), and gold (1.34) are of the most individual importance in determining the value of the first function, and hence contributing to the distinction between native and European copper. An increase in all elements but cobalt tended to provide more evidence that the artefact was of European origin. An increase in cobalt tended to provide more evidence that the artefact was of native copper.

In the second discriminant function, the most important contributions were made by antimony (0.56), gold (-0.36), copper (0.35) and arsenic (0.30). To a lesser extent, nickel (0.21) and cobalt (-0.21) contribute in distinguishing between Clusters 2 and 3. Silver is not important.

The categorization that explains the greatest percentage of the variation in the data is that between native and European coppers (94.92%), whereas the distinction between Clusters 2 and 3 only explains 5.08%. Therefore, the native-European copper distinction is the primary reason for the differences in elemental concentrations.

Analysis of the Second Set of Copper Samples

Eighty-nine samples, comprising most of the balance of the remaining Petun copper artefacts, were subsequently tested by INAA as above. The analyses run on the original artefact sample were rerun, first on the second sample only and then on the combined samples.

The cluster analysis of the second sample set created three clusters which could be identified as similar to the three clusters found in our initial cluster analysis. A fourth cluster was produced which was characterized by very high arsenic levels. This fourth cluster corresponded to a Sample 109, an outlier in the initial cluster analysis with a very high arsenic level.

The cluster analysis of both sample sets combined again created four clusters which could be identified as similar to the four clusters in the second cluster analysis.

Using the three clusters generated by the second cluster analysis, we assigned all artefacts to one of three groups, combining the fourth high arsenic group with the group corresponding to our initial Cluster 3. Using the initial sample of coppers as a training set for the creation of discriminant functions, we then classified all the coppers based on these functions. We then compared the two classification methods using a test of marginal homogeneity based on the exact distribution of the test statistic conditional on the marginal distributions of a misclassification table. There was no evidence that the two methods disagreed significantly as to how to classify the artefacts (MN (N=115) = 16, p = 0.234). Therefore we can conclude that the three clusters found in the second cluster analysis are consistent with the groups to which we assigned artefacts after the initial cluster analysis.

To investigate the stability of our discriminant function classification, we then produced discriminant functions based on the three clusters from the second cluster analysis for the entire 115 artefacts. These new discriminant functions were used to predict group membership, and the predictions were then compared with the predictions based on the original sample using a test of marginal homogeneity. There was no evidence that the classifications were different (MN (N = 115) = 14, p = 1.00). Therefore we can conclude that the procedure of predicting group membership using discriminant functions is stable over repeated sampling.

Three-Dimensional Studies of the Relationships between the Clusters

To better understand the relationship between Cluster 1 and Clusters 2 and 3 in the first dataset, a 3-dimensional scatterplot of antimony, silver and gold (the elements identified by the discriminant analysis as of the most individual importance in making this distinction), was constructed (Figure 6).

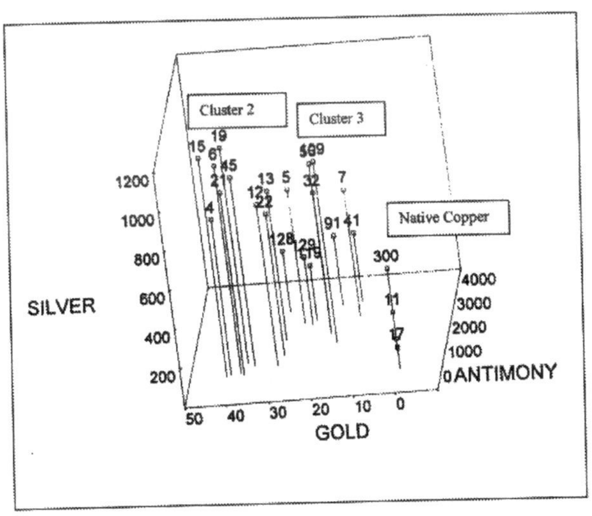

Figure 6. Scatterplot of silver, gold and antimony in all copper artifacts.

The native coppers, Cluster 1, are in the bottom right. Cluster 2, with high silver/gold and low antimony levels, is to the left. Cluster 3, with its high antimony levels, is more widely dispersed, in the centre. Samples 119, 129. 128, 41 and 91, with their low gold and silver levels, are seen below the others in the cluster. A scatterplot of gold, antimony and arsenic similarly illustrated the distinctions between Clusters 2 and 3 in the European coppers.

Differing trace element concentrations may be attributed to ore sources, smelting techniques and success rates, and metallurgy and manufacturing techniques. Although some trace elements found here are at concentrations high enough to have had at least some marginal effect on the physical or chemical properties of the metal, it is most likely that they are residual impurities, not deliberate additives.

In 16th century metallurgy, gold and silver were recovered as much as possible, although their differing commercial values and physical properties led to uneven results. Whilst it is reasonable to hypothesize that extraction abilities improved over time, lower gold or silver levels may simply indicate different mining or smelting locations. A hypothetical 30 ppm threshold for gold removal, between clusters 1 and 2, is suggested by Figure 6. Similarly, there may have been two success rates for silver, evident at 500 ppm. Whilst it appears that high gold is a characteristic of "earlier" coppers, Sample 7, from the earliest site, Sidey-Mackay, has the lowest gold content of any European sample, indicating a different early source. The linear relationship between quantities of gold and of silver is not strong ($R^2 = 0.7$) and gold is not a reliable predictor of silver.

Quantities of arsenic and antimony as small as 500 ppm will strengthen copper. If they are additives in these samples, their increase may be evidence of more sophisticated technology. On the other hand, low levels may simply indicate better copper smelting techniques. A linear model of the relationship between arsenic and antimony levels in these samples, at a 95% c.i., excluding Sample 109, yielded $R^2 = 0.9$. The general, but not invariable, tendency in these samples was for arsenic and antimony levels to be highest in samples from later sites, and for their levels to vary inversely with gold. In these samples, despite its lower melting/sublimation point, antimony levels are higher than arsenic. Wayman (personal communication) suggests that an arsenic-antimony sulfide was used as a copper source and that their levels are related to their ratios in the ore.

When Sample 109, with its very high arsenic content, was removed from the dataset and the discriminant analysis re-run, the main values for Function 1 were, as before, silver (1.37), gold (1.07) and antimony (1.09). For Function 2, antimony (-0.53) was again most important, followed by gold (0.36). Arsenic was no longer important.

A review of the literature suggests that the Petun samples are consistent with samples from other Ontario sites. Crawford Lake Neutral site Group D is similar to Petun Cluster 2; Neutral Group A to Petun Groups V and VIb; Neutral Group B, to Petun Group Va (Hancock *et al.* 1995:285ff, their Table 4).

Links Between Petun Sites as Suggested by Metal Groups

The copper groups suggested by the cluster analysis may contain artefacts from i) the same site, ii) contemporaneous sites that may have traded; or iii) sites aligned along migratory routes. Figure 2 was revised to indicate only those links suggested by the metal groupings.

In clusters produced by hierarchical methods, the level at which objects combine is more important than adjacency. The strength of the evidence for the arrows drawn for Figure 2 is hypothesized to be inversely proportional to how long it took to cluster all or most of the artefacts from one site with those from another. The earlier this is achieved on the dendrogram (Figure 5), the stronger the association between the sites. All of the arrows for which it took 20 distance units to cluster their artefacts were deleted. The three large clusters form just above distance-unit 10.

The procedure was as follows:

Step 1: A tree diagram was constructed based on Figure 4, combining sites according to the similarity of their ordering, from left (strongest) to right (weakest), e.g., artefacts 3 and 17 are closely related, implying a strong relationship between MacMurchy and Glebe.

Step 2: The initial precedence ordering was read horizontally, introducing the time ordering hypothesis of Figure 2 for each of the major clusters of the tree diagram.

Step 3: The results were combined, suggesting that mainly trading relations occurred between contemporary sites and migration between ones that were not. Two-way arrows (<—>) indicate trade, and one-way (—>) migration. Three stages were suggested. Only the first relationship, of three early sites (one from each Nation, one from the unassigned sites) and a later site, Glebe, was strongly indicated. It involves native copper only. The stages are as follows:

Stage 1, Trading:
SM <—> McQ <—> McM <—> GL

A second, slightly weaker group included two GBP II sites, and the "early" or "low arsenic/antimony" European coppers. SM and McQ sites are not represented.

Stage 2, Trading:
McM <—> GL <—> HC, with McA very weakly related.

A third, weakly related group included sites from all periods. MacMurchy is still represented.

Stage 3, Trading and Migration:
McM <—> HC <—> HL —> PF, with LS very weakly related.

These relationships are demonstrated in Figure 7. Kettles arriving in the area seem to have been widely distributed. There seems to be little justification for identifying "nations" from this data as the clusters contain representatives from both of them. MacMurchy and Glebe seem to have been important villages, based on the number of metal groups and stages in which they participate.

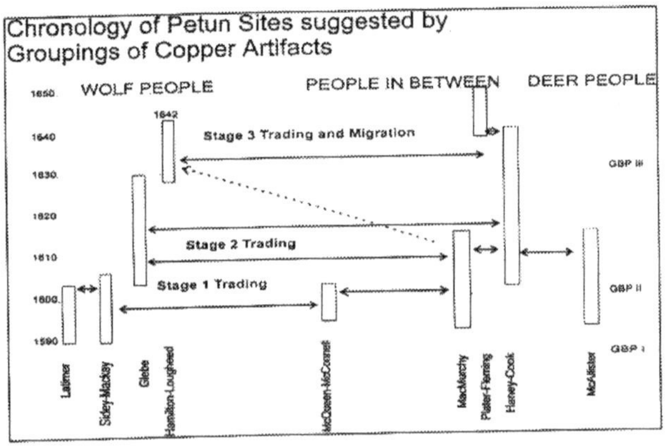

Figure 7. Revised version of figure 2, indicating trade and migration routes replicated by copper analysis.

Conclusions

The analysis of scratch-test results from copper-based metals, supplemented by chemical analysis and more sophisticated statistical methods, can provide useful information on contact-period chronologies. Cluster analysis of the sample sets gives results similar to the non-numerical method of Hancock *et al.* (1991). Discriminant analysis subsequently identified the most important elements (silver, gold, antimony, arsenic and nickel) in the formation of these clusters. Further discriminant analyses indicated that the classification scheme originally proposed was not falsified by the inclusion of over three times as many new artefacts. Thus the small sample is a reasonable predictor of the larger, and the initial conclusions are stable.

There is a great deal of noise inherent in the artefact element concentrations attained, as a result of variation in the sources of the smelted copper, the production of the kettles, the treatment of the artefacts after deposition and in the INAA itself. The several "metal groups" suggested by these procedures are mathematical summaries of the data, and are of use only insofar as they provide hypotheses for further research. The analyst must decide what constitutes a group, and on the number of groups expected, based on archaeological or other evidence.

Further investigation of the statistical treatment of elements at their detection levels is required. Substituting zero or a very low number is customary, but it can create artificial differences. In this test, substituting zero for levels of nickel below INAA detection limits may have been an important contributor to the clusters in which the artefacts eventually were grouped.

The detection limit for nickel, in the test here for instance, averages ≤200 ppm at a 68% c.i. (Table 3). At a 95% c.i., it is ≤400 ppm; at 99%, 600 ppm. A much large number of measurements of the same sample would give data below 400 ppm 95% of the time. However, the "real nickel levels" in Cluster 3 range from 300-800 ppm, overlapping the 95% c.i. The indium data, however, is not subject to this problem because of the low detection limits and the greater range in the data.

Dr. Hancock has suggested that substituting a number representing the upper boundary of a 95% confidence interval might eliminate the bias, which the use of zero may have induced in the cluster analysis. For data to be "good" for a particular element, a minimum criterion might be that the data must be >3 times the detection limit.

The hypotheses of Fitzgerald and Ramsden (1988) were not rejected: clearly there was a change in trade assemblages, at least by 1630. As yet it is impossible to source the metal groups back to European localities. However, as more research is done, further patterns may emerge which help us to better understand trading relations and the advance of European activities in North America.

Acknowledgments
Our thanks to Charles and Ella Garrad of the Petun Research Institute, to John Goffart of The Analytical Edge for help with the statistical tests, to Prof. Michael Wayman of the University of Alberta for assisting with the metallurgy, and to the University of Toronto Department of Anthropology for financially supporting the INAA.

References

BAXTER, M.J. 1989. Multivariate analysis of data on glass compositions: methodological note, *Archaeometry* 31: 45-53.

BRADLEY, J.W. & S. TERRY CHILDS. 1991. Basque earrings and panther's tails: the form of cross-cultural contact in sixteenth century Iroquoia, in R.M Ehrenreich (ed.), *Metals in society: theory beyond analysis*: 7-17. Philadelphia: MASCA Research Papers in Science and Archaeology 8, pt. 2.

CHERRY, J.F. 1978. Questions of efficiency and integration in assemblage sampling, in J.F. Cherry, C. Gamble and S. Shennan (eds.), *Contemporary British Archaeology*: 293-320. Oxford: British Archaeological Reports, British Series 50.

FITZGERALD, W.R. & P.G. RAMSDEN. 1988. Copper-based metal testing as an aid to understanding early European-Amerindian interaction: scratching the surface, *Canadian Journal of Archaeology* 12: 153-161.

FITZGERALD, W.R., L. TURGEON, R. HOLMES WHITEHEAD & J. BRADLEY. 1993. Late sixteenth-century Basque banded copper kettles, *Historical Archaeology* 27: 44-57.

GARRAD, C. 1989. Player-Fleming BdHb-2 site: a review, *ArchNotes* 89(3): 7-25.

GARRAD, C. & C.E. HEIDENREICH. 1978. Khionontateronon (Petun), in B.G. Trigger (ed.), *Handbook of North American Indians, Vol. 15, Northeast:* 394-397. Washington, DC: Smithsonian Institution Press.

HODSON, F.R. 1969. Searching for structure within multivariate archaeological data, *World Archaeology* 1: 90-105.

HAMELL, G.R. 1983. Trading in metaphors, in C. F. Hayes III (ed), *Proceedings of the 1982 Glass Trade Bead Conference*: 5-28. Rochester, NY: Rochester Museum and Science Centre Research Records.

HANCOCK, R.G.V., L.A. PAVLISH, R.M. FARQUHAR, & W.D. FINLAYSON. 1995. Analysis of copper-based metals from archaeological sites at Crawford Lake, south-central Ontario, Canada, in D.R. Hook & D.R.M. Gaimster (eds.), *Trade and discovery: the scientific study of artefacts from Post-Medieval Europe and beyond*: 283-297. London: British Museum Occasional Paper 109.

HANCOCK, R.G.V., L.A. PAVLISH, R.M. FARQUHAR, P.J. JULIG, W.A. FOX & G.C. WILSON. 1994. Chemical seriation of northeastern North American archaeological sites using copper artefacts, in D.A. Scott and P. Meyers (eds.), *Archaeometry of pre-Columbian sites and artefacts:* 255-265. Santa Monica, CA: Getty Conservation Institute.

HANCOCK, R.G.V., L.A. PAVLISH, R.M. FARQUHAR, R. SALLOUM, W.A. FOX & G.C. WILSON. 1991. Distinguishing European Trade copper and northeastern North American native copper, *Archaeometry* 33: 69-86.

MOREAU, J.-F. 1998. Traditions and cultural transformations: European copper-based kettles and Jesuit rings from 17th-Century Amerindian sites, *North American Archaeologist* 19:1-11.

POLLARD, A.M. & C. HERON. 1996. *Archaeological Chemistry.* Cambridge: Royal Society of Chemistry.

RAMSDEN, P.G. 1981. Rich man, poor man, dead man, thief: the dispersal of wealth in 17th Century society, *Ontario Archaeology* 35: 35-40.

WESTERMANN, E. 1981. Tendencies in the European copper market in the Fifteenth and Sixteenth Centuries, in H. Kellenbenz (ed.), *Precious metals in the Age of Expansion*: 71-77. Stuttgart: Klett-Cotta.

42

Faces of European copper alloy cauldrons from Québec and Ontario "contact" sites

J.F. MOREAU[1] & R.G.V. HANCOCK[2]

[1] *Laboratoire d'archéologie and Département des sciences humaines, Université du Québec à Chicoutimi, Chicoutimi, Québec, G7H 2B1, Canada*
[2] *SLOWPOKE Reactor Facility and Department of Chemical Engineering and Applied Chemistry, University of Toronto, Toronto, Ontario, M5S 3E5, Canada*

Since the late 16th century, copper alloy cauldrons crafted in Europe were traded for the furs obtained through hunting by the Amerindians of the New World. These cauldrons ultimately found their ways to archaeological sites where they are found whole or in parts in a variety of contexts (burial, habitation site, ...). INAA data from a series of twelve cauldrons spanning the ca. 1600 to ca. 1800 time period are used in order to assess the possibility of using elemental composition as a chronological indicator. With these twelve cauldrons as a basis of comparison, a specific aspect of cauldron's manufacture in Europe is addressed, namely that of riveting. The goal of such an exercise is to examine the rivets attaching the handle lugs to the bodies of the cauldrons to determine if they are of the same elemental composition as the bodies of these containers and to assess the influence of such a difference when these artefacts are spread in Amerindian sites in northeastern North America. Finally, elemental compositions are used to assess the importance to Amerindians of artefacts they shaped from fragments cut from copper alloy cauldrons. Such is the case for tinkling cones, typically small conical artefacts that were attached in series on clothes or on bags in order to obtain a soft tinkling noise as they hit one another.

Keywords: Contact Period; Northeastern North America; Instrumental Neutron Activation Analysis; European Trade Cauldrons.

Introduction

The year 1997 witnessed the quin-centennial of John Cabot's (Giovanni Caboto's) 1497 exploration of the northeastern section of the Atlantic coast of North America (see Biggar 1911, Litalien 1993). In the first decades of the 16th century, several other Europeans explored and reported on the same stretch of Atlantic coastline. It was not, however, until 1535 that the interior of the continent was explored, when Jacques Cartier, in his second voyage, explored the Saint-Lawrence River to Hochelaga (Montréal). All of these early explorations of northeastern North America involved rather incidental meetings with Amerindian groups, and irregular and unsystematic trading with them (Trigger & Swagerty 1996). More organized and systematic bartering between Amerindians and Europeans seems to have emerged during the 16th century, when the fishing for cod around New Foundland, and whaling for oil in the Saint-Lawrence River up to the Saguenay River mouth, became a regular endeavour of several European groups, among which the earliest were the Basques (Turgeon 1986; Proulx 1993). Historians often agree that the fishing and whaling activities were barely profitable most of the time. This may explain why, from early on, the bartering for furs took place with local Amerindian groups, who in turn sought several kinds of European goods. Of value to the Amerindians were iron knifes and axes, containers made of copper alloys, and glass beads (Fitzgerald 1995). These categories of goods are not only confirmed by description of the inventories of Basque ships sailing to northeastern North America, but are also the artefacts regularly recovered in the course of northeastern North American archaeological excavations. Furthermore, these goods continued to be traded regularly along with other types of trade goods such as "Jesuit" rings, guns, food and clothes during the 17th and 18th centuries, when fur trading was at its peak.

The scope of this paper is to illustrate aspects of the copper alloy cauldrons bartered between Amerindians and Europeans from the end of the 16th century to the end of the 18th century. From a theoretical point of view, this study is located at the crossroad of ethnoarchaeology, ethnohistory and historical archaeology. Strictly speaking, ethnoarchaeology is the study of cultural traits of ethnographic populations with the explicit goal of defining models - middle range theories - that could eventually be applied through the process of analogy to observations made from the material culture retrieved from prehistoric contexts in order to "reconstruct" past behaviours (Bahn 1992: 162; Gould 1978, 1996; Kramer 1979; Perlès 1997). This definition can not be applied strictly to the written records of the contact period - at least as this period presents itself in northeastern North America - since the records are not the product of works oriented by this kind of scientific goal, but were written centuries ago by writers who aimed at other objectives, such as "discovering new lands", occupying them and "civilizing" the people whom they encountered while proceeding to these "discoveries".

Although the ethnohistorical accounts certainly need to be critically considered before they could be used as a "scientific" record (Krech 1991, Wood 1990), the contact period does provide ethnoarchaeological information (Moreau 1992; Rogers & Wilson 1993). The contact period also provides information on some perspectives of historical archaeology, taking particularly into account that this archaeology is inscribed "in the growth and spread of mercantile capitalism" (Orser 1996). Copper alloy cauldrons illustrate this point well. Since they are manufactured in Europe, the contingencies of the different technological steps (smelting, shaping of the different parts, assembly of the parts, hammering) that take place in Europe to manufacture them may include peculiarities that must be understood in order to measure their effects on the processes of their dispersal in northeastern North America - whether this dispersal concerns the containers as a whole or the fragments into which they may eventually be transformed.

The three parts of this paper will attempt to deal with a) the chronological setting of the dispersal of copper alloy cauldrons in northeastern North America, b) the riveting of the lugs to the bodies of cauldrons as a specific aspect of the technology of their manufacture in Europe that could induce "distortions" in the interpretation of archaeological record from contact sites and c) "tinkling cones" as an example of transformation of cauldrons into an artefact made according to Amerindian cultural traditions.

Methods

These studies are based on instrumental neutron activation analysis performed since about 1990, following a modified methodology of Hancock *et al.* (1991a, 1991b). Specimens of the order of 8 mg to 40 mg are first manually abraded in order to remove the outside layer of "corrosion". A study comparing the corrosion's products to samples manually rubbed and samples etched with acid show that manual abrasion is almost as efficient as acid etching for cleaning of the samples, with the obvious gain of being an easier and faster process of cleaning samples before submission to INAA (Moreau & Hancock, n.d.).

The cleaned samples are individually stored in 1.2 mL polyethylene vials and analyzed using the INAA at the SLOWPOKE Reactor Facility of the University of Toronto. Following 1-5 minutes of neutron irradiation and a delay of about one hour, the samples are assayed for zinc (Zn), copper (Cu), manganese (Mn) and indium (In), using a hyperpure germanium detector-based gamma ray spectrometer. The elemental concentrations are calculated using the comparator method (Hancock 1976).

Medium and long half-life radioisotope-producing elements are analyzed by batch irradiation of twenty to forty samples for 16 hours at an neutron flux of 2.5×10^{11} neutrons per square centimetre per second. Five to seven days later, samples are serially assayed for 500 to 2000 seconds looking for gold (Au), copper (Cu), cadmium (Cd), arsenic (As), antimony (Sb), silver (Ag), zinc (Zn) and sodium (Na). A final counting is made ten to fourteen days after the irradiation, at which time the samples are counted for 4000 to 60,000 seconds (depending on sample mass and relative trace elemental contents) to determine the concentration of tin (Sn), selenium (Se), mercury (Hg), gold (Au), arsenic (As), antimony (Sb), silver (Ag), nickel (Ni), scandium (Sc), iron (Fe), zinc (Zn) and cobalt (Co). This procedure produces replicate measurements of enough elements to guarantee no sample mix-ups. Analytical precisions range from ±1% to detection limits. Results for selected elements are reported below.

Copper alloy cauldrons as a chronometric tool based on INAA

Copper alloy cauldrons found in northeastern North America have been found to be made of smelted copper, low-Sn bronze, gun metal, and low-Zn and high-Zn brass. As a baseline against which we could compare data from twelve copper alloy cauldrons, we used previously published data from eighteen analyzed samples from different native copper artefacts (see Table 1 (at end) and Fig. 1). These include two small beads sampled twice from a site in the Montréal area (see Fig. 2). From sites in Abitibi come a sample from each of two awls, single samples from three "butter knives", and one thick "cordiform" specimen. Also, from the Saguenay–Lac-Saint-Jean area are two "butter knives", sampled thrice, and two projectile points sampled once each. The mean and standard variation for the concentrations of Cu, Ag, Au, In, As, Sb, Sn, Zn in these 18 samples of native copper are grouped on the first line of the first section of Table 2 (at end).

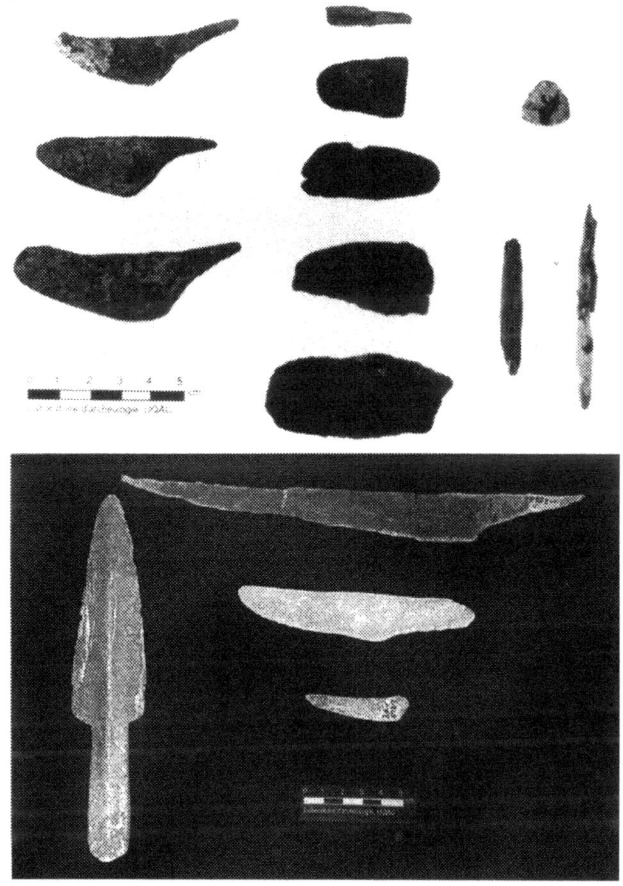

Figure 1. Examples of native artefacts.
A. *Native copper artefacts from Abitibi. Left: three examples of complete "butter knives", centre: undetermined fragment and four more or less complete blades of "butter knives", right: "hearth-like" object; bottom: left: undetermined fragment; left awl.*
B *Native copper artefacts from Saguenay-Luc-Saint-Jean. Left: socketed spear point; right: up long spear point; centre: "butter knife"; bottom: possible fragment of "butter knift".*

Next, the 12 copper alloy cauldrons are each represented by from 10 to 35 sub-samples, each which is either formally, or by its chemistry, assigned to a particular container. An example of formal assignment is the Kempt cauldron from the original analysis of which (Moreau & Hancock 1995a, 1995b) 16 samples are retained (see Table 3 and Fig. 3). Twelve of the original samples were excluded - 2 samples whose chemistries were affected by the rusting of the iron ring over which the copper sheet was folded to form the rim of the cauldron; a single tin-rich

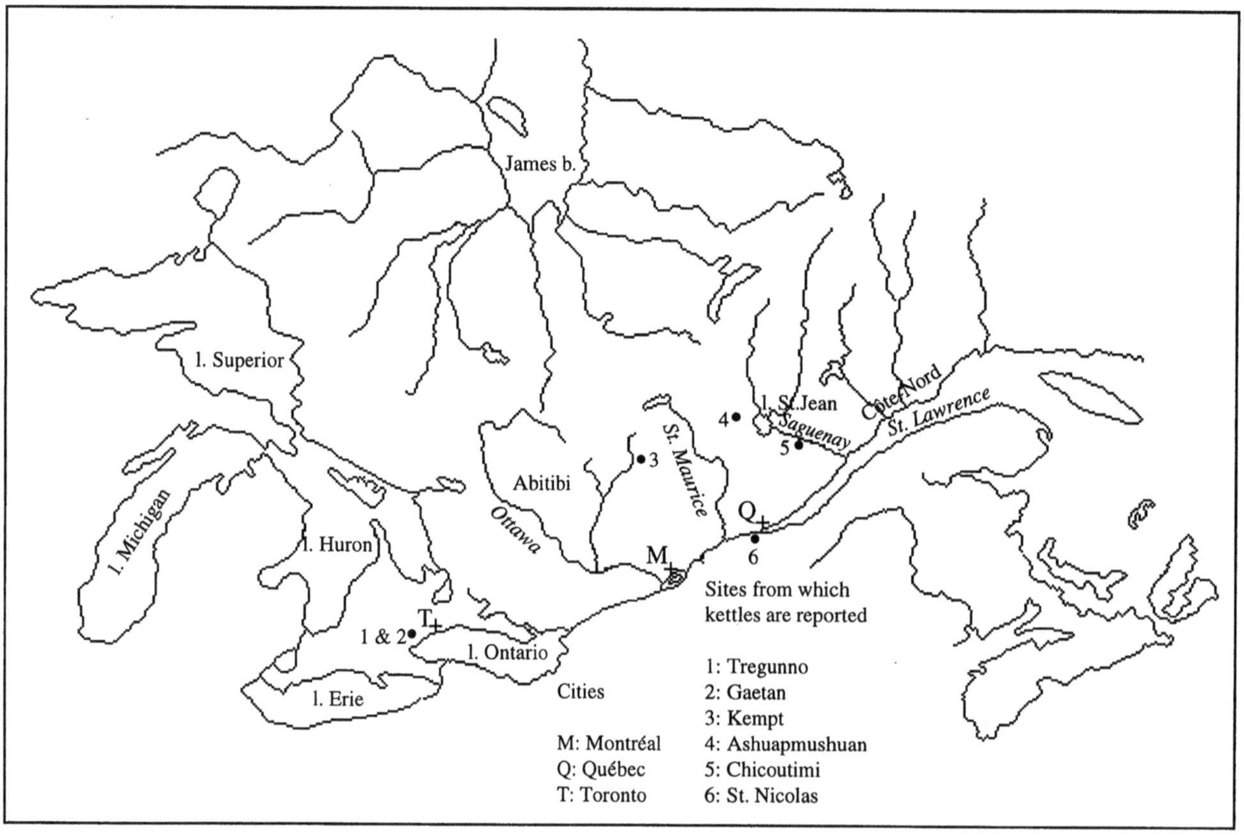

Figure 2. Northeastern North America: localisation of archaeology kettles from Ontario and Quebec.

sample from an obvious tin solder repair; 5 samples from the lugs; and 4 samples from the rivets attaching the lugs to the body - since all clearly exhibited different chemistries from the body of the cauldron.

The multiple samples from six cauldrons, five found in Ontario sites (G1, G2, G3 from the Gaetan site, Tr1 and Tr2 from the Treguno site: see Moreau & Hancock 1996b; and see Table 4) and one found in Québec (Ash.94, Ashuapmushuan site: data unpublished; see Table 3), are directly assignable to only one cauldron each, since each comes from a unique large fragment of copper sheet that has been sampled many times. The sizes of the Ontario specimens are of the order of 100 cm^2 (a square decimetre) while the Ash.94 specimen measured about twice that area.

The two St. Nicolas (St.Nic1 and St.Nic2) cauldrons were found in a burial near Québec city (Moreau & Hancock 1995c; see Table 5). In the course of their restoration, a series of tiny samples could not be reassembled and were thus spared for INAA, following a first sorting according to their thickness and colour. The elemental analyses confirmed the archaeologically-expected clustering of the samples into two chemistries (hence two cauldrons), except for a few samples that the INAA showed to be wrongly assigned on the basis of their physical characteristics and were, therefore, reassigned to the appropriate cauldron.

Finally, metal samples from Chicoutimi (Chic.451 and Chic.534; see Table 6) as well as from Ashuapmushuan (Ash.90; see Table 3 and Fig. 3) consist of samples presumably (from their chemistries) retrieved from individual cauldrons (Moreau & Hancock 1996a). These three cauldrons ended up as fragments, 1 to 2 cm wide and 5 to 10 cm long, that were found in very restricted volumes in the course of archaeological excavations. INAA showed that the chemistries of individual sets of samples from each of these clumped samples presented a high degree of homogeneity.

The distributions of the group of native copper artefacts and the twelve copper alloy cauldrons, as shown in Fig. 4, were established using the first two dimensions of a factor analysis for eight chemical elements. The first three dimensions total about 83% of the full variation, 68% of which is already comprised in the first two factors. The first dimension is strongly affected by Cu and Zn, with Au contributing significantly. The second factor is mainly affected by Ag and somewhat less significantly by As.

The two St. Nicolas clusters seem to produce a larger variation than any of the others. This might be either a consequence corrosion or the inclusion of fragments from the lugs. Since the tiny fragments could not be reassembled into one or the other cauldron, they were assigned to the former or latter cauldron on the basis of some of their morphological aspects. Their very reduced dimensions were taken into account. The somewhat larger cluster of Ashuapmushuan 1990 may also be accounted for by this same phenomenon: lugs could have been included among the series of fragments found in the archaeological concentration.

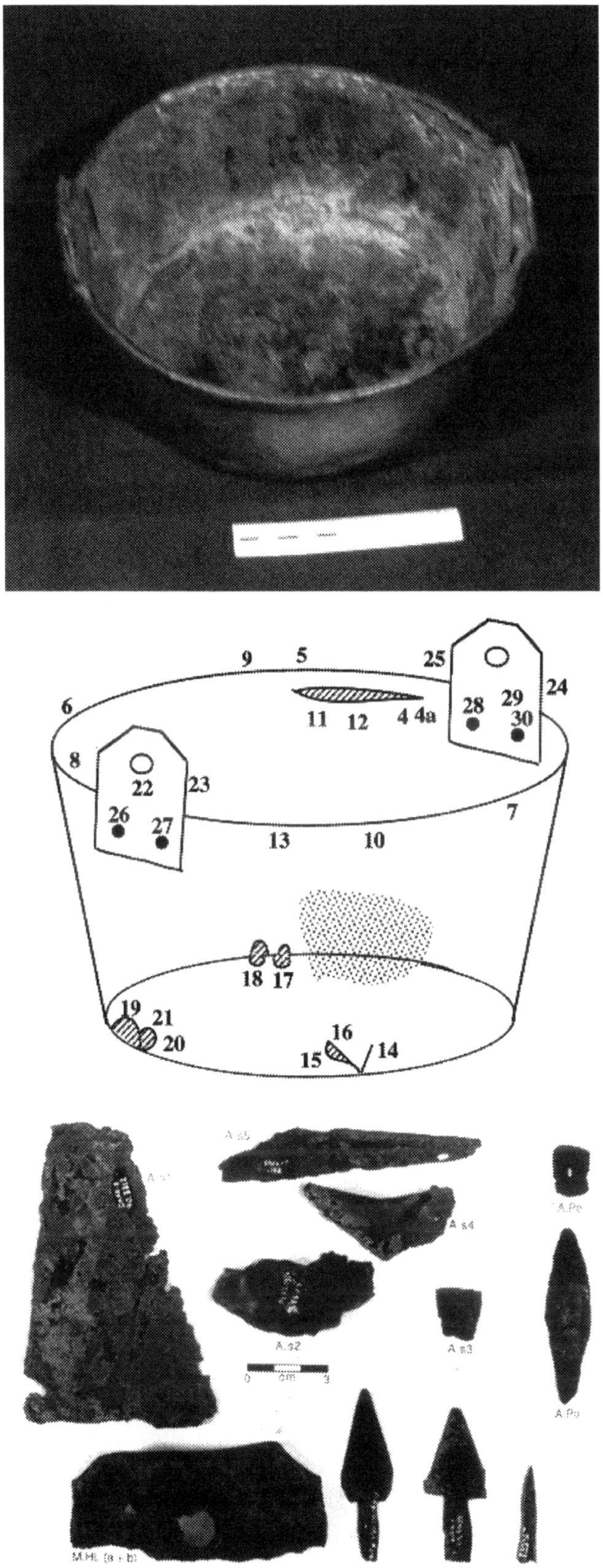

Figure 3. Examples of copper alloy kettles.

A. the complete cauldron from Kempt; note the missing handle, probably made of iron; note also that the cauldron is composed of the body to which are riveted two handle lugs.

B. Locatiob of samples on the Kempt cauldron.

C. Artefacts made of parts of copper alloy cauldrons from sites in Saguenay-Luc-Saint-Jean. As1, As5. As4, and As2: undetermined fragments. Ape: possible pendent. As3, Apo, Mpo1, Mpo2: tanged projectile point. Mpic: point object made of copper alloy sheet rolled over itself. MHL (a+b): broken handle lug.

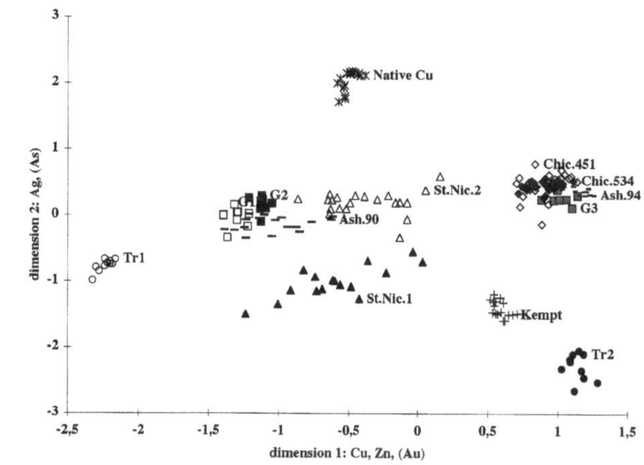

Figure 4. Native Cu artefacts and copper alloy kettles: factor analysis of INAA data.

The following points should be stressed about relative heterogeneity of the cauldron clusters. First, the position of the cluster of native copper artefacts is far from any cauldrons. Second, Tr1 (copper) also occupies a somewhat singular position while G1 and G2 (both copper) blend in part with Ashuapmushuan 1990 (low-Zn brass). Another remarkable grouping of clusters is that of G3, Ashuapmushuan 1994 and the two cauldrons from Chicoutimi (all brasses). Although, this multivariate distribution illustrates quite well the level of intra-cluster homogeneity as well as inter-cluster distinction, question arise from some of the groupings in light, at least, of their known temporal context.

A summary of these contexts follows. The native copper artefacts found respectively in the Abitibi, Montréal and Saguenay-Lac-Saint-Jean regions all date to either the Woodland (immediately pre-contact) or to the early Contact, periods. Among the cauldrons, those from Treguno (Tr1 and Tr2) and Gaetan (G1, G2, and G3) are dated to ca. 1590-1610 on the basis of their association to trade glass beads, specific to that time span (Fitzgerald 1990). It is noteworthy that the cauldrons' multivariate distribution of Fig. 4 presents one case of aggregation of these cauldrons (G1 and G2), while all others are spread far, one from the other.

On the basis of associated trade glass beads, the Ashuapmushuan cauldrons may also be assigned to a specific time period. Hence, in the northern sector of site DhFk-7, trade glass beads were found dating either to the 1625-1650 time period or to the 19th century, by means of INAA (Hancock *et al.* 1996). Stratigraphical arguments suggest that the cauldron fragments in this area of the sire are more likely akin to the early set than to the later set of beads (Moreau 1996). The Ashuapmushuan 1994 copper alloy sheet was found in the area south of that excavated in 1990. In as much as the nearby large trade beads concentration dating to the 18th century (Moreau *et al.* 1997,1998) could also be applied to the Ashuapmushuan 1994 cauldron, this would fix this cauldron in the range of 1700 to 1800.

On the basis of trade glass beads studies, with both

typological observations and INAA results (Moreau 1993, 1994; Hancock *et al.* 1996), the Chicoutimi cauldrons found in the *couche indienne* (as it was called by the original excavators) fit in the 1625 to 1675 time period. The two St.Nicolas cauldrons can be ranged in the 1625 to 1675 time period, on the basis of associated artefacts in the burial in which they were found (Chrétien *et al.* 1995, Moreau & Hancock 1995c). The less well associated cauldron is that of Kempt: it is acknowledged that it was found in the immediate vicinity of the burial site CiFo-12, that is associated mainly with artefacts of the 18th century, although 19th century objects are also present (Larocque 1995). In this context, a tentative working range date of 1700 to 1800 for that cauldron is proposed.

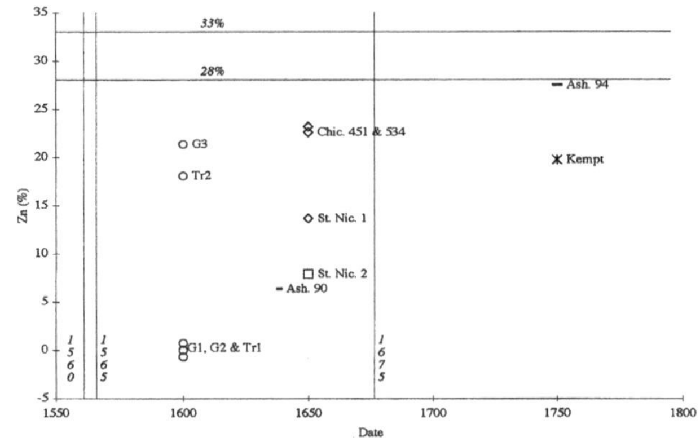

Figure 5. Copper alloy kettles: distribution of mean value of zinc vs. datation.

Given these temporal contexts, can some chronological sense from the INAA data retrieved from these cauldrons be found? Fig. 5 proposes such an interpretation. Here, the Zn concentration is presented in relation to the mid-range date for each of the twelve cauldrons, and trends from low to high Zn over time are also presented. Although arguments might be made that G1, G2 and Tr1 should be excluded given their trace levels of Zn, their inclusion is acceptable, because they are part of the full variation of cauldrons in the time range of 1590-1610 that is represented by all five cauldrons from Treguno and Gaetan. This argument is supported by the findings of Hancock *et al.* (1992), who analyzed copper-based metal samples from a number of sites. They found that long-used, and precious, native copper was replaced by European smelted copper, and then by brass, in 17th century Ontario.

A study of brass scientific instruments between the 15th and 19th centuries, as reported by Pollard & Heron (1996: 220-223), suggests that brass reached concentrations of 33%+ of Zn, depending on the country of origin, between the late 16th and the early 18th centuries. Assuming that these scientific instruments were relatively valuable, they probably represent the best quality of brass obtainable at a specific point in time (and for a specific brass manufacturing location.) Cauldrons, however, were potentially far less valuable than these European bass instruments, and may not appropriately be compared with them.

On the other hand, data for large collections of jettons (coins) show that Zn tends to increase through time from concentrations lower than 28% before 1560 to concentrations higher than 33% after 1675 (Pollard & Heron 1996: 217, Fig 6.5). Other data show that by the middle of the seventeenth century, concentrations of Zn in the jettons studied cover an extended range from almost none to over 33% with a modal concentration about 7%. In the later periods, the overall Zn concentration was limited to the 20-40% range (Pollard & Heron 1996: 219, Fig. 6.7). Although the influence of other elements such as Ni, Sn and Pb should not be underestimated in comparing the tendencies observed for jettons to those observed for copper alloy cauldrons, their comparison based mainly on Zn may be justified on the basis of two characteristics they share: they are the results of mass production and their use is common.

In summary, it may, therefore, be of interest to note in Fig. 5 that the same trends of variation that were noted for the jettons, are observed: a very large variation in Zn content from almost nothing to about 20% ca. 1650 tapering to a range from 20% to 27% ca. 1800, while the mid-range value increased from about 7% to about 28% in the same time period. In terms of chronometric assignment, this means that a low Zn (copper) cauldron may be assigned to an early context while, however, a high Zn content may imply either an early or late context. However, since this study is based on a limited series of single cauldrons among larger collections of fragments and if the steady increase in mean values of Zn through time is meaningful, it may be suggested that temporal assignment of a site might be calculated by developing a linear correlation of Zn through time on the basis of full range of Zn concentration presented by all the fragments located in a site, if these fragments can be shown to be coming from assemblages of short time range.

Riveting copper alloyed cauldrons

The previous study of the Kempt cauldron has shown, that at least one handle lug was not made from the same copper alloy sheet as that used for the body (Moreau & Hancock 1995a, 1995b). And so it was for the rivets. The gathering of data through many collections in Québec in February 1997, and the addition to previous studies, presented the opportunity to study forty rivet samples (see Table 7 as well as Table 2 for summarizing statistics and Fig. 1 for general location of the samples). These forty rivets include thirteen rivets, the data of which have been already reported. Among these, ten were found in sites located in Québec: five from the Kempt cauldron; three from the Chicoutimi site (Moreau & Hancock 1996a) and two the study of some artefacts from Abitibi (Moreau *et al.* 1994). Also included are data for three rivets from the Robitaille site in Ontario (Hancock *et al.* 1995). The remaining twenty-seven rivets comprise unpublished data from collections extensively studied in February 1997 in Abitibi (seven rivets), Centre de Conservation du Québec in Québec City (four rivets), Office of Parks Canada also in Québec City (eight rivets), and finally six rivets and two rivets from the collections stored respectively in the museums of Sept-Iles and Trois-Rivières.

Table 7 shows that, though many rivets are associated with large time ranges, some rivets represent more restricted time ranges (Chicoutimi, Maison Milot, Fort Témiscamingue, Kempt assuming that the 1700-1800 time range proposed above is correct), and so may be more useful in establishing chronological variabilities.

Following the same statistical procedure outlined above, a factor analysis was made for these rivets together with the group of native copper artefacts and the twelve copper alloy cauldrons. However, in order to maximize the differentiation between the clusters, a Varimax, instead of a Quartimax, variation was used (Anonymous 1994: 575). The first three factors total 77% of all the variation presented by the eight elements employed, the first two alone comprising 32% of that variation. The main contributors to dimension 1 are Cu and Zn, with a significant contribution from Sn, while the main contributor to dimension 2 is Au, with significant contributions from In and Ag.

Fig. 6 illustrates the distribution of the data from the forty rivet samples in comparison with the group of native copper artefacts and the twelve copper alloy cauldrons. An overwhelming number of these rivets (thirty-three) is made of relatively pure copper as shown by the association of these rivets with the group of native copper artefacts and the three copper alloy cauldrons. Among the seven low copper rivets, the five to the immediate left of the Kempt cauldron are those from this cauldron. The two other brass rivets are respectively from the Robitaille site, Ontario, an early Huron contact site of the first half of the seventeenth century, and the other rivet is a relatively modern (e.g., post 1800) rivet from Abitibi.

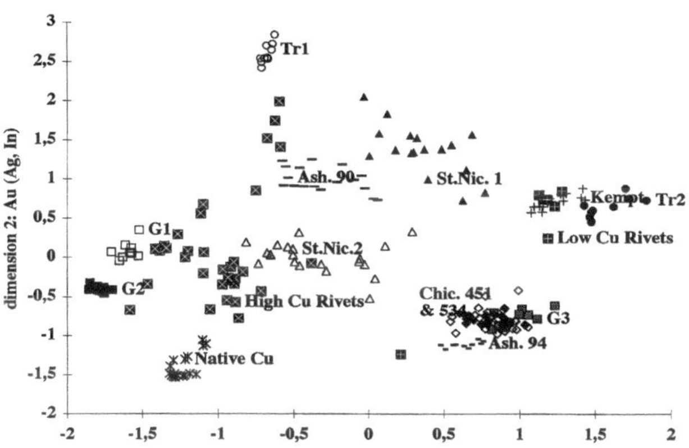

Figure 6. Rivets vs. Native Cu artefacts and copper alloy kettles: factor analysis of INAA data.

The thirty-three high copper rivets (97±2% copper) would be highly malleable and good for riveting purposes. This material is not "clean" copper, but an alloy with some Zn and Sn present in relatively low concentrations. In these rivets, arsenic averages ~3000 ppm (ranging from 200 ppm to 6100 ppm), relative to less than 13 ppm for the group of native copper artefacts and less than 1000 ppm for the three copper cauldrons (G1, G2 and Tr1). In contrast, the seven rivets with a Cu content of 71±4% are characterized by a concentration of Zn among the highest of all the data considered here. Given the possibility that the latest dates among all these rivets are for the Kempt cauldron, it may be that the process of riveting was also transformed, either as the concentration of Zn became less variable in the process of brass making, or as copper became too expensive relative to brass.

As shown by this riveting example, the chemistries of the various components used in European cauldron production, as seen in remains found both in North America and Europe, needs extensive attention. It was initially assumed that high malleability (high-Cu) would be an enduring characteristic of rivets. But this seems not to be so, since even from this small number of samples, it appears that well after the transition from the use of copper to brass and probably at the end of the transition from highly variable Zn to high Zn for cauldron brass bodies and lugs, brass rivets eventually swiftly supercede copper ones. Many reasons could be advocated to trade the malleability of copper for brass: lower cost of brass compared to copper; lower cost linked to producing only one type of metal (brass) instead of two (the high and low Zn varieties); uses of brass instead of copper rivets could also have improved the reliability of the cauldron's suspension system.

Combining the transition from copper to brass bodies of cauldrons to the observations retrieved from the rivets, the following chronological framework could be proposed: i) copper body with copper rivets, ii) brass body with copper rivets, iii) brass body with brass rivets. If such is the case, in the distribution of European cauldrons through northeastern North America, the long continuity of copper rivets may make them less efficient as a chronological tool than changes in the chemistries of the bodies of the cauldrons.

Musing with tinkling cones

In northeastern North America, copper alloy cauldrons were used by Amerindian groups as containers, both utilitarian and symbolic (as in burial contexts) and they were also cut into parts and transformed into a wide array of artefacts of traditional fashion - projectile points, beads, pendants (Martin 1975; Turgeon 1996; see Fig. 2). One such common category of artefacts is the so-called tinkling cone - a small object made from a sheet of copper rolled in conical shape of about 35±15 mm long with an average diameter aperture of about 6±2 mm. Tinkling cones were sewn in series onto clothes or small bags. One example of the latter includes tinkling cones (as well as white donut-like glass beads) sewn to the remnants of a bag found in Ontario and dated to the early 18th century (Ross *et al.* 1996).

Fifty-two tinkling cones were sampled in the course of the February 1997 sampling through Québec (see Table 8). None of these tinkling cones is, unfortunately, associated with a short time range. Nevertheless, following the same statistical procedure outlined above, a factor analysis was conducted for fifty-two tinkling cones, together with the group of native copper artefacts and the twelve copper alloy cauldrons (see Fig. 7; see also Table 2 for summarizing statistics). The first three factors total 69% of all the variation presented by the eight elements employed, the first two alone comprising 55% of that variation. The main load of dimension 1 is Cu and Zn with a significant contribution of Ag and Sn, while the main load of dimension 2 comprises Au, In and As.

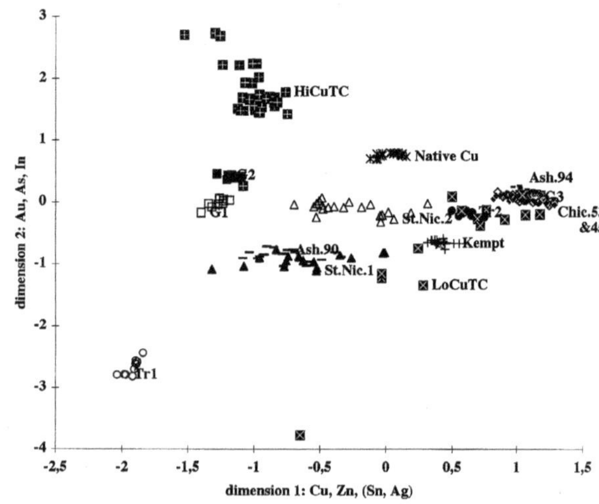

Figure 7. Tinkling cones vs. Native Cu artefacts and copper alloy kettles: factor analysis of INAA data.

The large majority (33 specimens) is made of relatively pure copper as shown by the closeness of these tinkling cones to the group of native copper artefacts and the G2 cauldron, made of high copper alloy sheet. Most of the fourteen low copper alloy tinkling cones are spread in the vicinity of the lowest copper alloy cauldrons (G3, Tr2, Kempt, Ash.94, Chic.451 and Chic.534). Finally, this collection of tinkling cones included three rare iron specimens.

Thirty-five of the tinkling cones are characterized by a mean of 98±1% copper with a high arsenic concentration at 5500±1700 ppm (Table 2). In contrast, the fourteen tinkling cones in the range of 74±4% copper concentration are characterized by a concentration of zinc of 26±3%, among the highest of all the Zn data considered here. Tin, Sb and As are all present in relatively low concentrations. Finally, the three iron tinkling cones with concentrations of Fe of 92%, 94% and 97% respectively, are peculiar in that their Sn concentration is the highest value with concentration of 4.9±1.7% along with relatively low concentrations for Zn, Sb and As. The iron tinkling cones were very difficult to sample. So, it must have been difficult to fold the sheet of iron over itself to form a conical shape.

Although, the change over time from high Cu (e.g., copper, low-Sn bronze) to low Cu (brass) could explain the variation in elemental composition of the copper-based tinkling cones, it should be noted that copper, brass and, in one instance, Fe tinkling cones are present in the same archaeological collection (see Table 8). Although, these collections could cover a long time span, the high homogeneity of physical characteristics among the tinkling cones in any of these series tends to suggest contemporaneity. If this is the case, the co-occurrence of high Cu (copper red) and low Cu (yellow brass) such a co-occurrence having been noted for cauldrons: see above, as well the Hancock *et al.* paper on Ball and Warminster sites in this volume could be suggestive of two avenues of research. First, combinations of the red and yellow colours could be researched from an aesthetic viewpoint, as could the distribution of colour on a string of tinkling cones to form a patterned motive. Second, do the high Cu and low Cu (and the high Fe as well?) tinkling cones produce differently pitched sounds? If so, is one pitch preferred to another or, again, is a blending of the two preferred? Albeit hypothetical, these suggestions represent potentially interesting avenues of research: just what are the sound wavelengths associated with tinkling cones?

Conclusions

The study of twelve cauldrons from Québec and Ontario parallels the general trend observed in the European change through time from highly fluctuating range of Zn fluxing by the 1600s to a narrow range in the high value of Zn by 1750s. Also noted is the progressive increase in the mean value of Zn through time during the same time range and follows Hancock *et al.* 1992 findings of native copper European copper brass.

In contrast, low variability seems to characterize high Cu rivets during most this period of time which, however, resulted in what seems to be a drastic change in rivet making (low Cu brass). If this is so, metal sheet and copper wire would be characterized by asynchronous transformations of their elemental contents, a conclusion of interest for cauldron manufacture in Europe. This, however, warns researchers in northeastern North America to distinguish sheet and wire samples from Amerindian sites.

If chronological trends could be suggested on the basis of copper alloy sheets and wire, other factors are also to be considered in order to understand the dynamics of the spreading of European goods into northeastern North America. Hence, it might be that aesthetic factors play a role in artefact uses, such as tinkling cones.

Acknowledgments

This study was made possible through the generous support of Université du Québec à Chicoutimi to J.-F.M. and by a Natural Sciences and Engineering Research Council infrastructure grant to the SLOWPOKE Reactor Facility of the University of Toronto. We thank the two anonymous reviewers who made suggestions that undoubtedly improved the original version of the text, including their suggestions to present all the analytical data, though most have been previously published elsewhere.

References

ANONYMOUS. 1994. *Statistica for Macintosh.* Tulsa, Statsoft.

BAHN P. 1992. *Collins Dictionary of archaeology.* Glasgow, Harper Collins.

BIGGAR, H.P. 1911. *The Precursors of Jacques Cartier 1497-1534.* Ottawa, Government Printing Bureau, *Publications of the Canadian Archives*, n° 5.

CHRÉTIEN Y, A. BERGERON & R. LAROCQUE. 1995. La sépulture historique ancienne du site Lambert (CeEu-12) à Saint-Nicolas in A.-M. Balac, C. Chapdelaine, N. Clermont & F. Duguay (dir.), *Archéologies québécoises*, Montréal, Recherches Améindiennes au Québec, coll. *Paléo-Québec*, n° 23, p. 203-226.

FITZGERALD W.R. 1990. *Chronology to Cultural Process: Lower Great Lakes Archaeology, 1500-1650*, Montréal, McGill University, Department of Anthropology, thèse de doctorat.

1995. A Late sixteenth-century European trade assemblage from North-Eastern North America, in Hook, D.R. & Gaimster, D.R.M. (eds.), Trade and Discovery: *The Scientific Study of Artefacts from Post-Medieval Europe and Beyond*, London, British Museum, *Occasional Paper* n° 109, p. 29-44.

GOULD R (ed.), 1978. *Explorations in Ethno-archaeology*. Albuquerque, University of New Mexico Press.

GOULD R. 1996. Ethnoarchaeology. in Fagan, B.M. (ed.), *The Oxford companion to archaeology*. New York, Oxford University Press, p. 207-208.

HANCOCK, R.G.V. 1976. Low flux multi-element instrumental neutron activation analysis in Archaeometry. *Analytical Chemistry* 48: 1443-1445.

HANCOCK R.G.V., S. AUFREITER, J.-F. MOREAU & I. KENYON. 1996. Chemical chronology of Turquoise Cluess Glass Beads from the Lac-Saint-Jean Region of Québec, in M.V. Orna (ed.), *Archaeological Chemistry. Organic, Inorganic and Biochemical Analysis*. Washington, American Chemical Society, Symposium Series 625, p.23-36.

HANCOCK, R.G.V., FARQUHAR, R.M., PAVLISH, L.A., SALLOUM, R., FOX, W.A. & WILSON, G.C. 1991a. North American Native Copper and European Trade Copper Analyses. In E. Pernicka, & G.A. Wagner (Eds), *Archaeometry-90*. Boston: Birkhaüser Verlag, 173-182.

1991b. Distinguishing European Trade Copper and North-Eastern North America Native Copper. *Archaeometry* 33(1): 69-86.

HANCOCK, R.G.V., PAVLISH, L.A., FARQUHAR, R.M., JULIG, P.J., FOX, W.A., & WILSON, G.C. 1994. Chemical seriation of northeastern North American archaeological sites using copper artefacts. in *Archaeometry of Pre-columbian Sites and Artefacts: Proceedings of a Symposium Organized by the UCLA Institute of Archaeology and the Getty Conservation Institute, Los Angeles, 1992* D.A. Scott & P. Meyers (eds.), Getty Conservation Institute, Santa Monica, California, U.S.A., 255-265.

HANCOCK, R.G.V., PAVLISH, L.A., FOX, W.A. & LATTA, M.A. 1995. Chemical Analysis of Copper Alloy Trade Metal from a Post-Contact Huron Site in Ontario, Canada. *Archaeometry* 37(2): 339-350.

KRAMER, C. (ed.) 1979. *Ethnoarchaeology. Implications of Ethnography for Archaeology*. New York, Columbia Press.

KRECH S. III. 1991. The State of ethnohistory, *Annual Review of Anthropology* 20: 345-275.

LANGEVIN, É., MCCAFFREY, M., MOREAU, J.-F. & HANCOCK, R.G.V. 1995. Le cuivre natif dans le Nord-Est québécois: contribution d'un site du lac Saint-Jean (Québec central). In *Archéologies québécoises*, A.-M. Balac, C. Chapdelaine, N. Clermont & F. Duguay (eds.). Montréal: Recherches Amérindiennes au Québec, coll. *Paléo-Québec*, n° 23: 307-320.

LAROCQUE, R. 1995. *Inventaire archéologique au lac Kempt*, Haute-Muricie, LaTuque, Conseil da la Nation Atikamekw, Unpublished Report.

LITALIEN, R. 1993. *Les Explorateurs de l'Amérique du Nord, 1492-1795*, Sillery (Québec), Septentrion.

MARTIN, C. 1975. The Four Lives of a Micmac Copper Pot. *Ethnohistory* 22(2): 111-133.

MOREAU, J.-F. 1992. L'apport de l'ethnohistorie à l'ethnoarchéologie. Le problème de la congruence des documents archéologiques et ethnologique, l'exemple du Québec méeridional, in A. Gallay, F. Audouze et V. Roux (dir.), *L'Ethnoarchéologie: justification, problèmes, limites*. XIIe rencontres internationales d'archéologie et d'histoire d'Antibes. Juan-les-Pins, Editions APDCA, p. 347-355.

1993. Histoires de perles ... d'avant Jean de Quen. *Saguenayensia* 35(2): 21-29.

1994. Des perles de la "protohistoire" au Saguenay-Lac-Saint-Jean? *Recherches Amérindiennes au Québec* 24(1-2): 31-48.

1996. Indices archéologiques de transferts culturels par la voie du Québec central in Turgeon, L., Delâge, D. & Ouellet, R. (eds.), *Transferts culturels et Métissage, Amérique/Europe, XVIe-XXe siècle*, Québec [Oaris], Presses de l'Université Laval [L'Harmattan], p.209-242.

MOREAU, J.-F. & HANCOCK, R.G.V. 1995a. Le chaudron en alliage de cuivre du lac Kempt, Annex B of R. Larocque *Inventaire archéologique au lac Kempt, Haute-Mauricie*, Research report submitted to Conseil de la Nation Attikamekw.

1995b. *L'anatomie d'un chaudron*. Communication presented in the Symposium L'Archéologie sous la loupe, held during the 63rd Annual Congress of the Association Canadienne Française pour l'Avancement des Sciences (ACFAS), Chicoutimi, 22-26 mai.

1995c. Analyse par activation neutronique des chaudrons de Saint-Nicolas (CeEu-12). In *Archéologies Québécoises*, A.-M. Balac, C. Chapdelaine, N. Clermont & F. Duguay (eds.) Montréal: Recherches Amérindiennes au Québec, coll. Paléo-Québec, n° 23: 227-236.

1996a. Intra-site and Inter-site Differentiations of Copper Based Artefacts from 'Early Contact' Period (1600-1650) Sites from the Eastern Subarctic of North America. In *Proceedings from the 6th Nordic Conference on the Application of Scientific Methods in Archaeology*, V. Mejdahl & P. Siemen (eds.). Esbjerg (Danmark): Esbjerg Museum, Arkaeologiske Rapporter no 1: 109-125.

1996b. Chrono-Cultural Technique Based on the Instrumental Neutron Activation Analysis of Copper-Based Artefacts from the "Contact" Period of Northeastern North America. In *Archaeological Chemistry. Organic, Inorganic and Biochemical Analysis*, M. V. Orna (ed.). Washington: American Chemical Society, 64-82.

1997. Analyse neutronique de deux perles en cuivre associées à une sépulture provenant de Rosemont, Annex 3 of R. Larocque *La sépulture du quartier Rosemont à Montréal (BjFj-98)*, Research report submitted to the Direction de Montréal of the Ministère de la Culture et des Communications du Québec.

n.d. The effects of corrosion on INAA characterizations of brass kettles of the early European contact period in Northeastern North America, *Journal of Archaeological Science* (in press).

MOREAU, J.-F., HANCOCK, R.G.V. & COTÉ, M. 1994. Analyse de la composition chimique d'objets en cuivre de l'Abitibi-Témiscamingue. *Recherches Amérindiennes au Québec* 24(1-2): 65-71.

MOREAU, J.-F., HANCOCK, R.G.V., AUFREITER, S. & KENYON, I. 1997. Taphonomical and chronological studies of a concentration of European glass trade beads from Ashuapmushuan, Central Québec (Canada), *Iskos* (Finska Forminnesföreningen) 11: 173-181.

1998. *Late French (1700-1750) to Early English (1750-1800) Regime white glass trade beads from a presumed decorated bag found at the Ashuapmushuan site (Eastern Central Québec)*, Canada, Paper presented at the 31st International Symposium on Archaeometry, Budapest, May 1998.

ORSER, C. E. 1996. Historical archaeology, in Fagan, B.M. (ed.), *The Oxford companion to archaeology*, New York, Oxford University Press, p.279-280.

PERLÈS, C. 1997. Ethnoarchéologie, in Leroi-Gourhan, A. (ed.), *Dictionnaire de la Préhistorie*, Paris, Presses universitaires de France, coll. *Quadrige* p.386.

POLLARD, A.M. & HERON, C. 1996. *Archaeological Chemistry*. London: Royal Society of Chemistry.

PROULX, J.-P. 1993. *Les Basques et la pêche de la baleine au Labrador au XVIe siècle*, Ottawa, Environnement Canada, Service des parcs, *Études en archéologie, architecture et histoire*.

ROGERS, J.D. & WILSON, S.M. 1993. *Ethnohistory and Archaeology. Approaches to Postcontact Change in the Americas*. New York, Plenum.

ROSS W.A., HINSHELWOOD, A. & CAMPBELL, P. 1996. The Wolf River Burial, (DeJc-2): a Preliminary Report, Ontario Heritage Foundation, *Annual Archaeological report, Ontario* 6: 25-31.

TRIGGER, B.G. & SWAGERTY, W.R. 1996. Entertaining Strangers: North America in the sixteenth century, in *Native Peoples of the Americas, Volume I. North America, part 1*. Trigger B.G. & Washburn, W.E. (eds.)., Cambridge, Cambridge University Press, pp. 325-398.

TURGEON, L. 1986. Pour Redécouvrir notre 16e siècle: les pêches à Terre-Neuve d'après les archives notariales de Bordeaux, *Revue d'Histoire de l'Amérique française* 39(4): 523-549.

1996. Le chaudron de cuivre en Amérique: parcours historique d'un objet interculturel. *Ethnologie française* XXVI: 58-73.

WOOD, W.R. 1990. Ethnohistory and historical method, *Archaeological Method and Theory* 2: 81-109.

Regions / Sites	Typological categories	Temporal assignment	Cu %	Ag ppm	Au ppm	In ppm	As ppm	Sb ppm	Sn ppm	Zn ppm
Rosemont (Montréal) *										
BjFj-98 bead A	Cylindrical bead	Late Woodland	96	390	≤ 0.036	≤ 1.5	≤ 4.7	≤ 0.44	≤ 500	≤ 52
BjFj-98 bead A	Cylindrical bead	Late Woodland	96	370	≤ 0.021	≤ 1.6	≤ 2.2	≤ 0.21	≤ 250	≤ 25
BjFj-98 bead B	Cylindrical bead	Late Woodland	98	280	≤ 0.033	≤ 1.3	≤ 3.0	≤ 0.38	≤ 460	≤ 32
BjFj-98 bead B	Cylindrical bead	Late Woodland	98	420	≤ 0.12	≤ 1.6	≤ 9.5	≤ 1.0	≤ 1400	≤ 99
Abitibi †										
DaGt-1	« Butter knife »	Late Woodland/Historic	98	79	≤ 0.096	≤ 1.2	≤ 8.5	≤ 1.1	≤ 2350	≤ 46
DcGt-4	« Hearth » like artifact	(Late) Woodland	98	260	≤ 0.089	≤ 0.92	≤ 3.8	≤ 0.24	≤ 270	≤ 37
DcGu-4	« Butter knife »	Late Woodland/Historic	99	130	≤ 0.088	≤ 1.2	≤ 3.9	≤ 0.23	≤ 290	≤ 37
DcGu4	« Butter knife »	Late Woodland/Historic	99	110	≤ 0.092	≤ 0.79	≤ 3.8	≤ 0.16	≤ 290	≤ 41
DdGt-5	awl	Late Woodland/Historic	99	83	≤ 0.093	≤ 0.97	15	≤ 0.44	≤ 1070	≤ 43
DdGt-9b	awl	(Late) Woodland	98	92	≤ 0.080	≤ 1.1	≤ 3.4	≤ 0.51	≤ 340	≤ 40
Saguenay-Lac-Saint-Jean ¥										
DdEw-10	Socketed point	Archaic	100	190	≤ 0.17	≤ 1.1	≤ 1.6	≤ 0.35	≤ 170	≤ 23
DdEw-26	«Bayonet » like point	?Archaic	100	250	≤ 0.04	≤ 1.2	≤ 0.81	≤ 0.23	≤ 77	≤ 11
DdEw-12 knife 1	« Butter knife »	(Archaic) Woodland	96	110	≤ 0.092	≤ 1.6	≤ 6.3	≤ 0.83	≤ 580	≤ 86
DdEw-12 knife 1	« Butter knife »	(Archaic) Woodland	99	120	≤ 0.085	≤ 1.3	≤ 5.6	≤ 2.2	≤ 470	≤ 67
DdEw-12 knife 1	« Butter knife »	(Archaic) Woodland	100	120	≤ 0.1	≤ 1.6	≤ 5.9	≤ 0.91	≤ 490	≤ 67
DdEw-12 knife 2	« Butter knife »	(Archaic) Woodland	98	100	≤ 0.13	≤ 1.4	≤ 2.6	≤ 0.32	≤ 220	≤ 57
DdEw-12 knife 2	« Butter knife »	(Archaic) Woodland	96	100	≤ 0.071	≤ 1.2	≤ 3.0	≤ 0.17	≤ 120	≤ 120
DdEw-12 knife 2	« Butter knife »	(Archaic) Woodland	94	100	≤ 0.092	≤ 1.7	≤ 5.0	≤ 0.82	≤ 400	≤ 53

Previously published data excerpted from
* Moreau and Hancock 1997
† Moreau et al 1994
¥ Langevin et al. 1995

Table 1. INAA data of Native Cu artefacts.

Cluster's names	N	Cu %	Ag ppm	Au ppm	In ppm	As ppm	Sb ppm	Sn % († : ppm)	Zn % († : ppm)
Native Cu	18	98 ± 2	180 ± 100	0.08 ± 0.04	1.3 ± 0.3	4.9 ± 3	0.59 ± 0.5	541 ± 600†	52 ± 30†
G2	10	99 ± 1	700 ± 30	16 ± 0.4	8 ± 0.5	720 ± 100	2300 ± 100	0.09 ± 0.06	0.03 ± 0.03
Tr1	10	99 ± 1	990 ± 50	38 ± 2	120 ± 7	140 ± 10	230 ± 10	0.08 ± 0.05	0.03 ± 0.02
G1	10	98 ± 1	770 ± 40	25 ± 2	11 ± 0.7	580 ± 90	2000 ± 200	0.08 ± 0.03	0.03 ± 0.02
Ash[uapmushuan 19]90	17	91 ± 4	960 ± 70	33 ± 3	15 ± 2	200 ± 20	61 ± 7	0.08 ± 0.05	6.4 ± 4
St.Nic[olas]2	21	89 ± 5	710 ± 80	17 ± 2	9.2 ± 1	630 ± 50	130 ± 10	0.2 ± 0.5	7.9 ± 4
St.Nic[olas]1	17	83 ± 4	1300 ± 200	31 ± 4	15 ± 2	610 ± 40	130 ± 40	0.2 ± 0.06	13.6 ± 5
Tr2	10	79 ± 2	1200 ± 70	11 ± 0.5	8 ± 0.7	1900 ± 40	410 ± 20	3.5 ± 0.4	18.1 ± 1
Kempt	16	77 ± 7	940 ± 40	21 ± 0.6	11 ± 0.8	1100 ± 50	320 ± 10	2.2 ± 0.3	20 ± 1
G3	10	76 ± 10	380 ± 300	2.9 ± 10	1.3 ± 30	580 ± 1000	150 ± 300	1.9 ± 1	21 ± 10
Chic[outimi]534	18	76 ± 2	410 ± 30	3.4 ± 0.3	2.1 ± 0.5	550 ± 30	200 ± 20	1.3 ± 0.2	23 ± 3
Chic[outimi]451	35	74 ± 3	400 ± 60	3.2 ± 0.4	2.2 ± 0.6	540 ± 50	180 ± 20	1.2 ± 0.4	23 ± 3
Ash[uapmushuan 19]94	10	71 ± 2	280 ± 20	2.2 ± 0.3	3.2 ± 0.4	1100 ± 30	190 ± 10	0.1 ± 0.06	28 ± 1
LoCuRivets	7	71 ± 4	760 ± 200	21 ± 9	17 ± 7	660 ± 500	330 ± 500	0.7 ± 0.9	26 ± 4
HiCuRivets	33	97 ± 2	900 ± 200	16 ± 14	9.5 ± 9	3000 ± 2000	730 ± 600	0.2 ± 0.3	1.2 ± 1
HiCuT[inkling]C[ones]	35	98 ± 1	1000 ± 100	4.3 ± 4	1.2 ± 0.3	5500 ± 2000	300 ± 200	0.1 ± 0.2	0.02 ± 0.01
LoCuT[inkling]C[ones]	14	74 ± 4	660 ± 300	20 ± 31	8.6 ± 6	440 ± 200	69 ± 20	0.1 ± 0.06	26 ± 3
HiFeT[inkling]C[ones]	3	0.13 ± 0.1	41 ± 50	0.06 ± 0.03	1.2 ± 0.2	92 ± 60	14 ± 20	4.9 ± 2	0.1 ± 0.07

Table 2. INAA data of Native Cu artefacts, 12 kettles, rivets and tinkling cones: mean +/- one standard deviation.

Sites (period) Kettles #	Archaeological information	Cu ppm	Ag ppm	Au ppm	In ppm	As ppm	Sb ppm	Sn ppm	Zn ppm
Ashuapmushuan (DhFk-7)									
(ca.1700-ca.1800)									
1994 single sheet †		70	250	1.9	3.8	1100	170	0.07	27
1994 single sheet †		70	310	2.6	3.6	1100	190	0.16	27
1994 single sheet †		70	300	2.6	2.5	1100	180	0.24	28
1994 single sheet †		72	250	1.9	3.2	1100	170	0.04	26
1994 single sheet †		74	280	2.3	3.1	1100	190	0.072	26
1994 single sheet †		72	270	2.1	3.5	1100	190	0.074	28
1994 single sheet †		70	290	2.3	2.9	1100	200	0.080	30
1994 single sheet †		71	290	2.3	2.7	1100	210	0.054	29
1994 single sheet †		73	290	2.0	3.2	1200	190	0.088	27
1994 single sheet †		74	290	2.0	3.5	1100	200	0.072	28
(ca.1625-ca.1650)									
1990 arch. conc. *	« scrap » 6	96	1100	35	14	210	61	0.12	2.2
1990 arch. conc. *	« scrap » 6	93	1100	36	15	210	70	0.18	2.9
1990 arch. conc. *	« scrap » 6	95	960	32	15	210	55	0.098	4.3
1990 arch. conc. *	« scrap » 7	93	950	31	16	220	60	0.11	6.6
1990 arch. conc. *	« scrap » 8	95	960	31	16	160	46	0.092	2.8
1990 arch. conc. *	« scrap » 9	84	830	31	14	200	63	0.1	12
1990 arch. conc. *	« scrap » 12	93	1000	35	17	180	61	0.1	2.1
1990 arch. conc. *	« scrap » 14	89	990	32	14	230	60	0.14	6.5
1990 arch. conc. *	« scrap »10	89	1000	38	15	210	77	≤ 0.036	7.7
1990 arch. conc. *	« scrap »10	90	930	31	14	200	66	0.098	5.0
1990 arch. conc. *	« scrap »10	87	900	33	13	190	66	≤ 0.031	12
1990 arch. conc. *	« scrap »11	88	930	35	14	200	66	≤ 0.033	9.7
1990 arch. conc. *	« scrap »12	93	900	33	16	170	57	≤ 0.029	5.4
1990 arch. conc. *	« scrap »13	96	1000	37	17	170	51	≤ 0.031	1.8
1990 arch. conc. *	« scrap »14	94	940	35	17	190	67	≤ 0.035	3.0
1990 arch. conc. *	« scrap »14	88	950	34	19	240	63	≤ 0.042	11
1990 arch. conc. *	« scrap »14	86	900	29	14	250	56	≤ 0.046	13
Kempt (CiFo-12) ¥									
(?1700-?1800)									
Whole kettle	body's sample	76	910	21	10	1100	320	2.0	18
Whole kettle	body's sample	76	940	20	12	1200	310	2.3	19
Whole kettle	body's sample	76	920	20	11	1100	280	2.7	19
Whole kettle	body's sample	77	960	21	13	1200	310	2.0	19
Whole kettle	body's sample	77	950	20	12	1100	310	2.1	19
Whole kettle	body's sample	77	980	20	12	1200	310	2.1	19
Whole kettle	body's sample	77	870	20	12	1100	310	1.8	20
Whole kettle	body's sample	77	880	20	10	1100	320	2.0	20
Whole kettle	body's sample	76	900	21	11	1100	320	2.0	21
Whole kettle	body's sample	77	890	21	12	1100	330	2.0	20
Whole kettle	body's sample	77	990	21	12	1200	330	2.3	20
Whole kettle	body's sample	77	950	20	11	1100	290	2.6	20
Whole kettle	body's sample	78	920	20	12	1100	320	1.9	20
Whole kettle	body's sample	78	980	21	12	1100	320	2.3	20
Whole kettle	body's sample	77	990	22	12	1100	330	2.5	21
Whole kettle	body's sample	78	930	21	12	1200	330	2.1	21

† Unpublished data

Previously published data excerpted from
* Moreau and Hancock 1996a
¥ Moreau and Hancock 1995a, 1995b

Table 3. INAA data of Kempt's and Ashuapmushuan's kettles.

Site (Period) Kettle #	Cu %	Ag ppm	Au ppm	In ppm	As ppm	Sb ppm	Sn %	Zn %
Gaétan (ca.1590-ca.1610)								
1	96	880	28	13	640	2100	≤ 0.05	≤ 0.0085
1	97	750	23	11	620	1800	≤ 0.04	≤ 0.0088
1	97	760	24	12	510	2000	≤ 0.09	≤ 0.05
1	98	770	26	11	620	1900	≤ 0.06	≤ 0.01
1	98	780	24	10	770	1900	≤ 0.05	≤ 0.01
1	98	770	23	11	530	2100	≤ 0.13	≤ 0.07
1	99	730	23	11	480	2000	≤ 0.11	≤ 0.05
1	99	790	26	12	480	2000	≤ 0.11	≤ 0.05
1	99	760	26	11	520	2200	≤ 0.1	≤ 0.05
1	100	750	24	12	600	1700	≤ 0.06	≤ 0.0087
2	97	690	16	7.6	730	2100	≤ 0.04	≤ 0.0082
2	97	690	16	8.3	630	2400	≤ 0.1	≤ 0.05
2	98	720	16	7.3	740	2200	≤ 0.04	≤ 0.0088
2	98	660	16	7.6	570	2300	≤ 0.21	≤ 0.09
2	99	690	16	8.6	730	2200	≤ 0.06	≤ 0.012
2	99	690	16	7.9	810	2200	≤ 0.06	≤ 0.012
2	99	680	17	8.6	580	2400	≤ 0.09	≤ 0.05
2	99	660	16	8.5	660	2200	≤ 0.18	≤ 0.06
2	100	750	17	8.3	940	2300	≤ 0.04	≤ 0.0089
2	100	720	16	8.0	770	2200	≤ 0.11	≤ 0.018
3	75	320	2.6	≤ 0.8	580	140	1.7	21
3	75	380	3.2	1.7	590	150	2.1	21
3	75	440	2.7	2.3	540	140	2.5	21
3	76	310	2.7	≤ 0.8	580	130	1.6	21
3	73	370	3.1	≤ 1.0	590	150	1.8	24
3	76	370	2.8	≤ 0.7	580	160	1.7	22
3	79	430	3.0	1.7	620	160	2.0	19
3	77	440	3.1	2.0	540	150	2.1	21
3	77	410	2.7	≤ 1.0	590	150	2.1	21
3	76	350	2.8	≤ 0.9	530	150	1.5	23
Treguno (ca.1590-ca.1610)								
1	97	940	39	130	150	220	≤ 0.05	≤ 0.037
1	98	1020	36	120	160	210	≤ 0.03	≤ 0.0046
1	98	980	38	120	150	220	≤ 0.06	≤ 0.0077
1	98	1090	43	120	150	250	≤ 0.04	≤ 0.037
1	99	960	38	120	160	220	≤ 0.08	≤ 0.035
1	99	990	38	110	130	230	≤ 0.08	≤ 0.04
1	99	930	36	130	130	230	≤ 0.09	≤ 0.04
1	99	1000	38	130	120	230	≤ 0.09	≤ 0.05
1	99	980	38	120	130	240	≤ 0.12	≤ 0.05
1	99	1010	38	130	140	230	≤ 0.19	≤ 0.04
2	77	1060	11	9.2	1900	420	4.2	17
2	77	960	11	8.2	1830	390	3.1	19
2	78	1010	11	6.8	1820	390	3.2	18
2	78	1120	11	8.0	1870	400	3.8	18
2	77	960	11	8.3	1910	390	3.0	20
2	79	1020	12	7.2	1850	390	3.3	18
2	79	1050	11	8.3	1860	390	3.3	18
2	79	1060	11	7.4	1880	420	3.7	18
2	81	1170	12	8.7	1940	460	4.1	17
2	83	1070	12	7.5	1920	400	3.5	18

Previously published data excerpted from Moreau and Hancock 1996b

Table 4. INAA data of Treguno's and Gaetan's kettles.

St.Nicolas (1625-1675) Kettle #	Cu %	Ag ppm	Au ppm	In ppm	As ppm	Sb ppm	Sn %	Zn %
1	83	1300	30	14	600	150	0.18	12
1	87	1300	29	15	560	130	≤ 0.091	8.3
1	85	1300	35	16	680	0	0.19	11
1	88	1400	33	15	620	150	0.24	7.9
1	84	1400	31	15	600	130	0.13	12
1	81	1300	31	14	640	140	0.22	15
1	83	1300	31	15	650	140	0.11	14
1	84	1300	30	14	640	150	0.11	13
1	81	1300	31	17	580	130	0.24	16
1	78	1200	27	13	590	120	0.18	19
1	79	1400	31	17	640	120	0.14	18
1	91	1600	39	16	650	170	0.13	6.4
1	76	1100	23	12	620	110	≤ 0.07	22
1	87	1300	31	14	550	140	0.24	11
1	82	1100	28	14	580	130	≤ 0.068	17
1	87	1500	36	16	630	150	0.19	7.5
1	77	980	25	12	600	120	≤ 0.064	21
2	85	650	14	9.2	660	120	≤ 0.051	8.8
2	93	740	18	9.1	630	130	≤ 0.052	3.7
2	94	680	17	7.9	720	150	≤ 0.044	3.5
2	98	810	19	9.5	530	120	0.11	2.3
2	80	480	11	6.5	550	100	≤ 0.056	14
2	89	680	15	9.1	660	110	≤ 0.093	6.2
2	92	710	18	9.4	570	110	≤ 0.057	3.3
2	87	670	16	8.7	650	120	≤ 0.083	8.7
2	90	660	17	11	620	140	≤ 0.047	6.5
2	83	680	16	9.6	610	130	≤ 0.051	13
2	84	700	16	9.5	570	130	0.14	12
2	82	780	17	11	680	130	≤ 0.069	15
2	81	570	14	8.6	570	120	≤ 0.061	15
2	92	790	19	8.8	680	140	≤ 0.049	4.2
2	90	730	17	10	660	140	≤ 0.039	6.3
2	93	760	17	7.7	630	130	≤ 0.093	3.6
2	91	760	19	9.6	680	140	≤ 0.048	5.5
2	91	760	18	9.9	700	130	≤ 0.069	6.3
2	84	690	15	7.8	620	120	0.1	14
2	92	830	21	11	600	130	≤ 0.086	6.7
2	91	740	18	9.2	560	110	2.2	6.7

Previously published data excerpted from Moreau and Hancock 1995c

Table 5. INAA data of St. Nicolas's kettles.

Chicoutimi (1625-1675) Kettles #	Cu %	Ag ppm	Au ppm	In ppm	As ppm	Sb ppm	Sn %	Zn %
451	72	350	3.0	≤ 1.0	510	170	≤ 0.99	22
451	71	390	3.0	2.0	510	170	1.2	24
451	78	470	3.8	2.3	540	210	1.8	17
451	71	300	2.8	2.2	490	160	0.87	25
451	71	320	2.7	2.0	500	160	0.99	25
451	70	330	2.8	2.0	500	160	1.0	26
451	72	370	2.6	2.2	510	170	1.1	24
451	76	410	3.3	2.2	530	190	1.2	20
451	74	380	3.0	1.8	530	190	1.1	23
451	74	380	3.3	2.3	530	180	1.2	23
451	73	390	3.1	≤ 1.3	560	190	1.3	24
451	77	430	3.3	2.4	530	190	1.3	20
451	76	420	3.4	≤ 1.8	550	200	1.3	21
451	74	390	3.4	2.6	550	180	1.4	23
451	77	420	3.8	3.0	570	200	1.4	20
451	78	410	3.8	2.9	540	210	1.5	19
451	78	590	3.8	2.8	590	220	1.6	19
451	71	380	2.9	2.1	520	180	0.72	27
451	77	410	3.5	3.0	540	200	0.87	21
451	77	430	3.9	2.2	670	190	1.9	20
451	72	380	2.7	2.0	480	160	1.0	26
451	74	390	3.1	2.5	510	180	1.1	24
451	73	380	3.1	0.9	510	170	0.12	26
451	78	570	4.4	3.4	720	200	2.2	19
451	74	400	3.1	2.2	520	180	1.3	24
451	74	400	3.1	1.9	540	180	1.3	24
451	76	470	3.3	2.6	530	210	1.3	22
451	71	350	2.7	2.1	520	170	1.1	28
451	73	400	2.9	2.0	510	170	1.1	26
451	74	360	2.9	2.8	520	170	1.2	25
451	73	400	3.0	2.0	510	190	1.2	26
451	78	410	3.4	≤ 0.99	530	190	1.3	21
451	72	350	2.9	1.7	530	170	0.91	28
451	74	370	3.0	2.2	520	170	1.0	26
451	80	460	3.5	2.4	560	210	0.95	23
534	75	410	3.3	2.4	550	190	1.1	21
534	77	430	4.1	2.3	540	200	1.3	19
534	79	450	3.8	≤ 1.1	600	230	1.6	17
534	74	370	3.3	1.9	500	180	1.1	23
534	74	400	3.3	2.5	540	180	1.1	23
534	78	420	3.6	2.2	580	220	1.5	19
534	76	420	3.5	2.5	610	220	1.6	21
534	70	380	2.7	2.4	510	160	0.9	28
534	75	370	3.6	1.7	550	190	1.1	23
534	73	390	3.6	2.0	540	180	1.1	25
534	75	390	3.3	2.7	500	180	1.2	23
534	78	400	3.5	2.6	590	230	1.4	20
534	76	380	3.2	2.1	560	210	1.3	23
534	75	440	3.4	≤ 1.2	540	200	1.3	24
534	73	390	3.0	1.8	550	180	1.0	27
534	78	390	3.3	≤ 1.2	550	210	1.2	22
534	77	430	3.6	2.4	520	190	1.3	23
534	77	490	3.6	2.3	560	220	1.5	24

Previously published data excerpted from Moreau and Hancock 1996a

Table 6. INAA data of Chicoutimi's kettles.

Region Site (period) Borden Code	Rivet #	Cu %	Ag ppm	Au ppm	In ppm	As ppm	Sb ppm	Sn %	Zn %
ONTARIO									
Robitaille (ca.1630)	Rivet†	100	890	31	14	230	670	0.028	0.21
BcHa-3	Rivet†	100	820	29	17	240	610	0.025	0.19
	Rivet†	**79**	**840**	**11**	**7.5**	**1700**	**360**	**2.7**	**18**
QUEBEC									
Saguenay-Lac-Saint-Jean									
Chicoutimi (ca.1625-ca.1675)	Rivet 533R†	100	610	7.4	13.8	2500	2000	0.65	1.1
DcEs-1	Rivet 1157R†	96	760	4.8	6.9	5400	370	0.82	2.3
	Rivet 21R†	99	890	7.6	7.1	6000	530	0.92	0.7
Québec									
Maison Milot (1640-1670)	Rivet C014	99	1250	42	24	320	150	0.06	1.3
1QU2150	Rivet C015	98	1100	35	42	210	81	0.074	0.39
	Rivet C032	92	1000	30	19	1500	1000	0.11	4.2
	Rivet C047	97	1000	36	31	220	130	0.25	1.1
Abitibi									
Fort Témiscamingue (?1675-1800)	Rivet P087	97	1000	0.25	5.1	4200	150	0.12	0.065
	Rivet P122	98	1200	1.1	0.7	4800	180	0.02	0.007
	Rivet P123	98	780	0.75	0.7	4500	120	0.05	0.009
	Rivet P141	94	640	2.5	1.4	4400	180	0.09	3.4
	Rivet P142	96	810	2.6	0.9	4400	210	0.05	3.3
	Rivet P143	94	950	7.7	4.1	5500	510	1.3	3.8
	Rivet P145	96	860	0.71	1.2	4100	130	0.07	0.11
	Rivet P146	98	1100	0.26	1.4	4800	160	0.13	0.069
DdGt-5 (historic)	Rivet 1375(RG)†	97	800	23	8.3	1000	1400	0.053	0.0095
	Rivet 1375(RD)†	97	820	24	7.3	910	1400	0.083	0.013
	Rivet A105	98	760	24	11	1000	1500	0.086	0.022
DdGt-9 (Historic)	Rivet A012	96	880	19	11	1800	1700	0.38	2.6
	Rivet A013	95	870	19	9.1	1700	1400	0.21	3.7
DeGu-4 (Historic)	Rivet A053	**70**	**230**	**4.4**	**6.2**	**680**	**1400**	**0.0035**	**28**
	Rivet A140	96	880	11	11	6100	550	0.28	1.9
	Rivet A141	97	950	8.3	8.5	7100	690	0.42	1.2
Gallichan (Historic)	JB1811, Rivet A219	97	790	6.3	12	5200	300	0.17	3.6
Mauricie									
Kempt (?1700-?1800)†	Rivet1(Lug1)†	**70**	**880**	**27**	**20**	**550**	**150**	**0.71**	**27**
	Rivet2(Lug1)†	**72**	**900**	**24**	**22**	**420**	**91**	**0.52**	**26**
	Rivet1(Lug2)†	**71**	**810**	**27**	**23**	**410**	**100**	**0.26**	**29**
	Rivet2(Lug2)†	**68**	**820**	**24**	**20**	**520**	**130**	**0.59**	**26**
	Rivet2(Lug2)†	**70**	**810**	**27**	**21**	**350**	**87**	**0.3**	**28**
Némiskachie (historic)	Rivet T009	97	870	22	5.1	1500	1400	0.04	0.037
CiFp-n	Rivet T010	97	1370	50	11	460	320	0.05	0.13
Côte-Nord									
Brador (historic)	Rivet S028	96	870	17	10	2600	1500	0.28	1.4
EiBh-n	Rivet S029	97	770	15	5.5	3800	1500	1.0	1.0
	Rivet S032	96	700	34	6.7	2000	1400	0.11	1.5
	Rivet S033	97	760	15	6.4	4300	1600	0.07	0.91
	Rivet S035	98	1000	1.6	1.2	3600	150	0.06	0.056
	Rivet S036	98	990	1.5	0.7	3800	170	0.06	0.011

Bold numbers: low Cu rivets

All data unpublished except † (see text)

Table 7. INAA data of the rivets.

	Region Site (period)	Tinkling cone #	Cu %	In ppm	Ag ppm	Au ppm	Fe %	As ppm	Sb ppm	Sn %	Zn %
HIGH Cu tinkling cones	Abitibi										
	Lac Abitibi (Historic)	A019	99	≤ 0.95	1400	2.0	≤ 0.15	9500	410	≤ 0.07	≤ 0.01
		A017	99	≤ 0.83	1100	1.3	≤ 0.18	9100	340	≤ 0.08	≤ 0.01
		A018	99	≤ 1.2	1100	1.2	≤ 0.15	9300	330	≤ 0.11	0.03
		C075	98	≤ 1.9	870	1.5	≤ 0.2	7200	310	≤ 0.14	0.03
		A020	97	≤ 1.1	950	0.74	≤ 0.15	3700	180	≤ 0.12	0.02
		C077	97	≤ 2.4	820	21	≤ 0.31	840	1620	≤ 0.19	0.06
	Gallichan (historic)	A207	97	≤ 1.5	1100	0.7	≤ 0.18	6500	350	0.71	≤ 0.01
	DdGt-5 (Historic)	A157	100	≤ 1.2	1100	5.5	≤ 0.16	5400	270	≤ 0.12	0.02
		A168	100	≤ 1.3	1200	5.5	≤ 0.16	4800	260	≤ 0.09	≤ 0.01
		A121	99	≤ 1.1	1200	1.4	≤ 0.15	6100	430	0.78	≤ 0.01
		A122	99	≤ 1.1	1200	1.4	≤ 0.13	6100	230	0.31	≤ 0.01
		A152	99	≤ 1.5	1000	5.0	≤ 0.24	4400	250	≤ 0.14	0.07
		A153	99	≤ 1.1	1000	5.1	≤ 0.15	4500	240	≤ 0.1	≤ 0.01
		A154	99	≤ 1.2	1100	6.6	≤ 0.14	4700	310	≤ 0.07	≤ 0.01
		A158	99	≤ 1.3	1000	6.3	≤ 0.16	4500	240	≤ 0.11	≤ 0.01
		A159	99	≤ 1.4	1200	5.5	≤ 0.21	4700	280	≤ 0.11	0.04
		A160	99	≤ 1.2	1000	4.9	≤ 0.2	4600	250	≤ 0.15	≤ 0.01
		A163	99	≤ 1.3	1000	4.7	≤ 0.17	5300	230	≤ 0.1	≤ 0.01
		A164	99	≤ 1.1	1000	4.9	≤ 0.14	4800	250	≤ 0.1	≤ 0.01
		A171	99	≤ 0.96	980	4.8	≤ 0.16	5100	220	≤ 0.1	≤ 0.01
		A172	99	≤ 1.2	860	4.6	≤ 0.15	4600	220	≤ 0.12	0.02
		A173	99	≤ 1.0	1100	4.8	≤ 0.18	4500	230	≤ 0.1	0.05
		A155	98	≤ 0.96	860	3.9	≤ 0.16	4800	180	≤ 0.09	≤ 0.01
		A156	98	≤ 1.4	890	3.8	≤ 0.17	5200	200	≤ 0.08	≤ 0.01
		A162	98	≤ 1.2	910	4.7	≤ 0.17	5200	220	≤ 0.1	≤ 0.01
		A165	98	≤ 1.0	850	4.0	≤ 0.16	5100	190	≤ 0.09	0.06
		A166	98	≤ 1.0	1100	5.0	≤ 0.17	5300	240	≤ 0.1	≤ 0.01
		A167	98	≤ 1.2	970	4.3	≤ 0.17	5500	210	≤ 0.08	≤ 0.01
		A169	98	≤ 0.97	890	4.0	≤ 0.16	4800	200	≤ 0.09	≤ 0.01
		A170	98	≤ 1.1	1000	4.9	≤ 0.16	4700	240	≤ 0.09	≤ 0.01
		A151	97	≤ 1.3	1000	5.7	≤ 0.17	4500	280	≤ 0.12	0.03
		A161	97	≤ 1.3	1200	5.2	≤ 0.17	7900	250	≤ 0.1	≤ 0.01
	DcGu-4 (Historic)	A148	98	≤ 1.1	1100	1.4	≤ 0.17	7100	370	≤ 0.12	0.03
		A023	97	≤ 1.2	810	≤ 0.27	≤ 0.16	4900	310	≤ 0.12	0.02
	St.Maurice Forge Saint-Maurice (1700-1800)	P002	98	≤ 1.2	950	1.8	≤ 0.15	7300	190	≤ 0.12	0.02
LOW Cu tinkling cones	Abitibi										
	Lac Abitibi (Historic)	C076	81	≤ 3.5	590	4.4	≤ 0.3	320	36	≤ 0.18	19
		C080	77	10	1100	33	≤ 0.29	300	92	≤ 0.12	26
		C078	68	7.5	890	6.9	≤ 0.34	530	56	≤ 0.19	28
	Gallichan (Historic)	A208	77	10	860	21	≤ 0.25	390	68	≤ 0.06	25
		A211	76	13	1000	31	≤ 0.18	260	76	0.17	22
		A212	76	10.1	580	7.3	≤ 0.24	430	42	≤ 0.06	27
		A209	75	2.9	540	7.2	≤ 0.22	430	63	≤ 0.04	26
		A210	74	7.3	410	120	≤ 0.22	430	110	≤ 0.06	26
		A206	72	25	810	28	≤ 0.28	300	80	≤ 0.07	28
	DdGt-5 (Historic)	A123	75	≤ 1.1	1100	2.0	≤ 0.34	260	83	≤ 0.09	29
	DcGu-4 (Historic)	A147	74	≤ 1.2	250	5.1	≤ 0.35	910	98	≤ 0.1	28
	Côte-Nord EbDo-1 (Historic)	S158	73	9.9	440	6.7	≤ 0.34	360	56	0.24	27
		S159	67	7.3	400	3.8	≤ 0.44	590	26	≤ 0.15	31
	Québec Nouvelles casernes	P001	70	11	320	5.4	≤ 0.26	690	81	≤ 0.06	30
HIGH Fe tinkling cones	Abitibi										
	DcGu-4 (Historic)	A050	0.2	≤ 0.92	≤ 14	0.075	94	57	4.2	3.02	0.2
		A022	0.1	1.4	≤ 7.0	0.082	97	61	4.7	6.3	0.05
		A051	0.1	1.3	100	≤ 0.025	92	157	34	5.3	0.093

All data unpublished

Table 8. INAA data of the tinkling cones.

43

The analysis of brass samples from the Ball and Warminster sites in southern Ontario, Canada

R.G.V. HANCOCK[1], L.A. PAVLISH[2], R.M. FARQUHAR[2] & D. KNIGHT[3]

[1]*SLOWPOKE Reactor Facility and Department of Chemical Engineering and Applied Chemistry, University of Toronto, Toronto, Canada M5S 3E5*
[2]*Department of Physics, University of Toronto, Toronto, Canada M5S 1A7*
[3]*Department of Anthropology, Wilfrid Laurier University, Waterloo, Canada N2L 3C5*

The Ball (circa AD 1616 - 1635) and Warminster (ca. 1625 - 1649) sites are located 0.8 km apart in southern Ontario. The chemical analyses of 218 brass samples from the Ball site split into three major groupings: low tin (~0.1%); medium tin (0.7 - 2.3%); high tin (2.7 - 6.3%). Each of these groupings were sub-dividable into tighter chemical sub-groups, based on their trace element contents, forming more than a dozen sub-groups. As expected from the proximity of the two sites, most of the 100 brass samples analyzed from the Warminster site match the Ball sub-groups, confirming that these sites were occupied by peoples who were closely related in time. Both sites contain red and yellow brass; and, from the relative abundances of yellow and red brass samples, and the overlapping compositions at Ball and Warminster, it may be argued that red brass kettles were traded into Huronia at a slightly earlier time than were yellow brass kettles. This observation appears to be consistent with the relative time difference associated with the introduction of copper and brass items into southern Ontario.

Keywords: Archaeometry; Brass; Copper; Contact Period; Huron; Iroquois; Ontario; Canada; Instrumental Neutron Activation Analysis; Seriation.

Introduction

The objective of this study is to obtain chemical data from historic brass samples that can be used in conjunction with archaeological evidence to construct a chronological picture of the Huron occupation of the Ball and Warminster sites. The frequency distributions of 'red brass' (defined here as copper alloy containing 5 - 20% zinc (Zn), 1 - 5.5% tin (Sn), 115 - 2330 ppm antimony (Sb) and 300 - 7200 ppm arsenic (As)) and 'yellow brass' (Zn 17 - 33%, Sn 0.1 - 2.2%, Sb 30-380 ppm and As 125-1150 ppm) on these two archaeological sites provides a clearer understanding of Huron occupational timing and both supports and augments an earlier study which focused on the seriational potential of the relative frequencies of copper and brass artefacts (Pavlish *et al.* 1996). The work reported here stems from a wider research project which has as its objective the chemical analysis, provenancing, and dating of copper and brass artefacts recovered from archaeological sites in eastern North America.

The Ball (*circa* AD 1616 - 1635) and Warminster (*ca.* 1625 - 1649) sites are located 0.8 km apart in the Huronia region of southern Ontario, 96 km north of Toronto and 13 km south of Georgian Bay. They were inhabited by the Huron people, who had established themselves there in the relatively recent past, having possibly stepwise-migrated through present day eastern Ontario from locations near the St. Lawrence River (Volterra 1994). Jesuit missions were established in Huronia in the 1630s and 1640s, the best known of these being Ste. Marie I near Georgian Bay, Ontario.

The presence of metal on Huron archaeological sites is a testament to their trade network which appears to have had the ability to acquire French, Dutch and English goods during protohistoric and early European contact periods. The Hurons appeared to operate as 'middlemen' in the trade between the Neutrals to the south and the Petun to the west.

The Iroquois Confederacy, whose own trade ambitions, perhaps helped by European geopolitics and commerce, continued to expand to the north from the western region of New York State during this period, became increasingly belligerent. This belligerence culminated in the massacre of the Huron peoples and the elimination of the Huron Nation in the wars of 1648 - 1650. During the summer of 1649, Ste. Marie II was built on Christian Island, in Georgian Bay, as a fallback position for those Huron who remained. In the spring of 1650 a small party of 300 Hurons and the few remaining Jesuit missionaries made the long trip east to Montreal. By the summer of that year little remained of the thousands who had sought refuge on the island. The fort, a national monument, stands today a grim reminder of this chapter in early Colonial history.

The Ball and Warminster sites are both spatially and chronologically close. In fact, there is every reason to believe from the archaeological evidence that these two villages, each with an expansion phase, represent slow movement and a transferral of the population from Ball to Warminster over the period of approximately a generation. The expansion phase at the Ball site suggests the assimilation of another group or a sudden increase in population. The Iroquois Confederacy activity in western New York State may well have precipitated this event. The within-site distribution of copper and red and yellow brass artefacts on the Ball site appears to distinguish older from newer areas of the site. The Warminster site also has two villages, termed the North Village and the South Village. Either the groups maintained their separate spatial identities on moving from the adjacent Ball site, or the South Village represents a very much later assimilation of peoples from neighbouring villages already devastated by the Iroquois. The presence of middens inside the South Village palisade would argue for a rather stressed and defensive disposition. The distribution of the copper, red brass and yellow brass on the Warminster site also appears to separate older from newer areas of the site. Furthermore, there are many chemically similar materials at both sites. The absolute copper to brass artefact ratio indicates that the Ball site is the earlier of the two. Chemical analysis of brass samples from both sites allows a re-assessment of the relationships between the inhabitants of the Ball and Warminster sites.

Analytical Procedure

Three hundred and eighteen manually-abraded analytical samples were taken from artefacts found at the Ball site (218 samples) and the Warminster site (100 samples). The samples were subjected to instrumental neutron activation analysis (INAA) at the SLOWPOKE Reactor Facility of the University of Toronto. Individual samples were stored in 1.2 ml polyethylene vials and were analyzed essentially as described by Hancock *et al.* (1991a; b).

Samples of 10-20 mg were first irradiated serially for five minutes at a neutron flux of 5.0×10^{11} $n.cm^{-2}.s^{-1}$ and assayed for five minutes after a delay time of approximately thirty minutes using germanium detector-based gamma ray spectrometers for zinc (Zn), copper (Cu), manganese (Mn) and indium (In). Elemental concentrations were calculated using the comparator method. Larger samples were irradiated at suitably lower neutron fluxes so that a relatively consistent total amount of radioactivity was produced in each sample.

Medium and long half-life radioisotope producing elements were analyzed by batch irradiating 10 to 20 samples per irradiation container for 16 hours at 2.5×10^{11} $n.cm^{-2}.s^{-1}$. After 5 to 9 days, these samples were serially assayed for 500 to 1000 seconds looking for samarium (Sm), gold (Au), lanthanum (La), copper (Cu), cadmium (Cd), arsenic (As), antimony (Sb), silver (Ag), zinc (Zn) and sodium (Na). A final counting was made after 9 to 18 days, at which time the samples were counted for 2,000 to 8,000 seconds to determine the concentrations of tin (Sn), selenium (Se), mercury (Hg), thorium (Th), chromium (Cr), gold (Au), arsenic (As), antimony (Sb), sliver (Ag), nickel (Ni), scandium (Sc), iron (Fe), zinc (Zn) and cobalt (Co).

This procedure produced replicate measurements of enough elements to guarantee no sample mix-ups. Analytical precisions ranged from ±1% to detection limits. Detection limits are consistent with those of 'dirth copper' (cf., Hancock *et al.* 1991: Table II). In the interests of brevity, the data of selected elements are presented in Tables 1 – 4 (at end). Complete analytical data may be obtained from the authors.

Results and Discussion

During the original sampling of artefacts, the colour of each sample was recorded to separate copper (red) from brass (light red-yellow). This process was rather subjective, because of colour variability, but it was a rapid and potentially valuable initial sorting method. "Yellow" was nearly always high-Zn brass, but "reddish" items were either the expected copper or a brass with lower than normal Zn content. (It is only the reddish brass that is reported here). By the end of this project, it was possible to define the terms 'red brass' and 'yellow brass' in chemical terms rather than by the original colour selection.

Selected elemental concentration data are presented in Tables 1 to 4, chosen to be representative of high, medium and low concentrations for the important elements of each chemical group. The first chemical criterion for sorting the samples was the Sn content, resulting inadvertently in a sorting by the colour of each sample, which in turn is related to the Zn/Sn content of the brass. Yellow brasses (Ball, Table 1 and Warminster Table 3) tended to have relatively high Zn contents (>20%) and low (<0.2%) to medium (<1%) Sn, whilst red brasses (Ball, Table 2 and Warminster, Table 4).tended to have relatively low Zn contents (<20%) and high Sn (>1%) contents.

Since the Ball brasses were in the majority, these were the first to be chemically classified. The yellow Ball brasses were formed into chemical groups primarily using their Au, As and Sn contents because these were the elements which exhibited maximum variation. Eight groups were initially identified, with five additional sub-groups being determined - three sub-groups (1a, 4a, 5a) were formed because of their very low Sn contents, and two (2a, 6a) because of higher than normal Au and Sn contents. In all, a total of thirteen Ball site yellow brass chemical groups were established (Table 1).

The yellow brass samples from Warminster were sorted in likewise manner and found to form ten chemical groups (Table 3). Where the Warminster chemical groups were found to be similar (within 10-20% relative concentrations for significant elements) to chemical groups already established for the yellow brass from the Ball site, a "B" designation was used for the group rather than a "W" designation.

Using the same criteria, the red brass samples from Ball were sorted into eleven groups, and one sub-group (16a) identified based on high As contents (Table 2), making a total of twelve chemical groups. The red brass samples from Warminster formed five chemical groups (Table 4).

Table 5 (at end) selected sample matches between Ball and Warminster material for five yellow brass and three red brass chemical groups. Given a potential relative homogeneity of the trace elements in the brass of ~10% to ~20%, these matches are convincing.

An overall comparison of the number of brass samples in each of the Ball and Warminster chemical groups (table 6 (at end) indicates contrasting commonality with respect to yellow and red brasses. The yellow brasses display a considerable overlapping of chemical groups between Ball (a total of 137 samples) and Warminster (80 samples), with ten out of thirteen groups found at Ball having similar a composition to groups found at Warminster, along with a possible high Sb variant sub-group (B6b) of the B6 group appearing at Warminster. The red brasses, however, with 80 samples in total from Ball and 20 samples from Warminster, share only two compositional groups out of Ball's twelve, with three different groups appearing only at Warminster. This difference in relative chemistries may arise from small sample numbers, but it is more likely a reflection of the small time difference between Ball (1616 - 1635) and Warminster (1625 - 1649), which copper seriation has shown to exist (Pavlish *et al.* 1996). If this is the case, then presumably the red brasses arrived first in Huronia and were used at Ball and later were carried to Warminster. On the other hand, it appears that most of the yellow brasses arrived later in the life of the Ball site and early in the life of Warminster.

Chronologically, separating the red and yellow brass using the current data is consistent with copper/(copper + brass) ratios derived for Ball (0.50) and Warminster (0.25). Also, if red and yellow brass sampling is representative of the total material at the respective sites, then the relative ratios of red brass to yellow brass (of all compositions) at each site - 0.59 at Ball and 0.25 at Warminster - may also contribute to this seriational chronology.

Conclusions

The chemically determined red and yellow brass from the Ball and Warminster sites in southern Ontario were subsequently placed in a number of chemical groups. From these chemical groups, ten similar yellow brass groups were found to be present at both the Ball and the Warminster site. This finding indicates strong ties between the inhabitants of both sites.The red brass chemical groups showed much less overlap, reinforcing the argument for a small chronological difference being present between the occupation of the two closely related sites - an observation that is consistent with the archaeological and historical record.

Acknowledgements

This research was supported by a grant from the Social Sciences and Humanities Research Council of Canada to RMF and by an infrastructure grant from the Natural Sciences and Engineering Council of Canada to the SLOWPOKE Reactor Facility.

References

HANCOCK, R.G.V., L.A. PAVLISH, R.M. FARQUHAR, R. SALLOUM, W.A. FOX & G.C. WILSON. 1991a. Distinguishing European trade copper and northeastern North American native copper, *Archaeometry* 33: 69-86.

HANCOCK, R.G.V., L.A. PAVLISH, R.M. FARQUHAR, R. SALLOUM, & G.C. WILSON. 1991b. North American native copper and European trade copper analysis, in E. Pernicka and G.A. Wagner (eds.), *Archaeometry '90*, Basel: Birkhauser Verlag, pp. 173-182.

PAVLISH, L.A., R.G.V. HANCOCK, R.M. FARQUHAR, R.P. BEUKENS, G.C. WILSON & A.C. D'ANDREA. 1996. Distinguishing different European trade opper and brass from archaeological sites in Ontario, in S. Demirci, A.M. Özer and G.D. Summers (eds.), *Archaeometry 1994*, Ankara: Tubitak, pp. 59-65.

VOLTERRA, V. 1994. *Dating and provenacing of sherds from five Balsam Lake sites*. Unpublished Masters Thesis, McMaster University.

	In ppm	As ppm	Ag ppm	Au ppb	Sb ppm	Sn %	Zn %
B1a							
599B	~0.5	567	335	1290	89	1.3	25
267W	~0.6	531	275	1254	80	1.2	27
B2a							
563B	1	826	362	4020	149	1.2	29
230W	3	869	392	3658	181	1.3	26
B3							
101B	8	970	490	8420	159	0.8	27
285W	6	822	494	9772	188	1.0	25
B5a							
032B	10	656	610	15770	148	1.1	25
175W	10	664	559	15680	147	1.1	24
B5							
673B	13	267	575	26100	70	1.4	20
261W	16	199	596	26210	81	1.9	28
B13							
103B	6	1790	847	8820	355	2.4	17
212W	7	2612	992	6418	374	3.4	14
249W	8	1968	906	6355	420	3.2	16
B16							
091B	11	2750	1266	13800	480	4.3	13
269W	10	2750	1136	13580	462	4.0	14
B16							
712B	16	3000	1440	16500	567	4.7	13
289W	13	2623	1218	15340	507	3.6	13

Table 5. Specific examples of matching samples for both yellow and red brasses from Ball and Warminster.

	In ppm	Mn ppm	As ppm	Cu %	Ag ppm	Au ppb	Co ppm	Ni ppm	Sb ppm	Sn ppm	Zn %
Group B1											
084B	~0.7	41	881	72	403	1400	3	1530	60	~0.1	28
275B	~0.7	37	509	71	291	970	~2	~530	122	~0.1	27
333B	~0.7	82	704	71	300	1440	10	~630	184	~0.1	29
Group B1a											
140B	~0.7	26	713	69	333	1030	~2	1990	209	0.6	27
168B	~0.7	17	645	78	355	1430	~2	~510	90	1.6	20
562B	~0.5	45	818	70	301	1800	5	~480	59	1.2	26
Group B2											
097B	1	~7	810	69	389	3370	9	1640	122	1.2	29
129B	2	11	683	73	402	2690	~2	1360	246	1.9	26
532B	1	13	669	71	369	2410	6	1120	119	2.2	27
Group B2a											
049B	2	22	784	71	365	3700	9	1010	142	1.2	27
112B	4	24	622	69	401	5470	9	1660	89	1.2	28
598B	3	33	724	70	385	4850	10	~440	265	2.0	26
Group B3											
088B	4	23	848	68	709	8660	46	~640	166	1.0	27
151B	5	21	1140	74	531	5730	12	830	182	1.4	21
187B	4	45	984	68	584	6240	11	1520	198	1.7	30
Group B7											
594B	5	30	522	75	642	9040	21	~370	74	0.9	20
Group B4											
078B	20	~10	170	76	790	27690	9	~470	70	0.8	21
200B	21	20	178	82	983	32500	10	~340	75	1.0	17
636B	20	17	173	74	818	27000	14	~470	64	0.8	23
Group B4a											
297B	17	21	167	74	843	26400	14	~460	63	~0.1	25
652B	20	18	169	73	835	27500	14	~470	68	~0.1	25
Group B5											
068B	12	~8	371	69	646	18630	14	~520	72	1.5	29
266B	29	~13	382	71	837	21090	25	~610	158	1.7	25
586B	14	28	200	71	612	25100	20	~500	46	1.9	26
592B	6	33	258	79	500	13900	18	820	110	0.9	25
Group B5a											
281B	14	27	298	71	503	21000	21	~500	79	~0.1	28
045B	14	14	383	71	607	21700	17	~480	94	~0.1	26
135B	5	45	375	66	427	11610	8	~520	79	~0.1	29
Group B6											
032B	10	23	656	70	610	15770	17	~620	148	1.1	25
064B	8	21	643	67	590	11840	25	~350	184	1.3	28
164B	12	~19	890	74	822	18180	15	~430	245	1.6	26
300B	13	22	642	73	658	20120	19	~440	138	1.9	23
Group B6a											
370B	17	33	674	73	665	26500	24	~440	104	1.7	23
789B	16	~14	699	75	711	24400	18	~420	153	2.2	21
Group B8											
222B	12	~9	1050	69	743	13800	31	1510	234	1.6	30

Table 1. Representative samples of yellow brass from the Ball site.

	In ppm	Mn ppm	As ppm	Cu %	Ag ppm	Au ppb	Co ppm	Ni ppm	Sb ppm	Sn ppm	Zn %
Group B11											
066B	4	26	2590	82	1630	4660	49	~470	118	3.4	13
034B	2	31	2896	79	813	3090	54	~400	127	1.1	17
Group B19											
308B	36	30	298	80	1712	67700	26	~850	213	1.9	14
Group B12											
124B	3	28	930	75	472	3120	10	1500	139	2.3	19
146B	3	45	1125	78	633	3740	13	~910	207	3.5	19
196B	3	64	1310	80	666	4260	17	~1030	174	4.1	14
Group B13											
040B	8	19	2540	80	940	8730	39	1090	575	3.8	14
103B	6	~7	1790	72	847	8820	36	1390	355	2.4	17
206B	8	~5	3890	79	1220	8780	56	1430	396	3.4	14
Group B14											
018B	9	~7	1630	72	994	17200	54	~1030	417	4.6	19
037B	6	~10	1280	73	638	10200	24	1110	328	2.9	20
102B	8	~9	1750	75	871	12600	38	1310	330	3.1	18
117B	6	~7	1340	72	693	10400	20	1590	354	3.4	18
Group B15											
016B	7	13	1610	81	1110	18400	52	1230	334	3.8	14
456B	12	~7	1640	78	819	18400	30	990	407	3.4	15
735B	12	~8	1880	82	1070	18000	55	~460	519	3.0	15
Group B16											
131B	6	11	2450	80	1070	10200	40	1280	637	4.2	12
180B	20	~8	3250	84	1560	23500	59	~450	509	4.7	12
192B	18	~6	2690	82	1320	18700	61	910	444	3.7	13
205B	17	~17	2940	80	1302	15200	58	960	511	4.0	11
Group B16a											
046B	22	~8	3460	79	1983	18900	65	1190	679	5.1	11
132B	9	~7	4380	82	1260	10200	58	2160	360	4.2	14
224B	17	32	4470	84	1710	14600	76	760	627	3.2	10
Group B17											
202B	28	13	4590	87	2030	19100	111	950	123	4.2	7
515B	30	24	4410	89	1840	18100	115	~310	115	4.0	7
Group B18											
106B	14	~16	5790	83	2325	12700	119	980	715	5.5	8
120B	16	18	5830	81	1983	12000	106	1310	697	5.4	8
Group B20											
134B	28	~12	7200	86	2371	12800	111	~300	1275	4.2	5
Group B21											
017B	46	~20	1760	86	1390	33900	150	~1070	1262	5.4	6

Table 2. Representative samples of red brass samples from the Ball site.

	In ppm	Mn ppm	As ppm	Cu %	Ag ppm	Au ppb	Co ppm	Ni ppm	Sb ppm	Sn ppm	Zn %
Group B1											
166W	~1.1	73	546	69	207	1444	~4	1896	381	~0.1	28
242W	~0.6	80	673	66	284	1771	~2	~640	160	0.5	33
267W	~0.6	20	531	71	275	1254	~1	~413	80	1.2	27
280W	~0.6	78	491	74	295	2064	3	833	86	~0.1	22
Group B2											
161W	9	31	585	67	246	2912	5	~433	95	0.4	28
178W	~0.5	51	709	69	307	2268	~2	~531	253	0.2	29
192W	~0.6	55	727	69	296	2297	~2	656	280	~0.1	30
272W	3	77	551	64	271	3086	6	1214	101	0.4	33
Group B2a											
262W	3	17	806	74	363	3600	5	1190	187	1.4	26
233W	3	25	850	71	370	4082	~2	1160	199	1.5	26
193W	9	18	519	72	522	5343	30	537	188	1.0	29
Group B7											
169W	24	22	270	67	159	7126	41	~583	41	0.2	30
236W	5	~22	270	70	174	7286	23	~729	39	0.4	30
237W	4	~13	288	73	204	8516	21	~613	47	~0.1	28
Group B3											
216W	8	20	1046	75	699	10600	18	~438	278	1.9	25
221W	5	18	698	69	501	9473	29	~457	170	0.9	28
222W	6	22	1015	70	599	9679	20	~496	244	1.7	26
234W	6	~13	588	71	487	8058	29	~752	167	0.9	25
Group B5											
170W	30	23	125	73	565	24540	19	~431	33	1.0	22
261W	16	20	199	70	596	26210	13	~391	81	1.9	28
274W	22	~13	210	69	694	21250	30	~490	51	0.3	29
293W	10	34	416	67	649	16490	19	~493	70	0.6	31
Group 5a											
180W	15	27	302	70	718	24370	25	~443	123	1.0	30
186W	13	45	280	71	619	13140	39	~691	131	1.2	26
Group B6											
159W	16	42	470	69	610	14160	42	~464	207	1.3	28
175W	10	29	664	75	559	15680	15	~449	147	1.1	24
191W	3	73	442	67	520	11460	26	~511	214	1.3	27
218W	10	30	397	68	516	13240	13	~462	102	0.5	28
Group B6a											
163W	16	37	418	73	815	17230	41	~652	254	1.4	26
255W	19	27	262	68	826	26940	21	~455	159	0.2	30
273W	6	25	266	70	665	21630	20	~375	109	1.0	26
Group B6b											
177W	20	~14	730	68	951	18800	51	~479	321	1.8	27
185W	24	45	514	69	788	23840	46	~616	298	0.4	27
260W	28	56	525	72	908	26110	115	732	300	0.6	30

Table 3. Representative samples of yellow brass samples from the Warminster site.

	In ppm	Mn ppm	As ppm	Cu %	Ag ppm	Au ppb	Co ppm	Ni ppm	Sb ppm	Sn ppm	Zn %
Group B16											
185W	24	45	2650	82	1210	13100	45	~340	445	4.1	12
211W	10	19	2530	84	1180	16000	48	1050	430	3.7	12
269W	10	~12	2700	78	1130	13600	53	730	462	4.0	14
289W	13	~8	2620	83	1210	15300	58	930	507	3.6	13
Group B13											
212W	7	~9	2600	80	990	6410	51	~400	374	3.4	14
249W	8	~9	1960	79	900	6350	33	650	420	3.2	16
Group W2											
238W	46	36	2940	86	1130	18000	90	~580	2140	4.2	5
275W	49	50	4090	88	1210	18700	119	~340	2330	4.1	6

Table 4. Representative samples of red brass samples from the Warminster site.

Group Yellow	Ball	Warminster	Group Red	Ball	Warminster
B1	13	4	B11	2	-
B1a	25	4	B12	4	-
B2	7	5	B13	3	3
B2a	10	6	B14	14	-
B3	6	13	B15	4	-
B4	5	-	B16	39	13
B4a	2	-	B16a	8	-
B5	14	11	B17	2	-
B5a	2	1	B18	2	-
B6	49	15	B19	1	-
B6a	2	12	B20	1	-
B6b	-	6	B21	1	-
B7	1	3	W1	-	1
B8	1	-	W2	-	2
			W3	-	1
Total	**137**	**80**		**81**	**20**

Table 6. Summary of numbers of samples from Ball and Warminster with similar chemistries.

Editor's notes

Addition to chapter 9 (permission to publish the copyrighted image was received very late).

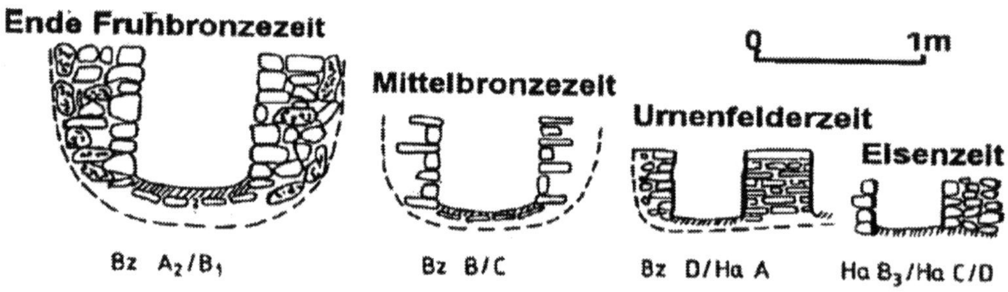

Figure 1. Eibner's Furnace typology showing the change in furnace dimensions through time. (Reproduced from Presslinger et al. 1988, by permission of Springer-Verlag).

All of the figures for chapter 7 and some of the figures for chapter 39 were never received. Permission to publish figure 3 of chapter 41 was never granted. If any of these images are needed for your research and understanding of the papers, please contact the authors of those chapters directly.

www.ingramcontent.com/pod-product-compliance
Lightning Source LLC
LaVergne TN
LVHW071203210926
848482LV00015B/716